Lecture Notes in Computer Science 8871

Commenced Publication in 1973
Founding and Former Series Editors:
Gerhard Goos, Juris Hartmanis, and Jan van Leeuwen

Christian Duncan   Antonios Symvonis (Eds.)

Graph Drawing

22nd International Symposium, GD 2014
Würzburg, Germany, September 24-26, 2014
Revised Selected Papers

 Springer

Volume Editors

Christian Duncan
Quinnipiac University
Hamden, CT, USA
E-mail: christian.duncan@quinnipiac.edu

Antonios Symvonis
National Technical University of Athens
Athens, Greece
E-mail: symvonis@math.ntua.gr

ISSN 0302-9743   e-ISSN 1611-3349
ISBN 978-3-662-45802-0   e-ISBN 978-3-662-45803-7
DOI 10.1007/978-3-662-45803-7
Springer Heidelberg New York Dordrecht London

Library of Congress Control Number: 2014956366

LNCS Sublibrary: SL 1 – Theoretical Computer Science and General Issues

Typesetting: Camera-ready by author, data conversion by Scientific Publishing Services, Chennai, India

Printed on acid-free paper

Springer is part of Springer Science+Business Media (www.springer.com)

Preface

This volume contains the papers that were presented at the 22nd International Symposium on Graph Drawing, which was held during September 24-26, 2014, in Würzburg, Germany. The symposium was hosted by the University of Würzburg and was attended by 106 participants from 16 countries. Fourteen of the participants were from industry. We thank Alexander Wolff, the local arrangements chair, and his team (Krzysztof Fleszar, Philipp Kindermann, Joachim Spoerhase, Mrs. Sigrid Keller, Fabian Lipp, Ben Morgan, and Wadim Reimche) for their warm hospitality.

Paper submissions were partitioned into two tracks with a separate poster track. Track 1 dealt with combinatorial and algorithmic aspects; track 2 with experimental, applied, and network visualization aspects. In total there were 84 submissions: 72 papers and 12 posters (each poster included a two-page description). Each submission was reviewed by at least three Program Committee members. The Program Committee decided to accept 41 papers and 11 posters. The acceptance rates were 33/52 in track 1, 8/20 in track 2, and 11/12 for posters.

We thank the Program Committee members and the additional reviewers for carefully reviewing the submitted papers and posters and for putting together a strong and interesting program. We also thank all authors for their hard work and for choosing GD 2014 as the publication venue for their research.

GD 2014 had two invited talks. Oswin Aichholzer from the University of Graz, Austria, gave a talk entitled "Good Drawings and Rotation Systems of Complete Graphs." Jean-Daniel Fekete from Inria, France, introduced the audience to the benefits of "Matrix-Based Visualization of Graphs." We thank both speakers for their excellent talks, which were very well received by the GD 2014 audience.

GD 2014 awarded prizes for Best Presentation and Best Poster. Most of the conference participants stayed until the last talk and voted for the winners. The Best Presentation award was split between two presenters: Vincent Kusters from ETH Zürich for his talk on the paper "Column Planarity and Partial Simultaneous Geometric Embedding," and Fidel Barrera-Cruz from the University of Waterloo for his talk on the paper "Morphing Schnyder Drawings of Planar Triangulations." The Best Poster award was won by Thomas Bläsius, Fabian Klute, Benjamin Niedermann and Martin Nöllenburg for their poster entitled "PIGRA — A Tool for Pixelated Graph Representations."

Following the community's well established tradition, the 21st Annual Graph Drawing Contest was held during the conference. It had two main categories: an off-line contest and an on-line challenge. This year's Contest Committee was chaired by Carsten Gutwenger (University of Dortmund). We thank the committee for preparing challenging problems and problem instances. A report of the contest is included in these proceedings.

We also wish to thank our sponsors: "diamond" sponsor German Research Foundation (DFG), "gold" sponsor Tom Sawyer Software, "silver" sponsors Microsoft and yWorks, and "bronze" sponsor Vis4. Their gracious support helps ensure the continued success of this conference.

In order to better reflect the diverse interests of the symposium both in theoretical aspects and in applications and systems design, the name of GD in 2015 will be extended to "International Symposium on Graph Drawing and Network Visualization." The 23rd International Symposium on Graph Drawing and Network Visualization will be held in Los Angeles, USA, during September 23-25, 2015. Emilio Di Giacomo and Anna Lubiw will be the Program Committee chairs. Csaba Tóth will be the Organizing Committee chair.

September 2014   Christian Duncan
                                                             Antonios Symvonis

Organization

Program Committee

Patrizio Angelini	Roma Tre University, Italy
Daniel Archambault	Swansea University, UK
David Auber	LaBRI , Université Bordeaux I, France
Michael Bekos	University of Tübingen, Germany
Anastasia Bezerianos	Université Paris-Sud, Inria and CNRS, France
Franz Brandenburg	University of Passau, Germany
Erin Chambers	St. Louis University, USA
Stephan Diehl	Universität Trier, Germany
Christian Duncan (Co-chair)	Quinnipiac University, USA
Tim Dwyer	Monash University, Australia
David Eppstein	University of California, Irvine, USA
Emden Gansner	AT&T Research
Michael Kaufmann	University of Tübingen, Germany
Stephen Kobourov	University of Arizona, USA
Jan Kratochvil	Charles University of Prague, Czech Republic
Giuseppe Liotta	University of Perugia, Italy
Maarten Löffler	Utrecht University, The Netherlands
Anna Lubiw	University of Waterloo, Canada
Petra Mutzel	TU Dortmund University, Germany
Lev Nachmanson	Microsoft Research
Antonios Symvonis (Co-chair)	National Technical University of Athens, Greece
Ioannis Tollis	University of Crete, Greece
Dorothea Wagner	Karlsruhe Institute of Technology (KIT), Germany
Hsu-Chun Yen	National Taiwan University, Taiwan

Organizing Committee

Krzysztof Fleszar	University of Würzburg, Germany
Philipp Kindermann	University of Würzburg, Germany
Joachim Spoerhase	University of Würzburg, Germany
Alexander Wolff (Chair)	University of Würzburg, Germany

Graph Drawing Contest Committee

Carsten Gutwenger (Chair)	TU Dortmund University, Germany
Maarten Löffler	Utrecht University, The Netherlands

Lev Nachmanson   Microsoft Research
Ignaz Rutter   Karlsruhe Institute of Technology (KIT),
                                       Germany

Additional Reviewers

Aerts, Nieke   Kostitsyna, Irina
Alam, Muhammad Jawaherul   Krug, Robert
Alamdari, Soroush   Lee, Bongshin
Argyriou, Evmorfia   Lu, Hsueh-I
Barba, Luis   Mchedlidze, Tamara
Bereg, Sergey   Mondal, Debajyoti
Binucci, Carla   Montecchiani, Fabrizio
Bläsius, Thomas   Morin, Pat
Borradaile, Glencora   Mumford, Elena
Bruckdorfer, Till   Niedermann, Benjamin
Chen, Ho-Lin   Nöllenburg, Martin
Da Lozzo, Giordano   Patrignani, Maurizio
Di Bartolomeo, Marco   Pedrosa, Lehilton L. C.
Di Donato, Valentino   Poon, Sheung-Hung
Di Giacomo, Emilio   Prutkin, Roman
Didimo, Walter   Pupyrev, Sergey
Durocher, Stephane   Raftopoulou, Chrysanthi
Evans, William   Riche, Nathalie
Felsner, Stefan   Roselli, Vincenzo
Fowler, J. Joseph   Rutter, Ignaz
Fox, Kyle   Schaefer, Marcus
Frati, Fabrizio   Schreiber, Falk
Fuchs, Fabian   Schulz, André
Fulek, Radoslav   Spisla, Christiane
Gemsa, Andreas   Squarcella, Claudio
Grilli, Luca   Staals, Frank
Gronemann, Martin   Strash, Darren
Gutwenger, Carsten   Toeniskoetter, Jackson
Heinsohn, Niklas   Tsiaras, Vassilis
Hlineny, Petr   Ueckerdt, Torsten
Holroyd, Alexander   van Goethem, Arthur
Kakoulis, Konstantinos   Zielke, Christian

Sponsors

Diamond Sponsor

Gold Sponsor

Silver Sponsors

Bronze Sponsor

Invited Talks

Good Drawings and Rotation Systems of Complete Graphs

Oswin Aichholzer

Institute for Software Technology,
Graz University of Technology, Austria.
oaich@ist.tugraz.at

Abstract. In a good drawing of a complete graph the vertices are drawn as distinct points in the plane, edges are drawn as non-self-intersecting continuous arcs connecting its two end points, but not passing through any other point representing a vertex. Moreover, any pair of edges intersects at most once, either in their interior or at a common endpoint, no tangencies are allowed and no three edges pass through a single crossing. These drawings are also called simple topological graphs.

A rotation system (of a good drawing of a complete graph) gives, for each vertex v of the graph, the circular ordering around v of all edges incident to v. In combinatorial mathematics, rotation systems were first used by Hefner in 1891 to encode embeddings of graphs onto orientable surfaces, determining its genus. In the plane (or equivalently on the sphere) the rotation system of a good drawing does not fully determine the drawing, but contains combinatorial information like all pairs of edges which intersect.

We present basic properties of these two concepts, as well as recent progress. This includes results on the number of realizable rotation systems, the crossing number of complete graphs (including the recent concept of shellability of a good drawing), relations to other systems like the order type of a point set, etc.

Matrix-Based Visualization of Graphs

Jean-Daniel Fekete

INRIA, France
Jean-Daniel.Fekete@inria.fr

Abstract. For decades, graph drawing has focused on the node-link representation, trying to address multiple important, difficult, and interesting issues related to 2D embeddings under some optimality criteria (planar drawing, minimizing crossings, optimizations, graph decompositions, and many more). Visualizing a graph structure using its adjancency matrix is much less common, although it has been shown to be more efficient than the node-link representation when the graph becomes dense, for important low-level tasks. The main question to address in matrix-based visualization is the computation of the vertices order. This problem is known with multiple names: linear ordering, seriation, reordering. With a proper ordering, a visualized matrix reveals important patterns and structures of the graph. We will briefly explain how the problem has been formalized in the past, some visual results sometimes revealing unexpected information. We then list some challenges to the community that could be used to motivate the graph drawing community to study the problem, and provide useful solutions for people who need to make sense of complex graphs.

Table of Contents

Contact Representations

k-Planar Graphs

Crossing Minimization

Level Drawings

Theory

Fixed Edge Directions

Drawing under Constraints

Clustered Planarity

Greedy Graphs

Graph Drawing Contest

Posters

Planar Induced Subgraphs of Sparse Graphs

Glencora Borradaile[1], David Eppstein[2], and Pingan Zhu[1]

[1] Oregon State University, Corvallis OR 97330, USA
[2] University of California, Irvine, Irvine CA 92697, USA

Abstract. We show that every graph has an induced pseudoforest of at least $n - m/4.5$ vertices, an induced partial 2-tree of at least $n - m/5$ vertices, and an induced planar subgraph of at least $n - m/5.2174$ vertices. These results are constructive, implying linear-time algorithms to find the respective induced subgraphs. We also show that the size of the largest K_h-minor-free graph in a given graph can sometimes be at most $n - m/6 + o(m)$.

1 Introduction

Planarization, a standard step in drawing non-planar graphs, involves replacing edge crossings with new vertices to form a planar graph with paths that represent the original graph's edges. *Incremental planarization*, does this by finding a large planar subgraph of the given graph, and then adding the remaining features of the input graph one at a time [6]. Thus, it is of interest to study the algorithmic problem of finding planar subgraphs that are as large as possible in a given graph. Unfortunately, this problem is NP-hard and, more strongly, MAX-SNP-hard [4]. A trivial algorithm, finding an arbitrary spanning tree, achieves an approximation ratio of $\frac{1}{3}$, and by instead searching for a partial 2-tree this ratio can be improved to $\frac{2}{5}$ [4]. The equivalent complementary problem, deleting a minimum number of edges to make the remaining subgraph planar, is fixed-parameter tractable and linear time for any fixed value of the parameter [12].

In this paper we study a standard variant of this problem: finding a large planar *induced* subgraph of a given graph. In the context of the planarization problem, one possible application of finding this type of planar subgraph would be to apply incremental planarization in a drawing style where edges are represented as straight-line segments. A planar induced subgraph can always be drawn without crossings in this style, by Fáry's theorem, after which the partial drawing could be used to guide the placement of the remaining vertices. As with the previous problem, the induced planar subgraph problem is NP-hard, but again there is a linear-time fixed-parameter tractable algorithm for the equivalent problem of finding the smallest number of vertices to delete so that the remaining induced subgraph is planar [11].

Because of the difficulty of finding an exact solution to this problem, we instead seek worst-case guarantees: what is the largest size of a planar induced subgraph that we can be guaranteed to find within a graph of a given size? In this we are inspired by a paper of Alon, Mubayi, and Thomas [1], who showed

C. Duncan and A. Symvonis (Eds.): GD 2014, LNCS 8871, pp. 1–12, 2014.

that every *triangle-free* input graph with n vertices and m edges contains an induced forest with at least $n - m/4$ vertices. This is tight, as shown by an input graph in the form of $n/4$ disjoint copies of a 4-cycle. Induced forests are a special case of induced planar subgraphs, and so this result guarantees the existence of an induced planar subgraph of $n - m/4$ vertices. As we show, an analogous improvement to the one in the approximation ratio for planar subgraphs can be obtained by seeking instead an induced partial 2-tree. Rossmanith [3] has posed the question: does every graph have an induced planar subgraph of size $n - m/6$? We shrink the gap on the worst-case bounds for the size of a planar induced subgraph by showing that every graph (not necessarily triangle-free) has an induced planar subgraph of $n - m/5.2174$ vertices and that there exist graphs for which the largest induced planar subgraph is not much larger than $n - m/6$ vertices.

1.1   New Results

We prove the following results:

Theorem 1. *Every graph with n vertices and m edges has an induced pseudo-forest with at least $n - \frac{2m}{9}$ vertices.*

Theorem 2. *Every graph with n vertices and m edges has an induced subgraph with treewidth at most 2 and with at least $n - \frac{m}{5}$ vertices.*

Theorem 3. *Every graph with n vertices and m edges has an induced planar subgraph with treewidth at most 3 and with at least $n - \frac{23m}{120}$ vertices.*

These three theorems can be implemented as algorithms which take linear time to find the induced subgraphs described by the theorems.

Theorem 4. *For every integer h, there is a family of graphs such that for any graph in this family with n vertices and m edges, the largest K_h-minor-free induced subgraph has at most $n - \frac{m}{6} + O(\frac{m}{\log m})$ vertices.*

The bounds of Theorems 1 and 2 are tight, even for larger classes of induced subgraphs. In particular, there exist graphs for which the largest induced outerplanar subgraph has size at most $n - m/4.5$, so the bound of Theorem 1 is tight for any class of graphs between the pseudoforests and the outerplanar graphs. There also exist graphs for which the largest induced K_4-free induced subgraph has $n - m/5$ vertices, so Theorem 2 is tight for every family of graphs between the treewidth 2 graphs and the graphs with no 4-clique.

1.2   Related Work

The worst-case size of the largest induced planar subgraph has been studied previously by Edwards and Farr [7], who proved a tight bound of $3n/(d+1)$ on its size as a function of the maximum degree d of the given graph. In contrast, by depending on the total number of edges rather than the maximum degree, our algorithms are sensitive to graphs with heterogeneous vertex degrees, and can construct larger induced subgraphs when the number of high-degree vertices is relatively small. Additionally, the algorithm given by Edwards and Farr is slower than ours, taking $O(mn)$ time. In a follow-up paper, Morgan and Farr [16] gave additional bounds on induced outerplanar subgraphs, and provided experimental results on the performance of their algorithms. A second paper by Edwards and Farr [8], like our Theorem 2, gives bounds on the size of the largest induced partial 2-tree in terms of n and m, which is $\frac{3n}{2m/n+1}$ for $m \geq 2n$. However, their bounds are asymptotically worse than Theorem 2 when $2n < m < 2.5n$ and require an additional assumption of connectivity for smaller values of m. A third paper by the same authors [9] gives improved bounds that are more difficult to state as a formula.

For some other graph classes than the ones we study, it is possible to prove trivial bounds on the size of the largest induced subgraph in the class, of a similar form to the bounds of Alon et al. and of our theorems. By repeatedly finding and removing a vertex of degree ≥ 1, one can obtain an independent set of at least $n - m$ vertices, and the example of a perfect matching shows this to be tight. By repeatedly finding and removing a vertex of degree ≥ 2, one can obtain a matching of at least $n - m/2$ vertices, and the example of the disjoint union of $n/3$ two-edge paths shows this to be tight. And by repeatedly finding and removing either a vertex of degree ≥ 3 or a vertex that is part of a 2-regular cycle, one can obtain a linear forest (forest with maximum degree 2) of $n - m/3$ vertices; the example of the disjoint union of $n/3$ triangles shows this to be tight even for more general forests.

Table 1 provides a comparison of these new results with previous known results on induced planar subgraphs of various types and with the trivial bounds for independent sets, matchings, and linear forests.

2   Preliminaries

For a graph G, we define $n(G)$ to be the number of vertices and $m(G)$ to be the number of edges in G. We drop the argument and write n and m when the choice of G is clear from context.

A subset S of the vertices of G corresponds to an *induced subgraph* $G[S]$, a graph having S as its vertices and having as edges every edge in G that has both endpoints in S. Equivalently, $G[S]$ may be constructed from G by deleting every vertex that is not in S and every edge that has at least one endpoint outside S.

A *pseudoforest* is an undirected graph in which every connected component has at most one cycle. Equivalently, the pseudoforests can be formed from forests (acyclic undirected graphs) by adding at most one edge per connected component. A *k-tree* is an undirected graph that can be constructed from a K_k graph by repeatedly picking a K_k subgraph and attaching its k vertices to a new vertex. A *partial k-tree* is a subgraph of a k-tree and is said to have *treewidth* at most k; the treewidth of a graph G is denoted $tw(G)$. Every pseudoforest is a partial

Table 1. Comparison of new and known results on induced subgraphs

Size of induced subgraph	Additional constraints	Type of subgraph	Reference
$n - m$		independent set	trivial
$n - \dfrac{m}{2}$		matching	trivial
$n - \dfrac{m}{3}$		linear forest	trivial
$n - \dfrac{m}{4}$	G is triangle-free	forest	[1]
$n - \dfrac{m+c}{4}$	max degree ≤ 3 $c = \#$ connected components	forest	[1]
$\dfrac{n}{\lceil (\Delta+1)/3 \rceil}$	max degree Δ	max degree 2	[10]
$\dfrac{3n}{\Delta + 5/3}$	max degree Δ	outerplanar	[16]
$\dfrac{3n}{\Delta + 1}$	max degree Δ	planar	[7]
$\dfrac{3n}{2m/n + 1}$	$m \geq 2n$ or connected and $m \geq n$	partial 2-tree	[8]
$\dfrac{5n}{6}$	claw-free subcubic	planar partial 4-tree	[5]
$n - \dfrac{m}{4.5}$		pseudoforest	Theorem 1
$n - \dfrac{m}{5}$		partial 2-tree	Theorem 2
$n - \dfrac{m}{5.2174}$		planar partial 3-tree	Theorem 3
$\leq n - \dfrac{m}{6} + o(m)$		any minor-closed property	Theorem 4

2-tree. A graph is a partial 2-tree if and only if every biconnected component is a series parallel graph. The operations of adding a vertex with two adjacent neighbors and of taking subgraphs preserve planarity, so every partial 2-tree and every pseudoforest is a planar graph.

When constructing induced subgraphs of size $n - m/k$, we will make the simplifying assumption that our graph G has maximum degree at most $\lceil k - 1 \rceil$.

Observation 1. *If every graph of maximum degree at most $\lceil k-1 \rceil$ contains an induced subgraph with property $\mathcal{P}$ and at least $n - m/k$ vertices, then the same is true for every graph.*

Proof. We use induction on n. Let G contain a vertex v of degree $\geq k$, and let G' be formed from G by removing v. By the induction hypothesis, G' has an induced subgraph H with the desired property $\mathcal{P}$ and at least $n(G') - \frac{m(G')}{k}$ vertices. Then H is an induced $\mathcal{P}$-subgraph of G with size at least

$$n(G) - 1 - \frac{m(G')}{k} \geq n(G) - 1 - \frac{m(G) - k}{k} = n(G) - \frac{m(G)}{k}.$$

3   Large Induced Pseudoforests

In this section, we prove Theorem 1 by showing that we can delete at most $\frac{m}{4.5}$ vertices from a graph G with m edges to leave a pseudoforest; by Observation 1, we assume G has degree at most 4.

We repeatedly perform the first applicable reduction in the following list of cases, until no edges are left. As we do, we construct a set S of vertices that will induce our desired subgraph. Initially S is empty and when no edges are left we add all remaining vertices to S. The steps of the reduction essentially identify "dangling trees" and contract these. After a series of vertex deletions (which identify vertices that will not belong the final induced subgraph) and edge contractions, if we create a component that consists of a single cycle (in fact a triangle, case Δ-a and *Vertex of degree 4* (c) (i) and (ii)), we "keep" this component; this triangle is the minor of an induced cycle in our final induced subgraph which will only be incident to the "dangling trees" which had been contracted into the cycle. This guarantees that the final induced subgraph has at most one cycle per component. To bound the size of the output S in terms of the number of edges, we use an amortized analysis, incurring a charge of -4.5 for every deleted vertex and a charge of $+1$ for every removed or contracted edge; we show that the net charge for every processing step is non-negative. Our cases are:

Leaf vertex. If there is a vertex a of degree 1, we add a to S and contract the edge incident to a. This incurs a charge of $+1$.

Vertex of degree 2 not in a triangle. If there is a vertex a of degree 2 that is not part of a triangle, we add a to S and contract an edge incident to a. This incurs a charge of $+1$.

Vertex of degree 2 in a triangle. If there is a vertex a of degree 2 in a triangle abc then we consider the four sub-cases illustrated below:

(Δ–a) If the triangle is isolated: add a, b and c to S, and remove the edges of the triangle from the graph. This incurs a charge of $+3$.

(Δ–b) If b has degree 2 and c has degree 3, then c is adjacent to a vertex d of degree at least 3 (otherwise d would be a degree 2 vertex not belonging to a triangle). Delete d, isolating triangle abc, and then apply case (Δ–a). This incurs a total charge of at least $+1.5$.

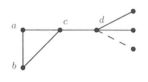

(Δ–c) If b has degree 3 and c has degree at least 3, then delete c, add a and b to S and contract the edges incident to b. This incurs a charge of at least $+0.5$.

(Δ–d) If b has degree 4, then delete b, add a to S and contract ac. This incurs a charge of $+0.5$.

Vertex of degree 3 adjacent to a vertex of degree 4. If there is a vertex a of degree 3 adjacent to a vertex b of degree 4, then we delete b. Deleting b incurs a net charge of -0.5, but reduces the degree of a to 2. Handling a as above incurs a charge of at least $+0.5$, for a net charge of at least 0.

Vertex of degree 3. If this is the first applicable case, then the graph must be 3-regular. Deleting any vertex a creates three vertices of degree 2 while incurring a charge of -1.5. Processing the three resulting degree-2 vertices as above incurs a charge of at least $+0.5$ per degree-2 vertex, for a net charge of at least 0.

Vertex of degree 4. If this is the first applicable case, then the graph must be 4-regular. We consider the following cases for the subgraph N_a induced by the neighbors b, c, d, e of a vertex a.

(a) If N_a has two non-adjacent pairs (b, c) and (d, e), then a, b, and c do not form a triangle, so we can delete d and e, keep b and c, add a to S, and contract ab. This removes nine edges from the graph and deletes two vertices, for a total charge of 0.

(b) If N_a is a star graph with center b, then consider the neighbors of c: a, b, f, g. Neither a nor b can be adjacent to f or g, because otherwise they would have too many neighbors. Thus we can process vertex c as in case (a) instead.

(c) If neither case (a) nor (b) applies, then N_a must contain a triangle. Without loss of generality the triangle is formed by vertices bcd, so a is a vertex of a tetrahedron (K_4) induced by vertices a, b, c, d. We may assume more strongly that every vertex in the graph belongs to a tetrahedron, for if not we may apply cases (a) or (b). We form four sub-subcases:

    (i) If any connected component is a complete graph K_5, then deleting two vertices leaves an isolated triangle and incurs a total charge of $+1$.

(ii) If two tetrahedra, a, b, c, d and b, c, d, e share triangle b, c, d without forming a K_5, then a and e are non-adjacent. Deleting a and e removes eight edges but leaves an isolated triangle (case Δ–a) for a net charge of $+2$. We illustrate this case:

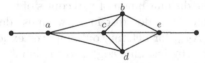

(iii) In the remaining cases all tetrahedra must be vertex-disjoint. If two tetrahedra $abcd$ and $efgh$ are connected to each other by at least two edges (be and dg), then we delete the two non-adjacent vertices d and e as illustrated here:

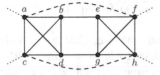   The dashed edges are possible connections from vertices a, c, f, h.

This incurs a charge of -1 but leaves two non-adjacent degree two vertices (b and g) each of which can be processed via case (Δ–c), adding charge $+0.5$ per vertex for a total charge of at least 0.

(iv) If every pair of tetrahedra are connected by at most one edge, then contracting every tetrahedron to a vertex reduces the input graph to a smaller 4-regular simple graph that necessarily contains a cycle of three or more edges. In the uncontracted graph, this gives a cycle of six or more edges that alternates between edges within tetrahedra and edges outside the tetrahedra:

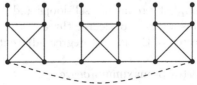

In this case, we choose one of the tetrahedra of the cycle, and delete the two vertices of this tetrahedron that do not belong to the cycle. This removes seven edges from the graph for a net charge of -2, but leaves two degree-2 vertices on a cycle of length at least 6. Each of these may be processed as a degree-2 vertex that does not belong to a triangle, giving a charge of $+1$ each and making the net charge be 0.

This case analysis concludes the proof of Theorem 1. The proof also gives the outline for an efficient algorithm for finding an induced pseudoforest of size at least $n - m/4.5$: after removing any high-degree vertices, form a data structure that lists the configurations of the graph obeying each of the cases in the analysis. Because the remaining graph has bounded degree, selecting the first applicable case, performing the reduction steps of the case, and updating the list of configurations for each case can all be done in constant time per case, leading to a linear overall time bound.

Theorem 1 is tight: there exist arbitrarily large graphs in which the largest induced pseudoforest has exactly $n - m/4.5$ vertices. In particular, let G be a graph formed by the disjoint union of $n/6$ copies of the complete bipartite graph $K_{3,3}$. Then, to form a pseudoforest in G, we must delete at least two vertices from each copy of $K_{3,3}$, for deleting only one vertex leaves $K_{2,3}$ which is not a pseudoforest. Each copy has nine edges, so the number of deleted vertices must be at least $m/4.5$. The same class of examples shows that even if we are searching for the broader class of induced outerplanar subgraphs, we may need to delete $m/4.5$ vertices.

4   Large Induced Treewidth Two Graphs

In this section, we prove Theorem 2 by showing that we can delete at most $\frac{m}{5}$ vertices from a graph G with m edges to leave a graph with treewidth at most 2. By Observation 1, we may assume without loss of generality G has degree at most 4. We prove the theorem algorithmically by arguing that the following procedure builds a vertex set S of size at least $n - \frac{m}{5}$ such that $G[S]$ has treewidth at most 2. The procedure modifies the graph by edge contractions but does not increase its degree over 4.

$S = \emptyset$
make G simple by removing self-loops and parallel edges
while G has more than 1 vertex:
    if there is a vertex v of degree one or two:
        contract an edge incident to v and add v to S:
        make G simple by removing self-loops and parallel edges
    else if G contains a vertex of degree three:
        delete a vertex of the largest degree adjacent to a degree-three vertex
    else:
        delete a vertex of maximum degree
add the last remaining vertex to S

Lemma 1. *The induced subgraph $G[S]$ produced by the algorithm above has treewidth at most 2.*

Proof. We use the following facts:

Fact 1. If H is a subdivision of G then $\mathrm{tw}(G) = \max\{1, \mathrm{tw}(H)\}$.
Fact 2. If H is a maximal simple subgraph of G then $\mathrm{tw}(G) = \max\{1, \mathrm{tw}(H)\}$.
Fact 3. If H is obtained from G by deleting a leaf vertex, $\mathrm{tw}(G) = \max\{1, \mathrm{tw}(H)\}$.

Let $s_1, s_2, \ldots, s_k$ be the vertices of S in the reverse of the order in which they were added to S. Let $S_i = \{s_1, s_2, \ldots, s_i\}$; let $S_0 = \emptyset$. Let G_i be the graph at the start of the iteration in which vertex s_i is added to S; let $G_{k+1} = G$. G_i is a minor of G, obtained by deletions of vertices and edge and contractions of edges. G_k is a minor of G and, for $i < k$, G_i is a minor of a graph obtained from G_{i+1} by contracting an edge incident to s_i.

For $i = 2, \ldots, k$:

- $G_i[S_i]$ is a maximal simple subgraph of $G_{i+1}[S_i]$; therefore $\text{tw}(G_{i+1}[S_i]) = \max\{1, \text{tw}(G_i[S_i])\}$ by Fact 2.
- $G_i[S_i]$ is obtained from $G_i[S_{i-1}]$ by the subdivision of an edge (by s_i) or the addition of a leaf vertex (s_i); therefore $\text{tw}(G_i[S_i]) = \max\{1, \text{tw}(G_i[S_{i-1}])\}$ by Facts 1 and 3.

By induction and the fact that a graph with a single vertex has treewidth 0, the lemma follows.   □

Lemma 2. $|S| \geq n - \frac{m}{5}$.

Proof. We show, equivalently, that the procedure deletes at most $\frac{m}{5}$ vertices by amortized analysis. For each vertex that we delete we incur a charge of -5. For each edge that we contract (incident to a degree-1 or -2 vertex), remove (as a self-loop or parallel edge) or delete (by way of deleting an adjacent vertex) we incur a charge of $+1$. Note that we distinguish between deleting and removing an edge for the purpose of this analysis. We will show that the net charge is positive, thus showing that for every 5 edges of the graph, we delete at most one vertex.

The first case of the algorithm, in which an edge is contracted, incurs only a positive charge. There are three remaining cases in which a vertex is deleted, according to the degree of the deleted vertex and whether it is adjacent to a degree 3 vertex.

Deleting a degree-3 vertex from a 3-regular graph. Deleting such a vertex incurs a charge of $+3 - 5 = -2$ but creates three degree-2 vertices. At least one edge incident to each of these will be contracted (or removed) for a total charge of $+3$. Therefore, before another vertex is deleted, the net charge for deleting this vertex is at least $+1$.

Deleting a degree-4 vertex adjacent to a degree-3 vertex. Deleting such a vertex v incurs a charge of $+4 - 5 = -1$ but creates at least one vertex u of degree 2; an edge incident to u will be contracted before another vertex is deleted. The net charge for deleting v is at least 0.

Deleting a degree-4 vertex from a 4-regular graph. Deleting such a vertex incurs a charge of $+4 - 5 = -1$. After this case happens, the remaining graph will have at least four degree-3 vertices, and will continue to have a nonzero number of both degree-3 and degree-4 vertices until either the graph becomes 4-regular again (by reducing the degree-3 vertices to degree-2 and contracting them away) or all degree-4 vertices have been removed. We consider the following sub-cases for the steps of the algorithm that either return to a 4-regular state or eliminate all degree-4 vertices:

(a) If the graph becomes 4-regular again, it can only be after removing a degree-4 vertex adjacent to four degree-3 vertices, followed by four contractions of the resulting degree-2 vertices. This deletion and contractions give a charge of $+3$.

(b) If the last degree-4 vertex is removed, and has no degree-four neighbors when it is removed, then again its removal causes its neighbors to have degree 2 and is followed by four contractions, for a charge of $+3$.

(c) If, in the last removal of a degree-4 vertex, the vertex has one or more degree-4 neighbors, then after this removal (and any ensuing edge contractions) the graph becomes 3-regular. The next deletion will incur a charge of $+1$.

Thus, in all cases, the negative charge for the removal of a degree-4 vertex from a 4-regular graph is balanced by a positive charge for a subsequent step of the algorithm.   □

The bound $n - m/5$ is tight, for arbitrarily large graphs. In particular, the graphs formed from $n/5$ disjoint copies of the complete graph K_5 can have at most $n - m/5$ vertices in any induced subgraph of treewidth at most 2, for otherwise one of the copies of K_5 would have only one of its vertices removed in the subgraph, leaving a K_4 subgraph which does not have treewidth 2.

The proof of Theorem 3 is similar in outline to this proof, but with a larger set of cases and a more complex system of charges. We defer the details to the full version of the paper [2].

5 No Very Large, Minor-Free Induced Subgraphs

In this section we prove Theorem 4. To prove this theorem, we begin with the well-known result that K_h-minor free graphs are sparse [13,14,17,18].

Lemma 3 (Theorem 1.1 [18]). *Every simple K_h-minor-free graph with n vertices has $O(nh\sqrt{\log h})$ edges.*

We will use this result to force the presence of a K_h minor even after deleting many vertices. Lemma 4 allows us to densify a graph in terms of its girth (allowing us to use Lemma 3 to argue the existence of a minor). We give a tighter bound on the number of edges in a K_h-minor free graph with girth g in Corollary 1. The proof of Theorem 4 may then be concluded by finding a family of graphs that have sufficiently large girth.

Lemma 4. *Let G be a graph with n vertices, m edges, and sufficiently large girth g. Then it has a minor G' that is a simple graph with $n' \leq \frac{5n}{g}$ vertices and $m - n + n'$ edges.*

Proof. Let T be an arbitrary rooted spanning tree of G, let r be the root of T, and let V_i be the set of vertices at i^{th} level of T. Let $\ell = \lfloor \frac{g-3}{4} \rfloor$. We choose an integer a such that

$$\mathcal{S} = r \cup \left\{ \bigcup_{k \geq 0} V_{a+k\ell} \right\} \tag{1}$$

contains at most $\frac{n}{\ell} \leq \frac{5n}{g}$ vertices. Set $\mathcal{S}$ is a collection of vertices at every ℓ^{th} level starting from level a along with root r.

Now we perform the following operation to obtain a minor G' of G: for every vertex $v \in G \setminus \mathcal{S}$ contract the edge uv where u is the parent of v in T. That is, for every $v \in V_i$, where $i \neq a + k\ell$, we contract v to its ancestor in V_{i-1}. Since the distance between two consecutive levels of vertices in $\mathcal{S}$ is ℓ and the girth of G is g, contracting these edges cannot result in self-loops or parallel edges. Therefore G' is simple.

Since we contract $n - |\mathcal{S}| = n - n'$ edges, the number of edges in G' is $m - (n - n') = m - n + n'$. □

Consider a graph G with n vertices, maximum degree 3, and girth g. If G has $n + \omega(\frac{n}{g} h \sqrt{\log h})$ edges, then, by Lemma 4, G has a minor G' with $O(\frac{n}{g})$ vertices and $\omega(\frac{n}{g} h \sqrt{\log h})$ edges. By Lemma 3, G' is dense enough to have a K_h minor. Therefore, we get:

Corollary 1. *Every simple K_h-minor-free graph G with n vertices, maximum degree 3, and girth g has $n + O(\frac{n}{g} h \sqrt{\log h})$ edges.*

Proof of Theorem 4. Let $G = (V, E)$ be a 3-regular graph with n vertices, $m = \frac{3n}{2}$ edges and girth $\Omega(\log n)$; for example, the Ramanujan graphs have this property [15]. In the following, we take h to be a constant. By Corollary 1, G has a K_h minor. Any subgraph G^* (with m^* edges and n^* vertices) of G also has girth $\Omega(\log n)$. By deleting k vertices, the best we can hope for is that we delete $3k$ edges. That is, $m^* \geq m - 3k$. To ensure that G^* does not have a K_h minor, we need

$$\frac{3n}{2} - 3k = m - 3k \leq m^* \leq n^* + O(n^*/g) = n - k + O\left(\frac{n-k}{\log n}\right)$$

Solving for k, we require that

$$k \geq \left(\frac{1}{4} - O(1/\log n)\right) n.$$

Substituting $2m/3$ for n gives the theorem. □

Acknowledgments. This material is based upon work supported by the National Science Foundation under Grant Nos. CCF-1252833 and CCF-1228639 and by the Office of Naval Research under Grant No. N00014-08-1-1015. The authors thank Amir Nayyeri for helpful discussions.

References

1. Alon, N., Mubayi, D., Thomas, R.: Large induced forests in sparse graphs. J. Graph Theory 113, 113–123 (2001) MR 1859785
2. Borradaile, G., Eppstein, D., Zhu, P.: Planar induced subgraphs of sparse graphs (2014), arXiv:1408.5939
3. Borradaile, G., Klein, P., Marx, D., Mathieu, C.: Algorithms for Optimization Problems in Planar Graphs (Dagstuhl Seminar 13421). Dagstuhl Reports 3(10), 36–57 (2014)
4. Călinescu, G., Fernandes, C.G., Finkler, U., Karloff, H.: A better approximation algorithm for finding planar subgraphs. J. Algorithms 27(2), 269–302 (1998) MR 1622397
5. Cheng, C., McDermid, E., Suzuki, I.: Planarization and acyclic colorings of subcubic claw-free graphs. In: Kolman, P., Kratochvíl, J. (eds.) WG 2011. LNCS, vol. 6986, pp. 107–118. Springer, Heidelberg (2011) MR 2914703
6. Di Battista, G., Eades, P., Tamassia, R., Tollis, I.G.: Graph Drawing: Algorithms for the Visualization of Graphs, 1st edn. Prentice-Hall (1998)
7. Edwards, K., Farr, G.: An algorithm for finding large induced planar subgraphs. In: Mutzel, P., Jünger, M., Leipert, S. (eds.) GD 2001. LNCS, vol. 2265, pp. 75–80. Springer, Heidelberg (2002) MR 2410446
8. Edwards, K., Farr, G.: Planarization and fragmentability of some classes of graphs. Discrete Math. 308(12), 2396–2406 (2008) MR 2410446
9. Edwards, K., Farr, G.: Improved upper bounds for planarization and series-parallelization of degree-bounded graphs. Electron. J. Combin. 19(2), P25 (2012) MR 2928640
10. Halldórsson, M.M., Lau, H.C.: Low-degree graph partitioning via local search with applications to constraint satisfaction, max cut, and coloring. J. Graph Algorithms Appl. 1(3), 1–13 (1997) MR 1600712
11. Kawarabayashi, K.-I.: Planarity allowing few error vertices in linear time. In: 50th Annual IEEE Symposium on Foundations of Computer Science (FOCS 2009), pp. 639–648 (2009) MR 2648441
12. Kawarabayashi, K.-I., Reed, B.: Computing crossing number in linear time. In: Proceedings of the Thirty-Ninth Annual ACM Symposium on Theory of Computing (STOC 2007), pp. 382–390 (2007) MR 2402463
13. Kostochka, A.V.: The minimum Hadwiger number for graphs with a given mean degree of vertices. Metody Diskret. Analiz 38, 37–58 (1982) MR 0713722
14. Kostochka, A.V.: Lower bound of the Hadwiger number of graphs by their average degree. Combinatorica 4(4), 307–316 (1984) MR 0779891
15. Lubotzky, A., Philips, R., Sarnak, R.: Ramanujan graphs. Combinatorica 8, 261–277 (1988) MR 0963118
16. Morgan, K., Farr, G.: Approximation algorithms for the maximum induced planar and outerplanar subgraph problems. J. Graph Algorithms Appl. 11(1), 165–193 (2007) MR 2354168
17. Thomason, A.: An extremal function for contractions of graphs. Math. Proc. Cambridge Philos. Soc. 95(2), 261–265 (1984) MR 0735367
18. Thomason, A.: The extremal function for complete minors. J. Combinatorial Theory, Series B 81(2), 318–338 (2001) MR 1814910

Picking Planar Edges; or, Drawing a Graph with a Planar Subgraph

Marcus Schaefer

School of Computing, DePaul University, Chicago, IL 60604, USA
`mschaefer@cdm.depaul.edu`

Abstract. Given a graph G and a subset $F \subseteq E(G)$ of its edges, is there a drawing of G in which all edges of F are free of crossings? We show that this question can be solved in polynomial time using a Hanani-Tutte style approach. If we require the drawing of G to be straight-line, but allow up to one crossing along each edge in F, the problem turns out to be as hard as the existential theory of the real numbers.

1 Introduction

Angelini, Binucci, Da Lozzo, Didimo, Grilli, Montecchiani, Patrignani, and Tollis [1] asked the following problem:

"Given a non-planar graph G and a planar subgraph S of G, decide whether G admits a drawing Γ such that the edges of S are not crossed in Γ, and compute Γ if it exists".

Their paper studies two variants of this problem: the unrestricted problem in which Γ is an arbitrary poly-line drawing, and the straight-line variant, in which Γ is restricted to straight-line drawings. Let us call these the *partial planarity* and the *geometric partial planarity* problem. It seems that these two problems are new to the literature. The closest previous variant may be the (also very recent) notion of partially embedded planarity [2], which differs in that a particular embedding of S is given, and the desired planar embedding of G has to extend the given embedding of S. For partially embedded planarity, a linear-time testing algorithm is known [2], as well as an obstruction set [13].

David Eppstein commented on the paper by Angelini et al. [1] in his blog [9]:

"If you're given a graph in which some edges are allowed to participate in crossings while others must remain uncrossed, how can you draw it, respecting these constraints? Unfortunately the authors were unable to determine the computational complexity of this problem, and leave it as an interesting open problem".

In other words, given a graph G and a subset of its edges $F \subseteq E(G)$, is there a (straight-line) drawing of G in which all edges of F are free of crossings? The subgraph and subset formulations are equivalent, of course, but we slightly prefer the second, since it emphasizes that we can specify for each edge whether it

C. Duncan and A. Symvonis (Eds.): GD 2014, LNCS 8871, pp. 13–24, 2014.

has to be planar (crossing-free) or not: we can pick the planar edges. Looking at planarity as a local requirement opens it up for combination with other properties; for example, what happens if we can specify a bound on the number of crossings along each edge, or on the number of bends?

Previous Research

Angelini, et al. [1] show that (G, S) is always partially planar if S is a spanning tree of G, even if the embedding of S is required to be a straight-line embedding. For geometric partial planarity, they show that (G, S) can always be realized if S is a spanning spider or caterpillar, even in polynomial area. However, they also exhibit examples of (G, S) where S is a spanning tree of G for which (G, S) has no geometric partial realization. There are further algorithms in the paper to test geometric partial planarity for various types of spanning trees S, though in some cases the layout algorithms require exponential area.

Our Contribution

In Section 2 we show that using a Hanani-Tutte style approach successfully settles the complexity of the poly-line variant of the problem: partial planarity can be solved in polynomial time. This is a further example of a planarity-style problem for which there is (as yet) no traditional polynomial-time algorithm for the problem, but the Hanani-Tutte approach leads to a solution. Other examples of this are surveyed in [21].

We have to leave the complexity of the straight-line variant open, but there is a good chance that it is as hard as the existential theory of the reals (see [20]). One indication for this is that the layout algorithm for geometric partial planarity suggested in [1] needs exponential area on some inputs. Secondly, the result is true if we replace planarity with 1-planarity: testing partial geometric 1-planarity is as hard as the existential theory of the reals, as we will see in Section 3. In comparison, the special case of geometric 1-planarity is **NP**-complete (this follows from known results in the literature, see Theorem 2).

2   Partial Planarity and Hanani-Tutte

We assume that the reader is somewhat familiar with the Hanani-Tutte characterization of planarity (see [21,22]). Briefly, Hanani [6] and Tutte [27] established the following algebraic characterization of planar graphs: a graph is planar if and only if it has a drawing in which every two independent edges cross evenly. This criterion can be rephrased as a linear system over GF(2): Create variables $x_{e,v}$ for every $e \in E(G)$ and $v \in V(G)$, and let $i_D(e, f)$ denote the number of times two edges e and f cross in a drawing D of G. Fix an arbitrary drawing D of G (e.g. a convex drawing). Let $P(D)$ be the system of linear equations over GF(2) containing:

$$i_D(uv, st) + x_{uv,s} + x_{uv,t} + x_{st,u} + x_{st,v} \equiv 0 \bmod 2,$$

for every pair of independent edges $uv, st \in E(G)$. Then G is planar if and only if $P(D)$ is solvable. The heart of the proof is showing that solvability of $P(D)$ leads to a planar drawing of G; we will not explain this part (see [21, Section 3] for a detailed discussion). The other direction is a consequence of the following well-known fact about drawings: as far as the crossing parity between pairs of independent edges is concerned, one can turn any drawing of a graph into any other drawing of the graph by performing a set of (e, v)-moves, where an (e, v)-*move* consists of taking a small piece of e, moving it close to v and then pushing it over v; the effect of an (e, v)-move is that the crossing parity between e and any edge incident to v changes. Imagining one drawing of a graph morphing into another, it is easy to believe that (e, v)-moves are sufficient to get from one drawing to another. We state this result without proof. For further details see [7, Section 4.6] or [22, Lemma 1.12].

Lemma 1. *If D and D' are two drawings of the same graph G, then there is a set of (e, v)-moves so that*

$$i_{D'}(uv, st) \equiv i_D(uv, st) + x_{uv,s} + x_{uv,t} + x_{st,u} + x_{st,v} \bmod 2,$$

for all edges $uv, st \in E(G)$, where $x_{e,v} = 1$ if an (e, v)-move is performed, and $x_{e,v} = 0$ otherwise.

For a graph G with a set of edges $F \subseteq E(G)$, fix an arbitrary drawing D of G, and let $P(D, F)$ be the following system of equations over GF(2):

$$i_D(uv, st) + x_{uv,s} + x_{uv,t} + x_{st,u} + x_{st,v} \equiv 0 \bmod 2,$$

for every pair of independent edges $uv \in F$ and $st \in E(G)$.

Lemma 2. *G has a drawing Γ in which F is free of crossings if and only if $P(D, F)$ is solvable.*

Since the solvability of a linear system of equations over a field (in this case GF(2)) can be decided in polynomial time, the following corollary is immediate.

Corollary 1. *Given a graph G with a set of edges $F \subseteq E(G)$, it can be decided in polynomial time whether G has a drawing in which all edges in F are free of crossings. In such a drawing we can assume that edges in F are straight-line, and each edge in $E(G) - F$ has at most $|E(G) - F| - 1$ bends.*

The running time of the algorithm is on the order $O((nm)^3)$, where $n = |V(G)|$ and $m = |E(G)|$, since systems of linear equations over a field can be solved in cubic time, and $P(D, F)$ can have as many as $O(nm)$ equations and $O(nm)$ variables (note that we can assume that $|F| = O(n)$: if the graph $(V(G), F)$ is not planar, then there is no drawing of G in which all edges of F are free of crossings; on the other hand, we cannot assume that G is planar). This may seem impractical at a first glance, but recent experiments with an algorithm of this type have been quite successful [11].

The hard direction in the proof of Lemma 2 is covered by the following result from an earlier paper on the independent odd crossing number [19]. We call an edge e in a drawing D *independently even* if it crosses every edge independent of it an even number of times. More formally, $i_D(e, f) \equiv 0 \bmod 2$ for every f which is independent of e.

Lemma 3 (Pelsmajer, Schaefer, Štefankovič [19]). *If D is a drawing of a graph G in the plane, then G has a drawing in which the independently even edges of D are crossing-free and every pair of edges crosses at most once.*

The proof of Lemma 3 is constructive in the sense that the new drawing of G can be found in polynomial time (there are no explicit time bounds in [19], but a running time quadratic in $O(|G|)$ seems achievable).

Proof (of Lemma 2). Suppose $P(D, F)$ is solvable, and fix a solution $x_{e,v} \in \{0, 1\}$, for $e \in E(G), v \in V(G)$, for some initial drawing D of G. Construct a drawing D' from D by performing an (e, v)-move for every $e \in E(G)$ and $v \in V(G)$ for which $x_{e,v} = 1$. Pick $uv \in F$ and let $st \in E(G)$ be an arbitrary edge independent of uv. Then

$$i_{D'}(uv, st) = i_D(uv, st) + x_{uv,s} + x_{uv,t} + x_{st,u} + x_{st,v} \equiv 0 \bmod 2,$$

since $x_{e,v}$ is a solution of the system $P(D, F)$. Thus uv is independently even. Since uv was arbitrary, all edges in F are independently even, and, by Lemma 3, there is a drawing of G in which all edges of F are free of crossings, and every pair of edges in $E(G) - F$ crosses at most once. Temporarily replace each crossing with a vertex, and take a planar straight-line embedding of the resulting graph. In that drawing, all edges of F are straight-line, and (after turning crossings into bends and perturbing them slightly), all remaining edges have at most $|E(G) - F| - 1$ bends.

For the other direction, assume G has a drawing D' in which all edges of F are free of crossings. By Lemma 1 we know that there is a set of (e, v)-moves so that

$$i_{D'}(uv, st) \equiv i_D(uv, st) + x_{uv,s} + x_{uv,t} + x_{st,u} + x_{st,v} \bmod 2$$

for all pairs of independent edges uv, $st \in E(G)$. Now if $uv \in F$, then $i_{D'}(uv, st) = 0$ for every edge $st \in E(G)$. In particular,

$$i_D(uv, st) + x_{uv,s} + x_{uv,t} + x_{st,u} + x_{st,v} \equiv i_{D'}(uv, st) \equiv 0 \bmod 2,$$

so $x_{e,v}$ is a solution to $P(D, F)$, which is what we had to show. □

3   Geometric Partial 1-Planarity

In the straight-line version of the partial planarity problem, we ask whether for a given G and $F \subseteq E(G)$, there is a straight-line drawing of G in which the edges of F are free of crossings. We cannot settle the complexity of this problem, but we have a suggestive result for a generalized version. Suppose we are allowed to specify sets $F_k \subseteq E(G)$, and ask whether G has a straight-line drawing in which all edges in F_k have at most k crossings, for every k. The problem posed by Angelini, Binucci, Da Lozzo, Didimo, Grilli, Montecchiani, Patrignani, and Tollis [1] corresponds to specifying a set F_0 of crossing-free edges. We will show that if instead we specify a set F_1 of edges that may be crossed at most once, the problem has the same complexity as deciding the truth of statements in the existential theory of the reals; in the terminology introduced in [24,20], it is $\exists\mathbb{R}$-complete. In analogy with the notion of 1-planarity (in which every edge may be crossed at most once), we call the problem *geometric partial 1-planarity*.

Remark 1 (Equivalent Drawings). Geometric 1-planarity was first studied by Eggleton [8] and Thomassen [26], and more recently in [12], and several other papers, but with one important difference: in these papers one is given an intial 1-planar drawing of G and asks whether there is an equivalent geometric 1-planar drawing, where two drawings are *equivalent* if they have the same facial structure (for this definition to make sense, we consider crossings to be vertices). With this stronger notion, Thomassen [26] was able to identify forbidden subconfigurations, which led to a linear-time testing algorithm [12]. Similarly, Nagamochi [18] shows that if we are given a drawing of G and a 2-connected, spanning subgraph S of G, one can test in linear time whether there is an equivalent drawing of G in which edges of S are free of crossings.   □

We will not give a formal definition of $\exists\mathbb{R}$ and $\exists\mathbb{R}$-completeness (that can be found in [24,20]), instead we will work with STRETCHABILITY, a complete problem for the class. This is just like working with SAT, the Boolean satisfiability problem, (or any other **NP**-complete problem) rather than the formal class **NP**.

An *arrangement of pseudolines* in the plane is a collection of x-monotone curves (that is, each pseudoline has exactly one crossing with every vertical line) so that every pair of pseudolines crosses exactly once. An arrangement of pseudolines is *stretchable* if all pseudolines can be replaced by straight lines so that the order of crossings along the lines remains the same. See Figure 1 for an example of a pseudoline arrangement, and an equivalent straight-line arrangement.

Mnëv [17] showed that STRETCHABILITY, the problem of deciding whether an arrangement of pseudolines is stretchable, is computationally equivalent to deciding the truth of a sentence in the existential theory of the real numbers (for an accessible treatment of Mnëv's proof, see Shor [25]).[1] This led to the introduction of the complexity class $\exists\mathbb{R}$, which contains all problems which can be translated in polynomial time to a sentence in the existential theory of the reals, see [24,20] for more details. Similar to the theory of **NP**-completeness, there are $\exists\mathbb{R}$-complete problems including stretchability, and truth in the existential theory of the reals, but many other problems as well, such as the rectilinear crossing

---

[1] Mnëv actually showed a stronger result, his universality theorem, here we are only interested in the computational aspects.

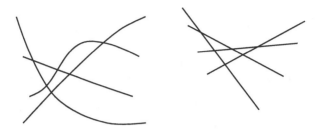

Fig. 1. *(Left)* A pseudoline arrangement. *(Right)* A straight-line arrangement equivalent to the pseudoline arrangement on the left.

number (there is a wikipedia page, for example [28]). We note that $\exists\mathbb{R}$ contains **NP**, since the existential theory of the real numbers easily encodes satisfiability, and in turn $\exists\mathbb{R}$ is contained in **PSPACE**, due to a famous result by Canny [5]. Therefore any $\exists\mathbb{R}$-complete problem, such as partial geometric 1-planarity is **NP**-hard, and can be solved in polynomial space.

Theorem 1. *Partial Geometric 1-Planarity is $\exists\mathbb{R}$-complete.*

In particular, we conclude that the problem is **NP**-hard, and lies in **PSPACE**. For the proof we make use of a simple gadget.

Lemma 4. *There is no drawing of a K_6 and a vertex-disjoint cycle C so that all edges in the K_6 have at most one crossing, and there is a crossing between an edge of K_6 and the cycle.*

Proof. Suppose there were a drawing as described in the lemma, in which a K_6-edge $e = uv$ crosses a cycle edge $f \in E(C)$. Then e cannot cross any of the edges in $E(C) - \{f\}$, since it has at most one crossing, and thus no edge incident to u can cross an edge incident to v: to have a common point, one of them would have to cross C, but then it would have two crossings, one with the cycle, and one with the other edge. Therefore, the edges adjacent to e do not cross each other at all. This implies that the drawing of the K_6 contains 4 triangles with a shared edge e whose other edges do not cross each other. On the sphere, there is only one such drawing: 4 nested triangles (with a common base). But this implies that two of the endpoints of those triangles are separated by the other two triangles, which means the original endpoints cannot be joined by an edge in a 1-planar drawing of the K_6, since it would have to cross the other two triangles (it cannot cross e, since e already has a crossing). □

Proof (of Theorem 1). The problem can easily be expressed using an existentially quantified statement over the real numbers: use the existential quantifiers to find the locations of the vertices of the graph; once the vertices are located, it is easy to express that each edge in F is crossed at most once. This shows that the problem lies in $\exists\mathbb{R}$.

Since stretchability of pseudoline arrangements is $\exists\mathbb{R}$-complete, it is sufficient to show that stretchability can be reduced to partial geometric 1-planarity to establish $\exists\mathbb{R}$-hardness of partial geometric 1-planarity. Let $\mathcal{A}$ be an arbitrary arrangement of pseudolines. We construct a graph $G_{\mathcal{A}}$ and a set of edges $F \subseteq G_{\mathcal{A}}$ so that $\mathcal{A}$ is stretchable if and only if $G_{\mathcal{A}}$ has a straight-line drawing in which every edge in F has at most one crossing.

Let R be a parabola-shaped region (boundary of the form $y = x^2 + c$ for some constant $c \in \mathbb{R}$) so that all crossings of pseudolines in $\mathcal{A}$ lie within the region R. Let $V_{\mathcal{A}}$ be the intersection points of pseudolines with the parabolic boundary of R (we can assume that all crossings of pseudolines lie in the convex hull of V_1). The region R is separated by $\mathcal{A}$ into faces, some of them adjacent to the boundary of R, and some of them inner faces of the arrangement. We choose a set of vertices V_I consisting of an interior vertex for each inner face of the arrangement; for each face on the boundary of R, we pick a vertex on the interior of a boundary arc of the face, let V_B of those boundary vertices; note that all faces except the infinite face, are incident to a unique boundary arc; the infinite face is incident to two boundary arcs, of which we pick one arbitrarily to place the V_B-vertex. Finally, pick a vertex p below R so that p can *see* all vertices of $V_{\mathcal{A}} \cup V_B$; that is, a straight-line segment between p and any vertex in $V_{\mathcal{A}} \cup V_B$ does not cross the boundary of R. Let $V = V_{\mathcal{A}} \cup V_R \cup V_I \cup \{p\}$.

For every two vertices in $V_{\mathcal{A}}$ belonging to the same pseudoline, add an edge between those vertices. Add a *frame* as follows: connect the vertices of $V_{\mathcal{A}} \cup V_B$ by a cycle that respects the order of those vertices along the boundary of R, and connect p to every vertex in $V_{\mathcal{A}} \cup V_B$ by an edge. Identify each edge of the frame with an edge in a (new) K_6. Finally, add the dual graph of the line arrangement to $V_I \cup V_B$. Let F consist of all edges, except for the edges corresponding to the original pseudolines. See Figure 2 for an example.

We first note that if $\mathcal{A}$ is stretchable, then $G_{\mathcal{A}}$ has a straight-line drawing in which every edge of F has at most one crossing. To see this, start with a straight-line realization of $\mathcal{A}$. Perform the construction of $G_{\mathcal{A}}$ as we described it above. Because of the convexity of R, we can draw the edges of the cycle on $V_{\mathcal{A}} \cup V_B$ as well as the straight-line edges to p. We can then add a straight-line drawing of each K_6 gadget to the frame so that the shared edge is free of crossings (and the remainder of K_6 does not participate in unnecessary crossings). Finally, the dual graph of the line arrangement can be added to $V_I \cup V_B$ since any edge connects two vertices in adjacent faces of the line arrangement which is always possible with a straight-line arrangement, unless the resulting edge coincides with the boundary of a cell. This cannot occur, however, since V_I vertices lie in the interior of faces, and the V_B vertices lie on the boundary of the convex region R. In this drawing, every edge in F has at most one crossing. Only edges corresponding to the original pseudolines are crossed more than once by dual edges.

For the other direction, start with a straight-line drawing of $G_{\mathcal{A}}$ in which all edges in F have at most one crossing. Suppose f is an edge of the frame and let e be another edge in $G_{\mathcal{A}}$ which does not belong to f's K_6 gadget. If e and

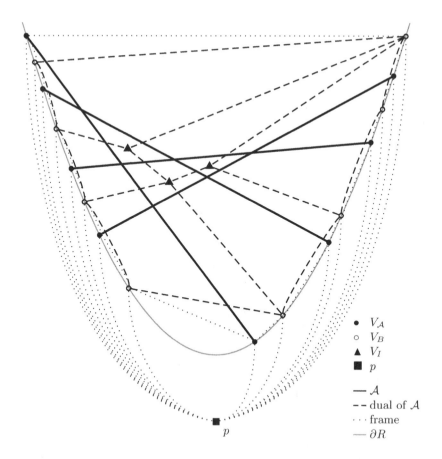

Fig. 2. The graph $G_{\mathcal{A}}$ corresponding to the pseudoline arrangement $\mathcal{A}$ shown in Figure 1. K_6-gadgets are not shown, and some edges are curved to improve readability.

f are adjacent, they cannot cross, since the drawing is straight-line. Hence e either belongs to another K_6-gadget or is one of the edges between vertices in $V_{\mathcal{A}} \cup V_B \cup V_I$. In either case, e belongs to a cycle which is vertex-disjoint from f's K_6-gadget, so Lemma 4 implies that e does not cross f. This means that after removal of all the K_6-gadgets, the frame is free of crossings. In particular, the cycle C on $V_{\mathcal{A}} \cup V_B$ is crossing-free, and hence its vertices occur in the order determined by the line arrangement $\mathcal{A}$. Let $\mathcal{A}'$ be the line arrangement obtained from $G_{\mathcal{A}}$ by erasing the frame (and its gadgets), the dual graph, and extending the edges corresponding to pseudolines to infinite lines. We claim that $\mathcal{A}'$ is equivalent to $\mathcal{A}$.

We just saw that the order of pseudolines along C is correct, and, since the frame does not cross edges corresponding to pseudolines, every two such edges have to cross inside the region bounded by C (since their endpoints along C alternate in $\mathcal{A}'$ just as they do in $\mathcal{A}$ (recall that every pair of pseudolines crosses once). We now show that the dual graph of $\mathcal{A}$ forces the facial structure of the line arrangement to be unique.

Let $v \in V_I \cup V_B$ be an arbitrary vertex representing a face of the line arrangement, and e an edge corresponding to some line in $\mathcal{A}$. We show that v lies on the same side of e (within the region bounded by the cycle C through $V_{\mathcal{A}} \cup V_B$) in both $\mathcal{A}$ and $\mathcal{A}'$, so the two line arrangements have to be equivalent. If $v \in V_B$ this is forced by the cycle C; if $v \in V_I$, we argue as follows: let s and t be the V_B-vertices closest to e (along C) and on the same side of e as v. We claim that there is an st-path of length $|\mathcal{A}| - 1$ on V_I vertices that passes through v. Clearly, any such path must have length at least $|\mathcal{A}| - 1$, so we only need to argue that a path of this length exists. To see this, start at s. Since v is an inner vertex, s and v do not lie in the same face of the line arrangement, hence there must be an edge f corresponding to a line of $\mathcal{A}$ so that s and v lie on opposite sides of f. More strongly, there must be such an edge f which contributes to the boundary of the cell v lies in (if the two vertices were on the same side of all lines contributing to the boundary of the cell, they would have to be in the same cell); in other words, there is a cell adjacent to the cell containing v (sharing f), which is closer to s (note that t and v have to lie on the same side of f, since otherwise s and t lie on the same side of both e and f, but then they cannot both be closest to e). By induction we can now show that there are paths sv and vt containing at most $|\mathcal{A}| - 1$ edges together (since e need never be crossed). But then the path svt in $G_{\mathcal{A}}$ on $|\mathcal{A}| - 1$ edges cannot cross e, since it has to cross all $|\mathcal{A}| - 1$ edges corresponding to pseudolines (other than e). Hence v lies on the same side of e in both $\mathcal{A}$ and $\mathcal{A}'$.

Since $\mathcal{A}'$ is a straight-line arrangement equivalent to $\mathcal{A}$, we conclude that $\mathcal{A}$ is stretchable, which is what we had to show.   □

In contrast, geometric 1-planarity is only **NP**-complete. This follows from two well-known results: 1-planarity is **NP**-complete [10,14,4], and geometric 1-planarity can be tested in linear time if a rotation system is given [26,12].

Theorem 2 (Folklore). *Testing geometric 1-planarity is **NP**-complete.*

Proof. The problem lies in **NP**, since we can guess the rotation system, and then use the linear time algorithm from [12] to check whether there is an obstruction to geometric 1-planarity with that rotation system. To see **NP**-hardness, we use the **NP**-hardness of testing 1-planarity. If a graph G is 1-planar, then it has a 1-planar drawing in which each edge has at most one bend: simply apply Fary's theorem to the graph obtained from G by replacing each crossing by a dummy vertex. To avoid that crossings and bends occur at the same location, we replace each edge in G with a path of length three to get a new graph G'. Then G is 1-planar if and only if G' has a geometric 1-planar embedding in which all edges incident to the original vertices of G are free of crossings. And that we can easily guarantee by identifying all of these edges with an edge of a K_6-gadget. Let H be this new graph. Then G is 1-planar if and only if H is geometrically 1-planar. Therefore, geometric 1-planarity is **NP**-hard.   □

4   Future Research

What can we say about traditional approaches to partial planarity? More specifically, can PQ-trees or $SPQR$-trees be used to solve this problem?

Recall that a *bridge* of S in G is either an edge in $E(G) - E(S)$ with both endpoints in S (a *trivial* bridge) or a connected component of $G - S$ together with its edges and vertices of attachment to S. Given an embedding of S, a group of vertices of S is *mutually visible* [2] if there is a face of S containing all vertices in the group on its boundary. The poly-line variant can be rephrased as follows: is there a poly-line embedding of S so that for every bridge of S in G, the vertices of attachment of the bridge are mutually visible? It seems quite likely that SPQR-trees could be used to decide that question, even in linear time, extending ideas for deciding partially embedded planarity developed in [2].

Another solution may come from progress on simultaneous embeddings, since partial planarity can be viewed as a special case of simultaneous planarity.[2] Two graphs G_1, G_2 are *simultaneously planar* if there is a drawing of $G_1 \cup G_2$ in which the induced drawings of G_1 and G_2 are (each by itself) planar. Given a graph G and S, we add edges in S to both G_1 and G_2. All other edges $E(G) - S$ we subdivide $2|E(G)|$ times, and assign the pieces along each subdivided edge of G to G_1 and G_2 alternatingly. If G has a drawing in which all edges in S are crossing-free, we can turn this into a simultaneous drawing of G_1 and G_2, since we can assume that any two edges in $E(G) - S$ cross at most once, so every edge has less than $|E(G)|$ crossings which we can now realize by matching up G_1 and G_2 pieces of the subdivided edges.

Weak Realizability

Before we leave partial planarity, we want to draw one more connection, to the *weak realizability* problem introduced by Kratochvíl [15,16]: Given a graph G and a symmetric relation R on $E(G)$, we can ask whether the *abstract topological graph* (G, R) is *weakly realizable*, that is, if there is a drawing in which only pairs of edges $(e, f) \in R$ are allowed to cross (but do not have to cross). The general problem is **NP**-complete [15,23], so one could ask whether there are special cases which are solvable. Let us shift the focus by viewing R itself as the edge set of a graph on the vertex set $E(G)$.

From this point of view, $R = \emptyset$ corresponds to the planarity problem for G, which can be solved in linear time. On the other hand, letting R be the complete graph on $E(G)$ leads to a trivial problem. What happens if we let R be a complete graph on a subset $E' \subseteq E(G)$ of all edges of G? It turns out that this captures partial planarity: (G, R) is weakly realizable, if and only if (G, E') is partially planar.

This could be the starting point of an attack on weak realizability using structural properties of R, an approach from the intersection-graph point of view. We quickly get into uncharted waters: If R is a complete bipartite graph,

---

[2] This is based on a suggestion by Ignaz Rutter.

then weak realizability of (G, R) expresses a simultaneous planarity problem for two graphs: if R is a complete bipartite graph on $E_1(G)$ and $E_2(G)$, and $E_0(G)$ contains the remaining (isolated) vertices of $E(G)$, then (G, R) is weakly realizable, if and only if G_1 with edge set $E_0(G) \cup E_1(G)$ and G_2 with edge set $E_0(G) \cup E_2(G)$ have a simultaneous embedding with fixed edges. The complexity of this problem is famously open, and related to several other open problems in graph drawing (e.g. c-planarity [21]). If R is a complete k-partite graphs, then the weak realizability problem corresponds to the sunflower case of the SEFE problem for k graphs, which is **NP**-complete even for $k = 3$ [3,21].

If R consists of two disjoint complete graphs that together partition the vertex set $E(G)$, we get a problem which is the opposite of the SEFE problem: it asks whether we can draw the two graphs G_1 and G_2 simultaneously (so that shared edges are drawn the same way) so that edges belonging to the same graph may cross each other, but edges belonging to different graphs may not. As far as we know, nobody has investigated this version of the problem. Even the case where R is a tree (or even a matching) does not seem immediately obvious.

References

1. Angelini, P., Binucci, C., Da Lozzo, G., Didimo, W., Grilli, L., Montecchiani, F., Patrignani, M., Tollis, I.G.: Drawings of non-planar graphs with crossing-free subgraphs. In: Wismath, S., Wolff, A. (eds.) GD 2013. LNCS, vol. 8242, pp. 292–303. Springer, Heidelberg (2013)
2. Angelini, P., Di Battista, G., Frati, F., Jelínek, V., Kratochvíl, J., Patrignani, M., Rutter, I.: Testing planarity of partially embedded graphs. In: Proceedings of the Twenty-First Annual ACM-SIAM Symposium on Discrete Algorithms, SODA 2010, pp. 202–221. Society for Industrial and Applied Mathematics, Philadelphia (2010)
3. Angelini, P., Da Lozzo, G., Neuwirth, D.: On some $\mathcal{NP}$-complete SEFE problems. In: Pal, S.P., Sadakane, K. (eds.) WALCOM 2014. LNCS, vol. 8344, pp. 200–212. Springer, Heidelberg (2014)
4. Cabello, S., Mohar, B.: Adding one edge to planar graphs makes crossing number and 1-planarity hard. SIAM Journal on Computing 42(5), 1803–1829 (2013)
5. Canny, J.: Some algebraic and geometric computations in pspace. In: STOC 1988: Proceedings of the Twentieth Annual ACM Symposium on Theory of Computing, pp. 460–469. ACM, New York (1988)
6. Chojnacki, C. (Hanani, H.): Über wesentlich unplättbare Kurven im dreidimensionalen Raume. Fundamenta Mathematicae 23, 135–142 (1934)
7. de Longueville, M.: A course in topological combinatorics. Universitext. Springer, New York (2013)
8. Eggleton, R.B.: Rectilinear drawings of graphs. Utilitas Math. 29, 149–172 (1986)
9. Eppstein, D.: Big batch of graph drawing preprints, http://11011110. livejournal.com/275238.html (last accessed September 4, 2013)
10. Grigoriev, A., Bodlaender, H.L.: Algorithms for graphs embeddable with few crossings per edge. Algorithmica 49(1), 1–11 (2007)
11. Gutwenger, C., Mutzel, P., Schaefer, M.: Practical experience with Hanani-Tutte for testing c-planarity. In: McGeoch, C.C., Meyer, U. (eds.) 2014 Proceedings of the Sixteenth Workshop on Algorithm Engineering and Experiments (ALENEX), pp. 86–97. SIAM (2014)

12. Hong, S.-H., Eades, P., Liotta, G., Poon, S.-H.: Fáry's theorem for 1-planar graphs. In: Gudmundsson, J., Mestre, J., Viglas, T. (eds.) COCOON 2012. LNCS, vol. 7434, pp. 335–346. Springer, Heidelberg (2012)
13. Jelínek, V., Kratochvíl, J., Rutter, I.: A Kuratowski-type theorem for planarity of partially embedded graphs. Comput. Geom. 46(4), 466–492 (2013)
14. Korzhik, V.P., Mohar, B.: Minimal obstructions for 1-immersions and hardness of 1-planarity testing. In: Tollis, I.G., Patrignani, M. (eds.) GD 2008. LNCS, vol. 5417, pp. 302–312. Springer, Heidelberg (2009)
15. Kratochvíl, J.: String graphs. I. The number of critical nonstring graphs is infinite. J. Combin. Theory Ser. B 52(1), 53–66 (1991)
16. Kratochvíl, J.: Crossing number of abstract topological graphs. In: Whitesides, S.H. (ed.) GD 1998. LNCS, vol. 1547, pp. 238–245. Springer, Heidelberg (1999)
17. Mäkinen, E.: On circular layouts. International Journal of Computer Mathematics 24(1), 29–37 (1988)
18. Nagamochi, H.: Straight-line drawability of embedded graphs. Technical Report 2013-005, Kyoto University (2013)
19. Pelsmajer, M.J., Schaefer, M., Štefankovič, D.: Removing independently even crossings. SIAM Journal on Discrete Mathematics 24(2), 379–393 (2010)
20. Schaefer, M.: Complexity of some geometric and topological problems. In: Eppstein, D., Gansner, E.R. (eds.) GD 2009. LNCS, vol. 5849, pp. 334–344. Springer, Heidelberg (2010)
21. Schaefer, M.: Toward a theory of planarity: Hanani-Tutte and planarity variants. Journal of Graph Algortihms and Applications 17(4), 367–440 (2013)
22. Schaefer, M.: Hanani-Tutte and related results. In: Bárány, I., Böröczky, K.J., Fejes Tóth, G., Pach, J. (eds.) Geometry—Intuitive, Discrete, and Convex—A Tribute to László Fejes Tóth. Bolyai Society Mathematical Studies, vol. 24. Springer, Berlin (2014)
23. Schaefer, M., Sedgwick, E., Štefankovič, D.: Recognizing string graphs in NP. In: Proceedings of the 33th Annual ACM Symposium on Theory of Computing (STOC 2002) (2002)
24. Schaefer, M., Štefankovič, D.: Fixed points, Nash equilibria, and the existential theory of the reals. Unpublished manuscript (2009)
25. Shor, P.W.: Stretchability of pseudolines is NP-hard. In: Applied Geometry and Discrete Mathematics. DIMACS Ser. Discrete Math. Theoret. Comput. Sci., vol. 4, pp. 531–554. Amer. Math. Soc., Providence (1991)
26. Thomassen, C.: Rectilinear drawings of graphs. J. Graph Theory 12(3), 335–341 (1988)
27. Tutte, W.T.: Toward a theory of crossing numbers. J. Combinatorial Theory 8, 45–53 (1970)
28. Wikipedia. Existential theory of the reals (2012), http://en.wikipedia.org/wiki/Existential_theory_of_the_reals (Online; accessed July 17, 2013)

Drawing Partially Embedded
and Simultaneously Planar Graphs

Timothy M. Chan[1], Fabrizio Frati[2], Carsten Gutwenger[3], Anna Lubiw[1],
Petra Mutzel[3], and Marcus Schaefer[4]

[1] Cheriton School of Computer Science, University of Waterloo, Canada
{tmchan,alubiw}@uwaterloo.ca
[2] School of Information Technologies, The University of Sydney, Australia
fabrizio.frati@sydney.adu.au
[3] Technische Universität Dortmund, Dortmund, Germany
{carsten.gutwenger,petra.mutzel}@tu-dortmund.de
[4] DePaul University, Chicago, Illinois, USA
mschaefer@cdm.depaul.edu

Abstract. We investigate the problem of constructing planar drawings with few
bends for two related problems, the *partially embedded graph* (PEG) problem—
to extend a straight-line planar drawing of a subgraph to a planar drawing of the
whole graph—and the *simultaneous planarity* (SEFE) problem—to find planar
drawings of two graphs that coincide on shared vertices and edges. In both cases
we show that if the required planar drawings exist, then there are planar drawings
with a linear number of bends per edge and, in the case of simultaneous planarity,
a constant number of crossings between every pair of edges. Our proofs provide
efficient algorithms if the combinatorial embedding information about the drawing is given. Our result on partially embedded graph drawing generalizes a classic
result of Pach and Wenger showing that any planar graph can be drawn with fixed
locations for its vertices and with a linear number of bends per edge.

1 Introduction

In many practical applications we wish to draw a planar graph while satisfying some
geometric or topological constraints. One natural situation is that we have a drawing of
part of the graph and wish to extend it to a planar drawing of the whole graph. Pach and
Wenger [20] considered a special case of this problem. They showed that any planar
graph can be drawn with its vertices lying at pre-assigned points in the plane and with
a linear number of bends per edge. In this case the pre-drawn subgraph has no edges.

If the pre-drawn subgraph H has edges, a planar drawing of the whole graph G
extending the given drawing $\mathcal{H}$ of H might not exist. Angelini et al. [1] gave a linear-
time algorithm for the corresponding decision problem; the algorithm returns, for a
positive answer, a planar embedding of G that *extends* that of $\mathcal{H}$ (i.e., if we restrict the
embedding of G to the edges and vertices of H, we obtain the embedding corresponding
to $\mathcal{H}$). If one does not care about maintaining the actual planar drawing of H this is
the end of the story, since standard methods can be used to find a straight-line planar
drawing of G in which the drawing of H is topologically equivalent to the one of $\mathcal{H}$. In

C. Duncan and A. Symvonis (Eds.): GD 2014, LNCS 8871, pp. 25–39, 2014.
© Springer-Verlag Berlin Heidelberg 2014

this paper we show how to draw G while preserving the actual drawing $\mathcal{H}$ of H, so that each edge has a linear number of bends. This bound is worst-case optimal, as proved by Pach and Wenger [20] in the special case in which H has no edges.

A result analogous to ours was claimed by Fowler et al. [10] for the special case in which H has the same vertex set as G. Their algorithm draws the edges of G one by one in a certain order, and they claim a linear number of bends per edge. However, we give an example where their algorithm produces exponentially many bends, confirming a claim of Schaefer [23] that greedy extensions can in general give many bends.

We also address the *simultaneous planarity* problem [4], also known as "simultaneous embedding with fixed edges (SEFE)". The SEFE problem is strongly related to the partially embedded graph problem and—in a sense we will make precise later—generalizes it. We are given two planar graphs G_1 and G_2 that share a *common subgraph* G (i.e., G is composed of those vertices and edges that belong to both G_1 and G_2). We wish to find a *simultaneously planar drawing*, i.e., a planar drawing of G_1 and a planar drawing of G_2 that coincide on G. Graphs G_1 and G_2 are *simultaneously planar* if they admit such a drawing. Both G_1 and G_2 may have *private* edges that are not part of G. In a simultaneous planar drawing the private edges of G_1 may cross the private edges of G_2. The simultaneous planarity problem arises in information visualization when we wish to display two relationships on two overlapping element sets.

The decision version of the simultaneous planarity problem is not known to be **NP**-complete, nor solvable in polynomial time, though it is **NP**-complete if more than two graphs are given [11]. However, there is a combinatorial characterization of simultaneous planarity, based on the concept of a "compatible embedding", due to Jünger and Schulz [16] (see below for details). Erten and Kobourov [8], who first introduced the problem, gave an efficient drawing algorithm for the special case where the two graphs share vertices but no edges. In this case, a simultaneous planar drawing always exists, and they construct a drawing in which each edge has at most three bends and therefore any two edges cross (when they legally can) at most 16 times. In this paper we show that if two graphs have a simultaneous planar drawing, then there is a drawing in which every edge has a linear number of bends and in which any two edges cross at most 24 times. Our result is algorithmic, assuming a compatible embedding is given.

More formally, our paper addresses the following two problems:

- **Planarity of a partially embedded graph (PEG).** Given a planar graph G and a straight-line planar drawing $\mathcal{H}$ of a subgraph H of G, find a planar drawing of G that extends $\mathcal{H}$ (see [1,15]).
- **Simultaneous planarity (SEFE).** Given two planar graphs G_1 and G_2 that share a subgraph G, find planar drawings of G_1 and G_2 that are the same on the shared subgraph (see [4]).

We prove the following results:

Theorem 1. *Let G be an n-vertex planar graph, let H be a subgraph of G, and let $\mathcal{H}$ be a straight-line planar drawing of H. Suppose that G has a planar embedding $\mathcal{E}$ that extends $\mathcal{H}$. Then we can construct a planar drawing of G in $O(n^2)$-time which realizes $\mathcal{E}$, extends $\mathcal{H}$, and has at most $102|V(H)| + 12$ bends per edge.*

Theorem 2. *Let G_1 and G_2 be simultaneously planar graphs on a total of n vertices with a shared subgraph G. Then there is a simultaneous planar drawing in which any edge of $G_1 - G$ and any edge of $G_2 - G$ intersect at most 24 times, and one of the following properties holds:*

1. *each edge of G is straight, and each private edge of G_1 and of G_2 has at most $72n$ bends; also, vertices, bends, and crossings lie on an $O(n^2) \times O(n^2)$ grid; or*
2. *each edge of G_1 is straight and each private edge of G_2 has at most $102|V(H)|+12$ bends per edge.*

If we are given a compatible embedding of the two graphs, we can construct such drawings in $O(n^2)$ time.

Theorem 1 generalizes Pach and Wenger's result, which corresponds to the special case in which the pre-drawn subgraph has no edges. Observe that Theorem 1 directly provides a weak form of Theorem 2: If G_1 and G_2 are simultaneously planar, then they admit a compatible embedding. We can hence take any straight-line planar drawing of G_1 realizing the embedding and extend the induced drawing of G to a drawing of G_2. By Theorem 1, we obtain a simultaneous planar drawing where each edge of G_1 is straight and each private edge of G_2 has at most $102|V(H)| + 12$ bends per edge. Our stronger result of 24 crossings between any two edges is obtained by modifying the proof of Theorem 1, rather than applying that result directly.

We note that Grilli et al. [12] have a paper in this conference with a result similar to Theorem 2. They show, using different techniques, that two simultaneously planar graphs have a simultaneous planar drawing with at most 9 bends per edge, vastly better than our $72n$ bound. Our primary goal, however, was to reduce crossings rather than bends. We achieve 24 crossings per pair of edges. They do not address the number of crossings, but the obvious bound from their result is 100 crossings per pair of edges. We also achieve a polynomial-size grid, but the obvious way of forcing their drawing onto a polynomial-sized grid increases the number of bends per edge to $300n$.

1.1 Related Work

The decision version of simultaneous planarity generalizes partially embedded planarity: given an instance $(G, H, \mathcal{H})$ of the latter problem, we can augment $\mathcal{H}$ to a drawing of a 3-connected graph G_1 and let $G_2 = G$. Then G_1 and G_2 are simultaneously planar if and only if G has a planar embedding extending $\mathcal{H}$. In the other direction, the algorithm [1] for testing planarity of partially embedded graphs solves the special case of the simultaneous planarity problem in which the embedding of the common graph G is fixed (which happens, e.g., if G or one of the two graphs is 3-connected).

Several optimization versions of partially embedded planarity and simultaneous planarity are **NP**-hard. Patrignani showed that testing whether there is a straight-line drawing of a planar graph G extending a given drawing of a subgraph of G is **NP**-complete [21], so bend minimization in partial embedding extensions is **NP**-complete; Patrignani's result holds even if a combinatorial embedding of G is given.[1] Bend minimiza-

---

[1] Patrignani does not explicitly claim **NP**-completeness in the case in which the embedding of G is fixed, but that can be concluded by checking his construction; only the variable gadget, pictured in his Figure 3, needs minor adjustments.

tion in simultaneous planar drawings is **NP**-hard, since it is **NP**-hard to decide whether there is a straight-line simultaneous drawing [9]. Crossing minimization in simultaneous planar drawings is also **NP**-hard, as follows from an **NP**-hardness result on *anchored planar drawings* by Cabello and Mohar [5] (see Section 4).

As mentioned above, the special cases of PEG and SEFE in which there are no edges in the pre-drawn subgraph and in the common subgraph have been already studied.

Concerning PEG, Pach and Wenger [20] proved the following result: given an n-vertex planar graph G with fixed vertex locations, a planar drawing of G in which each edge has at most $120n$ bends can be constructed in $O(n^2)$ time. They also proved that such a bound is tight in the worst case. A $3n + 2$ upper bound improving upon the $120n$ upper bound of Pach and Wenger has been proved by Badent et al. [2].

Concerning SEFE, Erten and Kobourov [8] proved the following result: given two planar graphs G_1 and G_2 sharing some vertices and no edges with a total number of n vertices, there is an $O(n)$-time algorithm to construct a simultaneous planar drawing of G_1 and G_2 on a grid of size $O(n^2) \times O(n^2)$, with at most 3 bends per edge, hence at most 16 crossings between any edge of G_1 and any edge of G_2. Building on Kaufmann and Wiese's drawing algorithm [17], the number of bends per edge and the number of crossings per pair of edges can be reduced to 2 and 9, respectively, at the expense of an exponential increase in the area of the simultaneous drawing.

Haeupler et al. [13] showed that if two simultaneously planar graphs G_1 and G_2 share a subgraph G that is connected, then there is a simultaneous planar drawing in which any edge of $G_1 - G$ and any edge of $G_2 - G$ intersect at most once. Introducing vertices at crossing points yields a planar graph, and a straight-line drawing of that graph provides a simultaneous planar drawing with $O(n)$ bends per edge, $O(n)$ crossings per edge, and with vertices, bends, and crossings on an $O(n^2) \times O(n^2)$ grid. Our result generalizes this to the case where the common graph G is not necessarily connected.

1.2   Graph Drawing Terminology

A *rotation system* for a graph is a cyclic ordering of the edges incident to each vertex. A rotation system of a connected graph determines its *facial walks*—the closed walks in which each edge (u, v) is followed by the next edge (v, w) in the cyclic order at v. The facial walks are the boundaries of the *faces* in an embedding of the graph. The *size* $|W|$ of a facial walk W is the length of W (edge repetitions are counted). A rotation system is *planar* if it corresponds to a planar drawing; a *planar embedding* of a connected graph consists of a planar rotation system together with a specified outer face.

These definitions do not handle the situation in which the graph is not connected. Following Jünger and Schulz [16], we define a *topological embedding* of a (possibly non-connected) graph as follows: We specify a planar embedding for each connected component. This determines a set of inner faces. For each connected component we specify a "containing" face, which may be an inner face of some other component or the unique outer face. Furthermore, we forbid cycles of containment—in other words, if a connected component is contained in an inner face, which is contained in a component, etc., then this chain of containments must lead eventually to the unique outer face.

A *facial boundary* in a topological embedding of a graph is the collection of facial walks along the (not necessarily connected) boundary of a face. Each face (unless it is the outer face) has a distinguished facial walk we call the *outer* facial walk separating the remaining *inner* facial walks from the outer face of the embedding. The *size* of a facial boundary is the sum of the sizes of the facial walks part of the facial boundary.

A *compatible embedding* of two planar graphs G_1 and G_2 consists of topological embeddings of G_1 and G_2 such that the common subgraph G inherits the same topological embedding from G_1 as from G_2 (where a subgraph inherits a topological embedding in a straightforward way; in particular, if we remove an edge that disconnects the graph, the face containment is determined by the edge that was removed). Jünger and Schulz [16] proved that G_1 and G_2 are simultaneously planar if and only if they have a compatible embedding. For that proof, they construct a simultaneous planar drawing of G_1 and G_2 by extending a drawing of G (thus proving a form of our Theorem 1). However, their method does not yield any bounds on the number of bends or crossings.

2   Partially Embedded Graphs

In this section we prove Theorem 1. We will construct a planar drawing of G that extends $\mathcal{H}$, assuming that we are given a planar embedding of G that extends $\mathcal{H}$. It suffices to prove the result for a single face F of $\mathcal{H}$ and the connected components of G that lie inside or on the boundary of F and are connected to H.

Pach and Wenger [20] proved their upper bound on the number of bends needed to draw a graph with fixed vertex locations by drawing a tree with leaves at the fixed vertex locations, and "routing" all the edges close to the tree, sometimes crossing the tree but never crossing each other. We will adapt their method to our setting.

One important difference is that we have to deal with fixed facial boundaries instead of fixed vertex locations. The solution is natural: We contract each facial boundary W_i of F to a single vertex v_i, fix vertex v_i inside F near W_i, and then apply the Pach-Wenger method to draw the contracted graph on the fixed vertex locations v_i. This must be done while keeping the drawing inside F. We keep the drawing at a small distance from the boundary of F, inside a polygonal region F' that is an "inner approximation" of F. Inside F' we draw a tree T with its leaves v_i at the fixed vertex locations, suitably bounding the size of T in order to get our bound on the number of bends. We then route the edges of the contracted graph close to T as in Pach-Wenger. Finally, to get back our uncontracted graph, we route the edges incident to v_i to their true endpoint on the facial boundary W_i—these routes use the empty buffer zone between F and F'.

We now fill in further details. We use n_A and m_A for the number of vertices and edges in subgraph A. Let W_i, with $1 \le i \le b$, be the boundary walks of F.

We now introduce the concept of inner ε-approximations. The *Hausdorff distance* $d_H(A, B)$ of two sets (in a space with metric d) is defined as:[2]

$$\max \{ \sup_{a \in A} \inf_{b \in B} d(a, b), \sup_{b \in B} \inf_{a \in A} d(a, b) \}.$$

Intuitively, the Hausdorff distance measures how far a point in one set can be from the other set. Sets A and B are *ε-close* if $d_H(A, B) < \varepsilon$. Then A is an *inner ε-approximation of B* if they are ε-close and there is a $\delta > 0$ so that all the points δ-close to A are a subset of B. The next lemma deals with inner ε-approximations of F.

---

[2] The underlying metric d can be Euclidean or some other appropriate metric.

Lemma 1. *Let k be the size of the boundary of F. For any $\varepsilon > 0$ we can efficiently construct an inner ε-approximation F' of F whose boundary has size $3k$ (see Figure 1).*

We prove Lemma 1 using Lemma 2 in which, for every sufficiently small $\varepsilon > 0$ we construct a closed polygonal arc P_ε that is ε-close to the facial walk, does not have too many bends, and so that the simple polygon bounded by $P_{\varepsilon'}$ lies in the interior of the simple polygon bounded by P_ε for all $0 < \varepsilon' < \varepsilon$ (in particular, any two polygonal arcs are disjoint). There are various ways to achieve this. Pach and Wenger [20] use the Minkowski sum of the facial walk (in their case the facial walk of a tree) and a square diamond centered at 0. We use a slightly different construction, because it seems easier (both computationally and conceptually) and it gives a slightly better bound on the number of bends (which is what we are most interested in); namely for the facial walk of an n-vertex tree, Pach and Wenger construct a polygonal arc with $4n - 2$ vertices, while our polygonal arcs have $2n - 2$ vertices. Our construction does have one disadvantage: the resulting drawings will get rather tight for sharp (acute or obtuse) angles (the Minkowski-sum construction has the same problem for highly obtuse angles only).

Lemma 2. *Let W be a facial walk in a face F of a drawing of a graph G in the plane. We can efficiently construct a disjoint family of polygonal arcs P_ε so that P_ε is ε-close to W and each P_ε has at most $\max\{3, |W|\}$ vertices.*

Proof. Let e, v, f be a *corner* of W, that is, two consecutive edges e, f and their shared vertex v. At v erect the angle bisector of e and f of length ε (inside F), and let v' be the endpoint of the bisector different from v. For computational reasons, it may be better to use the ℓ_1-norm at this point (the Euclidean norm will lead to square root expressions in the coordinates). If $(v_i)_{i=1}^k$ is the sequence of vertices along W, with $k = |W|$, then $(v_i')_{i=1}^k$ defines a closed polygonal arc. If ε is sufficiently small, namely less than half the distance between any vertex of W and a non-adjacent edge on W, the arc is free of self-crossings, and therefore bounds a simple polygon with $|W|$ vertices. There are two special cases in which this argument does not work: if the boundary walk is a boundary walk on an isolated vertex or an isolated edge. In both of these cases, we can approximate W using a triangular shape.   □

Lemma 2 allows us to replace a facial boundary with a *simple polygon with holes*, that is, a collection of closed polygonal arcs that bound a face which is very close to the original boundary, has bounded complexity, and can be constructed efficiently. This leads to a proof of Lemma 1. Namely, approximate each facial walk of the facial boundary with an ε-close polygonal arc lying in F. The union of those arcs is a simple polygon with holes as long as ε is less than half the distance between any two non-adjacent vertices or edges. The upper bound of $3k$ will generally be a large overestimate, but allows for the possibility that all the inner walks are walks on isolated vertices.

We now return to the proof of Theorem 1. After constructing an inner ε-approximation F' of F by using Lemma 1, the next step is to construct tree T. Triangulate F' using at most $m_{F'} + 2(b - 2)$ triangles[3] and use a result of Bern and Gilbert [3] to construct a

---

[3] Every n-vertex polygon with b boundary components can be triangulated by inserting edges in $O(n \log n)$ time. The number of resulting triangles is $n + 2(b - 2)$ (see [19, Lemma 5.1]).

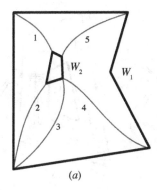
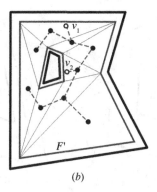

(a) (b)

Fig. 1. A face F with outer and inner and boundary walks W_1 and W_2. (a) The 5 edges of $G - H$. (b) The inner approximation F' (heavy blue lines), a triangulation of it (fine lines), and the dual spanning tree (dashed red) with extra vertices v_1 and v_2 close to W_1 and W_2, respectively.

straight-line drawing of the dual of the triangulation. Bern and Gilbert place a vertex at the *incenter* of each triangle (where the angle bisectors of the triangle meet) and prove that the straight-line edge joining two vertices in adjacent triangles lies within the union of the two triangles. Now take a spanning tree T of the dual. For each boundary walk W_i, we augment T with a new leaf v_i close to W_i and inside F'. This adds b vertices to T, so the number of vertices of T is now $n_T = m_{F'} + 3b - 4$.

Let G_F be the embedded multi-graph obtained by restricting G to vertices and edges lying inside or on the boundary of F and by contracting each boundary walk W_i of F to a single vertex v_i. We can now use the following result (extending ideas of Pach and Wenger) to embed G_F close to T.

Lemma 3. *Let G be a multi-graph with a given planar embedding and fixed locations for a subset $U \subseteq V(G)$ of its vertices. Suppose we are given a straight-line drawing of a tree T whose leaves include all the vertices in U at their fixed locations. Then for every $\varepsilon > 0$ there is a planar poly-line drawing of G that is ε-close to T, that realizes the given embedding, where the vertices in U are at their fixed locations, and where each edge has at most $12n_T$ bends. Moreover, each edge of G comes close to each vertex in U at most six times (where coming close means entering and leaving an ε-neighborhood of the vertex or terminating at the vertex).*

The proof of Lemma 3 is long and involved, hence we defer it to the end of the section, and we first proceed with the reminder of the proof of Theorem 1.

We use Lemma 3 to embed G_F along T so that vertices v_i are drawn at their fixed locations. Each edge of G_F has at most $12n_T$ bends.

We now want to connect edges in G_F to the boundary components they belong to. We will use the buffer between F' and F to do this. In fact, we need to split the buffer zone into two, so we apply Lemma 1 a second time to obtain an inner $\varepsilon/2$-approximation F'' of F, so that $F' \subseteq F'' \subseteq F$. See Figure 2. The size of the boundary of F'' is at most $3m_F$ (just like F'). Now for each walk W_i we extend the edges ending at v_i to their endpoint on W_i. Since we maintained the cyclic order of G_F-edges at v_i, we can simply

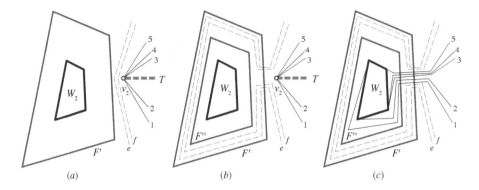

Fig. 2. A close-up of the situation near inner boundary walk W_2. (*a*) After drawing G_F around the tree T (heavy dashed line), edges $1, \ldots, 5$ are incident to v_2 in the correct cyclic order, but two other edges e and f pass by between v_2 and F'. (*b*) We add a second approximation F'' and route the edges e and f (in dashed red) around W_2 in the buffer zone between F'' and F'. (*c*) We route the edges incident to W_2 in the buffer zone between F and F''.

route these edges around W_i using approximations to W_i via Lemma 1, and we can do so in $F - F''$. This adds at most $m_{W_i} + 2$ bends to an edge with endpoint on W_i; the two additional bends are needed to separate edges at v_i, and turn to connect to W_i. There is one difficulty: there are edges of G_F that pass by v_i, separating it from the segment of F' close to v_i (which is our gate to W_i). To remedy this difficulty, we first route all of these edges around the whole obstacle W_i in the $F'' - F'$ part of the buffer, which adds $m_{W_i} + 2$ bends to an edge every time it passes v_i. Now we are free to route the G_F-edges incident to v_i to their endpoints along W_i. Since an edge can pass by and/or terminate at a vertex at most six times, the total number of additional bends in each edge caused by going around W_i is $6(m_{W_i} + 2) \leq 6(m_{F'} + 2) \leq 18m_F + 12$. Since each G_F edge started with $12n_T$ bends, each G_F edge now has at most $12n_T + 18m_F + 12$ bends. Using $m_F \leq m_H \leq 3n_H$, and $n_T \leq m_{F'} + 3b - 4 \leq 3m_F + 3b - 4 \leq 4n_H$ we conclude that each edge has at most $48n_H + 54n_H + 12 = 102n_H + 12$ bends.

Let us now analyze the running time of the algorithm. Most of the steps in the construction can be performed in linear time. Building the triangulation takes time $O(n_H \log n_H)$. The overall running time is thus bounded by the size of the resulting drawing which contains a linear number of edges each with a linear number of bends, yielding the quadratic running time.

We conclude the section by proving Lemma 3. Pach and Wenger's [20] algorithm to draw a planar graph G with vertices at fixed locations has three ingredients: (i) they show how to assume that G is Hamiltonian, (ii) they show how to draw the Hamiltonian cycle of G, and (iii), they show how to draw the remaining edges of G. In order to prove Lemma 3, we will follow their structure closely. We will use their result (i) directly:

Lemma 4 (Pach, Wenger [20]). *Given a planar graph G we can in linear time construct a Hamiltonian graph G' with $|E(G')| \leq 5|E(G)| - 10$ by adding and subdividing edges of G (each edge is subdivided by at most two new vertices).*

We will use a slightly stronger version of Lemma 4 in which G is allowed to be a mulitgraph. Pach and Wenger's proof of Lemma 4 works for this case.

For part (ii) Pach and Wenger show that a Hamiltonian cycle can be drawn at fixed vertex locations ε-close to a star connecting all the vertices. For our application, we replace their star with a straight-line drawing of a tree T whose leaves are the vertices v_i. Independently of our result, the generalization of part (ii) to trees has essentially been shown by Chan et al. [6]. Since their goal was the minimization of the edge lengths, they did not give an estimate on the number of bends. We now show how to draw the Hamiltonian cycle. We will later show how to draw the remaining edges.

Lemma 5. *Let C be a cycle with fixed vertex locations, and suppose we are given a straight-line planar drawing of a tree T, in which the vertices of C are leaves of T at their fixed locations. Then for every $\varepsilon > 0$ there is a planar poly-line drawing of C with at most $2|E(T)| - 1$ bends per edge and ε-close to T.*

Proof. Let $p_1, \ldots, p_n$ be the vertices of C in their order along the cycle. We build a planar poly-line drawing of C as follows. Let Θ_i be an $i\varepsilon/n$-approximation of T for $1 \leq i < n$ (which we can construct using Lemma 2). We start at p_1. Suppose we have already built the poly-line drawing of $p_1, \ldots, p_i$ and we want to add $p_i p_{i+1}$. Let Q_i be the unique path in T connecting p_i to p_{i+1}. Create Θ_i' from Θ_i by keeping only the vertices of Θ_i close to (approximating) vertices in $T_i := \bigcup_{j \leq i} Q_j$. This removes parts of the walk along Θ_i which we patch up as follows: suppose v is an interior vertex of T_i, and v is incident to e which does not lie on T_i. Then v is approximated by two vertices v_1 and v_2 which lie on bisectors formed by e with neighboring edges. Now v_1 and v_2 belong to Θ_i', but the path along Θ_i between them got removed (since e does not belong to T_i). We add $v_1 v_2$ to Θ_i' to connect them. Note that $v_1 v_2$ does not pass through v since v is incident to at least three edges (e and two edges of T_i), and it does not cross any edges of any Θ_j' with $j < i$, since T_i is monotone: if $e \notin E(\Theta_i)$, then $e \notin E(\Theta_j)$ for $j < i$. See Figure 3 for an illustration. Now both p_i and p_{i+1} correspond to unique vertices on Θ_i' (since they are leaves), so we can pick a facial walk $v_1, \ldots, v_k$ on Θ_i' which connects p_i to p_{i+1} and which avoids passing by p_1. We now add line segments $p_i v_2, v_2 v_3, \ldots, v_{k-2} v_{k-1}, v_{k-1} p_{i+1}$ to the poly-line drawing of C. We treat the final edge $p_n p_1$ similarly, except that we move along Θ_{n-1}' back to p_1 in the last step, which we can do, since none of the intermediate paths passed by p_1. Each edge of C is replaced by a polygonal arc with at most $2|E(T)| - 1$ bends. $\square$

As mentioned earlier, the following lemma is close to a result by Chan et al. [6], except for the claim about the number of bends, and the rotation system (which we require for our main result).

Lemma 6. *Let G be a Hamiltonian multi-graph with a given planar embedding and fixed vertex locations. Suppose we are given a straight-line drawing of a tree T whose leaves include all the vertices of G at their fixed locations. Then for every $\varepsilon > 0$ there is a planar poly-line drawing of G that is ε-close to T, that realizes the given embedding, where the vertices of G are at their fixed locations, where each edge has at most $4|E(T)| - 1$ bends, and where each edge comes close to any leaf of T at most twice.*

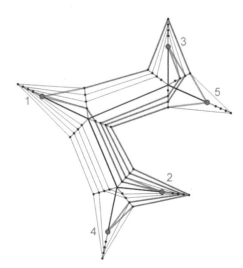

Fig. 3. The underlying tree T is in black (thick edges), angle bisectors in gray; the Θ'_i are drawn as thin black edges; to reduce clutter, we are not showing the remaining edges of Θ_i; the drawing of C is indicated by the green line.

The obvious idea—routing edges along the Hamiltonian cycle C—only gives a quadratic bound on the number of bends, since each edge would follow the path of a linear number of edges of C, and each edge of C has a linear number of bends. Pach and Wenger came up with an ingenious way to construct auxiliary curves with few bends based on the level curves Θ'_i which carry the cycle C in the proof of Lemma 5.

Proof. Let C be the Hamiltonian cycle of G and let G_1 and G_2 be the two outerplanar graphs composed of C and, respectively, of the edges of G outside and inside C. Using Lemma 5 we find a planar poly-line drawing of C on $V(G)$. We need to show how to draw G_1 and G_2 respecting the planar embeddings induced by the given embedding of G. Let $n = |V(G)|$ and $m_i = |E(G_i)|$. We only describe how to draw G_1, since G_2 can be handled analogously. Let $\Delta_{i,k}$, $1 \leq k \leq m_1$ be a $k\varepsilon/(nm_1)$-approximation of Θ'_i constructed using Lemma 2. For a fixed i, each $\Delta_{i,k}$ crosses C twice: when C moves from p_i to Θ'_{i+1}, and when it finally moves back from Θ'_n to p_1. As in Pach and Wenger, we can then split $\Delta_{i,k}$ at the crossings and connect their free ends to p_1 and p_i, resulting (for each k) in two curves $\Delta'_{i,k}$ and $\Delta''_{i,k}$ connecting p_1 to p_i, where $\Delta'_{i,k}$ lies outside C (these are the curves we use for G_1) and $\Delta''_{i,k}$ inside C (these are the curves we use for G_2). Each such curve has at most $2|E(T)| - 1$ bends. As in the proof of Pach and Wenger, we can create edges $p_i p_j \in E(G_1)$ by concatenating $\Delta'_{i,k}$ with $\Delta'_{j,k}$. Since we chose m_1 such approximations, we can do this for each edge in G_1. There are two problems remaining: edges $p_i p_j$ now all pass through p_1 and they could potentially cross (rather than just touch) there. Pach and Wenger show that any two edges touch, so the drawing can be modified close to p_1 so as to separate all edges $p_i p_j$ from each other. This introduces at most one more bend per edge, so that the resulting edges have

$2(2|E(T)|-1)+1 = 4|E(T)|-1$ bends. Finally, note that each edge p_ip_j comes close to each leaf of T (including p_1) at most twice, once for $\Delta'_{i,k}$ and once for $\Delta'_{j,k}$.   □

Now we are ready to finish the proof of Lemma 3. We show how to apply Lemma 6 in case G is not Hamiltonian, and not all its vertices are assigned fixed locations.

By Lemma 4, we can construct a graph G' with a Hamiltonian cycle C by subdividing each edge of G at most twice, and by adding some edges, where G' has a planar embedding extending the embedding of G. Traverse C: whenever we encounter an edge of C with at least one endpoint not in U, contract that edge. This yields a new Hamiltonian graph G'' with $V(G'') = U$ and a planar embedding induced by the planar embedding of G'. Use Lemma 6 to embed G'' at the fixed vertex locations, and ε-close to T, so that each edge of G'' has at most $4|E(T)| - 1$ bends. Each vertex $u \in U$ of G'' corresponds to a set of vertices $V_u \subseteq V(G')$ which was contracted to u, so the subgraph G'_u of G' induced by V_u is connected. Since we embedded G'' with the induced planar embedding of G', we can now do some surgery to turn u back into G'_u.

To this end, we define a graph G_u^+, which consists of G'_u, of a cycle C_u containing G'_u in its interior, and of some further edges. Each vertex of C_u corresponds to an edge of G' "incident to" G'_u, i.e., with an end-vertex in V_u and with an end-vertex not in V_u. Vertices appear in C_u in the same order as the corresponding edges incident to G'_u leave G'_u (this order also corresponds to the cyclic order of the edges incident to v in G''); each vertex of C_u corresponding to an edge e of G' is connected to the end-vertex of e in V_u. Finally, G_u^+ contains further edges that triangulate its internal faces.

Now consider a small disk δ around u. We erase the part of the drawing of G'' inside δ. We construct a straight-line convex drawing of G_u^+ in which each vertex of C_u is mapped to the point in which the corresponding edge crosses the boundary of δ. This drawing always exists (and can be constructed efficiently), given that G_u^+ is 2-connected and internally-triangulated. Removing the edges that triangulate the internal faces of G_u^+ completes the reintroduction of G'_u.

Overall, we added one bend to an edge with exactly one endpoint in V_u. Since an edge can have endpoints in at most two V_u, this process adds at most two bends per edge, so every edge has at most $4|E(T)|+1$ bends. Since each edge of G was subdivided at most twice to obtain G', each edge of G has at most $3(4|E(T)|+1)+2 = 12|E(T)|+5 < 12|V(T)|$ bends. Each edge of G' comes close to each leaf of T at most twice, so each edge of G comes close to each vertex of U at most six times. This concludes the proof of Lemma 3.

3   Extending Partial Straight-Line Planar Drawings Greedily

Let G be an n-vertex plane graph, let H be a spanning subgraph of G, let $\mathcal{H}$ be a straight-line planar drawing of H, and let $\sigma = [e_1, \ldots, e_m]$ be an ordering of the edges in $G \setminus H$. A drawing Γ of G *greedily extends* $\mathcal{H}$ *with respect to* σ if it is obtained by drawing edges $e_1, \ldots, e_m$ in this order, so that e_i is drawn as a polygonal curve that respects the embedding of G and with the minimum number of bends, for $i = 1, \ldots, m$.

Fowler *et al.* claimed in [10] that, for every ordering σ of the edges in $G \setminus H$ such that the edges between distinct connected components of H precede edges between vertices

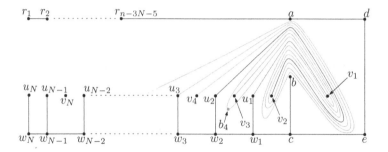

Fig. 4. A drawing Γ of G that greedily extends $\mathcal{H}$ with respect to σ. Drawing $\mathcal{H}$ consists of the black circles. The first edges $n - N - 1$ edges in σ are (black) straight-line segments. The last N edges (u_i, v_i) are (colored) polygonal lines whose bends have been made smooth to improve the readability. Only four of the latter edges are shown.

in the same connected component of H, there exists a drawing Γ of G greedily extending $\mathcal{H}$ with respect to σ where each edge has $O(n)$ bends. However, in the following we confirm a claim of Schaefer [23] stating that greedy extensions do not, in general, lead to drawings with a polynomial number of bends.

Theorem 3. *For every n, there exists an n-vertex plane graph G, a planar drawing $\mathcal{H}$ of the spanning empty subgraph H of G, and an order σ of the edges in G such that any drawing of G that greedily extends $\mathcal{H}$ with respect to σ has edges with $2^{\Omega(n)}$ bends.*

Proof. We adapt an example by Kratochvíl and Matoušek [18]. Refer to Fig. 4. Let $N = \lfloor \frac{n}{3} \rfloor - 6$, for any integer n. Graph H consists of n isolated vertices; namely vertices $u_1, \ldots, u_N, v_1, \ldots, v_N, w_1, \ldots, w_N, a, b, c, d, e, r_1, \ldots, r_{n-3N-5}$. The first $n - N - 1$ edges in σ are (u_i, w_i) for $i = 1, \ldots, N$, (w_i, w_{i+1}) for $i = 1, \ldots, N-1$, (r_i, r_{i+1}) for $i = 1, \ldots, n - 3N - 6$, (c, w_1), (b, c), (c, e), (e, d), (a, d), and (a, r_{n-3N-5}). All these edges are straight-line segments in any drawing Γ of G that greedily extends $\mathcal{H}$ with respect to σ. The last N edges in σ are $(u_1, v_1), \ldots, (u_N, v_N)$ in this order.

Consider any drawing Γ of G that greedily extends $\mathcal{H}$ with respect to σ. We claim that edge (u_i, v_i) has 2^{i-1} bends in Γ. In fact, it suffices to prove that (u_i, v_i) has 2^{i-1} intersections with the straight-line segment $\overline{ab}$ in Γ. Indeed, (u_1, v_1) has exactly one intersection with $\overline{ab}$ in Γ. Inductively assume that (u_i, v_i) has 2^{i-1} intersections with $\overline{ab}$ in Γ; we prove that (u_{i+1}, v_{i+1}) has 2^i intersections with $\overline{ab}$ in Γ. This proof is accomplished by citing Kratochvíl and Matoušek [18] almost *verbatim*. Since (u_{i+1}, v_{i+1}) does not cross (u_i, v_i), it has a bend b_{i+1} around v_i, i.e., inside the square defined by $u_{i-2}, w_{i-2}, w_{i-1}$, and u_{i-1}. Thus the polygonal curve representing (u_{i+1}, v_{i+1}) in Γ consists of two parts – one from u_{i+1} to b_{i+1}, the other from b_{i+1} to v_{i+1}. Both of these parts may be used as an edge joining u_i and v_i – after contracting u_{i+1} and v_{i+1} into u_i, and b_{i+1} into v_i. Hence, by induction, each of these two parts has 2^{i-1} intersections with $\overline{ab}$, and the whole edge (u_{i+1}, v_{i+1}) has 2^i intersections with $\overline{ab}$.

Hence, in any drawing Γ of G that greedily extends $\mathcal{H}$ with respect to σ, one edge has $2^{N-1} = 2^{\lfloor \frac{n}{3} \rfloor - 7} \in 2^{\Omega(n)}$ bends, which concludes the proof.

Note that the graph G in the proof of Theorem 3 is a tree, thus all of its edges connect vertices in distinct connected components of H. $\square$

4 Simultaneous Planarity

Before turning to our algorithm for drawing simultaneously planar graphs, we justify our claim that minimizing the number of crossings in a simultaneous planar drawing is **NP**-hard. This result follows from Cabello and Mohar's proof of **NP**-hardness for the *anchored planarity* problem [5, Theorem 2.1], but a more direct proof of a slightly stronger result is possible by reduction from the **NP**-complete crossing number problem. We briefly explain the reduction. Given a graph G with m edges, subdivide each edge $2m$ times. Let G_1 consist of all the edges incident to the original vertices of G together with every other edge along the paths connecting the original vertices. Let G_2 consist of the remaining edges. Note that G_1 and G_2 do not share any edges. It can be easily seen that the crossing number of G equals the smallest number of crossings between edges of G_1 and edges of G_2 in a simultaneous drawing of G_1 and G_2.[4] We now turn to the proof of Theorem 2.

Proof (of Theorem 2). We show how to find in $O(n^2)$ time a simultaneous planar drawing Γ such that any private edge of G_1 and any private edge of G_2 intersect at most 24 times, such that every edge of G_1 is straight, and such that every private edge of G_2 has at most $102|V(H)| + 12$ bends. In order to construct a simultaneous planar drawing Γ' on an $O(n^2) \times O(n^2)$ grid such that any private edge of G_1 and any private edge of G_2 intersect at most 24 times, such that each edge of G is straight, and such that every private edge has at most $72n$ bends, it suffices to introduce dummy vertces at the $O(n^2)$ crossing points in Γ, and then to construct a straight-line drawing of the resulting planar graph on a small grid. In particular, the number of bends per edge in Γ' is at most $72n$, since each edge in Γ crosses less than $3n$ edges, each at most 24 times.

We start by constructing any straight-line planar drawing Γ_1 of G_1. We now construct a drawing Γ_2 of G_2 by exploiting an approach analogous to the one of the proof of Theorem 1. Drawing Γ_1 induces a straight-line planar drawing Γ of G. Thus, in order to determine Γ_2, it remains to describe how to draw the private edges of G_2. We will accomplish this independently for each face F of G.

We construct a triangulation Σ of F by using all the vertices and edges of G_1 that lie inside F, as well as some extra edges. Next, we execute the same algorithm as for the proof of Theorem 2. Namely, we construct a straight-line drawing of the dual D of Σ and we take a spanning tree T of D. For each boundary walk W_i of F, we augment T with a leaf v_i close to W_i and inside F', where F' is an inner ε-approximation of F. Let G_2^F be the embedded multi-graph obtained by restricting G_2 to the vertices and edges inside or on the boundary of F, and by contracting each boundary walk W_i of F to a single vertex v_i. We use Lemma 3 to construct a planar poly-line drawing of G_2^F that realizes the given embedding, that is ε-close to T, and in which vertices v_i maintain their fixed locations. Finally, we reconnect edges in G_2^F to the boundary components they belong to. In order to do this, we first "wrap" the edges of G_2^F passing by a vertex

---

[4] Using the fact that crossing number is hard for cubic graphs [14], we can even show that minimizing the number of crossings in a simultaneous drawing of two graphs one of which is the disjoint union of paths of length at most two and the other is a matching is **NP**-hard. This is in some sense sharp, since the union of two matchings is always planar.

v_i around W_i, and we then extend the edges of G_2^F incident to v_i to their endpoint on W_i, by routing them around W_i.

By construction every edge of G_1 is straight. By Theorem 1 every private edge of G_2 has at most $102|V(H)| + 12$ bends. Also, the algorithmic steps are the same as for the proof of Theorem 1, hence the algorithm runs in $O(n^2)$ time. It remains to prove that any private edge of G_1 and any private edge of G_2 intersect at most 24 times.

Consider any private edge e of G_2 and any private edge e' of G_1. Recall that e' is an edge of Σ. Denote by W_i and W_j the boundary walks the end-vertices of e' belong to. Edge e intersects e' in two situations: when passing by v_i or v_j and when passing by the point p_T in which the edge of D dual to e' crosses e'. We prove that each of these two types of intersections happens at most 12 times.

For the first type of intersections, we have by Lemma 3 that edge e passes by each of v_i or v_j at most 6 times, hence at most 12 times in total. For the second type of intersections, we have by Lemma 4 that edge e is subdivided into at most three edges e_1, e_2, and e_3 in order to turn G_2^F into a Hamiltonian graph. For each $j = 1, 2, 3$, e_j either belongs to the Hamiltonian cycle of the subdivided G_2^F or not. In the former case, e_j is drawn as part of an $i\varepsilon/n$-approximation Θ_i of T, as in the proof of Lemma 5, hence it crosses e' at most twice. In the latter case, e_j is composed of two parts, denoted by $\Delta'_{p,k}$ and $\Delta'_{q,k}$, or by $\Delta''_{p,k}$ and $\Delta''_{q,k}$ in the proof of Lemma 6. Each of $\Delta'_{p,k}$, $\Delta'_{q,k}$, $\Delta''_{p,k}$ and $\Delta''_{q,k}$ is part of a $k\varepsilon/(nm_1)$-approximation of Θ'_i, which is part of Θ_i. Hence, each of $\Delta'_{p,k}$, $\Delta'_{q,k}$, $\Delta''_{p,k}$ and $\Delta''_{q,k}$ crosses e' at most twice; thus e_j crosses e' at most four times, and e crosses e' close to p_T at most 12 times.   □

5   Conclusions and Open Problems

We proved that if a graph has a planar drawing extending a straight-line planar drawing of a subgraph then there is such a drawing with at most $102n + O(1)$ bends per edge. This is asymptotically tight, but can the constant 102 be reduced? Our second result is that any two simultaneously planar graphs have a simultaneous planar drawing with at most 24 crossings per pair of edges and a linear number of bends per edge with a drawing on a polynomial-sized grid. The only lower bound on the number of crossings between two edges in a simultaneous planar drawing is 2 (see [7] or the figure in the margin for the entry "simultaneous crossing number" in [22]). There is a large gap between 2 and 24. Can two edges be forced to cross more than twice in a simultaneous planar drawing? Grilli et al. [12] showed that two simultaneously planar graphs have a drawing with at most 9 bends per edge, though with a larger constant for the number of crossings and not on a grid. Is it possible to achieve the best of both results: 9 bends per edge, 24 crossings per pair of edges, and a nice grid?

Acknowledgements. The University of Waterloo co-authors thank Vincenzo Roselli for contributions in the early stages of the work.

References

1. Angelini, P., Di Battista, G., Frati, F., Jelínek, V., Kratochvíl, J., Patrignani, M., Rutter, I.: Testing planarity of partially embedded graphs. In: Proc. Twenty-First Annual ACM-SIAM Symposium on Discrete Algorithms, SODA 2010, pp. 202–221. SIAM (2010)

2. Badent, M., Di Giacomo, E., Liotta, G.: Drawing colored graphs on colored points. Theor. Comput. Sci. 408(2-3), 129–142 (2008)
3. Bern, M., Gilbert, J.R.: Drawing the planar dual. Inform. Process. Lett. 43(1), 7–13 (1992)
4. Bläsius, T., Kobourov, S.G., Rutter, I.: Simultaneous embeddings of planar graphs. In: Tamassia, R. (ed.) Handbook of Graph Drawing and Visualization. Discrete Mathematics and Its Applications, ch. 11, pp. 349–382. Chapman and Hall/CRC (2013)
5. Cabello, S., Mohar, B.: Adding one edge to planar graphs makes crossing number and 1-planarity hard. SIAM Journal on Computing 42(5), 1803–1829 (2013)
6. Chan, T.M., Hoffmann, H.-F., Kiazyk, S., Lubiw, A.: Minimum length embedding of planar graphs at fixed vertex locations. In: Wismath, S., Wolff, A. (eds.) GD 2013. LNCS, vol. 8242, pp. 376–387. Springer, Heidelberg (2013)
7. Chimani, M., Jünger, M., Schulz, M.: Crossing minimization meets simultaneous drawing. In: PacificVis, pp. 33–40. IEEE (2008)
8. Erten, C., Kobourov, S.G.: Simultaneous embedding of planar graphs with few bends. J. Graph Algorithms and Appl. 9(3), 347–364 (2005)
9. Estrella-Balderrama, A., Gassner, E., Jünger, M., Percan, M., Schaefer, M., Schulz, M.: Simultaneous geometric graph embeddings. In: Hong, S.-H., Nishizeki, T., Quan, W. (eds.) GD 2007. LNCS, vol. 4875, pp. 280–290. Springer, Heidelberg (2008)
10. Fowler, J.J., Jünger, M., Kobourov, S.G., Schulz, M.: Characterizations of restricted pairs of planar graphs allowing simultaneous embedding with fixed edges. Comput. Geom. 44(8), 385–398 (2011)
11. Gassner, E., Jünger, M., Percan, M., Schaefer, M., Schulz, M.: Simultaneous graph embeddings with fixed edges. In: Fomin, F.V. (ed.) WG 2006. LNCS, vol. 4271, pp. 325–335. Springer, Heidelberg (2006)
12. Grilli, L., Hong, S.-H., Kratochvíl, J., Rutter, I.: Drawing simultaneously embedded graphs with few bends. In: Duncan, C., Symvonis, A. (eds.) GD 2014. LNCS, vol. 8871, pp. 40–51. Springer, Heidelberg (2014)
13. Haeupler, B., Jampani, K.R., Lubiw, A.: Testing simultaneous planarity when the common graph is 2-connected. J. Graph Algorithms and Appl. 17(3), 147–171 (2013)
14. Hliněný, P.: Crossing number is hard for cubic graphs. J. Combin. Theory Ser. B 96(4), 455–471 (2006)
15. Jelínek, V., Kratochvíl, J., Rutter, I.: A Kuratowski-type theorem for planarity of partially embedded graphs. Comput. Geom. 46(4), 466–492 (2013)
16. Jünger, M., Schulz, M.: Intersection graphs in simultaneous embedding with fixed edges. J. Graph Algorithms Appl. 13(2), 205–218 (2009)
17. Kaufmann, M., Wiese, R.: Embedding vertices at points: Few bends suffice for planar graphs. J. Graph Algorithms and Appl. 6(1), 115–129 (2002)
18. Kratochvíl, J., Matoušek, J.: String graphs requiring exponential representations. J. Comb. Theory, Ser. B 53(1), 1–4 (1991)
19. O'Rourke, J.: Art Gallery Theorems and Algorithms. Oxford University Press, NY (1987)
20. Pach, J., Wenger, R.: Embedding planar graphs at fixed vertex locations. Graphs Combin. 17(4), 717–728 (2001)
21. Patrignani, M.: On extending a partial straight-line drawing. Internat. J. Found. Comput. Sci. 17(5), 1061–1069 (2006)
22. Schaefer, M.: The graph crossing number and its variants: A survey. The Electronic Journal of Combinatorics 20, 1–90 (2013), Dynamic Survey, #DS21.
23. Schaefer, M.: Toward a theory of planarity: Hanani-Tutte and planarity variants. J. of Graph Algorthims and Appl. 17(4), 367–440 (2013)

Drawing Simultaneously Embedded Graphs with Few Bends[*]

Luca Grilli[1], Seok-Hee Hong[2], Jan Kratochvíl[3], and Ignaz Rutter[3,4]

[1] Dipartimento di Ingegneria, Università degli Studi di Perugia
luca.grilli@unipg.it
[2] School of Information Technologies, University of Sydney
shhong@it.usyd.edu.au
[3] Department of Applied Mathematics, Faculty of Mathematics and Physics,
Charles University in Prague
honza@kam.mff.cuni.cz
[4] Institute of Theoretical Informatics, Karlsruhe Institute of Technology
rutter@kit.edu

Abstract. We study the problem of drawing simultaneously embedded graphs with few bends. We show that for any simultaneous embedding with fixed edges (SEFE) of two graphs, there exists a corresponding drawing realizing this embedding such that common edges are drawn as straight-line segments and each exclusive edge has a constant number of bends. If the common graph is biconnected and induced, a straight-line drawing exists. This yields the first efficient testing algorithm for simultaneous geometric embedding (SGE) for a non-trivial class of graphs.

1 Introduction

Let $G_1 = (V_1, E_1)$ and $G_2 = (V_2, E_2)$ be two graphs sharing a *common graph* $G = (V, E) = (V_1 \cap V_2, E_1 \cap E_2)$. The vertices and edges in $V_i \setminus V$ and $E_i \setminus E$ are called *exclusive*. The problem of finding a simultaneous drawing of G_1 and G_2 such that each graph is drawn in a planar way and the subdrawing of G coincides in both drawings is a long-standing problem in Graph Drawing with applications to, e.g., dynamic graph drawing. The problem can be studied in a topological variant, SIMULTANEOUS EMBEDDING WITH FIXED EDGES (or SEFE for short), where edges are represented by arbitrary open Jordan curves between their endpoints or in the geometric variant, SIMULTANEOUS GEOMETRIC EMBEDDING (or SGE for short), where edges are represented by straight-line segments. Both problems naturally generalize to more than two input graphs. An important special case is the case of *sunflower intersection*, where one requires that the pairwise intersection of any two input graphs is the same.

[*] This work was started at the Bertinoro Workshop on Graph Drawing 2012. L. Grilli was partly supported by the MIUR project AMANDA "Algorithmics for MAssive and Networked DAta", prot. 2012C4E3KT_001. S. Hong was supported by ARC Future Fellowship and Humboldt Fellowship. Work by Jan Kratochvíl was supported by the grant no. 14-14179S of the Czech Science Foundation GAČR. Ignaz Rutter was supported by a fellowship within the Postdoc-Program of the German Academic Exchange Service (DAAD).

C. Duncan and A. Symvonis (Eds.): GD 2014, LNCS 8871, pp. 40–51, 2014.

The problem SGE is NP-hard [8] and, moreover, there are quite restricted graph classes that do not always admit an SGE, e.g., even a path and a tree do not always admit an SGE [3]. To date no efficient testing algorithms for a non-trivial class of restricted input instances is known.

In contrast, the complexity of testing the existence of a SEFE drawing for two input graphs is a long-standing open problem. Jünger and Schulz [13] showed that the problem is actually equivalent to determining planar embeddings of the two input graphs that induce the same embedding on the common graph. For three input graphs the problem is NP-complete [9]. In recent years considerable progress has been made, providing efficient testing algorithms for increasingly general sets of input instances. Most of the results revolve around assumptions on the connectivity or the maximum degree of the input graphs and the common subgraph. It is known that SEFE can be tested in polynomial time if the common subgraph has a fixed planar embedding [1], if the common graph is biconnected [2,10], if the two input graphs are biconnected and the common graph is connected or a forest [6], and if each connected component of the common graph is either biconnected or subcubic [18]. The last result can be improved to also allow connected components of the common graph that are outerplanar and whose cutvertices have degree at most 3 in the common graph [4]. See the recent survey by Bläsius et al. [5] for further details.

While the rephrasing of the original drawing problem SEFE as an embedding problem has certainly been fundamental in starting this evolution, it also comes at a disadvantage. Typically, the outputs of the above-mentioned algorithms are just planar embeddings of the input graphs, i.e., rotation systems and relative positions of the connected components, that coincide on the common graph. We call this a SEFE *embedding*. To obtain a visualization, it is necessary to transform this combinatorial description of a drawing into an actual drawing while preserving the given embedding. For clarity, we refer to such a drawing as a SEFE *drawing* and say that the SEFE drawing *realizes* the corresponding SEFE embedding. Although the complexity of SEFE is still open, the existing results allow efficient testing algorithms for a large range of instances, increasing the importance of the realization problem. The very first result on simultaneous drawings with few bends was obtained by Erten and Kobourov [7] in the context of SIMULTANEOUS EMBEDDING, where one only requires that common vertices are drawn the same, whereas shared edges may be drawn differently for different input graphs. They showed that three bends per edge suffice if the common graph does not contain any edges. Haeupler et al. [10] initiated the study of the realization problem for SEFE embeddings and showed that for any instance where the common graph is connected, it is always possible to find a SEFE drawing realizing a given SEFE embedding in such a way that one of the input graphs is drawn straight-line, whereas the other graph has at most as many bends per edge as the number of vertices in the common subgraph. We show that a constant number of bends per edge suffices if bends are allowed on the exclusive edges of both graphs, even if the common graph is disconnected.

A related notion is *partially embedded graphs*, where one seeks to extend a given drawing of a subgraph (partial drawing) into a planar drawing of the whole graph. Similar to the simultaneous drawing problem, the partial embedding problem has been studied both in the topological setting [1,12,18] and in the straight-line setting [15,17].

In fact, to obtain their result about SEFE realizations, Haeupler et al. [10] essentially show that k bends per edge suffice when extending a given drawing with k predrawn vertices provided that the predrawn graph is connected. Their result is then obtained by taking a straight-line drawing of G_1, considering the drawing it induces on the common graph G and extending it to a drawing of G_2 with at most $|V|$ bends per edge, where V is the vertex set of the common graph G. In the setting of partially embedded graphs their result is asymptotically tight; it is easy to construct an example that shows that $\Omega(k)$ bends per edge are necessary. To achieve our result, which only requires $O(1)$ bends on the exclusive edges, we allow bends on the exclusive edges of both graphs and make use of the fact that, in the SEFE realization problem, we can choose the drawing of the straight-line drawing of the common graph in such a way that it fits with both input graphs simultaneously.

Our Contribution. We study the problem of finding realizations of SEFE embeddings where the common graph is drawn without bends and the exclusive edges have few bends per edge. We refer to a drawing where the edges are represented by polygonal curves with at most c_1 bends per common edge and at most c_2 bends per exclusive edge as a (c_1, c_2)-*drawing*. In a SEFE drawing realizing a SEFE embedding, we require that the planar embeddings of the input graphs are preserved.

Our main result is that every SEFE embedding of two graphs admits a $(0, c)$-drawing with $c \le 9$; see Section 4. If the common graph is (bi)connected, we have $c \le 3$ and this even holds for an arbitrary number of input graphs intersecting in a sunflower-way (i.e., the pairwise intersections of the input graphs are identical); see Section 3. As a side result, we obtain the first efficient algorithm for testing SGE for a non-trivial class of graphs, namely for instances whose common graph is biconnected and an induced subgraph of the input graphs. Finally, we study lower bounds in Section 5 and show that some of the results from Section 3 are in fact tight. For the case of sunflower-intersection we show that there exist k-tuples of input graphs (with a disconnected common graph) that require $\Omega(\sqrt{2}^{-k}/k)$ bends per edge. We note that all proofs are constructive and can be turned into efficient drawing algorithms.

2 Preliminaries

A graph $G = (V, E)$ is *planar* if and only if it can be drawn in the Euclidean plane such that its vertices are represented by points and its edges are represented by internally disjoint open Jordan curves between their endpoints. If G is connected, a planar drawing can be combinatorially described by its *rotation system*, i.e., the circular ordering of the edges around each vertex and the choice of an outer face. We refer to this rotation system as the associated *(combinatorial) embedding*. For disconnected graphs, the embedding also encodes the relative positions of the connected components.

A *plane graph* H is a planar graph with an associated planar embedding. A drawing of a plane graph is a planar drawing of the graph with its given planar embedding. A *plane subgraph* of H is a subgraph G of H associated with the corresponding planar embedding induced by H. In this case we also say that H is a *plane supergraph* of G.

Consider a straight-line drawing Γ of a planar graph G. A face f is *star-shaped* if it contains a point p such that the straight-line segment from p to each vertex of f lies inside f. The set of all such points p is the *kernel* of the face. We say that Γ is *star-shaped* if all its faces are *star-shaped*. We will frequently use the following lemma, stating that any planar graph admits a star-shaped drawing. Of course it is always possible to find a drawing where the vertices are in general position. This implies that the kernel of each face has positive area.

Lemma 1. *Let $G = (V, E)$ be a plane graph. There exists a star-shaped planar straight-line drawing of G such that the kernel of the outer face contains points on or outside of the convex hull of the vertices in G.*

Proof. To construct such a drawing, we create for each face f of G a new vertex v_f that is connected to vertices incident to f and embed it inside f. Afterwards, we triangulate the graph in such a way that the vertex v_0 added for the outer face o is incident to the outer face. Call the resulting graph $G^\star$.

We then produce a straight-line drawing of $G^\star$. Since this drawing is planar and each vertex v_f is connected by straight-line segments to all vertices incident to f, removing v_f yields a drawing where face f is star-shaped. We remove all vertices v_f of all faces to obtain a star-shaped drawing. Note that the outer face o of $G^\star$ is a triangle, and hence v_o lies on or outside of the convex hull of G. $\qquad\square$

For an instance $G_1 = (V_1, E_1)$ and $G_2 = (V_2, E_2)$ of SEFE, we say that the common graph $G = (V, E)$ is *induced* if the induced subgraph of V in G_1 and G_2 is G. This is equivalent to the statement that each exclusive edge has at least one endpoint that is not in V. Assuming that G is induced will often simplify our arguments. Given a non-induced instance (G_1, G_2) of SEFE together with a SEFE embedding, we can construct its *associated induced instance* by subdividing each exclusive edge once (with an exclusive vertex). Note that this operation does not change the common graph G. By interpreting the subdivision vertices as bends in a drawing of the associated induced instance, we obtain the following lemma.

Lemma 2. *Let (G_1, G_2) be an instance of SEFE with a fixed SEFE embedding and let (G_1', G_2') be the associated induced instance. If (G_1', G_2') admits a $(0, c)$-drawing, then (G_1, G_2) admits a $(0, 2c + 1)$-drawing.*

Proof. Consider a $(0, c)$-drawing of (G_1', G_2'). We can interpret it as a SEFE drawing of (G_1, G_2) by interpreting the subdivision vertices as bends. Consider a subdivided edge e. By assumption, each half-edge into which e is subdivided has at most c bends. Together with the additional bend at the subdivision vertex this amounts to a total of $2c + 1$ bends per edge. $\qquad\square$

3 SEFE Drawing with (Bi-)Connected Common Graph

In this section we study realizations of SEFE embeddings where the common graph is biconnected or connected. The main result of this section is the following theorem.

Theorem 1. *Let* $(G_1, \ldots, G_k)$ *be a sunflower instance of* SEFE *with pairwise common graph G together with a* SEFE *embedding. Then the following* SEFE *drawings realizing the given* SEFE *embedding exist.*

(i) *If G is biconnected and induced, there exists a (0,0)-drawing.*

(ii) *If G is biconnected, there exists a (0,1)-drawing.*

(iii) *If G is connected and induced, there exists a (0,1)-drawing.*

(iv) *If G is connected, there exists a (0,3)-drawing.*

Before we proceed to prove Theorem 1, we first mention an interesting implication; the first efficient algorithm for testing the existence of SGE on a non-trivial class of graphs. By Theorem 1(i) two graphs whose common graph is biconnected and induced admit an SGE if and only if they admit a SEFE. The latter can be tested in linear time [2,10].

Corollary 1. *There is a linear-time algorithm for* SIMULTANEOUS GEOMETRIC EMBEDDING *if the common graph is biconnected and induced.*

The rest of this section is devoted to proving Theorem 1. The main tool for the proof is the existence of certain planar straight-line drawings of the common graph, so-called *universal* drawings, that can be extended to a drawing of any plane supergraph of G with few bends per edge. Theorem 1 is an immediate consequence of the existence of such drawings.

Let G be a planar graph and let Γ be a planar straight-line drawing of G. Let $H \supseteq G$ be a plane supergraph of G. We say that Γ is k-*extendable* for H if it can be extended to a drawing of H that has at most k bends per edge. The drawing Γ is k-*universal* if it is k-extendable to every plane supergraph H. The drawing Γ is *induced* k-*universal* if it is k-extendable for any plane supergraph H that contains G as an induced subgraph. Similar to Lemma 2, a drawing that is induced k-universal is $2k + 1$-universal.

Lemma 3. *Let G be a planar graph and let Γ be an induced k-universal drawing of G. Then Γ is $(2k + 1)$-universal.*

Proof. Let $H \supseteq G$ be an arbitrary plane supergraph of G. Let H' be the graph obtained from H by subdividing each edge of $H - G$ whose endpoints both belong to H. Since Γ is induced k-universal, we find a drawing Γ' of H' extending Γ that has at most k bends per edge. By interpreting the subdivision vertices as bends, the drawing Γ' can be seen as a drawing of H with at most $(2k + 1)$-bends per edge, extending Γ. This finishes the proof, since H is an arbitrary plane supergraph of G. □

We are now ready to prove the existence of universal drawings for biconnected planar graphs.

Theorem 2. *Every biconnected plane graph has an induced 0-universal drawing.*

Proof. Let G be a biconnected plane graph. We claim that a star-shaped drawing of G, which exists by Lemma 1 is induced 0-universal.

Let $H \supseteq G$ be a planar supergraph of G with a fixed embedding that extends that of G. Without loss of generality, we assume that H is a triangulation. If it is not, we add new vertices to triangulate its faces. Note that it may not be possible to triangulate H without adding new vertices as this might violate the property of G being an induced subgraph. All vertices and edges that are added during the triangulation can be removed after a drawing has been found.

The vertices and edges of $H - G$ are embedded in the faces of G. We show that, for each face f of G, the parts of $H - G$ embedded in f can be drawn inside f without any bends. To this end, let f be a face of G and let H' denote the subgraph of H consisting of the boundary of f and all edges and vertices of H that are embedded in f. Let further C denote the facial cycle bounding f. Note that C is a simple cycle since G is biconnected. Then the graph H' is a planar graph where one face is bounded by C and all remaining faces are triangles. Hence, considering C as the boundary of the outer face of H', it follows that H' is internally triconnected and, moreover, it is triconnected if and only if it does not contain a chord of C. This is, however, not the case since C is an induced subgraph of H'. It hence follows that H' is triconnected. Thus, H' is triconnected and its outer vertices have been fixed to a star-shaped polygon with positive kernel. A result of Hong and Nagamochi [11, Theorem 10] shows that the given drawing of C can be extended to one of H' without crossings and bends. Since this reasoning can be applied independently to all faces, we find the claimed extension.

For the outer face we use a technique similar to [10] and position the vertices of $H - G$ incident to G on the boundary of a small disk in the interior of the kernel of the outer face such that the edges between G and $H - G$ can be drawn in a planar way without intersecting the interior of the disk. Then, since the positioned exclusive vertices are in convex position, the remaining drawing can be completed without any bends by a Tutte drawing [19]. ☐

By applying Lemma 3, we immediately obtain the following corollary.

Corollary 2. *Every biconnected plane graph has a 1-universal drawing.*

Theorem 3. *Every connected plane graph has an induced 1-universal drawing.*

Proof. Let G be a connected planar graph with a fixed embedding. We show how to construct an induced 1-universal drawing of G. Let v be a cutvertex of G, let f be a face of G on whose boundary v occurs at least twice, and let vu and vw be two edges incident to f and v that are consecutive in the circular ordering around v. We call such a configuration an *angle* of G. We process this angle by adding to G a new vertex v' with neighbors u, v and w. We call v' the *representative vertex* of the angle; see Fig. 1. Call G' the graph resulting from G after processing all angles in this way. Note that G' is biconnected and hence has an induced 0-universal drawing Γ' by Theorem 2.

We claim that the restriction Γ of Γ' to G is induced 1-universal. Let $H \supseteq G$ be a plane graph that contains G as an induced subgraph. For each cutvertex v of G, the incident edges of H are embedded in one of the angles of v. Let u and w be the other two vertices of such an angle at v and let $vv_1, \ldots, vv_k$ denote the edges of H that are embedded inside this angle. We modify H by creating a new vertex v' that is adjacent to u, v and w. We further replace the edges $vv_1, \ldots, vv_k$ by $v'v_1, \ldots, v'v_k$, which is clearly possible in a planar way; see Fig. 1, where the edges of $E(H) \setminus E(G)$ are shown dashed. Let H' be the graph resulting from H by treating all angles of G in

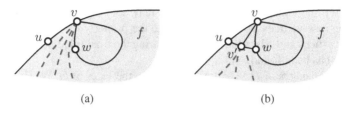

Fig. 1. Augmentation technique for removing cutvertices. (a) The graph G is drawn solid, the exclusive edges of H are dashed. (b) The modified graphs G' (solid) and H' (exclusive edges are dashed).

this way. Now H' contains G' as an induced subgraph. Since Γ' is induced 0-universal, there exists a straight-line drawing of H' that extends Γ'. To obtain a 1-bend drawing of H that extends Γ, we remove from this drawing the edges $v'u$ and $v'w$ for each angle of G. Now each edge xv of H, where v is a cutvertex of G, is drawn with one bend at the position of the representative vertex of the corresponding angle. We can remove any overlaps of segments by slightly moving the bend points apart from each other without creating any crossings. Thus Γ is induced 1-universal.   □

Corollary 3. *Every connected plane graph has a 3-universal drawing.*

We note that Theorem 1 follows easily by applying one of Theorems 2, 3 and Corollaries 2, 3 to the common graph of the SEFE embedding.

4   SEFE Drawing for General Graphs

Unfortunately, universal drawings cannot be used to prove the existence of SEFE drawings with few bends in general. Namely, Pach and Wenger [16] showed that drawing a planar graph with fixed vertex locations may require an edge with $\Theta(n)$ bends. Thus, even a graph consisting only of n isolated vertices has no (induced) $o(n)$-universal drawing.

Our goal is to show that any SEFE embedding of two graphs G_1 and G_2 with common graph $G = G_1 \cap G_2$ admits a $(0, c)$-drawing for some constant c. Unlike the previous constructions, this result does not generalize to an arbitrary number of graphs intersecting in a sunflower-fashion. In fact, we will see later, in Section 5, that for k graphs $\Omega(\sqrt{2}^k/k)$ bends per edge are necessary even in the case of sunflower-intersection.

As a first step, we show how to construct a $(0, 3)$-drawing of (G_1, G_2) if the common graph is an induced subgraph of G_1 and G_2, respectively, and where we require that each connected component of the common graph is biconnected. Afterwards, we apply the technique from Theorem 3 to treat cutvertices in the connected components of G, resulting in $(0, 4)$-drawings when G is an induced subgraph. Then Lemma 2 implies the existence of a $(0, 9)$-drawing for any SEFE embedding.

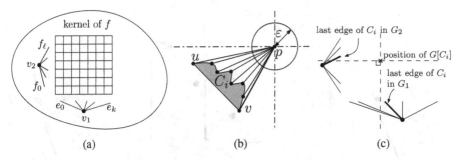

(a)   (b)   (c)

Fig. 2. Illustration of the placement of the $G[C_i]$ inside the face f for a $(0, 3)$-drawing

Theorem 4. *Let (G_1, G_2) be two planar graphs with a* SEFE *embedding. Assume further that $G = G_1 \cap G_2$ is an induced subgraph of G_1 and G_2, respectively. If each connected component of G is biconnected, there exists a $(0, 3)$-drawing of (G_1, G_2).*

Sketch of Proof. Without loss of generality, we can assume that G_1 and G_2 are internally triangulated and that the outer face of G does not contain any exclusive edges. Let C be a connected component of G and denote by $G[C]$ the subgraph of G consisting of all vertices and edges that either belong to C or are embedded inside some inner face of C. The graphs $G_1[C]$ and $G_2[C]$ are defined analogously. Note that if C' is a connected component of G with $C' \subseteq G[C]$, then $G[C'] \subseteq G[C]$. Hence, the relation $C' \prec C$ if and only if $G[C'] \subseteq G[C]$ defines a partial ordering of the components of G whose transitive reduction is a tree. We prove the following claim by induction on the depth of this tree; it implies the statement of the theorem.

Claim. Let C be a connected component of G with depth d. The induced SEFE of $G_1[C]$ and $G_2[C]$ admits a $(0, 3)$-drawing such that the outer face of $G[C]$ is star-shaped.

The base case, where $G[C] = C$ is biconnected, follows from Theorem 1(i). For the induction step, assume that C is a component such that $G[C]$ has depth d. The graph C is biconnected, and we take a star-shaped drawing with positive kernel area in all faces. Clearly, the outer face is star-shaped. We show that we can embed the remaining parts of $G_1[C]$ and $G_2[C]$ inside the inner faces of C using the given embedding and such that the result is a $(0, 3)$-drawing. The faces of C can be treated independently. In the following we fix an arbitrary internal face f and denote by $C_1, \ldots, C_c$ the connected components of G that are distinct from C and incident to f. Note that $G[C_i]$ has depth at most $d - 1$ and hence, by induction, we know that corresponding $(0, 3)$-drawings of $G_1[C_i]$ and $G_2[C_i]$ exist for $i = 1, \ldots, c$. We show how to arrange them in the interior of f and how to draw the exclusive edges embedded inside f to obtain a $(0, 3)$-drawing of $G_1[C]$ and $G_2[C]$. We assume that the only exclusive edges of G_1 and G_2 are the ones embedded inside f. This is not a restriction since there is no interaction between exclusive edges in distinct faces of G. We can thus treat all faces independently.

   Since G_1 and G_2 are triangulated, it follows that the subgraph induced by the exclusive vertices of G_i that are embedded inside f is connected, and we contract it into a single vertex v_i for $i = 1, 2$, preserving the edge ordering of the given embedding.

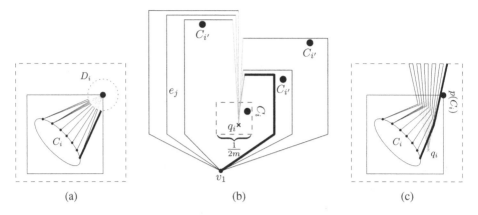

Fig. 3. Illustration of the drawing of exclusive internal edges

Note that we remove loops but not multiple edges. We will produce the largest part of the drawing inside a grid of size $2 \deg(v_1) \times 2 \deg(v_2)$, which we position inside the kernel of f. We distinguish the exclusive edges of G_1 and G_2 into two types. An exclusive edge is *internal* if its endpoint distinct from v_1 and v_2 belongs to one of the C_i for $i = 1, \ldots, c$. Otherwise the endpoint belongs to C and the edge is *external*.

The vertices v_1 and v_2 are positioned below and left of the grid, respectively; see Fig. 2a. By picking two external edges incident to v_1 and v_2 as the first edge, the circular ordering around these vertices determine corresponding linear orderings. Second, for each component C_i, we obtain a linear ordering of its incident internal edges by the ordering around a point in the kernel of the outer face; see Fig. 2b. Our goal is to draw the edges as shown in Fig. 3b. Since the linear ordering around v_1 and the linear ordering determined by C_i do not necessarily coincide, this requires that component C_i is positioned at a specific position, namely slightly left of the so-called last edge, which is shown bold in Fig. 3a and 3b. In this way, the ordering of the edges around v_1 and v_2 imply a certain ordering of the components C_i. Since these orderings generally differ, we use the ordering around v_1 to determine the x-coordinate and the ordering around v_2 to determine the y-coordinate on the grid. Thus, each of the two vertices "sees" the components in the expected ordering; see Fig. 2c.

We now sketch how to draw the edges of G_1. Graph G_2 is drawn analogously, but with exchanged x- and y-coordinates, which corresponds to mirroring the drawing along a diagonal. Let $e_0, \ldots, e_k$ denote the internal edges incident to v_1 linearly ordered from left to right. Let C_i be a component incident to edge e_j. We draw edge e_i first from v_1 to the grid point $(2j, 0)$, then vertically upwards up to some height $y(e_j)$, then horizontally above the position of C_i (see Fig. 3b), and from there down to the position of C_i and into its kernel. Finally, from there we route it to its endpoint in C_i (see Fig. 3c). The main point is the height $y(e_j)$, which we choose such that e_j passes above all components $C_{i'}$ whose x-coordinate lies between the x-coordinate of e_j, i.e., $2j$, and the x-coordinate of the target component C_i; see Fig. 3b. If e_j is an internal edge, then we draw it from v_1 to $(2j, 0)$ and from there vertically upwards through the hole grid. It is then not hard to reach the target vertex of C with one or two more bends.

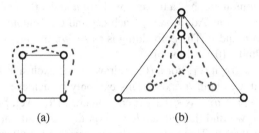

Fig. 4. SEFE instances that do not admit a $(0,0)$-drawing

Finally, we undo the contraction of v_1 and v_2 using a technique similar to Haeupler et al [10, Theorem 2]. □

If the connected components of the common graph are not biconnected, we use the same technique as in the proof of Theorem 3 to make them biconnected, showing that we obtain a $(0,4)$-drawing of the given SEFE embedding if the common graph is induced. Then, Lemma 2 implies the existence of a $(0,9)$-drawing for any SEFE embedding of two graphs.

Corollary 4. *Let* (G_1, G_2) *be two planar graphs whose common graph is induced with a* SEFE *embedding. Then* (G_1, G_2) *admits a* $(0,4)$-*drawing.*

Corollary 5. *Any* SEFE *embedding of two graphs admits a* $(0,9)$-*drawing.*

5 Lower Bounds

In this section we study lower bounds on c for $(0,c)$-drawings of SEFE embeddings. Since SEFE with an arbitrary number of input graphs is equivalent to the problem WEAK REALIZABILITY, which asks for a drawing of a graph specifying for each pair of edges whether they are allowed to cross, an example of Kratochvíl and Matoušek [14] shows that there are SEFE instances that require an exponential number of crossings between two edges. This implies that at least one of them must have an exponential number of bends. However, these graphs do not have sunflower-intersection.

We first show that the results from Theorem 1(ii) and (iii) are tight; examples are given in Fig. 4. Afterwards, we prove that for a SEFE embedding of k graphs with sunflower intersection $\Omega(\sqrt{2}^k/k)$ bends per edge are necessary.

Theorem 5. *There exist* SEFE *embeddings that do not admit a* $(0,0)$-*drawing even when (i) the common graph is biconnected or (ii) the common graph is connected and induced.*

Proof. For (i), consider a SEFE instance whose common graph is a cycle C of length 4. The exclusive edges are the two chords of C, which belong to different graphs and are embedded outside of C; see Fig. 4a. Clearly, at least one of the exclusive edges requires a bend.

For (ii), the common graph is a triangle with tip u and a path of length 2 attached to u. Let the path be u, v, w. Additionally, each G_1 and G_2 contain one exclusive vertex that is adjacent to u and v. The embedding is as shown in Fig. 4b. We claim that this SEFE does not admit a $(0, 0)$-drawing.

Consider the red graph G_1. For a $(0, 0)$-drawing, its exclusive vertex must be positioned such that it sees both u and v. This can only be achieved if the angle at v on the left side of the path uvw is strictly greater than π. However, by applying the same arguments for G_2, we find that the angle at v on the right side of the path uvw must be strictly greater than π. This is obviously not possible simultaneously, and hence a $(0, 0)$-drawing does not exist. $\qquad\square$

Theorem 6. *There exist* SEFE *embeddings of k graphs with sunflower intersection where any $(0, c)$-drawing has $c \in \Omega((\sqrt{2})^k/k)$.*

Proof. We define a graph as follows. Let C be a cycle of length four with vertices N, E, S, W in this clockwise ordering. In the interior, we embed 2^k vertices, each labeled with a distinct binary vector of length k. Fix k colors. We connect each vertex v in the interior of C by an edge of color i to either W or E. If the ith bit of the binary vector associated with v is 1, the edge of color i is vW, otherwise it is vE. Finally, we add edges of color i from S to N for $i = 1, \ldots, k$. We require that edges of the same color do not cross, whereas edges of different colors may cross and are drawn independently. We prove a lower bound on the number of bends in this model.

A SEFE instance can be obtained by interpreting each color as the exclusive edges of its own graph and subdividing each of the colored edges with an exclusive vertex of the corresponding color. In particular, the common graph consists of the cycle C and the 2^k vertices in the interior and each G_i additionally connects each of the 2^k vertices by a path of length 2 to either E or W. The lower bound of the colored instance then also yields a lower bound for this SEFE instance.

Consider an arbitrary admissible drawing of the colored graph. Since each of the 2^k vertices is attached by an edge of color i to either W or E, each of the edges from S to N partitions the points inside C into two sets of 2^{k-1} points. The k SN-edges then divide the interior of C into 2^k areas, each nonempty, as it contains exactly one vertex.

Now consider only the SN-edges. Let $f \geq 2^k$ be the number of such areas in the interior of C, let X be the number of crossings of the SN-edges in the drawing, and let e be the number of arcs on the SN-edges in the arrangement. We thus have a planar graph with $f + 1$ faces (including the outer face), $X + 4$ vertices (including the vertices of C), and $e = 2X + k + 4$ edges. By Euler's formula, we have $X + 4 - (2X + k + 4) + f + 1 = 2$, i.e., $-X - k2^k + 1 \geq 2$ or equivalently $X \geq 2^k + 1 - k - 1$. Since there are only $\binom{k}{2}$ pairs of edges, at least one pair crosses $\frac{2(2^k + 1 - k - 1)}{k(k-1)} = \Omega(2^k/k^2)$ times. Then at least one of them requires $\Omega((\sqrt{2})^k/k)$ bends. $\qquad\square$

6   Conclusion

We have studied the problem of constructing SEFE drawings with polygonal curves of low complexity. Our main result is that any SEFE embedding of two graphs can be drawn with at most nine bends per edge. Fewer bends suffice if the common graph is (bi)connected. Our main open questions concern lower bounds. What is the smallest c_0 such that every SEFE of two graphs admits a $(0, c_0)$-drawing? Is it possible to put some bends on the edges of G in order to save bends on the exclusive edges?

References

1. Angelini, P., Di Battista, G., Frati, F., Jelínek, V., Kratochvíl, J., Patrignani, M., Rutter, I.: Testing planarity of partially embedded graphs. In: Discrete Algorithms (SODA 2010), pp. 202–221. SIAM (2010)
2. Angelini, P., Di Battista, G., Frati, F., Patrignani, M., Rutter, I.: Testing the simultaneous embeddability of two graphs whose intersection is a biconnected or a connected graph. J. Discrete Alg. 14, 150–172 (2012)
3. Angelini, P., Geyer, M., Kaufmann, M., Neuwirth, D.: On a tree and a path with no geometric simultaneous embedding. J. Graph Algorithms Appl. 16(1), 37–83 (2012)
4. Bläsius, T., Karrer, A., Rutter, I.: Simultaneous embedding: Edge orderings, relative positions, cutvertices. In: Wismath, S., Wolff, A. (eds.) GD 2013. LNCS, vol. 8242, pp. 220–231. Springer, Heidelberg (2013)
5. Bläsius, T., Kobourov, S.G., Rutter, I.: Simultaneous embedding of planar graphs. In: Tamassia, R. (ed.) Handbook of Graph Drawing and Visualization. CRC Press (2013)
6. Bläsius, T., Rutter, I.: Simultaneous PQ-ordering with applications to constrained embedding problems. In: Discrete Algorithms (SODA 2013), pp. 1030–1043. SIAM (2013)
7. Erten, C., Kobourov, S.G.: Simultaneous embedding of planar graphs with few bends. J. Graph Algorithms Appl. 9(3), 347–364 (2005)
8. Estrella-Balderrama, A., Gassner, E., Jünger, M., Percan, M., Schaefer, M., Schulz, M.: Simultaneous geometric graph embeddings. In: Hong, S.-H., Nishizeki, T., Quan, W. (eds.) GD 2007. LNCS, vol. 4875, pp. 280–290. Springer, Heidelberg (2008)
9. Gassner, E., Jünger, M., Percan, M., Schaefer, M., Schulz, M.: Simultaneous graph embeddings with fixed edges. In: Fomin, F.V. (ed.) WG 2006. LNCS, vol. 4271, pp. 325–335. Springer, Heidelberg (2006)
10. Haeupler, B., Jampani, K.R., Lubiw, A.: Testing simultaneous planarity when the common graph is 2-connected. J. Graph Algorithms Appl. 17(3), 147–171 (2013)
11. Hong, S.H., Nagamochi, H.: Convex drawings of graphs with non-convex boundary constraints. Discrete Appl. Math. 156(12), 2368–2380 (2008)
12. Jelínek, V., Kratochvíl, J., Rutter, I.: A kuratowski-type theorem for planarity of partially embedded graphs. Computational Geometry Theory & Applications 46(4), 466–492 (2013)
13. Jünger, M., Schulz, M.: Intersection graphs in simultaneous embedding with fixed edges. J. Graph Algorithms Appl. 13(2), 205–218 (2009)
14. Kratochvíl, J., Matoušek, J.: String graphs requiring exponential representations. J. Comb. Theory, Ser. B 53(1), 1–4 (1991)
15. Mchedlidze, T., Nöllenburg, M., Rutter, I.: Drawing planar graphs with a prescribed inner face. In: Wismath, S., Wolff, A. (eds.) GD 2013. LNCS, vol. 8242, pp. 316–327. Springer, Heidelberg (2013)
16. Pach, J., Wenger, R.: Embedding planar graphs at fixed vertex locations. Graphs and Combinatorics 17, 717–728 (2001)
17. Patrignani, M.: On extending a partial straight-line drawing. International Journal of Foundations of Computer Science 17(5), 1061–1069 (2006)
18. Schaefer, M.: Toward a theory of planarity: Hanani-tutte and planarity variants. J. Graph Algorithms Appl. 17(4), 367–440 (2013)
19. Tutte, W.T.: How to draw a graph. London Math. Soc. s3-13(1), 743–767 (1963)

Planar and Quasi Planar
Simultaneous Geometric Embedding[*]

Emilio Di Giacomo[1], Walter Didimo[1], Giuseppe Liotta[1],
Henk Meijer[2], and Stephen Wismath[3]

[1] Dipartimento di Ingegneria, Università degli Studi di Perugia, Italy
{emilio.digiacomo,walter.didimo,giuseppe.liotta}@unipg.it
[2] University College Roosevelt, The Netherlands
h.meijer@ucr.nl
[3] Department of Mathematics and Computer Science, University of Lethbridge, Canada
wismath@uleth.ca

Abstract. A *simultaneous geometric embedding* (*SGE*) of two planar graphs G_1 and G_2 with the same vertex set is a pair of straight-line planar drawings Γ_1 of G_1 and Γ_2 of G_2 such that each vertex is drawn at the same point in Γ_1 and Γ_2. Many papers have been devoted to the study of which pairs of graphs admit a SGE, and both positive and negative results have been proved. We extend the study of SGE, by introducing and characterizing a new class of planar graphs that makes it possible to immediately extend several positive results that rely on the property of strictly monotone paths. Moreover, we introduce a relaxation of the SGE setting where Γ_1 and Γ_2 are required to be *quasi planar* (i.e., they can have crossings provided that there are no three mutually crossing edges). This relaxation allows for the simultaneous embedding of pairs of planar graphs that are not simultaneously embeddable in the classical SGE setting and opens up to several new interesting research questions.

1 Introduction

The *simultaneous embedding* (*SE*) problem is one of the most studied in Graph Drawing since the publication of the first seminal results on the subject by Braß *et al.* [5,6]. Given two planar graphs with the same vertex set $G_1 = (V, E_1)$ and $G_2 = (V, E_2)$, the SE problem asks whether a planar drawing Γ_1 of G_1 and a planar drawing Γ_2 of G_2 exist such that each vertex of V is drawn at the same point in Γ_1 and Γ_2. If so, pair $\langle \Gamma_1, \Gamma_2 \rangle$ is called a *simultaneous embedding* (*SE*) of $\langle G_1, G_2 \rangle$. Several variants and generalizations of the SE problem (e.g., extensions to $k > 2$ graphs) have also been studied. A comprehensive survey on SE can be found in [4].

In this paper we concentrate on the most desirable, but also the most restrictive, setting of the SE problem, namely the *simultaneous geometric embedding* (*SGE*) setting, where drawings Γ_1 and Γ_2 are required to have straight-line edges. A paper by Angelini *et al.* [3] establishes that there exist a tree of depth four and a path that do not admit a

---

[*] Research supported in part by the MIUR project AMANDA "Algorithmics for MAssive and Networked DAta", prot. 2012C4E3KT_001.

C. Duncan and A. Symvonis (Eds.): GD 2014, LNCS 8871, pp. 52–63, 2014.

SGE. On the positive side, a path can always be simultaneously embedded with a tree of depth at most two [3] and with other kinds of trees such as *caterpillars* (i.e., trees that become paths after the removal of the degree-one vertices) or *stars* (trees with at most one vertex of degree greater than one) and their extensions (see e.g., [6]). Other positive results involve simple types of cyclic graphs (see, e.g., [6,7]).

Most of the positive results about SGE rely on reducing one of the two graphs G_1 or G_2 to a path that is realized in a strictly monotone fashion. The fundamental property of a strictly monotone drawing of a path, say in the y-direction, is that it is planar independently of the x-coordinates given to the vertices. This property makes it possible to arbitrarily assign the x-coordinates of the other graph when looking for a SGE. Motivated by this observation, Fowler and Kobourov characterized the class of graphs that can be simultaneously embedded with any given path drawn in a strictly monotone way [11]; they call these graphs the *Unlabeled Level Planar* (*ULP*) graphs. A characterization for ULP trees was previously given in [9].

In this paper we extend the study of SGE in two different directions:

(i) We characterize the class of graphs that have the same property as strictly monotone paths; namely, the planar graphs such that there exists a suitable y-leveling of the vertices for which a planar drawing exists for any x-leveling of the vertices. We prove that this class, which we term *EAP* [1], is a proper sub-class of the ULP; every graph that can be simultaneously embedded with a strictly monotone path is also simultaneously embeddable with an EAP graph, hence our finding immediately enlarges the set of positive results on SGE. We also extend the result in [3], proving that EAP graphs can always be simultaneously embedded with a tree of depth at most two. We remark that the SGE technique in [3] does not rely on a strictly monotone drawing of the path.

(ii) Since SGE is a rather restricting setting, we study simultaneous geometric embeddings where both Γ_1 and Γ_2 are required to be "nearly planar" in some sense. Namely, we require that each of the two drawings is *quasi planar*, i.e., it does not contain three mutually crossing edges. We remark that quasi planar drawings have been widely studied in the literature (see, e.g., [1,2,12,13]); they fall into a line of research called "beyond planarity", which is receiving lots of interest in the graph drawing community. For *simultaneous geometric quasi planar embedding* (*SGQPE*) setting we generalize the ULP and the EAP graph classes, thus obtaining several positive results; for example, we prove that a tree and a path always admit a SGQPE in contrast to the negative result in [3] for the SGE setting. More in general we prove that every tree has a SGQPE with a meaningful subfamily of the outerplanar graphs.

The results of point (i) are presented in Sec. 3 and those of point (ii) are in Sec. 4. Preliminaries are in Sec. 2. Conclusions and open problems are in Sec. 5.

2 Preliminaries

Let $\langle G_1 = (V, E_1), G_2 = (V, E_2) \rangle$ be a pair of planar graphs with the same vertex set. A *simultaneous geometric embedding* (*SGE*) of $\langle G_1, G_2 \rangle$ is a pair of drawings $\langle \Gamma_1, \Gamma_2 \rangle$

---

[1] The explanation of this acronym is clarified in the paper. The same class of graphs is defined with the name of *column planar graphs* in [10].

such that: (i) Γ_i is a planar straight-line drawing of G_i for $i = 1, 2$; (ii) each vertex $v \in V$ is represented by the same point in Γ_1 and Γ_2.

Let G be a planar graph. An x-*leveling* of G is a mapping $\mathcal{X} : V \to \mathbb{R}$ that assigns to each vertex of G a distinct real x-coordinate. A y-*leveling* of G is a mapping $\mathcal{Y} : V \to \mathbb{R}$ that assigns to each vertex of G a distinct real y-coordinate. Let G be a planar graph with a given x-leveling $\mathcal{X}$ and a given y-leveling $\mathcal{Y}$. We denote by $\Gamma(\mathcal{X}, \mathcal{Y})$ a straight-line drawing of G obtained by drawing each vertex v at point $(\mathcal{X}(v), \mathcal{Y}(v))$ and each edge as a straight-line segment between its endpoints. If $\Gamma(\mathcal{X}, \mathcal{Y})$ has no three collinear vertices we say that $\mathcal{X}$ is *general with respect to* $\mathcal{Y}$, and $\mathcal{Y}$ is *general with respect to* $\mathcal{X}$.

Let G be a graph and let Γ be a straight-line drawing of G. Two edges of G are *independent* if they do not have an endvertex in common. The *independent horizontal stabbing number* of Γ, denoted by ihs(Γ), is the maximum number of independent edges of Γ intersected by a horizontal line. The *independent horizontal stabbing number* of G, denoted by ihs(G), is the minimum independent horizontal stabbing number over all straight-line drawings of G. A horizontal line is called a *stabber*; the stabber of equation $y = l$ is the *stabber at l*. Two drawings with the same top-to-bottom order of the vertices have the same independent horizontal stabbing number. The next lemmas are used to remove collinear points in a drawing.

Lemma 1. *Let G be a graph. Let $\mathcal{Y}$ be a y-leveling of G and let $\mathcal{X}$ be an x-leveling of G that is not general with respect to $\mathcal{Y}$. There exists an x-leveling $\mathcal{X}'$ that is general with respect to $\mathcal{Y}$ and such that if two edges cross in $\Gamma(\mathcal{X}', \mathcal{Y})$ then they cross in $\Gamma(\mathcal{X}, \mathcal{Y})$.*

Lemma 2. *Let G be a graph. Let $\mathcal{Y}$ be a y-leveling of G and let $\mathcal{X}$ be an x-leveling of G that is not general with respect to $\mathcal{Y}$. There exists a y-leveling $\mathcal{Y}'$ that is general with respect to $\mathcal{X}$ and such that ihs($\Gamma(\mathcal{X}, \mathcal{Y}')$) = ihs($\Gamma(\mathcal{X}, \mathcal{Y})$).*

3   EAP Graphs and Simultaneous Geometric Embedding

Before defining EAP graphs, we recall the definition of ULP graphs (introduced in [11]), which we rename as *AEP graphs*. Let G be a planar graph. G is an *AEP graph* if for any y-leveling $\mathcal{Y}$ of G, there exists an x-leveling $\mathcal{X}$ such that $\Gamma(\mathcal{X}, \mathcal{Y})$ is planar. G is an *EAP graph* if there exists a y-leveling $\mathcal{Y}$ of G, called *universal y-leveling*, such that for any x-leveling $\mathcal{X}$ that is general with respect to $\mathcal{Y}$, $\Gamma(\mathcal{X}, \mathcal{Y})$ is planar[2]. We denote by AEP the set of AEP graphs and by EAP the set of EAP graphs. Note that, in the definition of EAP graphs we consider only x-levelings that are general with respect to the y-leveling $\mathcal{Y}$. This restriction is necessary since otherwise no graphs other than paths could be EAP. Namely, if three collinear points are allowed, then for any given y-leveling there would exist an x-leveling such that the resulting drawing has edges that overlap each other. These trivial counterexamples are avoided by restricting the definition to those x-levelings that do not cause three collinear points. The interplay between AEP and

---

[2] The names AEP and EAP are acronyms coming from the definitions of the AEP and EAP graphs, respectively. Namely, a graph is an AEP graph if "for **A**ny y-leveling, there **E**xists an x-leveling such that the resulting drawing is **P**lanar", while a graph is EAP if "there **E**xists a y-leveling such that for **A**ny x-leveling the resulting drawing is **P**lanar".

EAP graphs makes it possible to extend several results about SGE previously described in the literature, thanks to the following result (see also Theorem 2 and Corollary 2).

Theorem 1. *Let $\langle G_1, G_2 \rangle$ be a pair of graphs such that $G_1 \in AEP$ and $G_2 \in EAP$. Then $\langle G_1, G_2 \rangle$ admits a SGE.*

Proof. Since G_2 is an EAP graph, there exists a universal y-leveling $\mathcal{Y}$ of G_2. Consider $\mathcal{Y}$ as a y-leveling of G_1; we aim at finding an x-leveling $\mathcal{X}$ of G_1 that is general with respect to $\mathcal{Y}$ and such that $\Gamma(\mathcal{X}, \mathcal{Y})$ is planar. Since $G_1 \in AEP$, then there exists an x-leveling $\mathcal{X}'$ such that $\Gamma_1(\mathcal{X}', \mathcal{Y})$ is a planar drawing of G_1. If $\mathcal{X}'$ is general with respect to $\mathcal{Y}$ we set $\mathcal{X} = \mathcal{X}'$; if $\mathcal{X}'$ is not general with respect to $\mathcal{Y}$, then, by Lemma 1, there exists an x-leveling $\mathcal{X}''$ that is general with respect to $\mathcal{Y}$ and such that $\Gamma(\mathcal{X}'', \mathcal{Y})$ is planar; in this case we set $\mathcal{X} = \mathcal{X}''$. In either case we have the desired x-leveling $\mathcal{X}$. Consider $\mathcal{X}$ as an x-leveling for G_2. Then $\Gamma_2(\mathcal{X}, \mathcal{Y})$ is a planar drawing of G_2 because G_2 is an EAP graph, $\mathcal{Y}$ is a universal y-leveling of G_2 and $\mathcal{X}$ is general with respect to $\mathcal{Y}$. Thus, $\langle \Gamma_1(\mathcal{X}, \mathcal{Y}), \Gamma_2(\mathcal{X}, \mathcal{Y}) \rangle$ is a SGE of $\langle G_1, G_2 \rangle$. $\square$

Characterization of EAP graphs. AEP graphs have been characterized by Fowler and Kobourov [11]. We first show that EAP graphs constitute a subfamily of AEP graphs with stronger properties (Lemma 3) and then we characterize EAP graphs (Theorem 2).

Lemma 3. *Let $G \in EAP$. Then $G \in AEP$.*

Proof. Let $\mathcal{Y}'$ be an arbitrary given y-leveling of G. In order to prove that $G \in AEP$ we must show that there exists an x-leveling $\mathcal{X}'$ such that $\Gamma(\mathcal{X}', \mathcal{Y}')$ is a planar drawing. Since G is an EAP graph, there exists a universal y-leveling $\mathcal{Y}$ of G. We define an x-leveling $\mathcal{X}$ for G as $\mathcal{X} = \mathcal{Y}'$. In other words we assign to each vertex v an x-coordinate $\mathcal{X}(v)$ that is equal to the y-coordinate $\mathcal{Y}'(v)$ assigned to v by $\mathcal{Y}'$. We have two cases:

Case 1: $\mathcal{X}$ is general with respect to $\mathcal{Y}$. In this case, since $G \in EAP$, $\Gamma(\mathcal{X}, \mathcal{Y})$ is a planar drawing. If we set $\mathcal{X}' = \mathcal{Y}$, we have that $\Gamma(\mathcal{X}', \mathcal{Y}')$ is the drawing $\Gamma(\mathcal{X}, \mathcal{Y})$ rotated by $90°$ and hence it is planar.

Case 2: $\mathcal{X}$ is not general with respect to $\mathcal{Y}$. By Lemma 2 there exists a y-leveling $\mathcal{Y}''$ that is general with respect to $\mathcal{X}$ such that $\mathsf{ihs}(\Gamma(\mathcal{X}, \mathcal{Y}'')) = \mathsf{ihs}(\Gamma(\mathcal{X}, \mathcal{Y}))$. Let $\mathcal{X}''$ be an x-leveling that is general with respect to $\mathcal{Y}$; we have $\mathsf{ihs}(\Gamma(\mathcal{X}, \mathcal{Y})) = \mathsf{ihs}(\Gamma(\mathcal{X}'', \mathcal{Y}))$ because $\Gamma(\mathcal{X}, \mathcal{Y})$ and $\Gamma(\mathcal{X}'', \mathcal{Y})$ have the same y-leveling. Since G is an EAP graph, $\mathsf{ihs}(\Gamma(\mathcal{X}'', \mathcal{Y})) = 1$ and hence $\mathsf{ihs}(\Gamma(\mathcal{X}, \mathcal{Y}'')) = 1$, which implies that $\Gamma(\mathcal{X}, \mathcal{Y}'')$ is planar (otherwise its independent horizontal stabbing number would be at least two). If we set $\mathcal{X}' = \mathcal{Y}''$ then $\Gamma(\mathcal{X}', \mathcal{Y}')$ is $\Gamma(\mathcal{X}, \mathcal{Y}'')$ rotated by $90°$, and hence is planar.

In both cases we have found an x-leveling $\mathcal{X}'$ such that $\Gamma(\mathcal{X}', \mathcal{Y}')$ is a planar drawing. Since $\mathcal{Y}'$ is arbitrary, we have that $G \in AEP$. $\square$

By Lemma 3 we have that $EAP \subseteq AEP$. By the characterization of Fowler and Kobourov [11] we know precisely the graphs in AEP; therefore in order to characterize the class EAP we can establish which graphs of the set AEP are EAP graphs. The family of AEP graphs is the union of the following three families: *radius-2 stars*, *extended degree-3 spiders*, and *generalized caterpillars*. In order to recall the definition of these families we start by defining four gadgets each having two vertices called *poles*.

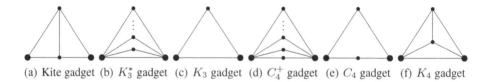

(a) Kite gadget (b) K_3^* gadget (c) K_3 gadget (d) C_4^+ gadget (e) C_4 gadget (f) K_4 gadget

Fig. 1. Illustration of the different gadgets. Bigger vertices are the poles of the gadget

Given an edge e of a graph, replacing e with one of the gadgets means to replace e with the gadget so that the end-vertices of e coincide with the two poles of the gadget. The *kite gadget* is a 4-cycle plus an edge connecting the two vertices that are not poles (Fig. 1(a)). A K_3^* *gadget* is a set of $k \geq 1$ 3-cycles sharing the two poles and the edge connecting them (Fig. 1(b)). A K_3^* gadget consisting of a single cycle is called a K_3 *gadget* (Fig. 1(c)). A C_4^+ *gadget* is a set of $k \geq 2$ paths with two edges, connecting the two poles (Fig. 1(d)). A C_4^+ gadget consisting of exactly two paths is called a C_4 *gadget* (Fig. 1(e)). A K_4 *gadget* is the complete graph on four vertices (Fig. 1(f)). The poles of a K_4 gadget are any two of its vertices. A *star* is a graph $K_{1,k}$ for some $k \geq 3$. A *radius-2 star* is a star in which at least one edge has been subdivided once (see Fig. 2(a)). A *degree-3 spider* is an arbitrary subdivision of $K_{1,3}$. A *1-connected extended degree-3 spider* is a degree-3 spider with two optional additional edges: an edge connecting two vertices adjacent to the unique degree-3 vertex and an edge connecting two leaves (see Fig. 2(b)). A *2-connected extended degree-3 spider* is either a cycle or a cycle where an edge is replaced by a K_3 gadget, a C_4 gadget, or a kite gadget (see Fig. 2(c)). An *extended degree-3 spider* is either a 1-connected extended degree-3 spider or a 2-connected extended degree-3 spider. A *caterpillar* is a tree such that removing all leaves we get a path, called the *spine* of the caterpillar. A *generalized caterpillar* is a caterpillar in which each edge of the spine can be replaced by a K_3^* gadget, or a C_4^+ gadget, or a kite gadget, and for each endvertex u of the spine, one edge connecting u to a leaf can be replaced by a K_3^* gadget, or a C_4^+ gadget, or a kite gadget, or a K_4 gadget (see Fig. 2(d)).

We define a new family of graphs that we call *fat caterpillars* and that is a subfamily of generalized caterpillars. A *fat caterpillar* is a graph obtained from a caterpillar by replacing some of the edges of the spine with a K_3 gadget (see Fig. 2(e)). Notice that fat caterpillars are exactly the generalized caterpillars with no cycle of length larger than three. Let G' be the subgraph of G obtained by removing all degree-one vertices; G' consists of a path $\Pi = (v_1, v_2, \ldots, v_k)$ plus a set of vertices each adjacent to two consecutive vertices of Π. The vertices of Π are called the *path vertices* of G; in particular v_1 and v_k are called *extreme path vertices*. The remaining vertices of G' are called the *tip vertices* of G; a tip vertex that is adjacent to v_i and v_{i+1} will be denoted by u_i. Each vertex of G of degree one is adjacent to a path vertex. The degree-one vertices adjacent to v_i will be denoted by $w_{i,j}$, with $j = 1, 2, \ldots, h_i$, where $h_i \geq 0$. We start with a technical lemma that will be used to characterize EAP graphs.

Lemma 4. *A graph G is an EAP graph if and only if* $\mathrm{ihs}(G) = 1$.

The next lemma can be proved by showing that, in each case, $\mathrm{ihs}(G) > 1$, which, by Lemma 4, implies that $G \notin EAP$.

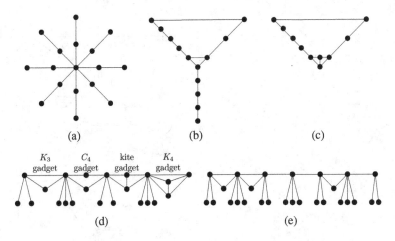

Fig. 2. (a) A radius-2 star. (b) A 1-connected extended degree-3 spider. (c) A 2-connected extended degree-3 spider (containing a kite gadget). (d) A generalized caterpillar. (e) A fat caterpillar.

Lemma 5. *Let G be a graph such that either: (i) G is a radius-2 star that is not a generalized caterpillar; or (ii) G contains a cycle of length at least 4; or (iii) G is an extended degree-3 spider that is not a generalized caterpillar. Then $G \notin EAP$.*

Theorem 2. *A planar graph G is an EAP graph if and only if it is a fat caterpillar.*

Proof. "Only if part". If G is an EAP graph, then by Lemma 3 it is also AEP and therefore it is either a radius-2 star, or an extended degree-3 spider, or a generalized caterpillar. By Lemma 5, G must be a generalized caterpillar and cannot contain a cycle of length four. Hence it must be a fat caterpillar.

"If part". Let G be a fat caterpillar. We prove that $\mathrm{ihs}(G) = 1$. We define a y-leveling $\mathcal{Y}$ of G as follows. For each path vertex v_i $(1 \leq i \leq k)$ of G we set $\mathcal{Y}(v_i) = 2i$. For each tip vertex u_i $(1 \leq i \leq k)$ of G we set $\mathcal{Y}(u_i) = \mathcal{Y}(v_i) + 1$. For each degree-one vertex $w_{i,j}$ we set $\mathcal{Y}(w_{i,j}) = \mathcal{Y}(v_i) + \frac{j}{h_i + 1}$ $(j = 1, 2, \ldots, h_i)$. Observe that the vertices of G have been assigned a different value and therefore $\mathcal{Y}$ is a valid y-leveling. Consider now an arbitrary x-leveling $\mathcal{X}$. We show that $\mathrm{ihs}(\Gamma(\mathcal{X}, \mathcal{Y})) = 1$, which implies that $\mathrm{ihs}(G) = 1$. Consider a stabber ℓ at l, $l \in \mathbb{R}$. If $l < 2$ or $l > 2k$, then ℓ does not intersect any edge of $\Gamma(\mathcal{X}, \mathcal{Y})$. If $l = 2i$ for $1 \leq i \leq k$, then ℓ passes through vertex v_i and it intersects only the edges incident to v_i, which are not independent. If $l = 2i + 1$ for $1 \leq i \leq k - 1$, then ℓ either intersects only the edge (v_i, v_{i+1}) (if the tip vertex u_i does not exist), or it intersects the three edges (v_i, v_{i+1}), (v_i, u_i), and (u_i, v_{i+1}) (if the tip vertex u_i exists) no two of which are independent. If $2i < l < 2i + 1$ for $1 \leq i \leq k - 1$, then ℓ intersects the edge (v_i, v_{i+1}), possibly the edge (v_i, u_i), and possibly some of the edges $(v_i, w_{i,j})$ $(j = 1, 2, \ldots, h_i)$. No two of these edges are independent. Finally, if $2i + 1 < l < 2i + 2$ for $1 \leq i \leq k - 1$, then ℓ intersects the edge (v_i, v_{i+1}) and possibly the edge (u_i, v_{i+1}), which are not independent. Thus, no stabber intersects two independent edges in $\Gamma(\mathcal{X}, \mathcal{Y})$ and therefore $\mathrm{ihs}(\Gamma(\mathcal{X}, \mathcal{Y})) = 1$. It follows that $\mathrm{ihs}(G) = 1$ and, by Lemma 4, $G \in EAP$. $\qquad\square$

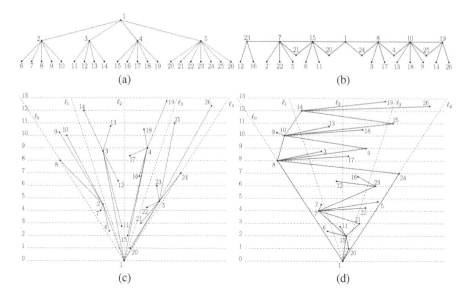

Fig. 3. (a)-(b) A pair of graphs $\langle G_1, G_2 \rangle$; G_1 is a tree of depth two while G_2 is a fat caterpillar. G_1 and G_2 have the same vertex set; each vertex has the same label in (a) and (b). (c)-(d) A simultaneous embedding of $\langle G_1, G_2 \rangle$.

An immediate consequence of Lemma 3 and Theorem 2 is that EAP is properly contained in AEP (Corollary 1). Theorems 1 and 2 imply Corollary 2.

Corollary 1. $EAP \subset AEP$.

Corollary 2. *Let $\langle G_1, G_2 \rangle$ be a pair of graphs such that G_1 is either a radius-2 star, an extended degree-3 spider or a generalized caterpillar, and G_2 is a fat caterpillar. Then $\langle G_1, G_2 \rangle$ admits a SGE.*

The characterization of Theorem 2 implies that all planar graphs that are known to be simultaneously embeddable with a strictly monotone path [4] are in fact simultaneously embeddable with a fat caterpillar (i.e., EAP graphs). Moreover, EAP graphs open up two further research directions. A first research direction is to extend to fat caterpillars other existing SGE results that involve paths that are not realized in a strictly monotone fashion. A second research direction is to extend the result of Theorem 1 to simultaneous drawings that are not planar but where some crossing configurations are forbidden. Regarding the first research direction, we prove below that fat caterpillars can be used to generalize a result by Angelini et al. [3] in which it is proved that a path and a tree of depth two always admit a SGE. As for "nearly planar" simultaneous geometric embeddings, we devote Sec. 4 to quasi planar drawings.

Theorem 3. *Let $\langle G_1, G_2 \rangle$ be a pair of graphs such that G_1 is a tree of depth 2 and G_2 is a fat caterpillar. Then $\langle G_1, G_2 \rangle$ admits a SGE.*

Sketch of Proof: The technique is inspired by the one described in [3]. Since G_1 is a tree of depth 2, removing the root of G_1 we obtain a set of stars $S_1, S_2, \ldots, S_h$. Let

$\ell_0, \ell_1, \ldots, \ell_h$ be a set of half-lines each having initial point $(0,0)$ and all extending in the upper half-plane, ordered as they are encountered rotating clockwise starting at the negative x-axis. Let C_i be the cone delimited by ℓ_{i-1} and ℓ_i. In order to obtain a planar drawing of a star it is sufficient to map its vertices to a set of points in general position, i.e., a set of points such that there are no three collinear points. So in order to draw G_1 we can map the vertices of each S_i ($i = 1, 2, \ldots, h$) to points in general position of a different cone C_j and the root of G_1 to point $(0,0)$ shared by all regions. In this way each star S_i is drawn planarly inside a different C_j (thus different stars do not cross each other) and the root vertex can be connected without crossings to its adjacent vertex of each star S_i (see Fig. 3(c)). It can be shown that G_2 admits a drawing such that the vertex corresponding to the root of G_1 is mapped to $(0,0)$ and the points corresponding to each star S_i are mapped to a different cone C_j (see Fig. 3(d)). $\quad\square$

4 Simultaneous Geometric Quasi Planar Embeddings

We introduce and study geometric simultaneous embeddings where each of the two drawings are not required to be planar but only quasi planar. We implicitly assume that all drawings are straight-lines. Let G be a graph and let Γ be a drawing of G. Γ is *quasi planar* if there are no three mutually crossing edges. A graph is *quasi planar* if it admits a quasi planar drawing. Let $\langle G_1 = (V, E_1), G_2 = (V, E_2) \rangle$ be a pair of quasi planar graphs with the same vertex set. A *simultaneous geometric quasi planar embedding* (*SGQPE*) of $\langle G_1, G_2 \rangle$ is a pair of drawings $\langle \Gamma_1, \Gamma_2 \rangle$ such that: (i) Γ_i is a quasi planar straight-line drawing of G_i for $i = 1, 2$; (ii) each vertex $v \in V$ is represented by the same point in Γ_1 and Γ_2. We extend the definition of AEP and EAP graphs to the quasi planar case as follows. A quasi planar graph G is an *AEQP graph* if for any y-leveling $\mathcal{Y}$ of G, there exists an x-leveling $\mathcal{X}$ such that $\Gamma(\mathcal{X}, \mathcal{Y})$ is quasi planar. G is an *EAQP graph* if there exists a y-leveling $\mathcal{Y}$ of G, called a *universal quasi planar leveling*, such that for any x-leveling $\mathcal{X}$ that is general with respect to $\mathcal{Y}$, $\Gamma(\mathcal{X}, \mathcal{Y})$ is quasi planar. Denote by $AEQP$ and $EAQP$ the set of AEQP and EAQP graphs, respectively. The next result generalizes Theorem 1.

Theorem 4. *Let* $\langle G_1, G_2 \rangle$ *be a pair of graphs such that* $G_1 \in AEQP$ *and* $G_2 \in EAQP$. *Then* $\langle G_1, G_2 \rangle$ *admits a SGQPE.*

Motivated by Theorem 4 we study the interplay between AEQP and EAQP graphs and also their relationships with AEP and EAP graphs, which are summarized by Fig. 4 and Theorem 6. We first show that any AEP graph is an EAQP graph (Lemma 7). The next technical lemma generalizes Lemma 4 (the proof is similar).

Lemma 6. *A graph G is an EAQP graph if and only if* ihs$(G) \leq 2$.

Lemma 7. $AEP \subset EAQP$.

Sketch of Proof: Let G be an AEP graph; it is either a radius-2 star, or a generalized degree-3 spider, or a generalized caterpillar. For each type of graph we describe a y-leveling that is universal quasi planar. The proof of correctness is omitted.

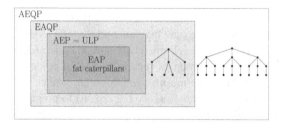

Fig. 4. Proper inclusions among the different classes of graphs studied in this paper

Radius-2 Star. (see Fig. 5(a)). Let G be a radius-2 star and let v be the unique vertex of G with degree larger than two. Denote by $u_1, u_2, \ldots, u_k$ the neighbors of v and denote by w_i the neighbor of u_i different from v (if it exists). We define a y-leveling $\mathcal{Y}$ of G as follows. We set $\mathcal{Y}(v) = 0$, $\mathcal{Y}(u_i) = 2i - 1$, and $\mathcal{Y}(w_i) = 2i$.

Extended Degree-3 Spider. Let G be an extended degree-3 spider. Suppose first that G is a 1-connected extended degree-3 spider (see Fig. 5(b)). We can assume that G is *full*, i.e., it has the two optional edges (if G is not full we can temporarily add the additional edges to G). Since G is full it consists of a cycle C, plus a vertex v connected to two consecutive vertices of C, plus a path P attached to v. Denote by $u_1, u_2, \ldots u_k$ the vertices of C (in the order they appear on C), where u_1 and u_2 are the vertices adjacent to v. Denote by $w_1, w_2, \ldots, w_h$ the vertices of P (in the order they appear on P), where w_1 is the vertex adjacent to v. We define a y-leveling $\mathcal{Y}$ of G as follows. We set $\mathcal{Y}(v) = 0$, $\mathcal{Y}(u_i) = i$, and $\mathcal{Y}(w_i) = -i$. Suppose now that G is a 2-connected extended degree-3 spider (see Fig. 5(c)). If G is a cycle where one edge has been replaced by a K_3 gadget, then it is a subgraph of a full 1-connected extended degree-3 spider and therefore a y-leveling for G can be defined as in the previous case. If G is a cycle where one edge has been replaced by a C_4 gadget, then G is the subgraph of a 2-connected extended degree-3 spider consisting of a cycle where one edge has been replaced by a kite gadget, and a y-leveling for G can be defined as in the next case. Thus, suppose that G is a cycle where one edge has been replaced by a kite gadget. In this case G consists of a cycle C plus a vertex v adjacent to three consecutive vertices of C. Denote by $u_1, u_2, \ldots u_k$ the vertices of C (in the order they appear on C), where u_1, u_2, and u_3 are the vertices adjacent to v. We define a y-leveling $\mathcal{Y}$ setting $\mathcal{Y}(v) = 0$ and $\mathcal{Y}(u_i) = i$.

Generalized Caterpillar. (see Fig. 5(d)). Let G be a generalized caterpillar. G consists of a caterpillar C where some edges of the spine and (possibly) two non spine edges have been replaced by a gadget. We first describe a y-leveling $\mathcal{Y}$ for C and then extend it to the vertices of G that are not in C. Let $u_1, u_2, \ldots, u_k$ be the vertices of the spine of C. If an edge connecting u_1 to a leaf is replaced by a gadget in G, then let this leaf be denoted by u_0; analogously, if an edge connecting u_k to a leaf is replaced by a gadget in G, then let this leaf be denoted by u_{k+1}. We set $\mathcal{Y}(u_i) = 2i$ ($i = 0, 1, \ldots, k + 1$). Denote by $w_{i,1}, w_{i,2}, \ldots w_{i,h}$ the leaves adjacent to u_i. We set $\mathcal{Y}(w_{i,j}) = \mathcal{Y}(u_i) + \frac{j}{h+1}$. Suppose that edge (u_i, u_{i+1}) ($i = 0, 1, \ldots, k$) is replaced by a gadget γ in G. Let $v_{i,1}, v_{i,2}, \ldots v_{i,h}$ ($h \geq 1$) be the vertices of γ other than u_i and u_{i+1}. We set $\mathcal{Y}(v_{i,j}) = 2i + 1 + \frac{j-1}{h}$ (for $j = 1, 2, \ldots, h$).

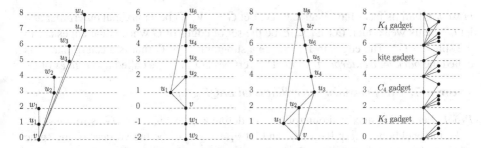

Fig. 5. Illustration of the y-levelings described in the proof of Lemma 7. (a) A radius-2 star. (b) A full 1-connected extended degree-3 spider. (c) A 2-connected extended degree-3 spider (with an edge replaced by a kite gadget). (d) A generalized caterpillar.

From the discussion above we have $AEP \subseteq EAQP$. Since the tree T in Fig. 6(a), which is not in AEP [11], is in $EAQP$ (see Fig. 6(b)), we have $AEP \neq EAQP$. $\square$

The proof of the next lemma is analogous to the one of Lemma 3.

Lemma 8. *Let* $G \in EAQP$. *Then* $G \in AEQP$.

It is natural to ask whether $AEQP$ and $EAQP$ coincide or not. Also, it is natural to study which families of graphs belong to $AEQP$ and to $EAQP$. Theorem 5 answers the first question and gives a result about the second research direction. Theorem 6 summarizes the relationships between EAP, AEP, $EAQP$, and $AEQP$.

Theorem 5. *All trees are AEQP graphs and there exist trees that are not EAQP graphs.*

Sketch of Proof: The fact that every tree is an EAQP graph follows easily from the result in [8]. A tree that is not in $EAQP$ is shown in Fig. 6(c). $\square$

Theorem 6. $EAP \subset AEP \subset EAQP \subset AEQP$.

In the next theorem we prove that *maximal outerpillars* are EAQP graphs. An outerplanar graph is called an *outerpillar* if its weak dual is a a *caterpillar*. An outerpillar is

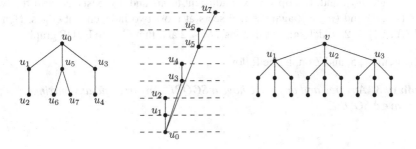

Fig. 6. (a) A tree T that is an EAQP graph but is not an AEP graph. (b) A universal quasi planar y-leveling of T. (c) A tree that is not an EAQP graph.

maximal if all its faces have degree three. Let $G = (V, E)$ be a maximal outerpillar and let G^* be its weak dual. Let $E_1 \subseteq E$ be the set of edges of G whose dual edges are in the spine of G^*. The edges in E_1 induce a subgraph C of G that is a caterpillar. C is called the *backbone caterpillar* of G. The vertices not in C are called *tip vertices* of G.

Theorem 7. *Every maximal outerpillar is an EAQP graph.*

Proof. Let G be a maximal outerpillar. We define a y-leveling $\mathcal{Y}$ of G as follows. Let C be the backbone caterpillar of G, and let $v_1, v_2, \ldots, v_h$ be the vertices of the spine of C in the order they appear along the spine (for a chosen walking direction). Let n_i (with $n_i \geq 0$) be the number of degree-one vertices of C adjacent to v_i (for $i = 1, 2, \ldots, h$). We set $\mathcal{Y}(v_1) = 0$ and for each vertex v_i $(1 < i \leq h)$ of G we set $\mathcal{Y}(v_i) = \sum_{j=1}^{i-1} 2(n_j + 1)$. Let $w_{i,1}, w_{i,2}, \ldots, w_{i,n_i}$ be the degree one nodes of C that are adjacent to v_i (for $i = 1, 2, \ldots, h$); we set $\mathcal{Y}(w_{i,j}) = \mathcal{Y}(v_i) + 2 \cdot j$. Finally, let u be a tip vertex of G; u is adjacent to two vertices u_1 and u_2 of C. Suppose that $\mathcal{Y}(u_1) < \mathcal{Y}(u_2)$, then we set $\mathcal{Y}(u) = \mathcal{Y}(u_2) - 1$.

Since the vertices of G have been assigned a different value, $\mathcal{Y}$ is a valid y-leveling. Consider now an arbitrary x-leveling $\mathcal{X}$. We show that $\mathrm{ihs}(\Gamma(\mathcal{X}, \mathcal{Y})) \leq 2$. Let ℓ be the stabber at l, with $l \in \mathbb{R}$. If $l < 0$ or $l > 2n_c - 2$ (where n_C is the number of vertices of C) then ℓ does not intersect any edge. If $l = 2i$ for some $0 \leq i \leq n_c - 1$, then ℓ passes through a vertex w of C. If w is a vertex of the spine of C, say v_j (for $1 \leq j \leq h$), then ℓ intersects all edges incident to v_j and at most two edges both incident to v_{j-1}. Thus ℓ intersects at most two independent édges. If w is a degree-one vertex of C, say $w_{i,j}$, then ℓ intersects all edges incident to $w_{i,j}$ and some of the edges incident to v_i. Also, in this case it intersects at most two independent edges. Consider now the case when $2i < l < 2i + 2$, and let w_1 and w_2 be the two vertices of C such that $\mathcal{Y}(w_1) = 2i$ and $\mathcal{Y}(w_2) = 2i + 2$. Suppose first that w_1 is a vertex of the spine of C, say v_j $(1 \leq j \leq h)$; in this case w_2 is either the next vertex of the spine of C, i.e. v_{j+1}, or a degree-one vertex adjacent to v_j, i.e. $w_{i,1}$. In both cases ℓ intersects some edges incident to v_j, edge (v_{j-1}, w_2), and, if a tip vertex u adjacent to v_{j-1} and w_2 exists, at most two edges (v_{j-1}, u) and (w_2, u); in any case ℓ intersects at most two independent edges. Suppose then that w_1 is a degree-one vertex of C, say $w_{j,k}$ $(1 \leq j \leq h, 1 \leq k \leq n_j)$; in this case w_2 is either another degree-one vertex adjacent to v_j, i.e. $w_{j,k+1}$, or the next vertex on the spine of C, i.e. v_{j+1}. In both cases ℓ intersects some edges incident to v_j, the edge (w_1, w_2), and, if a tip vertex u adjacent to w_1 and w_2 exists, at most the two edges (w_1, u) and (w_2, u); again, ℓ intersects at most two independent edges. Hence $\mathrm{ihs}(\Gamma(\mathcal{X}, \mathcal{Y})) \leq 2$, which implies that $\mathrm{ihs}(G) \leq 2$ and that G is an EAQP graph. $\square$

Theorems 4, 5, and 7 imply the following.

Corollary 3. *Any tree and any cycle have a SGQPE. Any tree and any maximal outerpillar have a SGQPE.*

5   Discussion and Open Problems

Several open problems arise from the study of the SGQPE:

Problem 1. Fowler and Kobourov [11] characterized AEP graphs, while in this paper we provided a characterization of EAP graphs. Thus the first obvious open problem is to characterize AEQP and EAQP graphs.

Problem 2. Does every pair of trees (or even every pair of planar graphs) admit a SGQPE? So far we were only able to prove the following.

Theorem 8. *There exists a pair of quasi planar graphs that does not admit a SGQPE.*

Problem 3. Extend the study of simultaneous embeddability to other families of drawings with forbidden crossing configurations, such as k-planar, RAC, LAC, fan-planar, fan-crossing-free-planar drawings.

References

1. Ackerman, E., Tardos, G.: On the maximum number of edges in quasi-planar graphs. J. of Combinatorial Theory, Series A 114(3), 563–571 (2007)
2. Agarwal, P.K., Aronov, B., Pach, J., Pollack, R., Sharir, M.: Quasi-planar graphs have a linear number of edges. Combinatorica 17(1), 1–9 (1997)
3. Angelini, P., Geyer, M., Kaufmann, M., Neuwirth, D.: On a tree and a path with no geometric simultaneous embedding. J. of Graph Algorithms and Applications 16(1), 37–83 (2012)
4. Bläsius, T., Kobourov, S.G., Rutter, I.: Simultaneous embedding of planar graphs. In: Tamassia, R. (ed.) Handbook of Graph Drawing and Visualization. CRC Press (2014)
5. Brass, P., Cenek, E., Duncan, C.A., Efrat, A., Erten, C., Ismailescu, D., Kobourov, S.G., Lubiw, A., Mitchell, J.S.B.: On simultaneous planar graph embeddings. In: Dehne, F., Sack, J.-R., Smid, M. (eds.) WADS 2003. LNCS, vol. 2748, pp. 243–255. Springer, Heidelberg (2003)
6. Braß, P., Cenek, E., Duncan, C.A., Efrat, A., Erten, C., Ismailescu, D., Kobourov, S.G., Lubiw, A., Mitchell, J.S.B.: On simultaneous planar graph embeddings. Comput. Geom. 36(2), 117–130 (2007)
7. Cabello, S., van Kreveld, M.J., Liotta, G., Meijer, H., Speckmann, B., Verbeek, K.: Geometric simultaneous embeddings of a graph and a matching. J. Graph Algorithms and Applications 15(1), 79–96 (2011)
8. Didimo, W., Kaufmann, M., Liotta, G., Okamoto, Y., Spillner, A.: Vertex angle and crossing angle resolution of leveled tree drawings. Inform. Process. Lett. 112(16), 630–635 (2012)
9. Estrella-Balderrama, A., Fowler, J.J., Kobourov, S.G.: Characterization of unlabeled level planar trees. Computational Geometry 42(6-7), 704–721 (2009)
10. Evans, W., Kusters, V., Saumell, M., Speckmann, B.: Column planarity and partial simultaneous geometric embedding. In: Duncan, C., Symvonis, A. (eds.) GD 2014. LNCS, vol. 8871, pp. 259–271. Springer, Heidelberg (2014)
11. Fowler, J.J., Kobourov, S.G.: Characterization of unlabeled level planar graphs. In: Hong, S.-H., Nishizeki, T., Quan, W. (eds.) GD 2007. LNCS, vol. 4875, pp. 37–49. Springer, Heidelberg (2008)
12. Fox, J., Pach, J., Suk, A.: The number of edges in k-quasi-planar graphs. SIAM J. on Discrete Mathematics 27(1), 550–561 (2013)
13. Valtr, P.: On geometric graphs with no k pairwise parallel edges. Discrete & Computational Geometry 19(3), 461–469 (1998)

Simultaneous Embeddability of Two Partitions

Jan Christoph Athenstädt[1], Tanja Hartmann[2], and Martin Nöllenburg[2]

[1] Department of Computer and Information Science, University of Konstanz, Germany
[2] Institute of Theoretical Informatics, Karlsruhe Institute of Technology (KIT), Germany

Abstract. We study the simultaneous embeddability of a pair of partitions of the same underlying set into disjoint blocks. Each element of the set is mapped to a point in the plane and each block of either of the two partitions is mapped to a region that contains exactly those points that belong to the elements in the block and that is bounded by a simple closed curve. We establish three main classes of simultaneous embeddability (*weak*, *strong*, and *full* embeddability) that differ by increasingly strict well-formedness conditions on how different block regions are allowed to intersect. We show that these simultaneous embeddability classes are closely related to different planarity concepts of hypergraphs. For each embeddability class we give a full characterization. We show that (i) every pair of partitions has a weak simultaneous embedding, (ii) it is NP-complete to decide the existence of a strong simultaneous embedding, and (iii) the existence of a full simultaneous embedding can be tested in linear time.

1 Introduction

Pairs of partitions of a given set of objects occur naturally when evaluating two alternative clusterings in the field of data analysis and data mining. A *clustering* partitions a set of objects into *blocks* or *clusters*, such that objects in the same cluster are more similar (according to some notion of similarity) than objects in different clusters. There are a multitude of clustering algorithms that use, e.g., an underlying graph structure or an attribute-based distance measure to define similarities. Many algorithms also provide configurable parameter settings. Consequently, different algorithms return different clusterings and judging which clustering is the most meaningful with respect to a certain interpretation of the data must be done by a human expert. For a structural comparison of two clusterings several numeric measures exist [20], however, a single numeric value hardly shows where the clusterings agree or disagree. Hence, a data analyst may want to compare different clusterings visually, which motivates the study of simultaneous embeddability of two partitions.

We provide fundamental characterizations and complexity results regarding the simultaneous embeddability of a pair of partitions. While simultaneous embeddability can generally be defined for any number $k \geq 2$ of partitions, we focus on the basic case of embedding *two* partitions, which is also the most relevant one in the data analysis application. We propose to embed two alternative partitions of the same set U into the plane $\mathbb{R}^2$ by mapping each element of U to a unique point and each block (of either of the two partitions) to a region bounded by a simple closed curve. Each block region must contain all points that belong to elements in that block and no point whose element belongs to a different block. Hence, in total, each point lies inside two block regions.

C. Duncan and A. Symvonis (Eds.): GD 2014, LNCS 8871, pp. 64–75, 2014.
© Springer-Verlag Berlin Heidelberg 2014

A simultaneous embedding of two partitions shares certain properties with set visualizations like Euler or Venn diagrams [8,12,19]. Its readability will be affected by well-formedness conditions for the intersections of the different block regions. Accordingly, we define a (strict) hierarchy of embeddability classes based on increasingly tight well-formedness conditions: *weak*, *strong*, and *full* embeddability. We show that (i) any two partitions are weakly embeddable, (ii) the decision problem for strong embeddability is NP-complete, and (iii) there is a linear-time decision algorithm for full embeddability. We fully characterize the embeddability classes in terms of the existence of a planar support (strong embeddability) or in terms of the planarity of the bipartite map (full embeddability). Interestingly, both concepts are closely related to hypergraph embeddings and different notions of hypergraph planarity. Our NP-completeness result implies that vertex-planarity testing of 2-regular hypergraphs is also NP-complete.

1.1 Related Work

In information visualization there are a large variety of techniques for visualizing clusters of objects, some of which simply map objects to (colored) points so that spatial proximity indicates object similarity [6, 16], others explicitly visualize clusters or general sets as regions in the plane [9, 19]. These approaches are visually similar to Euler diagrams [8, 12], however, they do not give hard guarantees on the final set layout, e.g., in terms of intersection regions or connectedness of regions, nor do they specifically consider the simultaneous embedding of two or more clusterings or partitions.

Clustered planarity is a concept in graph drawing that combines a planar graph layout with a drawing of the clusters of a single hierarchical clustering. Clusters are represented as regions bounded by simple closed and pairwise crossing-free curves. Such a layout is called *c-planar* if no edge crosses a region boundary more than once [11].

The simultaneous embedding of two planar graphs on the same vertex set is a topic that is well studied in the graph drawing literature, see the recent survey of Bläsius et al. [2]. In a simultaneous graph embedding each vertex is located at a unique position and edges contained in both graphs are represented by the same curve for both graphs. The remaining (non-shared) edges are embedded so that each graph layout by itself is crossing-free, but edges from the first graph may cross edges in the second graph.

Some of our results and concepts in this paper can be seen as a generalization of simultaneous graph embedding to simultaneous hypergraph embedding if we consider blocks as hyperedges: all vertices are mapped to unique points in the plane and two hyperedges, represented as regions bounded by simple closed curves, may only intersect if they belong to different hypergraphs or if they share common vertices. Several concepts for visualizing a single hypergraph are known [4, 5, 14, 15, 17], but to the best of our knowledge the simultaneous layout of two or more hypergraphs has not been studied.

1.2 Preliminaries

Let $U = \{u_1, \ldots, u_m\}$ be a finite universe. A *partition* $\mathcal{P} = \{B_1, \ldots, B_n\}$ of U groups the elements of U into disjoint *blocks*, i.e., every element $u \in U$ is contained in exactly one block $B_i \in \mathcal{P}$. In this paper, we consider pairs $\{\mathcal{P}_0, \mathcal{P}_1\}$ of partitions of the same

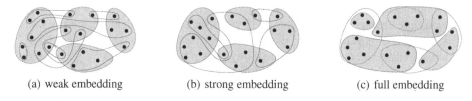

(a) weak embedding   (b) strong embedding   (c) full embedding

Fig. 1. Examples of simultaneous embeddings of two partitions

universe U, i.e., each element $u \in U$ is contained in one block of $\mathcal{P}_0$ and in one block of $\mathcal{P}_1$. In the following we often omit to mention U explicitly.

Let $\mathcal{S}$ be a collection of subsets of U. An *embedding* Γ of $\mathcal{S}$ maps every element $u \in U$ to a distinct point $\Gamma(u) \in \mathbb{R}^2$ and every set $S \in \mathcal{S}$ to a simple, bounded, and closed region $\Gamma(S) \subset \mathbb{R}^2$ such that $\Gamma(u) \in \Gamma(S)$ if and only if $u \in S$. Moreover, we require that each contiguous intersection between the boundaries of two regions is in fact a *crossing point* $p \in \mathbb{R}^2$, i.e., the local cyclic order of the boundaries alternates around p. A *simultaneous embedding* Γ of a pair of partitions $\{\mathcal{P}_0, \mathcal{P}_1\}$ is an embedding of the union $\mathcal{P}_0 \cup \mathcal{P}_1$ of the two partitions. We define $R_B = \Gamma(B)$ as the *block region* of a block B and denote its boundary by ∂R_B. Figure 1 shows examples of simultaneous embeddings in the three different embedding classes to be defined in Section 2.

A simultaneous embedding Γ induces a subdivision of the plane and we can derive a plane multigraph G_Γ by introducing a node for each intersection of two boundaries and an edge for each section of a boundary that lies between two intersections. Furthermore, a boundary without intersections is replaced by a node with a self loop nested inside its surrounding face. We call G_Γ the *contour graph* of Γ and its dual graph G_Γ^* the *dual graph* of Γ. The faces of G_Γ belong to zero, one, or two block regions. We call a face that belongs to no block region a *background face*, a face that belongs to a single block region a *linking face*, and a face that belongs to two block regions an *intersection face*. Only intersection faces contain points corresponding to elements in the universe, and no two faces of the same type are adjacent in the contour graph.

Alternatively, the union of the two partitions $\mathcal{P}_0 \cup \mathcal{P}_1$ can also be seen as a hypergraph $H = (U, \mathcal{P}_0 \cup \mathcal{P}_1)$, where every element $u \in U$ is a vertex and every block defines a hyperedge, i.e., a non-empty subset of U. The hypergraph H is *2-regular* since every vertex is contained in exactly two hyperedges. We denote $H = H(\mathcal{P}_0, \mathcal{P}_1)$ as the *corresponding hypergraph* of the pair of partitions $\{\mathcal{P}_0, \mathcal{P}_1\}$.

Hypergraph supports [15] play an important role in hypergraph embeddings and their planarity. A support of a hypergraph $H = (V, \mathcal{S})$ is a graph $G_p = (V, E)$ on the vertices of H, such that the *induced subgraph* $G_p[S]$ of every hyperedge $S \in \mathcal{S}$ is connected. We extend the concept of supports to pairs of partitions, i.e., we say that a graph $G_p = (V, E)$ is a support for $\{\mathcal{P}_0, \mathcal{P}_1\}$, if it is a support of $H(\mathcal{P}_0, \mathcal{P}_1)$.

We call a support *path based*, if the induced subgraphs of all hyperedges are paths,[1] and *tree based*, if all hyperedge-induced subgraphs are trees, i.e., they do not contain

---

[1] Brandes et al. [4] used a slightly different definition and called a support *path based* if the induced subgraph of each hyperedge has a Hamiltonian path.

any cycles. For any support G_p of a pair of partitions $\{\mathcal{P}_0, \mathcal{P}_1\}$ we can always create a tree-based support G_p' by removing edges from cycles: Suppose there exists a block $B \in \mathcal{P}_0$ such that $G_p[B]$ contains a cycle K. If the vertices in K are also contained in a common block of $\mathcal{P}_1$, we can just remove a random edge from K without destroying the support property. Otherwise, we can remove an edge from K that connects vertices in two different blocks of $\mathcal{P}_1$ without destroying the support property.

The *bipartite map* $G_b(H)$ of a hypergraph $H = (V, \mathcal{S})$ is defined as the bipartite graph $G_b(H) = (V \cup \mathcal{S}, E_b)$ that has a node for each vertex in V and for each hyperedge in $\mathcal{S}$ [21]. A node $v \in V$ is adjacent to a node $S \in \mathcal{S}$ if $v \in S$. We say that $G_b(H)$ is the bipartite map of a pair of partitions $\{\mathcal{P}_0, \mathcal{P}_1\}$ if $H = H(\mathcal{P}_0, \mathcal{P}_1)$.

Finally, we define the *block intersection graph* $G_s(\mathcal{P}_0, \mathcal{P}_1)$ as the graph with vertex set $V_s = \mathcal{P}_0 \cup \mathcal{P}_1$ and edge set $E_s = \{\{B, B'\} \mid B \cap B' \neq \emptyset\}$. Thus G_s has a vertex for each block and an edge between any two blocks that share a common element. Since only blocks of different partitions can intersect, we know that G_s is bipartite.

2   The Main Classes of Embeddability

We define three main concepts of simultaneous embeddability for pairs of partitions. We will see that these concepts induce a hierarchy of embeddability classes of pairs of partitions. We begin with *weak embeddability*, which is the most general concept.

Definition 1 (Weak Embeddability). *A simultaneous embedding of two partitions is weak if no two block regions of the same partition intersect. Two partitions are weakly embeddable if they have a weak simultaneous embedding.*

Prohibiting intersections of block regions of the same partition is our first well-formedness condition. A weak embedding emphasizes the fact that the blocks in each partition are disjoint. Since the blocks of any partition are disjoint by definition, it is not surprising that any pair of partitions is weakly embeddable (see Fig. 1(a) for an example).

Theorem 1. *Any two partitions of a common universe are weakly embeddable on any point set.*

Proof. A spanning forest (in fact, any planar graph) on n nodes can always be drawn in a planar way on any fixed set of n points in the plane [18]. Let now $\mathcal{P}$ be a partition. We choose arbitrary, but distinct points in the plane for the elements of U. We then generate a spanning tree on the elements in each block and embed the resulting forest in a planar way on the points. Slightly inflating the thickness of the edges of the trees yields simple bounded block regions. We can do this independently for a second partition on the same points and obtain a weak simultaneous embedding. $\square$

Although the concept of weak embedding does not seem to provide interesting insights into the structure of a given pair of partitions, it guarantees at least the existence of a simultaneous embedding for any pair of partitions that is more meaningful than an arbitrary embedding. An obvious drawback of weak embeddings is that the block regions of disjoint blocks are allowed to intersect, as long as both blocks belong to different partitions—even if they do not share common elements. Following the general idea of Euler diagrams [8], which do not show regions corresponding to empty intersections, we establish a stricter concept of embeddability. In a strong embedding block regions may only intersect if the corresponding blocks have at least one element in common, and even more, each intersection face of the contour graph must actually contain a point, see Fig. 1(b). This is our second well-formedness condition.

Definition 2 (Strong Embeddability). *A simultaneous embedding Γ of two partitions is* strong *if each intersection face of the corresponding contour graph contains a point $\Gamma(u)$ for some $u \in U$. Two partitions are* strongly embeddable *if they have a strong simultaneous embedding.*

Obviously, a strong embedding is also weak, since blocks of the same partition have no common elements, and thus, cannot form intersection faces. The class of strongly embeddable pairs of partitions is characterized by Theorem 2; we show in Section 3 that deciding the strong embeddability of a pair of partitions is NP-complete.

Theorem 2. *A pair of partitions of a common universe is strongly embeddable if and only if it has a planar support.*

Proof. Let $\{\mathcal{P}_0, \mathcal{P}_1\}$ be a pair of partitions and let G_Γ be the contour graph resulting from a strong embedding Γ of $\{\mathcal{P}_0, \mathcal{P}_1\}$. We construct a planar support of $\{\mathcal{P}_0, \mathcal{P}_1\}$ along G_Γ as follows. First recall that the elements of the universe, which correspond to the nodes in a support, are represented in Γ by points that are drawn inside intersection faces. Vice versa, since Γ is strong, each intersection face contains at least one point. Hence, we choose one point in each intersection face as the *center* of this face. We now create a dummy vertex for each linking face (observe that one block region may induce several linking faces) and link it to the centers of all adjacent intersection faces. The resulting graph is a subgraph of the dual graph of the contour graph G_Γ and therefore planar. We now connect all remaining vertices in a star-like fashion to the center of their intersection face, routing the edges in a non-crossing way. We finally remove the dummy vertices by merging them to an adjacent center, linking all adjacent vertices to that center. This graph remains planar. It also has the support property, since all intersection and linking faces of any block region are connected into a single component, and with them all vertices of that block region.

Now we construct a strong embedding from a planarly embedded support of $\{\mathcal{P}_0, \mathcal{P}_1\}$. To this end, we first construct a tree-based support by deleting edges from cycles as described in Section 1.2. Then, we simply inflate the edges of each block-induced subtree. Since the underlying support is embedded in a planar way, this yields a simple block region for every block in $\{\mathcal{P}_0, \mathcal{P}_1\}$ such that two block regions only intersect at the positions of the nodes. Hence, the constructed block regions together with the nodes of the support form a strong embedding of $\{\mathcal{P}_0, \mathcal{P}_1\}$. We note that the support graph as a planar graph can in fact be embedded on any point set [18]. Hence, a strongly embeddable pair of partitions can be strongly embedded on any point set. □

In a strong embedding, a single block region may still cross other block regions and intersect the same block regions several times forming distinct intersection faces—as long as each intersection face contains at least one common point. The last of our three embeddability classes prevents this behavior and requires that the block regions form a collection of *pseudo-disks*, i.e., the boundaries of every pair of regions intersect at most twice and the boundaries of two nested regions do not intersect. See Fig 1(c) for an example. This implies in particular that every block intersection is connected, which is a well-formedness condition widely used in the context of Euler diagrams [8], and that block regions do not cross and are thus more locally confined.

Definition 3 (Full Embeddability). *A simultaneous embedding of two partitions is* full *if it is a strong embedding and the regions form a collection of pseudo-disks. Two partitions are* fully embeddable *if they have a full simultaneous embedding.*

Using a linear-time algorithm for planarity testing [13], the following characterization of fully embeddable pairs of partitions directly implies a linear-time algorithm for deciding full embeddability. The proof of Theorem 3 constructs a bipartite map along a given full embedding, and vice versa. It uses similar techniques as the proofs of Theorems 1 and 2 and is found in the full version of this paper [1].

Theorem 3. *A pair of partitions of a common universe is fully embeddable if and only if its bipartite map is planar.*

A full embedding is strong by definition and we have seen above that a strong embedding is also weak. Hence, the three embeddability classes introduced in this section induce a hierarchy of embeddability classes. In the full version [1] we show that this hierarchy is strict. The weak embeddability class forms the basis of the hierarchy and contains all pairs of partitions. The strong embeddability class and the full embeddability class are characterized by the existence of a planar support and the planarity of the bipartite map of a pair of partitions, respectively, where the latter directly implies a linear time algorithm for the corresponding decision problem. Moreover, these characterizations reveal close relations to the hypergraph planarity concepts of *Zykov* and *vertex planarity*.

A hypergraph $H = (V, \mathcal{S})$ is Zykov-planar [22], if there exists a subdivision of the plane into faces, such that each hyperedge $S \in \mathcal{S}$ can be mapped to a face of the subdivision, and each vertex $v \in V$ can be mapped to a point on the boundary of all faces that represent a hyperedge containing v. Walsh [21] showed that a hypergraph is Zykov planar if and only if its bipartite map is planar.

In contrast, a hypergraph $H = (V, \mathcal{S})$ is vertex-planar [14] if there exists a subdivision of the plane into faces, such that every vertex $v \in V$ can be mapped to a face and for every hyperedge $S \in \mathcal{S}$, the interior of the union of all faces of the vertices in S is connected. Kaufmann et al. [15] showed that a hypergraph is vertex planar if and only if it has a planar support. This shows that the class of *fully* embeddable pairs of partitions is a subclass of Zykov planar hypergraphs, and the class of *strongly* embeddable pairs of partitions is a subclass of vertex planar hypergraphs.

3 Complexity of Deciding Strong Embeddability

In this section we show the NP-completeness of testing strong embeddability. As a consequence, testing whether the corresponding hypergraph of a pair of partitions has a planar support is also NP-complete by Theorem 2. This seems not very surprising considering the more general hardness results of Johnson and Pollak [14] and Buchin et

(a) Removing a linking face   (b) Removing a background face

Fig. 2. Two cases for transforming a strong embedding into a proper strong embedding

al. [5] who showed that deciding the existence of a planar support and a 2-outerplanar support in general hypergraphs is NP-hard. However, we consider a restricted subclass of 2-regular hypergraphs, thus, the NP-hardness of our problem does not directly follow from the previous results. Moreover, other special cases, e.g., finding path, cycle, tree, and cactus supports are known to be solvable in polynomial time [3, 5, 14]. Together with the characterization of Theorem 2, Theorem 4 immediately implies that testing the vertex planarity of a 2-regular hypergraph is NP-complete.

Theorem 4. *Deciding the strong embeddability of a pair of partitions is* NP-*complete.*

The existing hardness results [5, 14] rely on elements that are contained in more than two hyperedges and could not be adapted to our 2-regular setting. Instead we prove the hardness of deciding strong embeddability by a quite different reduction from the NP-complete problem MONOTONE PLANAR 3SAT [10]. A monotone planar 3Sat formula φ is a 3Sat formula whose clauses either contain only positive or only negated literals (we call these clauses *positive* and *negative*) and whose variable-clause graph H_φ is planar. A *monotone rectilinear representation* (MRR) of φ is a drawing of H_φ such that the variables correspond to axis-aligned rectangles on the x-axis and clauses correspond to non-crossing E-shaped "combs" above the x-axis if they contain only positive variables and below the x-axis otherwise; see Fig. 4(a).

An instance of MONOTONE PLANAR 3SAT is an MRR of a monotone planar 3Sat formula φ. In the proof of Theorem 4 we will construct a pair of partitions $\{\mathcal{P}_0, \mathcal{P}_1\}_\varphi$ that admits a strong embedding if and only if φ is satisfiable.

For the sake of simplicity, we restrict the class of strong embeddings to the subclass of *proper strong embeddings*, which is equivalent, as we can argue that a pair of partitions has a strong embedding if and only if it also has a proper one. A strong embedding is *proper* if the contour graph does not contain background or linking faces that are adjacent to only two other faces. Figure 2 illustrates how background or linking faces violating this condition can be removed, transforming a strong embedding into a proper one. We say that two proper strong embeddings are *equivalent* if the embeddings of their contour graphs are equivalent, i.e. if the cyclic order of the edges around each vertex is the same. A pair of partitions has a *unique strong embedding* if all proper strong embeddings are equivalent. Note that, analogously to the definition of equivalence of planar graph embeddings, two equivalent proper strong embeddings may have different unbounded outer background faces. Our construction in the hardness proof is independent of the choice of the outer face.

Next we define a special pair of partitions that has a unique grid-shaped embedding as a scaffold for the gadgets in the subsequent proof of Theorem 4. The first step is

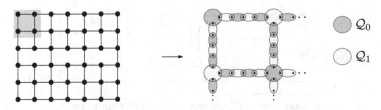

Fig. 3. Graph $G_{2,3}$ and the partitions $\{\mathcal{Q}_0, \mathcal{Q}_1\}$ sketched for the top-left grid cell marked in gray

to construct a base graph $G_{m,n}$ for two integers m and n. The graph $G_{m,n}$ is a grid with $mn + 1$ columns and $2m + 2$ rows of vertices with integer coordinates (i, j) for $0 \le i \le mn$ and $0 \le j \le 2m + 1$. Each vertex v with coordinates (i, j) is connected to the four vertices at coordinates $(i - 1, j), (i + 1, j), (i, j - 1), (i, j + 1)$ (if they exist). Between the middle rows m and $m + 1$ we remove all vertical edges except for those in columns $0, m, 2m, \ldots, nm$. This defines n larger grid cells of width m in this particular row. Figure 3 (left) shows an example.

From $G_{m,n}$ we construct a pair of partitions $\{\mathcal{Q}_0, \mathcal{Q}_1\}$ as follows (see Fig. 3). For each vertex v with coordinates (i, j) we create a *vertex block* B_v in partition $\mathcal{Q}_{(i+j) \pmod 2}$. For each edge (u, v) in $G_{m,n}$ we create a chain of four *edge blocks* $B_{u,v}^1, B_{u,v}^2, B_{u,v}^3, B_{u,v}^4$, such that $B_{u,v}^1$ and $B_{u,v}^3$ are in the same partition as B_v and $B_{u,v}^2$ and $B_{u,v}^4$ are in the same partition as B_u. We distribute five distinct elements among the edge blocks of (u, v) and the vertex blocks for u and v such that they form the desired chain pattern and each intersection face contains one common element. The pair $\{\mathcal{Q}_0, \mathcal{Q}_1\}$ is indeed a pair of partitions as every element belongs to exactly one block of each partition. Edge blocks contain two and vertex blocks up to four elements (depending on the degree of the corresponding vertex in $G_{m,n}$). Below we will add the gadgets of the reduction on top of $\{\mathcal{Q}_0, \mathcal{Q}_1\}$, for which it is required that there is an edge block in each partition that does not share any element with a vertex block. This explains why we link blocks of adjacent vertices by chains of four blocks.

The next lemma shows that $\{\mathcal{Q}_0, \mathcal{Q}_1\}$ has a unique embedding (proof in the full version [1]), which is a consequence of the fact that $G_{m,n}$ is a subdivision of a planar 3-connected graph (assuming $n \ge 2$) and thus it has a unique embedding. This property is inherited by $\{\mathcal{Q}_0, \mathcal{Q}_1\}$ in our construction.

Lemma 1. *The pair of partitions $\{\mathcal{Q}_0, \mathcal{Q}_1\}$ has a unique embedding.*

Now we have all the tools that we need to prove our main theorem in this section.

Proof (of Theorem 4). First we show that the problem is in NP. By Theorem 2 we know that a pair of partitions is strongly embeddable if and only if it has a planar support. Thus we can "guess" a graph on U and then test its planarity and support property in polynomial time. This shows membership in NP. It remains to describe the hardness reduction.

Let φ be a planar monotone 3Sat formula together with an MRR. First we construct the pair of partitions $\{\mathcal{Q}_0, \mathcal{Q}_1\}$ for the base graph $G_{m,n}$, where m is the number of clauses of φ and n is the number of variables of φ. By Lemma 1 $\{\mathcal{Q}_0, \mathcal{Q}_1\}$ has a unique

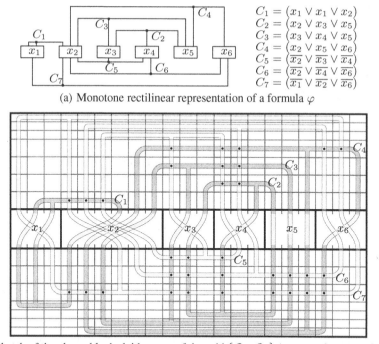

$$C_1 = (x_1 \lor x_1 \lor x_2)$$
$$C_2 = (x_2 \lor x_3 \lor x_5)$$
$$C_3 = (x_3 \lor x_4 \lor x_5)$$
$$C_4 = (x_2 \lor x_5 \lor x_6)$$
$$C_5 = (\overline{x_2} \lor \overline{x_3} \lor \overline{x_4})$$
$$C_6 = (\overline{x_2} \lor \overline{x_4} \lor \overline{x_6})$$
$$C_7 = (\overline{x_1} \lor \overline{x_2} \lor \overline{x_6})$$

(a) Monotone rectilinear representation of a formula φ

(b) Sketch of the clause blocks laid on top of the grid $\{\mathcal{Q}_0, \mathcal{Q}_1\}$ (empty columns omitted)

Fig. 4. Illustration of the NP-hardness reduction

proper grid-like embedding. We call $\{\mathcal{Q}_0, \mathcal{Q}_1\}$ the *base grid* and the n special cells between rows m and $m + 1$ the *variable cells* of the base grid.

Next we augment the pair of partitions $\{\mathcal{Q}_0, \mathcal{Q}_1\}$ by additional blocks, one for each clause, where positive clauses are added to $\mathcal{Q}_0$ and negative clauses to $\mathcal{Q}_1$. The definition of these *clause blocks* closely follows the layout of the given MRR, see Fig. 4(a). Let $C_1, C_2, \ldots, C_l$ be the positive clauses of φ ordered so that if C_i is nested inside the E-shape of C_j in the given MRR then $i < j$. Analogously let $C_{l+1}, \ldots, C_m$ be the ordered negative clauses. We describe the definition of the block B_i for a positive clause C_i $(1 \leq i \leq l)$; blocks for negative clauses are defined symmetrically. We create an *intermediate embedding* of B_i (which is not yet strong but serves as a template for a later strong embedding) by putting B_i on top of the base grid[2] and adding new elements to B_i and to certain edge blocks in $\mathcal{Q}_1$. This fixes B_i to run through two mirrored E-shaped sets of grid cells of our choice (Fig. 4(b)). In the upper half of the base grid, B_i is assigned to run between rows $m - i$ and $m - i + 1$. Furthermore, B_i is assigned to three columns leading towards the variable cells from the top. Let x_j be a variable contained in C_i and assume that C_i is the k-th positive clause from the right connecting to x_j in the embedding of the given MRR. Then B_i runs between columns $jm - k$ and

---

[2] The idea of fixing paths to an underlying grid is inspired by Chaplick et al. [7].

$jm - k + 1$. In the lower half of the base grid we translate and mirror the resulting E-shape as follows. We let B_i occupy the cells between rows $2m + 2 - l + i - 1$ and $2m + 2 - l + i$ and the three columns are shifted to the left by the number of occurrences of the respective variable in negative clauses (Fig. 4(b)). Since each variable cell is m columns wide, we can always assign each clause to a unique column of x_j in the top and bottom half of the grid in this way.

We actually fix B_i to the base grid by adding one shared element for each crossed edge of a grid cell to both B_i and the respective edge block of $\mathcal{Q}_1$ that does not share an element with a vertex block in $\mathcal{Q}_0$ (recall that $\{\mathcal{Q}_0, \mathcal{Q}_1\}$ contains such a block in each partition and for each grid edge). No two blocks of the same clause type (positive or negative) intersect, but blocks of different type do intersect in certain grid cells. For each grid cell shared between a positive and negative block (except for the n variable cells) we add one shared element (black dots in Fig. 4(b)) and call the respective grid cell the *home cell* for this element. Recall that the orders of the incoming blocks from the top and the bottom of each variable cell are inverted. Thus, within each variable cell the blocks of each pair of a positive and negative clause using the corresponding variable intersect, but no shared element is added. We denote the resulting new pair of partitions as $\{\mathcal{P}_0, \mathcal{P}_1\}_\varphi$ and observe that its size is polynomial in the size of φ.

Next we argue about the strong embedding options in contrast to the immediate embedding for a clause block B_i in $\{\mathcal{P}_0, \mathcal{P}_1\}_\varphi$. In the intermediate embedding each block has three connections through variable cells linking the upper E-shape with the lower E-shape. Any element shared with an edge block of the uniquely embedded base grid must obviously be reached by the block region of B_i. Since the block region must be simple, any strong embedding of B_i results from opening the intermediate embedding of B_i in exactly two grid cells so that the resulting block region of B_i is connected and has no holes. Additionally, a shared element must be placed in any intersection of the block region of B_i with block regions of other clause blocks.

First we assume that φ is a satisfiable formula and a satisfying variable assignment is given. We need to show that $\{\mathcal{P}_0, \mathcal{P}_1\}_\varphi$ has a strong embedding. If a variable x_j has the value *true* in the given assignment we open all blocks of negative clauses using x_j in the corresponding variable cell; if x_j is *false* we open all blocks of positive clauses using x_j. Thus no blocks intersect in variable cells any more. If a clause contains more than one *true* literal, we open all but one connection in its variable cells of *true* literals. Since the assignment satisfies φ, we know that each clause block is opened exactly twice in its variable cells and thus forms a valid simple block region. Moreover, we place all shared elements in their home cells so that every block intersection contains an element and the embedding is strong. We call a strong embedding of $\{\mathcal{P}_0, \mathcal{P}_1\}_\varphi$ with the above properties a *canonical embedding*.

Now assume that $\{\mathcal{P}_0, \mathcal{P}_1\}_\varphi$ has a strong embedding. We know that the base grid has its unique embedding and that each block is embedded as a simple region that results from opening the intermediate embedding (with its two E-shapes linked through three variable cells) in exactly two cells. If the embedding is already canonical, we can immediately construct a satisfying variable assignment for φ: if a variable cell is crossed by clause blocks in $\mathcal{Q}_0$ we set the variable to *true*, otherwise we set it to *false*. Since every clause block is connected we know that this assignment satisfies all clauses.

If the embedding is not canonical we show that it can be transformed into a canonical embedding as follows. In a non-canonical embedding it is possible that two blocks B_i and B_j intersect in a variable cell x_k and have a shared element in their intersection face in the cell of x_k rather than in the home cell of that element. This means, on the other hand, that in some shared home cell γ of B_i and B_j, say in the upper half, at least one of the two blocks is opened (as there is no more shared element to put into an intersection face). Thus the grid cell γ splits the E-shaped block region of one or both blocks in the upper half into two disconnected components, meaning that each opened block crosses at least two variable cells in order to connect both components via the lower half. Hence we can safely split any block that is opened in γ in the cell of variable x_k, re-connect it inside γ, and place the shared element of B_i and B_j into its home cell γ. This removes the block intersection in the cell of x_k. Once all block crossings within variable cells are removed, the resulting embedding is a canonical embedding and we can derive the corresponding satisfying variable assignment.   □

4   Extensions and Conclusion

We have characterized three main embeddability classes for pairs of partitions, which in fact form a strict hierarchy (see full version [1]), and we have shown NP-completeness of deciding strong embeddability. From a practical point of view the class of strong embeddings is of particular interest: it guarantees that every intersection between block regions is meaningful as it contains at least one element, but, in contrast to full embeddings, it allows multiple disconnected intersection regions between the same two blocks and it allows two blocks to cross.

There are interesting subclasses of strong embeddings that further structure the space between strong and full embeddability. They are discussed in more detail in the full version [1]. In *single-intersection strong embeddings* we adapt the *unique intersection region* condition of full embeddings, but still permit that two blocks cross in the embedding. This new class is a true subclass of strong embeddings. It is open whether the corresponding decision problem is still NP-complete since the proof of Theorem 4 is based on the existence of multiple intersection regions between pairs of blocks. In *strong grid embeddings*, a true subclass of single-intersection strong embeddings, the blocks of $\mathcal{P}_0$ and $\mathcal{P}_1$ are embedded as horizontal and vertical ribbons, respectively, which intersect in a matrix-like fashion.

It is an interesting direction for future work to generalize our embeddability concepts to $k > 2$ partitions. While weak embeddability and its properties extend readily to any number of partitions, it is less obvious how to generalize strong and full embeddability. One possibility is to require the properties in a pairwise sense; otherwise constraints for new types of faces in the contour graph belonging to more than one but less than k block regions might be necessary. On the practical side, future work could be the design of algorithms that find visually appealing simultaneous embeddings of two or more partitions. Finally, if the partitions are clusterings on a graph, one would ideally want to simultaneously draw both the partitions and the underlying graphs.

Acknowledgments. We thank the anonymous reviewers for helpful comments.

References

1. Athenstädt, J.C., Hartmann, T., Nöllenburg, M.: Simultaneous embeddability of two partitions. CoRR, abs/1408.6019 (August 2014)
2. Bläsius, T., Kobourov, S.G., Rutter, I.: Simultaneous embedding of planar graphs. In: Tamassia, R. (ed.) Handbook of Graph Drawing and Visualization, ch. 11, pp. 349–381. CRC Press (2013)
3. Brandes, U., Cornelsen, S., Pampel, B., Sallaberry, A.: Blocks of hypergraphs applied to hypergraphs and outerplanarity. In: Iliopoulos, C.S., Smyth, W.F. (eds.) IWOCA 2010. LNCS, vol. 6460, pp. 201–211. Springer, Heidelberg (2011)
4. Brandes, U., Cornelsen, S., Pampel, B., Sallaberry, A.: Path-based supports for hypergraphs. J. Discrete Algorithms 14, 248–261 (2012)
5. Buchin, K., van Kreveld, M., Meijer, H., Speckmann, B., Verbeek, K.: On planar supports for hypergraphs. In: Eppstein, D., Gansner, E.R. (eds.) GD 2009. LNCS, vol. 5849, pp. 345–356. Springer, Heidelberg (2010), see also Tech. Rep. UU-CS-2009-035, Utrecht University (2009)
6. Buja, A., Swayne, D.F., Littman, M.L., Dean, N., Hofmann, H., Chen, L.: Data visualization with multidimensional scaling. J. Comput. Graphical Statistics 17(2), 444–472 (2008)
7. Chaplick, S., Jelínek, V., Kratochvíl, J., Vyskočil, T.: Bend-bounded path intersection graphs: Sausages, noodles, and waffles on a grill. In: Golumbic, M.C., Stern, M., Levy, A., Morgenstern, G. (eds.) WG 2012. LNCS, vol. 7551, pp. 274–285. Springer, Heidelberg (2012)
8. Chow, S.: Generating and Drawing Area-Proportional Euler and Venn Diagrams. PhD thesis, University of Victoria (2007)
9. Collins, C., Penn, G., Carpendale, S.: Bubble sets: Revealing set relations with isocontours over existing visualizations. IEEE TVCG 15(6), 1009–1016 (2009)
10. de Berg, M., Khosravi, A.: Optimal binary space partitions in the plane. In: Thai, M.T., Sahni, S. (eds.) COCOON 2010. LNCS, vol. 6196, pp. 216–225. Springer, Heidelberg (2010)
11. Feng, Q.-W., Cohen, R., Eades, P.: Planarity for clustered graphs. In: Spirakis, P.G. (ed.) ESA 1995. LNCS, vol. 979, pp. 213–226. Springer, Heidelberg (1995)
12. Flower, J., Fish, A., Howse, J.: Euler diagram generation. J. Visual Languages and Computing 19(6), 675–694 (2008)
13. Hopcroft, J., Tarjan, R.: Efficient planarity testing. J. ACM 21(4), 549–568 (1974)
14. Johnson, D.S., Pollak, H.O.: Hypergraph planarity and the complexity of drawing Venn diagrams. J. Graph Theory 11(3), 309–325 (1987)
15. Kaufmann, M., van Kreveld, M., Speckmann, B.: Subdivision drawings of hypergraphs. In: Tollis, I.G., Patrignani, M. (eds.) GD 2008. LNCS, vol. 5417, pp. 396–407. Springer, Heidelberg (2009)
16. Kohonen, T.: Self-Organizing Maps, 3rd edn. Springer (2001)
17. Mäkinen, E.: How to draw a hypergraph. Int. J. Computer Math. 34(3-4), 177–185 (1990)
18. Pach, J., Wenger, R.: Embedding planar graphs at fixed vertex locations. Graphs and Combinatorics 17(4), 717–728 (2001)
19. Simonetto, P., Auber, D., Archambault, D.: Fully automatic visualisation of overlapping sets. Computer Graphics Forum 28(3), 967–974 (2009)
20. S. Wagner and D. Wagner. Comparing Clusterings – An Overview. Tech. Rep. 2006-04, Department of Informatics, Universität Karlsruhe, 2007.
21. Walsh, T.R.: Hypermaps Versus Bipartite Maps. J. Combinatorial Theory Series B 18(2), 155–163 (1975)
22. Zykov, A.A.: Hypergraphs. Russian Mathematical Surveys 29(6), 89–156 (1974)

Luatodonotes:
Boundary Labeling for Annotations in Texts

Philipp Kindermann, Fabian Lipp, and Alexander Wolff

Lehrstuhl für Informatik I, Universität Würzburg, Germany
http://www1.informatik.uni-wuerzburg.de/en/staff

Abstract. We present a tool for annotating Latex documents with comments. Our annotations are placed in the left, right, or both margins, and connected to the corresponding positions in the text with arrows (so-called *leaders*). Problems of this type have been studied under the name *boundary labeling*. We consider various leader types (straight-line, rectilinear, and Bézier) and modify existing algorithms to allow for annotations of varying height. We have implemented our algorithms in Lua; they are available for download as an easy-to-use Luatex package.

1 Introduction

Many word processing systems support annotations for the text. The most common case for this annotations are comments, which can be inserted in arbitrary positions inside the text. The comments themselves are placed as *labels* in the margin next to the text and connected to the corresponding position, called *site*, by a line called *leader*. The endpoint of a leader at a label is called a *port*. Such comments are available, for example, in LibreOffice (see Fig. 1) and Microsoft Word. This task can be expressed in the boundary labeling notion introduced by Bekos et al. [5]: the sites to be annotated lie inside the text area and the labels are to be placed outside the text area. They describe several types of leaders, such as straight-line leaders (s-leaders), rectilinear leaders with one bend (po-leaders) and rectilinear leaders with two bends (opo-leaders).

Previous work. Boundary labeling has been extensively investigated in the last few years, see a survey on the interaction between cartography and graph drawing [17]. For labels of uniform size, the problem is well-studied. Most algorithms try to minimize the total leader length. For s-leaders, it suffices to compute a minimum-weight perfect matching, which can be done in $O(n^{2+\varepsilon})$ time [1]. For opo-leaders, Bekos et al. [5] gave three different algorithms for the number of sides used by the labels, with running times $O(n \log n)$ (one-sided), $O(n^2)$ (two-sided), and $O(n^2 \log^3 n)$ (four-sided). Further, they presented an $O(n^2)$-time algorithm for po-leaders that lie on one side or on two opposite sides of the text. The result for po-leaders was improved by Benkert et al. [6] for the one-sided case. They gave an $O(n \log n)$-time algorithm for length minimization and an $O(n^3)$-time algorithm for a very general class of objective functions, including, for example,

C. Duncan and A. Symvonis (Eds.): GD 2014, LNCS 8871, pp. 76–88, 2014.
© Springer-Verlag Berlin Heidelberg 2014

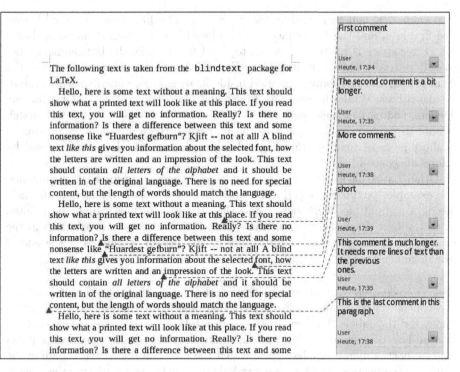

Fig. 1. Screenshot of comments in a document in LibreOffice 4.1.5

bend minimization. They also studied leaders that contain a diagonal part and gave an $O(n^2)$-time algorithm for the one-sided case. This result was extended by Bekos et al. [3] to more than one side. Recently, Kindermann et al. [10] gave the first efficient algorithms for po-leaders that decide whether an instance with labels on two adjacent, three, or four sides has a crossing-free solution (and, if yes, compute one).

Boundary labeling for non-uniform labels is still largely unexplored. Bekos et al. [4] showed that it is NP-hard to find a crossing-free labeling if the labels have to be placed on two sides (or two stacks on the same side). Huang et al. [9] considered a version of the problem that is always feasible: labels are placed into the right margin or into both margins, which are not bounded from below or above. For this model, opo-leaders, and labels of non-uniform size, they gave an $O(n^3)$-time algorithm that minimizes the total leader length in the one-sided case. For the two-sided case, they showed NP-hardness.

In this paper, we focus on comments for Latex documents. There are some packages that support the placement of textual comments in the margin, namely `todonotes` [12], `fixme` [16] and `fixmetodonotes` [2]. They have in common that they use Latex's `\marginpar` command to print the note as soon as the corresponding command is encountered in the source of the document. The drawback of this approach is that the positions of the following comments are not known and cannot be considered when placing a note. The first label is placed beside the first site, and the following ones are placed below. Often it happens that a lot of free space is wasted above the topmost label, while the bottommost label is only partially visible (if at all), see Fig. 4a. Another disadvantage is that the \marginpar method cannot be used inside floating environments such as tables or algorithms. While the packages fixme and fixmetodonotes do not draw any leaders, todonotes uses *opo*-leaders. With this leader style it is hard to match a note to its corresponding site in the text when there are many comments in a short piece of text. A similar problem occurs with the leader style used by LibreOffice; see Fig. 1.

Other Latex packages support annotations as metadata for PDF documents, for example, pdfcomment [11]. The drawback of this package is that the user needs a compatible PDF viewer and that the annotations cannot be printed with the text. Packages such as easy-todo [14] don't place annotations in the margins, but insert a marker into the text and list all comments at the end of the document.

Our contribution. Our approach is different from all those listed above in that we collect the comments for a whole page and then compute a good placement for the labels. Of course, this computation needs more resources than the ad-hoc placement of the existing packages. Additionally, our Latex package supports different leader types, which the user can select when loading the package; see Section 2. We give several algorithms for non-uniform labels, most of which are extensions of existing algorithms for the one-sided case; see Section 3. We improve upon these basic algorithms by considering label clustering and the two-sided case; see Section 4. We have implemented all of our algorithms and have evaluated them experimentally; see Section 5. We conclude with some open problems; see Section 6. The package is available on CTAN:

http://ctan.org/pkg/luatodonotes

2   Implementation

We have implemented the algorithms in Lua and have bundled them into a Luatex package, which we call luatodonotes. The package requires the modern Tex-processor Luatex [8], which allows us to embed Lua code inside our Tex sources. This gives us access to a high-level programming language for implementing our label-placement algorithms. From the user's point of view, this does not change much. Luatex is part of every modern Tex installation, for example, Tex Live. Assuming such an installation, the difference in usage is simply that instead of calling (pdf)latex, the user calls lualatex.

Our package is based on the todonotes package (see Section 1). It is downward compatible as it provides the same commands to the user as the original package. Usage is quite simple: the user loads the package with the command \usepackage{luatodonotes} and inserts a comment into the text with the command \todo{comment text}.

Now, we describe how our package works. Wherever the user inserts a `\todo` command in the text, we store its position and its argument (that is, the comment) in a Lua list, but we do not print anything at this moment. When a page is finished ("shipped out" in Tex terminology), we compute the position of the labels and draw them. Before calling our label-placement algorithm, we have Tex determine the label heights. To determine the absolute positions of the sites, we use PGF/TikZ [15], a widely used Tex package for producing vector graphics. This package can locate the position of a site on the page where the `\todo` command was inserted, even when the command occurs inside a floating environment (such as a figure or a table).

For each label, the placement algorithm computes the absolute coordinates on the page on which the label is to be placed. Then, we use TikZ to draw the labels and the leaders that connect the labels with their corresponding sites in text. Finally, a mark is placed at each site. This modular design simplifies the implementation of new algorithms and makes the package extensible.

The size and position of the rectangles that contain the label texts depend on the current page layout. We provide options to control the distances between the labels and the text (`distanceNotesText`) and between the labels and the border of the page (`distanceNotesPageBorder`). The algorithms can place labels in the left and in the right margin (see Section 4), but a margin is used only if it is wide enough to accomodate a label, that is, if the label can be at least of width `minNoteWidth`.

When loading the package with `\usepackage{luatodonotes}`, optional arguments can be specified in square brackets. The most relevant options are (a) the algorithm for label placement (`positioning`) and (b) the leader type (`leadertype`). Other options control the layout: the minimum vertical distance of the labels (`interNoteSpace`), the distance from the contents of the label to its border (`noteInnerSep`) and the color of the leaders (`linecolor`).

3   Algorithms for Label Placement

In the following, the algorithms are categorized by the leader type that they support. In principle, our package allows the user to combine any label-placement algorithm with any leader type. Still, some algorithms have been designed with certain leader types in mind. Other combinations will probably yield unwanted results, such as label overlap or crossing leaders.

In the descriptions of our algorithms below, we assume that labels are placed on the left side of the text, but this is not a restriction of our actual implementations. Additionally, we try to place the labels without gaps between them, while in reality we want to preserve a certain minimum distance between them. Clearly, this is easy to achieve.

3.1 s-Leaders

Our algorithms designed for s-leaders have a common property: they draw the leaders without crossing each other. Their common objective is to place the labels one below the other on the boundary while avoiding gaps between them. They differ in the position of the ports, that is, the position on the label boundary to which the leader is attached. A pleasant position for the port would be the center of the right side of the label. Unfortunately, we don't have an algorithm that can place the labels without gaps using this port position. We don't even know whether every instance of site positions and label heights is feasible w.r.t. these criteria; see Section 6.

We don't give algorithms that minimize the total leader length here, but concentrate on drawings without crossings. The clustering approach described in Section 4 can decrease the leader length as labels are placed closer to their corresponding sites.

NorthEast. We use an algorithm of Bekos et al. [5] for fixed labels, which can easily be adopted to our problem with labels of non-uniform heights: The upper right corner of each label is used as its port. The labels are placed consecutively from the top of the page to the bottom. In each step, we emit a ray from the port of the next label vertically to the top and rotate it clockwise until the first unlabeled site is hit. Obviously, by connecting this site to a label at the current position, we don't hide any other sites and can label the remaining sites without crossings.

NorthEastBelow. This algorithm is based on the preceding one. The difference is that we lower the port from the corner by a constant offset. In our opinion the result looks better when the leader is not attached directly at the corner. A good value for this offset is half of the height of the smallest label. As we know the position for each port while placing the label, we can still use the ray construction of the preceding algorithm to place the labels without spaces between them.

East. In this algorithm the port of every label is located at the center of its right side. When we try to find the next unlabeled site to be labeled, we do not know the port position as it depends on the height of the label. Therefore, we cannot use the ray construction from the previous algorithms. Algorithm 1 is a heuristic that guarantees crossing-free leaders while trying to avoid gaps between the labels. It can usually handle real-world inputs without additional gaps.

An instance that is not handled optimally by the heuristic is depicted in Fig. 2. The sites can be labeled without gaps when placing the labels in the order 2, 1, 3. As mentioned above it is an open question if this is possible for all instances.

3.2 Bézier Curves as Leaders

We base our Bézier curves on s-leaders using a force-directed algorithm described by Fink et al. [7]. We use cubic Bézier curves that are required to enter the port at the label horizontally. This means that the first control point has to stay on the same horizontal line as the port and can only be moved to the left or the

---

Algorithm 1. Placing labels using east anchors

Input: $p_1, \ldots, p_n$ are the sites in the text
Output: y-coordinate $y_1, \ldots, y_n$ of the top edge of each label

1   $P \leftarrow \{p_1, \ldots, p_n\}$
2   $L \leftarrow [\,]$ // list contains labels in the order in which they are placed
3   $lastY \leftarrow 0$
4   **while** $P \neq \emptyset$ **do**
5     $H \leftarrow \{height(p_j) \mid j = 1, \ldots, n\}$ // H is in ascending order
6   **foreach** $h \in H$ **do**
7   place a label of height h directly below the last label
8   emit a ray from the port of the newly placed label
9       $i \leftarrow$ index of first point in P that is hit by the ray (rotated clockwise)
10   **if** $height(p_i) \leq h$ **then**
11   **break**
12     $y_i \leftarrow lastY - (h - height(p_i))/2$
13     $L.\text{add}(p_i)$
14     $P \leftarrow P - \{p_i\}$
15     $lastY \leftarrow y_i - height(p_i)$

   // Postprocessing: try to shrink gaps
16 **foreach** $l \in L$ **do**
17   **if** there is a gap above l **then**
18   move l up as far as possible without creating any new intersection between leaders

---

right. The second control point is always placed in the center between the first control point and the site.

In the first iteration of the algorithm, the control points are placed on the endpoints of the leader, that is, it starts as a straight line. Later, the first control point of each curve is moved by applying forces to it. We use a force that pulls the control point to its optimal point, which is computed beforehand and usually yields a good-looking curve. Other forces try to increase the distance between curves. In every iteration the forces on every point are limited by the distance to the nearest curve to inhibit new intersections between leaders. Therefore, the algorithm guarantees crossing-free Bézier curves when starting with straight-line leaders without intersections.

The runtime of this algorithm is dominated by the calculation of the distances between each pair of curves. This calculation is done by an approximation of the curves. We need the distances to update the forces in every iteration.

3.3 *opo*-Leaders and *os*-Leaders

Positioning the labels for crossing-free *opo*-leaders is simple as Bekos et al. [5] show: we place the labels in the order given by the y-coordinates of their sites.

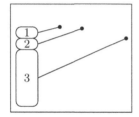 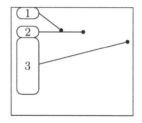

Fig. 2. An instance where the East algorithm does not yield a drawing without gaps. Left: label positions before postprocessing; Right: after postprocessing.

Sites with identical y-coordinates are processed from left to right. The vertical parts of the leaders are drawn in the *track routing area*, that is, the vertical strip between text and labels. The width of this track routing area is specified using the option `routingAreaWidth` of the package. We split the labels into groups, with labels sharing a common vertical segment being put in the same group. This can be done by a simple linear-time algorithm. Thus the vertical segments of the leaders in each group must be placed side by side. We draw the vertical segments in one group with equal distances between them, using the whole width of the track routing area.

The algorithm is even easier for *os*-leaders, a leader style that was not discussed until now. We list it here because this is the style that, for example, LibreOffice uses (see Fig. 1). Labels are placed in the same order as for *opo*-leaders. For the leaders, we connect the site with a horizontal line segment that extends to a fixed x-coordinate inside the margin. Then we connect the end of the horizontal segment to the label's port with a straight-line segment.

3.4 *po*-Leaders

Benkert et al. [6] developed an algorithm to compute an optimal crossing-free labeling using *po*-leaders with respect to an arbitrary badness function. This algorithm, which uses a dynamic programming approach, is designed for uniform labels only. It needs $O(n^3)$ running time and $O(n^2)$ space.

For our application, we extend the algorithm of Benkert et al. to non-uniform labels. To be able to work with the arbitrary heights of the labels, we need to raster the page, that is, we define the y-coordinates on which labels may be placed. Our algorithm yields a labeling respecting this raster with minimum total leader length. The height of the raster can be chosen using the parameter `rasterHeight` of the Latex package. The port for each label can be chosen arbitrarily. In the following, the ports are fixed to the center of the right side of the labels.

Let $p_1, \ldots, p_n$ denote the sites from top to bottom and let $r_1, \ldots, r_m$ be the slots obtained by rasterizing the page from top to bottom. We use a 5-dimensional table in our dynamic program. The entry $T[t, b, \tau, \beta, k]$ represents the minimum length of a labeling of the k leftmost sites in $\{p_t, \ldots, p_b\}$ using

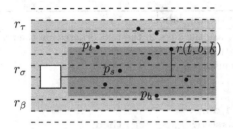

Fig. 3. The labeling problem for $T[t, b, \tau, \beta, k]$ is split into two independent subproblems by fixing the label position of $r(t, b, k)$. The dashed lines show the raster slots. The light gray area indicates the slots from r_τ to r_β. The dark gray area shows the sites between p_t and p_b.

only the raster slots $r_\tau, \ldots, r_\beta$. The labels must lie completely inside the given slots.

Let $r(t, b, k)$ the k-th point from the left in the set $\{p_t, \ldots, p_b\}$. The length of the shortest po-leader from the site p to its corresponding label beginning in slot r_σ is denoted by $l^*(p, \sigma)$. The entries of the table are computed using the following decomposition (illustrated in Fig. 3):

$$T[t, b, \tau, \beta, k] = \min_{\text{feasible } \sigma \in \{\tau, \ldots, \beta\}} l^*(r(t, b, k), \sigma) + T[t, s, \tau, \sigma - 1, k_1]$$
$$+ T[s + 1, b, \sigma + h, \beta, k_2]$$

In this formula p_s is the lowest point that lies above the leader arm (the horizontal part of the leader), when the label for $r(t, b, k)$ is placed at slot r_σ. Let h the height of this label. The number of sites from $\{p_t, \ldots, p_b\}$ lying left of $r(t, b, k)$ and above resp. below the leader arm is denoted by k_1 resp. k_2.

A position for the label is feasible, if both partial solutions (above and below the leader arm) are feasible, that is, there are enough slots to label the contained sites.

Clearly, $T[1, n, 1, m, n]$ is the optimal labeling of the whole instance. With this algorithm we can compute an optimal solution in $O(n^4 m^3)$ time with $O(n^3 m^2)$ space, where n is the number of sites to be labeled and m is the number of slots in the raster on the page.

Avoid overlappings with text lines. The algorithm described above does not take the position of the text lines of the document into account. Thus it can happen that a line gets striked out by the horizontal segment of a leader. We modified the algorithm to move the port up or down by a small offset to avoid such overlappings and place the leader into the gap between the lines.

It is quite hard to determine the positions of the lines in Tex because they are not fixed until the document is written to the output file. But in Luatex we can modify the linebreaking algorithm such that it inserts special nodes into the data structures of Tex that write the position of every line into a text file when

---

Algorithm 2. Clustering labels

Input: $p_1, \ldots, p_n$ are the sites in the text ordered by their y-coordinate from top to bottom

Output: list of clusters S

1 $S \leftarrow [\{p_1\}, \{p_2\}, \ldots, \{p_n\}]$
2 $i \leftarrow 1$
3 **while** $i \leq \#S - 1$ **do**
4 | **if** clustersIntersect($S[i]$, $S[i+1]$) **then**
5 | | $S[i] \leftarrow S[i] \cup S[i+1]$
6 | | S.delete($i+1$)
   | | `// as the size of stack i has increased we check again for`
   | | `   intersection with the previous stack in next iteration`
7 | | $i \leftarrow \max\{1, i - 1\}$
8 | **else**
9 | | $i \leftarrow i + 1$
10 **return** S

---

typesetting the page. In a second Tex run we can read the line positions from this file and use them for our algorithm.

4 Improvements

In this section we discuss some general improvements implemented in our package that can be used by every algorithm described in the previous section.

Label clustering. Most of the algorithms described in the previous sections place labels in a single stack (that is, without gaps between them) beginning at the upper margin of the page. This can produce unnecessarily long leaders, for example when the text contains a single site near the end of the page. We split the labels into separate clusters and place each of them near the corresponding sites in the text. An algorithm for clustered labeling is also described by Nöllenburg et al. [13]. Our approach is simpler but slower.

To group the labels into clusters we use Algorithm 2. It repeatedly joins adjacent clusters as long as they intersect each other. To test if two clusters intersect we place the contained labels as a stack each beneath the arithmetic mean of the sites in the cluster. The clusters intersect if their corresponding stacks overlap.

The positioning algorithm is executed independently for each of the identified clusters. The intended position is passed to the algorithm as a parameter.

Two-sided label placement. On some page layouts there is enough space to place labels in the margins on the left *and* the right side of the text. We have to decide for each label on which side of the text it should be placed. Our approach is to split the sites by a vertical line through the text. The sites which are left of this split line are labeled on the left side, those right of the split line are labeled in the right margin. There are several ways to determine the position of this split line. We use a weighted median to split the sites such that the sum of the label heights on the left side is approximately equal to that of the right side. With this algorithm it is not an issue if the widths of the two margins are different (which means that the height of a label depends on the side on which it is placed).

5 Experimental Results

We compare the leader styles presented in the previous sections on an example document with nine comments in it. This document stays the same, only the options of our package are modified to switch between the available algorithms. We used the label clustering approach described in the previous section for all examples except for that of the *po*-leader algorithm. For comparison, we also processed the document with the todonotes package (see Fig. 4a).

The NorthEastBelow algorithm for *s*-leaders (Fig. 4b) is straight-forward and fast. It is easy for the reader to match the sites to their corresponding label. Using Bézier curves (Fig. 4c) instead of the straight-line leaders yields a more aesthetic result with the disadvantage of a significantly higher runtime caused by the iterations of the force-directed algorithm. Using two-sided label placement with the same leader type produces shorter leaders because the labels can be placed closer to their site. Especially in text segments with a lot of comments this makes the relationship between sites and their labels clearer.

Our algorithm for *po*-leaders (Fig. 4d) has a high asymptotic runtime and space consumption. But in practice when there are only few comments per page this is not an issue. Among the algorithms we implemented, this is the only algorithm minimizing the total leader length.

The *opo*-leaders and *os*-leaders are available mainly for comparison. Clearly, it gets hard for the reader to match sites to their labels on pages with many comments. In particular, if several sites are in the same line it is hard to tell the matching between sites and labels. On the other hand the leaders only run between the lines and in the track routing area and thus don't disturb the text.

The running times of Luatex with the different leader types for some example documents are shown in Table 1. Note that Documents 2 and 3 with 15 resp. 25 comments on one page are quite unrealistic. When using two-sided label placement both sides are processed independently and thus the algorithm for *po*-leaders becomes feasible again. The measured times are for a single run of Tex only. When the absolute position of a site of a label changes, a second run is needed. When we deactivate our package, processing still needs 1.4 seconds. This means that *s*- and *opo*-leaders cause only small extra cost compared to a standard Latex run. With the classical todonotes package processing needs about 1.8 seconds, too.

We would have liked to give a numerical comparison of the drawing quality of the different algorithms, but it is not obvious how to find an appropriate indicator for the quality that is suitable for all of the available leader types. So we ask the reader to inspect Fig. 4 visually.

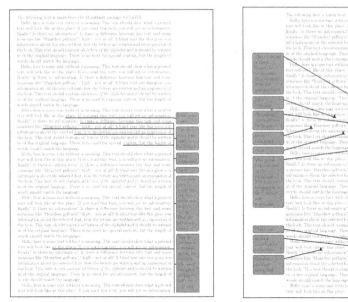

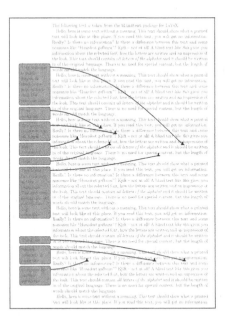

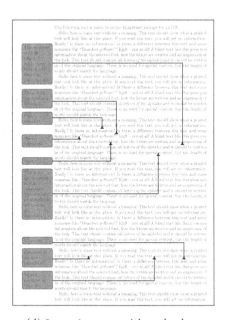

(a) `todonotes` with *opo*-leaders   (b) `luatodonotes` with *s*-leaders

(c) `luatodonotes` with Bézier leaders   (d) `luatodonotes` with *po*-leaders

Fig. 4. An example document with notes produced by the `todonotes` package (a) and the `luatodonotes` package (b–d) with different leader styles

Table 1. Running times of the different label styles on three one-page documents D1, D2, and D3 (in CPU seconds). The times were measured using a Intel Core 2 Duo E8400 with 3.0 GHz. D1 is the instance with 9 comments shown in the figures above. D2 has 15 comments, D3 has 25. For each document, we report two running times; for label placement into one margin vs. both margins. We use a raster height of 1 cm for *po*-leaders, resulting in 28 horizontal strips. We couldn't use *po*-leaders for D3 with one margin because the algorithm needed too much memory. For comparison we also give the running times for the classical **todonotes** package (which does not support placing labels in both margins) and the running times for the document without loading the **luatodonotes** package.

Document	D1		D2		D3	
Number of margins	1	2	1	2	1	2
s-leaders	1.8	1.7	1.9	1.9	2.2	2.2
Bézier leaders	5.7	5.4	33.2	11.1	322.9	116.3
po-leaders	4.8	3.0	17.7	6.2	—	27.6
po-leaders avoiding text lines	7.0	4.0	26.8	9.5	—	42.4
opo-leaders	1.8	1.7	1.9	1.9	2.2	2.2
classical **todonotes**	1.9		2.2		2.6	
without **luatodonotes**	1.4		1.4		1.3	

6 Conclusion and Open Problems

All our algorithms turned out to work well in practice—some of them cannot process too many labels on a single page. Using both margins helps in terms of speed. By visual inspection we reached the conclusion that *s*-leaders or Bézier leaders work better than the *os*-leaders used by other type-setting programs. The reason may be that the reader's eye can follow leaders without bends more easily. It would be interesting to verify this in a user study. With the modular design of our Latex package it is easy to improve the label-placement algorithms or add additional ones.

An interesting theoretical problem remains open: Given an instance with non-uniform label heights, is it always possible to place the labels without gaps so that *s*-leaders do not cross each other even if we insist that the ports are centered vertically at each label?

We have some ideas for further improvements of our package. The force-directed Bézier curve algorithm is quite slow at the moment. We think that we could speed up the computation of the distances between curves by doing a rough estimate first and computing the fine approximation only when needed. It would be interesting to transform the *po*-leaders into Bézier curves. As our algorithm yields a length-minimal *po*-labeling this could produce a shorter leader length than our approach with *s*-leaders. But it is not clear how to inhibit intersections between the curves.

Admittedly, our dynamic program for *po*-leaders is quite slow. Can we save time by computing labelings that are just feasible rather than length-minimal? For the other leader types, on the contrary, it would be interesting to minimize the total leader length. Such algorithms are known only for the case of uniform labels. When minimizing the leader length in the two-sided case, one could also try to improve the approach for partitioning the labels.

References

1. Agarwal, P.K., Efrat, A., Sharir, M.: Vertical decomposition of shallow levels in 3-dimensional arrangements and its applications. SIAM J. Comput. 29(3), 912–953 (1999)
2. Barabucci, G.: fixmetodonotes (2013), http://www.ctan.org/pkg/fixmetodonotes
3. Bekos, M.A., Kaufmann, M., Nöllenburg, M., Symvonis, A.: Boundary labeling with octilinear leaders. Algorithmica 57(3), 436–461 (2010)
4. Bekos, M.A., Kaufmann, M., Potika, K., Symvonis, A.: Multi-stack boundary labeling problems. In: Arun-Kumar, S., Garg, N. (eds.) FSTTCS 2006. LNCS, vol. 4337, pp. 81–92. Springer, Heidelberg (2006)
5. Bekos, M.A., Kaufmann, M., Symvonis, A., Wolff, A.: Boundary labeling: Models and efficient algorithms for rectangular maps. Comput. Geom. Theory Appl. 36(3), 215–236 (2007)
6. Benkert, M., Haverkort, H.J., Kroll, M., Nöllenburg, M.: Algorithms for multi-criteria boundary labeling. J. Graph Algorithms Appl. 13(3), 289–317 (2009)
7. Fink, M., Haunert, J.H., Schulz, A., Spoerhase, J., Wolff, A.: Algorithms for labeling focus regions. IEEE Trans. Vis. Comput. Graphics 18(12), 2583–2592 (2012)
8. Hagen, H., Henkel, H., Hoekwater, T.: Luatex (2007), http://www.luatex.org
9. Huang, Z.-D., Poon, S.-H., Lin, C.-C.: Boundary labeling with flexible label positions. In: Pal, S.P., Sadakane, K. (eds.) WALCOM 2014. LNCS, vol. 8344, pp. 44–55. Springer, Heidelberg (2014)
10. Kindermann, P., Niedermann, B., Rutter, I., Schaefer, M., Schulz, A., Wolff, A.: Two-sided boundary labeling with adjacent sides. In: Dehne, F., Solis-Oba, R., Sack, J.-R. (eds.) WADS 2013. LNCS, vol. 8037, pp. 463–474. Springer, Heidelberg (2013)
11. Kleber, J.: pdfcomment (2012), http://www.ctan.org/pkg/pdfcomment
12. Midtiby, H.S.: todonotes (2012), http://www.ctan.org/pkg/todonotes
13. Nöllenburg, M., Polishchuk, V., Sysikaski, M.: Dynamic one-sided boundary labeling. In: Proc. 18th SIGSPATIAL Int. Conf. Adv. Geogr. Inform. Syst. (ACM-GIS), pp. 310–319. ACM (2010)
14. Rada-Vilela, J.: easy-todo (2014), http://www.ctan.org/pkg/easy-todo
15. Tantau, T.: PGF and TikZ – Graphic systems for TeX, http://www.sourceforge.net/projects/pgf (accessed April 2, 2014)
16. Verna, D.: Fixme (2013), http://www.ctan.org/pkg/fixme
17. Wolff, A.: Graph drawing and cartography. In: Tamassia, R. (ed.) Handbook of Graph Drawing and Visualization, ch. 23. CRC Press (2013)

A Coloring Algorithm for Disambiguating Graph and Map Drawings

Yifan Hu[1] and Lei Shi[2,*]

[1] Yahoo Labs, 111 W 40th St, New York, NY 10018, USA
yifanhu@yahoo.com
[2] SKLCS, Institute of Software, Chinese Academy of Sciences, China
shil@ios.ac.cn

Abstract. Drawings of non-planar graphs always result in edge crossings. When there are many edges crossing at small angles, it is often difficult to follow these edges, because of the multiple visual paths resulted from the crossings that slow down eye movements. In this paper we propose an algorithm that disambiguates the edges with automatic selection of distinctive colors. Our proposed algorithm computes a near optimal color assignment of a dual collision graph, using a novel branch-and-bound procedure applied to a space decomposition of the color gamut. We conduct a user study to establish the effectiveness and limitations of this approach in clarifying drawings of real world graphs and maps.

Keywords: graph drawing, maps, edge coloring, branch-and-bound algorithm.

1 Introduction

Graphs are widely used for depicting relational information among objects. Typically, graphs are visualized as node-link diagrams [1]. In such a representation, edges are shown as straight lines, polylines or splines. Graphs that appear in real world applications are usually non-planar. For such graphs, edge crossings in the layout are unavoidable. It is a commonly accepted principle that the number of edge crossings should be minimized whenever possible, this principle was confirmed by user evaluations which showed that human performance in path-following is negatively correlated to the number of edge crossings [18,21]. Later studies found that the effect of edge crossings varies with the crossing angle. In particular, the task response time decreases as the crossing angle increases, and the rate of decrease levels off when the angle is close to 90 degree [14,15]. This implies that it is important not only to minimize the number of edge crossings, but also to maximize the angle of the crossings. Consequently, generating drawings that give large crossing angles, or even right crossing angles, became an active area of research (e.g., [6]). Nevertheless, for general non-planar graphs, there is no known algorithm that can guarantee large crossing angles for straight line drawings. Therefore, techniques to mitigate the adverse visual effect of small angle crossings are important in practice.

In this paper we propose to use colors to help differentiate edges. Our starting point is an existing layout, and our working assumption is that the graph is to be displayed as

* Supported by China National 973 project 2014CB340301 and NSFC grant 61379088.

a static image on paper, or on screen. The motivation comes from users of the Graphviz [10] software. These users were generally happy with the layouts of their graphs, but were asking whether there was any visual instrument that can help them follow edges better. Examining their layouts, we realized that because edges were drawn using the same color (e.g., black), when there were a lot of edge crossings, it was difficult to visually follow these edges. Thus the feedback from our users, and our own observation, echo the findings by Huang *et al.* [14,15]. When explaining why small crossing angles are detrimental to the task of following a path, they found, with the help of an eye tracking device, that *"when edges cross at small angles, crossings cause confusion, slowing down and triggering extra eye movements."* and that *"in many cases, it is crossings that cause confusion, making all the paths between two nodes, and branches along these paths, unforeseeable. Due to the geometric-path tendency, human eyes can easily slip into the edges that are close to the geometric path but not part of the target path.".*

Edge crossing is not the only hindrance to the visual clarity of a graph drawing. An additional problem is that when an edge from node u passes underneath the label of a node v and connects to a node w, it is impossible to tell visually whether there is one edge $u \leftrightarrow w$, or two edges $u \leftrightarrow v$ and $v \leftrightarrow w$, when all edges are of the same color (e.g., Fig. 3(b)). While these problems can be solved with user interactions by clicking on an edge of interest, or on a node to bring its neighbors closer (see, e.g., [17]), this involves an extra step for the user that may not be necessary if edges can be differentiated with a proper visual cue. Furthermore, there are situations where interaction is not possible, e.g., when looking at a static image of a graph on screen, or in print. These are the situations that are of particular interest in this paper.

We believe all the above mentioned problems of visually distinguishing and following edges can be greatly alleviated by choosing appropriate colors or line styles to differentiate edges. We first identify edge pairs that need to be differentiated (the *colliding edges*), and represent them as nodes of a dual collision graph. We then propose an algorithm to assign colors to the nodes of this collision graph, in a way that maximizes the color difference between nodes that share an edge. Thus our main contributions are:

- A novel branch-and-bound graph coloring algorithm that finds the globally optimal color embedding of each node with regard to its neighbors, and that works with both continuous color spaces and discrete color palettes.
- A user study that establishes the effectiveness/limitations of the coloring approach.

2   Related Work

Graph coloring is a classic problem in algorithmic graph theory. Traditionally the problem is studied in a combinatorial sense. For example, finding the smallest number of k colors on the vertices of a graph so that no two vertices sharing an edge have the same color. The difference between this and our work is that in $k-$colorability problem, a solution is valid as long as any pairs of vertices that share an edge have different colors, no consideration is given to maximizing the actual color differences. So in essence, the distance between colors is binary – either 0, or 1. For our problem we assume that even among distinctive colors, the differences are not equal, and are measured by color distances. In the special case when only k colors are allowed, our algorithm degenerates to find the optimal color assignment among all solutions of the $k-$colorability problem.

This last problem of optimal color assignment was also studied by Gansner *et al.* [9] and by Hu *et al.* [12], in the context of coloring virtual maps to maximize the color difference between neighboring regions. In these works, a set of k distinctive colors are assumed to be given, with k the number of countries in the map. Maps were then colored by an optimal permutation of the k colors. On the other hand, in this paper we assume that the color space can be either continuous or discrete, and we select among all colors in the color space to increase color differences. When applied to map coloring, our algorithm produces k distinctive colors as a side product.

Dillencourt *et al.* [7] studied the problem of coloring geometric graphs so that colors on nodes are as different as possible. The problem they studied is very related to ours, except that in their case the application is the coloring of geometric regions, while we are also interested in coloring edges of a graph. Dillencourt *et al.* used a force-directed gradient descent algorithm to find a *locally optimal* coloring of each node with regard to its neighbors. We propose a new algorithm based on a branch-and-bound process over an octree decomposition of the color space, that finds a *globally optimal* coloring for each node with regard to its neighbors. Furthermore, our approach is more flexible and works for discrete color palettes, in addition to continuous color spaces.

The angular resolution of a drawing is the sharpest angle formed by any two edges that meet at a common vertex.In addition to maximizing crossing angles (e.g., [6]), for the same reason of visual clarity, there have been researches to maximize the angular resolution of the drawing. Most recently, Lombardi Drawing of graphs was proposed [8,3], in which edges are drawn as arcs with perfect angular resolution. However, Purchase *et al.* [20] found that even though users prefer the Lombardi style drawings, straight-line drawings created by a spring-embedder gives better performance for path following and neighbor finding tasks. For straight-line drawings, while it is possible to adjust the layout to improve the angular resolution (e.g., [5,11]), the extent to which this can be done is limited. Although a previous study by Purchase *et al.* [19] did not find sufficient support for maximizing angular resolution, we do find that when two edges connected to the same node are almost on top to each other, it is difficult to tell whether these are two edges or one. For this reason we consider such edges as in collision too.

We note that a nice way to follow an edge, or to find the neighbors of a node, is to use interactive techniques such as "link sliding" and "bring & go" [17]. The algorithm we propose is primarily aimed at disambiguating a static drawing displayed on screen or on paper, it can nevertheless be used in conjunction with such interactive techniques.

Finally, we were made aware of the work of Jianu *et al.* [16] after the completion of this work. Jianu *et al.* [16] proposed a similar idea of using colors to differentiate edges. However there are multiple important differences between that work and ours. The construction of dual collision graph is different: Jianu *et al.* set the edge weights among all edges to be the inverse of either the intersection angle, or the edge distance if the edges do not intersect, which is not optimal since it is perfectly harmless to color edges that have no conflict with the same color. In fact, their method always results in a complete collision graph, making it more expensive for relatively large graphs. Furthermore, because of the complete collision graph, all edges of the original graph must have different colors. Therefore the drawings in [16], which are all of very small graphs, always contain a multitude of colors, which is unnecessary. Our collision graph almost always contains disconnected components. This decomposes the coloring problem into smaller ones, and allows us to use the same (black) colors for many edges. Jianu et al. [16] solved the coloring problem using a force-directed algorithm, motivated by Dillencourt *et al.* [7]. We were kindly given the source code for [16] from one of the authors. Based on reading the code, we found that it applies force directed algorithm to nodes of the collision graph in the 2D subspace of the LAB color space (the AB subspace). It then sets a fixed L value of 75 (L is the lightness, between 0 to 100). This observation is consistent with the drawings in [16], where black background is used for all drawings due to the high lightness value (see also Fig. 5(d)). This makes the algorithm limited to a small subset of all possible colors. Finally, the force-directed algorithms of Dillencourt *et al.* [7] and Jianu et al. [16] can only be applied to continuous color space in 2D or 3D. Neither works for user specified color palettes, or 1D colors. Our algorithm works for both continuous or discrete color spaces. Overall, we believe that the idea of using colors for disambiguating edges are quite natural to think of. It is how to design appropriate algorithms to make the idea work effectively in practice that is crucial and that differentiates our work and [16]. Furthermore, we present a first user study to evaluate results of our algorithm with real users. The results suggest possible scenarios when our edge coloring approach is effective.

3   The Edge Coloring Problem and a Coloring Algorithm

Appropriate coloring can help greatly in differentiating edges that cross at a small angle. Fig. 1 (left) illustrates such a situation. With many crossing edges, it is difficult to follow the edge from node 19 (top-middle, blue) to node 16 (lower-right, blue). In comparison, in Fig. 1 (right), it is easier to see that 19 is connected to 16 by a blue edge. The objective of this section is to identify situations where ambiguities in following edges can occur, and propose an edge coloring algorithm to resolve such ambiguities.

3.1   Edge Collisions

Two edges are considered in collision if an ambiguity arises when they are drawn using the same color. The following are four conditions for edge collision:

 – *C1: they cross at a small angle.*
 – *C2: they are connected to the same node at a small angle.*
 – *C3 (optional): they are connected to the same node at an angle close to 180 degree.*
 – *C4: they do not cross or share a node, but are very close to each other and are almost parallel.*

We now explain the rationale for considering each of these four conditions as being in collision. C1 is considered a collision following the user studies described in Section 1 by Huang *et al.* [14,15]. When eyes try to follow an edge to its destination, small crossing angles between this edge and other edges create multiple paths along the direction of the eye movement, either taking eyes to the wrong path, or slowing down the eye movement. C2 creates a situation where one edge is almost on top of the other, making it difficult to visually follow one of these edges.

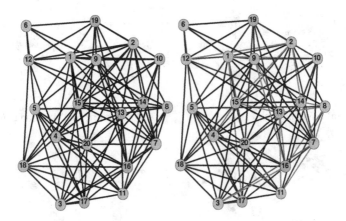

Fig. 1. Left: a graph with 20 nodes and 100 edges. It is difficult to follow some of the edges. For example, is node 19 (blue) connected to node 16 (blue)? Is node 19 connected to 17 (blue)? Right: the same graph, with the edges colored using our algorithm. Now it is easier to see that 19 and 16 are connected by a blue edge, but 19 and 17 are not connected.

C3 could create confusion as to whether the two edges connected at close to 180 degree are one edge, or two edges, when node labels are drawn. For example in Fig. 1 (left), it is difficult to tell whether nodes 19 and 17 are connected, or whether 19 is connected to 20 and 20 is connected to 17. When edges are properly colored (Fig. 1 (right)), it is clear that the latter is true. Note that if edges are allowed to be drawn on top of nodes, then an edge between 19 and 17 would be seen over the label of 20, thus this kind of confusion can be eliminated. Therefore we consider C3 as optional. But drawing edges over the label of nodes introduce extra clutter and make the node labels harder to read.

C4 causes a problem because when two edges are very close and almost parallel, it is difficult to differentiate between them. In addition, it can cause confusion when node labels are drawn. Fig. 3(a) shows two lines very close and almost parallel. While it is possible to differentiate between the two edges, when node labels are added (Fig. 3(b)), it is difficult to tell whether there are two edges ($1 \leftrightarrow 2$ and $3 \leftrightarrow 4$), or three edges ($1 \leftrightarrow 2$, $1 \leftrightarrow 4$ and $1 \leftrightarrow 3$), or whether there even exists an edge $3 \leftrightarrow 2$. This confusion can be avoided if suitable edge coloring is applied (Fig. 3(c)).

To resolve these collisions, we propose to color the edges so that any two edges in collision have as different colors as possible. We first construct a dual edge collision graph.

3.2   Constructing the Dual Collision Graph

Let the original graph be $G = \{V, E\}$. Denote by $N(v)$ the set of neighbors of a node v. The *dual collision graph* is $G_c = \{V_c, E_c\}$, where each node in V_c corresponds to an edge in the original graph. In other words, there is a one-to-one mapping $e : V_c \rightarrow E$. Two nodes of the collision graph i and j are connected if $e(i)$ and $e(j)$ collide in the original graph.

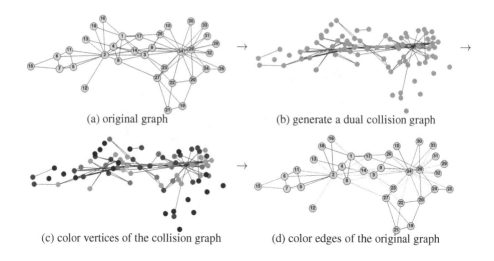

(a) original graph   (b) generate a dual collision graph

(c) color vertices of the collision graph   (d) color edges of the original graph

Fig. 2. The proposed pipeline for coloring the edges of the Zachary's Karate Club Graph: (a) the original graph; (b) the dual collision graph, with each node representing an edge of the original graph, and positioned at the center of that edge; (c) the collision graph, with nodes colored to maximize color differences along the edges; (d) the original graph, with edges colored using the node coloring in (c).

The problem of coloring the edges of G then becomes that of coloring nodes of the collision graph G_c. Let $\mathscr{C}$ be the color space, and $c(i) \in \mathscr{C}$ be the color of a node $i \in V_c$, we want to find a coloring scheme such that the color of each node in the collision graph is as different to its neighbors as possible. This task can be posed as a MaxMin optimization problem:

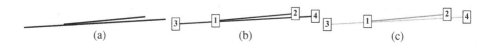

(a)   (b)   (c)

Fig. 3. An illustration of the rationale for collision condition C4. (a) Two edges that do not cross. (b) When nodes are shown, it is difficult to tell if there are two edges ($1 \leftrightarrow 2$ and $3 \leftrightarrow 4$), or three edges ($1 \leftrightarrow 2$, $1 \leftrightarrow 4$ and $1 \leftrightarrow 3$), or whether there even exists an edge $3 \leftrightarrow 2$. (c) After coloring each edges with a distinctive color, it is clear that there are two edges, $1 \leftrightarrow 2$ and $3 \leftrightarrow 4$

$$\arg\max_{c:V_c \to \mathscr{C}} \min_{\{i,j\} \in E_c} w_{ij}\|c(i) - c(j)\|, \tag{1}$$

where $w_{ij} > 0$ is a weight inversely proportional to how important it is to differentiate colors of nodes i and j, and $\|c(i) - c(j)\|$ is a measure of the difference between the colors assigned to the two nodes.

Note that (1) is stated rather generally: $\mathscr{C}$ could be a discrete, or continuous, color space. This is intentional since we are interested in both scenarios. All we assume is that $\mathscr{C}$ sits in a Euclidean space of dimension d.

Once we colored the collision graph, we can use the same coloring scheme for the edges of the original graph. The complete pipeline of our proposed approach is illustrated in Fig. 2. Notice that the collision graph in Fig. 2(b) is disconnected. We apply our algorithm on each component of the collision graph.

3.3 A Color Optimization Algorithm

Dillencourt et al. [7] proposed a force-directed algorithm in a Euclidean color space. They wanted *all* pairs of nodes to have distinctively different colors. Consequently their algorithm used a force model where repulsive forces exist among all pairs of nodes.

Because in our case edges can have the same color as long as they do not collide, there is no need to push all pairs of nodes of the collision graph apart in the color space. Therefore we can not use the algorithm of Dillencourt et al. [7] as is. Although it is possible to adapt their algorithm, we opt to propose an alternative algorithm. One reason is that we like to be able to use not only continuous color spaces, but also discrete color palettes.Another reason is due to the fact even when deciding the optimal color for one node of the collision graph with regard to all its neighbors, this seemingly simple problem can have many local maxima.

We give an example to illustrate this point. For simplicity of illustration, within this example, we assume for that our color space is 2D, and that the color distance is the Euclidean distance. Suppose we want to find the best color embedding for a node u in the collision graph with six neighbors, and the six neighbors are currently embedded as shown in Fig. 4 (left). We want to place u as far away from the set of six points as possible. Fig. 4 (left) shows a color contour of the distance from the set of six points (the distance of a point to a set of point is defined as the minimum distance between this point and all the points in the set, assuming unit weighting factors). Color scale is given in the figure, with blue for low values and off-white for large. From the contour plot it is clear that there are seven or more local maxima. In 3D there could be even more local maxima. A force-directed algorithm such as [7], even with the random jumps and swaps, is likely to settle in one of the local maxima.

Instead we hope to find the global maximum. A naive way to find the global maximum position in the color space with regard to a set of points is to search exhaustively by imposing a fine grid over the color space, and calculating the distance from each mesh point to the set. However, given that the color space are typically of three dimensions, even at a resolution of 100 subdivisions along each dimension, we need 10^6 distance calculations. This is computationally too expensive, bear in mind that this computation needs to be performed for each and every node of the collision graph repeatedly until the overall embedding in the color space converges.

We propose a more efficient algorithm based on the octree data structure (quadtree for 2D) that does not require evaluations of the distance over all mesh points. Pseudo code for the algorithm is given in [13]. We give a high level description here. Using Fig. 4 (left) as an example, we want to find a point in the color space that is of maximal distance to a target set of points. Define the objective function value of a square to be

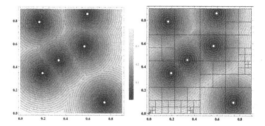

Fig. 4. Left: contour plot of the distance to a set of six (white) points in the space $[0,0.9] \times [0,0.9]$. There are seven or more local maxima. E.g., near $\{0,0.55\}$ and $\{0.4,0.7\}$. Right: an illustration of the quadtree structure generated during our algorithm for finding the global optimal embedding of a node that is farthest away from the set of six points. The final solution is $\{0,0\}$ (red point).

the distance from the center of the square to the target set. We start with a queue of one square covering the color space, and define the current optimal value as the maximal distance over all squares in the queue to the target set. Taking a square from the current queue, we subdivide it into four squares. If the distance of one of the four squares to the point set, plus the distance from the center of the square to a corner of the square, is less than the current optimal distance, this square is discarded. This is because no point in this square can have a larger distance to the target set than the current optimal distance. If the square is outside of the color space, it is also discarded. Otherwise the square is entered into the queue, and the optimal value updated. This continues until the half width of all squares in the queue is smaller than a preset threshold ε. The point that achieves the current optimal value is taken as the optimum. We know that the current optimal value should be within a value $\delta = d^{1/2}\varepsilon$ to the global optimal value, where δ is the half diagonal of the final square in d-dimensional space.

This algorithm is in essence a branch-and-bound algorithm operating on the octree (quadtree for 2D) decomposition of the color space. When applied to the problem in Fig. 4 (right), we can see that in the top-left quadrant, the quadtree branched twice and stopped, because the function values are relatively small in that quadrant. The top-right and bottom-right quadrants branched 3 and 4 times, respectively. The final optimal point is found in the bottom-left quadrant. Initially the algorithm homed in on two regions, one around $\{0.375,0\}$ and the other around $\{0,0\}$, eventually settled around the latter.

Of course this branch-and-bound algorithm only finds the global optimal embedding for one node. After applying the algorithm to every node of the collision graph once (one outer iteration), we repeated if the minimal color difference increases, or if it does not change, but the total sum of color difference across all nodes increases.

We name the algorithm CLARIFY (Edge Coloring for *CLARIFY*ing a Graph Layout) and formally state it in Algorithm 1 in the technical report [13].

4   Implementation and Results

CLARIFY works for both continuous color spaces (RGB and LAB), as well as discrete color space, including a fixed list of colors. Fig. 5 shows examples of applying CLAR-

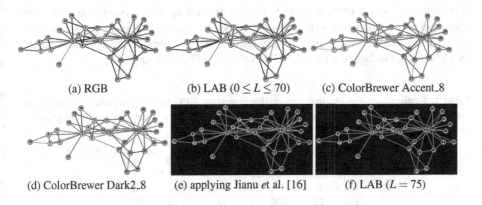

| (a) RGB | (b) LAB $(0 \leq L \leq 70)$ | (c) ColorBrewer Accent_8 |

| (d) ColorBrewer Dark2_8 | (e) applying Jianu et al. [16] | (f) LAB $(L = 75)$ |

Fig. 5. Applying CLARIFY on the Karate graph in RGB and LAB color spaces (a-b), and with two ColorBrewer palettes (c-d). For comparison we include the result of applying the algorithm of Jianu et al. [16], vs CLARIFY in LAB color space with fixed lightness of 75 (e-f).

IFY in the RGB color space, the LAB space with intensity $0 \leq L \leq 70$, and using two ColorBrewer [2] color palettes. In addition it shows how CLARIFY compared favorably with the result of [16]. CLARIFY can also be applied for coloring of virtual maps. Fig. 6 shows an author collaboration map (see [9]) colored using CLARIFY with two color palettes.

Detailed information on implementation, including how CLARIFY is made to work with both continuous and discrete color spaces, are given in the technical report [13]. CLARIFY is now available from Graphviz [10] as edgepaint (for edge coloring only; map coloring will be made available soon as part of gvmap).

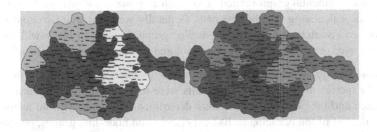

Fig. 6. CLARIFY on a virtual map with two ColorBrewer palettes: left: Accent_8, right: Dark2_8.

We now apply CLARIFY to graphs from real applications. Table 1 gives results on six of the graphs we tested, including running time and objective function (1) (color diff) achieved in LAB color space. These are from [4] or [10]. We intentionally avoided choosing mesh-like graphs – such graphs are easy to layout aesthetically. Their layouts also tend to exhibit a low perceptual complexity, making it relatively easy to follow edges and paths. Compared with a non-mesh-like graph, a mesh-like graph is easier for our algorithm because there are typically fewer colliding edges. We ran the experiment on a Macbook Pro laptop with a 2.3 GHz Intel Core i7 processor.

Table 1. Statistics on the original and dual test graphs, CPU time (in second) and objective function (cdiff) for CLARIFY. The time in bracket is for constructing the dual collision graph.

graph	$|V|$	$|E|$	$|E_c|$	CPU	cdiff
ngk_4	50	100	54	0.6 (0.)	122.69
NotreDame_yeast	1458	1948	1685	1.3 (0.2)	67.9
GD00_c	638	1020	1847	1.7 (0.1)	64.32
Erdos971	429	1312	4427	2.1 (0.1)	59.3
Harvard500	500	2043	11972	2.3 (0.3)	35.0
extr1	5670	11405	34696	14.5 (7.9)	47.1

It can be seen from Table 1 that for graphs of up to a few thousand nodes and edges, CLARIFY runs quickly. The majority of the CPU time is spent on color assignment, while the construction of the collision graph takes relatively little time even with the naive dual graph construction algorithm. The Harvard500 graph gives a large $|E_c|$ (number of edges in the collision graph) in comparison to the number of edges, because it has a few almost complete subgraphs, which results in a lot of crossings at small angles.

5   User Study

We conducted a controlled experiment to study the effect of edge coloring on user's performance in fundamental graph-related tasks, such as visually following edges, finding neighbors and calculating the shortest path. Generally we compared two approaches, defined as two visualization types: the baseline graph drawing in black-white (B/W) and the improved graph drawing with edges colored by our algorithm (Color).

Experiment Design. We recruited 12 participants (8 male, 4 female) for this paper-and-pencil experiment. 10 of the participants were graduate students majoring computer science and the other 2 of them were department assistants with no technology background. Half of the participants had experiences on node-link graphs, one student was even an expert on graphs. The other half did not have previous knowledge with the node-link graph. The experiment followed a within-subject design with every participant doing all tasks with both visualization types. To eliminate the learning effect over the same task, we used two different layouts of the same graph data. We had a full factorial deign on the choice of two visualization types and two graph layouts. Each participant entered the same task four times in total. The experiment order was randomized across participants. Half of them completed the tasks first with the B/W approach and then with the Color approach. Another half adopted the opposite order. Further, in half of the time when participants were given the colored drawing, the algorithm is fixed to use the LAB palette. In another half, the participants selected their favorite palette and completed tasks with the colored drawings generated by this palette.

Data and Task. Two layouts of the Zachary's Karate Club Graph were used. One was exactly the layout in Fig. 2. Another was rotated and re-labeled. Three types of graph-related tasks were designed:

T1 (Connectivity): Determine whether two nodes are connected by a direct edge;
T2 (Neighbor): Estimate the number of nodes a particular node connects directly;
T3 (Path): Estimate the minimum number of hops from a particular node to another, including the source and destination.

On each type, four tasks were selected on each graph layout with similar difficulty levels. To eliminate user's visual node querying time from their task completion time, we annotated the related nodes in each task on the corresponding graph layout before participants took the task.

Result. Results were analyzed separately on each task type. Detailed analysis and error bar charts are given in the technical report [13]. The major findings are that on connectivity tasks, the average task error of the Color group is less than 30% of the B/W group, and is statistically significant. Performance difference on neighbor/path tasks and color palettes were not statistically significant.

6 Conclusions

Edge crossings, particularly those at small crossing angles, are known to be detrimental to the visual understanding of graph drawings. This paper proposes an edge coloring algorithm for disambiguating edges that are in collision because of small crossing angles or partial overlaps. The algorithm, based on a branch-and-bound procedure applied to a space decomposition of the color gamut, generates color assignments that maximize color differences of the colliding edges, and works for both continuous color space and discrete color palettes. The algorithm can also be applied to generate coloring for disambiguating virtual maps. Our user study found that coloring edges in graph drawings helped user's performance in 1-hop graph connectivity task significantly. Consequently we have made the CLARIFY code available as part of Graphviz open source software.

The approach of coloring edges for disambiguating drawings has its limitations. Our working assumption is that the drawing is to be displayed as a static image on paper or screen. When an interactive environment is available, techniques such as "link slid-ing" and "bring & go" [17] could be more effective. In such a situation, the algorithms proposed here can be used as an additional visual aid to the interaction.

While the algorithm proposed here can run on relatively large graphs, our experience is that for graphs with a lot of edges, a static image is insufficient to allow the user to clearly see and follow each edge. Therefore our approach is best suited for small- to medium- sized graphs. Typical usage scenarios are illustrations of diagrams, such as computer or biological networks.

Finally, we note that sometimes edge colors are used to encode attributes on the edges. To apply our approach without interfering with the need to display such attributes, edges can be differentiated using dashed lines of different style and/or thickness, using CLARIFY through mapping different line styles to 1D or 2D spaces.

References

1. Battista, G.D., Eades, P., Tamassia, R., Tollis, I.G.: Algorithms for the Visualization of Graphs. Prentice-Hall (1999)
2. Brewer, C.: ColorBrewer - Color Advice for Maps, http://www.colorbrewer2.org
3. Chernobelskiy, R., Cunningham, K.I., Goodrich, M.T., Kobourov, S.G., Trott, L.: Force-directed lombardi-style graph drawing. In: Speckmann, B. (ed.) GD 2011. LNCS, vol. 7034, pp. 320–331. Springer, Heidelberg (2011)
4. Davis, T.A., Hu, Y.: University of Florida Sparse Matrix Collection. ACM Transaction on Mathematical Software 38, 1–18 (2011)
5. Di Battista, G., Vismara, L.: Angles of planar triangular graphs. In: Proceedings of the Twenty-Fifth Annual ACM Symposium on Theory of Computing, STOC 1993, pp. 431–437. ACM, New York (1993)
6. Didimo, W., Eades, P., Liotta, G.: Drawing graphs with right angle crossings. Theor. Comput. Sci. 412(39), 5156–5166 (2011)
7. Dillencourt, M.B., Eppstein, D., Goodrich, M.T.: Choosing colors for geometric graphs via color space embeddings. In: Kaufmann, M., Wagner, D. (eds.) GD 2006. LNCS, vol. 4372, pp. 294–305. Springer, Heidelberg (2007)
8. Duncan, C., Eppstein, D., Goodrich, M.T., Kobourov, S., Nöllenburg, M.: Lombardi drawings of graphs. J. Graph Algorithms and Applications 16, 85–108 (2012)
9. Gansner, E.R., Hu, Y., Kobourov, S.: Visualizing Graphs and Clusters as Maps. IEEE Computer Graphics and Applications 30, 54–66 (2010)
10. Gansner, E.R., North, S.: An open graph visualization system and its applications to software engineering. Software - Practice & Experience 30, 1203–1233 (2000)
11. Garg, A., Tamassia, R.: Planar drawings and angular resolution: Algorithms and bounds. In: van Leeuwen, J. (ed.) ESA 1994. LNCS, vol. 855, pp. 12–23. Springer, Heidelberg (1994)
12. Hu, Y., Kobourov, S., Veeramoni, S.: On maximum differential graph coloring. In: Brandes, U., Cornelsen, S. (eds.) GD 2010. LNCS, vol. 6502, pp. 274–286. Springer, Heidelberg (2011)
13. Hu, Y., Shi, L. (2014), http://arxiv.org/abs/1409.0436
14. Huang, W.: Using eye tracking to investigate graph layout effects. In: 2007 6th International Asia-Pacific Symposium on Visualization, APVIS 2007, pp. 97–100 (2007)
15. Huang, W., Hong, S.-H., Eades, P.: Effects of crossing angles. In: Proceedings of IEEE Pacific Visualization Symposium, pp. 41–46. IEEE (2008)
16. Jianu, R., Rusu, A., Fabian, A.J., Laidlaw, D.H.: A coloring solution to the edge crossing problem. In: Proceedings of the 13th International Conference in Information Visualization (iV 2009), pp. 691–696. IEEE Computer Society (2009)
17. Moscovich, T., Chevalier, F., Henry, N., Pietriga, E., Fekete, J.: Topology-aware navigation in large networks. In: CHI 2009: Proceedings of the 27th International Conference on Human Factors in Computing Systems, pp. 2319–2328. ACM, New York (2009)
18. Purchase, H.C.: Which aesthetic has the greatest effect on human understanding? In: DiBattista, G. (ed.) GD 1997. LNCS, vol. 1353, pp. 248–261. Springer, Heidelberg (1997)
19. Purchase, H.C., Carrington, D., Allder, J.-A.: Experimenting with aesthetics-based graph layout. In: Anderson, M., Cheng, P., Haarslev, V. (eds.) Diagrams 2000. LNCS (LNAI), vol. 1889, pp. 498–501. Springer, Heidelberg (2000)
20. Purchase, H.C., Hamer, J., Nöllenburg, M., Kobourov, S.G.: On the usability of lombardi graph drawings. In: Didimo, W., Patrignani, M. (eds.) GD 2012. LNCS, vol. 7704, pp. 451–462. Springer, Heidelberg (2013)
21. Ware, C., Purchase, H., Colpoys, L., McGill, M.: Cognitive measurements of graph aesthetics. Information Visualization 1(2), 103–110 (2002)

Untangling Hairballs*
From 3 to 14 Degrees of Separation

Arlind Nocaj, Mark Ortmann, and Ulrik Brandes

Computer & Information Science, University of Konstanz, Germany

Abstract. Small-world graphs have characteristically low average distance and thus cause force-directed methods to generate drawings that look like hairballs. This is by design as the inherent objective of these methods is a globally uniform edge length or, more generally, accurate distance representation. The problem arises in graphs of high density or high conductance, and in the presence of high-degree vertices, all of which tend to pull vertices together and thus clutter variation in local density.

We here propose a method to draw online social networks, a special class of hairball graphs. The method is based on a spanning subgraph that is sparse but connected and consists of strong ties holding together communities. To identify these ties we propose a novel measure of embeddedness. It is based on a weighted accumulation of triangles in quadrangles and can be determined efficiently. An evaluation on empirical and generated networks indicates that our approach improves upon previous methods using other edge indices. Although primarily designed to achieve more informative drawings, our spanning subgraph may also serve as a sparsifier that trims a hairball graph before the application of a clustering algorithm.

1 Introduction

Online social networks such as Facebook friendship graphs are an amalgamation of a variety of social relations. The existence of a friendship tie might be due to shared interests, spatial proximity, kinship, or professional relations to name but a few. When such a multitude of relations is conflated in the same network, any two nodes are likely to be connected via at most a few links – thus leading to a *small world* effect [21]. Visualizations of these graphs using standard layout methods such as force-directed placement produce drawings in which variation in local structure is hidden in a densely-looking, overlap-ridden *hairball*. An example is given in Fig. 1(a).

Various approaches to reduce the clutter in drawings of small worlds and other hairball graphs have been proposed [12], most notably *edge bundling* [10], *edge lensing* [11], modified *layout algorithms* or *representations* [1,7,29], and *graph simplification* [2,17,18,20,23,30]. The idea of graph simplification is to identify

---

* We gratefully acknowledge financial support from DFG under grants GRK 1042 and Br 2158/6-1. The proposed method is available in visone.

C. Duncan and A. Symvonis (Eds.): GD 2014, LNCS 8871, pp. 101–112, 2014.

a subset of edges such that only the resulting graph, the so-called backbone, needs to be laid out. We adopt this approach and propose a new method to trim hairballs.

Problem formulations in graph simplification include the preservation of properties such as cuts [2], spectra [23,24], connectivity [30], collapsing substructures into supernodes [18], and emphasizing deeply embedded connections [17,20]. As graph invariants such as cuts are more easily affected by noise in empirical networks, we opt for locally defined graph simplification criteria.

Substantiated by the sociological work of Simmel [22], Nick et al. [17] define the strength of an edge by the number of triangles it is contained in, and then determine the degree of structural embeddedness for an edge by comparing the ranked neighborhoods of its two vertices. The purpose is to identify edges that are more likely to be inside of cohesive groups than between such groups. If the initial edge strengths are uniform, the Simmelian backbone reduces to a backbone proposed by Satuliri et al. [20]. In both methods, the backbone is obtained by finally removing all edges with weights below a specified nodal or network wide threshold.

These filtering techniques are related to, but not the same as graph partitioning. Since we want to use them for graph drawing, the difference is even greater because maintaining connectedness becomes a crucial constraint. Otherwise, the layout algorithm is oblivious to edges of the original graph connecting vertices in different components of the backbone as can be inferred from Fig. 1(b). When connected components happen to be placed far apart, these edges will run across the drawing and produce even worse clutter.

We present an efficient preprocessing technique that allows to draw a certain class of small-world social networks with standard layout algorithms that would produce hairball layouts otherwise. Our main contributions are:

- a novel method to identify strong ties,
- the use of the union of all maximum spanning trees as a sparsifier that maintains connectedness and avoids subtree-ordering ambivalence, and
- an evaluation on observed and generated networks.

We outline our overall method for drawing hairball graphs in the next section and describe our edge embeddedness metric in Sect. 3. Different metrics are evaluated in Sect. 4 and we conclude in Sect. 5.

2   Drawing Algorithm

The main challenges in drawing hairball graphs are their high density, low diameter and noisy group structure. Therefore, our goal is to find a backbone of the graph that retains deeply embedded edges and thus can be used to draw the original graph, e.g., by a force-directed method [13] to reveal the actual variation in cohesiveness.

Since most drawing methods cannot put vertices of different graph components into a meaningful spatial relation, cf. Fig. 1(b), we need to maintain the graph connectivity to retain the global context.

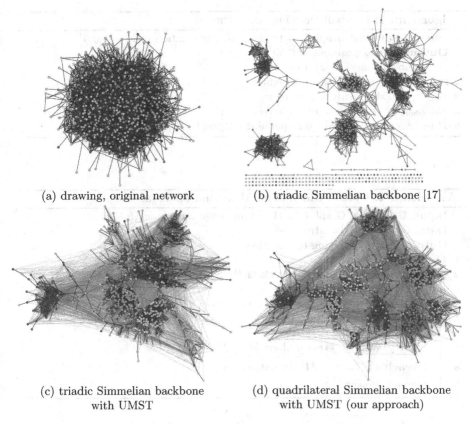

(a) drawing, original network

(b) triadic Simmelian backbone [17]

(c) triadic Simmelian backbone
with UMST

(d) quadrilateral Simmelian backbone
with UMST (our approach)

Fig. 1. Facebook friendships at California Institute of Technology (Caltech36). Vertex color corresponds to dormitory (gray for missing values), but has not been utilized in the layout algorithm. The layout in (a) is based on the entire hairball graph, whereas (b)-(d) use edge embeddedness, which spreads the graph while keeping cohesive groups together. Embeddedness mapped to edge color; backbone edges dark gray.

This leads to the following requirements on our backbone:

(i) Edges should be favored based on their structural embeddedness only.
(ii) Connectedness has to be maintained.

Two common approaches to simplify a graph $G = (V, E, w)$ with vertex set V, edge set E, and edge weight $w : E \to \mathbb{R}_{\geq 0}$, are *sampling* [2,23] and *thresholding* [1,17,20]. Note that we assume that w reflects the embeddedness of an edge and a higher value corresponds to stronger embeddedness. Although sampling can be used for sparsification purposes the random selection of edges violates both of our requirements. In contrast thresholding guarantees that edges are favored by their weights and consequently their structural properties, as it retains only the top k percent of edges with respect to w. Nevertheless, neither nodal nor network wide thresholding can ensure that the simplified graph stays connected.

Algorithm 1: Hairball Drawing Algorithm

    Input: Undirected Graph $G = (V, E)$ and sparsification ratio $s \in [0, 1]$.
    Output: Vertex positions $P \in \mathbb{R}^{|V| \times 2}$

1 $w \leftarrow$ embeddedness weights of edges
2   sort edges by non-increasing weight
3 $E_{\text{union}} \leftarrow$ UMST with respect to w
4 $E_{\text{threshold}} \leftarrow \{e \in E : w(e) \geq w(e_{\lceil (1-s)|E| \rceil})\}$
5 $P \leftarrow$ layout determined from spanning subgraph $(V, E_{\text{union}} \cup E_{\text{threshold}})$

Algorithm 2: UMST: Union of all Maximum Spanning Trees

    Input: Undirected Graph $G = (V, E)$ and edge weights $w : E \to \mathbb{R}_{\geq 0}$.
    Data: Union-Find datastructure
    Output: Edges belonging to any MST

1 $E_{\text{union}} \leftarrow \emptyset$
2   partition edges by weight into buckets $B_1, \ldots, B_k$
3   sort buckets by decreasing weight
4   **for** $i \leftarrow 1$ **to** k **do**
5 $M \leftarrow \emptyset$
6   **foreach** $e = (u, v) \in B_i$ **do**
7   **if** *find(u)* $\neq$ *find(v)* **then** $M \leftarrow M \cup \{e\}$
8   **foreach** $e = (u, v) \in M$ **do** union(u, v)
9 $E_{\text{union}} \leftarrow E_{\text{union}} \cup M$

Sparse connected subgraphs of edges not likely to be between cohesive groups have been proposed, e.g., by van Ham and Wattenberg [28] (planar graphs) and Tumminello et al. [27] (graph of bounded genus). A minimally connected subgraph of edges with high weights is a *maximum spanning tree* (MST), and Mantegna [14] proposed these as a backbone. Trees, however, have severe drawbacks: firstly, they do not maintain any local variation in density and, secondly, they introduce a subtree ordering ambiguity. While the first also means that arbitrary choices must be made when edges have equal embeddedness, the second creates a degree of freedom that is almost as bad as disconnected components.

We combine thresholding (to maintain local variation) with the union of all maximum spanning trees (UMST; to maintain connectedness). The UMST does not only solve the problem of tie breaks but also reduces the ordering problem by resulting in higher connectivity (Fig. 1(b)-(d)).

The complete algorithm to compute the layout of a hairball graph is presented in Alg. 1. Note that the UMST only contributes the (strongest) edges necessary to connect the components that result from the thresholding process.

Kruskal's algorithm for minimum spanning trees is easily adapted to determine the union of all maximum spanning trees. Since every edge of maximum weight that has not been processed yet could be chosen next, we batch-process them before components are merged; cf. Algorithm 2.

The final layout emphasizes variation in local density by considering only deeply embedded edges as expressed by the weights introduced in the next section.

3 Edge Embeddedness by Accumulating Triadic Effects

Real world networks are often aggregates of different relations, which can hamper the detection of subgroups or clusters. Our goal is to determine strong embedded edges, which are likely to be in dense groups, so that we can use them to emphasize the inherent structure. The assumption here is that vertices in the same subgroup of a network are connected stronger with each other than to members outside of the group.

Satuliri et al. [20] propose to capture the embeddedness of an edge $e = (u, v)$ by the Jaccard coefficient over u's and v's neighborhood. Nick et al. [17] suggest a more general framework, consisting of the following main steps:

1. For each edge, determine its strength
2. For each vertex, rank all its neighbors according to the edge strength
3. For each edge, determine its redundancy

The approach of Satuliri et al. can be seen as using a uniform edge strength for step one and the Jaccard coefficient for the redundancy in step three. Contrary to this, Nick et al. use the number of triangles an edge is embedded in (Simmelian strength) for step one and the best prefix Jaccard coefficient for step three. The latter chooses k such that the Jaccard coefficient of the first top k ranked neighbors of u and v is maximized. The effect of this ranking measure is that the highly ranked neighbors have more importance attached, since fewer common vertices are needed to get a high coefficient.

A more intuitive interpretation of this framework is that for an edge $e = (u, v)$ the edge strength allows us to determine the most important neighbors of u and v. If these most important neighbors are *the same*, e is strongly embedded; otherwise e is connecting two vertices, which are likely to be in different groups.

We follow the main idea, but propose a different edge strength than the number of triangles.

Consider the setting in Fig. 2. Clearly, edge e is strongly embedded. Compared to all other edges it closes many triangles resulting in an increase of the *group performance* [5] by introducing mediator effects. Similar to this, an edge (s, t) connecting two triangles at e introduces additional mediator effects on the triangles, which in turn increases the importance of e. We call these edges *mediator edges* on e.

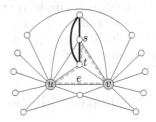

Fig. 2. Triangles at edge e [17,20] do not capture *mediator edges* (bold), while quadrangles do.

Counting the number of triangles at e does not capture the importance of mediator edges. But since each mediator edge creates two quadrangles at e, cf. dashed-contour in Fig. 2, we can use the number of quadrangles containing e

to capture this mediator effect. While there can be additional quadrangles at e, they will be counted only once from e's perspective, which makes their influence rather low. Furthermore, counting the two different types of quadrangles at e would be too time consuming and therefore we will not distinguish between them.

Using the absolute number of quadrangles poses difficulties, when the network contains subgroups of different densities. Hence, we normalize this absolute value by putting it into relation to all edges at vertex u and v. Let $q(u, v)$ be the number of quadrangles containing edge $(u, v) \in E$. We define the *quadrilateral edge embeddedness* as

$$Q(u, v) = \frac{q(u, v)}{\sqrt{q(u) \cdot q(v)}},$$

where $q(v) = \sum_{w \in N(v)} q(v, w)$, for $v \in V$, and $N(v)$ the neighborhood of v. We use the geometric mean over the arithmetic mean, since it takes the dependency of two variables into stronger consideration. Note that edge-metrics using quadrangles have already been proposed by Auber et al. [1] and Radicchi et al. [19], but are different from our method as they focus on density. For a comparison of different edge metrics we refer the reader to Melançon and Sallaberry [16].

Computation and Time Complexity

The quadrangles of a graph G can be listed in $\mathcal{O}(m\alpha(G))$ [6], where m is the number of edges and $\alpha(G)$, the *arboricity* of G, is the minimum number of edge-disjoint forests necessary to cover all edges of G. While the arboricity can be as large as $\sqrt{m}$, it is bounded from above by the h-index of a graph which in turn is found to be very small in social networks [8]. Together with the normalization, the computation of the edge strengths takes $\mathcal{O}(m\alpha(G))$ time.

Neighbors can be ranked in $\mathcal{O}(m \log \triangle(G))$ time and redundancy can be computed in $\mathcal{O}(m\triangle(G))$, where $\triangle(G)$ is the maximum vertex degree. For example, the overall backbone computation took 0.2s on a network with 762 vertices and 16k edges (Caltech65) and 2.3s on a network with 2970 vertices and 100k edges (Smith60) with our Java 7 implementation and an Intel Core i7-2600K CPU@3.40GHz. The approach thus scales to large networks and we turn to the evaluation of its effectiveness in the next section.

4   Evaluating Methods for Edge Embeddedness

In this section we introduce the dataset and a graph model, from which we generate artificial hairball graphs. Then we explain our output quality indicators and the different edge embeddedness methods. For each graph and edge embeddedness method, we iteratively increase the sparsification ratio by 10% and compute the corresponding backbone. Layouts are computed using stress majorization [9] initialized by PivotMDS [3] as suggested in [4].

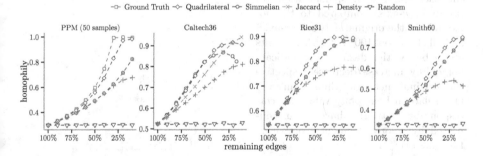

Fig. 3. Homophily (y-axis) is plotted against the number of remaining edges (x-axis) for the synthetic model (PPM) and three of the Facebook networks. Overall Quadrilateral performs better than the others. For the synthetic networks it comes very close to the ground truth.

4.1 Dataset and Model

As real world samples, we use the *Facebook100* dataset [26], which contains social relations of 100 higher educational institutes in the US. The network size varies from 762 to 41K vertices and from 16K to 1.6M edges. The dataset is directly from Facebook, not sampled, and thus very complete in terms of capturing the social relations according to a widely used service at that time. Additional attributes obtained from the Facebook profiles are gender, year of graduation, dormitory, etc. Due to incomplete profiles, a number of attribute values are missing. We will use the dormitory attribute for our evaluation, because it has been argued to be important for the creation of social relations in many of the networks [26].

Note that, in spite of a strong empirical association with homophilous attribute values, no ground-truth group structure is available for Facebook networks. Therefore, we also generated artificial networks from a model that represents the idealized version of the networks we are considering in this application.

A simple model generating random graphs with cohesive groups that are connected into a small world is the *planted partition model* (PPM) [15]. Let $\mathcal{C} = \{C_1, \ldots, C_k\}$ be a partition of V for a graph $G = (V, E)$. Then $\mathcal{C}$ is called a clustering of G with class $c(v) \in \mathcal{C}$ for a vertex $v \in V$. The probability of an edge (u, v) is p_{in} if $c(u) = c(v)$ and p_{out} if $c(u) \neq c(v)$. We generated 50 graphs from a PPM with 500 vertices, $k = 9$, $p_{in} = 0.3$, and $p_{out} = 0.01$. On top of that, we ran a random noise model with $p_{in} = p_{out} = 0.1$ to obfuscate the underlying group structure. The resulting graphs are very dense, have a low diameter, and are real hairballs without any visible structure when laid out using force-directed methods. The presented results of our model are averaged over 50 samples.

Fig. 4. Dormitory-homophily of different backbones, with sparsification ratio 70%, (*y*-axis) compared to the homophily in the original network (*x*-axis)
for all Facebook100 networks. Points
above/below the dashed line indicate
homophily increase/decrease respective
the original network. Simmelian and
Quadrilateral homophily values for corresponding networks have been connected by colored segments comparing
their performance.

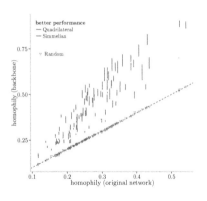

4.2   Edge Embeddedness Methods

We compare different methods which assign a weight $w : E \to \mathbb{R}_{\geq 0}$ to each edge
$e = (u, v) \in E$ depicting its embeddedness. All these methods are then extended
using our UMST approach to guarantee the connectivity, such that a layout can
be computed from the resulting graph. We use the following approaches to assign
a weight to the edges.

Random: Assigns uniform random weights, as base line.

Jaccard: Jaccard coefficient, $\frac{|N(u) \cap N(v)|}{|N(u) \cup N(v)|}$, as proposed by Satuliri et al. [20].

Simmelian: Triadic Simmelian backbone, as proposed by Nick et al. [17].

Quadrilateral: Quadrilateral Simmelian backbone, based on our embeddedness
method, which accumulates triadic effects at an edge with quadrangles (Sect. 3).

Density: Metric by Auber et al. [1] accumulating densities of different subgroups
in the local neighborhood.

Ground Truth: Knowledge of class membership in the synthetic network is
used to assign directly a low value to inter-cluster edges and a big value to
intra-cluster edges.

4.3   Quality Metrics

In contrast to the synthetic networks there is no ground truth available for the
Facebook networks. This makes it hard to evaluate outcomes of the different
methods. Nevertheless, it was found that for many of the Facebook networks,
the housing structure (dormitory attribute) is very relevant for the underlying
formation of social relations [17,26]. We, therefore, use the dormitory attribute
as a reference for evaluation.

Assume that we know the ground truth, meaning the class membership $c(v)$ of
each vertex. A *perfect* algorithm, for example, would first remove all inter-cluster
edges before starting to remove intra-cluster edges while obeying the required
sparsification ratio. Since inter-cluster edges are removed priorly, this increases
the ratio between intra-cluster or homophily edges and the total number of edges.

If the edge embeddedness methods perform similar to this, the ratio of homophily edges

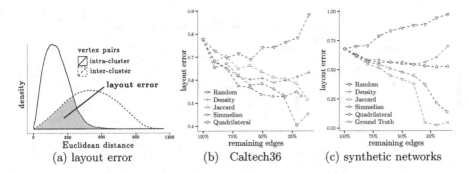

(a) layout error   (b)   Caltech36   (c) synthetic networks

Fig. 5. Layout error of different edge embedding methods combined with our UMST for (b) a real world network and (c) synthetic networks. (a) shows the layout error for a single point of the line chart in (b).

$$\text{homophily}(G) = \frac{\#\text{homophily edges}}{\#\text{homophily edges} + \#\text{heterophily edges}}$$

should monotonically increase, while gradually removing edges from the network according to their weight. Edges for which the class membership (attribute) of at least one vertex is missing are neglected.

Additionally, we would like to see how well this class membership is reflected in the layouts. Vertex pairs of the same class should have a small Euclidean distance, while pairs of different classes should have a large Euclidean distance. Looking at the curve of the Euclidean distance distribution of the intra-cluster and inter-cluster vertex pairs in Fig. 5(a), we define the layout error as the intersection area of these two curves. The layout error can also be interpreted as the percentage of vertex pairs, where the distinction whether they are in the same cluster or not cannot clearly be made based on the Euclidean distance. Since the computation of this quality metric is very time intensive, it was not feasible to analyze all 100 Facebook networks with it.

4.4   Results and Discussion

An interesting observation from Fig. 3 is that Jaccard and Simmelian perform very similar for most Facebook networks. Our method (Quadrilateral) clearly manages to distinguish between the different types of edges better than the other methods, especially in earlier phases of the sparsification.

For all 100 Facebook networks, the difference in homophily between Simmelian and Quadrilateral is shown by the length of a vertical segment in Fig. 4. While both approaches increase the percentage of homophily edges (all segments above the diagonal dashed line), Quadrilateral clearly performs better, especially for networks with higher percentage of homophily edges.

Although the homophily of Jaccard and Quadrilateral is nearly the same for the last but one step of the Caltech network (Fig. 3) the Quadrilateral embedding

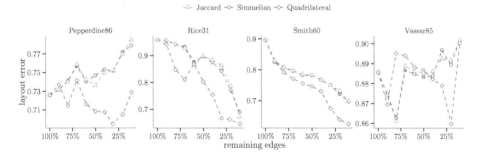

Fig. 6. Layout error of Facebook networks w.r.t. the dormitory attribute. While improvement is not clear for Pepperdine86 and Vassar85, the layout is improved a lot for the networks with high homophily (Rice31 and Smith60).

creates the superior layout (Fig. 5). Furthermore, for the synthetic networks, Quadrilateral comes very close to the ground truth (Fig. 5).

Figure 6 shows the layout error for four Facebook networks and the three best performing edge metrics (according to homophily). The layout clearly improves for the Rice and Smith network, but not much for the other two. One possible explanation for this could be that the dormitory attribute is not the explanatory variable for the formation of social relations in these two networks.

The effectiveness of our layout quality metric can also be verified, by looking at the final drawings in Fig. 1(c) and 1(d). In the latter many clusters, as light green and light blue, are more clearly visible. For the synthetic networks our method comes also very close to the ground truth, in terms of layout error (Fig. 5(c)). Again, the drawings of a synthetic network (Fig. 7) support this conclusion.

5   Conclusion

We proposed a sparsification approach to draw hairball graphs as encountered in online social networks. It is based on the idea that pairwise distances (the "degrees of separation") need to be increased without disrupting tightly-knit groups. The deeply embedded edges such groups are made of are identified using a suitably modified Simmelian backbone [17], and overall layout organization is stabilized by maintaining connectedness via the union of all maximum spanning trees.

An evaluation with empirical and generated networks showed that our novel metric manages to reveal relations deeply embedded in latent primary groups. In the resulting drawings such groups are separated from each other but still positioned in their global context. On the Facebook100 dataset, average distances increased from about 3 in the original friendship networks to about 14 in the backbone, thus easing the layout task for force-directed algorithms.

Our proposed edge embeddedness metric proved to be more effective than previous approaches with respect to improving layout quality by way of amplifying homophily. It is thus likely to be useful as a preprocessing step for graph clustering algorithms as well.

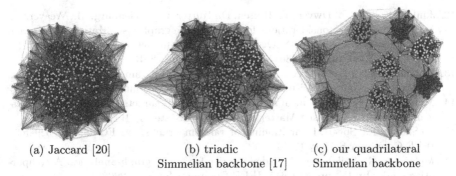

(a) Jaccard [20]	(b) triadic Simmelian backbone [17]	(c) our quadrilateral Simmelian backbone

Fig. 7. Layouts of the same synthetic network determined by different edge embeddedness methods combined with our UMST (sparsification ratio of 80%). Colors encode groups – ground truth.

By design, our technique appears to be best suited for small-world networks with multiple centers. While these are common, especially in social media, it will be interesting to identify variants for hierarchically clustered graphs and single-centered core-periphery structures such as the network of world trade [25].

References

1. Auber, D., Chiricota, Y., Jourdan, F., Melançon, G.: Multiscale visualization of small world networks. In: INFOVIS. IEEE Computer Society (2003)
2. Benczúr, A.A., Karger, D.R.: Approximating s-t minimum cuts in $õ(n^2)$ time. In: Miller, G.L. (ed.) STOC, pp. 47–55. ACM (1996)
3. Brandes, U., Pich, C.: Eigensolver methods for progressive multidimensional scaling of large data. In: Kaufmann, M., Wagner, D. (eds.) GD 2006. LNCS, vol. 4372, pp. 42–53. Springer, Heidelberg (2007)
4. Brandes, U., Pich, C.: An experimental study on distance-based graph drawing. In: Tollis, I.G., Patrignani, M. (eds.) GD 2008. LNCS, vol. 5417, pp. 218–229. Springer, Heidelberg (2009)
5. Burt, R.S.: Structural holes versus network closure as social capital. Social Capital: Theory and Research, pp. 31–56 (2001)
6. Chiba, N., Nishizeki, T.: Arboricity and subgraph listing algorithms. SIAM J. Comput. 14(1), 210–223 (1985)
7. Cohen, J.D.: Drawing graphs to convey proximity: An incremental arrangement method. ACM Trans. Comput.-Hum. Interact. 4(3), 197–229 (1997)
8. Eppstein, D., Spiro, E.S.: The h-index of a graph and its application to dynamic subgraph statistics. J. Graph Algorithms Appl. 16(2), 543–567 (2012)
9. Gansner, E.R., Koren, Y., North, S.C.: Graph drawing by stress majorization. In: Pach, J. (ed.) GD 2004. LNCS, vol. 3383, pp. 239–250. Springer, Heidelberg (2005)
10. Holten, D., van Wijk, J.J.: Force-directed edge bundling for graph visualization. Comput. Graph. Forum 28(3), 983–990 (2009)
11. Hurter, C., Telea, A., Ersoy, O.: Moleview: An attribute and structure-based semantic lens for large element-based plots. IEEE TVCG 17(12), 2600–2609 (2011)

12. Jankun-Kelly, T.J., Dwyer, T., Holten, D., Hurter, C., Nöllenburg, M., Weaver, C., Xu, K.: Scalability considerations for multivariate graph visualization. In: Kerren, A., Purchase, H.C., Ward, M.O. (eds.) Multivariate Network Visualization 2013. LNCS, vol. 8380, pp. 208–236. Springer, Heidelberg (2013)
13. Kobourov, S.G.: Force-directed drawing algorithms. In: Tamassia, R. (ed.) Handbk. of Graph Drawing and Visualization, pp. 383–408. Chapman & Hall/CRC (2013)
14. Mantegna, R.N.: Hierarchical structure in financial markets. The European Physical Journal B-Condensed Matter and Complex Systems 11(1), 193–197 (1999)
15. McSherry, F.: Spectral partitioning of random graphs. In: FOCS, pp. 529–537. IEEE Computer Society (2001)
16. Melançon, G., Sallaberry, A.: Edge metrics for visual graph analytics: A comparative study. In: IV, pp. 610–615. IEEE Computer Society (2008)
17. Nick, B., Lee, C., Cunningham, P., Brandes, U.: Simmelian backbones: amplifying hidden homophily in facebook networks. In: Rokne, J.G., Faloutsos, C. (eds.) ASONAM, pp. 525–532. ACM (2013)
18. Pfaltz, J.L.: The irreducible spine(s) of undirected networks. In: Lin, X., Manolopoulos, Y., Srivastava, D., Huang, G. (eds.) WISE 2013, Part II. LNCS, vol. 8181, pp. 104–117. Springer, Heidelberg (2013)
19. Radicchi, F., Castellano, C., Cecconi, F., Loreto, V., Parisi, D.: Defining and identifying communities in networks. Proc. Natl. Acad. Sci. 101(9), 2658–2663 (2004)
20. Satuluri, V., Parthasarathy, S., Ruan, Y.: Local graph sparsification for scalable clustering. In: Sellis, T.K., Miller, R.J., Kementsietsidis, A., Velegrakis, Y. (eds.) SIGMOD Conference, pp. 721–732. ACM (2011)
21. Schnettler, S.: A structured overview of 50 years of small-world research. Social Networks 31(3), 165–178 (2009)
22. Simmel, G.: The sociology of Georg Simmel, vol. 92892. Simon and Schuster (1950)
23. Spielman, D.A., Srivastava, N.: Graph sparsification by effective resistances. SIAM J. Comput. 40(6), 1913–1926 (2011)
24. Spielman, D.A., Teng, S.H.: Nearly-linear time algorithms for graph partitioning, graph sparsification, and solving linear systems. In: Babai, L. (ed.) STOC, pp. 81–90. ACM (2004)
25. Subramanian, A., Wei, S.J.: The WTO promotes trade, strongly but unevenly. Journal of International Economics 72(1), 151–175 (2007)
26. Traud, A.L., Kelsic, E.D., Mucha, P.J., Porter, M.A.: Comparing community structure to characteristics in online collegiate social networks. SIAM Review 53(3), 526–543 (2011)
27. Tumminello, M., Aste, T., Di Matteo, T., Mantegna, R.: A tool for filtering information in complex systems. Proc. Natl. Acad. Sci. 102(30), 10421–10426 (2005)
28. Van Ham, F., Wattenberg, M.: Centrality based visualization of small world graphs. Computer Graphics Forum 27(3), 975–982 (2008)
29. Zaidi, F., Sallaberry, A., Melançon, G.: Revealing hidden community structures and identifying bridges in complex networks: An application to analyzing contents of web pages for browsing. In: Web Intelligence, pp. 198–205. IEEE (2009)
30. Zhou, F., Mahler, S., Toivonen, H.: Network simplification with minimal loss of connectivity. In: Webb, G.I., Liu, B., Zhang, C., Gunopulos, D., Wu, X. (eds.) ICDM, pp. 659–668. IEEE Computer Society (2010)

GION: Interactively Untangling Large Graphs on Wall-Sized Displays

Michael R. Marner[1], Ross T. Smith[1], Bruce H. Thomas[1], Karsten Klein[2], Peter Eades[2], and Seok-Hee Hong[2]

[1] University of South Australia, Adelaide, Australia
{michael.marner,ross.smith,bruce.thomas}@unisa.edu.au
[2] The University of Sydney, Sydney, Australia
{karsten.klein,peter.d.eades,seokhee.hong}@sydney.edu.au

Abstract. Data sets of very large graphs are now commonplace; the scale of these graphs presents considerable difficulties for graph visualization methods. The use of interactive techniques and large screens have been proposed as two possible avenues to address these difficulties.This paper presents GION, a new skeletal animation technique for interacting with large graphs on wall-sized displays. Our technique is based on a physical simulation, and aims to enhance the users' ability to efficiently interact with the graph visualization for exploratory analysis. We conducted a user study to evaluate our technique against standard operations available in most graph layout editors, and the study shows that the new technique produces layouts with less stress, and fewer edge crossings. GION is preferred by users, and requires significantly less mouse movement.

1 Introduction

Graphs provide a versatile model for data from a large variety of application domains, including biology, finance, telecommunication, software engineering, and social sciences. Graph visualization helps scientists and engineers to understand critical issues in these domains. However, the depth of understanding depends on the quality of the drawing. Automatic graph layout methods are developed for computational efficiency and quality, i.e. readability. These methods however can only optimize a few criteria in combination, and it is impossible to define a quality measure that allows to create optimal layouts for all graphs, tasks, and observers. Moreover, the size of relevant data sets for analysis has grown exponentially over the last years. For example, data from social networks, biology, and finace continue to grow at a rate that is not accommodated by current methodologies.

While some layout algorithms are capable of laying out graphs with hundreds of thousands of nodes in a few seconds [4], data sizes from practice are still a challenge in a number of ways:

– There is a trade-off between computational resources and layout quality. For example, algebraic methods run quickly but in many cases give poor results [6], while stress minimization [5] gives good quality layouts but is too slow for interactive work on large graphs.

C. Duncan and A. Symvonis (Eds.): GD 2014, LNCS 8871, pp. 113–124, 2014.

- Existing methods do not scale well visually. A standard screen with a few megapixels cannot faithfully display graphs of a few million edges.
- An underlying problem lies in the optimisation criteria for layout algorithms for large graphs. For small scale graphs, criteria such as the number of edge crossings have been successfully used and validated [13, 14]. However, all commonly used algorithms for large graph layout ignore these criteria and generally use an optimisation criterion based on a notion of energy or stress [4].
- Readable large graphs might not be sufficient for understanding, as requirements differ based on the application and the task at hand. Moreover, the user can interactively explore different regions of the graph which are not known in advance, e.g. to compare or investigate the local structures, while keeping their global context.

This paper makes the contribution of a new interaction technique, GION, for manipulating layouts of large graphs. GION is novel by employing a physics engine to simplify the process of interacting with large graphs, treating the graph as a set of connected rigid bodies. The physics engine provides smooth animation for the user while interactively laying out the graph, and this animation improves the understanding for the user of the graphs underlying structure. When the user moves a cluster, the connected clusters are also moved, as if they were connected in a chain. Our contribution is validated with a user study conducted to evaluate the effectiveness of the technique in a *graph untangling* task.

The remainder of this paper is structured as follows. Section 2 describes previous research related to this paper. Section 3 describes the details of the new graph interaction technique. Section 4 outlines the user study conducted to evaluate the new interaction technique, with the results presented in Section 5 followed by a discussion of these results in Section 6. Finally, the paper concludes with a discussion of future work.

2 Background

While readability of graph drawings has been a topic of research for decades, and there is a wealth of papers on the evaluation of drawing quality, e.g. [7, 13], the research has focused mainly on task based performance for diagrams of small to medium size. Several well established quality criteria for graph layouts exist. The most prominent one is the number of crossings, which was verified to be an impediment for the human understanding of small graphs in empirical experiments [13]. Further well established quality criteria are angular resolution, edge length deviation, and stress. Recently, Huang et al. [7] suggested that it is often better to make compromises between aesthetics, instead of trying to satisfy one or two of them to the fullest. It has however not yet been investigated if the results obtained for small graphs can be extended to huge graphs or if different quality criteria have to be employed.

Recently, Dwyer et al. [3] compared user-generated and automatic graph layouts, where users were asked to optimize the layout for aesthetics and social network analysis tasks. In their study, users that were asked to optimize a graph layout for an analysis task used the term "untangling" to describe their process. In contrast the term "untangle" here is not task related but used in a relatively informal sense: a user "untangles" a graph drawing when they improve the layout, in the subjective opinion of the user. In particular, we did not ask the participants of our user study to optimize any pre-specified quality metric. Indeed, our long-term goals include discovery of users' metrics.

Our new technique employs animation techniques found in modern computers games, and many applications apply animation techniques to enrich the look and feel of the user interface. Animations to the interface smooth the rough edges and sudden transitions common in many current graphical interfaces, and strengthen the illusion of direct manipulation that many interfaces strive to present [16]. Animation improves a user's understanding of the direct manipulation of the data by better portraying such concepts as constraints, relationships, and connectivity. These are powerful cues for the direct manipulation of large graph structures.

Ball, North, and Bowman [1] evaluated interaction techiques for large display visualisations. They tracked physical navigation in 3D space via the participants head with a VICON system. Their experiments found with increased size of the display, there was more physical navigation. When combined with the reduced performance time on large displays, they found a compelling suggestion that physical navigation was also more efficient. They also found that physical navigation was preferred over virtual navigation.

Peck, North, and Bowman [12] defined a new 3D interaction technique, *multiscale interaction*, which associates the user's scale of perception to their scale of interaction. Multiscale interactions exploit the user's physical navigation in front of a large display to directly control the scale of interaction, while adjusting their scale of perception. Overall, they found evidence that multiscale interaction is a natural behavior, and this technique can be useful in interaction design for large high-resolution displays.

Skeletal animation (see, for example, [8]) is a well established technique in graphics. An object is modelled as a mesh, with "bones" as links between "joints". Movement of the mesh in between keyframes can be computed using methods of *inverse kinematics* [17]. The skeletal animation technique is mainly used to animate people and animals, but Merrick et al. [9, 10] investigate application to graph interaction. Their work, however, is limited to very small graphs.

3   GION: Graph Interaction Operation for Nodes

To untangle a graph G, the user has to rearrange nodes by dragging them to new positions. Moving nodes one by one is time consuming. With thousands to millions of nodes, the human resources required are too large both for untangling in practice and for evaluation experiments like the one in this paper. Large graphs thus need interaction methods for untangling that move more than one node at a time. The *GION* technique uses ideas and off-the-shelf software from skeletal animation to simplify the process of interacting with large graphs. More specifically, GION adapts a physics engine to move many nodes at a time. GION treats the graph as a skeleton, where bones simulate edges, and joints simulate nodes. However, the simple approach of representing every edge as a bone and each node as a joint does not scale to handle large graphs. Thus large graphs are clustered to enable the user to move large numbers of nodes at once. The physics engine treats clusters as rigid bodies connected by joints. The effect of this approach is that connected clusters move as a chain. This section describes the interaction technique in detail and provides rationale for design choices.

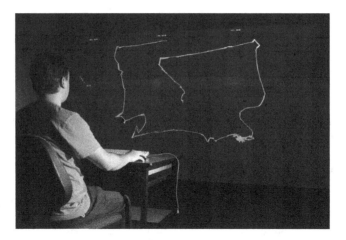

Fig. 1. Untangling a graph layout on the tiled wall display

3.1   Graph Clustering

Allowing the user to move larger chunks of the graph at a time can help to reduce this effort in case these chunks are specified in a way that supports the user's untangling process. Graph clustering partitions the nodes of the graph into clusters, that is, disjoint node sets, aiming to have high *cohesion* (that is, many intra-cluster edges), and low *coupling* (that is, few inter-cluster edges).

Many different clustering algorithms are available. We use a fast and simple clustering method based on random walks [2]. This algorithm aims to detect dense local substructures using a random walk based graph traversal. Roughly speaking, random walks tend to stay within a highly cohesive substructure with a high probability [2]. This algorithm allows to influence the number of clusters over parameter settings; this property is helpful for tuning the interaction fidelity. As the cluster boundaries created by the random walk approach can be somewhat fuzzy, we apply a Kernigan-Lin style postprocessing technique [18] that flips the cluster affiliation of single nodes that are connected more strongly to a different cluster than to the cluster they are affiliated with.

Clusters as Rigid Bodies: Each cluster in the graph is represented as a polygonal rigid body. The polygon shape is calculated by taking the convex hull of the vertices in the cluster, and then simplifying the polygon down to a maximum of eight vertices. This simplification greatly improves the performance of the simulation at runtime. The polygon is given physical properties that drive the simulation: **1) Damping** reduces the velocity of rigid bodies when in motion. **2) Density** determines the mass of the polygon, and thus its momentum when in motion. **3) Static Friction** prevents rigid bodies from moving unless a minimum threshold force is applied. This is important in limiting the number of nodes that move in response to the movement of a node. **4) Collision** In most physics simulations, rigid bodies can collide. For our purposes, collisions are disabled and bodies can pass through each other. Physics engines such as that provided by Box2D allow specification of control parameters for the above properties; for GION, we chose values for these parameters from experience.

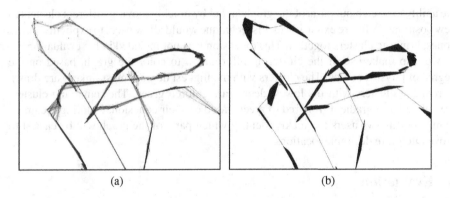

       (a)   (b)

Fig. 2. (a) A section of a graph, and (b) physics system representation

Edges, clustered edges, and bones: If there is at least one edge between two clusters C_1 and C_2, then we say that C_1 and C_2 are *linked*. If two clusters are linked, then we place an elastic *bone* between them; physically this acts to approximately maintain a specified distance between the two clusters. The bone has a *length* that defines this distance. In GION, the length of a bone is held constant during a single user operation, but varies from one operation to the next, as explained below. The physics engine provides two parameters to define the elasticity of a bone: *frequency* and *damping ratio*. In GION, the elasticity is held constant throughout.

3.2 User Operations

Users interact with the graph using the mouse to drag clusters into new positions. Each GION user operation consists of three steps: 1) The user selects a node with a button-down mouse event. 2) The user moves the mouse. Mouse movement is applied to the physics simulation as a force that acts on the cluster containing the node selected by the user. Using a force provides intuitive feedback about the connections from the selected cluster; heavily connected clusters are more difficult (require more mouse movement) to move than clusters with only a few connections. 3) The user releases (button-up) the mouse. GION then re-sets the length of each bone to the distance between its endpoints.

GION differs significantly from previous skeletal animation methods in that the length of each bone can vary from one user operation to the next. In a classical skeletal animation (say of a human walking), bones have constant length; but GION introduces some elasticity to the bone length. This provides more information to the user, as they can more easily see how the graph is connected. However, we found through informal pilot studies that users felt as though they did not have enough control over the layout of the graph. Users became frustrated when clusters, connected by constant length bones, would rebound back towards their initial positions.

To improve this behaviour, GION resets all bone lengths when the user releases the mouse. The distance between two clusters at that point in time becomes the new defined distance for the bone: thus the clusters stay in their positions when the user releases the mouse. This improves the control that users have over the layout. It also gives the users the ability to stretch out parts of the graph layout by quickly moving different clusters to new positions. A future extension to this technique would allow users to explicitly shrink bones, bringing clusters together. This extension was not included in the evaluation.

We also make use of the clustering information to color the graph, based on the degree of physics bodies. The clusters with the highest degree (most linked) are drawn in red, and clusters with the lowest degree are colored green. The remaining clusters are colored as a gradient from red to green based on their cohesion. Coloring the graph in this way allows users to quickly identify which parts of the graph will be easiest to move into more desirable locations.

4 Evaluation

We conducted a user study to evaluate the benefits of the physics-based graph interaction technique. We chose an *untangling* task for the experiment. Participants were shown a series of graph layouts, and were asked to 'untangle' the layout to better show the overall structure. Untangling was chosen as the task because it would require many mouse operations to complete, and resulting graphs could be compared to the initial layout. The experiment is a 2x2, within participant, repeated measures design. The conditions tested were interaction mode: *physics* or *normal*, and coloring: *colored* or *plain*. Participants used the GION technique for the *physics* condition. The *normal* interaction allowed users to move nodes to new locations one cluster at a time, emulating movement operations commonly used in graph layout software.

The hypotheses tested in the experiment are as follows:

H1 *Physics* interaction leads to lower stress than *normal* interaction.
H2 *Physics* interaction leads to fewer edge crossings than *normal* interaction.
H3 *Colored* graphs would have lower stress than the *plain* graphs.
H4 *Colored* graphs would have fewer edge crossings than the *plain* graphs.
H5 *Physics* interaction requires less mouse movement than *normal* interaction.
H6 *Physics* interaction requires fewer clicks than *normal* interaction.
H7 *Physics* interaction leads to lower stress than the starting layout.
H8 *Physics* interaction leads to fewer edge crossings than the starting layout.
H9 *Physics* interaction is preferred by users.

The following data were collected during each trial of the experiment: mouse movement, in millimetres on the videowall, mouse clicks, and snapshots of the graph layout (captured every five seconds). From the graph layout snapshots other properties of the layout could be calculated and analysed. Participants were asked to fill out a questionnaire at the end of the user study session. Participants answered questions 1-4 for both the physics and normal interaction conditions, and questions 5-6 for the color condition. All questions were answered using a visual analogue scale. The participants were asked to rank the interaction conditions in order of preference, and comment on strategies used for untangling the graphs.

1. Moving graph clusters into new positions was {very easy - very hard}
2. Untangling the graphs was {very easy - very hard}

3. The interaction mode made the graphs {more understandable - less understandable}
4. With the results of untangling the graphs, I was {very happy - very unhappy}
5. When deciding which clusters to move first, the coloring made it {easy - hard}
6. Coloring the graph made untangling {fast - slow}

4.1 Graphs

We selected graphs that come from a real world application where graph visualization is used for data analysis and the graph size and layout quality requirements pose a challenge for state-of-the-art layout methods. Our graph set consists of RNA sequence graphs that are used for the analysis of repetitive sequences in sequencing data [11]. They have been created by running pairwise alignment on genomic sequence reads, and to represent reads as nodes and large overlaps between reads as edges. Eight graphs were chosen for the experiment. Participants untangled all graphs, with the conditions randomised for each graph.

We applied the Fruchterman-Reingold algorithm FR to obtain the initial layouts used in the experiments. Our goal here was to start with a layout that did not reveal the overall graph structure completely. A completely random layout might pose a too difficult challenge. Using a layout generated by FR allows the user to identify starting points for untangling while leaving enough space for improvement based on individual preferences. Graph properties are provided in Table 1. The set of graphs can be downloaded from http://wcl.ml.unisa.edu.au/graph-untangling/graphs.zip. All graphs have high local density, and a sparse global structure that allows to create layouts far from hairballs that are showing the structure well.

Table 1. Overview on the graph set used for the experiment

Graph	# nodes	# edges	density	avg deg.	clus. coeff.	avg sh. path	diameter
A	1159	6424	5.5	11.1	0.65	19.5	59
B	1748	13957	8	16	0.64	17.9	63
C	1785	20459	11.5	22.9	0.61	10.7	41
D	3010	41757	13.9	27.7	0.67	26.4	77
E	4924	52502	10.7	21.3	0.65	36	121
F	5452	118404	21.7	43.4	0.73	46.6	216
G	5953	186279	31.3	62.6	0.72	56.2	163
H	7885	427406	54.2	108.4	0.69	24.4	55

4.2 Experimental Procedure

Each participant completed the experiment in a single session. Participants were first asked to complete a graph theory quiz. This quiz asked simple graph structure questions and was designed to allow results from participants with different levels of knowledge to be compared, and did not affect the rest of the experiment. Participants were given instructions on how to interact with the system, and that they would have two minutes to untangle each graph. Specifically, participants were shown an example starting layout and untangled layout (shown in Fig. 3) and told "to untangle the graph. This involves moving parts of the graph to new locations in order to make the underlying structure of the graph clear". Following this, the untangling trials began. The display provided instructions informing the participant of the conditions of the next trial (for example, *physics colored*). The participant clicked the mouse to begin the trial and the display changed to show the initial graph layout. The participant then had two minutes to best untangle the graph. After two minutes, the video wall went blank while the next trial was loaded. After eight trials the participants were asked to complete the subjective questionnaire and the session was concluded.

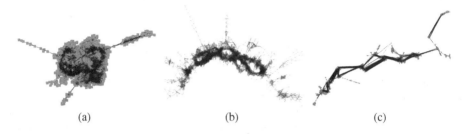

(a)   (b)   (c)

Fig. 3. (a) The starting layout for graph B. (b) Layout for graph B using stress minimization. (c) Final layout created by one of the participants.

5   Results

Sixteen participants completed the experiment, recruited from staff and students from the University of South Australia, and the general public. Participant ages ranged from 23 to 57, with a mean age of 33. Five of the participants were female, and all but one of the participants were right handed. The mean score for the graph theory quiz was 74.22% (std. dev 25.19). Pearson Correlation analysis was performed to see how quiz score affected results. A significant correlation was found between quiz score and change in edge crossings for the normal/color condition, $r = -.505[-.876, .453], p < 0.05$. Quiz results did not affect any other conditions. All other quantitative results were analysed using a 2x2 repeated measures ANOVA.

5.1   Mouse Usage

There was a significant main effect on interaction mode, $F(1, 15) = 36.586$, $p < 0.001$. There was significantly less mouse travel in the *physics* condition compared to *normal*, with 59294.661mm (SE 5001.180) of mouse travel for the *physics* condition compared to 76227.881mm (SE 4153.657) for the *normal* condition. The graph coloring did not produce a significant effect on mouse travel, and there was no significant interaction between interaction mode and graph coloring.

There was a significant main effect on the interaction mode, $F(1, 15) = 6.279$, $p < 0.05$. Participants made fewer mouse clicks for the *normal* condition than *physics*, with mean mouse clicks of 38.563 (SE 3.235) for the *normal* condition compared to 43.109 (SE 3.036) for the *physics* condition.

There was also a significant main effect on the graph coloring, $F(1,15) = 11.614$, $p <$ 0.01. Participants made significantly fewer mouse clicks for the *plain* condition than *colored*, with mean mouse clicks of 39.125 (SE 2.685) for the *plain* condition compared to 42.547 (SE 3.367) for the *colored* condition.

5.2   Graph Layout Analysis

Layouts were analysed for stress and edge crossings. The results presented here are represented as a ratio of change in values compared to the starting values of the initial graph layout, i.e. *Result* = (*EndValue* − *StartValue*)/*StartValue*. Using a ratio of change allows comparisons between graph layouts of different sizes, numbers of clusters, and vastly different initial stress and edge crossing values. In all conditions, stress was higher after participants interacted with the graph. Of the 128 trials conducted in the experiment, only two resulted in lower stress values than the initial conditions. However, there was a significant main effect on interaction mode, $F(1,15)$, $p < 0.05$. The physics based interaction produced layouts with significantly less stress than the normal interaction mode, with mean stress change of 22.884 (SE 2.772) for the *physics* condition compared to 95.763 (SE 14.955) for the *normal* condition. Graph coloring did not produce a significant effect on graph stress, and no significant interaction between interaction mode and graph coloring was found. Edge crossings were also higher after participants interacted with the graph. Sixteen of the 128 trials showed a reduction in edge crossings.

There was a significant main effect on interaction mode, $F(1,15)$, $p < 0.05$. The physics based interaction produced layouts with significantly fewer edge crossings than the normal interaction mode, with mean change in edge crossings of 0.578 (SE 0.087) compared to 2.811 (SE 0.538) for the normal interaction mode. Graph coloring did not significantly affect edge crossings, and no significant interaction between interaction mode and graph coloring was found.

5.3   Questionnaire Results

The results of the questionnaire comparing the physics interaction to normal interaction are summarised in Fig. 4. Significant results ($p < .05$) were found for questions 1, 3, and 4, with participants giving higher scores for the physics condition in all three questions. Participants overwhelmingly preferred the physics interaction, with 87.5% of participants choosing physics as the preferred mode. Participants also responded favourably to the graph coloring. Results for Q5 were 71.69 (SD 18.095), and Q6 68.19 (SD 15.753).

6   Discussion

The results of the user study show that the GION interaction technique is better in the untangling task than existing interactive methods. Specifically, hypotheses H1, H2, H5,

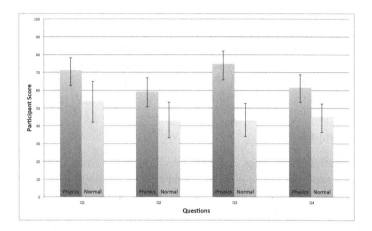

Fig. 4. Questionnaire Results for physics and normal interaction modes. Error bars show 95% CI

and H9 were confirmed in the experiment. An interesting result is that users moved the mouse less in the GION condition, but made more mouse clicks. This suggests that participants spent more time fine-tuning the layout. Another unexpected result is that while GION produced layouts with lower stress and fewer edge crossings than the normal interaction, those values were worse than for the starting graph layouts. Using human interaction to untangle a large graph raises the question of what actually makes a good layout in this context. In particular, classical metrics like stress or edge crossings do not take into account the dynamics of the process and the mental map that the user creates during interaction. Intermediate layouts might not be good with respect to such metrics, but the user's interactive operations to create them can increase the users insight in the graph structure, and their value may depend on the preceding untangling process. Trying to measure such effects is however beyond the scope of this paper.

We deliberately restricted the user interface for the sake of a robust evaluation. For example, participants were not able to zoom or pan the graph drawing. A more complete system would also allow users to temporarily disable the physics engine in order to precisely control a single cluster. Multiple levels of clustering would also improve the technique, by allowing users to switch from coarse to fine grained interaction.

7   Implementation Details

Our video-wall consists of six NEC NP510W projectors each with a resolution of 1280x800 arranged in a 3x2 configuration. A camera based technique, as described by Raskar et al. [15], is used for geometric calibration of the projectors and for producing blending masks for smooth transitions between projectors. The computer used in this work consists of 2x Quad Core Xeon processors, 12GB of RAM, and 2x Nvidia Geforce GTX 780 GPUs.

The system presented in this paper consists of a custom built application written in C++ with OpenGL for rendering. The physics simulation was developed using Box2D[1],

---

[1] http://box2d.org

an open source 2D physics library popular for game development. The Open Graph Drawing Framework (OGDF)[2] is used to provide graph data structures used by the application, saving and loading graphs at runtime, as well as for layout metrics. Vertex data is stored in an OpenGL Vertex Buffer Object (VBO). Edge data references the VBO and is rendered using an Index array. This reduces the amount of data transferred to the graphics card each frame and improves rendering performance by reducing the number of draw calls needed. Further enhancements were needed to improve rendering times on on a multi-projector display. A naive approach would simply involve rendering the graph in its entirety for each projector. Instead, we use a deferred rendering technique. The entire display is first rendered to an off-screen Framebuffer Object (FBO). Following this step, a portion of the FBO is rendered to each projector. This approach scales much better as the number of projectors increases, as the cost of each projector is just a single textured polygon.

8 Conclusion

In this paper we presented GION, a new interactive graph layout technique of large graph structures. GION is based on a physics engine to provide smooth and understandable animations to update the graph layout while the user moves a cluster. The results of a user study comparing GION with moving a single cluster at a time found the use of physics engine produced graphs with less stress, fewer edge crossings, and less mouse movement. Participants preferred the GION technique to moving a single cluster during the experiment.

We applied two standard quality layout metrics: stress and crossings. With GION, users constructed graph layouts that did not show significantly less stress or significantly fewer edge crossings, in comparison with the Fruchterman-Reingold algorithm. These results from our experiments lead us to question the validity of these two standard metrics for large graphs in the context of human layout improvement, and our work raises the question as to what quality metrics should be applied instead. We conjecture that measures like the precision of neighborhood preservation [5] will be better suited in this context than standard metrics for small graphs.

Acknowledgment. This work was supported in part by a grant from the Australian Research Council - Discovery Grant DP120100248, Linkage Grant H2814 A4421, Tom Sawyer Software, and NewtonGreen Technologies.

References

1. Ball, R., North, C., Bowman, D.: Move to improve: promoting physical navigation to increase user performance with large displays. In: Proc. of the SIGCHI Conference on Human Factors in Computing Systems, pp. 191–200. ACM (2007)
2. Catherine, R., Sudarshan, S.: Graph clustering for keyword search. In: Chawla, S., Karlapalem, K., Pudi, V. (eds.) COMAD. Computer Society of India (2009)

---

[2] http://www.ogdf.net

3. Dwyer, T., Lee, B., Fisher, D., Quinn, K.I., Isenberg, P., Robertson, G.G., North, C.: A comparison of user-generated and automatic graph layouts. IEEE Trans. Vis. Comput. Graph. 15(6), 961–968 (2009)
4. Gansner, E.R., Hu, Y., Krishnan, S.: Coast: A convex optimization approach to stress-based embedding. In: Wismath, S., Wolff, A. (eds.) GD 2013. LNCS, vol. 8242, pp. 268–279. Springer, Heidelberg (2013)
5. Gansner, E.R., Hu, Y., North, S.C.: A maxent-stress model for graph layout. IEEE Trans. Vis. Comput. Graph. 19(6), 927–940 (2013)
6. Hachul, S., Jünger, M.: Large-graph layout algorithms at work: An experimental study. J. Graph Algorithms Appl. 11(2), 345–369 (2007)
7. Huang, W., Eades, P., Hong, S.H., Lin, C.C.: Improving force-directed graph drawings by making compromises between aesthetics. In: VL/HCC, pp. 176–183 (2010)
8. Lewis, J.P., Cordner, M., Fong, N.: Pose space deformation: A unified approach to shape interpolation and skeleton-driven deformation. In: Proc. of the 27th Annual Conference on Computer Graphics and Interactive Techniques, SIGGRAPH 2000, pp. 165–172. ACM Press/Addison-Wesley Publishing Co., New York (2000)
9. Merrick, D., Dwyer, T.: Skeletal animation for the exploration of graphs. In: Australasian Symposium on Information Visualisation, InVis.au, pp. 61–70. Christchurch, New Zealand (2004)
10. Murray, C., Merrick, D., Takatsuka, M.: Graph interaction through force-based skeletal animation. In: Australasian Symposium on Information Visualisation, InVis, pp. 81–90. Christchurch, New Zealand (2004)
11. Novák, P., Neumann, P., Macas, J.: Graph-based clustering and characterization of repetitive sequences in next-generation sequencing data. BMC Bioinformatics 11, 378 (2010)
12. Peck, S., North, C., Bowman, D.: A multiscale interaction technique for large, high-resolution displays. In: IEEE Symposium on 3D User Interfaces, 3DUI 2009, pp. 31–38. IEEE (2009)
13. Purchase, H.C.: Which aesthetic has the greatest effect on human understanding? In: DiBattista, G. (ed.) GD 1997. LNCS, vol. 1353, pp. 248–261. Springer, Heidelberg (1997)
14. Purchase, H.C.: Metrics for graph drawing aesthetics. Journal of Visual Languages & Computing 13(5), 501–516 (2002)
15. Raskar, R., Brown, M.S., Yang, R., Chen, W.C., Welch, G., Towles, H., Seales, B., Fuchs, H.: Multi-projector displays using camera-based registration. In: Proc. of the Conference on Visualization 1999: Celebrating Ten Years, pp. 161–168. IEEE Computer Society Press, San Francisco (1999)
16. Thomas, B.H., Calder, P.: Applying cartoon animation techniques to graphical user interfaces. ACM Transactions on Computer-Human Interaction (TOCHI) 8(3), 198–222 (2001)
17. Welman, C.: Inverse kinematics and geometric constraints for articulated figure movement. Master's thesis, Simon Fraser University (1993)
18. Kernighan, B.W., Lin, S.: An efficient heuristic procedure for partitioning graphs. Bell System Technical Journal 49, 291–307 (1970)

Balanced Circle Packings for Planar Graphs

Md. Jawaherul Alam[1], David Eppstein[2], Michael T. Goodrich[2],
Stephen G. Kobourov[1], and Sergey Pupyrev[1,3]

[1] Department of Computer Science, University of Arizona, Tucson, Arizona, USA
[2] Department of Computer Science, University of California, Irvine, California, USA
[3] Institute of Mathematics and Computer Science, Ural Federal University, Russia

Abstract. We study *balanced* circle packings and circle-contact representations
for planar graphs, where the ratio of the largest circle's diameter to the smallest
circle's diameter is polynomial in the number of circles. We provide a number of
positive and negative results for the existence of such balanced configurations.

1 Introduction

Circle packings are a frequently used and important tool in graph drawing [3,5,10,11,
18]. In this application, they can be formalized using the notion of a *circle-contact representation* for a planar graph; this is a collection of interior-disjoint circles in $\mathbb{R}^2$, corresponding one-for-one with the vertices of the graph, such that two vertices are adjacent
if and only if their corresponding two circles are tangent to each other [15]. In a classic
paper, Koebe [16] proved that every triangulated planar graph has a circle-contact representation, and this has been subsequently re-proved several times. Generalizing this,
every planar graph has a circle-contact representation: we can triangulate the graph by
adding "dummy" vertices connected to the existing vertices within each face, produce
a circle-contact representation for this augmented graph, and then remove the circles
corresponding to dummy vertices. It is not always possible to describe a circle-contact
representation for a given graph by a symbolic formula involving radicals [2,5], but
they can nevertheless be constructed numerically and efficiently by polynomial-time
iterative schemes [7,19].

One of the drawbacks of some of these constructions, however, is that the sizes of
the circles in some of these configurations may vary exponentially, leading to drawings
with very high area or with portions that are so small that they are below the resolution of the display. For this reason, we are interested in *balanced* circle packings and
circle-contact representations for planar graphs, where the ratio of the maximum and
minimum diameters for the set of circles is polynomial in the number of vertices in the
graph; see Fig. 1.

Related Work. There is a large body of work about representing planar graphs as contact graphs, where vertices are represented by geometrical objects and edges correspond
to two objects touching in some pre-specified fashion. For example, Hliněný [15] studies contact representations using curves and line segments as objects. Several authors
have considered contact graphs of triangles of various types. For instance, de Fraysseix *et al.* [12] show that every planar graph has a triangle-contact representation, and

C. Duncan and A. Symvonis (Eds.): GD 2014, LNCS 8871, pp. 125–136, 2014.

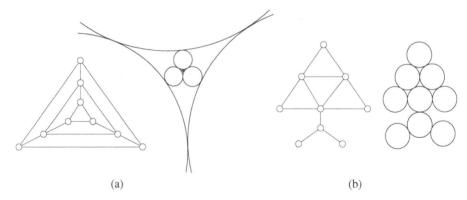

Fig. 1. Two planar graphs with possible circle-contact representations: (a) a representation that is not optimally balanced; (b) a perfectly-balanced representation

Gonçalves *et al.* [14] prove that every 3-connected planar graph and its dual can be simultaneously represented by touching triangles (and they point out that 4-connected planar graphs also have contact representations with homothetic triangles). Also, Duncan *et al.* [9] show that every planar graph has a contact representation with convex hexagons all of whose sides have one of three possible slopes, and that hexagons are necessary for some graphs, if convexity is required. With respect to balanced circle-contact representations, Breu and Kirkpatrick [4] show that it is NP-complete to test whether a graph has a perfectly-balanced circle-contact representation, in which every circle is the same size.

New Results. In this paper, we provide a number of positive and negative results regarding balanced circle-contact representations for planar graphs:

- Every planar graph with bounded maximum vertex degree and logarithmic outerplanarity admits a balanced circle-contact representation.
- There exist planar graphs with bounded maximum degree and linear outerplanarity, or with linear maximum degree and bounded outerplanarity, that do not admit a balanced circle-contact representation.
- Every tree admits a balanced circle-contact representation.
- Every outerpath admits a balanced circle-contact representation.
- Every cactus graph admits a balanced circle-contact representation.
- Every planar graph with bounded tree-depth admits a balanced circle-contact representation.

2   Bounded Degree and Logarithmic Outerplanarity

A plane graph (that is, a combinatorially fixed planar embedding of a planar graph) is *outerplanar* if all of its vertices are on the outer face. A *k-outerplanar graph* is defined recursively. As a base case, if a plane graph is outerplanar, then it is a 1-outerplanar graph. A plane graph is *k*-outerplanar, for $k > 1$, if the removal of all the outer vertices (and their incident edges) yields a graph such that each of the remaining components is $(k-1)$-outerplanar. The *outerplanarity* of a plane graph G is the minimum value for k such that G is k-outerplanar.

2.1 Balanced Circle-Contact Representations

Theorem 1. *Every n-vertex k-outerplanar graph with maximum degree Δ admits a circle-contact representation where the ratio of the maximum and the minimum diameter is at most $f(\Delta)^{k+\log n}$, for some positive function f. In particular, when Δ is a fixed constant and k is $\mathcal{O}(\log n)$, this ratio is polynomial in n.*

In order to prove the theorem, we need the following result from [18].

Lemma 1 (Malitz-Papakostas). *The vertices of every triangulated planar graph G with the maximum degree Δ can be represented by nonoverlapping disks in the plane so that two disks are tangent to each other if and only if the corresponding vertices are adjacent, and for each two disks that are tangent to each other, the ratio of the radii of the smaller to the larger disk is at least $\alpha^{\Delta-2}$ with $\alpha = \frac{1}{3+2\sqrt{3}} \approx 0.15$.*

As a direct corollary, every maximal planar graph with maximum degree $\Delta = \mathcal{O}(1)$ and diameter $d = \mathcal{O}(\log n)$ has a balanced circle-contact representation. Theorem 1 goes beyond this.

Proof of Theorem 1: To prove the claim, it is sufficient to show how to augment a given k-outerplanar graph into a maximal planar graph with additional vertices so that its maximum degree remains $\mathcal{O}(\Delta)$ and its diameter becomes $\mathcal{O}(k + \log n)$. By Lemma 1, the resulting graph admits a circular contact representation with the given bounds on the ratio of radii. Removing the circles corresponding to the added vertices yields the desired balanced representation of the original graph.

Let G be an n-vertex k-outerplanar graph with the maximum degree Δ. If the outerplanarity k of G is bounded by a constant, we can easily augment G to logarithmic diameter, preserving its constant maximum degree, as follows. Inside each non-triangular face f of G, insert a balanced binary tree with $\lceil \log |f| \rceil$ levels and $|f|$ leaves and then triangulate the remaining non-triangular faces by inserting an *outerpath* (an outerplanar graph whose weak dual is a path) with constant maximum degree; see Fig. 2. However, such an augmentation results in a maximal planar graph with the diameter

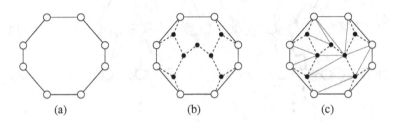

(a)   (b)   (c)

Fig. 2. (a) A face, (b) augmentation with a balanced binary tree, (c) triangulation with grey edges

$d = \mathcal{O}(k \log n)$, which does not yield a balanced circle-contact representation when k is non-constant. For $k = \Omega(\log n)$, we present a different augmentation to achieve the diameter $d = \mathcal{O}(k + \log n)$ in the resulting graph.

We augment the graph using *weight-balanced binary trees*. Let T be a binary tree with leaves $l_1, l_2, \ldots, l_{|f|}$ and a prespecified weight w_i assigned to each leaf l_i. The tree T is *weight-balanced* if the depth of each leaf l_i in T is $\mathcal{O}(\lceil \log(W/w_i) \rceil)$, where $W = \sum_{i=1}^{f} w_i$. There exist several algorithms for producing a weight-balanced binary tree with positive integer weights defined on its leaves [13, 21].

To augment G, we label each vertex v of G with the number $l + 1$, where l is the number of outer cycles that need to be removed before v becomes an outer vertex. By our assumption that the outerplanarity of G is k, the label of every vertex is at most k. It follows from this labeling that, for each vertex v of G with label $l > 1$, there exists a face f containing v such that f has at least one vertex of label $l - 1$ and such that all the vertices on f have label either l or $l - 1$. We insert a weight-balanced binary tree inside f; we choose an arbitrary vertex of f with label $l - 1$ as the root of the tree, and a subset of vertices with label l as the leaves; see Fig. 3. We construct these trees inside the different faces in such a way that each vertex of G with label $l > 1$ becomes a leaf in exactly one of the trees. Finally, we insert another weight-balanced tree T_0 on the outer face containing all the outer vertices as the leaves. Note that we have yet to specify the weights we assign to these leaves for producing the weight-balanced trees. By the construction, the union of all these trees forms a connected spanning tree of G; we can consider the root of T_0 to be the root of the whole spanning tree.

Let us now specify the weights assigned to the leaves of the different weight-balanced trees. We label each tree with the label of its root, and define the weights for the leaves of each tree in a bottom-up ordering, by decreasing order of the labels of the trees. In a tree T with label $l = (k - 1)$, all the leaves have label k and are not the root of any other tree; we assign each of these leave the weight 1. In this case, the total weight of T is the number of its leaves. Similarly, for a tree with label $l < k - 1$, we assign a weight

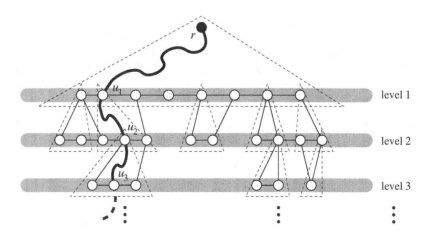

Fig. 3. Augmentation of G with a weight-balanced binary trees

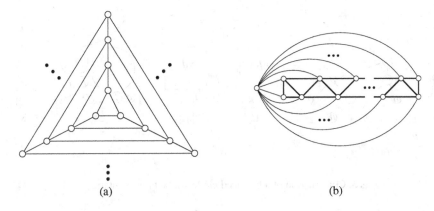

(a)   (b)

Fig. 4. Planar graphs with no balanced circle-contact representation: (a) the nested-triangles graph [8]; (b) a 2-outerplanar graph

of 1 to those leaves v that do not have any tree rooted at them; otherwise, if v is the root of a tree T_v with label $l + 1$, the weight of v is the total weight of T_v. The total weight of T is defined as the summation of the weights of all its leaves.

Now, for each vertex v of G, the distance to v from the root r of T_0 is $\mathcal{O}(k + \log n)$. Indeed, assume that $v = u_l$ is a vertex with label l and $u_{l-1}, \ldots, u_1, u_0 = r$ are the root vertices of the successive weight-balanced trees $T_{u_{l-1}}, \ldots, T_{u_1}, T_0$ with labels $l - 1, \ldots, 1, 0$, respectively on the way from v to r; see Fig. 3. Then the distance from v to r is $\mathcal{O}(\lceil \log w(r)/w(u_1) \rceil) + \mathcal{O}(\lceil \log w(u_1)/w(u_2) \rceil) + \ldots + \mathcal{O}(\lceil \log w(u_{l-1})/w(v) \rceil) = \mathcal{O}(k + \log w(r))$. Here $w(u_i)$ denotes the weight of vertex u_i as the root; $w(r)$ is the weight of the root of T_0, which is equal to the total number of vertices, n, in G. Therefore, the diameter of the augmented graph is $\mathcal{O}(k + \log n)$, where the first term, k, comes from the ceilings in the summation. Finally, we triangulate the graph by inserting outerpaths with constant maximum degree inside each non-triangular face to obtain a maximal planar graph with constant maximum degree and $\mathcal{O}(k + \log n)$ diameter. The result follows from Lemma 1.   □

2.2 Negative Results

Next we show that, for a graph with unbounded maximum degree or unbounded outerplanarity, there might not be a balanced circle-contact representation with circles.

Lemma 2. *There is no balanced circle-contact representation for the graphs in Fig. 4.*

Lemma 2, which we prove in the full version of this paper [1], shows the tightness of the two conditions for balanced circle-contact representations in Theorem 1. Note that the example of the graph in Fig. 4(b) can be extended for any specified maximum degree, by adding a simple path to the high-degree vertex. Furthermore, the example is a 2-outerplanar graph with no balanced circle-contact representation.

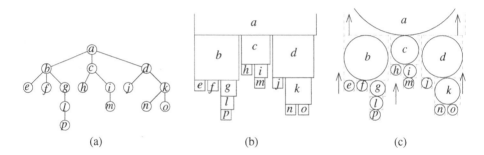

Fig. 5. Construction of a balanced circle-contact representation

3   Trees and Outerplanar Graphs

Theorem 2. *Every tree has a balanced circle-contact representation. Such a representation can be found in linear time.*

Proof: We first find a contact representation Γ of a given tree T with squares such that the ratio of the maximum and the minimum sizes for the squares is polynomial in the number of vertices n in T. To this end, we consider T as a rooted tree with an arbitrary vertex r as the root. Then we construct a contact representation of T with squares where each vertex v of T is represented by a square $R(v)$ such that $R(v)$ touches the square for its parent by its top side and it touches all the squares for its children by its bottom side; see Figs. 5(a) and 5(b). We choose the size of $R(v)$ as $l(v) + \varepsilon(n(v) - 1)$, where $\varepsilon > 0$ is a small positive constant and $n(v)$ and $l(v)$ denote the number of vertices and the number of leaves in the subtree of T rooted at v. In particular, the size of $R(v)$ is 1 when v is a leaf. If v is not a leaf, then suppose $v_1, \ldots, v_d$ are the children of v in the counterclockwise order around v. Then we place the squares $R(v_1), \ldots, R(v_d)$ from left-to-right touching the bottom side of $R(v)$ such that for each $i \in \{1, \ldots, d - 1\}$, $R(v_{i+1})$ is placed ε unit to the right of $R(v_i)$; see Fig. 5(b). There is sufficient space to place all these squares in the bottom side of $R(v)$, since $n(v) = (\sum_{i=1}^{d} n(v_i)) - 1$ and $l(v) = \sum_{i=1}^{d} l(v_i)$. The representation contains no crossings or unwanted contacts since for each vertex v, the representation of the subtrees rooted at v is bounded in the left and right side by the two sides of $R(v)$, and all the subtrees rooted at the children of v are in disjoint regions ε unit away from each other. The size of the smallest square is 1, while the size of the largest square (for the root) has size $l(T) + \varepsilon(n - 1) = \mathcal{O}(n)$, where $l(T)$ is the number of leaves in T.

Using Γ, we find a balanced circle-contact representation of T as follows. We replace each square $R(v)$, representing vertex v, by an inscribed circle of $R(v)$; see Fig. 5(c). The operation removes some contacts from the representation. We re-create these contacts by a top-down traversal of T and moving each circle upward until it touches its parent. Note that a given circle will not touch or intersect any circle other than the circles for its parent and its children, as for every vertex in the infinite strip between its leftmost and rightmost point for its circle, the closest circle in the upward direction is its parent's one. Thus, we obtain a contact representation of T with circles. The representation is balanced since the diameter for every circle is equal to the side-length for its square and we started with a balanced representation Γ.

The linear running time can be achieved by a linear-time traversal of T. First, by a bottom-up traversal of T, we compute the values $n(v)$ and $l(v)$ for each vertex v of T. Using the values for each vertex, we compute the square-contact representation for T by a linear-time top-down traversal of T. Finally, in another top-down traversal of T, for each vertex v of T, we can compute the exact translation required for the inscribed circles of $R(v)$ to touch the parent circle.   □

Let us now describe how to compute a balanced circle-contact representation for a *cactus* graph, which is a connected graph in which every biconnected component is either an edge or a cycle. We use the algorithm described in the proof of Theorem 2, and we call it **Draw_Tree**.

Let T be a rooted tree with a plane embedding. For each vertex v of T, add an edge between every pair of children of v that are consecutive in the clockwise order around v. Call the resulting graph an *augmented fan-tree* for T. Clearly for any rooted tree T, the augmented fan-tree is outerplanar. We call an outerplanar graph a *fan-tree* graph if it is an augmented fan-tree for some rooted tree. A *star* is the complete bipartite graph $K_{1,n-1}$. The center of a star is the vertex that is adjacent to every other vertex. An augmented fan-tree for a star is obtained by taking the center as the root. Thus, an augmented fan-tree for a star is a *fan*. The *center* of a fan is again the vertex adjacent to all the other vertices.

Lemma 3. *Every subgraph of a fan admits a contact representation with circles in which, for each circle $c(v)$ representing a vertex v other than the center, the vertical strip containing $c(v)$ is empty above $c(v)$.*

Proof: Let G be a subgraph of a fan and let T be the star contained in the fan. We now use the contact representation Γ of T obtained by **Draw_Tree** to compute a representation for G. Consider the square-contact representation computed for T in the algorithm. This defines a vertical strip for each circle $c(v)$ in Γ representing a vertex v, and for all the vertices other than the center, these strips are disjoint; see Fig. 6(a). Call the left and right boundary of this strip the *left-* and *right-line* for c, respectively.

We now consider a set S of circles, one for each vertex of G other than the center, with the following properties:

(P1) The circles are interior-disjoint.
(P2) Each circle $c'(v)$ representing a vertex v spans the entire width of the vertical strip for v, and the vertical strip above $c'(v)$ is empty.
(P3) For each vertex v, the circle $c'(v)$ touches the circle c_0 representing the center in Γ if v is adjacent to the center; otherwise, $c'(v)$ is exactly ε distance away from c_0, for some fixed constant $\varepsilon > 0$.
(P4) If a vertex v is not adjacent to the vertex on its left (or if v is the leftmost vertex), then the leftmost point of $c'(v)$ is on the left-line of v; similarly, if v is not adjacent to the vertex on its right (or if v is the rightmost vertex), then the rightmost point of $c'(v)$ is on the right-line of v.

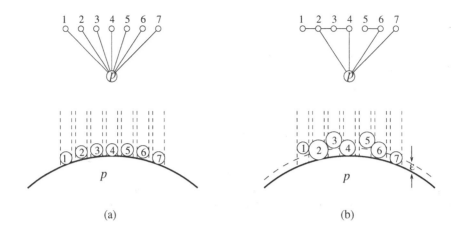

Fig. 6. (a) A star T and a contact representation of T with circles; (b) a subgraph of the fan for T and its contact representation with circles

(P5) The sizes for the circles are maximal with respect to the above properties.

Note that there exists a set of circles with the properties (P1)–(P4); in particular, the set of circles in Γ representing the vertices of T other than the center is such a set. We now claim that the set S of circles with properties (P1)–(P5) together with the circle c_0 gives a contact representation for G; see Fig. 6(b). First note that a circle $c'(v)$ cannot touch any circle other than c_0 and the two circles $c(v_l)$ and $c(v_r)$ representing the vertices v_l and v_r on its left and right, respectively. Indeed, it cannot pass the vertical strip for v_l and v_r above them due to (P2) and behind them due to (P3). Furthermore, the ε distance between c_0 and the circles for vertices non-adjacent to the center and the restriction on the left and right side in (P4) ensures that there is no extra contact. Hence, it is sufficient to show that for each edge in G, we have the contact between the corresponding circles.

Since each circle $c'(v)$ is maximal in size, it must touch at least three objects. One of them is either the circle c_0 or the ε offset line for c_0. Thus, if v_l and v_r are the left and right neighbors of v (if any), then $c'(v)$ must touch two of the followings: (i) $c'(v_l)$ (or the left line of v if v_l does not exists), (ii) the right line for v_l, (iii) $c'(v_r)$ (or the right line of v if v_r does not exists), and (iv) the left line for v_r. Assume without loss of generality that both v_l and v_r exist for v. Then if $c'(v)$ touches both $c'(v_l)$ and $c'(v_r)$, we have the desired contacts for v. Therefore, for a desired contact of $c'(v)$ to be absent, either $c'(v)$ touches both $c'(v_l)$ and the right-line of v_l (and misses the contact with $c'(v_r)$), or it touches both $c'(v_r)$ and the left-line of v_r (and misses the contact with $c'(v_r)$).

Assume, for the sake of a contradiction, that there are two consecutive vertices x and y that are adjacent in G but $c'(x)$ and $c'(y)$ do not touch each other. Let l and r be the vertices to the left of x and to the right of y, respectively. Then it must be the case that x touches both $c'(l)$ and the right line for l and y touches both $c'(r)$ and the left line of r; see Fig. 7(a). One can then increase the size of either $c'(x)$ or $c'(y)$ (say $c'(y)$) such

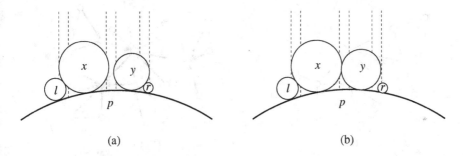

Fig. 7. Illustration for the proof of Lemma 3: if the circles for x and y do not touch each other, at least one can be increased in size

that it now touches $c'(x)$ and the left-line for r (but not $c'(r)$), a contradiction to the maximality for the circles; see Fig. 7(b). □

Using the lemma, we can obtain a quadratic-time algorithm as follows. Given a subgraph G of a fan, compute the balanced circle-contact representation Γ for the corresponding star T using **Draw_Tree**. Then pick the vertices of T other than the center in an arbitrary order and for each vertex v, replace the circle $c(v)$ in Γ by a circle of maximum size that does not violate any of the properties (P1)–(P4) in the proof of Lemma 3. This takes a linear time. Now for every edge (x, y) for which $c(x)$ and $c(y)$ do not touch, replace one of the two circles (say, $c(y)$) with a circle that touches $c(x)$ as in Fig. 7(b). Note that this may result in a loss of a contact between $c(y)$ and the circle to its right. We perform a similar operation for the circle to the right of $c(y)$, then possibly for the circle on its right and so on, until all missing contact are repaired. This process requires linear time per edge; hence, the total running time to compute the desired contact representation is quadratic. The contact representation is balanced since the representation obtained by **Draw_Tree** is balanced and afterwards we only increase the size of circles that are not of the largest size.

Theorem 3. *Every n-vertex fan-tree graph has a balanced circle-contact representation. Such a representation can be found in $\mathcal{O}(n^2)$ time.*

Proof: Let G be a fan-tree graph and let T be the corresponding tree for which G is the augmented fan-tree. Using **Draw_Tree**, we first obtain a balanced circle-contact representation of T. As in the proof of Lemma 3, this defines a vertical strip for each vertex in T. In a top-down traversal of T, we can find a contact representation of G with circles by repeating the quadratic-time algorithm for the subgraphs of fans. Hence, the total complexity is $\sum_{v \in V(T)} \deg_T^2(v) = \mathcal{O}(n^2)$. □

As a corollary of Theorem 3, we obtain an algorithm for creating balanced circle-contact representation of a cactus graph.

Corollary 1. *Every n-vertex cactus graph has a balanced circle-contact representation. Such a representation can be found in $\mathcal{O}(n^2)$ time.*

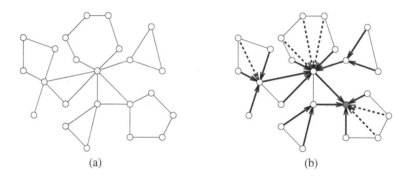

Fig. 8. (a) A cactus graph G; (b) augmenting G to a fan-tree so that the directed edges form a rooted tree and are oriented towards the root

Proof: Given cactus graph G, choose a root vertex v arbitrarily. For each cycle C of G, add an edge from each vertex of C to the (unique) closest vertex to v in C (Fig. 8). The resulting supergraph of G is a fan-tree; the result follows by Theorem 3.   □

In the full version of this paper [1], we provide a linear-time algorithm for balanced circle-contact representation of outerpaths. The main idea of this construction is to partition a given outerpath into a sequence of fans, use unit circles to represent the zigzag outerpath formed by the vertices at the ends of each fan, and then perturb these circles by small rotations to make room for the other circles that should go between them.

Theorem 4. *Every outerpath has a balanced circle-contact representation. Such a representation can be found in linear time.*

4   Bounded Tree-Depth

A graph G has *tree-depth* t if there exists a supergraph of G, and a depth-first search tree T of the supergraph, with at most t vertices on every root–leaf path in T. A family of graphs has bounded tree-depth if and only if there is a constant bound on the length of the longest path that can be found in any of its graphs [20].

Theorem 5. *For every constant bound d, every planar graph with tree-depth at most d has a balanced circle-contact representation.*

We sketch the proof from the full version of this paper [1]. The first step characterizes the planar graphs with bounded tree-depth, using block-cut trees and SPQR trees to represent the 2-vertex-connected and 3-vertex-connected components of a graph. We show that a family of planar graphs has bounded tree-depth if and only if the block-cut trees of graphs in the family have bounded depth, the SPQR trees of 2-connected components of these graphs have bounded depth, and each 3-connected component has a bounded number of vertices. If all three conditions are true, the longest path length can be bounded by a recursion of bounded height and branching factor. Conversely, if any one of these conditions is violated, then there exist paths of unbounded length: a long path in one of the trees leads directly to a long path in the graph and large 3-connected components have long paths by results of Chen and Yu [6].

Because each 3-connected component must have bounded size, the circle packing theorem gives it a balanced circle packing. Next, we construct a contact representation for a supergraph of the given graph, by using Möbius transformations to glue together these packings. The virtual edge representing two adjacent components in an SPQR tree should be represented by a pair of tangent circles shared by the packings for the two components; two tangent circles may be shared by an unbounded number of components. We find a family of Möbius transformations that pack all these components into the space surrounding the two shared tangent circles, so that the components are otherwise disjoint from each other, and each is distorted by a polynomial factor. By using this method to combine adjacent nodes of the block-cut and SPQR trees, we obtain a balanced circle packing for the whole graph in which each component is transformed a constant number of times with polynomial distortion per transformation. However, we may have additional unwanted tangencies between circles, coming from virtual edges in an SPQR tree node that do not correspond to graph edges.

The final part of our proof of Theorem 5 shows how to perturb these glued-together packings, in a controlled way, to eliminate the contacts between pairs of vertices that are connected by virtual edges but not by edges of the input graph while still allowing the Möbius gluing to work correctly. The existence of a Möbius transformation from one pair of circles to another is controlled by an invariant of pairs of circles called their *inversive distance* that equals 1 for tangent circles, is less than 1 for crossing circles, and is greater than 1 for disjoint circles. The theory of inversive distance circle packings is not as well-developed as the theory of tangent circle packings, but a theorem of Luo [17] implies that, for a maximal planar graph with specified positions for the centers of the three circles representing the outer face of the graph and specified inversive distances on each edge of the graph, a circle packing of this type is unique when it exists. By combining this fact with Brouwer's theorem of invariance of domain, we show that for any fixed maximal planar graph (and fixed three outer circle centers) the space of feasible assignments of inversive distances to edges of the graph forms an open set. Therefore, for all sufficiently small $\varepsilon > 0$, there exist packings for which all virtual-but-not-actual edges have inversive distance $1 + \varepsilon$ and all actual edges have inversive distance 1. Choosing ε to be inverse-polynomially small allows the same gluing method to complete the construction and the proof.

5   Conclusion

We studied balanced circle packings for planar graphs, showing that several rich classes of graphs have balanced circle packings. One interesting open problem is whether or not every outerplanar graph has a balanced circle packing representation. While we identified several subclasses of outerplanar graphs that admit such representations, the question remains open for general outerplanar graphs.

Acknowledgments. This work is supported in part by the National Science Foundation under grants CCF-1228639, CCF-1115971, DEB 1053573, and by the Office of Naval Research under Grant No. N00014-08-1-1015.

References

1. Alam, J., Eppstein, D., Goodrich, M.T., Kobourov, S.G., Pupyrev, S.: Balanced circle packings for planar graphs. Arxiv report arxiv.org/abs/1408.4902 (2014)
2. Bannister, M.J., Devanny, W.E., Eppstein, D., Goodrich, M.T.: The Galois complexity of graph drawing: Why numerical solutions are ubiquitous for force-directed, spectral, and circle packing drawings. In: Duncan, C., Symvonis, A. (eds.) GD 2014. LNCS, vol. 8871, pp. 149–161. Springer, Heidelberg (2014)
3. Bern, M., Eppstein, D.: Optimal Möbius transformations for information visualization and meshing. In: Dehne, F., Sack, J.-R., Tamassia, R. (eds.) WADS 2001. LNCS, vol. 2125, pp. 14–25. Springer, Heidelberg (2001)
4. Breu, H., Kirkpatrick, D.G.: Unit disk graph recognition is NP-hard. Comput. Geom. Th. Appl. 9(1-2), 3–24 (1998)
5. Brightwell, G., Scheinerman, E.: Representations of planar graphs. SIAM J. Discrete Math. 6(2), 214–229 (1993)
6. Chen, G., Yu, X.: Long cycles in 3-connected graphs. J. Comb. Theory B 86(1), 80–99 (2002)
7. Collins, C.R., Stephenson, K.: A circle packing algorithm. Comput. Geom. Th. Appl. 25(3), 233–256 (2003)
8. Dolev, D., Leighton, T., Trickey, H.: Planar embedding of planar graphs. Advances in Computing Research 2, 147–161 (1984)
9. Duncan, C.A., Gansner, E.R., Hu, Y.F., Kaufmann, M., Kobourov, S.G.: Optimal polygonal representation of planar graphs. Algorithmica 63(3), 672–691 (2012)
10. Eppstein, D.: Planar Lombardi drawings for subcubic graphs. In: Didimo, W., Patrignani, M. (eds.) GD 2012. LNCS, vol. 7704, pp. 126–137. Springer, Heidelberg (2013)
11. Eppstein, D., Holten, D., Löffler, M., Nöllenburg, M., Speckmann, B., Verbeek, K.: Strict confluent drawing. In: Wismath, S., Wolff, A. (eds.) GD 2013. LNCS, vol. 8242, pp. 352–363. Springer, Heidelberg (2013)
12. de Fraysseix, H., de Mendez, P.O., Rosenstiehl, P.: On triangle contact graphs. Combinatorics, Probability & Computing 3(2), 233–246 (1994)
13. Gilbert, E.N., Moore, E.F.: Variable-length binary encodings. Bell System Technical Journal 38(4), 933–967 (1959)
14. Gonçalves, D., Lévêque, B., Pinlou, A.: Triangle contact representations and duality. Discrete Comput. Geom. 48(1), 239–254 (2012)
15. Hliněný, P.: Classes and recognition of curve contact graphs. J. Comb. Theory B 74(1), 87–103 (1998)
16. Koebe, P.: Kontaktprobleme der konformen Abbildung. Ber. Sächs. Akad. Wiss. Leipzig, Math.-Phys. Kl. 88, 141–164 (1936)
17. Luo, F.: Rigidity of polyhedral surfaces, III. Geometry & Topology 15(4), 2299–2319 (2011)
18. Malitz, S.M., Papakostas, A.: On the angular resolution of planar graphs. SIAM J. Discrete Math. 7(2), 172–183 (1994)
19. Mohar, B.: A polynomial time circle packing algorithm. Discrete Math. 117(1-3), 257–263 (1993)
20. Nešetřil, J., Ossona de Mendez, P.: Sparsity: Graphs, Structures, and Algorithms. Springer (2012)
21. Nievergelt, J., Reingold, E.M.: Binary search trees of bounded balance. SIAM J. Comput. 2, 33–43 (1973)

Unit Contact Representations of Grid Subgraphs with Regular Polytopes in 2D and 3D

Linda Kleist and Benjamin Rahman

Institut für Mathematik, Technische Universität Berlin, Berlin, Germany
{kleist,rahman}@math.tu-berlin.de

Abstract. We present a strategy to construct unit proper contact representations (UPCR) for subgraphs of certain highly symmetric grids. This strategy can be applied to obtain graphs admitting UPCRs with squares and cubes, whose recognition is NP-complete.

We show that subgraphs of the square grid allow for UPCR with squares which strengthens the previously known cube representation. Indeed, we give UPCR for subgraphs of a d-dimensional grid with d-cubes. Additionally, we show that subgraphs of the triangular grid admit a UPCR with cubes, implying that the same holds for each subgraph of an Archimedean grid. Considering further polygons, we construct UPCR with regular $3k$-gons of the hexagonal grid and UPCR with regular $4k$-gons of the square grid.

1 Introduction

In this paper, we study *unit contact representations* (UCR) which are *contact representations* (CR) with congruent objects. We are particularly interested in *proper contacts* (PCR), that is, contacts are realized by segments of non-zero length in 2D or polygons with non-zero area in 3D. Contacts not of this type are disregarded. Typical objects considered are regular polygons and cubes.

1.1 Related Work

Considering homothetic copies of disks, the celebrated circle packing theorem of Koebe [10] states that every planar graph has a CR with disks. Schramm [11] gives the following generalization: Assigning a convex set in 2D with smooth boundary to each vertex, every planar graph has a CR with non-degenerated homothetic copies of the prescribed sets. Allowing convex sets without smooth boundary, this results in CRs with possibly degenerated homothetic copies of the assigned sets [11, 12]. Gonçalves, Lévêque and Pinlou [8] observe that this result can be exploited for triangle CRs with non-degenerated homothetic triangles for 4-connected planar triangulations. Felsner and Francis [6] employ possibly degenerated homothetic triangle representations to show that all planar graphs have a CR with cubes, where contacts are not necessarily proper.

Studying PCRs with polygons, Gansner et al. [7] show that every planar graph has a PCR with hexagons, but not always with pentagons.

C. Duncan and A. Symvonis (Eds.): GD 2014, LNCS 8871, pp. 137–148, 2014.
© Springer-Verlag Berlin Heidelberg 2014

Considering UCRs with congruent objects, Breu and Kirkpatrick [4] prove that the recognition of unit disk graphs is NP-complete; indeed, this holds even for the recognition of bounded-ratio disk graphs. For a survey on recognition-complexity results with balls and disks we refer to [9].

Czyzowicz et al. [5] are interested in discrete versions of unit disk graphs and study UCRs of non-rotated copies of regular k-gons with two different contact types: vertex-to-vertex and whole edge contact. For even k, these graph classes coincide (for odd k the second class is empty). It turns out that these graphs are also unit disk graphs and that the recognition of graphs allowing for a representation with $4k$-gons is NP-complete.

In 3D, Bremner et al. [3], show that it is NP-complete to decide whether a graph admits a UPCR with cubes. In [1, 2] UPCR with cubes of subgraphs of 5 Archimedean grids are given. These are partly obtained by threshold colorings. However, threshold colorings cannot be used to find UPCR with cubes for all subgraphs of Archimedean grids; in particular, not for all subgraphs of the triangular grid [1].

1.2 Our Contributions

We develop a strategy to construct unit proper contact representations (UPCR) in 2D and 3D for subgraphs of certain highly symmetric grids. This strategy is used to show that subgraphs of the square grid allow for UPCR with squares. This is a strengthening of the unit cube representation of Alam et al. [2]. We generalize this result to $4k$-gons as well as to higher dimensions, namely, we prove that subgraphs of a d-dimensional grid have UPCR with d-cubes. Furthermore, we show that subgraphs of the triangular grid admit UPCRs with cubes, implying that the same holds for all subgraphs of Archimedean grids (grids originating from regular and semi-regular tilings of the plane). This solves an open problem posed by Alam et al. [2]. Considering other geometric objects, we construct UPCRs with regular $3k$-gons of the hexagonal grid.

Additionally, we observe that with the ideas of [3], we can show the NP-completeness of recognizing unit square proper contact graphs.

2 Definitions and Properties

Let $G = (V, E)$ be a graph. A function $\phi\colon V \to \mathcal{P}(\mathbb{R}^d), v \mapsto S_v$ is called a *proper contact representation* (PCR) in $\mathbb{R}^d$ of G if the sets in $\phi(V) := \bigcup_{v \in V} \phi(v)$ are pairwise interiorly disjoint and $(u, v) \in E \iff S_u \cap S_v$ is $(d-1)$-dimensional. Usually, the assigned sets are compact and path-connected and in this paper regular polytopes. A PCR $\mathcal{C}$ is called *unit* (UPCR), if all sets in $\mathcal{C}$ are congruent. This means for UPCR with regular n-gons, the polygons (in a component of G) can be transformed into one another by translations for even n, and by translations and rotation by π for odd n. In particular, we consider UPCRs of squares (USqPCR), cubes (UCuPCR) and triangles (UTriPCR). We define a *unit square* in $\mathbb{R}^2$ by its characteristic corner (x, y) as $S(x, y) := [x, x + 1] \times [y, y+1]$, and analogously, a *unit d-cube* in $\mathbb{R}^d$ as $Q(x) := [x_1, x_1+1] \times \ldots \times [x_d, x_d+1]$. Further, we define an upward and downward *unit triangle* with height $h := \sin(\frac{\pi}{3})$ as

$$\triangle(x,y) := \left\{ \begin{pmatrix} a \\ b \end{pmatrix} \in \mathbb{R}^2 \mid \begin{pmatrix} a \\ b \end{pmatrix} = \begin{pmatrix} x \\ y \end{pmatrix} + r\begin{pmatrix} 1 \\ 0 \end{pmatrix} + s\begin{pmatrix} 1/2 \\ h \end{pmatrix}; \ r,s \geq 0; \ r+s \leq 1 \right\},$$

$$\triangledown(x,y) := \left\{ \begin{pmatrix} a \\ b \end{pmatrix} \in \mathbb{R}^2 \mid \begin{pmatrix} a \\ b \end{pmatrix} = \begin{pmatrix} x \\ y \end{pmatrix} + r\begin{pmatrix} 1 \\ 0 \end{pmatrix} + s\begin{pmatrix} 1/2 \\ -h \end{pmatrix}; \ r,s \geq 0; \ r+s \leq 1 \right\}.$$

For two touching polygons S_u, S_v in the plane, we define the *size of contact* $\mathrm{cs}(S_u, S_v)$ by the length of its realizing segment; and for two touching d-dimensional polytopes S_u, S_v, we define $\mathrm{cs}(S_u, S_v)$ by the shortest edge length of the $(d-1)$-dimensional polytope realizing the contact. The *contact size* of a PCR $\mathcal{C}$ is given by $\mathrm{cs}(\mathcal{C}) := \min\{\mathrm{cs}(S_u, S_v) \mid (u,v) \in E\}$. The *translation* of a set $S \in \mathbb{R}^n$ by a vector $t \in \mathbb{R}^n$ is defined by the addition $S + t := \{(x + t) \in \mathbb{R}^n \mid x \in S\}$. The *space* $\mathrm{sp}(S_u, S_v, d)$ of two objects S_u, S_v in direction d is the maximum $\delta \in \mathbb{R}$ such that either object can be translated by δd and they remain interiorly disjoint. Finally, $[n] := \{0, \dots, n\}$.

2.1 Some Properties

We start with some basic properties of UPCRs. Clearly, PCRs in the plane may only exist for planar graphs. Due to the congruent objects, graphs with UPCR have bounded maximum degree and fulfill spatial constraints.

Observation 1. *Let G be a graph admitting a UPCR with regular n-gons. Then, the maximum degree is bounded by 6, i.e. $\Delta(G) \leq 6$.*

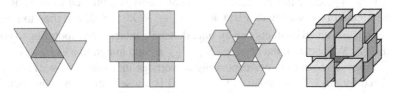

Fig. 1. Examples of polygons and a cube with maximum number of neighbors in a UPCR

For triangles and squares, it is easy to see that the maximum degree is 4 and 6, respectively, see Figure 1. A general upper bound for regular polygons is given by their *kissing number*, the maximal number of congruent copies a single polygon can be touched by. Regular n-gons with $n \geq 5$ have a kissing number of ≤ 6, see [13, 14] for further explanation.

Together with $\Omega(\sqrt{n})$ polygons at the boundary of any (component of a) representation, this results in a low edge density: If a graph G on n vertices has a UPCR with k-gons, then the number of edges is $\leq 3n - \Omega(\sqrt{n})$. The analogous result for unit cube graphs has already been shown in [3]: For $G = (V, E)$ with UCuPCR it holds that $\Delta(G) \leq 14$ and $|E| \leq 7n - \Omega(n^{\frac{2}{3}})$.

We formulate spatial restrictions, due to the congruent objects for d-cubes:

Observation 2. *Let G be a graph with a UPCR with d-cubes in $\mathbb{R}^d$. Then, every vertex has at most $(2r + 1)^d$ vertices in distance $\leq r$.*

This follows easily by comparing the available space in distance $\leq r$ and the needed space to place all congruent d-cubes. Consequently, there exist binary trees with no UPCR with squares nor cubes.

For cubes, Bremner et al. [3], show that it is NP-complete to decide whether a graph admits a UCuPCR. With the same ideas, we can show the analogous result for squares:

Theorem 1. *The recognition of graphs with USqPCR or UCuPCR is NP-complete.*

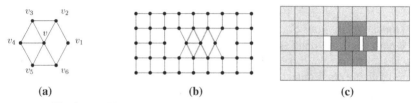

Fig. 2. (a) Aligning graph (b) Box graph (c) USqPCR of box graph

Due to the similarity, we only want to provide the strategy of the proof. The logic engine of the proof by Bremner et al. [3] is composed of copies of the boxgraph, see fig. 2. Since in any USqPCR of the boxgraph the rectangular shape of the USqPCR remains, see fig. 2, the recognition of a USqPCR of the composed graph is equivalent to the recognition of a non-overlapping layout of the corresponding logic engine.

3   The Strategy

Let $G = (V, E)$ be a subgraph of some grid $\mathbb{G} = (V_\mathbb{G}, E_\mathbb{G})$. In order to find a UPCR of G, we set $V := V_\mathbb{G}$, since independent vertices can easily be removed from a CR. The idea of the method is to start with a UPCR $\hat{\phi} : V \to \mathcal{P}(\mathbb{R}^n)$ of $\mathbb{G}$, and then to modify $\hat{\phi}$ by removing unwanted contacts one by one, that are contacts corresponding to edges in $\overline{E} := E_\mathbb{G} \backslash E$. We partition the edge set $E_\mathbb{G} = \bigcup_i E_i$ and assign to $e \in E_i$ a direction $d_i \in \mathbb{R}^n$. For $e \in \overline{E}$ objects corresponding to vertices in a *moving set* $M(e) \subset V$ are translated by δd_i. The translation step $\delta > 0$ is chosen small enough, such that apart from e all contacts remain and no additional ones occur. This strategy yields a straight-forward construction of a final UPCR ϕ of G:

$$\phi(v) = \hat{\phi}(v) + \sum_i |\{e \in \overline{E} \cap E_i \mid v \in M(e)\}| \cdot \delta d_i.$$

4   Representations with Unit Squares

The square grid $\mathbb{S}_{m,n} = (V_\mathbb{S}, E_\mathbb{S})$ of size $(m \times n)$ is defined as follows:

$$\mathbb{S}_{m,n} := \left(\{v_{i,j} \mid i \in [m], j \in [n]\}, \left\{ (v_{i,j}, v_{i',j'}) \mid \left\| \begin{pmatrix} i-i' \\ j-j' \end{pmatrix} \right\|_1 = 1 \right\} \right).$$

Theorem 2. *Every subgraph of the square grid $\mathbb{S}_{m,n}$ has a USqPCR.*

Proof. We consider subgraphs of type $G = (V_\mathbb{S}, E)$; for subgraphs with $V \subset V_\mathbb{S}$, the theorem follows by removing independent vertices. The proof applies the strategy described in Section 3. We fix an $\varepsilon \in (0, 1)$ and present a USqPCR $\hat{\phi}$ of $\mathbb{S}_{m,n}$, see also Figure 3:

$$\hat{\phi} \colon V \to \mathcal{P}(\mathbb{R}^2)$$
$$\hat{\phi}(v_{i,j}) = S(i - j\varepsilon, j + i\varepsilon).$$

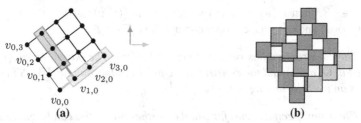

Fig. 3. (a) Square grid $\mathbb{S}_{3,3}$; for thick edges, the moving sets are indicated by the framed vertices (b) USqPCR $\hat{\phi}$ of $\mathbb{S}_{3,3}$ with illustration of modification step.

We claim that $\hat{\phi}$ is a USqPCR of $\mathbb{S}_{m,n}$

- $\mathrm{cs}\big(\hat{\phi}(u),\hat{\phi}(v)\big) = 1 - \varepsilon$ for all $(u,v) \in E_{\mathbb{S}}$ and
- $\mathrm{sp}\big(\hat{\phi}(u),\hat{\phi}(v),d\big) \geq \varepsilon$ for all $(u,v) \notin E_{\mathbb{S}}$ and the two directions d parallel to the square's sides: $d_1 := \binom{1}{0}$ and $d_2 := \binom{0}{1}$

Consider a vertex v and its four neighbors. Each contact is realized by one side of $\hat{\phi}(v)$. By definition of $\hat{\phi}$, the characteristic corners differ by $\pm\binom{1}{\varepsilon}, \pm\binom{-\varepsilon}{1}$. Hence, the squares are disjoint and $\mathrm{cs}\big(\hat{\phi}(u),\hat{\phi}(v)\big) = (1 - \varepsilon)$, for all $(u,v) \in E_{\mathbb{S}}$. Consider a non-edge $(u,v) \notin E_{\mathbb{S}}$. The characteristic corners of $\hat{\phi}(u)$ and $\hat{\phi}(v)$ differ by $a\binom{1}{\varepsilon} + b\binom{-\varepsilon}{1}$ with $a,b \in \mathbb{Z}$ and $|a| + |b| \geq 2$. This implies that $\mathrm{sp}\big(\hat{\phi}(u),\hat{\phi}(v),d_i\big) \geq \varepsilon$ for $i \in \{1,2\}$.

We now define the moving sets and translation vectors: For $e = (v_{i,j}, v_{i+1,j})$, we set $d(e) := d_1 = \binom{1}{0}$ and $M(e) := \{v_{k,j} \in V \mid k > i\}$, otherwise $e = (v_{i,j}, v_{i,j+1})$, we set $d(e) := d_2 = \binom{0}{1}$ and $M(e) := \{v_{i,k} \in V \mid k > j\}$. For simplification, we define $E_i := \{e \in E \mid d(e) = d_i\}$ for $i \in \{1,2\}$ and assume wlog $n \geq m$. The idea is to remove each contact of $e \in \overline{E}$ by translating $M(e)$ in direction $d(e)$. The integer $r_i(v)$ describes how often v is moved in direction d_i, that is $r_i(v) := |\{e \in \overline{E} \cap E_i \mid v \in M(e)\}|$. Observe that $r_i(v) \leq n$ for $i \in \{1,2\}$ and $v \in V$. Choosing $\delta < \frac{1}{n}\min\{\varepsilon, 1 - \varepsilon\}$ the following mapping is a USqPCR of $G \subseteq \mathbb{S}_{m,n}$:

$$\phi \colon V \to \mathcal{P}(\mathbb{R}^2)$$
$$\phi(v) = \hat{\phi}(v) + r_1(v) \cdot \delta d_1 + r_2(v) \cdot \delta d_2.$$

We verify that ϕ is a USqPCR of G by showing three properties: Let $(u,v) \notin E_{\mathbb{S}}$. Recall that $\mathrm{sp}\big(\hat{\phi}(u),\hat{\phi}(v),d_i\big)$ gives the maximum translation step λ such that $\hat{\phi}(u)$ and $\hat{\phi}(v)$ remain interiorly disjoint when either one of them is translated by λd_i. By construction $r_i(w)\delta \leq n\delta < \varepsilon \leq \mathrm{sp}\big(\hat{\phi}(u),\hat{\phi}(v),d_i\big)$ for $w \in \{u,v\}$ and $i \in \{1,2\}$. Consequently, $\phi(u)$ and $\phi(v)$ remain disjoint.

Let $(u,v) \in E_i$, then $\hat{\phi}(u)$ and $\hat{\phi}(v)$ have a contact segment parallel to d_{i+1} of length $\mathrm{cs}(\hat{\phi}(u),\hat{\phi}(v)) \geq 1 - \varepsilon$. Hence, translations in direction d_{i+1} have no effect on the contact: $\mathrm{cs}(\phi(u),\phi(v)) \geq \mathrm{cs}(\hat{\phi}(u),\hat{\phi}(v)) - r_{i+1}(v)\delta \geq 1 - \varepsilon - n\delta > 0$. Wlog we suppose that $\hat{\phi}(u)$ is left or below of $\hat{\phi}(v)$, then by definition of the moving sets: $r_i(u) \leq r_i(v)$. This implies that $\phi(u)$ and $\phi(v)$ remain interiorly disjoint.

Since the translation vector d_i is not parallel to the contact segment, the contact remains iff $r_i(u) = r_i(v)$. By definition, this is the case iff $(u,v) \in E$. Note therefore, that if $(u, v) \in \overline{E}$, then $r_i(u) < r_i(v)$ since $u \notin M((u,v))$ and $v \in M((u,v))$. Consequently, $\phi(u)$ and $\phi(v)$ have proper contact iff $(u, v) \in E$. $\qquad\square$

Remark 1. *The construction has running time of $O(|V|^2)$: The parameters $r_i(v)$ can be determined in $O(n|V|)$ and the construction can be produced in $O(|V|)$. Since $n \leq |V|$, this gives an overall running time of $O(|V|^2)$.*

Remark 2. *Choosing specific values for ε and δ, further properties can be guaranteed: With $\varepsilon = \frac{1}{2}$ and $\delta = \frac{1}{2n+2}$ one obtains a USqPCR ϕ of G where the proper contacts and non-contacts are guaranteed to be of size δ. As ε and δ approach 0, the constact sizes are arbitrarily close to 1. This is exploited in Section 6 for USqPCR with 4k-gons.*

5   Representations with Unit Cubes

In this section we investigate further subgraphs of grids for UPCR with cubes.

5.1 d-Dimensional Grid

Indeed, Theorem 2 can be generalized to all dimensions. As a generalization of the square grid, we define the d-dimensional grid $\mathbb{S}_n^d = (V_\mathbb{S}, E_\mathbb{S})$:

$$\mathbb{S}_n^d := \left(\{ v_x | x \in [n]^d \}, \{ (v_x, v_y) \mid \|x - y\|_1 = 1 \} \right)$$

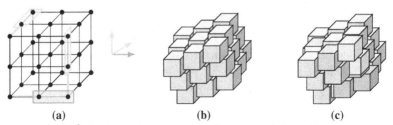

(a)   **(b)**   **(c)**

Fig. 4. (a) Cubic grid $\mathbb{S}_2^3$; for the thick edges, the moving sets are indicated by the framed vertices **(b)** UCuPCR $\hat{\phi}$ of $\mathbb{S}_2^3$ **(c)** Illustration of modification step.

Theorem 3. *Every subgraph of the grid $\mathbb{S}_n^d$ admits a UPCR with d-cubes.*

Proof. The proof applies the strategy presented in Section 3. First, we give a UPCR with d-cubes of $\mathbb{S}_n^d$, depicted in Figure 4. We choose $\varepsilon \in (0, 1)$, $\delta < \frac{1}{n} \min\{\varepsilon, 1 - \varepsilon\}$ and define the UPCR with the help of matrix $A \in \mathbb{R}^{d \times d}$:

$$\hat{\phi} \colon V \to \mathcal{P}(\mathbb{R}^d)$$
$$\hat{\phi}(v_x) = Q(A \cdot x)$$

$$A := \begin{pmatrix} 1 & \varepsilon & \dots & \varepsilon \\ -\varepsilon & 1 & \ddots & \vdots \\ \vdots & \ddots & \ddots & \varepsilon \\ -\varepsilon & \dots & -\varepsilon & 1 \end{pmatrix}$$

In order to prove that $\hat{\phi}$ is a UPCR of $\mathbb{S}_n^d$ with d-cubes, we note: Two axis-aligned unit d-cubes with characteristic corners x and y, have proper $(d-1)$-dimensional contact iff there exist $k \in [d]$ such that $|x_k - y_k| = 1$ and $|x_i - y_i| < 1$ for all $i \in [d] \setminus \{k\}$. It remains to show that two cubes $\hat{\phi}(v_x)$ and $\hat{\phi}(v_y)$ have proper contact iff $(v_x, v_y) \in E_{\mathbb{S}_n^d}$: Suppose $(v_x, v_y) \in E_{\mathbb{S}}$. Then, it holds $\|x - y\|_1 = 1$; that is $\exists! k \in [d]$ with $|x_k - y_k| = 1$ and $|x_i - y_i| = 0$ for all $i \in [d] \setminus \{k\}$. Wlog let $x \geq_{\text{dom}} y$. Consider the characteristic corners $A \cdot x$ and $A \cdot y$ of the cubes $\hat{\phi}(v_x)$ and $\hat{\phi}(v_y)$. Let e_k denote the kth standard basis vector of $\mathbb{R}^d$, then for the characteristic corners hold: $(A \cdot x - A \cdot y) = A \cdot (x - y) = A \cdot e_k$. This implies that the two cubes $\hat{\phi}(v_x)$ and $\hat{\phi}(v_y)$ have proper contact which is of size $cs(\hat{\phi}(v_x), \hat{\phi}(v_y)) = 1 - \varepsilon$.

Suppose $(v_x, v_y) \notin E_{\mathbb{S}}$, then $\|x - y\|_1 \geq 2$. Either there is a coordinate $k \in [d]$, such that $|x_k - y_k| \geq 2$ or there exist at least two coordinates $k, j \in [d]$ such that $|x_k - y_k| \geq 1$. It is easy to check, that in either case, there is a coordinate in $(A \cdot x - A \cdot y)$ which has an absolute value of $\geq 1 + \varepsilon$, and hence, the cubes $\hat{\phi}(v_x)$ and $\hat{\phi}(v_y)$ are disjoint. Thus for $(v_x, v_y) \notin E_{\mathbb{S}}$ the space is $\text{sp}(\hat{\phi}(v_x), \hat{\phi}(v_y), e_k) \geq \varepsilon$ for $k = 1, \ldots, d$.

We proceed with the translation vectors and moving sets: For $e = (v_x, v_y) \in E_{\mathbb{S}}$, there exists a unique $k \in [d]$ with $|x_k - y_k| = 1$ and $x_i = y_i$ for $i \neq k$. Wlog we assume $x \leq_{\text{dom}} y$. Then, the direction vector is defined as $d(e) := e_k$ and the moving set as $M(e) := \{v_z \mid z \in [n]^d, z_k > x_k, z_i = x_i \text{ for all } i \neq k\}$.

Observe that the translation vector $d(e)$ of edge e is parallel to each of the following crucial $(d-1)$-dimensional facets: These are facets realizing the contacts corresponding to edges of type $(u, v) \in E \setminus \{e\}$ with $u \in M(e)$ and $v \notin M(e)$. Additionally, it holds that $r_i(w)\delta \leq n\delta < \varepsilon \leq \text{sp}(\hat{\phi}(u), \hat{\phi}(v), d_i)$ for $w \in \{u, v\} \notin E_{\mathbb{S}}$ and $i \in \{1, \ldots, d\}$. Following the same lines as the proof of Theorem 2, this shows that the construction yields a UPCR with d-cubes for each subgraph of $\mathbb{S}_n^d$. □

5.2   Triangular Grid

Alam et al. [2] asked whether subgraphs of the triangular grid have UCuPCRs. In this section, we answer this question affirmatively.

We introduce an unconventional definition of the triangular grid $\mathbb{T}_{m,n} = (V_{\mathbb{T}}, E_{\mathbb{T}})$ of size $m \times n$. For $i \in [m]$, and $j \in [n]$, the vertex set is $V := \{b_{i,j}\} \cup \{t_{i,j}\}$. The set of edges is $E := \bigcup_{i=1}^{5} E_i$, where $E_1 := (x_{i,j}, x_{i,j+1})$ with $x \in \{b, t\}$, $E_2 := (b_{i,j}, t_{i,j})$, $E_3 := (b_{i,j}, t_{i,j-1})$, $E_4 := (b_{i,j}, t_{i-1,j-1})$, and $E_5 := (b_{i,j}, t_{i-1,j})$, see also Figure 5. For fixed j, we call the vertex set of type $\{x_{i,j}\}_{i \in [m]}$ a *line* and two neighboring lines a *level*. Edges between two lines are called *level edges*. Note that we have $2n$ lines.

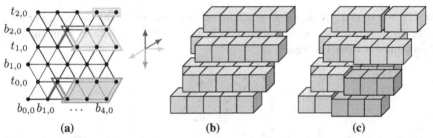

Fig. 5. (a) Triangular grid $\mathbb{T}_{4,2}$; for thick edges, the moving sets are indicated by framed vertices (b) UCuPCR $\hat{\phi}$ of $\mathbb{T}_{4,2}$ (c) Illustration of modification step.

Theorem 4. *Every subgraph of the triangular grid* $\mathbb{T}_{m,n}$ *has a UCuPCR.*

Proof. The proof uses the observation that subgraphs of $\mathbb{T}_{1,n}$ allow for USqPCR. Therefore, contacts corresponding to level edges are realized alternating in planes parallel to the xy and xz plane. Choosing $\varepsilon \in (0,1)$ and $\delta < \frac{1}{n+1}\min\{\varepsilon, 1-\varepsilon\}$, we define a UCuPCR of $\mathbb{T}_{m,n}$, see also Figure 5:

$$\hat{\phi}\colon V \to \mathcal{P}(\mathbb{R}^3)$$

$$\hat{\phi}(v) = \begin{cases} Q\big(i, & j(1+\varepsilon), & j\big) & \text{if } v = b_{i,j}, \\ Q\big(i+\varepsilon, & j(1+\varepsilon)+\varepsilon, & j+1\big) & \text{if } v = t_{i,j}. \end{cases}$$

It is easy to verify that $\hat{\phi}$ is a UCuPCR of $\mathbb{T}_{m,n}$ with contact size $cs(\hat{\phi}) \geq (1-\varepsilon)$. Additionally, it holds that $sp(\hat{\phi}(u), \hat{\phi}(v), d_i) \geq \varepsilon$ for the following direction vectors:

For an edge $e \in E_i$ we set the direction vector d_i to $d_1 := (1,0,0)$, $d_2 := (0,0,1)$, $d_3 := (0,0,-1)$, $d_4 := (0,1,0)$, $d_5 := (0,-1,0)$. The moving set belonging to an edge is slightly more involved. For a level edge e, $M(e)$ roughly consists of the vertices in the level with larger index; for a line edge, it consists of the line vertices with larger index:
for $e = (b_{i,j}, t_{k,l}) \in E_2 \cup E_4$, we define $M(e) := \{b_{i,r} \mid r > j\} \cup \{t_{k,r} \mid r \geq l\}$, and
for $e = (b_{i,j}, t_{k,l}) \in E_3 \cup E_5$, we define $M(e) := \{b_{i,r} \mid r \geq j\} \cup \{t_{k,r} \mid r > l\}$, and
for $e = (x_{i,j}, x_{i,j+1}) \in E_1$, we define $M(e) := \{x_{i,r} \mid r > j\}$. Figure 5 depicts the moving sets, direction vectors and modifications.

As the proof is analogous to the proof of Theorem 2, we give some useful observations and leave the rest to the reader: The translation vector $d(e)$ of an edge e is parallel to each of the crucial cube facets, realizing contacts corresponding to edges of type $(u,v) \in E\backslash\{e\}$ with $u \in M(e)$ and $v \notin M(e)$. Also note that $r_i(v) \leq n+1$ and hence $r_i(u) \cdot \delta \leq (n+1)\delta \leq \varepsilon \leq sp(\hat{\phi}(u), \hat{\phi}(v), d_i)$ for all $(u,v) \notin E_\mathbb{S}$ and all d_i. Combining this, the construction yields a UCuPCR for each subgraph of $\mathbb{T}_{m,n}$. $\quad\square$

5.3   Archimedean Grids

There exist eleven grids originating from regular and semi-regular tilings of the plane, so called *Archimedean grids*, which are depicted in Table 1. As mentioned before, UCuPCR for subgraphs of five Archimedean grids are known [1, 2]. With the results of the previous section, UCuPCR for subgraphs of all Archimedean grids follow directly:

Corollary 1. *Every subgraph of an Archimedean grid has a UCuPCR.*

Proof. Observe that Archimedean grids are subgraphs of the triangular grid. For proving this fact, we provide convincing pictures in Table 1. With this observation, the claim follows directly from Theorem 4. $\quad\square$

Fig. 6. The pufferfish graph and the star $K_{1,5}$

The remaining question is, whether subgraphs of Archimedean grids admit a representation with unit squares. In general, this is not the case since we find two forbidden subgraphs: $K_{1,5}$ and the *pufferfish graph*, which is a C_6 together with two private neighbors for all but one vertex, see Figure 6. Table 1 summarizes for which Archimedean grids, all subgraphs allow for a USqPCR.

Table 1. The table gives an overview for which Archimedean grids each subgraph allows for a USqPCR. By Corollary 1, each subgraph has a UCuPCR.

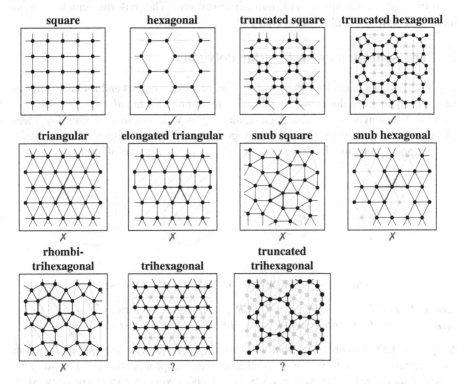

Lemma 1. *The pufferfish graph and $K_{1,5}$ have no USqPCR.*

Proof. We first note that for two touching squares, at least one corner of each square is involved in the segment of contact; two corners are involved iff the squares have full side contact. In a USqPCR of $K_{1,5}$, there must be a corner involved in two contacts which leads to full side contacts. Since a square has only four sides, it follows that $K_{1,5}$ has no USqPCR.

Assume the pufferfish graph has a USqPCR. We first observe that no two squares have full side contact: Suppose there are two such squares, then one of them represents a vertex v of the C_6 which has 4 independent neighbors. Due to its 4 independent neighbors, it follows that all of its contacts are full side contacts. It is easy to see that repeating the argument for a neighbor of v with degree 4 leads to a contradiction. Hence, no two squares have full side contact and each corner of the squares in the C_6 is involved in a contact segment.

Consequently, the inner face bounded by the squares of the C_6 is an orthogonal 8-gon (of T- or Z-shape) with side lengths ≤ 1. This implies that private neighbors cannot be placed in the inner face and the two convex corners belonging to different squares do not account for a contact: a contradiction. $\square$

Corollary 2. *Not all subgraphs of the triangular, elongated triangular, snub square, snub hexagonal, and rhombitrihexagonal grid have a USqPCR.*

Proof. $K_{1,5}$ (and the pufferfish graph) are induced subgraphs of the triangular, elongated triangular, snub square, and snub hexagonal grid. The pufferfish graph is an induced subgraph of rhombitrihexagonal grid. $\square$

6   Representations with Regular Polygons

In this section, we consider UPCR with further regular polygons and for ease, refer to them as polygons. To do so, we introduce the notion of *pseudo polygons*. A *pseudo n-gon* (with side length s) is a subset of a regular n-gon, which includes a central segment (of length $\geq s$) of each boundary edge. A segment of a boundary edge is called *central* if their midpoints coincide. It can be understood as a n-gon with cut-off corners, consider Figure 7.

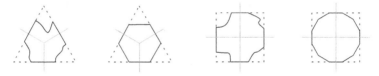

Fig. 7. Examples of pseudo-triangles and pseudo-squares

Lemma 1. *Let G be a graph with a UPCR ϕ with regular k-gons and $cs(\phi) > 1 - s$. Then, G has a UPCR with pseudo k-gons with side length at least s.*

Proof. This UPCR is obtained from ϕ by inscribing a pseudo k-gon into each k-gon: Consider two touching k-gons and the sides realizing the proper contact. The midpoints of these sides differ by $< s$ since $cs(\phi) > 1 - s$. Hence, every contact can be certified by two intersecting central segments of size s. Since the pseudo k-gons (with side length s) contain these segments, the contacts remain. Additionally, no new contacts occur, because each pseudo k-gon is a subsets of a k-gon. Hence, ϕ still serves as a UPCR. $\square$

6.1   Representations with Regular $4k$-gons

Corollary 3. *Every subgraph of the square grid $\mathbb{S}_{m,n}$ has a UPCR with regular $4k$-gons, $k \geq 1$.*

Proof. Note that $4k$-gons are pseudo-squares with side length $s_k := \sin(\frac{\pi}{4k})$. Choosing $\varepsilon < \frac{s_k}{2}$ in the proof of Theorem 2, we obtain a UPCR ϕ for every subgraph G of $\mathbb{S}_{m,n}$ with $cs(\phi) > 1 - s_k$. With this, the claim follows directly from Theorem 2 and Lemma 1. $\square$

6.2 Representations with Regular $3k$-gons

We define the hexagonal grid $\mathbb{H}_{m,n} = (V_{\mathbb{H}}, E_{\mathbb{H}})$ as a subgraph of the square grid $\mathbb{S}_{m,n}$:

$$\mathbb{H}_{m,n} := \left(V_{\mathbb{S}}, E_{\mathbb{S}} \setminus \{(v_{i,j}, v_{i+1,j}) \in E_{\mathbb{S}} \mid (i+j) \text{ odd}\} \right).$$

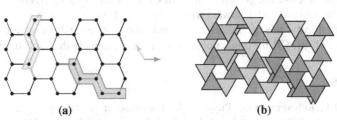

(a)   **(b)**

Fig. 8. (a) Hexagonal Grid $\mathbb{H}_{5,5}$; for thick edges, the moving sets are indicated by framed vertices (b) UTriPCR $\hat{\phi}$ of $\mathbb{H}_{5,5}$ with illustration of modification step

Theorem 5. *Every subgraph of the hexagonal grid $\mathbb{H}_{m,n}$ has a UTriPCR.*

Proof. For the proof, we apply the already known technique. With $\varepsilon \in (0,1)$, a UTriPCR of $\mathbb{H}_{m,n}$ is given by the following mapping and depicted in Figure 8:

$$\phi \colon V \to \mathcal{P}(\mathbb{R}^3)$$

$$\phi(v_{i,j}) = \begin{cases} \triangle\left(\frac{i}{4}\left[\binom{3}{2h} + \varepsilon\binom{3}{-6h}\right] + \frac{j}{4}\left[\binom{-1}{2h} + \varepsilon\binom{3}{2h}\right]\right) & \text{if } i+j \text{ even,} \\ \triangledown\left(\frac{i}{4}\left[\binom{3}{2h} + \varepsilon\binom{3}{-6h}\right] + \frac{j}{4}\left[\binom{-1}{2h} + \varepsilon\binom{3}{2h}\right] + \frac{1}{4}\left[\binom{-1}{h} + \varepsilon\binom{-1}{h}\right]\right) & \text{else.} \end{cases}$$

Analyzing the shifts of the characteristic corner, it is not hard to verify that $\hat{\phi}$ is a UTriPCR of the triangular grid with contact size $cs(\hat{\phi}) \geq (1-\varepsilon)$ and has moving space $sp(\phi(u), \phi(v), d) \geq \varepsilon$ for $(u,v) \in E_{\mathbb{H}}$ and direction vectors d parallel to the sides of the triangles.

Indeed, two types of edges and direction vectors suffice: For $e = (v_{i,j}, v_{i,j+1})$, we define direction $d(e) := (-\frac{1}{2}, h)$ and moving set $M(e) := \{v_{i,k} \in V \mid k > j\}$; for $e = (v_{i,j}, v_{i+1,j})$, we define direction $d(e) := (1,0)$ and moving set $M(e) := \{v_{i+1+k,j-k} \in V \mid k \in [m-i]\} \cup \{v_{i+1+k,j-1-k} \in V \mid k \in [m-i]\}$. A crucial property is that, again, the translation vector $d(e)$ of an edge e is parallel to each of the segments realizing contacts corresponding to edges of type $(u,v) \in E \setminus \{e\}$ with $u \in M(e)$ and $v \notin M(e)$. Moreover, $d(e)$ is not parallel to the segment realizing this contact in $\hat{\phi}$. Note also that each edge belongs to at most n moving sets with the same translation direction. Therefore, choosing $\delta < \frac{1}{n} \min\{\varepsilon, 1 - \varepsilon\}$ the construction analogous to Theorem 2 yields a UTriPCR for each subgraph G of $\mathbb{H}_{m,n}$. $\square$

Corollary 4. *Every subgraph of the hexagonal grid $\mathbb{H}_{m,n}$ has a UPCR with regular $3k$-gons, $k \geq 1$.*

Proof. Note that $3k$-gons are pseudo-triangles with side length $s_k := \tan(\frac{\pi}{3k})h$. Choosing $\varepsilon < \frac{s_k}{2}$ in the proof of Theorem 5, we obtain a UPCR ϕ for every subgraph G of $\mathbb{H}_{m,n}$ with $cs(\phi) > 1 - s_k$. With this, the claim follows directly from Theorem 5 and Lemma 1. $\square$

7   Open Questions

We want to conclude with a list of open questions:

- What is the complexity of recognizing graphs admitting UPCRs with regular polygons other than squares?
- Can we characterize the graphs with USqPCRs? Or with other polygons?
- Do the trihexagonal and truncated trihexagonal grid admit a USqPCR?
- Do subgraphs of duals of Archimedean grids not containing $K_{1,9}$ have UCuPCR? What about USqPCR for duals not containing $K_{1,5}$, namely the snubsquare grid?

References

[1] Alam, M.J., Kobourov, S.G., Pupyrev, S., Toeniskoetter, J.: Happy edges: Threshold-coloring of regular lattices. In: Ferro, A., Luccio, F., Widmayer, P. (eds.) FUN 2014. LNCS, vol. 8496, pp. 28–39. Springer, Heidelberg (2014)

[2] Alam, M.J., Chaplick, S., Fijavž, G., Kaufmann, M., Kobourov, S.G., Pupyrev, S.: Threshold-coloring and unit-cube contact representation of graphs. In: Brandstädt, A., Jansen, K., Reischuk, R. (eds.) WG 2013. LNCS, vol. 8165, pp. 26–37. Springer, Heidelberg (2013)

[3] Bremner, D., Evans, W., Frati, F., Heyer, L., Kobourov, S.G., Lenhart, W.J., Liotta, G., Rappaport, D., Whitesides, S.H.: On representing graphs by touching cuboids. In: Didimo, W., Patrignani, M. (eds.) GD 2012. LNCS, vol. 7704, pp. 187–198. Springer, Heidelberg (2013)

[4] Breu, H., Kirkpatrick, D.G.: On the complexity of recognizing intersection and touching graphs of disks. In: Brandenburg, F.J. (ed.) GD 1995. LNCS, vol. 1027, pp. 88–98. Springer, Heidelberg (1996)

[5] Czyzowicz, J., Kranakis, E., Krizanc, D., Urrutia, J.: Discrete realizations of contact and intersection graphs. International Journal of Pure and Applied Mathematics 13(4), 429 (2004)

[6] Felsner, S., Francis, M.C.: Contact representations of planar graphs with cubes. In: Proceedings of the 27th Annual ACM Symposium on Computational Geometry, pp. 315–320. ACM (2011)

[7] Gansner, E.R., Hu, Y.F., Kaufmann, M., Kobourov, S.G.: Optimal polygonal representation of planar graphs. In: López-Ortiz, A. (ed.) LATIN 2010. LNCS, vol. 6034, pp. 417–432. Springer, Heidelberg (2010)

[8] Gonçalves, D., Lévêque, B., Pinlou, A.: Triangle contact representations and duality. Discrete & Computational Geometry 48(1), 239–254 (2012)

[9] Hliněný, P., Kratochvíl, J.: Representing graphs by disks and balls (a survey of recognition-complexity results). Discrete Mathematics 229(1-3), 101–124 (2001)

[10] Koebe, P.: Kontaktprobleme der konformen abbildung. Berichte über die Verhandlungen der Sächsischen Akademie der Wissenschaften zu Leipzig, Math.-Phys. Kl. 88, 141–164 (1936)

[11] Schramm, O.: Combinatorially prescribed packings and applications to conformal and quasiconformal maps. Ph. D. thesis. Princeton University (1990)

[12] Schramm, O.: Square tilings with prescribed combinatorics. Israel Journal of Mathematics 84(1-2), 97–118 (1993)

[13] Zhao, L.: The kissing number of the regular polygon. Discrete Mathematics 188(1), 293–296 (1998)

[14] Zhao, L., Xu, J.: The kissing number of the regular pentagon. Discrete Mathematics 252(1), 293–298 (2002)

The Galois Complexity of Graph Drawing: Why Numerical Solutions Are Ubiquitous for Force-Directed, Spectral, and Circle Packing Drawings*

Michael J. Bannister, William E. Devanny,
David Eppstein, and Michael T. Goodrich

Department of Computer Science, University of California, Irvine

Abstract. Many well-known graph drawing techniques, including force directed drawings, spectral graph layouts, multidimensional scaling, and circle packings, have algebraic formulations. However, practical methods for producing such drawings ubiquitously use iterative numerical approximations rather than constructing and then solving algebraic expressions representing their exact solutions. To explain this phenomenon, we use Galois theory to show that many variants of these problems have solutions that cannot be expressed by nested radicals or nested roots of low-degree polynomials. Hence, such solutions cannot be computed exactly even in extended computational models that include such operations.

1 Introduction

One of the most powerful paradigms for drawing a graph is to construct an algebraic formulation for a suitably-defined optimal drawing of the graph and then solve this formulation to produce a drawing. Examples of this *algebraic graph drawing* approach include the force-directed, spectral, multidimensional scaling, and circle packing drawing techniques (which we review in the full version of the paper for readers unfamiliar with them).

Even though this paradigm starts from an algebraic formulation, the ubiquitous method for solving such formulations is to approximately optimize them numerically in an iterative fashion. That is, with a few exceptions for linear systems [1–3], approximate numerical solutions for algebraic graph drawing are overwhelmingly preferred over exact symbolic solutions. It is therefore natural to ask if this preference for numerical solutions over symbolic solutions is inherent in algebraic graph drawing or due to some other phenomena, such as laziness or lack of mathematical sophistication on the part of those who are producing the algebraic formulations.

In this paper, we introduce a framework for deciding whether certain algebraic graph drawing formulations have symbolic solutions, and we show that exact symbolic solutions are, in fact, impossible in several algebraic computation models, for some simple examples of common algebraic graph drawing formulations, including force-directed

---

* This research was supported in part by ONR MURI grant N00014-08-1-1015 and NSF grants 1217322, 1011840, and 1228639.

C. Duncan and A. Symvonis (Eds.): GD 2014, LNCS 8871, pp. 149–161, 2014.

graph drawings (in both the Fruchterman–Reingold [4] and Kamada–Kawai [5] approaches), spectral graph drawings [6], classical multidimensional scaling [7], and circle packings [8]. Note that these impossibility results go beyond saying that such symbolic solutions are computationally infeasible or undecidable to find—instead, we show that such solutions do not exist.

To prove our results, we use Galois theory, a connection between the theories of algebraic numbers and abstract groups. Two classical applications of Galois theory use it to prove the impossibility of the classical Greek problem of doubling the cube using compass and straightedge, and of solving fifth-degree polynomials by nested radicals. In our terms, these results concern quadratic computation trees and radical computation trees, respectively. Our proofs build on this theory by applying Galois theory to the algebraic numbers given by the vertex positions in different types of graph drawings. For force-directed and spectral drawing, we find small graphs (in one case as small as a length-three path) whose drawings directly generate unsolvable Galois groups. For circle packing, an additional argument involving the compass and straightedge constructibility of Möbius transformations allows us to transform arbitrary circle packings into a canonical form with two concentric circles, whose construction is equivalent to the calculation of certain algebraic numbers. Because of this mathematical foundation, we refer to this topic as the *Galois complexity* of graph drawing.

Related Work. The problems for which Galois theory has been used to prove unsolvability in simple algebraic computational models include shortest paths around polyhedral obstacles [9], shortest paths through weighted regions of the plane [10], the geometric median of planar points [11], computing structure from motion in computer vision [12], and finding polygons of maximal area with specified edge lengths [13]. In each of these cases, the non-existence of a nested radical formula for the solution is established by finding a Galois group containing a symmetric group of constant degree at least five. In our terminology, this shows that these problems cannot be solved by a radical computation tree. We are not aware of any previous non-constant lower bounds on the degree of the polynomial roots needed to solve a problem, comparable to our new bounds using the root computation tree model. Brightwell and Scheinerman [14] show that some circle packing graph representations cannot be constructed by compass and straightedge (what we call the quadratic computation tree model).

2   Preliminaries

Models of Computation. We define models of computation based on the *algebraic computation tree* [15, 16], in which each node computes a value or makes a decision using standard arithmetic functions of previously computed values. Specifically, we define the following variant models:

- A *quadratic computation tree* is an algebraic computation tree in which the set of allowable functions for each computation node is augmented with square roots and complex conjugation. These trees capture the geometric constructions that can be performed by compass and unmarked straightedge.

- A *radical computation tree* is an algebraic computation tree in which the set of allowable functions is augmented with the kth root operation, where k is an integer parameter to the operation, and with complex conjugation. These trees capture the calculations whose results can be expressed as nested radicals.
- A *root computation tree* is an algebraic computation tree in which the allowable functions include the ability to find complex roots of polynomials whose coefficients are integers or previously computed values, and to compute complex conjugates of previously computed values. For instance, this model can compute any algebraic number. As a measure of complexity in this model, we define the *degree* of a root computation tree as the maximum degree of any of its polynomials. A *bounded-degree root computation tree* has its degree bounded by some constant unrelated to the size of its input. Thus, a quadratic computation tree is exactly a bounded-degree root computation tree (of degree two).

Our impossibility results and degree lower bounds for these models imply the same results for algorithms in more realistic models of computation that use as a black box the corresponding primitives for constructing and representing algebraic numbers in symbolic computation systems. Because our results are lower bounds, they also apply *a fortiori* to weaker primitives, such as systems limited to *real* algebraic numbers, which don't include complex conjugation.

It is important to note that each of the above models can generate algebraic numbers of unbounded degree. For instance, even the quadratic computation tree (compass and straightedge model) can construct regular 2^k-gons, whose coordinates are algebraic numbers with degrees that are high powers of two. Thus, to prove lower bounds and impossibility results in these models, it is not sufficient to prove that a problem is described by a high-degree polynomial; additional structure is needed.

Algebraic Graph Theory. In algebraic graph theory, the properties of a graph are examined via the spectra of several matrices associated with the graph. The *adjacency matrix* $A = \text{adj}(G)$ of a graph G is the $n \times n$ matrix with $A_{i,j}$ equal to 1 if there is an edge between i and j and 0 otherwise. The *degree matrix* $D = \deg(G)$ of G is the $n \times n$ matrix with $D_{i,i} = \deg(v_i)$. From these two matrices we define the *Laplacian matrix*, $L = \text{lap}(G) = D - A$, and the *transition matrix*, $T = \text{tran}(G) = D^{-1}A$.

Lemma 1. *For a regular graph G, $\text{adj}(G)$, $\text{lap}(G)$, and $\text{tran}(G)$ have the same set of eigenvectors.*

Lemma 2. *For the cycle on n vertices, the eigenvalues of $\text{adj}(G)$ are $2\cos(2\pi k/n)$, for $0 \leq k < n$.*

Möbius Transformations. We may represent each point p in the plane by a complex number, z, whose real part represents p's x coordinate and whose imaginary part represents p's y coordinate. A *Möbius transformation* is a fractional linear transformation, $z \mapsto (az + b)/(cz + d)$, defined by a 4-tuple (a, b, c, d) of complex numbers, or the complex conjugate of such a transformation. We prove the following in the full version of the paper.

Lemma 3. *Given any two disjoint circles, a Möbius transformation mapping them to two concentric circles can be constructed using a quadratic computation tree.*

Number Theory. The *Euler totient function*, $\phi(n)$, counts the number of integers in the interval $[1, n-1]$ that are relatively prime to n. It can be calculated from the prime factorization $n = \prod p_i^{r_i}$ by the formula

$$\phi(n) = \prod p_i^{r_i-1}(p_i - 1).$$

A *Sophie Germain prime* is a prime number p such that $2p + 1$ is also prime [17]. It has been conjectured that there are infinitely many of them, but the conjecture remains unsolved. The significance of these primes for us is that, when p is a Sophie Germain prime, $\phi(2p + 1)$ has the large prime factor p. An easy construction gives a number n for which $\phi(n)$ has a prime factor of size $\Omega(\sqrt{n})$: simply let $n = p^2$ for a prime p, with $\phi(n) = p(p-1)$. Baker and Harman [18] proved the following stronger bound.

Lemma 4 (Baker and Harman [18]). *For infinitely many prime numbers p, the largest prime factor of $\phi(p)$ is at least $p^{0.677}$.*

Field Theory. A *field* is a system of values and arithmetic operations over them (addition, subtraction, multiplication, and division) obeying similar axioms to those of rational arithmetic, real number arithmetic, and complex number arithmetic: addition and multiplication are commutative and associative, multiplication distributes over addition, subtraction is inverse to addition, and division is inverse to multiplication by any value except zero. A field K is an *extension* of a field F, and F is a *subfield* of K (the *base field*), if the elements of F are a subset of those of K and the two fields' operations coincide for those values. K can be viewed as a vector space over F (values in K can be added to each other and multiplied by values in F) and the *degree* $[K : F]$ of the extension is its dimension as a vector space. For an element α of K the notation $F(\alpha)$ represents the set of values that can be obtained from rational functions (ratios of univariate polynomials) with coefficients in F by plugging in α as the value of the variable. $F(\alpha)$ is itself a field, intermediate between F and K. In particular, we will frequently consider field extensions $\mathbf{Q}(\alpha)$ where $\mathbf{Q}$ is the field of rational numbers and α is an *algebraic number*, the complex root of a polynomial with rational coefficients.

Lemma 5. *If α can be computed by a root computation tree of degree $f(n)$, then $[\mathbf{Q}(\alpha) : \mathbf{Q}]$ is $f(n)$-smooth, i.e., it has no prime factor $> f(n)$. In particular, if α can be computed by a quadratic computation tree, then $[\mathbf{Q}(\alpha) : \mathbf{Q}]$ is a power of two.*

Proof. See the full version of the paper.   □

A *primitive root of unity* ζ_n is a root of $x^n - 1$ whose powers give all other roots of the same polynomial. As a complex number we can take $\zeta_n = \exp(2i\pi/n)$.

Lemma 6 (Corollary 9.1.10 of [19], p. 235). $[\mathbf{Q}(\zeta_n) : \mathbf{Q}] = \phi(n)$.

Galois Theory. A *group* is a system of values and a single operation (written as multiplication) that is associative and in which every element has an inverse. The set of permutations of the set $[n] = \{1, 2, \ldots, n\}$, multiplied by function composition, is a standard example of a group and is denoted by S_n. A *permutation group* is a subgroup of S_n; i.e., it is a set of permutations that is closed under the group operation.

A *field automorphism* of the field F is a bijection $\sigma : F \to F$ that respects the field operations, i.e., $\sigma(xy) = \sigma(x)\sigma(y)$ and $\sigma(x + y) = \sigma(x) + \sigma(y)$. The set of all field automorphism of a field F forms a group denoted by $\text{Aut}(F)$. Given a field extension K of F, the subset of $\text{Aut}(K)$ that leaves F unchanged is itself a group, called the *Galois group* of the extension, and is denoted

$$\text{Gal}(K/F) = \{\sigma \in \text{Aut}(K) \mid \sigma(x) = x \text{ for all } x \in F\}.$$

The *splitting field* of a polynomial, p, with rational coefficients, denoted $\text{split}(p)$ is the smallest subfield of the complex numbers that contains all the roots of the polynomial. Each automorphism in $\text{Gal}(\text{split}(p)/\mathbf{Q})$ permutes the roots of the polynomial, no two automorphisms permute the roots in the same way, and these permutations form a group, so $\text{Gal}(\text{split}(p)/\mathbf{Q})$ can be thought of as a permutation group.

Lemma 7. *If α can be computed by a radical computation tree and K is the splitting field of an irreducible polynomial with α as one of its roots, then $\text{Gal}(K/\mathbf{Q})$ does not contain S_n as a subgroup for any $n \geq 5$.*

Proof. If α is computable by a radical computation tree, it can be written as an expression using nested radicals. If K is the splitting field of an irreducible polynomial with such an expression as a root, $\text{Gal}(K/\mathbf{Q})$ is a solvable group (Def. 8.1.1 of [19], p. 191 and Theorem 8.3.3 of [19], p. 204). But S_n is not solvable for $n \geq 5$ (Theorem 8.4.5 of [19], p. 213), and every subgroup of a solvable group is solvable (Proposition 8.1.3 of [19], p. 192). Thus, $\text{Gal}(K/\mathbf{Q})$ cannot contain S_n $(n \geq 5)$ as a subgroup. $\square$

The next lemma allows us to infer properties of a Galois group from the coefficients of a *monic* polynomial, that is, a polynomial with integer coefficients whose first coefficient is one. The *discriminant* of a monic polynomial is (up to sign) the product of the squared differences of all pairs of its roots; it can also be computed as a polynomial function of the coefficients. The lemma is due to Dedekind and proven in [19].

Lemma 8 (Dedekind's theorem). *Let $f(x)$ be an irreducible monic polynomial in $\mathbf{Z}[x]$ and p a prime not dividing the discriminant of f. If $f(x)$ factors into a product of irreducibles of degrees $d_0, d_1, \ldots d_r$ over $\mathbf{Z}/p\mathbf{Z}$, then $\text{Gal}(\text{split}(f)/\mathbf{Q})$ contains a permutation that is the composition of disjoint cycles of lengths $d_0, d_1, \ldots, d_r$.*

A permutation group is *transitive* if, for every two elements x and y of the elements being permuted, the group includes a permutation that maps x to y. If K is the splitting field of an irreducible polynomial of degree n, then $\text{Gal}(K/\mathbf{Q})$ (viewed as a permutation group on the roots) is necessarily transitive. The next lemma allows us to use Dedekind's theorem to prove that $\text{Gal}(K/\mathbf{Q})$ equals S_n. It is a standard exercise in abstract algebra (e.g., [20], Exercise 3, p. 305).

Lemma 9. *If a transitive subgroup G of S_n contains a transposition and an $(n-1)$-cycle, then $G = S_n$.*

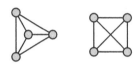

Fig. 1. Two stable drawings of K_4.

Fig. 2. A drawing whose coordinates cannot be computed by a quadratic computation tree.

3 Impossibility Results for Force Directed Graph Drawing

In the Fruchterman and Reingold [4] force-directed model, each vertex is pulled toward its neighbors with an attractive force, $f_a(d) = d^2/k$, and pushed away from all vertices with a repulsive force, $f_r(d) = k^2/d$. The parameter k is a constant that sets the scale of the drawing, and d is the distance between vertices. We say that a drawing is a Fruchterman and Reingold equilibrium when the total force at each vertex is zero.

In the Kamada and Kawai [5] force-directed model, every two vertices are connected by a spring with rest length and spring constant determined by the structure of the graph. The total energy of the graph is defined to be

$$E = \sum_i \sum_{j>i} \frac{1}{2} k_{ij} \big(\operatorname{dist}(p_i, p_j) - \ell_{ij} \big)^2,$$

where p_i = position of vertex v_i, d_{ij} = graph theoretic distance between v_i and v_j, L = a scaling constant, $\ell_{ij} = Ld_{ij}$, K = a scaling constant, and $k_{ij} = K/d_{i,j}^2$. We say that a drawing is a Kamada–Kawai equilibrium if E is at a local minimum. The necessary conditions for such a local minimum are as follows:

$$\frac{\partial E}{\partial x_j} = \sum_{i \neq j} k_{ji}(x_j - x_i) \left(1 - \frac{\ell_{ji}}{\operatorname{dist}(p_j, p_i)} \right) = 0 \qquad 1 \leq j \leq n$$

$$\frac{\partial E}{\partial y_j} = \sum_{i \neq j} k_{ji}(y_j - y_i) \left(1 - \frac{\ell_{ji}}{\operatorname{dist}(p_j, p_i)} \right) = 0 \qquad 1 \leq j \leq n.$$

For either of these approaches to force-directed graph drawing, a graph can have multiple equilibria (Figure 1). In such cases, typically, one equilibrium is the "expected" drawing of the graph and others represent undesired drawings that are not likely to be found by the drawing algorithm. To make the positions of the vertices in this drawing concrete, we assume that the constants k (Fruchterman–Reingold), L, and K (Kamada–Kawai) are all equal to 1. As we will demonstrate, there exist graphs whose expected drawings cannot be constructed in our models of computation. Interestingly, the graphs we use for these results are not complicated configurations unlikely to arise in practice, but are instead graphs so simple that they might at first be dismissed as insufficiently challenging even to be used for debugging purposes.

Root Computation Trees. Consider the cycle C_n with n vertices. When drawn with force directed algorithms, either Fruchterman and Reingold or Kamada and Kawai, the embedding typically places all vertices equally spaced on a circle, such that neighbors are placed next to each other, as shown in Figure 2. As an easy warm-up to our main results, we observe that this is not always possible using a quadratic computation tree.

Theorem 1. *There exist a graph with seven vertices such that it is not possible in a quadratic computation tree to compute the coordinates of every possible Fruchterman and Reingold equilibrium or every possible Kamada and Kawai equilibrium.*

Proof. Let G be the cycle C_7 on seven vertices. Both algorithms have the embedding shown in Figure 2 (suitably scaled) as an equilibrium. In this embedding let a and b be two neighboring vertices and α and β their corresponding complex coordinates. Then α/β is equal to $\pm\zeta_7$ the seventh root of unity. By Lemma 6

$$[\mathbf{Q}(\zeta_7) : \mathbf{Q}] = \phi(7) = 6.$$

Since 6 is not a power of two, Lemma 5 implies that ζ_7 cannot be constructed by a quadratic computation tree. Therefore, neither can this embedding. □

Theorem 2. *For arbitrarily large values of n, there are graphs on n vertices such that constructing the coordinates of all Fruchterman and Reingold equilibria on a root computation tree requires degree $\Omega(n^{0.677})$. If there exists infinitely many Sophie Germain primes, then there are graphs for which computing the coordinates of any Fruchterman and Reingold equilibria requires degree $\Omega(n)$. The same results with the same graphs hold for Kamada and Kawai equilibria.*

Proof. As in the previous theorem we consider embedding cycles with their canonical embedding, which is an equilibrium for both algorithms. The same argument used in the previous theorem shows we can construct ζ_n from the coordinates of the canonical embedding of the cycle on n vertices.

We consider cycles with p vertices where p is a prime number for which $\phi(p) = p-1$ has a large prime factor q. If arbitrarily large Sophie Germain primes exist we let q be such a prime and let $p = 2q + 1$. Otherwise, by Lemma 4 we choose p in such a way that its largest prime factor q is at least $p^{0.677}$. Now, by Lemma 6 we have:

$$[\mathbf{Q}(\zeta_p) : \mathbf{Q}] = \phi(p) = p - 1.$$

This extension is not D-smooth for any D smaller than q, and therefore every construction of it on a root computation tree requires degree at least q. □

Thus, such drawings are not possible on a bounded-degree root computation tree.

Radical Computation Trees. To show that the coordinates of a Fruchterman and Reingold equilibrium are in general not computable with a radical computation tree we consider embedding the path with three edges, shown in Figure 3. We assume that all of the vertices are embedded colinearly and without edge or vertex overlaps. These assumptions correspond to the equilibrium that is typically produced by the Fruchterman and Reingold algorithm.

Fig. 3. A graph whose Fruchterman–Reingold coordinates cannot be computed by a radical computation tree.

Fig. 4. A graph whose Kamada–Kawai coordinates cannot be computed by a radical computation tree.

Let $a > 0$ be the distance from v_0 to v_1 (equal by symmetry to the distance from v_2 to v_3) and let $b > 0$ be the distance from v_1 to v_2. We can then express the sum of all the forces at vertex v_0 by the equation

$$F_0 = a^2 - \frac{1}{a} - \frac{1}{a+b} - \frac{1}{2a+b} = \frac{2a^5 + 3a^4b + a^3b^2 - 5a^2 - 5ab - b^2}{2a^3 + 3a^2b + ab^2},$$

and the sum of all the forces at vertex v_1 by the equation

$$F_1 = -a^2 + \frac{1}{a} + b^2 - \frac{1}{b} - \frac{1}{a+b} = \frac{-a^4b - a^3b^2 + a^2b^3 - a^2 + ab^4 - ab + b^2}{a^2b + ab^2}.$$

In an equilibrium state we have $F_1 = F_2 = 0$. Equivalently, the numerator p of F_1 and the numerator q of F_2 are both zero, where

$$p(a,b) = 2a^5 + 3a^4b + a^3b^2 - 5a^2 - 5ab - b^2 = 0$$
$$q(a,b) = -a^4b - a^3b^2 + a^2b^3 - a^2 + ab^4 - ab + b^2 = 0.$$

To solve this system of two equations and two unknowns we can eliminate variable a and produce the following polynomial, shown as a product of irreducible polynomials, whose roots give the values of b that lead to a solution.

$$\frac{1}{3}b^2(3b^{15} - 48b^{12} + 336b^9 - 1196b^6 + 1440b^3 + 144).$$

The factor b^2 corresponds to degenerate drawings and may safely be eliminated. Let f be the degree-fifteen factor; then $f(x) = g(x^3)$ for a quintic polynomial g. A radical computation tree can compute the roots of f from the roots of g, so we need only show that the roots of g cannot be computed in a radical computation tree. To do this, we convert g to a monic polynomial h with the same splitting field, via the transformation

$$h(x) = \frac{x^5}{144}g(6/x) = x^5 + 60x^4 - 299x^3 + 504x^2 - 432x + 162.$$

The polynomial h can be shown to be irreducible by manually verifying that it has no linear or quadratic factors. Its discriminant is $-2^6 \cdot 3^9 \cdot 2341^2 \cdot 2749$, and h factors modulo primes 5 and 7 (which do not divide the discriminant) into irreducibles:

$$h(x) \equiv (x+1)(x^4 + 3x^3 + 6x^2 + x + 1) \pmod 7$$
$$h(x) \equiv (x^2 + 3x + 4)(x^3 + 2x^2 + x + 3) \pmod 5.$$

By Dedekind's theorem, the factorization modulo 7 implies the existence of a 4-cycle in $\mathrm{Gal}(\mathrm{split}(h)/\mathbf{Q})$, and the factorization modulo 5 implies the existence of a permutation that is the composition of a transposition and a 3-cycle. Raising the second permutation to the power 3 yields a transposition. By Lemma 9, $\mathrm{Gal}(\mathrm{split}(h)/\mathbf{Q}) = S_5$. So by Lemma 7 the value of b cannot be computed by a radical computation tree. Thus, we cannot compute the equilibrium coordinates of the path with three edges under the assumptions that the vertices are collinear and there are no vertex or edge overlaps.

Theorem 3. *There exists a graph on four vertices such that it is not possible on a radical computation tree to construct the coordinates of every possible Fruchterman and Reingold equilibrium.*

To show that the coordinates of a Kamada and Kawai equilibrium are in general not computable with a radical computation tree we consider the graph depicted in Figure 4.

Theorem 4. *There exists a graph on four vertices such that it is not possible on a radical computation tree to construct the coordinates of every possible Kamada and Kawai equilibrium.*

Proof. See the full version of the paper. □

4   Impossibility Results for Spectral Graph Drawing

Root computation trees. We begin with the following result for root computation trees.

Theorem 5. *For arbitrarily large values of n, there are graphs on n vertices such that constructing spectral graph drawings based on the adjacency, Laplacian, relaxed Laplacian, or transition matrix requires a root computation tree of degree $\Omega(n^{0.677})$. If there exist infinitely many Sophie Germain primes, then there are graphs for which computing these drawings requires degree $\Omega(n)$.*

Proof. Since all of the referenced matrices have rational entries, it suffices to consider the computability of their eigenvalues. Further, if we restrict our attention to regular graphs it suffices to consider the eigenvalues of just the adjacency matrix, $M = \mathrm{adj}(G)$, by Lemma 1. Let p be a prime and G the cycle on p vertices. By Lemma 2 the eigenvalues of $A = \mathrm{adj}(G)$ are given by $2\cos(2\pi k/p)$ for $0 \leq k \leq p-1$. In a root computation tree of degree at least 2 the primitive root of unity $\zeta_p = \exp(2i\pi/p)$ can be computed from $2\cos(2\pi k/p)$ for all $k \neq 0$. Therefore, from the proof of Theorem 2, for arbitrarily large n, there are graphs on n vertices such that M has one rational eigenvector (for $k = 0$) and the computation of any other eigenvector on a root computation tree requires degree $\Omega(n^{0.667})$. If infinitely many Sophie Germain primes exist, there are graphs for which computing these eigenvectors requires degree $\Omega(n)$. □

Thus, such drawings are not possible on a bounded-degree root computation tree.

Radical Computation Trees. To show that in general the eigenvectors associated with a graph are not constructible with a radical tree we consider the graph, Y, on nine vertices in Figure 5 for the Laplacian and relaxed Laplacian matrices, and in the

Fig. 5. A graph Y whose Laplacian eigenvectors are uncomputable by a radical tree

full version of the paper we consider another graph for the adjacency and transition matrices.

The characteristic polynomial, $p(x) = \det(M - xI)$, for the Laplacian matrix for Y, can be computed to be

$$p(x) = \text{char}(\text{lap}(Y))$$
$$= x(x^8 - 16x^7 + 104x^6 - 354x^5 + 678x^4 - 730x^3 + 417x^2 - 110x + 9).$$

Lemma 10 (Stäckel [21]). *If $f(x)$ is a polynomial of degree n with integer coefficients and $|f(k)|$ is prime for $2n + 1$ values of k, then $f(x)$ is irreducible.*

Let $q = p(x)/x$. The polynomial q is irreducible by Lemma 10, as it produces a prime number for 17 integer inputs from 0 to 90. The discriminant of q is $2^8 \cdot 9931583$ and we have the following factorizations of q modulo the primes 31 and 41.

$$p_1(x) \equiv (x + 27)(x^7 + 19x^6 + 25x^5 + 25x^4 + 3x^3 + 26x^2 + 25x + 21) \quad (\text{mod } 31)$$
$$p_1(x) \equiv (x + 1)(x^2 + 15x + 39)(x^5 + 9x^4 + 29x^3 + 10x^2 + 36x + 16) \quad (\text{mod } 41).$$

By Dedekind's theorem, the factorization modulo 31 implies the existence of a 7-cycle, and the factorization modulo 41 implies the existence of a permutation that is the composition of a transposition and a 5-cycle. The second permutation raised to the fifth power produces a transposition. Thus, Lemma 9 implies $\text{Gal}(\text{split}(p_1)/\mathbf{Q}) = S_8$. So by Lemma 7 the only eigenvalue of $\text{lap}(Y)$ computable in a radical computation tree is 0. For the relaxed Laplacian we consider the two variable polynomial $f(x, \rho) = \text{char}(\text{lap}_\rho(Y))$. Since setting ρ equal to 1 produces a polynomial with Galois group S_8, Hilbert's irreducibility theorem tells us that the set of ρ for which the Galois group of $f(x, \rho)$ is S_8 is dense in $\mathbf{Q}$.

Theorem 6. *There exists a graph on nine vertices such that it is not possible to construct a spectral graph drawing based on the Laplacian matrix in a radical computation tree. For this graph there exists a dense subset A of $\mathbf{Q}$ such that it is not possible to construct a spectral graph drawing based on the relaxed Laplacian with $\rho \in A$ in a radical computation tree.*

In the full version of the paper we similarly prove that spectral drawings based on the adjacency matrix and the transition matrix cannot be constructed by a radical computation tree. In the full version of the paper we similarly prove that drawings produced by classical multidimensional scaling cannot be constructed by a radical computation tree.

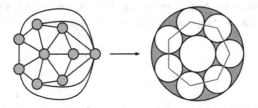

Fig. 6. The graph Bipyramid(7) and its associated concentric circle packing

5   Impossibility Results for Circle Packings

Root Computation Trees. A given graph may be represented by infinitely many circle packings, related to each other by Möbius transformations. But as we now show, if one particular packing cannot be constructed in our model, then there is no other packing for the same graph that the model can construct.

Lemma 11. *Suppose that a circle packing P contains two concentric circles. Suppose also that at least one radius of a circle or distance between two circle centers, at least one center of a circle, and the slope of at least one line connecting two centers of circles in P can all be constructed by one of our computation models, but that P itself cannot be constructed. Then the same model cannot construct any circle packing that represents the same underlying graph as P.*

Proof. Suppose for a contradiction that the model could construct a circle packing Q representing the same graph as P. By Lemma 3 it could transform Q to make the two circles concentric, giving a packing that is similar either to P or to the inversion of P through the center of the concentric circles. By one more transformation it can be made similar to P. The model could then rotate the packing so the slope of the line connecting two centers matches the corresponding slope in P, scale it so the radius of one of its circles matches the corresponding radius in P, and translate the center of one of its circles to the corresponding center in P, resulting in P itself. This gives a construction of P, contradicting the assumption.   □

We define Bipyramid(k) to be the graph formed by the vertices and edges of a $(k + 2)$-vertex bipyramid (a polyhedron formed from two pyramids over a k-gon by gluing them together on their bases). In graph-theoretic terms, it consists of a k-cycle and two additional vertices, with both of these vertices connected by edges to every vertex of the k-cycle. The example of Bipyramid(7) can be seen in Figure 6, left.

Theorem 7. *There exists a graph whose circle packings cannot be constructed by a quadratic computation tree.*

Proof. Consider the circle packing of Bipyramid(7) in which the two hubs are represented by concentric circles, centered at the origin, with the other circle centers all on the unit circle and with one of them on the x axis. One of the centers of this packing is at the root of unity ζ_7. By Lemma 6, $[\mathbf{Q}(\zeta_7) : \mathbf{Q}] = \phi(7) = 6$. 6 is not a power of two, so by Lemma 5 ζ_7 cannot be constructed by a quadratic computation tree. By Lemma 11, neither can any other packing for the same graph.   □

In the full version of the paper, we prove that certain circle packings also cannot be constructed by radical computation trees nor by bounded-degree root computation trees.

6   Conclusion

We have shown that several types of graph drawing cannot be constructed by models of computation that allow computation of arbitrary-degree radicals, nor by models that allow computation of the roots of bounded-degree polynomials. Whether the degree of these polynomials must grow linearly as a function of the input size, or only proportionally to a sublinear power, remains subject to an open number-theoretic conjecture.

It is natural to ask whether these drawings might be computable in a model of computation that allows both arbitrary-degree radicals and bounded-degree roots. We leave this as open for future research.

Acknowledgements. We used the Sage software package to perform preliminary calculations of the Galois groups of many drawings. Additionally, we thank Ricky Demer on MathOverflow for guiding us to research on large factors of $\phi(n)$.

References

[1] Chrobak, M., Goodrich, M.T., Tamassia, R.: Convex drawings of graphs in two and three dimensions. In: 12th Symp. on Computational Geometry (SoCG), pp. 319–328 (1996)

[2] Hopcroft, J.E., Kahn, P.J.: A paradigm for robust geometric algorithms. Algorithmica 7, 339–380 (1992)

[3] Tutte, W.T.: How to draw a graph. Proc. London Math. Soc. 3, 743–767 (1963)

[4] Fruchterman, T.M.J., Reingold, E.M.: Graph drawing by force-directed placement. Software: Practice and Experience 21, 1129–1164 (1991)

[5] Kamada, T., Kawai, S.: An algorithm for drawing general undirected graphs. Information Processing Letters 31, 7–15 (1989)

[6] Koren, Y.: Drawing graphs by eigenvectors: theory and practice. Computers & Mathematics with Applications 49, 1867–1888 (2005)

[7] Kruskal, J.B., Seery, J.B.: Designing network diagrams. In: Proc. First General Conf. on Social Graphics, pp. 22–50 (1980)

[8] Koebe, P.: Kontaktprobleme der Konformen Abbildung. Ber. Sächs. Akad. Wiss. Leipzig, Math.-Phys. Kl. 88, 141–164 (1936)

[9] Bajaj, C.: The algebraic complexity of shortest paths in polyhedral spaces. In: Proc. 23rd Allerton Conf. on Communication, Control and Computing, pp. 510–517 (1985)

[10] Carufel, J.L.D., Grimm, C., Maheshwari, A., Owen, M., Smid, M.: A Note on the unsolvability of the weighted region shortest path problem. In: Booklet of Abstracts of the 28th European Workshop on Computational Geometry, pp. 65–68 (2013)

[11] Bajaj, C.: The algebraic degree of geometric optimization problems. Discrete Comput. Geom. 3, 177–191 (1988)

[12] Nister, D., Hartley, R., Stewenius, H.: Using Galois theory to prove structure from motion algorithms are optimal. In: IEEE Conf. Computer Vision & Pattern Recog., pp. 1–8 (2007)

[13] Varfolomeev, V.V.: Galois groups of the Heron–Sabitov polynomials for inscribed pentagons. Mat. Sb. 195, 3–16 (2004); Translation in Sb. Math. 195, 149–162 (2004)

[14] Brightwell, G., Scheinerman, E.: Representations of planar graphs. SIAM J. Discrete Math. 6, 214–229 (1993)
[15] Ben-Or, M.: Lower bounds for algebraic computation trees. In: Proc. 15th Annu. Symp. Theory of Computing, pp. 80–86 (1983)
[16] Yao, A.C.: Lower bounds for algebraic computation trees of functions with finite domains. SIAM J. Comput. 20, 655–668 (1991)
[17] Shoup, V.: A Computational Introduction to Number Theory and Algebra. Cambridge Univ. Press (2009)
[18] Baker, R.C., Harman, G.: Shifted primes without large prime factors. Acta Arith. 83, 331–361 (1998)
[19] Cox, D.A.: Galois Theory. 2nd edn. Pure and Applied Mathematics. Wiley (2012)
[20] Jacobson, N.: Basic Algebra I, 2nd edn. Dover Books on Mathematics. Dover (2012)
[21] Stäckel, P.: Arithmetische Eigenschaften ganzer Funktionen (Fortsetzung.). J. Reine Angew. Math. 148, 101–112 (1918)

Bitonic st-orderings of Biconnected Planar Graphs

Martin Gronemann

Institut für Informatik, Universität zu Köln, Germany
gronemann@informatik.uni-koeln.de

Abstract. Vertex orderings play an important role in the design of graph drawing algorithms. Compared to canonical orderings, st-orderings lack a certain property that is required by many drawing methods. In this paper, we propose a new type of st-ordering for biconnected planar graphs that relates the ordering to the embedding. We describe a linear-time algorithm to obtain such an ordering and demonstrate its capabilities with two applications.

1 Introduction

Being a fundamental part of incremental drawing procedures, various types of orderings have been developed and improved over the years. De Fraysseix, Pach and Pollack [3] introduced the canonical ordering to create straight-line drawings of maximal planar graphs. Afterwards, Kant [10] extended this concept to the triconnected case. Obtainable in linear-time, both have been used in the graph drawing literature extensively. A few attempts have been made to generalize them to the biconnected case by relaxing their properties [7,9]. However, an alternative that in nature works for biconnected graphs and that can be computed in linear time, are st-orderings [6]. In the field of graph drawing, they have been used in several methods, reaching from the construction of visibility representations to drawings of non-planar graphs, see e.g. [4,12]. Although canonical and st-orderings share some properties in the planar case, it seems that they are usually not used in the same context.

In the following, we investigate these differences in more detail, especially one property of canonical orderings that is used implicitly in many drawing algorithms. Consider the successors of a single vertex in the clockwise ordering as implied by the embedding. Then their ranks in the canonical ordering form an increasing and then decreasing sequence, i.e., a *bitonic* sequence. Common st-orderings do not necessarily have this property, rendering them unsuitable for some applications.

We counteract by introducing a new type of st-ordering for biconnected planar graphs: the *bitonic st-ordering*, an st-ordering in which the successors of every vertex appear in the aforementioned pattern. We show that every biconnected planar graph admits such an ordering. The proof is constructive and yields a linear-time algorithm that computes the ordering and a corresponding embedding. For the case where a fixed embedding is given, we prove that one cannot always find a bitonic st-order. In order to further support our idea, we briefly describe two applications. In the first one, we extend the straight-line algorithm of de Fraysseix, Pach and Pollack [3] to bitonic st-orderings. In the second one, we describe how to obtain a special visibility representation and then transform it into a rectilinear T-shaped polygon contact representation.

C. Duncan and A. Symvonis (Eds.): GD 2014, LNCS 8871, pp. 162–173, 2014.

2   Preliminaries

In the following, we first introduce some notations and definitions that are used throughout this work. If not stated otherwise, we consider only simple, planar biconnected graphs. One exception is the following definition of *st*-orderings that does not require planarity.

Definition 1. *Let* $G = (V, E)$ *be a biconnected graph with* $s, t \in V$, $s \neq t$ *and let* $\pi : V \rightarrow \{1, \ldots, |V|\}$ *be the rank of the vertices in an ordering* $s = v_1, v_2, \ldots, v_n = t$, *i.e.,* $\pi(v_i) = i$ *with* $1 \leq i \leq n$. π *is called an st-ordering, if for all vertices* $v \in V$ *with* $1 < \pi(v) < n$, *there exists* $(u, v), (v, w) \in E$ *with* $\pi(u) < \pi(v) < \pi(w)$.

From now on we assume that a graph is planar and a corresponding combinatorial embedding is given. In that case an *st*-ordering π of G has a nice property which has been used in the graph drawing literature extensively [4]: When considering the circular order induced by the embedding, the set of predecessors and successors form a consecutive sequence in the circular order of the embedding at a vertex. We denote this ordered sequence of successors of a vertex v by $S(v) = \{w_1, \ldots, w_m\}$ such that for $1 \leq i < m$, w_i precedes w_{i+1} in the circular clockwise order around v and $\pi(v) < \pi(w_i)$ holds for all $1 \leq i \leq m$. This property is particularly useful in an incremental drawing procedure. However, one has no control over which successor is placed when.

Consider a simple example where a vertex v has been placed that has three successors, let us say $S(v) = \{w_1, w_2, w_3\}$. Then, π may be chosen such that w_2 must be placed before w_1 and w_3, i.e., $\pi(w_2) < \pi(w_1)$ and $\pi(w_2) < \pi(w_3)$. This may cause problems when attaching the edges (v, w_1) and (v, w_3), since (v, w_2) has already been attached. This lack of control is avoided by the canonical ordering that is limited to triconnected planar graphs:

Definition 2 ([10]). *Let* $G = (V, E)$ *be a triconnected plane graph and* (v_1, v_2) *an edge on the outer face. Let* $V_1 \cup \cdots \cup V_K$ *be an ordered partition of* V *and* G_k $(1 \leq k \leq K)$ *the subgraph induced by* $V_1 \cup \cdots \cup V_k$ *with outer face* C_k. $V_1 \cup \cdots \cup V_K$ *is a canonical ordering of* G *if:*

- $V_1 = \{v_1, v_2\}$ *and* $V_K = \{v_n\}$, *where* v_n *lies on the outer face and is adjacent to* v_1.
- *Each* C_k $(k > 1)$ *is a cycle containing* (v_1, v_2).
- *Each* G_k *is biconnected and internally triconnected.*
- *For* $1 < k < K$ *one of the two following conditions holds:*
  1. $V_k = \{z\}$ *is a singleton where* z *belongs to* C_k *and has at least one neighbor in* $G - G_k$.
  2. $V_k = \{z_1, \ldots, z_m\}$ *where each* z_i $(1 \leq i \leq m)$ *has at least one neighbor in* $G - G_k$, *and where* z_1 *and* z_m *each have one neighbor in* C_{k-1}, *and these are the only two neighbors of* V_k *in* G_{k-1}.

Clearly, a situation as in the small example cannot occur with canonical orderings, because of the biconnectivity of G_k. In fact one can go one step further and claim (as we did in the introduction) that the partition indices of the successors when considered in the clockwise ordering as implied by the embedding, form an increasing and then

Table 1. Comparison of the features of various orderings.

	biconnected	successor	bitonic
st-ordering	yes	yes	no
biconnected shelling- & canonical ordering	yes	no	yes
canonical ordering	no	yes	yes
bitonic st-ordering	yes	yes	yes

decreasing sequence. We will prove this for canonical orderings as an intermediate step in the main section of this paper. For now we refer to this as the bitonic property.

The concept of canonical ordering has been generalized to the biconnected case. Gutwenger and Mutzel [7] use an ordered partition of the vertices, referred to as *biconnected shelling order*, to create poly-line drawings in an incremental manner. A similar but more vertex ordering-based concept is used by Harel and Sardas [9]. They introduce the so called *biconnected canonical ordering* for drawing planar graphs in a straight-line style. In both definitions, the constraints of the triconnected version have been relaxed. But this generalization sacrifices an important property that is required for some applications. In the triconnected case, every vertex $v \in V_k$, except for $k = K$, has a neighbor in $G - G_k$. We are not aware of any canonical ordering-like approach for the biconnected case, where this is guaranteed. In order to draw a connection to st-orderings, we refer to this property as the successor property. Table 1 summarizes the orderings and their features including our contribution (*bitonic st-ordering*).

Another common technique for the biconnected case that can be found in the literature is to first develop an algorithm using the canonical ordering and is therefore limited to triconnected graphs. Afterwards, the algorithm is extended to the biconnected case using *SPQR-trees*. An SPQR-tree $\mathcal{T}$ reflects the decomposition of a biconnected graph $G = (V, E)$ into its *triconnected components* and their relationships [5,8]. In fact, every triconnected component $G_\mu = (V_\mu, E_\mu)$ is represented by a tree node μ in $\mathcal{T}$ where G_μ itself is called the *skeleton* of μ. The interrelationship between two triconnected components is described by a pair of so called *virtual edges*. Both virtual edges share the same endpoints that correspond to a *split pair* $\{s, t\}$. A split pair $\{s, t\}$ is either a pair of adjacent nodes in G or a *separation pair*, i.e., the removal of $\{s, t\}$ disconnects G. Every G_μ can be categorized to be one of four types based on its structure. A bundle of at least three parallel edges is referred to as *P-node*. In case G_μ is a simple cycle of length at least three, it classifies as an *S-node*, whereas if the skeleton is a simple triconnected graph, we call it an *R-node*. The leaves of $\mathcal{T}$ are formed by *Q-nodes* that are bundles of two edges, one being a virtual edge while the other corresponds to an edge of G. Usually it is convenient to root $\mathcal{T}$, hence, inducing a hierarchy on the triconnected components. Except for the root, every skeleton G_μ contains then a virtual edge $(s, t) \in E_\mu$ that represents a link to μ's parent. We refer to (s, t) as the *reference edge* of μ and to its endpoints $\{s, t\}$ as the *poles* of μ. When considering a node μ in a rooted SPQR-tree $\mathcal{T}$, μ induces a subgraph of G referred to as the *pertinent graph* of μ.

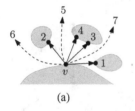   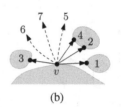

(a)   (b)

Fig. 1. (a) Example in which seven successors of a vertex v are placed in a non-bitonic manner. The last three edges to be attached to v (dashed) are separated by previously attached ones (solid). In (b), when using a bitonic ordering, they appear consecutively in the embedding around v.

The main task, when extending a triconnected drawing procedure to a biconnected one using SPQR-trees, can be sketched as follows. The original algorithm serves as a basis for the case in which μ is an R-node. It is then modified such that each (virtual) edge in the drawing can be replaced recursively by a drawing of the corresponding pertinent graph. Usually a drawing has to match certain invariant properties. For S- and P-nodes alternative methods are used. Finding a good invariant and presenting a clear proof can be tedious work and its complexity may outweigh the description of the original triconnected algorithm. We offer a different approach by defining a new type of st-ordering whose successor lists have the aforementioned property of being bitonic.

3   The Bitonic st-ordering

A sequence is said to be *bitonic*, if it can be partitioned into two subsequences such that one is monotonically increasing while the other is decreasing. More specifically:

Definition 3. *An ordered sequence $A = \{a_1, \ldots, a_n\}$ is **bitonic increasing**, if there exists $1 \le k \le n$ such that $a_1 \le \cdots \le a_k \ge \cdots \ge a_n$ holds and **bitonic decreasing** if $a_1 \ge \cdots \ge a_k \le \cdots \le a_n$. Moreover, we say A is bitonic increasing (decreasing) with respect to a function f if $A' = \{f(a_1), \ldots, f(a_n)\}$ is bitonic increasing (decreasing).*

One property of bitonic sequences that is very useful in our context is the following:

Property 1. If a sequence $A = \{a_1, \ldots, a_n\}$ is bitonic increasing (decreasing), then the reversed sequence $A' = \{a_n, \ldots, a_1\}$ is bitonic increasing (decreasing) as well.

In the following, we restrict ourselves to bitonic increasing sequences. Thus, we abbreviate it by just referring to it as being bitonic.

Definition 4. *Let $G = (V, E)$ be a biconnected planar graph with a fixed embedding and $(s, t) \in E$. We say an st-ordering π is a **bitonic st-ordering**, if at every vertex $v \in V$ the ordered sequence of successors $S(v) = \{w_1, \ldots, w_m\}$ as implied by the embedding is bitonic with respect to π.*

An ordering with this additional property is particularly useful in an incremental algorithm; the edges that correspond to those successors of a vertex v that have not been placed yet, appear consecutively in the embedding around v. See Figure 1 for an example. Next, we describe how to obtain such a bitonic st-ordering.

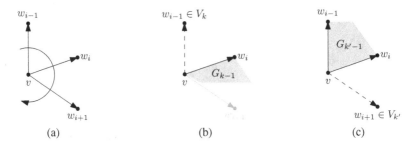

(a)   (b)   (c)

Fig. 2. (a) The initial situation at v with $S(v) = \{\ldots, w_{i-1}, w_i, w_{i+1}, \ldots\}$. (b) G_{k-1} with $k = \pi'(w_{i-1})$ where w_{i-1} has to be in the outer face of G_{k-1}. (c) $G_{k'-1}$ with $k < k' = \pi'(w_{i-1})$ where w_{i+1} has to be in the outer face of $G_{k'-1}$.

Lemma 1. *Every triconnected planar graph $G = (V, E)$ admits a bitonic st-ordering for every $(s, t) \in E$.*

Proof. From its definition it is easy to see that a canonical ordering $V_1 \cup \cdots \cup V_K$ can be transformed into an st-ordering π. We start by describing the construction of π and then show that it is indeed bitonic with respect to π. Given an edge $(s, t) \in E$, we compute a canonical ordering $V_1 \cup \cdots \cup V_K$ of G by choosing $V_1 = \{s, s'\}$ and $V_K = \{t\}$ with s' being the vertex that precedes t in the clockwise order around s. Notice that by definition of the canonical ordering, the edges (s, t) and (s, s') are on the outer face. For the st-ordering π we follow a simple principle that is sometimes referred to as the vertex ordering of a canonical ordering: Regardless of $V_k = \{v_1, \ldots, v_m\}$ with $1 \leq k \leq K$ being a chain or singleton, we choose π for $1 \leq i \leq m$ such that $\pi(v_i) = |V_1| + \cdots + |V_{k-1}| + i$.

For the sake of notation we may refer to the partition of a vertex $v \in V_k$ with $\pi'(v) = k$. Notice that by construction of π for all $u, v \in V$ with $\pi'(u) < \pi'(v)$, it holds that $\pi(u) < \pi(v)$. By definition of the canonical ordering, every $v \in V_k$ with $k < K$ has at least one neighbor w in $V_{k+1} \cup \cdots \cup V_K$. It holds then that $\pi(w) > \pi(v)$ and as a result every $v \neq t$ has at least one successor. In case $V_k = \{v\}$ ($1 < k \leq K$) is a singleton, v has at least two neighbors, say c_l and c_r, in $V_1 \cup \cdots \cup V_{k-1}$ with $\pi(c_l) < \pi(v)$ and $\pi(c_r) < \pi(v)$, thus v has at least two predecessors. The other case, i.e., $V_k = \{v_1, \ldots, v_m\}$ ($k > 1$) is a chain, only v_1 and v_m have one neighbor each, let us say c_l and c_r, in $V_1 \cup \cdots \cup V_{k-1}$. However, for every $v_i \in V_k$ with $i > 1$ it holds that $\pi(v_{i-1}) < \pi(v_i)$. Hence, every v_i with $i < m$ has exactly one predecessor while v_m has even two. Special attention must be paid to $V_1 = \{s, s'\}$ since for this chain no c_l and c_r exist. However, the predecessor of s' is s and s itself does not require a predecessor for π being an st-ordering. Since all vertices $v \neq s$ have predecessors the order in $S(v)$ is well-defined by considering them clockwise. For s we have to break the cyclic order and set $S(s) = \{t = w_1, w_2, \ldots, w_{m-1}, w_m = s'\}$.

In order to prove that π is a bitonic st-ordering, we first show that every successor list obtained from π is bitonic with respect to π' instead of π. To do so, assume to the contrary that there exists a successor list $S(v) = \{w_1, \ldots, w_i, \ldots, w_m\}$ of some vertex v that is not bitonic with respect to π', i.e., there is a $w_i \in S(v)$ with $1 <$

$i < m$ for which $\pi'(w_{i-1}) > \pi'(w_i)$ and $\pi'(w_{i+1}) > \pi'(w_i)$ holds. Furthermore, let w.l.o.g. $\pi'(w_{i-1}) < \pi'(w_{i+1})$. Notice that by construction of π and $S(v)$, it follows that $\pi'(w_{i-1}) \neq \pi'(w_{i+1})$. See Figure 2a for the initial situation at v. Now we set $k = \pi'(w_{i-1})$ and $k' = \pi'(w_{i+1})$ and argue that in a canonical ordering this can only occur for $k = 2$. By definition of the canonical ordering, $w_{i-1} \in V_k$ has to be in the outer face of G_{k-1} as displayed in Figure 2b. Similarly, $w_{i+1} \in V_{k'}$ has to be in the outer face of $G_{k'-1}$ (see Figure 2c). As a result, the outer face of G_{k-1} must be on both sides of the edge (v, w_i) and there is only one such G_{k-1} for which this is the case, namely G_1. Hence, $k = 2$, $v = s$, $w_i = s'$ and $w_{i+1} = t$. However, we defined $S(s)$ such that it ends with $w_m = s'$ which is a contradiction.

It remains to show that all $S(v)$ are not only bitonic with respect to π', but also for π. As aforementioned, by construction of π from π', for two vertices $u, v \in V$ with $\pi'(u) < \pi'(v)$ it follows that $\pi(u) < \pi(v)$. And since we have just shown for the successor list $S(v) = \{w_1, \ldots, w_i, \ldots, w_m\}$ of every vertex $v \in V$ it holds that $\pi'(w_{i-1}) < \pi'(w_i)$ or $\pi'(w_{i+1}) < \pi'(w_i)$, we may deduce that $\pi(w_{i-1}) < \pi(w_i)$ or $\pi(w_{i+1}) < \pi(w_i)$. Hence, every $S(v)$ is bitonic with respect to π.   □

The proof is constructive and reveals one additional property: The successor list of s is a special case, because it contains s' and t. Furthermore, s is the only vertex with $\pi(s) < \pi(s')$ and for every vertex $v \in V$ with $v \neq t$, $\pi(v) < \pi(t)$ holds. Since the successor list of s starts with t, ends with s' by our construction, and is bitonic with respect to π, we can state the following:

Corollary 1. *The successor list of s starts with t, ends with s' and is sorted decreasingly with respect to π, i.e., $S(s) = \{t, w_2, \ldots, w_{m-1}, s'\}$ such that $\pi(t) > \pi(w_2) > \cdots > \pi(w_{m-1}) > \pi(s')$.*

While the above results follow the intuition of canonical orderings, they hold only for the case where the input is triconnected. Next, we extend this result to the biconnected case using SPQR-trees. Corollary 1 provides us with the necessary ingredient for an invariant. More details are given in the proof of the main result of this section:

Theorem 1. *Every biconnected planar graph $G = (V, E)$ has a bitonic st-ordering π for any given st-edge $e^* \in E$. The ordering π and a corresponding embedding can be computed in time $\mathcal{O}(|V|)$.*

Proof. The overall challenge is to recursively compose a bitonic *st*-ordering along an SPQR-tree. For a subtree, we assume that we have already constructed a bitonic *st*-ordering that complies with an invariant. Then we show that we can combine it in the skeleton of the parent node with the solutions of other subtrees.

Invariant: For the assignment of an index in π, we maintain a single global counter that we use to label the vertices in an incremental manner. The poles $\{s, t\}$ of a tree node μ are labeled by the parent. Moreover, s has already been labeled such that we may assume that the global counter has a value greater than $\pi(s)$. Furthermore, π is a bitonic *st*-ordering for the subgraph induced by μ when assigning t the current value of the counter. Additionally, the successor list of s is sorted decreasingly with respect to π. We start by embedding G, creating the SPQR-tree $\mathcal{T}$ and rooting it at the Q-node representing the given *st*-edge $e^* = (s^*, t^*)$. Then we initialize the global counter, label s^*, and recurse on the only child of the root. Following standard practice, we now distinguish the different types of tree nodes.

Serial case: Let the skeleton of the S-node μ be the simple cycle $s, v_1, \ldots, v_{m-1}, t, s$, where (s, t) is the reference edge representing the parent of μ. The remaining edges $(s, v_1), \ldots, (v_{m-1}, t)$ correspond to the children $\mu_1, \ldots, \mu_m$ of μ. We recurse on μ_1, label v_1, recurse on μ_2, and so on, until μ_m. Notice that we do not label t. Clearly, the result is an st-ordering when assigning t the current value of the counter. The successor lists of $s, v_1, \ldots, v_{m-1}$ are all sorted decreasingly due to our invariant, thus, are bitonic.

Parallel case: We first check if one of the children $\mu_1, \ldots, \mu_m$ of the P-node μ is a Q-node. In that case we change the order of the children such that μ_1 is the Q-node. Notice that this implies a change in the embedding of G. Then we recurse on the children in their reverse order, i.e. $\mu_m, \ldots, \mu_1$. Consider now the successor list $S(s)$ of s: The neighbors $w_1^i, \ldots, w_{k'}^i$ with $1 \leq i \leq m$ that are located in the induced subgraph of μ_i form a consecutive sequence in $S(s)$:

$$S(s) = \{\ldots, \underbrace{w_1^i, \ldots, w_k^i}_{\text{neighbors in } \mu_i}, \underbrace{w_1^{i+1}, \ldots, w_{k'}^{i+1}}_{\text{neighbors in } \mu_{i+1}}, \ldots\}$$

By our invariant, it follows that $\pi(w_j^i) > \pi(w_{j+1}^i)$ and since we recursed on $\mu_1, \ldots, \mu_m$ in reverse order, $\pi(w_k^i) > \pi(w_1^{i+1})$ holds. Hence, the sequence is decreasing.

Rigid case: We start by constructing a temporary ordering π' for the triconnected skeleton $G_\mu = (V_\mu, E_\mu)$ of the R-node μ using Lemma 1 and choosing the reference edge (s, t) as input. Then we traverse the vertices of V_μ in the ordering as given by π'. At a vertex $v \in V_\mu$, we recurse on the incident edges $(u, v) \in E_\mu$ with $\pi'(u) < \pi'(v)$, i.e., the incoming edges of v with respect to π'. Afterwards, we label v unless $v = t$. The resulting ordering is not necessarily a bitonic st-ordering. We proceed in two steps: First we derive some useful properties of π and narrow down the problem. Then we argue that mirroring the embedding of some children of μ changes the successor lists such that they become bitonic with respect to π.

Let us take a closer look at the properties of π: Since we labeled all $v \in V_\mu$ in the order as provided by π', for any two vertices $u, v \in V_\mu$ with $u \neq v$, it holds that $\pi'(u) < \pi'(v)$ if and only if $\pi(u) < \pi(v)$. Hence, π is a feasible bitonic st-ordering for G_μ. Recall that we recursed on the children in a special way. Consider a vertex v' in the induced subgraph of a child μ_{uv} represented by the virtual edge $(u, v) \in E_\mu$ with $\pi(u) < \pi(v)$. Furthermore, assume that v' is not a pole of μ_{uv}, i.e., $u \neq v' \neq v$. Then v' has been labeled before v and after any $w \in V_\mu$ with $\pi(w) < \pi(v)$, thus $\pi(w) < \pi(v') < \pi(v)$. When now considering a fourth vertex, say w', that is defined similar as v', i.e., a non-pole vertex located in the subgraph induced by a virtual edge $(x, w) \in E_\mu$ with $\pi(x) < \pi(w)$, then we may deduce the implication $\pi(w) < \pi(v) \Rightarrow \pi(w') < \pi(v')$. Stemming from the special traversal of the edges, this property is of particular interest when considering the successor lists.

Let $S'(v) = \{w_1', \ldots, w_h', \ldots, w_m'\} \subset V_\mu$ be the successor list of $v \in V_\mu$. See Figure 3a for an example. Notice that $\pi(w_1') < \cdots < \pi(w_h') > \cdots > \pi(w_m')$ holds. Furthermore, let $\mu_1, \ldots, \mu_m$ be the corresponding children of μ that are represented by the virtual edges $(v, w_1'), \ldots, (v, w_m')$ with $\pi(v) < \pi(w_i')$ for $1 \leq i \leq m$. Similar

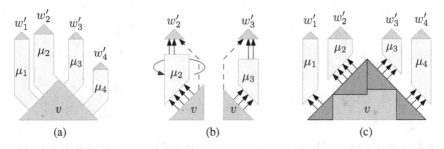

Fig. 3. (a) Example of virtual edges $(v, w_1'), \ldots, (v, w_4')$ in an R-node representing the tree nodes $\mu_1, \ldots, \mu_4$. (b) Mirroring the embedding of the subgraph induced by μ_2 turning the decreasing sequence into an increasing sequence. (c) The bitonic successor list at v after mirroring the embedding of μ_1 and μ_2.

to the P-node case, we refer to the neighbors of v that are contained in the subgraph induced by μ_i as $w_1^i, \ldots, w_{k_i}^i$. These form a consecutive sequence in $S(v)$, hence, we may write $S(v)$ as

$$S(v) = \{\underbrace{w_1^1, \ldots, w_{k_1}^1}_{\text{neighbors in } \mu_1}, \ldots, \underbrace{w_1^h, \ldots, w_{k_h}^h}_{\text{neighbors in } \mu_h}, \ldots, \underbrace{w_1^m, \ldots, w_{k_m}^m}_{\text{neighbors in } \mu_m}\}.$$

The idea now is to distinguish between two cases, depending on if either $i < h$ or $i \geq h$ holds, i.e., w_i' is in either the increasing or decreasing partition of $S'(v)$.

Let us first consider the case $h \leq i$: Since $\pi(w_i') > \pi(w_{i+1}')$ for $h \leq i < m$, it follows that $\pi(w_{k_i}^i) > \pi(w_1^{i+1})$ for all $h \leq i < m$, i.e., the last neighbor in the subgraph induced by μ_i has a greater label than the one in μ_{i+1}. By our invariant we may assume that $\pi(w_1^i) > \cdots > \pi(w_{k_i}^i)$ for all $h \leq i \leq m$ holds, i.e., with respect to π, we have a decreasing subsequence in $S(v)$. Hence, the sequence $w_1^h, \ldots, w_{k_m}^m$ is decreasing with respect to π.

In the second case where $1 \leq i < h$ holds, an increasing sequence is required. We mirror the embedding of every subgraph induced by μ_i with $1 \leq i < h$ along its poles (v, w_i'). As a result the decreasing subsequences in $S(v)$ turn into increasing ones, i.e., $\pi(w_1^i) < \cdots < \pi(w_{k_i}^i)$ for all $1 \leq i < h$ (μ_2 in Figure 3b). Notice that by Property 1 the successor list of every vertex in the mirrored subgraph remains bitonic. Now similar to the first case, we argue that from $\pi(w_i') < \pi(w_{i+1}')$ it follows that $\pi(w_{k_i}^i) < \pi(w_1^{i+1})$ for all $1 \leq i < h$. Thus, the sequence $w_1^1, \ldots, w_{k_{h-1}}^{h-1}$ is increasing with respect to π. And as a result, the sequence $w_1^1, \ldots, w_{k_{h-1}}^{h-1}, w_1^h, \ldots, w_{k_m}^m$ is bitonic with respect to π (Figure 3c). Notice that for $v = s$, there exists no i with $\pi(w_i') < \pi(w_{i+1}')$, thus, $S(s)$ is sorted decreasingly with respect to π as required by the invariant.

The case where μ is a Q-node is trivial. Both, the canonical ordering and the SPQR-tree, can be computed in linear time, thus, the runtime follows immediately.   □

In the proof of the main theorem, we changed the embedding of G in two places. At first in the P-node case, we had to ensure that a possible Q-node follows the reference edge in clockwise order around s. Afterwards in the R-node case, we mirrored the embedding

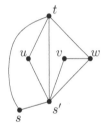

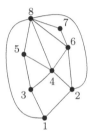

Fig. 4. A graph for which no bitonic st-ordering exists for the given embedding

Fig. 5. Running example with its bitonic st-ordering and corresponding embedding

along the poles to turn a decreasing sequence into an increasing one. The latter change is caused by our invariant that only provides a decreasing sequence at s for the sake of an easier maintainable invariant. In an actual implementation, this can easily be avoided by mirroring the embedding twice, once before recursing on the corresponding child and then afterwards. Thus, the resulting embedding is equivalent to the initial one. However, for the P-node case it is not trivial and the question may arise if it is necessary in general, or if one may always find a bitonic st-ordering for every edge when a fixed embedding is given. To answer this question, we give a small counterexample.

Lemma 2. *Given a fixed embedding, there exist biconnected planar graphs that do not admit a bitonic st-ordering for every edge.*

Proof. Consider the graph in Figure 4 and its embedding. The triangle consisting of s',t and w is attached to the source s via s'. Clearly, in any feasible st-ordering $\pi(u) < \pi(t)$ and $\pi(v) < \pi(w) < \pi(t)$ must hold. Thus, the successor list $S(s') = \{u, t, v, w\}$ of s' as implied by the illustrated embedding is not bitonic with respect to π, because it follows that $\pi(u) < \pi(t) > \pi(v) < \pi(w)$, which is neither bitonic increasing nor decreasing.   □

Although this is a drawback, it is worth mentioning that in many approaches that employ SPQR-trees for drawing purposes, implicit changes to the embedding are made anyway.

4   Applications

In the following, we present two simple applications of bitonic st-orderings. The results are not new, but we believe that the bitonic st-ordering simplify things. By its nature, it works out of the box for biconnected planar graphs and therefore no augmentation of the input is required. For both applications, we assume that a biconnected planar graph $G = (V, E)$ with a bitonic st-ordering π and the corresponding embedding is given. The graph, its embedding and ordering displayed in Figure 5 serves as a running example.

We start with a classic problem: Straight-line drawings of biconnected graphs by borrowing some ideas from Harel and Sardas [9]. They first describe an algorithm to obtain a biconnected canonical ordering. Then a modification of the classic algorithm of de Fraysseix, Pach and Pollack [3] is used to obtain a planar straight-line layout. We

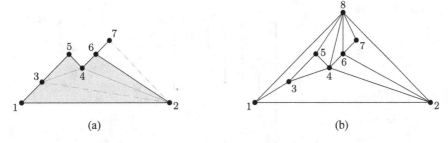

Fig. 6. (a) Adding v_7 which has only one predecessor but right support. (b) The final straight-line drawing produced by the modified de Fraysseix-Pach-Pollack algorithm.

only outline the approach here: during every step k, the algorithm maintains a straight-line drawing for the already placed vertices, $v_1, \ldots, v_{k-1}$ of the biconnected canonical ordering. Similar to the original algorithm, they maintain for the contour of the outer face of G_{k-1} the property that it consists only of segments with slopes $+1$ or -1. Adding a new vertex v_k with leftmost neighbor c_l and rightmost neighbor c_r in G_{k-1} results in a stretch of the drawing such that the edges (v_k, c_l) and (v_k, c_r) have slope $+1$ and -1, respectively. Of course, this works only for $c_l \neq c_r$. In the other case, where $c_l = c_r$ holds, i.e., v_k has only one predecessor, say $u = c_l = c_r$, one has to decide if v_k is placed to the right or to the left of u. Harel and Sardas [9] introduce for those vertices the property of having *left* or *right support*. Their ordering guarantees that either the successor or predecessor of v_k in the clockwise ordering around u has already been placed. Since π has by definition the same property, we may proceed similar. Avoiding sub cases, we always try to place v_k to the left, i.e., choosing a new c_l such that c_l is the predecessor of $u = c_r$ on the contour of G_{k-1}. However, in the case where there exists a w that precedes v_k in $S(u)$ and for which $\pi(w) > \pi(v_k)$ holds, we have to place v_k to the right by choosing $c_l = u$ and c_r to be the successor of u on the contour. Figure 6b shows an example generated by our implementation. Notice that in difference to the ordering as proposed in [9], in an st-ordering every vertex except of t has a successor, hence the faces of the drawing are y-monotone.

Next, we turn our attention to the second application: *contact representations* using rectilinear T-shaped polygons. Alam et al. [1] recently used these as an intermediate step to create cartograms. The idea is to represent a planar graph by touching sides of simple interior-disjoint polygons, in this case upside-down oriented T-shaped polygons. Their approach employs *Schnyder realizer* and their close relationship to canonical orderings. For more details see [1]. However, we choose a different approach and consider instead a special *visibility representation* of G. We assume that the reader is familiar with the basics of visibility representations. For an introduction, see e.g. [4]. The common way to obtain such a visibility representation can be summarized as follows: The y-coordinates $y(v)$ of the horizontal segments that represent the vertices $v \in V$ of G are computed by an optimal topological ordering of a planar st-graph induced by an st-ordering. For the x-coordinate $x(e)$ of a vertical segment that represents an edge $e \in E$, the same procedure is repeated but on the dual planar st-graph. We skip the first step and choose π itself for the y-coordinates, i.e., $y(v) = \pi(v)$. As a result every vertex has now its own

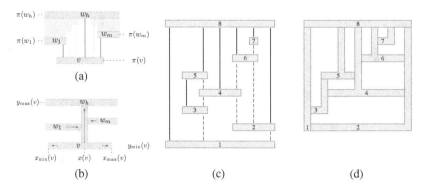

Fig. 7. (a) Successors $w_1, \ldots, w_h, \ldots, w_m$ of v whose ordering in the embedding is bitonic with respect to the y-coordinates. (b) Creating a pole at w_h, i.e., the highest successor, and pulling the bars of the remaining towards it. (c) Visibility representation for the running example. The edges to the highest successor are drawn solid. (d) The resulting T-shaped contact representation.

row that corresponds to its rank in π. See Figure 7c for such a visibility representation for the running example. Although a visibility representation can be derived this way for any st-ordering, we may now benefit from the property that π is a bitonic st-order. Since for every $v \in V$, $S(v)$ is bitonic with respect to π, by construction it is also bitonic with respect to the y-coordinates, i.e., the successors are located above v in an increasing and then a decreasing staircase pattern. See Figure 7a for an illustration.

By using a simple trick, we now transform this wedge-like structure into a rectilinear T-shaped polygon. The idea is straightforward: We create a vertical segment on top of the horizontal bar that reaches all the way up to w_h, i.e., the highest successor of v. Afterwards we pull the bars of the remaining successors towards this pole. See the arrows in Figure 7b for a sketch of the idea. Notice that in case of a non-bitonic st-ordering, a single pole is not sufficient. More specifically, let $x_{\min}(v)$ ($x_{\max}(v)$) denote the left (right) border of the upside-down T representing v and $x(v)$ the horizontal offset of the pole. Furthermore, let $y_{\min}(v)$ and $y_{\max}(v)$ denote the vertical offset of the horizontal bar and the upper border of the pole, respectively. Then, for every $v \in V$ with $S(v) = \{w_1, \ldots, w_h, \ldots, w_m\}$ in which $y(w_1) < \cdots < y(w_h) > \cdots > y(w_m)$ holds, we create the vertical segment by choosing $x(v) = x((v, w_h))$, where $x(v, w_h)$ denotes the x-coordinate of (v, w_h) in the visibility representation. Furthermore, we set $y_{\max}(v) = y_{\min}(w_h)$. For the remaining successors w_i with $1 \leq i < h$, i.e., those located to the left of the pole, we establish contact with the pole from the left by choosing $x_{\max}(w_i) = x(v)$. In a symmetric manner, we set $x_{\min}(w_i) = x(v)$ with $h < i \leq m$ for those successors that are located on the right. Notice that $x_{\min}(v)$ and $x_{\max}(v)$ are only defined in the case where there exists such a pole on both sides. Otherwise, we have to ensure that the horizontal bar of v covers at least the attaching poles from below. Hence, for every u with $v \in S(u)$ and $\pi(v) = \max_{w \in S(u)}\{\pi(w)\}$, i.e., all u for which v is the highest successor, we set $x_{\max}(v) = \max\{x(v), x(u)\}$ and $x_{\min}(v) = \min\{x(v), x(u)\}$. See v_3 and v_5 in Figure 7c. The final contact representation for our running example is shown in Figure 7d.

5   Implementation Details

The presented work has been implemented in C++ using the *Open Graph Drawing Framework (OGDF)* [11]. For the canonical ordering, we implemented the *leftist canonical ordering* algorithm as described by Badent et al. [2]. The linear-time implementation of Gutwenger and Mutzel [8] is used for the SPQR-tree that is required for Theorem 1. It is part of the OGDF, publicly available and provides a convenient interface to navigate the tree and the skeletons.

6   Conclusion

We have shown that every biconnected planar graph has a bitonic st-order that can be obtained in linear time. Moreover, two applications have been presented, both requiring the property of being bitonic. We believe that the bitonic st-ordering is a useful addition to the set of existing tools. Besides having potentially a broad range of applications, it may simplify existing methods considerably.

References

1. Alam, M.J., Biedl, T., Felsner, S., Kaufmann, M., Kobourov, S.G., Ueckerdt, T.: Computing cartograms with optimal complexity. In: Proceedings of the Twenty-Eighth Annual Symposium on Computational Geometry, SoCG 2012, pp. 21–30. ACM (2012)
2. Badent, M., Brandes, U., Cornelsen, S.: More canonical ordering. Journal of Graph Algorithms and Applications 15(1), 97–126 (2011)
3. de Fraysseix, H., Pach, J., Pollack, R.: How to draw a planar graph on a grid. Combinatorica 10(1), 41–51 (1990)
4. Di Battista, G., Eades, P., Tamassia, R., Tollis, I.G.: Graph Drawing: Algorithms for the Visualization of Graphs. Prentice Hall, Englewood Cliffs (1999)
5. Di Battista, G., Tamassia, R.: Incremental planarity testing. In: 30th Annual Symposium on Foundations of Computer Science, pp. 436–441 (1989)
6. Even, S., Tarjan, R.E.: Computing an st-numbering. Theoretical Computer Science 2(3), 339–344 (1976)
7. Gutwenger, C., Mutzel, P.: Planar polyline drawings with good angular resolution. In: Whitesides, S.H. (ed.) GD 1998. LNCS, vol. 1547, pp. 167–182. Springer, Heidelberg (1999)
8. Gutwenger, C., Mutzel, P.: A linear time implementation of SPQR-trees. In: Marks, J. (ed.) GD 2000. LNCS, vol. 1984, pp. 77–90. Springer, Heidelberg (2001)
9. Harel, D., Sardas, M.: An algorithm for straight-line drawing of planar graphs. Algorithmica 20(2), 119–135 (1998)
10. Kant, G.: Drawing planar graphs using the canonical ordering. Algorithmica 16, 4–32 (1996)
11. OGDF - Open Graph Drawing Framework, http://www.ogdf.net/
12. Tamassia, R.: Handbook of Graph Drawing and Visualization (Discrete Mathematics and Its Applications). Chapman & Hall/CRC (2007)

Drawing Outer 1-planar Graphs with Few Slopes*

Emilio Di Giacomo, Giuseppe Liotta, and Fabrizio Montecchiani

Dip. di Ingegneria, Università degli Studi di Perugia, Italy
{emilio.digiacomo,giuseppe.liotta,fabrizio.montecchiani}@unipg.it

Abstract. A graph is outer 1-planar if it admits a drawing where each vertex is on the outer face and each edge is crossed by at most another edge. Outer 1-planar graphs are a superclass of the outerplanar graphs and a subclass of the partial 3-trees. We show that an outer 1-planar graph G of bounded degree Δ admits an outer 1-planar straight-line drawing that uses $O(\Delta)$ different slopes, which extends a previous result by Knauer *et al.* about the planar slope number of outerplanar graphs (CGTA, 2014). We also show that $O(\Delta^2)$ slopes suffice to construct a crossing-free straight-line drawing of G; the best known upper bound on the planar slope number of planar partial 3-trees of bounded degree Δ is $O(\Delta^5)$ and is proved by Jelínek *et al.* (Graphs and Combinatorics, 2013).

1 Introduction

The *slope number* of a graph G is defined as the minimum number of distinct edge slopes required to construct a straight-line drawing of G. Minimizing the number of slopes used in a straight-line graph drawing is a desirable aesthetic requirement and an interesting theoretical problem which has received considerable attention since its first definition by Wade and Chu [21]. Let Δ be the maximum degree of a graph G and let m be the number of edges of G, clearly the slope number of G is at least $\frac{\Delta}{2}$ and at most m.

For non-planar graphs, there exist graphs with $\Delta \geq 5$ whose slope number is unbounded (with respect to Δ) [3,19], while the slope number of graphs with $\Delta = 4$ is unknown, and the slope number of graphs with $\Delta = 3$ is four [18].

Concerning planar graphs, the *planar slope number* of a planar graph G is defined as the minimum number of distinct slopes required by any planar straight-line drawing of G (see, e.g., [9]). Keszegh, Pach and Pálvölgyi [14] prove that $O(2^{O(\Delta)})$ is an upper bound and that $3\Delta - 6$ is a lower bound for the planar graphs of bounded degree Δ. The gap between upper and lower bound has been reduced for special families of planar graphs with bounded degree. Knauer, Micek and Walczak [15] prove that an outerplanar graph of bounded degree $\Delta \geq 4$ admits an outerplanar straight-line drawing that uses at most $\Delta - 1$ distinct edge slopes, and this bound is tight. Jelínek *et al.* [13] prove that the slope number of the planar partial 3-trees of bounded degree Δ is $O(\Delta^5)$, while in [17] it is proved that all partial 2-trees of bounded degree Δ have $O(\Delta)$ slope number. Di Giacomo *et al.* [7] show that planar graphs of bounded degree $\Delta \leq 3$ and at least five vertices have planar slope number four, which is worst case optimal.

---

* Research supported in part by the MIUR project AMANDA "Algorithmics for MAssive and Networked DAta", prot. 2012C4E3KT_001.

C. Duncan and A. Symvonis (Eds.): GD 2014, LNCS 8871, pp. 174–185, 2014.

The research in this paper is motivated by the following observations. The fact that the best known upper bound on the slope number is $O(\Delta^5)$ for planar partial 3-trees while it is $O(\Delta)$ for partial 2-trees suggests to further investigate the planar slope number of those planar graphs whose treewidth is at most three. Also, the fact that non-planar drawings may require a number of slopes that is unbounded in Δ while the planar slope number of planar graphs is bounded in Δ, suggests to study how many slopes may be needed to construct straight-line drawings that are "nearly-planar" in some sense, i.e. where only some types of edge crossing are allowed.

We study *outer 1-planar graphs* that are graphs which admit drawings where each edge is crossed at most once and each vertex is on the boundary of the outer face (see, e.g., [2,5,11]). In 2013, Auer *et al.* [2], and independently Hong *et al.* [11], presented a linear-time algorithm to test outer 1-planarity. Both algorithms produce an outer 1-planar embedding of the graph if it exists. Given an outer 1-planar graph G, we define the *outer 1-planar slope number* of G, as the minimum number of distinct slopes required by any outer 1-planar straight-line drawing of G. We prove the following results.

1. The outer 1-planar slope number of outer 1-planar graphs with maximum degree Δ is at most $6\Delta + 12$ (Section 3). Since outerplanar drawings are a special case of the outer 1-planar drawings, this result extends the above mentioned upper bound on the planar slope number of outerplanar graphs [15].
2. Outer 1-planar drawings are known to be planar graphs and they have treewidth at most three [2]. We study crossing-free straight-line drawings of outer 1-planar graphs of bounded degree Δ and show an $O(\Delta^2)$ upper bound to the planar slope number (Section 4). Hence, for this special family, we are able to reduce the general $O(\Delta^5)$ upper bound [13].

Our results are constructive and give rise to linear-time drawing algorithms. Also, it may be worth recalling that the study of the 1-planar graphs, i.e. those graphs that can be drawn with at most one crossing per edge, has received a lot of interest in the recent graph drawing literature (see, e.g., [1,4,8,10,12,16,20]).

In Section 2 we introduce preliminaries. Section 5 lists some open problems. For reasons of space some proofs are sketched or omitted.

2   Preliminaries and Basic Definitions

A *drawing* Γ of a graph $G = (V,E)$ is a mapping of the vertices in V to points of the plane and of the edges in E to Jordan arcs connecting their corresponding endpoints but not passing through any other vertex. Also, no two edges that share an endpoint cross. Γ is a *straight-line drawing* if every edge is mapped to a straight-line segment. Γ is a *planar drawing* if no edge is crossed; it is a *1-planar drawing* if each edge is crossed at most once. A *planar graph* is a graph that admits a planar drawing; a *1-planar graph* is a graph that admits a 1-planar drawing.

A planar drawing of a graph partitions the plane into topologically connected regions, called *faces*. The unbounded region is called the *outer face*. A *planar embedding* of a planar graph is an equivalence class of planar drawings that define the same set of faces. The concept of planar embedding can be extended to 1-planar drawings as follows. In a

$$\text{(a)} \qquad\qquad \text{(b)} \qquad\qquad \text{(c)} \qquad\qquad \text{(d)}$$

Fig. 1. Illustration of Properties 2– 4. The pertinent graph of: (a) an R-node μ; (b) a P-node μ (case (ii) of Property 3);. (c) a P-node μ that is AOS with respect to s_μ; (d) An S-node ν with a child μ that is AOS with respect to s_μ. Dashed edges cross in the embedding of the graph.

1-planar drawing Γ of a graph G each crossed edge is divided into two *edge fragments*. Also in this case, Γ partitions the plane into topologically connected regions, which we call faces. A 1-*planar embedding* of a 1-planar graph is an equivalence class of 1-planar drawings that define the same set of faces. An *outer 1-planar drawing* is a 1-planar drawing with all vertices on the outer face. An *outer 1-plane graph* G is a graph with a given outer 1-planar embedding.

The slope s of a line ℓ is the angle that an horizontal line needs to be rotated counterclockwise in order to make it overlap with ℓ. The slope of a segment representing an edge in a straight-line drawing is the slope of the supporting line containing the segment.

Our drawing techniques use $SPQR$-trees, whose definition can be found in [6].

Properties of Outer 1-planar Graphs. The structural properties of outer 1-planar graphs have been studied in [2,11]. In this paragraph we derive properties that hold in the fixed outer 1-planar embedding setting and that easily follow from the results in [11]. In Section 4 we will use the same properties explaining how to adapt them to the planar embedding setting. The following property can be found as Lemma 1 in [11].

Property 1. Let G be an outer 1-plane graph. If G is triconnected, then it is isomorphic to K_4 and it has exactly one crossing.

In what follows we consider a biconnected outer 1-plane graph G and its $SPQR$-tree T. Let μ be a node of T, the *pertinent graph* G_μ of μ is the subgraph of G whose $SPQR$-tree (with respect to the reference edge e of μ) is the subtree of T rooted at μ. Notice that the edge e is not part of G_μ. From now on we assume G_μ to be an outer 1-plane graph using the embedding induced from G. We give the following definition [11].

Definition 1. *A node μ of T is one sided with respect to its poles s_μ and t_μ, or simply OS, if the edge (s_μ, t_μ) is on the outer face of G_μ.*

Furthermore, we consider T to be rooted at a Q-node ρ whose (only) child is denoted by ξ. In particular, we choose ρ to be associated with an edge that is not crossed and that belongs to the boundary of the outer face of G. It can be shown that such an edge always exists. This choice implies that ξ is OS by definition. The next property derives from Lemma 5 in [11] and defines the structure of the skeleton of R-nodes, see also Figure 1(a).

Property 2. Let μ be an R-node of T. Then: (i) The skeleton $\sigma(\mu)$ is isomorphic to K_4 and it has one crossing; (ii) The children of μ are all OS; (iii) Two children of μ are Q-nodes whose associated edges cross each other in G_μ.

Observe that if μ is an R-node of T, then it is always OS. In order to handle P-nodes, we first need to define a special kind of S-nodes [11].

Definition 2. *Let μ be an S-node of T. Let η be the unique child of μ having s_μ as a pole, and let η' be the unique child of μ having t_μ as a pole. Node μ has a tail at s_μ (t_μ), if η (η') is a Q-node.*

The next property derives from Lemma 6 in [11], see also Figure 1(b).

Property 3. Let μ be an OS P-node of T. One of the following cases holds: (i) μ has two children one of which is a Q-node and the other one is OS; (ii) μ has two children and none of them is a Q-node. Then both are OS S-nodes, one of them has a tail at s_μ, and the other one has a tail at t_μ. Also, the two edges associated with these two tails cross each other in G; (iii) μ has three children and one of them is a Q-node. For the remaining two children case (ii) applies.

Property 3 is restricted to P-nodes that are OS. However, an internal P-node μ (different from ξ) might not have the edge (s_μ, t_μ) on the outer face of G_μ [11], see also Figure 1(c) for an illustration.

Definition 3. *Let μ be a P-node of T different from ξ. Node μ is almost one sided with respect to s_μ (t_μ), or simply AOS with respect to s_μ (t_μ), if μ has $2 \leq k \leq 4$ children, one of them is an S-node with a tail at s_μ (t_μ), and for the remaining children one of the following cases applies: (i) If $k = 2$, then the other child is OS; (ii) If $k > 2$, all and only the cases in Property 3 can apply for the remaining $k - 1$ children.*

Let μ be AOS with respect to s_μ (t_μ), then, in order to guarantee that the graph is outer 1-planar, the edge associated with the tail at s_μ (t_μ) crosses another edge, represented by a Q-node ψ in T, having t_μ (s_μ) as an end-vertex. This implies that in fact, μ and ψ are two children of an S-node v in T [11] (see also Figure 1(d)). This observation will be used in Section 3 and in the next property, that is derived from Lemma 7 in [11].

Property 4. Let μ be an S-node of T. Let $\eta_1, \eta_2, \ldots, \eta_k$ be the k children of μ in T, such that $t_{\eta_{i-1}} = s_{\eta_i}$, for $i = 2, \ldots, k$. For each $1 \leq i \leq k$, one of the following cases applies: (i) η_i is OS; (ii) η_i is AOS with respect to s_{η_i} and η_{i+1} ($i < k$) is a Q-node; (iii) η_i is AOS with respect to t_{η_i} and η_{i-1} ($i > 1$) is a Q-node.

An immediate observation from these properties is that every node μ of T different from ξ is OS if it is an S- or R-node, while it is either OS or AOS if it is a P-node.

3   The Outer 1-planar Slope Number

In this section we first present an algorithm, called BO1P-DRAWER, that takes as input a biconnected outer 1-plane graph G with maximum degree Δ, and returns a straight-line drawing Γ of G that uses at most 6Δ slopes. This result is then extended to simply connected graphs with a number of slopes equal to $6\Delta + 12$.

A Universal Set of Slopes. We define a universal set of slopes used by algorithm BO1P-DRAWER to draw every biconnected outer 1-plane graph G with maximum degree Δ. Let $\alpha = \frac{\pi}{2\Delta}$ and observe that $0 < \alpha \le \frac{\pi}{6}$ when $\Delta \ge 3$. We call *blue slopes* the set of slopes defined as $b_i = (i-1)\alpha$, for $i = 1, 2, \ldots, 2\Delta$. For each of the 2Δ blue slopes, we also define two *red slopes* as $r_i^- = b_i - \varepsilon$ and $r_i^+ = b_i + \varepsilon$, for $i = 1, 2, \ldots, 2\Delta$, where the value of ε only depends on Δ. The union of the blue and red slopes defines the universal set of slopes $\mathscr{S}_\Delta$ of size 6Δ. We choose ε as follows: $\varepsilon = \alpha - \arctan\left(\frac{\tan(\alpha)}{1 + 2\tan(2\alpha)\tan(\alpha) - 2\tan(\alpha)\tan(\alpha)}\right)$. The reason of this choice will be clarified in the proof of Lemma 3. Clearly, ε depends only on Δ and it is possible to see that it is a positive value.

Algorithm Overview. Algorithm BO1P-DRAWER exploits *SPQR*-trees and the structural properties presented in Section 2. It takes as input a biconnected outer 1-plane graph G with maximum degree Δ and returns a straight-line drawing Γ of G that uses only slopes in $\mathscr{S}_\Delta$. We first construct the *SPQR*-tree T rooted at a Q-node ρ, whose (only) child is denoted by ξ. Moreover, the edge associated with ρ is not crossed and belongs to the boundary of the outer face of G. Then we draw G by visiting T bottom-up, handling ρ and ξ together as a special case. At each step we process an internal node μ of T and compute a drawing Γ_μ of its pertinent graph G_μ by properly combining the already computed drawings of the pertinent graphs of the children of μ. Let s_μ and t_μ be the poles of μ. With a slight overload of notation for the symbol Δ, we denote by $\Delta(s_\mu)$ and $\Delta(t_\mu)$ the degree of s_μ and t_μ in G_μ, respectively. For each drawing Γ_μ we aim at maintaining the following three invariants. **I1.** Γ_μ is outer 1-plane with respect to the embedding of G_μ. **I2.** Γ_μ uses only slopes in $\mathscr{S}_\Delta$. **I3.** Γ_μ is contained in a triangle τ_μ such that s_μ and t_μ are placed at the corners of its base. Also, $\beta_\mu < (\Delta(s_\mu)+1)\alpha$ and $\gamma_\mu < (\Delta(t_\mu)+1)\alpha$, where β_μ and γ_μ are the internal angles of τ_μ at s_μ and t_μ.

We now explain how to compute a drawing Γ_μ of G_μ, by combining the drawings $\Gamma_{\eta_1}, \Gamma_{\eta_2}, \ldots, \Gamma_{\eta_h}$ of the pertinent graphs $G_{\eta_1}, G_{\eta_2}, \ldots, G_{\eta_h}$ of the children $\eta_1, \eta_2, \ldots, \eta_h$ of μ. To this aim, the drawings $\Gamma_{\eta_1}, \Gamma_{\eta_2}, \ldots, \Gamma_{\eta_h}$ are possibly manipulated. First, observe that the triangle τ_{η_j} ($1 \le j \le h$) can be arbitrarily scaled without modifying the slopes used in Γ_{η_j}. Furthermore, due to the symmetric choice of the blue and red slopes, if we rotate τ_{η_j} by an angle $c \cdot \alpha$, with c integer, the resulting drawing maintains invariant *I2*. Namely each blue slope b_i, for $i = 1, 2, \ldots, 2\Delta$, used in τ_{η_j} will be transformed in another blue slope $b_{i+c} = b_i + c \cdot \alpha = (i-1+c)\alpha$, where $i+c$ is considered modulo 2Δ. Similarly, any red slope will be transformed into another red slope. Moreover, let η_1 and η_2 be two children of μ. When we draw G_{η_1} and G_{η_2}, although they may share one or both the poles, we consider each graph to have its own copy of its poles. Then, when computing Γ_μ, we say that we *attach* Γ_{η_1} to Γ_{η_2} if they share either two poles (this is always true when μ is a P-node) or one pole (this may happen when μ is either an S- or R-node), meaning that we may scale, shift and rotate Γ_{η_1} or Γ_{η_2} in such a way that the points representing the shared poles on the drawing coincide.

As observed in Section 2, all the internal nodes of T are OS except for some P-nodes which are AOS. Let μ be any of these P-nodes, we know that μ is one of the children of an S-node, say v, and it shares a pole with a Q-node, denoted by η (also a children of v). We replace μ and η in T with a new node φ, that, for the sake of description, is called an S^*-*node*. Also, the children of μ become children of φ. If μ and η were

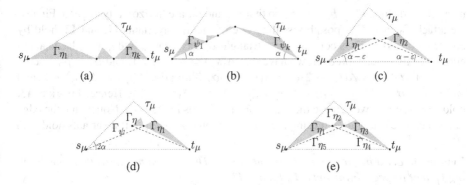

Fig. 2. The drawing of the pertinent graph of: (a) an S-node; (b) a P-node with two children such that one is a Q-node and the other one is an S-node; (c) a P-node with two children such that none of them is a Q-node; (d) an S^*-node; (e) an R-node. Edges drawn with red slopes are dashed.

the only two children of v, then we also replace v with φ. The pertinent graph of φ is $G_\varphi = G_\mu \cup G_\eta$, while the reference edge of φ is (s_μ, t_μ), if μ is AOS with respect to s_μ, or (s_η, t_μ), if μ is AOS with respect to t_μ. It is easy to see that φ is OS. By means of this transformation we can consider only P-nodes that are OS. Similarly we can handle just S-nodes whose children are OS. In what follows we distinguish between S-, P-, S^*-, and R-nodes different from ξ.

Lemma 1. *Let μ be an S-node different from ξ. Then G_μ admits a straight-line drawing Γ_μ that respects Invariants I1., I2. and I3.*

Proof sketch: The drawings of the pertinent graphs of the children $\eta_1, \eta_2, \ldots, \eta_k$ of μ are attached to each other as shown in Figure 2(a). Clearly all invariants hold. $\qquad\square$

Lemma 2. *Let μ be a P-node different from ξ. Then G_μ admits a straight-line drawing Γ_μ that respects Invariants I1., I2. and I3.*

Proof sketch: Recall that, thanks to the definition of S^*-nodes, here we need to only handle only P-nodes that are OS. By Property 3, one of the following cases applies: (i) μ has two children one of which is a Q-node and the other one is OS. (ii) μ has two children and none of them is a Q-node. Then both are OS S-nodes, one of them has a tail at s_μ, and the other one has a tail at t_μ. Also, the two edges associated with these two tails cross each other in G. (iii) μ has three children and one of them is a Q-node. For the remaining two children case (ii) applies.

Case (i) can be easily handled as shown in Figure 2(b). Consider case (ii) and let η_1 be the child of μ that is an S-node with a tail at t_μ, and η_2 be the child of μ that is an S-node with a tail at s_μ. Refer to Figure 2(c). Recall that $s_{\eta_1} = s_{\eta_2} = s_\mu$ and $t_{\eta_1} = t_{\eta_2} = t_\mu$. We modify the drawing Γ_{η_1} as follows. We first rotate Γ_{η_1} so that the segment $\overline{s_{\eta_1} t_{\eta_1}}$ uses the blue slope b_2. Then we redraw the tail of η_1 using the red slope $r_{2\Delta}^+ = b_{2\Delta} + \varepsilon$ and so that s_{η_1} and t_{η_1} are horizontally aligned. Similarly, we modify the drawing Γ_{η_2}. We rotate Γ_{η_2} so that the segment $\overline{s_{\eta_2} t_{\eta_2}}$ uses the blue slope $b_{2\Delta}$ and redraw the tail of η_2

using the red slope $r_2^- = b_2 - \varepsilon$ and so that s_{η_2} and t_{η_2} are horizontally aligned. Finally, we attach Γ_{η_1} and Γ_{η_2} (possibly scaling one of them). Invariants **I1.** and **I2.** hold by construction. Also, Γ_μ is contained in a triangle τ_μ such that s_μ and t_μ are placed at the corners of its base. Moreover, we have that $\Delta(s_\mu) = \Delta(s_{\eta_1}) + 1$, and $\beta_\mu = \beta_{\eta_1} + \alpha <$ $\Delta(s_{\eta_1} + 1)\alpha + \alpha = \Delta(s_{\eta_1} + 2)\alpha = \Delta(s_\mu + 1)\alpha$. Similarly, $\Delta(t_\mu) = \Delta(t_{\eta_2}) + 1$, and $\gamma_\mu = \gamma_{\eta_2} + \alpha < \Delta(t_{\eta_2} + 1)\alpha + \alpha = \Delta(t_{\eta_2} + 2)\alpha = \Delta(t_\mu + 1)\alpha$. Hence, Invariant **I3.** holds. In case (*iii*) we can use the same construction as in case (*ii*). Notice that the edge (s_μ, t_μ) can be safely drawn using the horizontal blue slope b_1. All invariants hold.   □

Lemma 3. *Let μ be an S^*-node different from ξ. Then G_μ admits a straight-line drawing Γ_μ that respects Invariants I1., I2. and I3.*

Proof. Refer to Figure 2(d). Denote by η the child of μ that is an S-node with a tail at either s_μ or t_μ. Suppose that η has a tail at t_μ (the case when the tail is at s_μ is symmetric). Denote by ψ the child of μ that is a Q-node having $t_\psi = s_\eta$ and $s_\psi = s_\mu$ as poles. Finally denote by $\eta_1, \eta_2, \ldots, \eta_k$ the remaining children of μ. Recall that $s_{\eta_1} = s_{\eta_i} = s_{\eta_k}$ and that $t_{\eta_1} = t_{\eta_i} = t_{\eta_k}$. If $k = 1$, we first rotate Γ_{η_1} so that the segment $\overline{s_{\eta_1} t_{\eta_1}}$ uses the blue slope $b_{2\Delta}$. If $k > 1$, we combine the drawings $\Gamma_{\eta_1}, \Gamma_{\eta_2}, \ldots, \Gamma_{\eta_k}$ with the same technique described for P-nodes (recall that indeed they were children of a P-node before the creation of the S^*-node), and, again, we rotate the resulting drawing so that the base of its bounding triangle uses the blue slope $b_{2\Delta}$. Then we attach Γ_η to Γ_{η_1} (after Γ_η has been horizontally flipped). Also, we scale Γ_η so that its tail can be redrawn by using the red slope $r_{2\Delta}^+$ and such that $t_\eta = t_\mu$ coincides with $t_{\eta_1} = t_{\eta_k}$. Finally, we redraw the edge associated with ψ, starting from the point representing $t_\psi = s_\eta$, using the red slope r_2^- and stretch it enough that $s_\psi = s_\mu$ and t_μ are horizontally aligned. See also Figure 2(d) for an illustration. Invariants **I1.** and **I2.** hold by construction. Consider now Invariant **I3.**. By construction Γ_μ is contained in a triangle τ_μ such that s_μ and t_μ are placed at the corners of its base. For the sake of description, in what follow we still denote by Γ_η the drawing of G_η minus the tail of η (i.e., minus an edge), and as τ_η the surrounding triangle of Γ_η. To prove the second part of Invariant **I3.**, we should prove that the line ℓ passing through s_μ with slope $b_3 = 2\alpha$ does not cross the drawing of Γ_η, i.e., is such that Γ_η is placed in the half-plane $\mathcal{H}$ defined by ℓ and containing the segment $\overline{s_\mu t_\mu}$. Denote by δx the horizontal distance between the point where s_μ is drawn and the leftmost endpoint of τ_η. Also, denote by h_η the height of τ_η. Our condition is satisfied if the following inequality holds $\tan(2\alpha)\delta x \geq \tan(\alpha)\delta x + h_\eta$. Let w_η be the length of the base of τ_η, in the worst case (the case that maximizes h_η), we have that $h_\eta = \frac{w_\eta}{2} \frac{1}{\tan(\alpha)}$, which means that the degree of the two vertices placed as endpoints of the base of τ_η is Δ. Moreover, it is possible to see that $w_\eta = \frac{\tan(\alpha)\delta x - \tan(\alpha - \varepsilon)\delta x}{\tan(\alpha - \varepsilon)}$. Substituting w_η in h_η and h_η in the above inequality we have: $\tan(2\alpha) \geq \tan(\alpha) + \frac{\tan(\alpha) - \tan(\alpha - \varepsilon)}{2\tan(\alpha - \varepsilon)\tan(\alpha)}$. With some manipulation we get: $\tan(\alpha - \varepsilon) \geq \frac{\tan(\alpha)}{2\tan(2\alpha)\tan(\alpha) - 2\tan(\alpha)\tan(\alpha) + 1}$. Now, since the tangent function is strictly increasing in $(-\frac{\pi}{2}, \frac{\pi}{2})$, we have: $\varepsilon \leq \alpha - \arctan\left(\frac{\tan(\alpha)}{2\tan(2\alpha)\tan(\alpha) - 2\tan(\alpha)\tan(\alpha) + 1}\right)$. Since the value of ε has been chosen equal to the right-hand side of the above inequality, the inequality holds. Hence, $\beta_\mu < 2\alpha = (\Delta(s_\mu) + 1)\alpha$ (since $\Delta(s_\mu) = 1$). With a symmetric argument one can prove that the line ℓ' passing through t_μ with slope $b_{2\Delta-1} = \frac{(\Delta-1)\pi}{\Delta}$ does not cross the drawing of Γ_η. Since $\Delta(t_\mu) = \Delta(t_{\eta_k}) + 1$, and $\gamma_\mu = \gamma_{\eta_k} + \alpha < (\Delta(t_{\eta_k})+1)\alpha + \alpha = (\Delta(t_{\eta_k})+2)\alpha = (\Delta(t_\mu)+1)\alpha$, Invariant **I3.** holds.   □

Lemma 4. *Let μ be an R-node different from ξ. Then G_μ admits a straight-line drawing Γ_μ that respects Invariants I1., I2. and I3.*

Proof. Refer to Figure 2(e). Recall that, by Property 2, (i) the skeleton $\sigma(\mu)$ is isomorphic to K_4 and it has one crossing; (ii) the children of μ are all OS; (iii) two children of μ are Q-nodes whose associated edges cross each other in G_μ. Hence, denote by η_1, η_2, η_3 the three children of μ whose associated virtual edges lie on the boundary of the outer face of $\sigma(\mu)$ with $s_\mu = s_{\eta_1}$, $t_{\eta_1} = s_{\eta_2}$, $t_{\eta_2} = s_{\eta_3}$, and $t_{\eta_3} = t_\mu$. Also, denote by η_4 and η_5 the two children of μ that are Q-nodes whose associated edges cross each other in G_μ, and so that the poles of η_4 coincides with t_{η_1} and t_{η_3}, while the poles of η_5 coincides with t_{η_2} and s_{η_1}. We rotate Γ_{η_1} in such a way that the segment $\overline{s_{v_1}t_{v_1}}$ uses the blue slope b_2. Similarly, we rotate Γ_{η_3} in such a way that the segment $\overline{s_{\eta_3}t_{\eta_3}}$ uses the blue slope $b_{2\Delta}$. Furthermore, we scale one of the two drawings so that t_{η_1} and s_{η_3} are horizontally aligned. Moreover, we redraw the edge associated with η_4 by using the red slope $r_{2\Delta}^+$ and we redraw the edge associated with η_5 by using the red slope r_2^-. Observe that, attaching η_4 and η_5 to η_1 and η_3, the length of the segment $\overline{t_{\eta_1}s_{\eta_3}}$ is determined. Thus, we attach Γ_{η_2} so that s_{η_2} coincides with t_{η_1} and that t_{η_2} coincides with s_{η_3}.

It is easy to see that Invariant **I1.** and **I2.** are respected by construction. Concerning Invariant **I3.**, again by construction Γ_μ is contained in a triangle τ_μ such that s_μ and t_μ are placed at the corners of its base. Moreover, with the same argument used in the proof of Lemma 3, one can show that $\beta_\mu = \beta_{\eta_1} + \alpha$ and that $\gamma_\mu = \gamma_{\eta_3} + \alpha$. Since $\Delta(s_\mu) = \Delta(\eta_1) + 1$ and $\Delta(t_\mu) = \Delta(\eta_3) + 1$, Invariant **I3.** holds.   □

Lemma 5. *Let ρ be the root of T and let ξ be its unique child. Graph $G = G_\rho \cup G_\xi$ admits a straight-line drawing Γ that respects Invariants I1., I2. and I3.*

Proof sketch: It is possible to prove that at least one edge (s,t) of the outer face of G is not crossed. If we root T at the Q-node associated with (s,t), the root's child ξ is OS and a drawing of $G_\rho \cup G_\xi$ can be computed as in Lemmas 1, 2, 3, and 4.   □

Lemma 6. *Let G be a biconnected outer 1-plane graph with n vertices and with maximum degree Δ. G admits an outer 1-planar straight-line drawing that maintains the given outer 1-planar embedding, and that uses at most 6Δ slopes. Also, this drawing can be computed in $O(n)$ time.*

Proof sketch: By Lemmas 1, 2, 3, 4, and 5, G has an outer 1-planar straight-line drawing that maintains the embedding, with at most 6Δ slopes.   □

A simply connected outerplane graph can be augmented (in linear time) into a biconnected outerplane graph by adding edges so that the maximum degree is increased by at most two. This technique can be directly applied also to outer 1-plane graphs.

Theorem 1. *Let G be an outer 1-plane graph with n vertices and with maximum degree Δ. G admits an outer 1-planar straight-line drawing that maintains the given outer 1-planar embedding, and that uses at most $6\Delta + 12$ slopes. Also, this drawing can be computed in $O(n)$ time.*

4   The Planar Slope Number

In this section we describe an algorithm, called BP-DRAWER, that computes a planar drawing of an outer 1-planar graph G, using at most $4\Delta^2 - 4\Delta$ slopes. This result is then extended to simply connected graphs with a number of slopes equal to $4\Delta^2 + 12\Delta + 8$.

A Universal Set of Slopes. We start by defining a universal set of slopes that are used by algorithm BP-DRAWER. Let $\theta = \frac{\pi}{4\Delta}$ and observe that $0 < \theta \le \frac{\pi}{12}$ when $\Delta \ge 3$. We call *green slopes* the set of slopes defined as $g_i = (i-1)\theta$, for $i = 1, 2, \ldots, 4\Delta$. For each green slope g_i, we define $\Delta - 1$ *yellow slopes* as $y_{i,j} = g_i + \arctan\left(\frac{\tan(g_{4\Delta})\tan(g_3)}{\tan(g_j)}\right)$ with $j = 3\Delta, \ldots, 4\Delta - 2$. The reason of this choice will be clarified in the proof of Lemma 10. The union of the green and yellow slopes defines the universal set of slopes $\mathscr{T}_\Delta$. It is possible to see that $g_i < y_{i,j} < g_{i+1}$, for each $1 \le i < 4\Delta$ and $3\Delta \le j \le 4\Delta - 2$.

Algorithm Overview. Algorithm BP-DRAWER takes as input a biconnected outer 1-plane graph G with maximum degree Δ and returns a planar straight-line drawing Γ of G that uses only slopes in $\mathscr{T}_\Delta$. As in Section 3 we construct the *SPQR*-tree T of G rooted at a Q-node associated with an edge that is not crossed and belongs to the boundary of the outer face of G in the outer 1-planar embedding of G. Then we draw G by visiting T bottom-up. At each internal node μ of T we compute a drawing Γ_μ of G_μ by combining the already computed drawings of the pertinent graphs of the children of μ. For each drawing Γ_μ we maintain the following three invariants: **Ia.** Γ_μ is planar. **Ib.** Γ_μ uses only slopes in $\mathscr{T}_\Delta$. **Ic.** Γ_μ is contained in a triangle τ_μ such that s_μ and t_μ are placed at the corners of its base. Also, $\beta_\mu < (\Delta(s_\mu) - 1)\theta$ and $\gamma_\mu < (\Delta(t_\mu) - 1)\theta$, where β_μ and γ_μ are the internal angles of τ_μ at s_μ and t_μ, respectively.

As in Section 3 the root ρ of T and its unique child ξ will be handled in a special way. Also, in order to construct Γ_μ we may shift, scale and rotate the drawings of the pertinent graphs of the children of μ. We observe that if we rotate τ_μ by an angle $c \cdot \theta$, with c integer, the resulting drawing maintains invariant *Ib*. Namely each green slope g_i, for $i = 1, 2, \ldots, 4\Delta$, used in τ_μ will be transformed in another green slope $g_{i+c} = g_i + c \cdot \theta = (i - 1 + c)\theta$, where $i + c$ is considered modulo 4Δ. Similarly, any yellow slope $y_{i,j}$ will be transformed into another yellow slope $y_{i+c,j}$.

Before describing how the drawing of the pertinent graph of each node μ is obtained by combining the drawing of the pertinent graphs of its children, we observe that the structural properties described in Properties 2, 3, or 4 hold, depending on the type of μ. However, since we want to produce a planar drawing, our algorithm embeds each pertinent graph in a planar way. One of the consequence of this fact is that we no longer need to introduce S^*-nodes; namely, the P-nodes that are *AOS* in the outer 1-planar embedding must be embedded in a planar way and therefore they do not need to be handled in a special way anymore. On the other hand, we need to distinguish between R-nodes whose poles are adjacent in G and R-nodes whose poles are not adjacent in G. For this reason we introduce R^*-nodes. Let μ be an R-node; if the poles s_μ and t_μ of μ are adjacent in G, then the parent ν of μ is a P-node that has (at least) another child η that is a Q-node (the edge associated with η is (s_μ, t_μ)). We replace μ and η in T with a new node φ, that, for the sake of description, is called an R^*-node. Also, the children of μ become children of φ. If μ and η were the only two children of ν, then we also

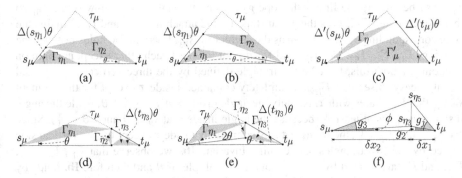

Fig. 3. The planar drawing of the pertinent graph of: (a) a P-node with two children such that none of them is a Q-node; (b) a P-node with three children, one of which is a Q-node; (c) a P-node that is *AOS* in the outer 1-planar embedding of G; (d) an R-node; (e) an R^*-node. (f) Illustration for the proof of Lemma 10.

replace v with φ. The pertinent graph of φ is $G_\varphi = G_\mu \cup G_\eta$, and the reference edge of φ is (s_μ, t_μ). We now explain how the different types of node are handled.

The proof of next lemmas are omitted. An illustration of how Γ_μ is constructed is shown in Figures 2(a) and 3.

Lemma 7. *Let μ be an S-node different from ξ. Then G_μ admits a straight-line drawing Γ_μ that respects Invariants **Ia.**, **Ib.** and **Ic.***

Lemma 8. *Let μ be a P-node different from ξ. Then G_μ admits a straight-line drawing Γ_μ that respects Invariants **Ia.**, **Ib.** and **Ic.***

Lemma 9. *Let μ be an R-node different from ξ. Then G_μ admits a straight-line drawing Γ_μ that respects Invariants **Ia.**, **Ib.** and **Ic.***

Lemma 10. *Let μ be an R^*-node different from ξ. Then G_μ admits a straight-line drawing Γ_μ that respects Invariants **Ia.**, **Ib.** and **Ic.***

Proof. Since μ is an R^*-node, it is obtained by merging an R-node μ' and a Q-node representing the edge $(s_{\mu'}, t_{\mu'})$. By Property 2, the skeleton $\sigma(\mu')$ of μ' is isomorphic to K_4 and two children of μ' are Q-nodes. The two edges corresponding to these Q-nodes do not share an end vertex and each one of them is incident to a distinct pole of μ. Let $\eta_1, \eta_2, \eta_3, \eta_4$, and η_5 be the children of μ'; we assume that η_4 and η_5 are the two Q-nodes. Also, μ has a sixth child η_6 that is a Q-node corresponding to the edge (s_μ, t_μ). We assume that $s_\mu = s_{\eta_1} = s_{\eta_4}$, $t_\mu = t_{\eta_3} = t_{\eta_5}$, $t_{\eta_1} = t_{\eta_2} = s_{\eta_5}$, and $t_{\eta_4} = s_{\eta_2} = s_{\eta_3}$. We construct a drawing of G_μ as follows (see Figure 3(e)). We rotate Γ_{η_3} so that the segment $\overline{s_{\eta_3} t_{\eta_3}}$ uses the green slope $g_{4\Delta}$, and draw the edge associated with η_5 as a segment whose slope is the green slope $(4\Delta - \Delta(t_{\eta_3}))\theta$ and whose length is such that s_{η_5} is vertically aligned with s_{η_3}. We rotate Γ_{η_2} so that the segment $\overline{s_{\eta_2} t_{\eta_2}}$ uses the green slope $g_{2\Delta+1} = \frac{\pi}{2}$. We then attach $\Gamma_{\eta_2}, \Gamma_{\eta_3}$, and Γ_{η_5} (possibly scaling some of them). We draw the edge corresponding to η_6 with the horizontal slope g_1 and stretch it so that

$s_{\eta_6} = s_\mu$ belongs to the line with slope g_2 passing through s_{η_5}. We now rotate Γ_{η_1} so that the segment $\overline{s_{\eta_1} t_{\eta_1}}$ uses the green slope g_2 and attach it to Γ_{η_5} and Γ_{η_6}. Finally, the edge corresponding to η_4 is drawn as the segment $\overline{s_\mu s_{\eta_3}}$. Invariant **Ia.** holds because the drawings $\Gamma_{\eta_1}, \Gamma_{\eta_2}, \Gamma_{\eta_3}, \Gamma_{\eta_4}, \Gamma_{\eta_5}$, and Γ_{η_6} do not intersect each other except at common endpoints. About this, let τ be the triangle defined by the three vertices s_μ, s_{η_3}, and s_{η_5}; it is easy to see that Γ_{η_2} is completely contained inside τ except for the segment $\overline{s_{\eta_3} s_{\eta_5}}$ that Γ_{η_2} shares with τ. Namely the angle inside τ at s_{η_3} is $\frac{\pi}{2} + \theta$, while the angle inside τ at s_{η_5} is at least $\frac{\pi}{4}$ (because the angle inside τ at s_μ is θ and $2\theta < \frac{\pi}{4}$). Since $\beta_{\eta_2} < \frac{\pi}{4}$ and $\gamma_{\eta_2} < \frac{\pi}{4}$, the triangle τ_{η_2} is completely inside τ except for the vertical side shared by the two triangles. Concerning Invariant **Ib.**, we observe that $\Gamma_{\eta_1}, \Gamma_{\eta_2}, \Gamma_{\eta_3}$, Γ_{η_4}, and Γ_{η_5} are rotated by an angle that is a multiple of θ and therefore **Ib.** holds by construction for each of them. We now show that the slope ϕ of the edge corresponding to η_4 is in fact either a green slope or a yellow one (refer to Figure 3(f)). Let δx_1 be the horizontal distance between s_{η_3} and t_μ and let δx_2 be the horizontal distance between s_μ and s_{η_3}. By simple trigonometry we have $\delta x_1 \tan(g_{4\Delta}) = \delta x_2 \tan(\phi)$ and $\delta x_1 \tan(g_j) = \delta x_2 \tan(g_3)$, where g_j is the slope of the segment representing the edge corresponding to η_5 (and therefore $j = 4\Delta - \Delta(t_{\eta_3})$). From the two previous equations we obtain $\tan(\phi) = \frac{\tan(g_{4\Delta})\tan(g_3)}{\tan(g_j)}$. Notice that $1 \le \Delta(t_{\eta_3}) \le \Delta$ and therefore $3\Delta \le j \le 4\Delta - 1$. If $j = 4\Delta - 1$, then $\tan(g_3) = -\tan(g_j)$ and $\tan(\phi) = -\tan(g_{4\Delta}) = \tan(g_2)$, hence $\phi = g_2$, i.e., ϕ is a green slope. Otherwise $\phi = \arctan\left(\frac{\tan(g_{4\Delta})\tan(g_3)}{\tan(g_j)}\right) = $ and therefore ϕ is the yellow slope $y_{1,j}$ (recall that $g_1 = 0$). Concerning Invariant **Ic.**, we have that $\Delta(s_\mu) = \Delta(s_{\eta_1}) + 2$ and $\Delta(t_\mu) = \Delta(t_{\eta_3}) + 2$. Moreover, $\beta_\mu = \beta_{\eta_1} + 2\theta \le (\Delta(s_{\eta_1}) - 1)\theta + 2\theta = (\Delta(s_\mu) - 1)\theta$. Finally, $\gamma_\mu = \gamma_{\eta_3} + 2\theta \le (\Delta(t_{\eta_3}) - 1)\theta + 2\theta = (\Delta(t_\mu) - 1)\theta$.   □

Lemma 11. *Let ρ be the root of T and let ξ be its unique child. Graph $G = G_\rho \cup G_\xi$ admits a straight-line drawing Γ that respects Invariants **Ia.**, **Ib.** and **Ic.***

By Lemmas 7, 8, 9, 10, and 11, we can prove the following lemma.

Lemma 12. *Let G be a biconnected outer 1-plane graph with n vertices and with maximum degree Δ. G admits a planar straight-line drawing that uses at most $4\Delta^2 - 4\Delta$ slopes. Also, this drawing can be computed in $O(n)$ time.*

The result above can be extended to simply connected outer 1-planar graph with the same technique described in Section 3. We obtain the following theorem.

Theorem 2. *Let G be an outer 1-plane graph with n vertices and with maximum degree Δ. G admits a planar straight-line drawing that uses at most $4\Delta^2 + 12\Delta + 8$ slopes. Also, this drawing can be computed in $O(n)$ time.*

5 Open Problems

An interesting open problem motivated by our result of Section 3 is whether the 1-*planar slope number* of 1-planar straight-line drawable graphs (not all 1-planar graphs admit a 1-planar straight-line drawing [12]), is bounded in Δ or not. A second problem is whether the quadratic upper bound of Section 4 is tight or not. Finally, it could be interesting to further explore trade-offs between slopes and crossings, e.g., can we draw planar partial 3-trees with $o(\Delta^5)$ slopes and a constant number of crossings per edge?

References

1. Alam, M.J., Brandenburg, F.J., Kobourov, S.G.: Straight-line grid drawings of 3-connected 1-planar graphs. In: Wismath, S., Wolff, A. (eds.) GD 2013. LNCS, vol. 8242, pp. 83–94. Springer, Heidelberg (2013)
2. Auer, C., Bachmaier, C., Brandenburg, F.J., Gleißner, A., Hanauer, K., Neuwirth, D., Reislhuber, J.: Recognizing outer 1-planar graphs in linear time. In: Wismath, S., Wolff, A. (eds.) GD 2013. LNCS, vol. 8242, pp. 107–118. Springer, Heidelberg (2013)
3. Barát, J., Matousek, J., Wood, D.R.: Bounded-degree graphs have arbitrarily large geometric thickness. The Electronic Journal of Combinatorics 13(1) (2006)
4. Brandenburg, F.-J., Eppstein, D., Gleißner, A., Goodrich, M.T., Hanauer, K., Reislhuber, J.: On the density of maximal 1-planar graphs. In: Didimo, W., Patrignani, M. (eds.) GD 2012. LNCS, vol. 7704, pp. 327–338. Springer, Heidelberg (2013)
5. Dehkordi, H.R., Eades, P.: Every outer-1-plane graph has a right angle crossing drawing. International Journal on Computational Geometry and Applications 22(6), 543–558 (2012)
6. Di Battista, G., Tamassia, R.: On-line planarity testing. SIAM Journal on Computing 25(5), 956–997 (1996)
7. Di Giacomo, E., Liotta, G., Montecchiani, F.: The planar slope number of subcubic graphs. In: Pardo, A., Viola, A. (eds.) LATIN 2014. LNCS, vol. 8392, pp. 132–143. Springer, Heidelberg (2014)
8. Didimo, W.: Density of straight-line 1-planar graph drawings. IPL 113(7), 236–240 (2013)
9. Dujmović, V., Suderman, M., Wood, D.R.: Graph drawings with few slopes. Computational Geometry 38(3), 181–193 (2007)
10. Eades, P., Hong, S.-H., Katoh, N., Liotta, G., Schweitzer, P., Suzuki, Y.: Testing maximal 1-planarity of graphs with a rotation system in linear time. In: Didimo, W., Patrignani, M. (eds.) GD 2012. LNCS, vol. 7704, pp. 339–345. Springer, Heidelberg (2013)
11. Hong, S.-H., Eades, P., Katoh, N., Liotta, G., Schweitzer, P., Suzuki, Y.: A linear-time algorithm for testing outer-1-planarity. In: Wismath, S., Wolff, A. (eds.) GD 2013. LNCS, vol. 8242, pp. 71–82. Springer, Heidelberg (2013)
12. Hong, S.-H., Eades, P., Liotta, G., Poon, S.-H.: Fáry's theorem for 1-planar graphs. In: Gudmundsson, J., Mestre, J., Viglas, T. (eds.) COCOON 2012. LNCS, vol. 7434, pp. 335–346. Springer, Heidelberg (2012)
13. Jelínek, V., Jelínková, E., Kratochvíl, J., Lidický, B., Tesar, M., Vyskocil, T.: The planar slope number of planar partial 3-trees of bounded degree. Graphs and Combinatorics 29(4), 981–1005 (2013)
14. Keszegh, B., Pach, J., Pálvölgyi, D.: Drawing planar graphs of bounded degree with few slopes. SIAM Journal on Discrete Mathematics 27(2), 1171–1183 (2013)
15. Knauer, K.B., Micek, P., Walczak, B.: Outerplanar graph drawings with few slopes. Computational Geometry 47(5), 614–624 (2014)
16. Korzhik, V.P., Mohar, B.: Minimal obstructions for 1-immersions and hardness of 1-planarity testing. Journal of Graph Theory 72(1), 30–71 (2013)
17. Lenhart, W., Liotta, G., Mondal, D., Nishat, R.: Planar and plane slope number of partial 2-trees. In: Wismath, S., Wolff, A. (eds.) GD 2013. LNCS, vol. 8242, pp. 412–423. Springer, Heidelberg (2013)
18. Mukkamala, P., Pálvölgyi, D.: Drawing cubic graphs with the four basic slopes. In: van Kreveld, M.J., Speckmann, B. (eds.) GD 2011. LNCS, vol. 7034, pp. 254–265. Springer, Heidelberg (2011)
19. Pach, J., Pálvölgyi, D.: Bounded-degree graphs can have arbitrarily large slope numbers. The Electronic Journal of Combinatorics 13(1) (2006)
20. Pach, J., Tóth, G.: Graphs drawn with few crossings per edge. Comb. 17(3), 427–439 (1997)
21. Wade, G.A., Chu, J.-H.: Drawability of complete graphs using a minimal slope set. The Computer Journal 37(2), 139–142 (1994)

Fan-Planar Graphs: Combinatorial Properties and Complexity Results[*]

Carla Binucci[1], Emilio Di Giacomo[1], Walter Didimo[1], Fabrizio Montecchiani[1],
Maurizio Patrignani[2], and Ioannis G. Tollis[3]

[1] Università degli Studi di Perugia, Italy
{carla.binucci,emilio.digiacomo,
walter.didimo,fabrizio.montecchiani}@unipg.it
[2] Università Roma Tre, Italy
patrigna@dia.uniroma3.it
[3] Univ. of Crete and Institute of Computer Science-FORTH, Greece
tollis@ics.forth.gr

Abstract. In a *fan-planar drawing* of a graph an edge can cross only edges with a common end-vertex. Fan-planar drawings have been recently introduced by Kaufmann and Ueckerdt, who proved that every n-vertex fan-planar drawing has at most $5n - 10$ edges, and that this bound is tight for $n \geq 20$. We extend their result from both the combinatorial and the algorithmic point of view. We prove tight bounds on the density of constrained versions of fan-planar drawings and study the relationship between fan-planarity and k-planarity. Also, we prove that testing fan-planarity in the variable embedding setting is NP-complete.

1 Introduction

There is a growing interest in the study of non-planar drawings of graphs with forbidden crossing configurations. The idea is to relax the planarity constraint by allowing edge crossings that do not affect too much the drawing readability. Among the most popular types of non-planar drawings studied so far we recall: *k-planar drawings*, where an edge can have at most k crossings (see, e.g., [5,8,9,10,16,18,22,26,27,29,30,32]); *k-quasi-planar drawings*, which do not contain k mutually crossing edges (see, e.g., [1,3,4,15,24,33]); *RAC drawings*, where edges can cross only at right angles (see, e.g., [19] and [20] for a survey); *ACE$_\alpha$ drawings* [2] and *ACL$_\alpha$ drawings* [6,14,21], which are generalizations of RAC drawings; namely, in an ACE$_\alpha$ drawing edges can cross only at an angle that is *exactly* α ($\alpha \in (0, \pi/2)$); in an ACL$_\alpha$ drawing edges can cross only at angles that are *at least* α (see also [20]); *fan-crossing free drawings*, where there cannot be an edge that crosses two other edges with a common end-vertex [11].

Given a desired type T of non-planar drawing with forbidden crossing configurations, a classical combinatorial problem is to establish bounds on the maximum

[*] Research supported in part by the MIUR project AMANDA "Algorithmics for MAssive and Networked DAta", prot. 2012C4E3KT_001. This work started at the Bertinoro Workshop on Graph Drawing 2014. We thank Michael Kaufmann and Torsten Ueckerdt for suggesting the study of fan-planar graphs during the workshop. We also thank all the participants of the workshop for the useful discussions on this topic.

C. Duncan and A. Symvonis (Eds.): GD 2014, LNCS 8871, pp. 186–197, 2014.
© Springer-Verlag Berlin Heidelberg 2014

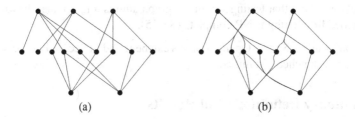

Fig. 1. (a) A fan-planar drawing of a graph G with 12 crossings; (b) A confluent drawing of G with 3 crossings

number of edges that a drawing of type T can have; this problem is usually dubbed a Turán-type problem, and several tight bounds have been proved for the types of drawings mentioned above, both for straight-line and for polyline edges (see, e.g., [1,2,4,10,11,18,19,21,24,30,33]). From the algorithmic point of view, the complexity of testing whether a graph G admits a drawing of type T is one of the most interesting. Also for this problem several results have been shown, both in the variable and in the fixed embedding setting (see, e.g., [8,12,13,25,26,29]).

In this paper we investigate *fan-planar drawings* of graphs, in which an edge cannot cross two independent edges, i.e., an edge can cross several edges provided that they have a common end-vertex. Fan-planar drawings have been recently introduced by Kaufmann and Ueckerdt [28]; they proved that every n-vertex graph without loops and multiple edges that admits a fan-planar drawing has at most $5n - 10$ edges, and that this bound is tight for $n \geq 20$. Fan-planar drawings are on the opposite side of fan-crossing free drawings mentioned above. Besides its intrinsic theoretical interest, we observe that fan-planarity can be also used for creating drawings with few edge crossings in a confluent drawing style (see, e.g., [17,23]). For example, Fig. 1(a) shows a fan-planar drawing Γ with 12 crossings; Fig. 1(b) shows a new drawing with just 3 crossings obtained from Γ by bundling crossing "fans".

We prove both combinatorial properties and complexity results related to fan-planar drawings of graphs. The main contributions of our work are as follows:

(i) We study the density of constrained versions of fan-planar drawings (Sec. 3), namely *outer fan-planar drawings*, where all vertices must lie on the external boundary of the drawing, and 2-*layer fan-planar drawings*, where vertices are placed on two distinct horizontal lines and edges are vertically monotone lines. We prove tight bounds for the edge density of these drawings. Namely, we show that n-vertex outer fan-planar drawings have at most $3n - 5$ edges (a tight bound for $n \geq 5$), and that n-vertex 2-layer fan-planar drawings have at most $2n - 4$ edges (a tight bound for $n \geq 3$). We remark that outer and 2-layer non-planar drawings have been previously studied in the 1-planarity setting [8,18,26] and in the RAC planarity setting [12,13].

(ii) Since general fan-planar drawable graphs have at most $5n - 10$ edges and the same bound holds for 2-planar drawable graphs [30], we investigate the relationship between these two graph classes. More in general, we are able to prove that in fact for any $k \geq 2$ there exist fan-planar drawable graphs that are not k-planar, and vice versa (Sec. 4).

(iii) Finally, we show that testing whether a graph admits a fan-planar drawing in the variable embedding setting is NP-complete (Sec. 5).

Preliminaries are in Sec. 2. Open problems can be found in Sec. 6. For space reasons some proofs are sketched or omitted.

2 Preliminary Definitions and Results

A *drawing* Γ of a graph G maps each vertex to a distinct point of the plane and each edge to a simple Jordan arc between the points corresponding to the end-vertices of the edge. For a subgraph G' of G, we denote by $\Gamma[G']$ the restriction of Γ to G'. Throughout the paper we consider only *simple graphs*, i.e., graphs with neither multiple edges nor self-loops; also we only consider *simple drawings*, i.e., drawings such that the arcs representing two edges have at most one point in common, which is either a common end-vertex or a common interior point where the two arcs properly cross each other.

For each vertex v of G, the set of edges incident to v is called the *fan of v*. Clearly, each edge (u, v) of G belongs to the fan of u and to the fan of v at the same time. Two edges that do not share a vertex are called *independent edges*; two independent edges always belong to distinct fans. A *fan-planar drawing* Γ of G, is a drawing of G such that: (a) no edge is crossed by two independent edges; (b) there are not two adjacent edges (u, v), (u, w) that cross an edge e from different "sides" while moving from u to v and from u to w. The forbidden configurations (a) and (b) are depicted in Fig. 2(a) and Fig. 2(b), respectively. Figures 2(c) and 2(d) show two allowed configurations of a fan-planar drawing. A *fan-planar graph* is a graph that admits a fan-planar drawing.

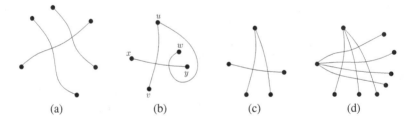

(a)   (b)   (c)   (d)

Fig. 2. (a)-(b) Forbidden configurations in a fan-planar drawing: (c)-(d) Allowed configurations in a fan-planar drawing

The next property immediately follows from the definition of fan-planar drawings.

Property 1. A fan-planar drawing does not contain 3-mutually crossing edges.

Let Γ be a non-planar drawing of G; the *planar enhancement* Γ' of Γ is the drawing obtained from Γ by replacing each crossing point with a dummy vertex. The boundary of each face f' of Γ' consists of a sequence of real and dummy vertices; the connected region f of the plane that corresponds to f' in Γ consists of a sequence of vertices and crossing points. For simplicity we call f a *face* of Γ. The *outer face* of Γ is the face corresponding to the outer face of Γ'. A fan-planar drawing of G with all vertices on the outer face is called an *outer fan-planar drawing* of G. Observe that the configuration in Fig. 2(b) cannot occur in a drawing with all vertices on the outer face; hence, a drawing is outer fan-planar if and only if all vertices are on the outer face and it does not contain an edge crossed by two independent edges. An *outer fan-planar graph* is a graph that admits an outer fan-planar drawing. An outer fan-planar graph G is *maximal*, if no edge can be added to G without loosing the property that G remains outer fan-planar. An outer fan-planar graph G with n vertices is *maximally dense* if it has the maximum number of edges among all outer fan-planar graphs with n vertices. If G is maximally dense then it is also maximal, but not vice versa. The following property holds.

Lemma 1. *Let $G = (V, E)$ be a maximal outer fan-planar graph and let Γ be an outer fan-planar drawing of G. The outer face of Γ does not contain crossing points, i.e., it consists of $|V|$ uncrossed edges.*

Given an outer fan-planar drawing Γ of a maximal outer fan-planar graph G, the edges of G on the external boundary of Γ will be also called the *outer edges* of Γ.

A *2-layer fan-planar drawing* is a fan-planar drawing such that: (i) each vertex is drawn on one of two distinct horizontal lines, called *layers*; (ii) each edge connects vertices of different layers and it is drawn as a vertical monotone curve. By definition, a 2-layer fan-planar drawing is also an outer fan-planar drawing. A *2-layer fan-planar graph* is a graph that admits a 2-layer fan planar drawing.

3   Density of Outer and 2-layer Fan-Planar Graphs

We first prove that an n-vertex outer fan-planar graph G has at most $3n - 5$ edges. Then we describe how to construct outer fan-planar graphs with n vertices and $3n - 5$ edges.

Let G be a graph and let Γ be a drawing of G. The *crossing graph* of Γ, denoted as $\mathrm{CR}(\Gamma)$, is a graph having a vertex for each edge of G and an edge between any two vertices whose corresponding edges cross in Γ. A cycle of $\mathrm{CR}(\Gamma)$ of odd length will be called an *odd cycle* of $\mathrm{CR}(\Gamma)$; similarly, an *even cycle* of $\mathrm{CR}(\Gamma)$ is a cycle of even length. In order to prove the $3n - 5$ upper bound, we can assume that G is a maximally dense outer fan-planar graph. We start by proving some interesting combinatorial properties of G related to the cycles of the crossing graph of G.

Lemma 2. *Let $G = (V, E)$ be a maximal outer fan-planar graph with $n = |V|$ vertices and $m = |E|$ edges. Let Γ be an outer fan-planar drawing of G. If $\mathrm{CR}(\Gamma)$ does not have odd cycles then $m \leq 3n - 6$.*

Proof. If $\mathrm{CR}(\Gamma)$ does not contain odd cycles, then it is bipartite and its vertices can be partitioned into two independent sets W_1 and W_2. Since by Lemma 1 the outer edges of Γ are not crossed, they correspond to n isolated vertices in $\mathrm{CR}(\Gamma)$. We can arbitrarily assign all these vertices to the same set, say W_1. Denote by E_i the set of edges of G corresponding to the vertices of W_i ($i \in \{1, 2\}$). Clearly, E_1 and E_2 partition the set E. Since no two edges of E_i cross in Γ, then the two subgraphs $G_1 = (V, E_1)$ and $G_2 = (V, E_2)$ are outerplanar graphs, where $|E_1| \leq 2n - 3$ and $|E_2| \leq 2n - 3 - n$. Thus, $m = |E| = |E_1| + |E_2| \leq 3n - 6$. □

The next lemma shows that the length of any odd cycle of $\mathrm{CR}(\Gamma)$ is at most 5.

Lemma 3. *Let G be a maximally dense outer fan-planar graph with n vertices and let Γ be an outer fan-planar drawing of G. CR(Γ) does not contain odd cycles of length greater than 5.*

Proof. Let C be an odd cycle of length ℓ in CR(Γ). Let $E(C) = \{e_0 = (u_0, v_0), \ldots, e_{\ell-1} = (u_{\ell-1}, v_{\ell-1})\}$ be the set of ℓ edges of G corresponding to the vertices of C, such that e_i crosses e_{i+1} for $i = 0, \ldots, l - 1$, where indices are taken modulo ℓ. Recall that all vertices of G are on the outer face of Γ, which implies that the end-vertices of the edges in $E(C)$ are encountered in the following order when walking clockwise on the boundary of the outer face of Γ: u_i precedes v_{i-1} and v_i precedes u_{i+2} (see, e.g., Fig. 3(a)). Furthermore, vertices v_i and u_{i+2} must coincide, for $i = 0, \ldots, \ell - 1$. Indeed, if v_i and u_{i+2} are distinct, for some $i = 0, \ldots, \ell - 1$, then edge e_{i+1} is crossed by two independent edges (i.e., e_i and e_{i+2}), which contradicts the hypothesis that Γ is fan-planar. See also Fig. 3(a). Thus, we have that u_i precedes u_{i+1} while walking clockwise on the boundary of the outer face of Γ, for $i = 0, \ldots, \ell - 1$, as shown in Fig. 3(b). Moreover, it can be seen that the edges in $E(C)$ are not crossed by any edge not in $E(C)$, as otherwise the drawing would not be fan-planar.

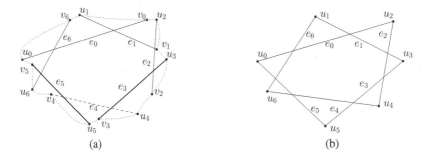

Fig. 3. Illustration for the proof of Lemma 3. (a) An edge set $E(C)$ with $\ell = 7$. If v_3 and u_5 do not coincide, e_4 (dashed) is crossed by the two independent edges e_3 and e_5 (in bold). (b) $E(C)$ with $\ell = 7$, where v_i coincides with u_{i+2}, for $i = 0, \ldots, 7$.

Now, suppose by contradiction that ℓ is odd and greater than 5 (see Fig. 3(b)). Consider a vertex u_i, for some $i = 0, \ldots, \ell-1$, and denote by $\overline{V}$ the set of vertices encountered between u_{i+3} and u_{i-3} while walking clockwise on the boundary of the outer face of Γ (including u_{i+3} and u_{i-3}). Vertex u_i cannot be adjacent to any vertex in $\overline{V}$. Namely, if an edge $e = (u_i, u_j)$ existed, for some $u_j \in \overline{V}$, then it would be crossed by the two independent edges e_{i-1} and e_{j-1}. Thus, removing e_{i-1} from Γ, one can suitably connect u_i to all the vertices in $\overline{V}$, still obtaining a fan-planar drawing Γ^* with n vertices. Since the size of $\overline{V}$ is $\ell - 5$, and $\ell \geq 7$ by assumption, then Γ^* has at least two edges more than Γ, which contradicts the hypothesis that G is maximally dense.   □

The following corollary is a consequence of Lemma 3 and Property 1.

Corollary 1. *Let G be a maximally dense outer fan-planar graph. Any odd cycle in the crossing graph of a fan-planar drawing of G has exactly length 5.*

The next lemma claims that odd cycles in the crossing graph correspond to K_5 (the proof uses similar arguments as the proof of Lemma 3.

Lemma 4. *Let G be a maximally dense outer fan-planar graph and let Γ be an outer fan-planar drawing of G. If $\mathrm{CR}(\Gamma)$ contains a cycle C of length 5, then the subgraph of G induced by the end-vertices of the edges corresponding to the vertices of C is K_5.*

We now prove the upper bound on the density of outer fan-planar graphs.

Lemma 5. *Let G be a maximally dense outer fan-planar graph with n vertices and m edges. Then $m \leq 3n - 5$ edges.*

Proof. Let Γ be an outer fan-planar drawing of G. We first claim that G is biconnected. Suppose by contradiction that G is not biconnected, and let C_1 and C_2 be two distinct biconnected components of G that share a cut-vertex v. Let u be the first vertex of G encountered while moving from v clockwise on the external boundary of $\Gamma[C_1]$, and let w be the first vertex encountered while moving from v counterclockwise on the external boundary of $\Gamma[C_2]$. One can suitably add edge (u, w) in Γ, still getting an outer fan-planar drawing, which contradicts the hypothesis that G is maximally dense.

Now, by Corollary 1, $\mathrm{CR}(\Gamma)$ can only have either even cycles or cycles of length 5. Also, by Lemma 4, every cycle of length 5 in $\mathrm{CR}(\Gamma)$ corresponds to a subset of edges whose end-vertices induce K_5. We prove the statement by induction on the number h of K_5 subgraphs in G.

Base Case. If $h = 0$ then, by Lemma 2, G has at most $3n - 6$ edges.

Inductive Case. Suppose by induction that the claim is true for $h \geq 0$, and suppose G contains $h + 1$ subgraphs that are K_5. Let G^* be one of these $h + 1$ subgraphs. Let $e = (u, v)$ be an edge on the outer face of $\Gamma[G^*]$ that is not on the outer face of Γ. Vertices u and v are a separation pair of G, otherwise there would be a vertex of G that is not on the outer face of Γ, which is impossible because Γ is an outer fan-planar drawing by hypothesis. Hence, we can split G into two biconnected subgraphs that share only edge e, one of them containing G^*. Let $G_1, G_2, \ldots, G_k$ ($k \leq 5$) be the biconnected subgraphs of G distinct from G^* such that each G_i shares exactly one edge with G^*. Each G_i ($i = 1, 2, \ldots, k$) contains at most h subgraphs that are K_5, and therefore it has at most $3n_i - 5$ edges by induction, where n_i denotes the number of vertices of G_i. On the other hand, G^* has $3n^* - 5 = 10$ edges, where $n^* = 5$ is the number of vertices of G^*. It follows that $m \leq 3(n^* + n_1 + \cdots + n_k) - 5(k + 1) - k$ ($k \leq 5$). Since $n^* + n_1 + \cdots + n_k \leq n + 2k$ we have $m \leq 3(n + 2k) - 5(k + 1) - k = 3n - 5$. □

The existence of an infinite family of outer fan-planar graphs that match the $3n - 5$ bound is proved in the next lemma. Refer to Fig. 4 for an illustration.

Lemma 6. *For any integer $h \geq 1$ there exists an outer fan-planar graph G with $n = 3h + 2$ vertices and $m = 3n - 5$ edges.*

Proof. Consider h graphs $X_1, \ldots, X_h$, such that each X_i is a K_5 graph, for $i = 1, \ldots, h$. We now describe how to construct G. The idea is to "glue" $X_1, \ldots, X_h$ together in such a way that they share single edges one to another. The proof is by induction on the number of merged graphs. Denote by G_i the graph obtained after merging

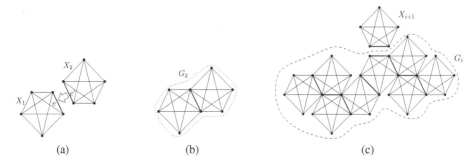

Fig. 4. Illustration for the proof of Lemma 6. (a) X_1 and X_2 before being merged. (b) Merging X_1 and X_2 into G_2. (c) G_i and X_{i+1}, the bold edges are used for merging.

$X_1, \ldots, X_i$, for $1 < i \leq h$. We prove by induction that G_i respects the following invariants: (I1) it is an outer fan-planar graph; (I2) it has $n_i = 3i + 2$ vertices and $m_i = 3n_i - 5$ edges. In the base case $i = 2$, we merge $G_1 = X_1$ to X_2 as follows. Pick an edge e on the outer face of X_1 and an edge e' on the outer face of X_2. Merge X_1 and X_2 by identifying e with e', see also Figs. 4(a) and 4(b). The new graph G_2 is clearly an outer fan-planar graph with $n_2 = 5 + 5 - 2 = 8$ vertices and $m_2 = 10 + 10 - 1 = 19$ edges. Thus, the two invariants hold. In the inductive case, suppose we constructed G_i for $2 < i < h$ and we want to attach X_{i+1} (see also Fig. 4(c)). Pick any edge e on the outer face of G_i and any edge e' on the outer face of X_{i+1}. Merge the two graphs in the same way as done in the base case. It is immediate to see that (I1) holds. Also, $n_{i+1} = n_i + 3$ and $m_{i+1} = m_i + 9$. Since by induction $m_i = 3n_i - 5$, then $m_{i+1} = 3n_i - 5 + 9 = 3n_{i+1} - 5$. □

Lemmas 5 and 6 imply the following theorem.

Theorem 1. *An outer fan-planar graph with n vertices has at most $3n - 5$ edges, and this bound is tight for $n \geq 5$.*

An obvious consequence of Theorem 1 and of the definition of outer fan-planar graphs that are maximally dense is the following fact.

Corollary 2. *Every maximally dense outer fan-planar graph with $n = 3h + 2$ vertices ($h \geq 1$) has $3n - 5$ edges.*

Concerning 2-layer fan planar graphs, we already observed that a 2-layer fan planar graph G is an outer fan-planar graph. Also, since all vertices on the same layer form an independent set, G is bipartite.

Theorem 2. *A 2-layer fan-planar graph with n vertices has at most $2n - 4$ edges, and this bound is tight for $n \geq 3$.*

Proof. Let G be a maximally dense 2-layer fan-planar graph with n vertices and m edges, and let Γ be a 2-layer fan-planar drawing of G. Let $V_1 = \{v_1, \ldots, v_{n_1}\}$ and

$V_2 = \{v_{n_1+1}, \ldots, v_n\}$ the two independent sets of vertices of G. Suppose that in Γ v_i precedes v_{i+1} along the layer of V_1 (for $i = 1, \ldots, n_1 - 1$), and v_j precedes v_{j+1} along the layer of V_2 (for $j = n_1 + 1, \ldots, n - 1$). Construct from G a super-graph G^*, by adding an edge (v_i, v_{i+1}), for $i = 1, \ldots, n_1 - 1$, and an edge (v_j, v_{j+1}), for $j = n_1 + 1, \ldots, n$. Graph G^* is still outer fan-planar. Moreover, since G does not contain a K_5 subgraph (because it is bipartite), also G^* does not contain a K_5 subgraph, as otherwise at least three vertices of the same layer in G should form a 3-cycle in G^* (which does not happen by construction). Thus, by Lemma 3 and Property 1, the crossing graph of any outer fan-planar drawing of G^* contains only even cycles. Hence, denoted as m^* the number of edges of G^*, by Lemma 2 we have $m^* \leq 3n - 6$, and therefore $m = m^* - (n - 2) \leq 2n - 4$. A family of 2-layer fan-planar graphs with $2n - 4$ edges (for $n \geq 3$) are the complete bipartite graphs $K_{2,n-2}$. □

4 Fan-Planar and k-planar Graphs

A k-*planar drawing* is a drawing where each edge is crossed at most k times, and a k-*planar graph* is a graph that admits a k-planar drawing. Clearly, every 1-planar graph is also a fan-planar graph. Also, both the maximum number of edges of fan-planar graphs [28] and the maximum number of edges of 2-planar graphs [30] have been shown to be $5n - 10$. Thus it is natural to ask what is the relationship between fan-planar and 2-planar graphs. More in general, we prove that there are fan-planar graphs that are not k-planar, for any $k \geq 1$, and that there are k-planar graphs (for $k > 1$) that are not fan-planar. The existence of fan-planar graphs that are not k-planar is proved with a counting argument on the minimum number of crossings of graph drawings. The *crossing number* $cr(G)$ of G is the smallest number of crossings required in any drawing of G. We give the following.

Theorem 3. *For any integer $k \geq 1$ there is a graph that is fan-planar but not k-planar.*

Proof. Consider the complete 3-partite graph $K_{1,3,h}$. This graph is fan-planar for every $h \geq 1$ (see Fig. 5(a)). It is known that $cr(K_{1,3,h}) = 2 \lfloor \frac{h}{2} \rfloor \lfloor \frac{h-1}{2} \rfloor + \lceil \frac{h}{2} \rceil$ [7,31]. For $h = 4k+2$, we have $cr(K_{1,3,4k+2}) = 2 \lfloor \frac{4k+2}{2} \rfloor \lfloor \frac{4k+1}{2} \rfloor + \lceil \frac{4k+2}{2} \rceil = 4k(2k+1)+2k+1 = 8k^2 + 6k + 1$. Thus, in every drawing of $K_{1,3,4k+2}$ there are at least $8k^2 + 6k + 1$ crossings. On the other hand, in a k-planar drawing there can be at most $\frac{km}{2}$ crossings, where m is the number of edges in the drawing. Since $K_{1,3,4k+2}$ has $16k + 11$ edges, to be k-planar it should admit a drawing with at most $\frac{km}{2} = \frac{k(16k+11)}{2} = 8k^2 + \frac{11}{2}k$ crossings. Since $6k + 1 > \frac{11}{2}k$ for every $k \geq 1$, $K_{1,3,4k+2}$ is not k-planar. □

To prove that for any $k > 1$ there are k-planar graphs that are not fan-planar (Theorem 4), we first give a technical result (Lemma 7), which will be also reused in Sec. 5. Let Γ be a fan-planar drawing of a graph. We may regard crossed edges of Γ as composed by *fragments*, where a fragment is the portion of the edge that is between two consecutive crossings or between one of the two end-vertices of the edge and the first crossing encountered while moving along the edge towards the other end-vertex. An edge that is not crossed does not have any fragment. Figure 5(b) shows a fan-planar drawing of the K_7 graph and Fig. 5(c) shows the fragments of the drawing in Fig. 5(b).

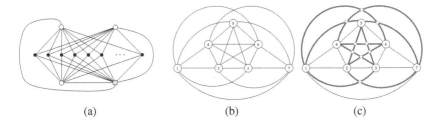

Fig. 5. (a) A fan-planar drawing of $K_{1,3,h}$. (b) A fan-planar drawing of the K_7 graph. (c) The fragments of the fan-planar drawing in (b) are the thicker lines.

We consider two fragments *adjacent* if they share a common crossing or a common end-vertex. The next lemma provides an interesting and useful property.

Lemma 7. *In any fan-planar drawing of the K_7 graph, any pair of vertices is joined by a sequence of adjacent fragments.*

Theorem 4. *For any integer $k > 1$ there is a graph that is k-planar but not fan-planar.*

Proof sketch: Since 2-planar graphs are also k-planar graphs, for $k > 1$, we can prove that there is a 2-planar graph that is not fan-planar. Let G' be a graph consisting of a cycle $C = (v_1, v_2, \ldots, v_{10})$ and of the edges (v_1, v_4), (v_5, v_{10}), (v_6, v_9) (see Fig. 6(a)). Let G'' be the graph obtained from G' by replacing each edge (v_i, v_j) $(1 \le i, j \le 10)$ with a copy of K_7, whose vertices are denoted as $u_1, u_2, \ldots, u_7$, so that $v_i = u_1$ and $v_j = u_7$ (see Fig. 6(b)). The copy of K_7 that replaces (v_i, v_j) is denoted as $K_7^{i,j}$. Let G be the graph obtained from G'' by adding the edges (v_1, v_7), (v_2, v_6), (v_3, v_9), (v_4, v_8) (see Fig. 6(c)). G is 2-planar (planarly embed G' as shown in Fig. 6(a)). Construct a drawing Γ of G by replacing each edge of G' with a drawing of $K_7^{i,j}$ like that in Fig. 5(b) (see Fig. 6(b)), and draw the edges (v_1, v_7), (v_2, v_6), (v_3, v_9), (v_4, v_8) inside C as in Fig. 6(c). Γ is 2-planar. To prove that G is not fan-planar note that, by Lemma 7, each $K_7^{i,j}$ $(1 \le i, j \le 10)$ has a sequence of fragments leading from $v_i = u_1$ to $v_j = u_7$, which we call *spine*. In any fan-planar drawing of G, this spine cannot be crossed by edges that do not belong to $K_7^{i,j}$ (otherwise there would be an edge crossed by two independent edges). Thus, every $K_7^{i,j}$ is not "traversed" by external edges, and this forces (v_1, v_7), (v_2, v_6), (v_3, v_9), (v_4, v_8) to violate fan-planarity, as in Fig. 6(c).   □

5   Complexity of the Fan-Planarity Testing Problem

We exploit the results of Secs. 3 and 4 to prove that testing whether a graph is fan-planar in the variable embedding setting is NP-complete. We call this problem the *fan-planarity testing*. We use a reduction from the *1-planarity testing*, which is NP-complete in the variable embedding setting [25,29]. The 1-planarity testing asks whether a given graph admits a 1-planar drawing. We prove the following.

Theorem 5. *Fan-planarity testing is NP-complete.*

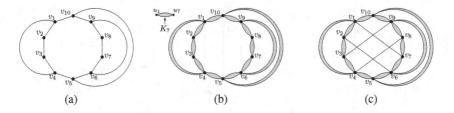

Fig. 6. (a)–(c) Illustration for the proof of Theorem 4: (a) the graph G'; (b) the graph G''; (c) the graph G

Proof. Similarly to [34], a non-deterministic algorithm to test if a graph admits a fan-planar drawing with k crossings considers all possible k pairs of edges that cross (and the order of the crossings along the edges), discards the configurations where an edge crosses more than one fan, replaces crossings with dummy vertices, and tests the obtained graph for planarity. Then the problem belongs to NP.

We now prove the hardness. Given an instance $G = (V, E)$ of the 1-planarity testing we build an instance $G_f = (V_f, E_f)$ of the fan-planarity testing by replacing each edge $(u, v) \in E$ with two K_7 graphs with vertices $u = u_1, u_2, \ldots, u_7$ and $v = v_1, v_2, \ldots, v_7$, called *attachment gadgets* and joined by a *spanning* edge (u_7, v_7) (see Fig. 7 for an illustration). $G_f = (V_f, E_f)$ can be constructed in polynomial time, having $|V_f| = |V| + |E| \times 12$ vertices and $|E_f| = |E| \times 43$ edges, where $|E| \times 42$ of them belong to the attachment gadgets and the remaining $|E|$ are spanning edges that join different attachment gadgets. We show that G is 1-planar if and only if G_f is fan-planar. If G admits a 1-planar drawing, replace each edge (u, v) of G with two fan-planar drawings of K_7 like those depicted in Fig. 5(b) and with edge (u_7, v_7), in such a way that the possible crossing of (u, v) occurs on (u_7, v_7). The obtained drawing of G_f is fan-planar since each attachment gadget has a fan-planar drawing and each spanning edge has at most one crossing. Conversely, suppose G_f admits a fan-planar drawing Γ_f. By Lemma 7, for any attachment gadget of G_f attached to vertex u, there is at least a sequence of fragments leading from $u = u_1$ to u_7. As in the proof of Theorem 4, call such a sequence of fragments the *spine* of the attachment gadget. Delete from Γ_f all fragments except those in the spines. Delete from Γ_f all uncrossed edges except the spanning edges. Remove also isolated vertices. A drawing Γ of G is obtained, where the drawing of edge (u, v) is given by the spine from $u = u_1$ to u_7, the spanning edge (u_7, v_7), and the spine from v_7 to $v_1 = v$. Observe that, $u \neq v$, as otherwise there would be a self-loop in G. We claim that Γ is a 1-planar drawing of G. Indeed, fragments in the spines can not be crossed by any other fragment or spanning edge of Γ_f. It follows that spanning edges can cross only among themselves in Γ_f. However, they can cross only once, as they are a matching of G_f and Γ_f is fan-planar. Hence, Γ is a 1-planar drawing, but not necessarily simple; indeed, it may happen that two crossing edges (u, v) and (w, z) in Γ share an end-vertex, say $u = w$ (this happens when in Γ_f there are two crossing spanning edges of two K_7 attached to u). The crossing between (u, v) and (u, z) in Γ can be easily removed by rerouting the edges (see Fig. 7(c)). □

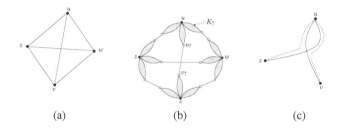

(a)   (b)   (c)

Fig. 7. Illustration of the reduction used in Theorem 5. (a) An instance G of 1-planarity testing; (b) The reduced instance G_f of fan-planarity testing. (c) Two adjacent edges of G that cross one to another in Γ; the crossing can be removed by rerouting the two edges as shown by the dashed lines.

6   Open Problems

We suggest two questions: (i) What is the minimum number of edges of maximal fan-planar graphs? (ii) Can we efficiently recognize maximally dense fan-planar graphs?

References

1. Ackerman, E.: On the maximum number of edges in topological graphs with no four pairwise crossing edges. Discrete & Computational Geometry 41(3), 365–375 (2009)
2. Ackerman, E., Fulek, R., Tóth, C.D.: Graphs that admit polyline drawings with few crossing angles. SIAM J. on Discrete Mathematics 26(1), 305–320 (2012)
3. Ackerman, E., Tardos, G.: On the maximum number of edges in quasi-planar graphs. J. of Combinatorial Theory, Series A 114(3), 563–571 (2007)
4. Agarwal, P.K., Aronov, B., Pach, J., Pollack, R., Sharir, M.: Quasi-planar graphs have a linear number of edges. Combinatorica 17(1), 1–9 (1997)
5. Alam, M.J., Brandenburg, F.J., Kobourov, S.G.: Straight-line grid drawings of 3-connected 1-planar graphs. In: Wismath, S., Wolff, A. (eds.) GD 2013. LNCS, vol. 8242, pp. 83–94. Springer, Heidelberg (2013)
6. Angelini, P., Di Battista, G., Didimo, W., Frati, F., Hong, S.H., Kaufmann, M., Liotta, G., Lubiw, A.: Large angle crossing drawings of planar graphs in subquadratic area. In: Márquez, A., Ramos, P., Urrutia, J. (eds.) EGC 2011. LNCS, vol. 7579, pp. 200–209. Springer, Heidelberg (2012)
7. Asano, K.: The crossing number of $K_{1,3,n}$ and $K_{2,3,n}$. J. of Graph Theory 10(1), 1–8 (1986)
8. Auer, C., Bachmaier, C., Brandenburg, F.J., Gleißner, A., Hanauer, K., Neuwirth, D., Reislhuber, J.: Recognizing outer 1-planar graphs in linear time. In: Wismath, S., Wolff, A. (eds.) GD 2013. LNCS, vol. 8242, pp. 107–118. Springer, Heidelberg (2013)
9. Auer, C., Brandenburg, F.J., Gleißner, A., Hanauer, K.: On sparse maximal 2-planar graphs. In: Didimo, W., Patrignani, M. (eds.) GD 2012. LNCS, vol. 7704, pp. 555–556. Springer, Heidelberg (2013)
10. Brandenburg, F.J., Eppstein, D., Gleißner, A., Goodrich, M.T., Hanauer, K., Reislhuber, J.: On the density of maximal 1-planar graphs. In: Didimo, W., Patrignani, M. (eds.) GD 2012. LNCS, vol. 7704, pp. 327–338. Springer, Heidelberg (2013)
11. Cheong, O., Har-Peled, S., Kim, H., Kim, H.S.: On the number of edges of fan-crossing free graphs. In: Cai, L., Cheng, S.-W., Lam, T.-W. (eds.) ISAAC 2013. LNCS, vol. 8283, pp. 163–173. Springer, Heidelberg (2013)

12. Dehkordi, H.R., Eades, P.: Every outer-1-plane graph has a right angle crossing drawing. International J. on Computational Geometry and Appl. 22(6), 543–558 (2012)
13. Di Giacomo, E., Didimo, W., Eades, P., Liotta, G.: 2-layer right angle crossing drawings. Algorithmica 68(4), 954–997 (2014)
14. Di Giacomo, E., Didimo, W., Liotta, G., Meijer, H.: Area, curve complexity, and crossing resolution of non-planar graph drawings. Theory of Computing Syst. 49(3), 565–575 (2011)
15. Di Giacomo, E., Didimo, W., Liotta, G., Montecchiani, F.: h-quasi planar drawings of bounded treewidth graphs in linear area. In: Golumbic, M.C., Stern, M., Levy, A., Morgenstern, G. (eds.) WG 2012. LNCS, vol. 7551, pp. 91–102. Springer, Heidelberg (2012)
16. Di Giacomo, E., Didimo, W., Liotta, G., Montecchiani, F.: Area requirement of graph drawings with few crossings per edge. Computational Geometry 46(8), 909–916 (2013)
17. Dickerson, M., Eppstein, D., Goodrich, M.T., Meng, J.Y.: Confluent drawings: Visualizing non-planar diagrams in a planar way. J. of Graph Algorithms and Appl. 9(1), 31–52 (2005)
18. Didimo, W.: Density of straight-line 1-planar graph drawings. Information Processing Letters 113(7), 236–240 (2013)
19. Didimo, W., Eades, P., Liotta, G.: Drawing graphs with right angle crossings. Theor. Comput. Sci. 412(39), 5156–5166 (2011)
20. Didimo, W., Liotta, G.: The crossing angle resolution in graph drawing. In: Pach, J. (ed.) Thirty Essays on Geometric Graph Theory. Springer (2012)
21. Dujmović, V., Gudmundsson, J., Morin, P., Wolle, T.: Notes on large angle crossing graphs. Chicago J. on Theoretical Computer Science 2011 (2011)
22. Eades, P., Liotta, G.: Right angle crossing graphs and 1-planarity. Discrete Applied Mathematics 161(7-8), 961–969 (2013)
23. Eppstein, D., Goodrich, M.T., Meng, J.Y.: Confluent layered drawings. Algorithmica 47(4), 439–452 (2007)
24. Fox, J., Pach, J., Suk, A.: The number of edges in k-quasi-planar graphs. SIAM J. on Discrete Mathematics 27(1), 550–561 (2013)
25. Grigoriev, A., Bodlaender, H.L.: Algorithms for graphs embeddable with few crossings per edge. Algorithmica 49(1), 1–11 (2007)
26. Hong, S.H., Eades, P., Katoh, N., Liotta, G., Schweitzer, P., Suzuki, Y.: A linear-time algorithm for testing outer-1-planarity. In: Wismath, S., Wolff, A. (eds.) GD 2013. LNCS, vol. 8242, pp. 71–82. Springer, Heidelberg (2013)
27. Hong, S.-H., Eades, P., Liotta, G., Poon, S.-H.: Fáry's theorem for 1-planar graphs. In: Gudmundsson, J., Mestre, J., Viglas, T. (eds.) COCOON 2012. LNCS, vol. 7434, pp. 335–346. Springer, Heidelberg (2012)
28. Kaufmann, M., Ueckerdt, T.: The density of fan-planar graphs. CoRR abs/1403.6184 (2014), http://arxiv.org/abs/1403.6184
29. Korzhik, V.P., Mohar, B.: Minimal obstructions for 1-immersions and hardness of 1-planarity testing. J. of Graph Theory 72(1), 30–71 (2013)
30. Pach, J., Tóth, G.: Graphs drawn with few crossings per edge. Combinatorica 17(3), 427–439 (1997)
31. Schaefer, M.: The graph crossing number and its variants: A survey. Electronic J. of Combinatorics 20(2) (2013)
32. Suzuki, Y.: Re-embeddings of maximum 1-planar graphs. SIAM J. on Discrete Mathematics 24(4), 1527–1540 (2010)
33. Valtr, P.: On geometric graphs with no k pairwise parallel edges. Discrete & Computational Geometry 19(3), 461–469 (1998)
34. Garey, M., Johnson, D.: Crossing Number is NP-Complete. SIAM Journal on Algebraic Discrete Methods 4(3), 312–316 (1983), doi:10.1137/0604033

On the Recognition of Fan-Planar
and Maximal Outer-Fan-Planar Graphs [*]

Michael A. Bekos[1], Sabine Cornelsen[2], Luca Grilli[3],
Seok-Hee Hong[4], and Michael Kaufmann[1]

[1] Wilhelm-Schickard-Institut für Informatik, Universität Tübingen, Germany
{bekos,mk}@informatik.uni-tuebingen.de
[2] Dept. of Computer and Information Science, University of Konstanz, Germany
sabine.cornelsen@uni-konstanz.de
[3] Dipartimento di Ingegneria, Università degli Studi di Perugia, Italy
luca.grilli@unipg.it
[4] School of Information Technologies, University of Sydney, Australia
shhong@it.usyd.edu.au

Abstract. *Fan-planar* graphs were recently introduced as a generalization of
1-*planar* graphs. A graph is *fan-planar* if it can be embedded in the plane, such
that each edge that is crossed more than once, is crossed by a bundle of two or
more edges incident to a common vertex. A graph is *outer-fan-planar* if it has
a fan-planar embedding in which every vertex is on the outer face. If, in addition, the insertion of an edge destroys its outer-fan-planarity, then it is *maximal
outer-fan-planar*.

In this paper, we present a polynomial-time algorithm to test whether a given
graph is *maximal outer-fan-planar*. The algorithm can also be employed to produce an outer-fan-planar embedding, if one exists. On the negative side, we show
that testing fan-planarity of a graph is NP-hard, for the case where the *rotation
system* (i.e., the cyclic order of the edges around each vertex) is given.

1 Introduction

A *simple drawing* of a graph is a representation of a graph in the plane, where each
vertex is represented by a point and each edge is a Jordan curve connecting its endpoints
such that no edge contains a vertex in its interior, no two edges incident to a common
end-vertex cross, no edge crosses itself, no two edges meet tangentially, and no two
edges cross more than once.

An important subclass of drawn graphs is the class of planar graphs, in which there
exist no crossings between edges. Although planarity is one of the most desirable properties when drawing a graph, many real-world graphs are in fact non-planar.

---

[*] This work started at the *Bertinoro Workshop on Graph Drawing 2014*. The work of Bekos is
implemented within the framework of the Action "Supporting Postdoctoral Researchers" of
the Operational Program "Education and Lifelong Learning" (Action's Beneficiary: General
Secretariat for Research and Technology), and is co-financed by the European Social Fund
(ESF) and the Greek State. Grilli was partly supported by the MIUR project AMANDA "Algorithmics for MAssive and Networked DAta", prot. 2012C4E3KT_001. Hong was supported
by ARC Future Fellowship and Humboldt Fellowship.

C. Duncan and A. Symvonis (Eds.): GD 2014, LNCS 8871, pp. 198–209, 2014.

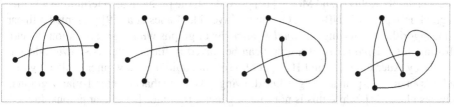

(a) Fan-crossing   (b) Forbidden pattern I (c) Forbidden pattern II   (d) Triangle crossing

Fig. 1. (taken from [18]) (a) A fan-crossing. (b) Forbidden crossing pattern I: An edge cannot be crossed by two independent edges. (c) Forbidden crossing pattern II: An edge cannot be crossed by two edges having their common end-point on different sides of it. (d) Forbidden crossing pattern II implies that an edge cannot be crossed by three edges forming a triangle.

On the other hand, it is accepted that edge crossings have negative impact on the human understanding of a graph drawing [22] and simultaneously it is NP-complete to find drawings with minimum number of edge crossings [13]. This motivated the study of "almost planar" graphs which may contain crossings as long as they do not violate some prescribed forbidden crossing patterns. Typical examples include k-planar graphs [23], k-quasi planar graphs [2], RAC graphs [8] and fan-crossing free graphs [6].

Fan-planar graphs were recently introduced in the same context [18]. A *fan-planar drawing* of graph $G = (V, E)$ is a simple drawing which allows for more than one crossing on an edge $e \in E$ iff the edges that cross e are incident to a common vertex on the same side of e. Such a crossing is called *fan-crossing*. An equivalent definition can be stated by means of forbidden crossing patterns; see Fig. 1. A graph is *fan-planar* if it admits a fan-planar drawing. So, the class of fan-planar graphs is in a sense the complement of the class of fan-crossing free graphs [6], which simply forbid fan-crossings.

Kaufmann and Ueckerdt [18] showed that a fan-planar graph on n vertices cannot have more than $5n - 10$ edges; a tight bound. An *outer-fan-planar drawing* is a fan-planar drawing in which all vertices are on the outer-face. A graph is *outer-fan-planar* if it admits an outer-fan-planar drawing. An outer-fan-planar graph is *maximal outer-fan-planar* if adding any edge to it yields a graph that is not outer-fan-planar. Note that the forbidden pattern II is irrelevant for outer-fan-planarity.

Our main contribution is a polynomial time algorithm for the recognition of maximal outer-fan-planar graphs (Section 2). We also prove that the general fan-planar problem is NP-hard, for the case where the *rotation system* (i.e., the circular order of the edges around each vertex) is given (Section 3). Due to space restrictions, some proofs are omitted or only sketched in the text; full proofs for all results can be found in [5].

Related Work. As already stated, k-planar graphs [23], k-quasi planar graphs [2], RAC graphs [8] and fan-crossing free graphs [6] are closely related to the class of graphs that we study.

A graph is k-*planar*, if it can be embedded in the plane with at most k crossings per edge. Obviously, 1-planar graphs are also fan-planar. A 1-planar graph with n vertices has at most $4n - 8$ edges and this bound is tight [21]. Grigoriev and Bodlaender [14], and independently Kohrzik and Mohar [19] proved that the problem of determining whether a graph is 1-planar is NP-hard. On the positive side, Eades et al. [9] presented a linear time algorithm for testing *maximal 1-planarity* of graphs with a given rotation system. Testing *outer-1-planarity* of a graph can be solved in linear time, as shown independently by Auer et al. [4] and Hong et al. [16]. In addition, every outer-1-planar graph admits an outer-1-planar straight-line drawing [11]. Note that an outer-1-planar graph is always planar [4], while this is not true in general for outer-fan-planar graphs. Indeed, the complete graph K_5 is outer-fan-planar, but not planar. Recently, it was shown that testing *full outer-2-planarity* (i.e., outer-2-planar embedding with no crossing on the outer face) of graphs can be solved in linear time [17].

A drawn graph is k-*quasi planar* if it has no k mutually crossing edges. It is conjectured that the number of edges of a k-quasi planar graph is linear in the number of its vertices. Pach et al. [20] and Ackerman [1] affirmatively answered this conjecture for 3- and 4-quasi planar graphs. Fox and Pach [12] showed that a k-quasi-planar n-vertex graph has at most $O(n \log^{1+o(1)} n)$ edges. Fan planar graphs are 3-quasi planar [18].

A different forbidden crossing pattern arises in RAC drawings where two edges are allowed to cross, if the crossings edges form right angles. Graphs that admit such drawings (with straight-line edges) are called *RAC graphs*. Didimo et al. [8] showed that a RAC graph with n vertices has no more than $4n - 10$ edges; a tight bound. RAC graphs are quasi planar [8], while *maximally dense* (i.e., exactly $4n - 10$ edges) RAC graphs are 1-planar [10]. Testing whether a graph is RAC is NP-hard [3], while testing outer-RAC graphs with a given vertex ordering and a rotation system can be solved in linear time [7].

Preliminaries. Unless otherwise specified, we consider finite, undirected, simple graphs. We also assume basic familiarity with SPQR-trees [15] (a short introduction is given in [5]). The *rotation system* of a drawing is the counterclockwise order of the incident edges around each vertex. The *embedding* of a drawn graph consists of its rotation system and for each edge the sequence of edges crossing it. For a graph G and a vertex $v \in V[G]$, we denote by $G - \{v\}$ the graph that results from G by removing v.

Lemma 1. *A biconnected graph G is outer-fan-planar if and only if it admits a straight-line outer-fan-planar drawing in which the vertices of G are restricted on a circle C.*

Sketch of Proof. Let G be an outer-fan-planar graph and let Γ be an outer-fan-planar drawing of G. We will only show that G has a straight-line outer-fan-planar drawing whose vertices lie on a circle C (the other direction is trivial). The order of the vertices along the outer face of Γ completely determines whether two edges cross, as in a simple drawing no two incident edges can cross and any two edges can cross at most once. Now, assume that two edges cross another edge in Γ. Then, both edges have to be incident to the same vertex; hence, cannot cross each other. So, the order of the crossings on an edge is also determined by the order of the vertices on the outer face. Therefore, we can construct a drawing Γ_C by placing the vertices of G on a circle C preserving their order in the outer face of Γ and draw the edges as straight-line segments.   □

2   Recognizing and Drawing Maximal Outer-Fan-Planar Graphs

In this section, we prove that given a graph $G = (V, E)$ on n vertices, there is a poly-nomial time algorithm to decide whether G is maximal outer-fan-planar and if so a corresponding straight-line drawing can be computed in linear time. By Lemma 1, we only have to check, whether G has a straight-line drawing on a circle C that is fan-planar. Note that such a drawing is determined by the cyclic order of the vertices on C. Since fan-planar graphs with n vertices have at most $5n - 10$ edges [18], we may assume that the number of edges is linear in the number of vertices. We first consider the case that G is 3-connected and then using SPQR-trees we show how the problem can be solved for biconnected graphs. Observe that biconnectivity is a necessary condition for maximal outer-fan-planarity. Indeed, if an outer-fan-planar drawing has a cut-vertex c, it is easy to see that it is always possible to draw an edge connecting two neighbors of c while preserving the outer-fan-planarity.

The 3-Connected Case. Assume that a straight-line drawing of a 3-connected graph G with n vertices on a circle C is given. Let $v_1, \ldots, v_n$ be the order of the vertices around C. An edge $\{v_i, v_j\}$ is an *outer edge*, if $i - j \equiv \pm 1 \pmod{n}$, a *2-hop*, if $i - j \equiv \pm 2 \pmod{n}$, and a *long edge* otherwise. G is a *complete 2-hop graph*, if there are all outer edges and all 2-hops, but no long edges. Two crossing long edges are a *scissor* if their end-points form two consecutive pairs of vertices on C. We say that a triangle is an *outer triangle* if two of its three edges are outer edges. We call an outer-fan-planar drawing *maximal*, if adding any edge to it yields a drawing that is not outer-fan-planar.

Our algorithm is based on the observation that if a graph is 3-connected maximal outer-fan-planar, then it is a complete 2-hop graph, or we can repeatedly remove any degree-3 vertex from any 4-clique until only a triangle is left. In a second step, we reinsert the vertices maintaining outer-fan-planarity (if possible). It turns out that we have to check a constant number of possible embeddings. In the following, we prove some necessary properties. The first three lemmas are used in the proof of Lemma 5. Their proofs are based on the 3-connectivity of the input graph; see Fig. 2a, 2b and 2c.

Lemma 2. *Let G be a 3-connected outer-fan-planar graph embedded on a circle C. If two long edges cross, then two of its end-points are consecutive on C.*

Lemma 3. *Let G be a 3-connected outer-fan-planar graph embedded on a circle C. If there are two long crossing edges, then there is a scissor, as well.*

Lemma 4. *Let G be a 3-connected graph embedded on a circle C with a maximal outer-fan-planar drawing. If G contains a scissor, then its end vertices induce a K_4.*

Lemma 5. *Let G be a 3-connected graph with a maximal outer-fan-planar drawing and assume that the drawing contains at least one long edge. Then, G contains a K_4 with all four vertices drawn consecutively on the circle.*

Proof. First consider the case where the graph contains at least two crossing long edges and, thus, by Lemma 3 a scissor. Removing the vertices of a scissor, splits G into two connected components. Assume that we have chosen the scissor such that the smaller

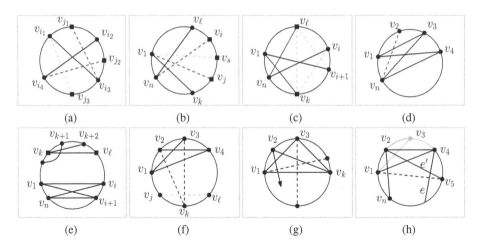

(a)   (b)   (c)   (d)

(e)   (f)   (g)   (h)

Fig. 2. Different configurations used in: (a) Lemma 2, (b) Lemma 3, (c, d) Lemma 4, (e) Lemma 5, (f) Lemma 6, (g) Lemma 7, (h) Lemma 9

of the two components is as small as possible (thus, scissor-free) and that the vertices around C are labeled such that this scissor is $\{v_1, v_{i+1}\}$, $\{v_i, v_n\}$ with $i \leq n - i$, i.e., the component induced by $v_2, \ldots, v_{i-1}$ is the smaller one. Recall that by Lemma 4 a scissor induces a K_4.

If $i = 3$, i.e., if $\{v_1, v_3\}$ is a 2-hop, then G should contain either $\{v_2, v_n\}$ or $\{v_2, v_4\}$, as otherwise v_1 and v_3 is a separation pair; see Fig. 2d. Say w.l.o.g. $\{v_2, v_n\}$. Then, v_1, v_2, v_3 together with v_n induce a K_4 with all vertices consecutive on circle C.

If $i > 3$, let $\{v_k, v_\ell\}$, $1 \leq k < \ell \leq i$ be a long edge such that there is no long edge $\{v_{k'}, v_{\ell'}\} \neq \{v_k, v_\ell\}$ with $k \leq k' < \ell' \leq \ell$; see Fig. 2e. Then, no long edge is crossing the edge $\{v_k, v_\ell\}$, as otherwise by Lemma 3 such a crossing would yield a new scissor, contradicting the choice of $\{v_1, v_{i+1}\}$ and $\{v_i, v_n\}$. Since $\{v_k, v_\ell\}$ is not crossed by a long edge, it must be crossed by exactly one 2-hop, say $\{v_{k-1}, v_{k+1}\}$. Now, $\ell - k > 3$ is not possible, since we could add the edge $\{v_{k+1}, v_\ell\}$, which is long. Hence, $\ell - k = 3$ and by maximality of the outer-fan-planar drawing, $v_k, v_{k+1}, v_{k+2}, v_\ell$ induces a K_4 with all vertices consecutive on C. Finally, if G contains no two crossing long edges, let $\{v_k, v_\ell\}$, $1 \leq k < \ell \leq n$ be a long edge such that there is no long edge $\{v_{k'}, v_{\ell'}\} \neq \{v_k, v_\ell\}$ with $k \leq k' < \ell' \leq \ell$. By the same argumentation as above, we obtain that $v_k, v_{k+1}, v_{k+2}, v_\ell$ induces a K_4 with all vertices consecutive on C.   □

Lemma 6. *Let G be a 3-connected outer-fan-planar graph with at least six vertices. If G contains a K_4 with all vertices drawn consecutively on circle C, then this K_4 contains exactly one vertex of degree three and this vertex is neither the first nor the last of the four vertices.*

Proof. Let the vertices around circle C be labeled so that v_1, v_2, v_3, v_4 induce a K_4. Since v_1 and v_4 is not a separation pair, there is an edge between v_2 or v_3 and a vertex, say v_k, among $v_5, \ldots, v_n$. Hence, three out of the four vertices v_1, v_2, v_3 and v_4 have degree at least four; see Fig. 2f. If v_3 had a neighbor in $v_5, \ldots, v_n$, then this could only be v_k, as otherwise $\{v_1, v_4\}$ would be crossed by two independent edges. Since G has at least 6 vertices, we assume w.l.o.g. that $k > 5$. Since v_4 and v_k is not a separation pair, there has to be an edge $\{v_\ell, v_m\}$ for some $4 < \ell < k$ and a $j \notin \{4, \ldots, k\}$. But such an edge would not be possible in an outer-fan-planar drawing. $\qquad\square$

Lemma 7. *Let G be a 3-connected outer-fan-planar graph with at least six vertices. If G contains a K_4 with a vertex of degree 3, then this K_4 has to be drawn consecutively on circle $\mathcal{C}$ in any outer-fan-planar drawing of G.*

Proof. Observe that any outer-fan-planar drawing of a K_4 contains exactly one pair of crossing edges. If two 2-hops cross, then all vertices of the K_4 are consecutive. If the K_4 contains two crossing long edges, then each of the vertices of the K_4 is incident to an outer edge not contained in the K_4; thus, has degree at least four. If a long edge and a 2-hop cross, assume that the vertices around $\mathcal{C}$ are labeled such that v_1, v_2, v_3, v_k induce a K_4 for some $5 \leq k < n$; see Fig. 2g. Since v_1, v_3 and v_k are incident to an outer edge not contained in the K_4, they have degree at least four. We claim that v_2 has degree at least four. Since v_3 and v_k is not a separation pair, there is an edge between a vertex among $v_4, \ldots, v_{k-1}$ and v_2 or v_1 and an edge between a vertex among $v_{k+1}, \ldots, v_n$ and v_2 or v_3. Choosing v_1 and v_3 in the first and second case respectively, yields two independent edges crossing $\{v_2, v_k\}$. So, v_2 is connected to a vertex outside K_4. $\qquad\square$

Lemma 8. *Let G be a 3-connected graph with $n \geq 5$ vertices and let $v \in V[G]$ be a vertex of degree three that is contained in a K_4. Then, $G - \{v\}$ is 3-connected.*

Proof. Let a, b, c and d be four arbitrary vertices of $G - \{v\}$. Since G was 3-connected, there was a path P from a to b in $G - \{c, d\}$. Assume that P contains v. Since v is only connected to vertices that are connected to each other, there is also another path from a to b in $G - \{c, d\}$ not containing v. Hence, a and b cannot be a separation pair in $G - \{v\}$. Since a and b were arbitrarily selected, $G - \{v\}$ is 3-connected. $\qquad\square$

Lemma 9. *Let G be a 3-connected graph with $n > 6$ vertices, let v_1, v_2, v_3 and v_4 be four vertices that induce a K_4, such that the degree of v_3 is three. Then, $G - \{v_3\}$ has a maximal outer-fan-planar drawing if G has a maximal outer-fan-planar drawing.*

Proof. Consider a maximal outer-fan-planar drawing of G on a circle $\mathcal{C}$ and let $v_1, v_2, v_3, v_4, \ldots, v_n$ be the order of the vertices on $\mathcal{C}$ (recall Lemma 7). Assume to the contrary that after removing v_3, we could add an edge e to the drawing; see Fig. 2h. By Lemma 6, $\{v_3, v_1\}$ is the only edge incident to v_3 that crosses some edges of $G - \{v_3\}$. Hence, there must be an edge e' that is crossed by e and $\{v_3, v_1\}$. Since $\{v_3, v_1\}$ crosses only edges incident to v_2 that also cross $\{v_1, v_4\}$, it follows that e' has to be incident to v_2. Further, since $G - \{v_3\}$ plus e is outer-fan-planar it follows that e is incident to v_1 or v_4. Moreover, since G plus e is not outer-fan-planar it follows that e is incident to v_4.

Let i be maximal so that there is an edge $\{v_2, v_i\}$. If $i \neq n$, then v_1 and v_i is a separation pair: Any edge connecting $\{v_{i+1}, \ldots, v_{n-1}\}$ to $\{v_2, v_3, \ldots, v_{i-1}\}$ and not being incident to v_2 crosses $\{v_2, v_i\}$. But edges crossing $\{v_2, v_i\}$ can only be incident to v_1, a contradiction. Now, let $j > 4$ be minimum such that there is an edge $\{v_2, v_j\}$.

We claim that $j = 5$. If this is not the case, then similarly to the previous case v_4 and v_j would be a separation pair in $G - \{v_3\}$ plus e, which is not possible due to Lemma 8.

It follows that G has to contain edge $\{v_1, v_5\}$: Since G is outer-fan-planar, in G there cannot be an edge $\{v_4, v_k\}$ for some $k = 6, \ldots, n$, since it would cross $\{v_2, v_5\}$ which is crossed by $\{v_3, v_1\}$. So, $\{v_1, v_5\}$ crosses only edges incident to v_2 that are already crossed by $\{v_3, v_1\}$ and $\{v_4, v_1\}$. Hence, $\{v_1, v_5\}$ could be added to G without violating outer-fan-planarity; a clear contradiction. Since e and $\{v_2, v_n\}$ both cross $\{v_1, v_5\}$ it follows that $e = \{v_4, v_n\}$. But now, v_5 and v_n has to be a separation pair.   □

Remark 1. Let G be a graph with 6 vertices containing a vertex v of degree three. Then G is maximal outer-fan-planar if and only if $G - \{v\}$ is a K_5 missing one of the edges that connects a neighbor of v to one of the other two vertices.

Lemma 10. *It can be tested in linear time whether a graph is a complete 2-hop graph. Moreover, if a graph is a complete 2-hop graph, then it has a constant number of outer-fan-planar embeddings and these can be constructed in linear time.*

Proof. Let G be an n-vertex graph. We test whether G is a complete 2-hop as follows. If $n \in \{4, 5\}$, then G is either K_4 or K_5. Otherwise, check first whether all vertices have degree four. If so, pick one vertex as v_1, choose a neighbor as v_2 and a common neighbor of v_1 and v_2 as v_3 (if no such common neighbor exists then G is not a complete 2-hop). Assume now that we have already fixed $v_1, \ldots, v_i$, $3 \leq i < n$. Test whether there is a unique vertex $v \in V \setminus \{v_1, \ldots, v_i\}$ that is adjacent to v_i and v_{i-1}. If so, set $v_{i+1} = v$. Otherwise reject. If we have fixed the order of all vertices check whether there are only outer edges and 2-hops. Do this for any possible choices of v_2 and v_3, i.e., for totally at most 6 choices.   □

Remark 2. No degree 3 vertex can be added to an n-vertex complete 2-hop with $n \geq 5$.

We are now ready to describe our algorithm. If the graph is not a complete 2-hop graph, recursively try to remove a vertex of degree 3 which is contained in a K_4. If G is maximal outer-fan-planar, Lemmas 5 and 6 guarantee that such a vertex always exists in the beginning. Remark 2 guarantees that also in subsequent steps there is a long edge and, thus, Lemmas 8 and 9 guarantee that also in subsequent steps, we can apply Lemma 5 as long as we have at least six vertices. Remark 1 guarantees that we can also remove two more vertices of degree 3 ending with a triangle.

At this stage, we already know that if the graph is outer-fan-planar, it is indeed maximal outer-fan-planar. Either, we started with a complete 2-hop graph or we iteratively removed vertices of degree three yielding a triangle. Note that in the latter case we must have started with $3n - 6$ edges. On the other hand, if we apply the above procedure to an n-vertex 3-connected maximal outer-fan-planar graph, we get that the number of its edges is exactly $2n$ or $3n - 6$.

Finally, we try to reinsert the vertices in the reversed order in which we have deleted them. By Lemma 7, we can insert the vertex of degree three only between its neighbor, that is, there are at most two possibilities where we could insert the vertex. Lemma 11 guarantees that in total, we have to check at most four possible drawings for G.

Lemma 11. *When reinserting a sequence of degree 3 vertices starting from a triangle, at most the first two vertices have two choices where they could be inserted.*

Proof. Let H be a outer-fan-planar graph and let three consecutive vertices v_1, v_2, v_3 induce a triangle. Assume, we want to insert a vertex v adjacent to v_1, v_2, v_3. By Lemma 6, we have to insert v between v_1 and v_2 or between v_2 and v_3. Note that the edges that are incident to v_2 and cross $\{v_1, v_3\}$ are also crossed by an edge e incident to v. So, if there is an edge incident to v_2 that was already crossed twice before inserting v, this would uniquely determine whether e is incident to v_1 or v_3 and, thus, where to insert v.

We will now show that after the first insertion each relevant vertex is incident to an edge that is crossed at least twice. When we insert the first vertex we create a K_4. From the second vertex on, whenever we insert a new vertex, it is incident to an edge that is crossed at least twice. Also, after inserting the second degree 3 vertex, three among the four vertices of the initial K_4 are also incident to an edge that is crossed at least twice. The forth vertex of the initial K_4 is not the middle vertex of a triangle consisting of three consecutive vertices. It can only become such a vertex if its incident inner edges are crossed by a 2-hop. But then these inner edges are all crossed at least twice.   □

Summarizing, we obtain the following theorem; in order to exploit this result in the biconnected case, it is also tested whether a prescribed subset (possibly empty) of edges can be drawn as outer edges.

Theorem 1. *Given a 3-connected graph G with a subset E' of its edge set, it can be tested in linear time whether G is maximal outer-fan-planar and has an outer-fan-planar drawing such that the edges in E' are outer edges. Moreover if such a drawing exists, it can be constructed in linear time.*

Sketch of Proof. Let n be the number of vertices. By Lemma 10, a complete 2-hop graph has only a constant number of outer-fan-planar embeddings which can be computed in linear time. Whenever we remove a vertex from the graph, we append it to a queue. Any vertex that was removed from the queue will never be appended again. Hence, there are at most n iterations.

To check whether the degree three vertices can be reinserted back in the graph, we only have to consider in total four different embeddings. Assume that we want to insert a vertex v into an outer triangle v_1, v_2, v_3. Then we just have to check whether v_1 or v_3 are incident to edges other than the edge $\{v_1, v_3\}$ that cross an edge incident to v_2. This can be done in constant time by checking only two pairs of edges.   □

The Biconnected Case. We now sketch how to test outer-fan-planar maximality on a biconnected graph.

Lemma 12. *Let $v_1, \ldots, v_n$ be the order of the vertices around the circle in an outer-fan-planar drawing of a 3-connected graph G. If we can add a vertex v between v_1 and v_n with an edge $\{v, v_i\}$ for some $i = 2, \ldots, n-1$, then $i = 2$ or $i = n-1$.*

Proof. Otherwise, since v_1, v_i cannot be a separation pair of G, there has to be an edge from a v_k for some $k = 2, \ldots, i-1$ that crosses $\{v, v_i\}$ and hence an edge $\{v_k, v_n\}$. Since v_n, v_i cannot be a separation pair of G, there has to be an edge $\{v_1, v_\ell\}$ for some $\ell = i+1, \ldots, n-1$. But now there are three independent edges crossing.   □

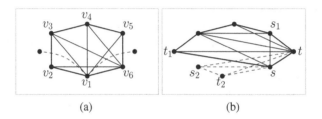

(a)   (b)

Fig. 3. (a) In the solid graph, edge $\{v_2, v_3\}$ ($\{v_5, v_6\}$) is porous around v_2 (v_6, resp.). (b) Illustration of Case 2 of Theorem 2.

We say that an outer edge $\{v_1, v_n\}$ is *porous* around v_1 if we could add a vertex v between v_1 and v_n and an edge $\{v, v_2\}$ maintaining outer-fan-planarity. Note that any edge of a simple cycle, i.e., of the skeleton of an S-node is porous around any of its end vertices. Any outer edge of a K_4 is porous around any of its end vertices; see Fig. 3.

We use the SPQR-tree of a biconnected graph to characterize whether it is maximal outer-fan-planar; see [5] for a proof of this theorem.

Theorem 2. *A biconnected graph is maximal outer-fan-planar iff the following hold:*

1) *The skeleton of any R-node is maximal outer-fan-planar and has an outer-fan-planar drawing in which all virtual edges are outer edges,*
2) *No R-node is adjacent to an R-node or an S-node,*
3) *All S-nodes have degree three,*
4) *All P-nodes have degree three and are adjacent to a Q-node, and*
5) *Let G_1 and G_2 be the skeleton of the two neighbors of a P-node other than the Q-node and let $\{s, t\}$ be the common virtual edge of G_1 and G_2. Then $G_i, i = 1, 2$ must not admit an outer-fan-planar drawing with t_i, s, t, s_i being consecutive around the circle and*
   (a) *edge $\{s, t\}$ is porous in both G_1 and G_2 around the same vertex, or*
   (b) *edge $\{t_1, s\}$ ($\{s_2, t\}$) is real and porous around s (t, resp.), or*
   (c) *edge $\{s_1, t\}$ ($\{t_2, s\}$) is real and porous around t (s, resp.).*

As the number of outer-fan-planar embeddings of a 3-connected graph is bounded by a constant, the conditions of Thm. 2 can be tested in polynomial time. If the conditions are fulfilled, then an outer-fan-planar drawing can be constructed in linear time.

3   Recognizing Fan-Planar Graphs with Rotation System

In this section, we study the FAN-PLANARITY WITH FIXED ROTATION SYSTEM problem (FP-FRS), that is, the problem of deciding whether a graph $G = (V, E)$ with a fixed rotation system $\mathcal{R}$ admits a fan-planar drawing preserving $\mathcal{R}$.

Theorem 3. FAN-PLANARITY WITH FIXED ROTATION SYSTEM *is NP-hard.*

Proof. We prove the statement by using a reduction from 3-PARTITION (3P). An instance of 3P is a multi-set $A = \{a_1, a_2, \ldots, a_{3m}\}$ of $3m$ positive integers in the range $(B/4, B/2)$, where B is an integer such that $\sum_{i=1}^{3m} a_i = mB$. 3P asks whether A can be partitioned into m subsets $A_1, A_2, \ldots, A_m$, each of cardinality 3, such that the sum of the numbers in each subset is B. As 3P is *strongly* NP-hard [13], it is not restrictive to assume that B is bounded by a polynomial in m.

Before describing our transformation, we need to introduce the concept of *barrier gadget*. An n-vertex *barrier gadget* is a graph consisting of a cycle of $n \geq 5$ vertices plus all its 2-hop edges; a barrier gadget is therefore a maximal outer-2-planar graph. We make use of barrier gadgets in order to constraint the routes of some specific paths of G_A. Indeed, in a fan-planar drawing of a biconnected graph containing an outer-2-planar drawing Γ_b of a barrier gadget, no path can enter inside the boundary cycle of Γ_b and cross a 2-hop edge. Also, if a path enters in Γ_b without crossing any 2-hop edge, then it must immediately exit from Γ_b forming a fan-crossing with an outer edge of Γ_b.

Now, we are ready to describe how to transform an instance A of 3P into an instance $\langle G_A, \mathcal{R}_A \rangle$ of FP-FRS. We start from the construction of graph G_A which will be always biconnected. First of all, we create a *global ring barrier* by attaching four barrier gadgets G_t, G_r, G_b and G_l as depicted in Fig. 4. G_t is called the *top beam* and contains exactly $3mK$ vertices, where $K = \lceil B/2 \rceil + 1$. G_r is the *right wall* and has only five vertices. G_b and G_r are called the *bottom beam* and the *left wall*, respectively, and they are defined in a specular way. Observe that G_t, G_r, G_b and G_l can be embedded so that all their vertices are linkable to points within the closed region delimited by the global ring barrier. Then, we connect the top and bottom beams by a set of $3m$ *columns*, see Fig. 4 for an illustration of the case $m = 3$. Each *column* consists of a stack of $2m - 1$ *cells*; a *cell* consists of a set of pairwise disjoint edges, called the *vertical edges* of that cell. In particular, there are $m - 1$ *bottommost cells*, one *central cell* and $m - 1$ *topmost cells*. Cells of a same column are separated by $2m - 2$ barrier gadgets, called *floors*. Central cells (that are $3m$ in total) have a number of vertical edges depending on the elements of A. Precisely, the central cell C_i of the i-th column contains a_i vertical edges connecting its delimiting floors ($i \in \{1, 2, \ldots, 3m\}$). Instead, all the remaining cells have, each one, K vertical edges. Hence, a non-central cell contains more edges than any central cell. Further, the number of vertices of a floor is given by the number of its incident vertical edges minus two. Let u and v be the "central" vertices of the left and right walls, respectively (see also Fig. 4). We conclude the construction of graph G_A by connecting vertices u and v with m pairwise internally disjoint paths, called the *transversal paths* of G_A; each transversal path has exactly $(3m - 3)K + B$ edges.

Concerning the choice of a rotation system $\mathcal{R}_A$, we define a cyclic order of edges around each vertex that is compatible with the one depicted in Fig. 4. From what said, it is straightforward to see that an instance of 3P can be transformed into an instance of FP-FRS in polynomial time in m.

Let A be a *Yes*-instance of 3P, we show that $\langle G_A, \mathcal{R}_A \rangle$ admits a fan-planar drawing Γ_A preserving $\mathcal{R}_A$. We observe that such a drawing is easy to compute if one omits all the transversal paths. It is essentially a drawing like that one depicted in Fig. 4, where columns are one next to the other within the closed region delimited by the global ring barrier. However, by exploiting a solution $\{A_1, A_2, \ldots, A_m\}$ of 3P for the

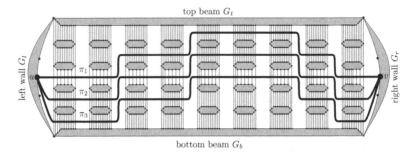

Fig. 4. Illustration of the reduction of FP-FRS from 3P, where $m = 3$, $B = 24$ and $A = \{7, 7, 7, 8, 8, 8, 8, 9, 10\}$. Transversal paths are routed according to the following solution of 3P: $A_1 = \{7, 7, 10\}$, $A_2 = \{7, 8, 9\}$ and $A_3 = \{8, 8, 8\}$.

instance A, also the transversal paths can be easily embedded without violating the fan-planarity. The idea is to route these paths in such a way that: (R.1) they do not cross each other; (R.2) they do not cross any barrier; (R.3) each path passes through exactly 3 central cells and $3m - 3$ non-central cells; (R.4) each cell is traversed by at most one path. Eventually, each transversal path crosses exactly $(3m - 3)K + B$ vertical edges, which is the same number of its edges. Therefore, it is possible to draw these paths by ensuring that each of their edges crosses exactly one vertical edge, which preserves the fan-planarity. Hence, eventually we get a fan-planar drawing Γ_A preserving the rotation system $\mathcal{R}_A$.

We conclude the proof by showing that if $\langle G_A, \mathcal{R}_A \rangle$ is a *Yes*-instance of FP-FRS, then A is a *Yes*-instance of 3P. Let Γ_A be a fan-planar drawing of G_A preserving the rotation system $\mathcal{R}_A$. We first observe that the top beam and the bottom beam are disjoint, otherwise there would be at least a 2-hope edge in one beam that is crossed by another edge of the other beam, thus violating the fan-planarity. We also note that columns can partially cross each other, but this does not actually affect the validity of the proof. Indeed, an edge e of a column L might cross an edge e' of another column L' only if e is incident to a vertex in the rightmost (leftmost) side of L, e' is a leftmost (rightmost) vertical edge of L', and L and L' are two consecutive columns. With a similar argument, it is immediate to see that vertices u and v must be separated by all the columns. Therefore, every transversal path satisfies conditions R.1, R.2 and it must pass through at least three central cells, if not it would cross a number of pairwise disjoint edges that is greater than the number of its edges, hence Γ_A would not be fan-planar. On the other hand, because of condition R.4, which is obviously satisfied, there cannot be any transversal path passing through more than three central cells. Otherwise, there would be some other transversal path that traverses a number of central cells that is strictly less than three. Hence, also condition R.3 is satisfied. In conclusion, every transversal path π_j ($j \in \{1, 2, \dots, m\}$) crosses $(3m-3)K+B$ vertical edges and traverses exactly three central cells C_{1j}, C_{2j} and C_{3j}. If $m(C_{1j})$, $m(C_{2j})$ and $m(C_{3j})$ denote the number of edges of these cells, then $m(C_{1j}) + m(C_{2j}) + m(C_{3j}) = B$, because each non-central cell has K edges. Therefore, the partitioning of A defined by $A_1, A_2, \dots, A_m$, where $A_j = \{m(C_{1j}), m(C_{2j}), m(C_{3j})\}$, is a solution of 3P for the instance A. □

References

1. Ackerman, E.: On the maximum number of edges in topological graphs with no four pairwise crossing edges. Discrete & Computational Geometry 41(3), 365–375 (2009)
2. Agarwal, P.K., Aronov, B., Pach, J., Pollack, R., Sharir, M.: Quasi-planar graphs have a linear number of edges. Combinatorica 17(1), 1–9 (1997)
3. Argyriou, E.N., Bekos, M.A., Symvonis, A.: The straight-line RAC drawing problem is NP-hard. J. Graph Algorithms Appl. 16(2), 569–597 (2012)
4. Auer, C., Bachmaier, C., Brandenburg, F.J., Gleißner, A., Hanauer, K., Neuwirth, D., Reislhuber, J.: Recognizing outer 1-planar graphs in linear time. In: Wismath, S., Wolff, A. (eds.) GD 2013. LNCS, vol. 8242, pp. 107–118. Springer, Heidelberg (2013)
5. Bekos, M.A., Cornelsen, S., Grilli, L., Hong, S.H., Kaufmann, M.: On the recognition of fan-planar and maximal outer-fan-planar graphs. CoRR abs/1409.0461 (September 2014)
6. Cheong, O., Har-Peled, S., Kim, H., Kim, H.S.: On the number of edges of fan-crossing free graphs. In: Cai, L., Cheng, S.-W., Lam, T.-W. (eds.) ISAAC 2013. LNCS, vol. 8283, pp. 163–173. Springer, Heidelberg (2013)
7. Reisi Dehkordi, H., Nguyen, Q., Eades, P., Hong, S.-H.: Circular graph drawings with large crossing angles. In: Ghosh, S.K., Tokuyama, T. (eds.) WALCOM 2013. LNCS, vol. 7748, pp. 298–309. Springer, Heidelberg (2013)
8. Didimo, W., Eades, P., Liotta, G.: Drawing graphs with right angle crossings. Theor. Comput. Sci. 412(39), 5156–5166 (2011)
9. Eades, P., Hong, S.H., Katoh, N., Liotta, G., Schweitzer, P., Suzuki, Y.: A linear time algorithm for testing maximal 1-planarity of graphs with a rotation system. Theor. Comput. Sci. 513, 65–76 (2013)
10. Eades, P., Liotta, G.: Right angle crossing graphs and 1-planarity. Discrete Applied Mathematics 161(7-8), 961–969 (2013)
11. Eggleton, R.: Rectilinear drawings of graphs. Utilitas Mathematica 29, 149–172 (1986)
12. Fox, J., Pach, J., Suk, A.: The number of edges in k-quasi-planar graphs. SIAM J. Discrete Math. 27(1), 550–561 (2013)
13. Garey, M.R., Johnson, D.S.: Computers and Intractability: A Guide to the Theory of NP-Completeness. W. H. Freeman & Co., New York (1979)
14. Grigoriev, A., Bodlaender, H.L.: Algorithms for graphs embeddable with few crossings per edge. Algorithmica 49(1), 1–11 (2007)
15. Gutwenger, C., Mutzel, P.: A linear time implementation of SPQR-trees. In: Marks, J. (ed.) GD 2000. LNCS, vol. 1984, pp. 77–90. Springer, Heidelberg (2001)
16. Hong, S.H., Eades, P., Katoh, N., Liotta, G., Schweitzer, P., Suzuki, Y.: A linear-time algorithm for testing outer-1-planarity. In: Wismath, S., Wolff, A. (eds.) GD 2013. LNCS, vol. 8242, pp. 71–82. Springer, Heidelberg (2013)
17. Hong, S.H., Nagamochi, H.: Testing full outer-2-planarity in linear time. Technical Report 2014-003, Department of Applied Mathematics and Physics, Kyoto University (2014)
18. Kaufmann, M., Ueckerdt, T.: The density of fan-planar graphs. CoRR abs/1403.6184 (2014)
19. Korzhik, V.P., Mohar, B.: Minimal obstructions for 1-immersions and hardness of 1-planarity testing. Journal of Graph Theory 72(1), 30–71 (2013)
20. Pach, J., Radoičić, R., Tóth, G.: Relaxing planarity for topological graphs. In: Akiyama, J., Kano, M. (eds.) JCDCG 2002. LNCS, vol. 2866, pp. 221–232. Springer, Heidelberg (2003)
21. Pach, J., Tóth, G.: Graphs drawn with few crossings per edge. Combinatorica 17(3), 427–439 (1997)
22. Purchase, H.C.: Effective information visualisation: a study of graph drawing aesthetics and algorithms. Interacting with Computers 13(2), 147–162 (2000)
23. Ringel, G.: Ein Sechsfarbenproblem auf der Kugel. Abh. Math. Sem. Univ. Hamburg 29, 107–117 (1965)

Crossing Minimization for 1-page and 2-page Drawings of Graphs with Bounded Treewidth

Michael J. Bannister and David Eppstein

Department of Computer Science, University of California, Irvine

Abstract. We investigate crossing minimization for 1-page and 2-page book drawings. We show that computing the 1-page crossing number is fixed-parameter tractable with respect to the number of crossings, that testing 2-page planarity is fixed-parameter tractable with respect to treewidth, and that computing the 2-page crossing number is fixed-parameter tractable with respect to the sum of the number of crossings and the treewidth of the input graph. We prove these results via Courcelle's theorem on the fixed-parameter tractability of properties expressible in monadic second order logic for graphs of bounded treewidth.

1   Introduction

A *k-page book embedding* of a graph G is a drawing that places the vertices of G on a line (the *spine* of the book) and draws each edge, without crossings, inside one of k half-planes bounded by the line (the *pages* of the book) [16,19]. In one common drawing style, an *arc diagram*, the edges in each page are drawn as circular arcs perpendicular to the spine [24], but the exact shape of the edges is unimportant for the existence of book embeddings. These embeddings can be generalized to *k-page book drawings*: as before, we place each vertex on the spine and each edge within a single page, but with crossings allowed. The *crossing number* of such a drawing is defined to be the sum of the numbers of crossings within each page, and the *k-page crossing number* $\mathrm{cr}_k(G)$ is the minimum number of crossings in any k-page book drawing [22]. In an optimal drawing, two edges in the same page cross if and only if their endpoints form interleaved intervals on the spine, so the problem of finding an optimal drawing may be solved by finding a permutation of the vertices and an assignment of edges to pages minimizing the number of pairs of edges with interleaved intervals on the same page.

As with most crossing minimization problems, k-page crossing minimization is NP-hard; even the simple special case of testing whether the 2-page crossing number is zero is NP-complete [8]. However, it may still be possible to solve these problems in polynomial time for restricted families of graphs and restricted values of k. For instance, recently Bannister, Eppstein and Simons [3] showed the computation of $\mathrm{cr}_1(G)$ and $\mathrm{cr}_2(G)$ to be fixed-parameter tractable in the almost-tree parameter; here, a graph G has almost-tree parameter k if every biconnected component of G can be reduced to a tree by removing at most k edges. In this paper we improve these results by finding fixed-parameter tractable algorithms

C. Duncan and A. Symvonis (Eds.): GD 2014, LNCS 8871, pp. 210–221, 2014.

for stronger parameters, allowing k-page crossing minimization to be performed in polynomial time for a much wider class of graphs.

New Results. We design fixed-parameter algorithms for computing the minimum number of crossings $cr_1(G)$ in a 1-page drawing of a graph G, and the minimum number of crossings $cr_2(G)$ in a 2-page drawing of G. Ideally, fixed-parameter algorithms for crossing minimization should be parameterized by their *natural parameter*, the optimal number of crossings. We achieve this ideal bound, for the first time, for $cr_1(G)$. However, for $cr_2(G)$, even testing whether a given graph is 2-page planar (that is, whether $cr_2(G) = 0$) is NP-complete [8]. Therefore, unless P = NP, there can be no fixed-parameter-tractable algorithm parameterized by the crossing number. Instead, we show that $cr_2(G)$ is fixed-parameter tractable in the sum of the natural parameter and the treewidth of G. One consequence of our result on $cr_2(G)$ is that it is possible to test whether a given graph is 2-page planar, in time that is fixed-parameter tractable with respect to treewidth.

We construct these algorithms via Courcelle's theorem [9,10], which connects the expressibility of graph properties in monadic second order logic with the fixed-parameter tractability of these properties with respect to treewidth. Recall that second order logic extends first order logic by allowing the quantification of k-ary relations in addition to quantification over individual elements. In monadic second order logic we are restricted to quantification over unary relations (equivalently subsets) of vertices and edges. The property of having a 2-page book embedding is easy to express in (full) second-order logic, via the known characterization that a graph has such an embedding if and only if it is a subgraph of a Hamiltonian planar graph [4]. However, this expression is not allowed in monadic second-order logic because the extra edges needed to make the input graph Hamiltonian cannot be described by a subset of the existing vertices and edges of the graph. Instead, we prove a new structural description of 2-page planarity that is more easily expressed in monadic second order logic.

Related Work. As well as the previous work on crossing minimization for almost-trees [3], related results in fixed-parameter optimization of crossing number include a proof by Grohe, using Courcelle's theorem, that the topological crossing number of a graph is fixed-parameter tractable in its natural parameter [15]. This result was later improved by Kawarabayashi and Reed [17]. Based on these results the crossing number itself was also shown to be fixed-parameter tractable; Pelsmajer et al. showed a similar result for the odd crossing number [20]. In *layered graph drawing*, Dujmović et al. showed that finding a drawing with k crossings and h layers is fixed-parameter tractable in the sum of these two parameters; this result depends on a bound on the pathwidth of such a drawing, a parameter closely related to its treewidth [11].

Like many of these earlier algorithms, our algorithms have a high dependence on their parameter, rendering them impractical. For this reason we have not attempted an exact analysis of their complexity nor have we searched for optimizations to our logical formulae that would improve this complexity.

2   Preliminaries

Bridges vs Flaps and Isthmuses. There is an unfortunate terminological confusion in graph theory: two different concepts, a maximal subgraph that is internally connected by paths that avoid a given cycle, and an edge whose removal disconnects the graph, are both commonly called *bridges*. We need both concepts in our algorithms. To avoid confusion, we call the subgraph-type bridges *flaps* and the edge-type bridges *isthmuses*. To be more precise, given a graph G and a cycle C, we define an equivalence relation on the edges of $G \setminus C$ in which two edges are equivalent if they belong to a path that has no interior vertices in C, and we define a *flap* of C to be the subgraph formed by an equivalence class of this relation. (Different cycles may give rise to different flaps.) Given a graph G, we define an *isthmus* of G to be an edge of G that does not belong to any simple cycles in G.

Treewidth and Graph Minors. The *treewidth* of G can be defined to be one less than the number of vertices in the largest clique in a chordal supergraph of G that (among possible chordal supergraphs) is chosen to minimize this clique size [6]. The problem of computing the treewidth of a general graph is NP-hard [1], but it is fixed-parameter tractable in its natural parameter [5].

A graph H is said to be a *minor* of a graph G if H can be constructed from G via a sequence edge contractions, edge deletions, and vertex deletions. It can be determined whether a graph H is a minor of a graph G, in time that is polynomial in the size of G and fixed-parameter tractable in the size of H [21].

Logic of Graphs. We will be expressing graph properties in *extended monadic second-order logic* ($\mathrm{MSO_2}$). This is a fragment of second-order logic that includes:

- variables for vertices, sets of vertices, edges, and sets of edges;
- binary relations for equality ($=$), inclusion of an element in a set ($\in$) and edge-vertex incidence (I);
- the standard propositional logic operations: $\neg, \wedge, \vee, \rightarrow$;
- the universal quantifier ($\forall$) and the existential quantifier ($\exists$), both which may be applied to variables of any of the four variable types.

To distinguish the variables of different types, we will use $u, v, w, \ldots$ for vertices, $e, f, g, \ldots$ for edges, and capital letters for sets of vertices or edges (with context making clear which type of set). Given a graph G and an $\mathrm{MSO_2}$ formula ϕ we write $G \models \phi$ ("G models ϕ") to express the statement that ϕ is true for the vertices, and sets of vertices and edges in G, with the semantics of this relation defined in the obvious way. $\mathrm{MSO_2}$ differs from full second order logic in that it allows quantification over sets, but not over higher order relations, such as sets of pairs of vertices that are not subsets of the given edges.

The reason we care about expressing graph properties in $\mathrm{MSO_2}$ is the following powerful algorithmic meta-theorem due to Courcelle.

Lemma 1 (Courcelle's theorem [9, 10]). *Given an integer $k \geq 0$ and an MSO_2-formula ϕ of length ℓ, an algorithm can be constructed that takes as input a graph G of treewidth at most k and decides in $O\big(f(k, \ell) \cdot (n + m)\big)$ time whether $G \models \phi$, where the function f appearing in the time bound is a computable function of the treewidth k and formula length ℓ.*

Combinatorial Enumeration of Crossing Diagrams. In order to show that the properties we study can be represented by logical formulas of finite length, we need to bound the number of combinatorially distinct ways that a subset of edges in a k-page graph drawing can cross each other.

We define a 1-*page crossing diagram* to be a placement of some points on the circumference of a circle, together with some straight line segments connecting the points such that each point is incident to a segment, no segment is uncrossed and no three segments cross at the same point. Two crossing diagrams are *combinatorially equivalent* if they have the same numbers of points and line segments and there exists a cyclic-order-preserving bijection of their points that takes line segments to line segments. The *crossing number* of a 1-page crossing diagram is the number of pairs of its line segments that cross each other.

We define a 2-*page crossing diagram* to be a 1-page crossing diagram together with a labeling of its line segments by two colors. For a 2-page crossing diagram we define the *crossing number* to be the total number of crossing pairs of line segments that have the same color as each other.

Lemma 2. *There are $2^{O(k^2)}$ 1-page crossing diagrams with k crossings, and there are $2^{O(k^2)}$ 2-page crossing diagrams with k crossings.*

Proof. Place $4k$ points around a circle. Then every 1-page crossing diagram with k or fewer crossings can be represented by choosing a subset of the points and a set of line segments connecting a subset of pairs of the points. There are $4k$ points and $4k(4k - 1)/2$ pairs of points, so $2^{O(k^2)}$ possible subsets to choose.

Similarly, every 2-page crossing diagram can be represented by a subset of the same $4k$ points, and two disjoint subsets of pairs of points, which again can be bounded by $2^{O(k^2)}$. $\qquad\Box$

Two combinatorially equivalent crossing diagrams, as defined above, may have a topology that differs from each other, or from combinatorially equivalent diagrams with curved edges. This is because, for an edge with multiple crossings, the order of the crossings along this edge may differ from one diagram to another, but this ordering is not considered as part of the definition of combinatorial equivalence. For our purposes such differences are unimportant, as we are concerned only with the total number of crossings. So we consider two crossing diagrams to be equivalent if they have the same crossing pairs of edges, regardless of whether the crossings occur in the same order.

3   1-page Crossing Minimization

Outerplanarity. Recall that a graph is *outerplanar* if there exists a placement of its vertices on the circumference of a circle such that when its edges are drawn as straight line segments they do not cross. Topologically, the circle and the half-plane are equivalent, so a graph is outerplanar if and only if it has a crossing-free 1-page drawing. For incorporating a test of outerplanarity into methods using Courcelle's theorem, it is convenient to use a standard characterization of the outerplanar graphs by forbidden minors:

Lemma 3 (Chartrand and Harary [7]). *A graph G is outerplanar (1-page planar) if and only if it contains neither K_4 nor $K_{2,3}$ as a minor.*

Lemma 4 (Corollary 1.15 in [10]). *Given any fixed graph H there exists a MSO_2-formula ϕ such that, for all graphs G, $G \models \phi$ if and only if G contains H as a minor. We will write* MINOR$_H$ *for ϕ.*

Let OUTERPLANAR be the formula $\neg$ MINOR$_{K_4}$ $\wedge \neg$ MINOR$_{K_{2,3}}$. Then Lemma 3 implies that, for all graphs G, $G \models$ OUTERPLANAR if and only if G is outerplanar. Because outerplanar graphs have bounded treewidth (at most two), Courcelle's theorem together with Lemma 4 guarantee the existence of a linear time algorithm for testing outerplanarity. There are of course much simpler linear time algorithms for testing outerplanarity [18, 25].

Crossings vs Treewidth. Next, we relate the natural parameter for 1-page crossing minimization (the number of crossings) to the parameter for Courcelle's theorem (the treewidth). This relation will allow us to construct a fixed-parameter-tractable algorithm for the natural parameter.

Lemma 5. *Every graph G has treewidth $O(\sqrt{cr_1(G)})$.*

See the full version of this paper (arXiv:1408.6321) for the proof.

Logical Characterization. Let G be a graph with bounded 1-page crossing number, and consider a drawing of G achieving this crossing number. Then the set of crossing edges of the drawing partitions the halfplane into an arrangement of curves, and we can partition G itself into the subgraphs that lie within each face of this arrangement. Each of these subgraphs is itself outerplanar, because it lies within a subset of the halfplane (with its vertices on the boundary of the subset) and has no more crossing edges; see Figure 1. This intuitive idea forms the basis for the following characterization of the 1-page crossing number, which we will use to construct an MSO_2-formula for the property of having a drawing with low crossing number.

Lemma 6. *A graph $G = (V, E)$ has $cr_1(G) \leq k$ if and only if there exist edges $F = \{e_0, \ldots, e_r\}$ with $r = O(k)$, vertices $W = \{v_0, \ldots, v_\ell\}$ with $\ell = O(k)$, and a partition $U_0, \ldots, U_\ell$ of $V \setminus W$ into (possibly empty) subsets, satisfying the following properties:*

1. *W is the set of vertices incident to edges in F.*
2. *F contains all edges in the induced subgraph on W.*
3. *There are no edges between U_i and U_j for $i \neq j$.*

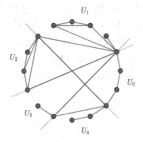

Fig. 1. A 1-page drawing of a graph with two crossings and five outerplanar subgraphs.

Fig. 2. A 2-page planar graph with its edges partitioned into the six sets A_b (green edges), A_c (blue edges), A_i (red edges), B_b (yellow edges), B_c (purple edges), and B_i (gray edges).

4. There is an outerplanar embedding of the induced subgraph on $U_i \cup \{v_i, v_{i+1}\}$ with v_i and v_{i+1} adjacent for all $0 \leq i < \ell$.
5. The edges in F produce at most k crossings when their endpoints (the vertices in W) are placed in order according to their indices.

We now construct a formula ONEPAGE$_k$, based on Lemma 6, such that $G \models$ ONEPAGE$_k$ if and only if $\mathrm{cr}_1(G) \leq k$. The formula ONEPAGE$_k$ will have the overall form of a disjunction, over all crossing configurations, of a conjunction of sub-formulas representing Properties 1–4 in Lemma 6. Property 5 will be represented implicitly, by the enumeration of crossing configurations. The first three properties are easy to express directly: the formulas

$$\theta_1(W, F) \equiv (\forall v)[v \in W \to (\exists e)[e \in F \wedge I(e, v)]]$$
$$\theta_2(F, W) \equiv (\forall e)[(\forall v)[I(e, v) \to v \in W] \to e \in F]$$
$$\theta_3(U_i, U_j) \equiv \neg(\exists e)(\exists u, v)[I(e, u) \wedge I(e, v) \wedge u \in U_i \wedge v \in U_j]$$

express in MSO$_2$ Properties 1, 2, and 3 of Lemma 6 respectively.

To express Property 4 we first observe that it is equivalent to the property that the induced subgraph on $U_i \cup \{v_i, v_{i+1}\}$ with v_i and v_{i+1} identified (merged) to form a single supervertex is outerpalanar. That is, the requirement in Property 4 that vertices v_i and v_{i+1} be adjacent in the outerplanar embedding can be enforced by identifying the vertices. To express this property we need the following lemma, which can be proved in straightforward manner using the method of syntactic interpretations. (For details on this method see [13, 15].)

Lemma 7. For every MSO$_2$-formula ϕ there exists an MSO$_2$-formula $\phi^*(v_1, v_2)$ such that $G \models \phi^*(a, b)$ if and only if $G/a \sim b \models \phi$, where $G/a \sim b$ is the graph constructed from G by identifying vertices a and b.

Now, to construct $\theta_4(U_i, v_i, v_j)$ we first modify the formula OUTERPLANAR by restricting its quantifiers to only quantify over vertices (and sets of vertices) in $U_i \cup \{v_i, v_j\}$ and edges (and sets of edges) between these vertices. This modified formula describes the outerplanarity of $U_i \cup \{v_i, v_j\}$. We then apply the transformation of Lemma 7 to produce the formula $\theta_4(U_i, v_i, v_j)$, expressing the outerplanarity of the induced graph on $U_i \cup \{v_i, v_j\}$ with v_i and v_j identified.

Lemma 2 tells us that there are $2^{O(k^2)}$ ways of satisfying Property 5 of Lemma 6. For each crossing diagram D with k crossings we can construct a formula $\alpha_D(v_0, \ldots, v_\ell, e_0, \ldots, e_r)$ specifying that the vertices $v_0, \ldots, v_\ell$ and edges $e_0, \ldots, e_r$ are in configuration D. We then construct the formula

$$\beta_D \equiv (\exists v_0, \ldots v_\ell)(\exists e_0, \ldots, e_r)(\exists U_0, \ldots, U_\ell)$$

$$\left[\alpha_D(v_0, \ldots, v_\ell, e_0, \ldots, e_r) \wedge \bigcup_0^\ell U_i = V \setminus \{v_0, \ldots, v_\ell\} \wedge \bigwedge_{i \neq j} U_i \cap U_j = \emptyset \right.$$

$$\wedge \, \theta_1(v_0, \ldots, v_\ell; e_0, \ldots, e_r) \wedge \theta_2(e_0, \ldots, e_r; v_0, \ldots, v_\ell)$$

$$\left. \wedge \bigwedge_{i \neq j} \theta_3(U_i, U_j) \wedge \bigwedge_{i=0}^\ell \theta_4(U_i, v_i, v_{i+1}) \right]$$

of length $O(k^2)$. This formula expresses the property that, in the given graph G, we can construct a crossing diagram of type D, and a corresponding partition of the vertices into subsets U_i, that obeys Properties 1–4 of Lemma 6. By Lemma 6, this is equivalent to the property that G has a 1-page drawing with k crossings in configuration D. Finally, we construct ONEPAGE$_k$ by taking the disjunction of the β_D where D ranges over all crossing diagrams with $\leq k$ crossings. Thus, ONEPAGE$_k$ is a formula of length $2^{O(k^2)}$, expressing the property that $\mathrm{cr}_1(G) \leq k$.

Theorem 1. *There exists a computable function f such that $\mathrm{cr}_1(G)$ can be computed in $O(f(k)n)$ time for a graph G with n vertices and with $k = \mathrm{cr}_1(G)$.*

Proof. We have shown the existence of a formula ONEPAGE$_k$ such that a graph $G \models$ ONEPAGE$_k$ if and only if $\mathrm{cr}_1(G) \leq k$. By Lemma 5, the treewidth of any graph with crossing number k is $O(k)$. Applying Courcelle's theorem with the formula ONEPAGE$_k$ and the $O(k)$ treewidth bound, it follows that computing $\mathrm{cr}_1(G)$ is fixed-parameter tractable in k . □

4   2-page Planarity

A classical characterization of the graphs with planar 2-page drawings is that they are exactly the subhamiltonian planar graphs:

Lemma 8 (Bernhart and Kainen [4]). *A graph is 2-page planar if and only if it is the subgraph of planar Hamiltonian graph.*

However, this characterization does not directly help us to construct an MSO$_2$-formula expressing the 2-page planarity of a graph, as we do not know how to construct a formula that asserts the existence of a supergraph with the given property. Hamiltonicity and planarity are both straightforward to express in MSO$_2$, but there is no obvious way to describe a set of edges that may be of more than constant size, is not a subset of the existing edges, and can be used to augment the given graph to form a planar Hamiltonian graph.

For this reason we provide a new characterization, which we model on a standard characterization of planar graphs: a graph is planar if and only if, for every cycle C, the flaps of C can be partitioned into two subsets (the interior and exterior of C) such that no two flaps in the same subset cross each other. For instance, this characterization has been used as the basis for a cubic-time divide and conquer algorithm for planarity testing, which recursively subdivides the graph into cycles and non-crossing subsets of flaps [2,14,23]. In our characterization of 2-page graphs, we apply this idea to a special set of cycles, the boundaries of maximal regions within each halfplane that are separated from the spine of a 2-page book embedding by the edges of the embedding. The cycles of this type are edge-disjoint, and if a single cycle of this type has been identified then its interior flaps can also be identified easily: each interior flap is a single edge, and an edge forms an interior flap if and only if it belongs to the same page as the cycle in the book embedding and has both its endpoints on the cycle. As well as identifying which of the two pages each edge of a given graph is assigned to, our MSO_2 formula will partition the edges into three different types of edge: the ones that belong to these special cycles, the ones that form interior flaps of these special cycles, and the remaining *isthmus* edges that, if deleted, would disconnect parts of their page.

Suppose we are given a graph $G = (V, E)$ and a partition of its edges into two subsets A, B, intended to represent the two pages of a 2-page drawing of G. We define the graph separate$(G; A, B)$ that splits each vertex of G into two vertices, one in each page, with a new edge connecting them. Thus, separate$(G; A, B)$ has $2n$ vertices, which can be labeled by pairs of the form (v, X) where v is a vertex in V and X is one of the two sets in A, B. It has an edge between (x, X) and (y, Y) if either of two conditions is met: (1) $x = y$ and $X \neq Y$, or (2) $X = Y$ and there is an edge between x and y in X.

Lemma 9. *A graph $G = (V, E)$ is 2-page planar if and only if there exists a partition $A_b, A_c, A_i, B_b, B_c, B_i$ of E into six subsets such that, for each of the two choices of $X = A$ and $X = B$, these subsets satisfy the following properties:*

1. *X_c is a union of edge-disjoint cycles.*
2. *$X_c \cup X_b$ does not contain any additional cycles that involve edges in X_b.*
3. *For every edge e in X_i there exists a cycle in X_c containing both endpoints of e.*
4. *The graph formed by the edges $X_i \cup X_c \cup X_b$ is outerplanar.*
5. *For each cycle C in X_c it is not possible to find two vertex-disjoint paths P_1 and P_2 in E such that neither path is a single edge in X_i, all four path endpoints are distinct vertices of C, neither path contains a vertex of C in its interior, and the two pairs of path endpoints are in crossing position on C.*
6. *The subdivision separate$(G; A_b \cup A_c \cup A_i, B_b \cup B_c \cup B_i)$ is planar.*

Figure 2 illustrates the division of edge into six subsets described in Lemma 9. For the proof of Lemma 9, see the full version of this paper.

We construct a formula TWOPAGE based on Lemma 9 with the property that $G \models$ TWOPAGE if and only if G is 2-page planar. First, we construct formulas $\theta_1, \ldots, \theta_5$ expressing Properties 1 through 5 in Lemma 9, as we did for 1-page crossing; each of these properties has a straightforward expression in MSO_2. To express Property 6 we will need the following technical lemma, which can be proved using the method of syntactic interpretations.

Lemma 10. *For every MSO_2-formula ϕ there exists an MSO_2-formula $\phi^*(A, B)$ such that $G \models \phi^*(A, B)$ if and only if $\text{separate}(G; A, B) \models \phi$.*

Now, we can express Property 6 as an MSO_2-formula θ_6 using Lemma 10, as planarity is expressible by Lemma 4 and the fact that planar graphs are the graph that avoid K_5 and $K_{3,3}$ as minors. Thus, we define TWOPAGE to be the formula expressing the existence of $A_b, A_c, A_i, B_b, B_c, B_i$ satisfying $\theta_1, \ldots \theta_6$.

Theorem 2. *There exists a computable function f and an algorithm that can decide whether a given graph with treewidth k is 2-page planar in $O(f(k)n)$ time.*

Proof. The result follows from Courcelle's theorem together with the construction of the MSO_2 formula TWOPAGE representing the existence of a two-page planar embedding. $\square$

5   2-page Crossing Minimization

We now extend the results of the previous section from 2-page planarity to 2-page crossing minimization. As in the 1-page case, we will use a formula that involves a disjunction over crossing diagrams. Given a crossing diagram D with k crossings and $r+1$ edges, whose graph is G, we define the *planarization* of G with respect to D to be the graph in which each edge e_i is replaced by a path of degree four vertices, such that two of these replacement paths share a vertex if and only if the original two edges cross in D. As explained earlier, we do not care about the order of crossings along each edge (two crossing diagrams with the same sets of crossing pairs but with different crossing orders are considered equivalent. Nevertheless, we do preserve the order of crossings from (one representative of an equivalence class of) crossing diagrams to their planarizations, in order to ensure that the planarizations form planar graphs.

Lemma 11. *A graph $G = (V, E)$ has $\text{cr}_2(G) = k$ if and only if there exists edges $e_0, e_1, \cdots, e_r$ with $r < 2k$ and a 2-page crossing diagram D with k crossings on these edges such that when G is planarized with respect to D the resulting graph $G_D = (V_D, E_D)$ has a partition of E_D into $A_b, A_c, A_i, B_b, B_c, B_i$ such that, for $X = A, B$:*

1. *X_c is a union of edge disjoint cycles.*
2. *None of the cycles $X_c \cup X_b$ contains an edge in X_b.*
3. *If e is an edge introduced in the planarization, then $e \in A_b \cup A_c \cup A_i$ if e is in the first page of D, and $e \in B_b \cup B_c \cup B_i$ if it is in the second page of D.*

4. For every edge e in X_i, there exists a subgraph P containing e and a cycle C in X_c such that P consists only of vertices of C and of degree-four vertices introduced in the planarization, P contains at least two vertices of C, and P includes all four edges incident to each of its planarization vertices.

5. For each two edges e and f in X_i, the two subgraphs P_e and P_f satisfying Property 4 do not each have a pair of endpoints in crossing position on the same cycle C.

6. For each cycle C in X_c there do not exist two paths in E, such that neither path uses edges of X_i or interior vertices of C, with four distinct endpoints on C in crossing position.

7. the subdivision separate$(G; A_b \cup A_c \cup A_i, B_b \cup B_c \cup B_i)$ is planar.

Now, we construct a MSO$_2$-formula ζ_k based on Lemma 11 such that $G \models \zeta_k$ if and only if $\mathrm{cr}_2(G) = k$. To handle the planarization process we use the following lemma. In the lemma, the notation $G^{e_1 \times e_2}$ describes the graph obtained from a graph G by deleting two edges e_1 and e_2 that do not share a common endpoint, and adding a new degree-4 vertex connected to the endpoints of e_1 and e_2.

Lemma 12 (Grohe [15]). *For every MSO$_2$-formula ϕ there exists an MSO-formula $\phi^*(x_1, x_2)$ such that $G \models \phi^*(e_1, e_2)$ if and only if $G^{e_1 \times e_2} \models \phi$.*

Given any MSO$_2$-formula ϕ and crossing diagram D, we can repeatedly apply the lemma above to construct a formula ϕ^D such that $G \models \phi^D(e_0, \dots, e_r)$ if and only if $G_D \models \phi$. With this tool in hand it is straightforward to construct a formula γ_D, expressing the property that, in a given graph G we can build a crossing diagram with the structure of D, and partition the planarization G_D into six sets, satisfying Lemma 11. So we can define ζ_k to be the disjunction of the γ_D ranging over all 2-page crossing diagrams with k-crossings.

Theorem 3. *There exists a computable function f such that $\mathrm{cr}_2(G)$ can be computed in $O(f(k,t)n)$ time for a graph G with n vertices, $k = \mathrm{cr}_2(G)$, and $t = \mathrm{tw}(G)$.*

6   Conclusion

We have provided new fixed-parameter algorithms for computing the crossing numbers for 1-page and 2-page drawings of graphs with bounded treewidth. The use of monadic second order logic and Courcelle's theorem in our solutions causes the running times of our algorithms to have an impractically high dependence on their parameters. We believe that it should be possible to achieve a better dependence by directly designing dynamic programming algorithms that use tree-decompositions of the given graphs, rather than by relying on Courcelle's theorem to prove the existence of these algorithms. Can this dependency be reduced to the point of producing practical algorithms? For 2-page crossing minimization the runtime is parameterized by both the treewidth and the crossing number. Is 2-page crossing minimization NP-hard for graphs of fixed treewidth? We leave these questions open for future research.

Dujmović and Wood asked [?], "is there a polynomial-time algorithm for computing the book thickness of graphs with bounded treewidth?" Our Theorem 2 provides a partial solution to this question for book thickness 2. Can the graph property of having book thickness k be expressed in MSO_2, answering the question of Dujmović and Wood? The special case of $k = 3$ is of particular interest, to provide a computational attack on the still-open problem of whether there exist planar graphs that require four pages [12, 26]. Heath has shown that every planar graph of treewidth three has a planar 3-page drawing [?], but recognizing three-page graphs of higher treewidth efficiently remains open.

Acknowledgments. This material is based upon work supported by the National Science Foundation under Grant CCF-1228639 and by the Office of Naval Research under Grant No. N00014-08-1-1015.

References

[1] Arnborg, S., Corneil, D., Proskurowski, A.: Complexity of finding embeddings in a k-tree. SIAM J. Alg. Disc. Meth. 8(2), 277–284 (1987), doi:10.1137/0608024

[2] Auslander, L., Parter, S.V.: On imbedding graphs in the sphere. Journal of Mathematics and Mechanics 10(3), 517–523 (1961)

[3] Bannister, M.J., Eppstein, D., Simons, J.A.: Fixed parameter tractability of crossing minimization of almost-trees. In: Wismath, S., Wolff, A. (eds.) GD 2013. LNCS, vol. 8242, pp. 340–351. Springer, Heidelberg (2013), doi:10.1007/978-3-319-03841-4_30

[4] Bernhart, F., Kainen, P.C.: The book thickness of a graph. Journal of Combinatorial Theory, Series B 27(3), 320–331 (1979), doi:10.1016/0095-8956(79)90021-2

[5] Bodlaender, H.L.: A linear time algorithm for finding tree-decompositions of small treewidth. In: Proceedings of the Twenty-fifth Annual ACM Symposium on Theory of Computing, STOC 1993, pp. 226–234. ACM (1993), doi:10.1145/167088.167161

[6] Bodlaender, H.L.: A partial k-arboretum of graphs with bounded treewidth. Theoretical Computer Science 209(1-2), 1–45 (1998), doi:10.1016/S0304-3975(97)00228-4

[7] Chartrand, G., Harary, F.: Planar permutation graphs. Annales de l'institut Henri Poincaré (B) Probabilités et Statistiques 3(4), 433–438 (1967), http://eudml.org/doc/76875

[8] Chung, F.R.K., Leighton, F.T., Rosenberg, A.L.: Embedding graphs in books: A layout problem with applications to VLSI design. SIAM J. Alg. Disc. Meth. 8(1), 33–58 (1987), doi:10.1137/0608002

[9] Courcelle, B.: The monadic second-order logic of graphs. I. Recognizable sets of finite graphs. Information and Computation 85(1), 12–75 (1990), doi:10.1016/0890-5401(90)90043-H

[10] Courcelle, B., Engelfriet, J.: Graph Structure and Monadic Second-Order Logic: A Language-Theoretic Approach. Cambridge University Press (2012)

[11] Dujmović, V., Fellows, M.R., Kitching, M., Liotta, G., McCartin, C., Nishimura, N., Ragde, P., Rosamond, F., Whitesides, S., Wood, D.R.: On the parameterized complexity of layered graph drawing. Algorithmica 52(2), 267–292 (2008), doi:10.1007/s00453-007-9151-1

[12] Dujmović, V., Wood, D.R.: Graph treewidth and geometric thickness parameters. Discrete Comput. Geom. 37(4), 641–670 (2007), doi:10.1007/s00454-007-1318-7

[13] Ebbinghaus, H.-D., Flum, J., Thomas, W.: Mathematical logic, 2nd edn. Undergraduate Texts in Mathematics. Springer (1994), doi:10.1007/978-1-4757-2355-7; Translated from the German by Margit Meßmer

[14] Goldstein, A.J.: An efficient and constructive algorithm for testing whether a graph can be embedded in a plane. In: Graph and Combinatorics Conference (1963)

[15] Grohe, M.: Computing crossing numbers in quadratic time. Journal of Computer and System Sciences 68(2), 285–302 (2004), doi:10.1016/j.jcss.2003.07.008

[16] Kainen, P.C.: Some recent results in topological graph theory. In: Graphs and Combinatorics. Lecture Notes in Mathematics, vol. 406, pp. 76–108. Springer (1974), doi:10.1007/BFb0066436

[17] Kawarabayashi, K., Reed, B.: Computing crossing number in linear time. In: ACM Symp. Theory of Computing (STOC 2007), pp. 382–390 (2007), doi:10.1145/1250790.1250848

[18] Mitchell, S.L.: Linear algorithms to recognize outerplanar and maximal outerplanar graphs. Information Processing Letters 9(5), 229–232 (1979), doi:10.1016/0020-0190(79)90075-9

[19] Ollmann, L.T.: On the book thicknesses of various graphs. In: Proc. 4th Southeastern Conference on Combinatorics, Graph Theory and Computing, vol. 8, p. 459 (1973)

[20] Pelsmajer, M.J., Schaefer, M., Štefankovič, D.: Crossing numbers and parameterized complexity. In: Hong, S.-H., Nishizeki, T., Quan, W. (eds.) GD 2007. LNCS, vol. 4875, pp. 31–36. Springer, Heidelberg (2008), doi:10.1007/978-3-540-77537-9_6.

[21] Robertson, N., Seymour, P.D.: Graph minors. XIII. The disjoint paths problem. Journal of Combinatorial Theory, Series B 63(1), 65–110 (1995), doi:10.1006/jctb.1995.1006

[22] Shahrokhi, F., Sýkora, O., Székely, L.A., Vrťo, I.: Book embeddings and crossing numbers. In: Mayr, E.W., Schmidt, G., Tinhofer, G. (eds.) WG 1994. LNCS, vol. 903, pp. 256–268. Springer, Heidelberg (1995)

[23] Shirey, R.W.: Implementation and Analysis of Efficient Graph Planarity Testing Algorithms. Ph.D. thesis, The University of Wisconsin – Madison (1969)

[24] Wattenberg, M.: Arc diagrams: visualizing structure in strings. In: IEEE Symposium on Information Visualization (INFOVIS 2002), pp. 110–116 (2002), doi:10.1109/INFVIS.2002.1173155

[25] Wiegers, M.: Recognizing outerplanar graphs in linear time. In: Tinhofer, G., Schmidt, G. (eds.) WG 1986. LNCS, vol. 246, pp. 165–176. Springer, Heidelberg (1987)

[26] Yannakakis, M.: Four pages are necessary and sufficient for planar graphs. In: Proc. 18th ACM Symp. on Theory of Computing (STOC 1986), pp. 104–108 (1986), doi:10.1145/12130.12141

A Crossing Lemma for the Pair-Crossing Number

Eyal Ackerman[1] and Marcus Schaefer[2]

Dept. Math., Physics, and Comp. Sci., University of Haifa at Oranim, Tivon, Israel
`ackerman@sci.haifa.ac.il`
School of Computing, DePaul University, Chicago, Illinois 60604, USA
`mschaefer@cdm.depaul.edu`

Abstract. The *pair-crossing number* of a graph G, $\mathrm{pcr}(G)$, is the minimum possible number of pairs of edges that cross each other (possibly several times) in a drawing of G. It is known that there is a constant $c \geq 1/64$ such that for every (not too sparse) graph G with n vertices and m edges $\mathrm{pcr}(G) \geq c\frac{m^3}{n^2}$. This bound is tight, up to the constant c. Here we show that $c \geq 1/34.2$ if G is drawn without adjacent crossings.

1 Introduction

Throughout this paper we consider graphs with no loops or parallel edges. A *topological graph* is a graph drawn in the plane with its vertices as distinct points and its edges as Jordan arcs that connect the corresponding points and do not contain any other vertex as an interior point. Every pair of edges in a topological graph has a finite number of intersection points. If every pair of its edges intersect at most once, then a topological graph is called *simple*. The intersection point of two edges is either a vertex that is common to both edges, or a crossing point at which one edge passes from one side of the other edge to its other side.

A *crossing* in a topological graph consists of a pair of crossing edges and a point in which they cross. The *crossing number* of a graph G, $\mathrm{cr}(G)$, is the minimum possible number of crossings in a drawing of G as a topological graph in the plane. The *pair-crossing number* of a graph G, $\mathrm{pcr}(G)$, is the minimum possible number of *pairs* of crossing edges in a drawing of G as a topological graph in the plane. There has been some confusion between these two notions in the literature, probably due to the fact that in a drawing with the least number of crossings no pair of edges intersects more than once. Perhaps for the same reason there has also been some confusion as to whether adjacent crossings are allowed or counted.[1] For examples and history of this confusion and other variants of the crossing number, see the recent survey of Schaefer [14] and the paper titled "Which crossing number is it anyway?" by Pach and Tóth [12].

Considering adjacent crossings, Pach and Tóth [11] introduced the following notation:

---

[1] By adjacent crossings we mean crossings between edges that share a common vertex.

C. Duncan and A. Symvonis (Eds.): GD 2014, LNCS 8871, pp. 222–233, 2014.
© Springer-Verlag Berlin Heidelberg 2014

Rule +: adjacent crossings are not allowed.
Rule -: adjacent crossings are allowed but not counted.
Rule 0: adjacent crossings are allowed and counted (this is the default rule).

Clearly, $pcr_-(G) \le pcr(G) \le pcr_+(G) \le cr_+(G) = cr(G)$ for every graph G. On the other hand, it is known [7] that $cr(G) = O(pcr(G)^{3/2} \log^2 pcr(G))$, and it follows from the results in [13] that $cr(G) \le \binom{2pcr_-(G)}{2}$. Perhaps the main related open problem is to determine whether there is a graph G for which $pcr(G) < cr(G)$.

The following lower bound on the crossing number was proved by Ajtai, Chvátal, Newborn, Szemerédi [4] and, independently, by Leighton [6].

Theorem 1 ([4,6]). *There is an absolute constant $c > 0$ such that for every graph G with n vertices and $m \ge 4n$ edges we have $cr(G) \ge c\frac{m^3}{n^2}$.*

This celebrated result is known as the *Crossing Lemma* and has numerous applications in combinatorial and computational geometry, number theory, and other fields of mathematics. The Crossing Lemma is tight, apart from the multiplicative constant c. This constant was originally small, and later was shown to be at least $1/64 \approx 0.0156$, by a very elegant probabilistic argument due to Chazelle, Sharir, and Welzl [3]. Pach and Tóth [10] proved that $0.0296 \approx 1/33.75 \le c \le 0.09$ (the lower bound applies for $m \ge 7.5n$). Their lower bound was later improved by Pach, Radoičić, Tardos, and Tóth [9] to $c \ge 1024/31827 \approx 1/31.1 \approx 0.0321$ (when $m \ge \frac{103}{16}n$). Recently, Ackerman [1] further improved the lower bound to $c \ge \frac{1}{29}$ (when $m \ge 6.95n$).

Pach et al. [9] pointed out that the original proofs of the Crossing Lemma generalize to the pair-crossing number, yielding $pcr(G) \ge \frac{1}{64}\frac{|E(G)|^3}{|V(G)|^2}$ when $|E(G)| \ge 4|V(G)|$. They also remarked that they were unable to extend their lower bound on the crossing number to the pair-crossing number. Our main result is the following.

Theorem 2. *For every graph G with n vertices and $m \ge 6.75n$ edges we have $pcr_+(G) \ge \frac{2^6}{3^7}\frac{m^3}{n^2} \ge \frac{1}{34.2}\frac{m^3}{n^2}$.*

All the above-mentioned improvements for the crossing number were obtained using the same approach, namely, by showing that a sparse graph has an edge that is involved in several crossings. Denote by $e_k(n)$ the maximum number of edges in a topological graph with $n > 2$ vertices in which every edge is involved in at most k crossings. Let $e_k^*(n)$ denote the same quantity for *simple* topological graphs. It follows from Euler's Polyhedral Formula that $e_0(n) \le 3n-6$. Pach and Tóth showed that $e_k^*(n) \le 4.108\sqrt{k}n$ and also gave the following better bounds for $k \le 4$.

Theorem 3 ([10]). $e_k^*(n) \le (k+3)(n-2)$ *for $0 \le k \le 4$. Moreover, these bounds are tight when $0 \le k \le 2$ for infinitely many values of n.*

Pach et al. [9] observed that the upper bound in Theorem 3 applies also for not necessarily simple topological graphs when $k \le 3$, and proved a better bound for $k = 3$, namely, $e_3(n) \leq 5.5n - 11$. Ackerman [1] proved that $e_4^*(n) \leq 6n - 12$. The last two bounds are tight up to an additive constant.

The bounds $e_k(n)$ are used to get a weak lower bound on the crossing number of the form $\mathrm{cr}(G) \geq \alpha |E(G)| - \beta |V(G)|$. This linear bound is then used instead of the trivial bound $\mathrm{cr}(G) \geq |E(G)| - 3|V(G)|$ in the well-known probabilistic proof of the Crossing Lemma. The same approach would work to get a better lower bound for the pair-crossing number (and its variant pcr_+), if one can show that a sparse graph has an edge that crosses several other edges (each of them possibly many times).

Denote by $e_k''(n)$ the maximum number of edges in a topological graph with $n > 2$ vertices in which every edge crosses at most k other edges (each of them possibly more than once).[2] Clearly, $e_k''(n) \geq e_k(n) \geq e_k^*(n)$. For $0 \leq k \leq 3$ we have the following upper bounds on $e_k''(n)$.

Theorem 4. *Let G be a graph with $n \geq 3$ vertices that can be drawn as a topological graph in which every edge crosses at most k other edges (each of them possibly more than once). If $0 \leq k \leq 3$, then G has at most $(k+3)(n-2)$ edges.*

Note that for $0 \leq k \leq 2$ it is known that there are infinitely many values of n for which one can draw a (simple) topological graph with n vertices and $(k+3)(n-2)$ edges such that every edge is crossed at most k times. Therefore, the bounds for $0 \leq k \leq 2$ in Theorem 4 are tight. On the other hand, our upper bound for $e_3''(n)$ is inferior to the known bound on $e_3(n)$.

Organization. In Section 2 we collect some useful facts towards proving Theorem 4. A sketch of the proof of this theorem is then presented in Section 3. In Section 4 we recall how such a result can be used to get a better bound for the pair-crossing number when adjacent crossings are not allowed. Due to space limitation, some of the proofs are omitted or only sketched. The missing details can be found in the full version of the paper.

2   Preliminaries

In this section we collect some useful facts towards proving Theorem 4. Since we will be interested in the number of crossing pairs and the number of edges crossing a single edge, we may assume henceforth that the topological graphs that we consider have no three edges crossing at a single point. Indeed, if more than two edges cross at a point p, then we can redraw these edges in a small neighborhood of p such that no three of them cross at a point without changing the set of edges that each of these edges cross.

Recall that Pach et al. [9] proved that $e_k^*(n) = e_k(n)$, for $0 \leq k \leq 3$. This is implied by the following lemma.

---

[2]   Think of the double prime symbol as a *pair* of prime symbols. Note that adjacent crossings are allowed and counted although for improving the crossing lemma for pcr_+ it would suffice to consider drawings without adjacent crossings.

Lemma 1 ([9]). *For every $0 \leq k \leq 3$, if a graph can be drawn as a topological graph such that each of its edges is crossed at most k times, then in a drawing with that property and the least number of crossings every pair of edges intersects at most once.*

Let G be a graph and let D be a drawing of G as a topological graph. Let D' be a drawing of G as a topological graph with the least number of crossings such that every pair of crossing edges in D' are also crossing in D. The following is implied by the proof of Theorem 3.2 in [15].

Lemma 2. *There is no pair of edges e and e' in D' such that there are two crossing points between them that are consecutive along e.*

Lemma 3. *If every edge in D (and hence in D') is crossed by at most k edges, then every edge in D' contains fewer than 2^k crossing points with other edges.*

3   Proof of Theorem 4

Recall that we wish to show that $e_k''(n) \leq (k+3)(n-2)$, for $0 \leq k \leq 3$. That is, for every $0 \leq k \leq 3$, if a graph G with $n > 2$ vertices can be drawn such that each of its edges crosses at most k edges (possibly several times), then G has at most $(k+3)(n-2)$ edges. Theorem 3 (due to Pach and Tóth) yields this edge bound for $k \leq 3$, but under the stronger assumption that each edge is crossed at most k times. As we see in the next section, for $k \leq 2$, the assumption is not really stronger, so that we can use Theorem 3 for these cases (though for $k = 0$ and $k = 1$ direct proofs are easier). For $k = 3$ we need a more sophisticated discharging argument, detailed in Section 3.2.

3.1   The Local Pair-Crossing Number and Bounding e_k'' for $k \leq 2$

The *local crossing number*, $\mathrm{lcr}(G)$, of a graph G is the smallest k so that G can be drawn with at most k crossings per edge. The *local pair-crossing number*, $\mathrm{lpcr}(G)$, is the smallest k so that G has a drawing in which each edge is crossed by at most k other edges. By definition, $\mathrm{lpcr}(G) \leq \mathrm{lcr}(G)$. Following the hints in [14], it is possible to construct a graph G for which $\mathrm{lpcr}(G) = 4$ and $\mathrm{lcr}(G) = 5$, therefore, the two local crossing numbers differ. This is in marked contrast to the pair crossing number, which we cannot at this point separate from the standard crossing number.

Theorem 5. *If $\mathrm{lpcr}(G) \leq 2$, then $\mathrm{lpcr}(G) = \mathrm{lcr}(G)$.*

This leaves open the question whether equality holds for $\mathrm{lpcr}(G) = 3$. We believe a counterexample is possible, implying that we cannot take the easy route via the local crossing number to establish the bound of e_3''.

Proof. The statement is immediate for $\mathrm{lpcr}(G) = 0$ (by definition), and follows from Lemma 2 for $\mathrm{lpcr}(G) \leq 1$. Therefore, suppose that G is a graph with

$\mathrm{lpcr}(G) = 2$. Fix a drawing D of G in which every edge crosses at most two other edges and (under this condition) the least possible number of crossings.

It follows from Lemma 3 that every edge in D is crossed at most three times. We claim that every edge in D is crossed at most twice. Suppose for the sake of contradiction that there is an edge e that is crossed exactly three times. Orient e arbitrarily and let x_1, x_2, x_3 be the crossing points on e, in the order they appear on e according to its orientation. Denote by e_1, e_2, e_3 the edges that cross e at x_1, x_2, x_3, respectively. It follows from Lemma 2 that $e_1 = e_3$ and $e_1 \neq e_2$. Moreover, by the same argument, the segment of e_1 between x_1 and x_3 must contain a crossing point of e_1 with an edge e'. Note also that e' crosses e_1 once and e crosses e_2 once, since e and e_1 cross each other twice and there are at most three crossing points on every edge. Denote by D' the topological graph we obtain by swapping the segments of e and e_1 between x_1 and x_3 and redrawing them at small neighborhoods of x_1 and x_3 such that these segments are disjoint (see Fig. 1 for an illustration). Note that in D' every edge still crosses at most

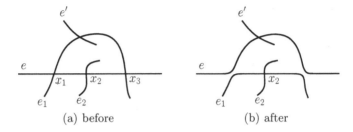

(a) before   (b) after

Fig. 1. Decreasing the number of crossings in case of an edge that is crossed three times

two other edges, however, D' has fewer crossing points than D, contradicting the minimality of D.

Thus, every edge in D is crossed at most twice, showing that $\mathrm{lcr}(G) \leq 2$. □

For $k \leq 2$, we can now conclude that $e_k''(n) \leq (k+3)(n-2)$ as follows: By Theorem 5, if a graph G can be drawn so that each of its edges crosses at most $k \leq 2$ other edges, then G has a drawing in which each of its edges is crossed at most k times; by Lemma 1, we can assume that G has such a drawing which is simple. Now Theorem 4 yields the desired bound $e_k''(n) \leq e^*(n) = (k+3)(n-2)$.

The local pair-crossing number seems to be a useful tool for approaching arguments about the pair-crossing number, so we would like to know more about its properties. For example, one can ask whether lcr can be bounded in lpcr? This is true, by Lemma 3, which yields the exponential bound $\mathrm{lcr}(G) < 2^{\mathrm{lpcr}(G)}$, however, we think that much better bounds should be true, since we have much more flexibility with the local pair-crossing number than with string graphs.

In particular, it would bear investigating whether the upper bounds achieved by Tóth and Matoušek (bounding the crossing number in a function of the pair-crossing number), aren't really arguments about the local version of these crossing numbers.

3.2   A Proof that $e_3''(n) \leq 6(n-2)$

Suppose that G is a graph with $n > 2$ vertices that can be drawn such that every edge crosses at most three other edges (possibly several times), and let D be a drawing of G as a topological graph with that property and the least possible number of crossings.

We prove that G has at most $6n - 12$ edges by induction on n. For $n \leq 10$ we have $6n - 12 \leq \binom{n}{2}$ and thus the theorem trivially holds. Therefore, we assume that $n \geq 11$. Furthermore, we may assume that the degree of every vertex in G is at least 7, for otherwise the theorem easily follows by removing a vertex of small degree and applying induction.

We denote by $M(D)$ the plane map induced by D. That is, the vertices of $M(D)$ are the vertices and crossing points in D, and the edges of $M(D)$ are the crossing-free segments of the edges of D (where each such edge-segment connects two vertices of $M(D)$). If a certain vertex of $M(D)$ is known to be a vertex of D we will denote it by a capital letter, otherwise, we will use small letters. Unless the context is clear, we will refer to edges of D as D-edges and to edges of $M(D)$ as M-edges. An edge-segment will be denoted by a concatenation of its endpoints (no comma in between), where round or square brackets will indicate whether an endpoint is included in the edge-segment. E.g., $[xy)$ is an edge-segment whose endpoints are x and y, such that x is included in $[xy)$ and y is not. In all the following figures bold edge-segments mark M-edges.

Proposition 1. *If $M(D)$ is not 2-connected, then D has at most $6n-12$ edges.*

By Proposition 1, we may assume henceforth that $M(D)$ is 2-connected. The *boundary* of a face f in $M(D)$ consists of all the M-edges that are incident to f. Since $M(D)$ is 2-connected, the boundary of every face in $M(D)$ is a simple cycle. Thus, we can define the *size* of a face f, $|f|$, as the number of M-edges on its boundary. It can be shown that if $M(D)$ contains a face of size two, then there is a drawing of G as a topological with fewer crossings than D and such that every edge crosses at most three other edges. Therefore, the size of every face in $M(D)$ is at least three.

We use the *Discharging Method* (see, e.g., [5]) to prove that $|E(D)| \leq 6(n-2)$. We begin by assigning a *charge* to every face of the planar map $M(D)$ such that the total charge is $4n - 8$. Then, we redistribute the charge in several steps such that eventually the charge of every face is nonnegative and the charge of every vertex $A \in V(D)$ is $\frac{1}{3}\deg(A)$. Hence, $\frac{2}{3}|E(D)| = \sum_{A \in V(D)} \frac{1}{3}\deg(A) \leq 4n - 8$ and we get the desired bound on $|E(D)|$. Next we describe the proof in details.

Charging. Let V', E', and F' denote the vertex, edge, and face sets of $M(D)$, respectively. For a face $f \in F'$ we denote by $V(f)$ the set of vertices of D that are incident to f. It is easy to see that $\sum_{f \in F'} |V(f)| = \sum_{A \in V(D)} \deg(A)$ and that $\sum_{f \in F'} |f| = 2|E'| = \sum_{u \in V'} \deg(u)$. Note also that every vertex in $V' \setminus V(D)$ is a crossing point in D and therefore its degree in $M(D)$ is four. Hence,

$$\sum_{f \in F'} |V(f)| = \sum_{A \in V(D)} \deg(A) = \sum_{u \in V'} \deg(u) - \sum_{u \in V' \setminus V(D)} \deg(u) = 2|E'| - 4\left(|V'| - n\right).$$

Assigning every face $f \in F'$ a charge of $|f| + |V(f)| - 4$, we get that total charge over all faces is

$$\sum_{f \in F'} (|f| + |V(f)| - 4) = 2|E'| + 2|E'| - 4(|V'| - n) - 4|F'| = 4n - 8,$$

where the last equality follows from Euler's Polyhedral Formula by which $|V'| + |F'| - |E'| = 2$ (recall that $M(D)$ is connected).

Discharging. We will redistribute the charges in several steps. We denote by $ch_i(x)$ the charge of an element x (either a face in F' or a vertex in $V(D)$) after the ith step, where $ch_0(\cdot)$ represents the initial charge function. We will use the terms *triangles*, *quadrilaterals* and *pentagons* to refer to faces of size 3, 4 and 5, respectively. An integer before the name of a face denotes the number of original vertices it is incident to. For example, a 2-triangle is a face of size 3 that is incident to 2 original vertices.

Step 1: Charging the Vertices of D. In this step every vertex of D takes $1/3$ units of charge from each face it is incident to.   ⤳

After Step 1 the charge of every vertex $A \in V(D)$ is $\frac{1}{3} \deg(A)$. Next, we need to make sure that the charge of every face is nonnegative. Let $f \in F'$ be a face. Note that $ch_1(f) \geq |f| + \frac{2}{3}|V(f)| - 4$ and therefore $ch_1(f) \geq 0$ if $|f| \geq 4$. Recall that $M(D)$ has no faces of size two. Thus, it remains to consider the case that f is a triangle: if f is a 3-triangle, then $ch_1(f) = 1$; if f is a 2-triangle, then $ch_1(f) = \frac{1}{3}$; if f is a 1-triangle, then $ch_1(f) = -\frac{1}{3}$; and if f is a 0-triangle, then $ch_1(f) = -1$.

In order to describe the way the charge of 0- and 1-triangles becomes nonnegative, we will need the following definitions. Let f be a face, let e be one of its edges, and let f' be the other face that shares e with f. We say that f' is the *immediate neighbor* of f at e.

Let f_0 be a face in $M(D)$ and let x_1 and y_1 be two vertices of f_0 that are consecutive on its boundary and are crossing points in D. Denote by e_1 (resp., e_2) the D-edge that crosses the D-edge that contains the edge-segment $[x_1 y_1]$ at x (resp., y). Notice that it follows from Lemma 2 that e_1 and e_2 are distinct D-edges. Let f_1 be the immediate neighbor of f_0 at $[x_1 y_1]$. For $i \geq 1$, if f_i is a 0-quadrilateral, then denote by $[x_{i+1} y_{i+1}]$ the edge opposite to $[x_i y_i]$ in f_i, such that e_1 contains x_{i+1} and e_2 contains y_{i+1}, and let f_{i+1} be immediate neighbor of f_i at $[x_{i+1} y_{i+1}]$ (see Fig. 2 for an illustration).

Clearly, it is impossible that $x_j = x_k$ or $y_j = y_k$ for $j \neq k$, since e_1 and e_2 do not cross themselves. Suppose that $x_j = y_k$ for some j and k. Assume without loss of generality that $j \leq k$. It cannot happen that $j = k$ for then f_{j-1} is not a 0-quadrilateral. Since $x_j = y_k$, e_2 crosses e_1 at x_j. The M-edges $[x_{j-1} x_j]$ and $[x_j x_{j+1}]$ are contained in e_1 (note that $[x_j x_{j+1}]$ exists since $j < k$). Therefore, e_2 contains $[x_j y_j]$. However, this implies that e_2 crosses itself at y_j (see Fig. 2 for an illustration).

It follows that it cannot happen that there are two 0-quadrilaterals f_i and f_j such that $i \neq j$ and $f_i = f_j$. By Lemma 3 every edge in D contains at most

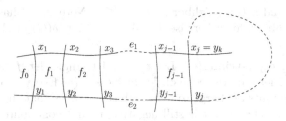

Fig. 2. If $x_j = y_k$, then e_2 crosses itself

seven crossing points. Therefore there must be an index $1 \le k \le 7$ such that f_k is not a 0-quadrilateral (notice that if f_i is not a 0-quadrilateral, then f_{i+1} is not defined). We say that f_k is the *distant neighbor* of f_0 at $[x_1 y_1]$, and that the M-edge $[x_k y_k]$ is the edge of f_k that *faces* $[x_1 y_1]$. Note that x_k and y_k must be crossing points, since they belong to the 0-quadrilateral f_{k-1} or coincide with x_1 and y_1, if $k = 1$. It is also important to note that if f_0 is not a 0-quadrilateral, then f_0 is the distant neighbor of f_k at $[x_k y_k]$ and $[x_1 y_1]$ is the edge of f_0 that faces $[x_k y_k]$. Indeed, this follows from the definition of a distant neighbor and from the fact that the relation immediate neighbor at a certain M-edge and the relation opposite edge in a 0-quadrilateral are one-to-one.

Proposition 2. *Let t be a 0- or 1-triangle whose vertices are x, y and z, such that x and y are crossing points in D, and let e_1 (resp., e_2) be the D-edge that contains $[zx]$ (resp., $[zy]$). Suppose that f is the distant neighbor of t at $[xy]$ and e' is the edge of f that faces $[xy]$. Then:*

1. *the endpoints of e' are crossing points in D;*
2. *one endpoint of e' (denote it by p) lies on e_1 and the other endpoint of e' (denote it by q) lies on e_2;*
3. *t is the distant neighbor of f at $[pq]$, and $[xy]$ is the edge of t that faces $[pq]$;*
4. *the edge-segment $(zp]$ of e_1 (resp., $(zq]$ of e_2) does not intersect e_2 (resp., e_1); and*
5. *$|f| \ge 5$ or $|f| = 4$ and $|V(f)| \ge 1$.*

Step 2: Charging 0-triangles. Let t be a 0-triangle, let e be one of its edges, let f be the distant neighbor of t at e, and let e' be the edge of f that faces t. We move $1/3$ units of charge from f to t, and say that f *contributed* $1/3$ units of charge to t *through* e'.

In a similar way t obtains $1/3$ units of charge from each of its distant neighbors at its other edges.

After the second discharging step the charge of every 0-triangle becomes zero. It remains to deal with 1-triangles, and then to make sure that the charge of every face did not become negative after the discharging steps. Let t be a 1-triangle and let $A \in V(D)$ be the vertex of D that is incident to t. Let g be an immediate neighbor of t that is incident to A. We call g a *good* neighbor of

t if $ch_2(g) > 0$, and a *bad* neighbor if $ch_2(g) \leq 0$. Note that t has two distinct (good/bad) neighbors, for otherwise $\deg(A) = 2 < 7 \leq \delta(G)$ or A is cut vertex in $M(D)$.

Step 3: Charging 1-triangles. Let t be a 1-triangle, let f be the distant neighbor of t and let e be the edge of f that faces t. (a) Every good neighbor of t contributes $1/6$ units of charge to t through the M-edge that they share. (b) If after Step 3(a) the charge of t is still negative, then f contributes $1/6$ units of charge to t through e.   ↜⤳

See Fig. 3 for an illustrations of the discharging steps.

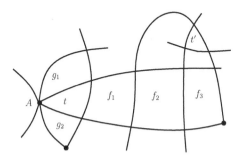

Fig. 3. Discharging steps: in Step 1 each of t, g_1 and g_2 contributes $1/3$ units of charge to A; in Step 2 f_3 contributes $1/3$ units of charge to t'; in Step 3(a) g_2 contributes $1/6$ units of charge to t; and in Step 3(b) f_3 contributes $1/6$ units of charge to t.

Considering Steps 2 and 3 and Proposition 2, we have the following observations.

Observation 3. *Let f be a face that contributes charge through one of its M-edges $[xy]$, and let z (resp., w) be the other vertex of f that is adjacent to x (resp., y). If x (resp., y) is a crossing point, then denote by (X, X') (resp., (Y, Y')) the D-edge that contains xz (resp., yw) such that $x \in [Xz]$ (resp., $y \in [Yw]$). We have:*

1. *f contributes charge through $[xy]$ exactly once;*
2. *if f contributes charge through $[xy]$ in Step 3(a), then one of x and y is a crossing point and the other is a vertex of D; and*
3. *if f contributes charge through $[xy]$ in Step 2 or Step 3(b), then both x and y are crossings points and f is a distant neighbor of a 0- or 1-triangle t. Furthermore, $[Xx]$ and $[Yy]$ intersect at a point q that is a vertex of t, (qx) and (qy) do not intersect, and $q = X = Y$ if t is a 1-triangle (otherwise, q is a crossing point).*

Recall that our plan was to distribute the initial charge such that the charge of every original vertex is one third of its degree and the charge of every face is nonnegative.

Lemma 4. *For every vertex* $A \in V(D)$ *we have* $ch_3(A) = \frac{1}{3}\deg(A)$ *and for every face* $f \in F'$ *we have* $ch_3(f) \geq 0$.

Proof. (sketch) The first part of the claim follows from the first discharging step. Let f be a face in $M(D)$. Since f contributes at most $1/3$ units of charge through each of its edges, we have $ch_3(f) \geq \frac{2}{3}|f| + \frac{2}{3}|V(f)| - 4$. Therefore, if $|f| \geq 6$, then $ch_3(f) \geq 0$. Recall that there are no faces of size two in $M(D)$, thus, it remains to consider faces of size three, four and five, i.e., triangles, quadrilaterals and pentagons.

Triangles. Suppose that $|f| = 3$. It is easy to show that $ch_3(f) \geq 0$ if f is not a 1-triangle. If f is a 1-triangle, then we show that after Step 3(a) its charge is at least $-1/6$, since it cannot have two bad neighbors (for otherwise, there is an edge of D crossing four other edges). It then follows from Step 3(b) that the final charge of f is zero.

Quadrilaterals. Suppose that $|f| = 4$. A 0-quadrilateral does not contribute charge and therefore if $|V(f)| = 0$ we have $ch_3(f) = 0$. If $|V(f)| \geq 2$, then it is easy to see that $ch_3(f) \geq 0$. It remains to consider the case that f is a 1-quadrilateral. Observe that if f is a 1-quadrilateral and $ch_3(f) < 0$, it must be that f contribute $1/3$ units of charge through precisely one of its edges in Step 2, and $1/6$ units of charge through each of its other edges. However, this case implies that there is an edge of D that crosses four other edges (see Fig. 4 for an illustration).

Pentagons. Suppose that $|f| = 5$. Recall that $ch_3(f) \geq \frac{2}{3}|f| + \frac{2}{3}|V(f)| - 4$. Therefore, if $|V(f)| \geq 1$, then $ch_3(f) \geq 0$, and it remains to consider the case that f is a 0-pentagon. Recall that we may assume that the boundary of f consists of a simple 5-cycle. It follows from Lemma 2 that all the D-edges that contain the M-edges of f are distinct. From this fact and Observation 3 one concludes that it is impossible that f contributes charge through two of its edges that are not consecutive on its boundary. Thus, if $ch_3(f) < 0$, then f must contribute $1/3$ units of charge through two (consecutive) edges in Step 2 and

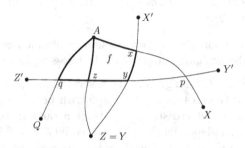

Fig. 4. If f is a 1-quadrilateral that contributes $1/3$ units of charge through $[xy]$ and $1/6$ units of charge through each of $[Az]$ and $[yz]$, then (Y', Z') crosses four edges

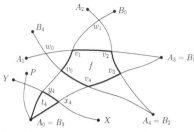
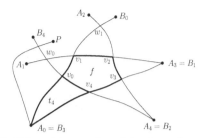
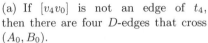

(a) If $[v_4v_0]$ is not an edge of t_4, then there are four D-edges that cross (A_0, B_0).

(b) If $[v_4v_0]$ is an edge of t_4, then there are four D-edges that cross (A_4, B_4).

Fig. 5. The 0-pentagon f contributes charge through $[v_0v_1]$ and $[v_1v_2]$ in Step 2 and through $[v_2v_3]$, $[v_3v_4]$ and $[v_4v_0]$ in Step 3(b)

1/6 units of charge through each of its other three edges in Step 3(b). A simple case-analysis shows that this scenario is impossible, as it implies that there is an edge of D that crosses four other edges (see Fig. 5 for illustrations).

It follows from Lemma 4 that the final charge of every face in $M(D)$ is nonnegative and that the charge of every vertex of D equals to one third of its degree. Recall that the total charge is $4n-8$. Therefore, $\frac{2}{3}|E(D)| = \sum_{A \in V(D)} \frac{1}{3} \deg(A) \leq 4n - 8$ and thus $|E(G)| = |E(D)| \leq 6n - 12$. This concludes the proof of Theorem 4. □

4   A Better Lower Bound on pcr$_+$

Recall that $e_k''(n)$ is the maximum number of edges in a topological graph with $n > 2$ vertices in which every edge crosses at most k other edges (each of them possibly more than once). Using the bounds on $e_k''(n)$ from Theorem 4, one can prove:

Theorem 6. *For every graph G with $n > 2$ vertices and m edges we have:* $pcr(G) \geq m - 3(n - 2)$; $pcr(G) \geq 2m - 7(n - 2)$; $pcr(G) \geq 3m - 12(n - 2)$; *and* $pcr(G) \geq 4m - 18(n - 2)$.

Using the new linear bounds it is now possible to obtain a better lower bound for pcr$_+$, following the probabilistic proof of the Crossing Lemma, as in [8,9,10].

Proof of Theorem 2: Let G be a graph with n vertices and $m \geq 6.75n$ edges and consider a drawing of G as a topological graph with pcr$_+(G)$ pairs of crossings edges and without adjacent crossings. Construct a random subgraph of G by selecting every vertex independently with probability $p = 6.75n/m \leq 1$. Let G' be the subgraph of G that is induced by the selected vertices. Denote by n' and m' the number of vertices and edges in G', respectively. Clearly, $\mathbb{E}[n'] = pn$ and $\mathbb{E}[m'] = p^2m$. Denote by x' the number of pairs of crossing edges in the drawing of G' inherited from the drawing of G. Then $\mathbb{E}[\mathrm{pcr}_+(G')] \leq \mathbb{E}[x'] = p^4 \cdot \mathrm{pcr}_+(G)$.[3] It follows from Theorem 6 that $\mathrm{pcr}_+(G') \geq \mathrm{pcr}(G') \geq 4m' - 18n'$ (note that this it true for any $n' \geq 0$), and this holds also for the expected values: $\mathbb{E}[\mathrm{pcr}_+(G')] \geq 4\mathbb{E}[m'] - 18\mathbb{E}[n']$. Plugging in the expected values we get that $\mathrm{pcr}_+(G) \geq \frac{2^6}{3^7} \frac{m^3}{n^2} \geq \frac{1}{34.2} \frac{m^3}{n^2}$. $\qquad\qquad\square$

References

1. Ackerman, E.: On topological graphs with at most four crossings per edge (manuscript)
2. Ackerman, E., Tardos, G.: On the maximum number of edges in quasi-planar graphs. J. Combinatorial Theory, Ser. A 114(3), 563–571 (2007)
3. Aigner, M., Ziegler, G.: Proofs from the Book. Springer, Heidelberg (2004)
4. Ajtai, M., Chvátal, V., Newborn, M., Szemerédi, E.: Crossing-free subgraphs. In: Theory and Practice of Combinatorics. North-Holland Math. Stud., vol. 60, pp. 9–12. North-Holland, Amsterdam (1982)
5. Cranston, D.W., West, D.B.: A guide to discharging (manuscript)
6. Leighton, F.T.: Complexity Issues in VLSI: Optimal Layouts for the Shuffle-Exchange Graph and Other Networks. MIT Press, Cambridge (1983)
7. Matoušsek, J.: Near-optimal separators in string graphs, ArXiv (February 2013)
8. Montaron, B.: An improvement of the crossing number bound. J. Graph Theory 50(1), 43–54 (2005)
9. Pach, J., Radoičić, R., Tardos, G., Tóth, G.: Improving the crossing lemma by finding more crossings in sparse graphs. Disc. Compu. Geometry 36(4), 527–552 (2006)
10. Pach, J., Tóth, G.: Graphs drawn with few crossings per edge. Combinatorica 17(3), 427–439 (1997)
11. Pach, J., Tóth, G.: Thirteen problems on crossing numbers. Geombinatorics 9(4), 194–207 (2000)
12. Pach, J., Tóth, G.: Which crossing number is it anyway? J. Combin. Theory Ser. B 80(2), 225–246 (2000)
13. Pelsmajer, M.J., Schaefer, M., Štefankovič, D.: Removing independently even crossings. SIAM J. on Disc. Math. 24(2), 379–393 (2010)
14. Schaefer, M.: The graph crossing number and its variants: A survey. Electronic Journal of Combinatorics, Dynamic Survey 21 (2013)
15. Schaefer, M., Štefankovič, D.: Decidability of string graphs. J. Comput. System Sci. 68(2), 319–334 (2004)

---

[3] Here we need the restriction that there are no adjacent crossings: an adjacent crossing would survive with probability p^3 instead of p^4. This was overlooked in a preliminary version of this paper when this theorem was stated for pcr instead of pcr_+.

Are Crossings Important for Drawing Large Graphs?

Stephen G. Kobourov[1], Sergey Pupyrev[1,2], and Bahador Saket[1]

[1] Department of Computer Science, University of Arizona, Tucson, Arizona, USA
[2] Institute of Mathematics and Computer Science, Ural Federal University, Russia

Abstract. Reducing the number of edge crossings is considered one of the most important graph drawing aesthetics. While real-world graphs tend to be large and dense, most of the earlier work on evaluating the impact of edge crossings utilizes relatively small graphs that are manually generated and manipulated. We study the effect on task performance of increased edge crossings in automatically generated layouts for graphs, from different datasets, with different sizes, and with different densities. The results indicate that increasing the number of crossings negatively impacts accuracy and performance time and that impact is significant for small graphs but not significant for large graphs. We also quantitatively evaluate the impact of edge crossings on crossing angles and stress in automatically constructed graph layouts. We find a moderate correlation between minimizing stress and the minimizing the number of crossings.

1 Introduction

Graphs are often used to model a set of entities and their relationships. They are usually visualized with node-link diagrams, where vertices are depicted as points and edges as line-segments connecting the corresponding points. Many different methods for drawing graphs have been developed and they typically aim to optimize one or more *aesthetic criteria*. According to the seminal work of Purchase [20], aesthetic criteria include: number of edge crossings, number of edge bends, symmetry of the drawing, angular resolution, crossing angles, and vertex distribution. Such criteria are often proposed based on human intuition and the personal judgement of algorithm designers, and therefore the task of validating graph drawing aesthetics is of high importance.

A great deal of the prior experimental evaluations of graph drawing aesthetics utilize relatively small and nearly planar graphs and networks. For example, Purchase et al. [21] conduct a user study with graphs on 16 vertices and $18 - 28$ edges. Huang et al. [13, 14] generate graphs having between 10 and 40 vertices. Larger graphs with 50 vertices are used by Dwyer et al. [5] but the number of edges is only 75, which results in graphs with almost tree-like structure. Real-world graphs, however, tend to be large, dense, and non-planar.

There are several of-the-shelf methods for drawing large graphs. Classical force-directed methods such as Fruchterman-Reingold [7] and Kamada-Kawai [17], and more recent multiscale variants [10, 12], define and minimize the "energy" of the layout; layouts with minimal energy tend to be aesthetically pleasing and to exhibit symmetries. Similarly, methods based on multidimensional scaling (MDS) minimize a particular energy function of the layout, called "stress" [8]. Note that the classical methods

C. Duncan and A. Symvonis (Eds.): GD 2014, LNCS 8871, pp. 234–245, 2014.

are not designed to directly optimize a specific graph drawing aesthetic criterion. Yet minimizing edge crossings remains the most cited and the most commonly used aesthetic [13,15,20,21,23]. With this in mind, we consider *the impact of edge crossings on the readability of graphs in automatically generated straight-line layouts of real-world large graphs*.

Many real-world graphs (e.g., biological networks, social networks, research citation graphs) have tens of thousands or even millions of vertices. Such graphs are not usually explored with static node-link diagrams, but rather with alternative visualization methods based on interaction, abstraction, overview-detail views, etc [1, 16]. Still, static node-link diagrams with more than a hundred vertices are common today. We would like to determine a reasonable upper limit on the size of a graph, for which typical tasks can be performed using a static node-link diagram. In order to empirically define the notion of a "large graph" in this setting, we run a preliminary experiment with graphs on 100-150 vertices. For graphs with 150 vertices and density (the number of edges divided by the number of vertices) of 3.5, task accuracy is steadily below 39%, even in the most advantageous setting (e.g., high resolution display, unlimited time, the simplest path-finding tasks, graph layouts with close-to-optimal number of edge crossings, etc). The results of this preliminary experiment helped us determine useful ranges of size and density of the graphs used in our main experiment. In the main experiment, we consider *small* (40 vertices) and *large* (120 vertices) graphs. The graphs are constructed from two real-world datasets and drawn with the classical force-directed and MDS-based algorithms. We vary edge density (from 1.5 to 2.5) and the number of crossings (by a factor of two), and analyze accuracy and completion time for four tasks, frequently utilized in prior experiments. We also quantitatively evaluate the relationship between edge crossings and several other layout quality measures. Thus our contributions are two-fold:

1. We measure accuracy and completion time for four graph tasks to evaluate the effect of edge crossings on small and large graphs with varying densities. The experiments indicate that increasing the number of crossings has a negative impact, but the change is not significant for large graphs.
2. We quantitatively evaluate the impact of edge crossings on crossing angles and stress in automatically constructed graph layouts. We find a moderate correlation between minimizing stress and minimizing the number of edge crossings.

2 Related Work

Several empirical studies aim to determine the impact of various aesthetic criteria on human understanding of graph visualizations. A series of experiments by Purchase shows that many of the aesthetics are indeed important [20]. The experiments indicate that the number of edge crossings is by far the most important aesthetic, while the number of edge bends and the local symmetry displayed have a lesser impact. These results are confirmed by Huang et al. [15], who found that edge crossings significantly impact user preference and task performance. Overall, it is a common belief that minimizing the number of edge crossings is one of the most important goals in drawing graphs.

These findings have made the area of crossing minimization one of the most active research topics in the graph drawing community; see [3] for an excellent survey. However, the problem of crossing minimization is computationally hard [9], and it remains hard even when restricted to special graphs [11]. In fact, one cannot even compute in polynomial time a crossing-optimal solution for a graph obtained from a planar one by adding a single edge [4]. Given that the problem is difficult, several heuristics have been designed. The heuristics are usually hard to implement and they do not scale well with the size of a graph [3]. Hence, it is a reasonable question to ask to what extent one should try to minimize edge crossings to justify the cost.

Other graph aesthetics have also been considered. Huang et al. [14] study crossing angles (the minimum angle between pairs of crossing edges) and conclude that larger crossing angles make graphs easier to read. This motivates the research area of right-angle-crossing (RAC) drawings, where the goal is to make all crossing angles close to 90 degrees. Several studies consider the relative importance of various aesthetic criteria, which is relevant as some of them can be conflicting (e.g., minimizing crossings in planar graph drawings usually results in poor angular resolution). Huang and Huang [13] argue that the number of edge crossings is relatively more important than the crossing angles.

Alternative representations of large graphs and networks have also been considered. Archambault et al. [1] show that coarsening graph representations, in which several interconnected vertices are merged into metanodes, does not result in significant improvements over node-link diagrams. However, such representations might be beneficial for specific tasks in very dense graphs. Jianu et al. [16], and Saket et al. [22] investigate several methods of representing cluster information in large graphs. Their results indicate that classical node-link diagrams are not the most efficient way to visualize large clustered datasets.

3   Experiments

Objectives. We conduct a controlled experiment to explore how edge crossings affect the understandability of graph layouts. Although several studies assess the impact of crossings, a number of important questions remain open. Our specific objectives are:

1. to confirm the results of prior studies that increasing the number of edge crossings negatively impacts the usability of node-link diagrams for **small graphs**;
2. to verify whether increasing the number of edge crossings also negatively impacts the usability of node-link diagrams for **large graphs**;
3. to explore the impact of edge crossings while varying the **edge density** for both large and small graphs;
4. to analyze the impact of edge crossings on **different tasks**.

Controlled experiments in graph drawing often involve manually creating different layouts of the same graph, by varying only one aesthetic, while the others are kept unchanged. However, due to the computational hardness of the crossing minimization problem, and the use of larger graphs than those in previous studies, it is almost impossible to do this in our setting. Instead we use a different approach to accomplish a similar

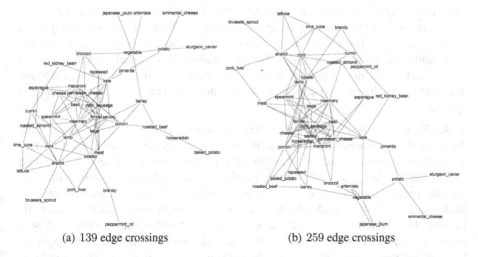

(a) 139 edge crossings   (b) 259 edge crossings

Fig. 1. A *small dense* graph with 40 vertices and 100 edges constructed from the **Recipes** dataset with (a) the *low* number of crossings and (b) the *high* number of crossings.

result by automatically generating all our drawings, without any manual postprocessing, as suggested in [13, 23]. We emphasize here that unlike most previous studies, we work only with real-world graphs and automatically computed layouts.

Our study involves a two-phase evaluation. In the first step (Experiment 1), the participant perform simple tasks on several graphs with different sizes (number of vertices) and densities (ratio of number of edges to number of vertices). This is how we determine the size of the largest graphs for which task accuracy is steadily above 50%. We use the information to design the main experiment (Experiment 2) in which we record performance, in terms of accuracy and completion time for our four tasks.

Datasets and Visualization. In order to minimize potential bias, we use two different datasets in our evaluation. The **Recipes** dataset contains 381 unique ingredients extracted from cooking recipes. The edges correspond to co-occurrence of the ingredients in the recipes. The **GD** dataset models co-authorship in the Graph Drawing conference. The vertices represent 506 authors and an edge between two vertices indicates that this pair of authors have co-authored a paper. For each dataset, we randomly sample vertices and edges creating graphs with different sizes and densities. The number of vertices is 40 (*small*) and 120 (*large*), and the edge density is 1.5 (*sparse*) and 2.5 (*dense*), making a total of 4 unweighted undirected graphs per dataset. Section 3.1 explains why we choose these sizes and densities.

We use two classical straight-line drawing algorithms implemented in GRAPHVIZ [6]. The **Recipes** graphs are embedded using the multidimensional scaling layout algorithm; for this purpose, we utilize the neato tool in GRAPHVIZ. For drawing the **GD** graphs, we use the force-directed placement algorithm, fdp in GRAPHVIZ. In order to perform our experiments, we need to have layouts of the same graph with different number of crossings. To this end, we run the layout algorithms 10, 000 times on the same graph, varying the initial positions of the vertices. Since both algorithms are sensitive to the initial embedding, the resulting layouts are different. We choose two layouts of the same graph: the one with the minimum number of crossings and one with approximately twice as many crossings. These two layouts are referred to as the drawings with the *low* and *high* number of crossings; see Fig. 1. Note that neither MDS-based nor force-directed algorithms provide any guarantees about the number of crossings. However, due to the many runs for each graph, we expect that the *low* number of crossings is not too far from optimal.

Tasks. We choose the tasks for our experiments based on several considerations. First, the tasks should represent standard problems, commonly encountered when analyzing relational data. Second, the number of edge crossings in a graph visualization should likely affect task performance. Finally, the tasks should be present in existing graph task taxonomies and often utilized in other graph drawing user evaluations. With this in mind, we consider the task taxonomy for graph visualization suggested by Lee et al. [19], which categorizes the tasks into groups: topology-based, attribute-based, browsing, and overview tasks. Each of the categories specifies different subcategories. Previous studies clearly indicate that the number of edges crossings affects tasks in the topology-based category, while tasks in the other three categories are less likely to be significantly impacted by the number of crossings or do not fit in our experimental setup. The graphs in our experiments do not contain special attributes (e.g., color or shape), and hence the attribute-based tasks are not suitable. The browsing category deals with navigational tasks that do not require a specific answer, making it difficult to measure the task performance. Overview tasks are related to compound tasks (e.g., identifying changes over time, comparing the relative size of a pair of graphs) are also not suitable to our setting and less likely to be affected by the number of edge crossings. Therefore, we focus on topology-based tasks, grouped into four subcategories: connectivity, accessibility, adjacency, and common connections. For each subcategory, we choose a task that is frequently used in prior user studies on graph visualization.

Task 1: *How many edges are in a shortest path between two given nodes?*
Task 2: *What is the node with the highest degree?*
Task 3: *What nodes are all adjacent to the given node?*
Task 4: *Which of the following nodes are adjacent to both given nodes?*

The vertices for each question were randomly selected (in the case of Task 1, additionally ensuring that the pair of vertices is at most 5 edges away).

Participants and Apparatus. For the first experiment we recruited 6 participants (3 male, 3 female) aged 21–27 years (mean 23) with normal vision. For the second experiment we recruited 16 new participants (12 male, 4 female) aged 21–30 years (mean 25) with normal vision. All the participants were undergraduate and graduate science and engineering students familiar with graphs and networks. Both experiments were conducted on a computer with i7 CPU 860 @ 2.80GHz processor and 24 inch screen with 1600x900 resolution. The participants interacted with a standard mouse to complete the tasks. We used custom-built software to guide the users through the experiment by providing instructions and collecting data about time and accuracy.

3.1 Procedure: Experiment 1

Real-world graphs are typically large and non-planar. In drawings of such graphs there could be many edge crossings, which likely makes the drawings difficult to understand. To evaluate the impact of the number of crossings for different sizes and densities of graphs, while keeping the experiment to a reasonable length and complexity, we want to choose the graphs so that the average completion time is below 120 seconds and the average accuracy for a single task is higher than 50%.

To determine reasonable upper limits for the main experiment, we generated different graphs with 100-150 vertices, in increments of 10, and densities ranging from 1.5 to 3.5, in increments of 1. For every graph, we used the layout with the smallest number of crossings and for each of these layouts the participants performed the four tasks described above. The resulting completion time ranges from 63 seconds for a 100-vertex graph to 184 seconds for a 150-vertex graph. The accuracy (the number of correct answers divided by the total number of questions) ranges from 85% for 100-vertex graphs with 1.5 density to 39% for 150-vertex graphs with 3.5 density. Based on these results, we choose 120 vertices as the maximum number of vertices and 2.5 as the maximum density value for our main experiment.

3.2 Procedure: Experiment 2

An experimental system was implemented to present the 64 (2 *sizes* × 2 *number of crossings* × 2 *densities* × 2 *datasets* × 4 *tasks*) stimuli and questions for this within-subjects experiment, and to collect the participant answers and response times.

Before the controlled experiment, the participants were briefed about the purpose of the study. Although all participants were familiar with graphs, we explained all the required definitions (e.g., graphs, edges, paths). The participants then answered 8 training questions (two for each of the tasks) as quickly and as accurately as possible. The participants were encouraged to ask questions during this stage and we did not record time and accuracy for the training questions.

The main experiment consisted of the 64 tasks, presented in a reduced Latin square to counterbalance learning and order effects (to prevent participants from extrapolating new judgements from previous ones). The participants were able to zoom and pan the diagram on the screen (if needed) and were required to select one of the provided multiple choices. We recorded time and accuracy for each task. After every 12 questions, there was a break and the participants could continue when they were ready.

Hypotheses. Based on prior work and results from our preliminary experiment, we hypothesize that:

H1 Increasing the number of crossings negatively impacts accuracy and performance time and that impact is significant for small graphs but not significant for large graphs.

H2 The negative impact of increasing the number of crossings on performance is significant for both small sparse and small dense graphs.

H3 The negative impact of increasing the number of crossings on performance is not significant for both large sparse and large dense graphs.

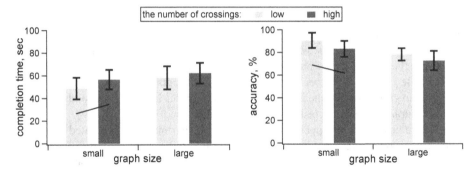

Fig. 2. Mean and standard deviation for time and accuracy in *small* and *large* graphs with different number of crossings. The differences are significant (indicated by the diagonal line segments) only for small graphs.

3.3   Results

We used a Shapiro-Wilk test to check normality of the collected data. The p-values for graphs with low/high number of crossings were 0.15 and 0.42, respectively. This, together with Q-Q plots, indicates that the data has close to normal distribution. With this in mind, we use the within-subjects t-test to analyze the results. Accuracy is measured using the number of correct trials divided by the total number of trials, thus showing a percentage. Time is measured in seconds.

Completion Time. We exclude incorrect answers, about 11% of the total, and analyze the completion time data only for the correct answers. Otherwise, the measurements of performance time might not be fair (e.g., a participant might quickly give up and give a random answer). Exclusion of incorrect answers does not decrease our sample size significantly since the average number of wrong answers per participant was 7 out of 64 questions.

Increasing the number of edge crossings for small graphs results in statistically significant reduction in performance time. For large graphs there is also a negative impact on performance time, but the results are not statistically significant; see Fig. 2. These results support H1.

Looking at the breakdown into large and small and dense and sparse provides further information. The data are summarized in Table 1, where the small (large) category refers to the average results computed for small (large) sparse and dense graphs.

Increasing the number of edge crossings results in statistically significant reduction in performance time for both small sparse and small dense graphs. This supports H2.

Increasing the number of edge crossings does not result in statistically significant reduction in performance time for large dense graphs (but the reduction is statistically significant for large sparse graphs). This partially supports H3.

Further breakdown by task, reveals more interesting results. For small graphs the main contributors to the statistically significant impacts observed earlier are Tasks 2 and 3 . For large graphs, there is a statistically significant impact for Task 1, although

Table 1. Mean (μ) and standard deviation (σ) of *Completion Time* (in seconds). Statistically significant differences between performance time in layouts with the low and high number of edge crossings are highlighted.

graphs	the number of crossings		t-test results	
	low	high	p-value	t-value
small	$\mu = 48.8 \; \sigma = 9.4$	$\mu = 56.6 \; \sigma = 8.4$	$p < .05$	$t(15) = 2.9$
large	$\mu = 58.0 \; \sigma = 10.1$	$\mu = 62.2 \; \sigma = 9.0$	$p = .24$	$t(15) = 2.0$
small sparse	$\mu = 44.2 \; \sigma = 11.0$	$\mu = 51.3 \; \sigma = 6.7$	$p < .05$	$t(15) = 2.4$
small dense	$\mu = 53.4 \; \sigma = 11.9$	$\mu = 62.0 \; \sigma = 11.9$	$p < .05$	$t(15) = 2.3$
large sparse	$\mu = 53.6 \; \sigma = 12.7$	$\mu = 59.8 \; \sigma = 9.6$	$p = .13$	$t(15) = 1.6$
large dense	$\mu = 62.5 \; \sigma = 11.2$	$\mu = 64.7 \; \sigma = 16.0$	$p = .61$	$t(15) = 0.5$

over all tasks the impact is not significant. Surprisingly, increasing the crossings in large graphs improved the performance time of Task 3 by 10 seconds.

Accuracy. Increasing the number of edge crossings for small graphs results in statistically significant reduction in performance accuracy. For large graphs there is also a negative impact on performance accuracy, but the results are not statistically significant; see Fig. 2. These results support H1.

Looking at the breakdown into large and small and dense and sparse provides further information; see Table 2.

Increasing the number of edge crossings results in statistically significant reduction in accuracy for small dense graphs (but the reduction is not statistically significant for small sparse graphs). This partially supports H2.

Increasing the number of edge crossings results in statistically significant reduction in accuracy for large dense graphs (but the reduction is not statistically significant for large sparse graphs). This partially supports H3.

Further breakdown by task shows that for small graphs Tasks 2 and 4 contribute to the statistically significant impacts observed earlier. Although over all tasks the impact is not significant for large graphs, there is statistically significant difference in accuracy of Tasks 1 and 2. This is counterbalanced with a statistically significant difference in accuracy in opposite direction for Task 4; see more about this below.

3.4 Discussion

Our first hypothesis (H1) is confirmed: increasing the number of edge crossings significantly affects performance time and accuracy for small graphs and the impact is not statistically significant for large graphs. The second hypothesis (H2) is partially confirmed: crossings have a statistically significant impact on time for both sparse and dense small graphs. However, the effect is not statistically significant for accuracy in both sparse and dense small graphs. The third hypothesis (H3) is also only partially confirmed: increasing the number of edge crossings has no significant impact on completion time for large graphs. However, there is statistically significant impact on accuracy for large dense graphs.

Table 2. Mean (μ) and standard deviation (σ) of *Accuracy* (in percentage). Statistically significant differences between completion time in layouts with the low and high number of edge crossings are highlighted.

graphs	the number of crossings		t-test results	
	low	*high*	p-value	t-value
small	$\mu = 94.1\%\ \sigma = 4.3$	$\mu = 89.4\%\ \sigma = 4.4$	$p < .05$	$t(15) = 2.8$
large	$\mu = 86.3\%\ \sigma = 3.4$	$\mu = 83.1\%\ \sigma = 4.0$	$p = .06$	$t(15) = 2.0$
small sparse	$\mu = 93.7\%\ \sigma = 6.4$	$\mu = 92.9\%\ \sigma = 6.3$	$p = .77$	$t(15) = 0.2$
small dense	$\mu = 94.5\%\ \sigma = 7.8$	$\mu = 85.9\%\ \sigma = 13.5$	$p < .05$	$t(15) = 2.2$
large sparse	$\mu = 89.1\%\ \sigma = 11.1$	$\mu = 89.0\%\ \sigma = 9.0$	$p = .81$	$t(15) = 0.2$
large dense	$\mu = 83.5\%\ \sigma = 7.5$	$\mu = 77.3\%\ \sigma = 13.1$	$p < .05$	$t(15) = 2.4$

It is somewhat surprising to see that increasing the crossings affects different task in markedly different ways. It is particularly unexpected to see a statistically significant positive impact on accuracy, with the increase of edge crossings, for Task 4 in large graphs! It is also worth noting that with the increase of edge crossings, the average accuracy increases for Task 3 in small graphs for Tasks 3 and 4 in large graphs. This might be due to participants paying more attention in the cases where the problem was more difficult, possibly related to the "chart junk" effect [2]. But it is also possible that edge crossings may not be as bad as we normally think, as indicated by Huang et al. [15], who found that crossings have negative effect only on some of their tasks.

There are good indications that density plays a possibly independent role, especially on accuracy. Note that we only considered two density settings (1.5 and 2.5), both of which are relatively low. Yet, together with increased number of crossings, the high density settings resulted in statistically significant decrease in accuracy both for small and large graphs. It is probably worth exploring further the nature of the interactions between size (number of vertices), density (ratio of number of edges to number of vertices) and edge crossings upper limit of density.

4   Edge Crossings and Other Aesthetic Criteria

As mentioned earlier, several traditional methods for drawing large undirected graphs are based on the assumption that minimizing a suitably-defined energy function of the graph layout results in aesthetically pleasant drawing. But do such methods also (possibly indirectly) optimize some of the standard aesthetic criteria? Next we qualitatively analyze layouts produced by fdp (force-directed) and neato (MDS-based), with respect to three commonly used and well-defined quality measures: the energy of the layout, the number of crossings, and the angles between pairs of crossing edges.

In a number of studies, the energy of a layout is defined as the variance of edge lengths in the drawing, known as *stress* [18]. Assume a graph $G = (V, E)$ is drawn with p_i being the position of vertex $i \in V$. Denote the distance between two vertices

Table 3. Correlations between three aesthetics: $r(\mathrm{En}, \mathrm{Cr})$, $r(\mathrm{En}, \mathrm{Ang})$, $r(\mathrm{Cr}, \mathrm{Ang})$ stand for the correlation coefficients r between the layout energy En, the number of crossings Cr, and the average crossing angle Ang. Absolute values between 0.7 and 1.0 indicate a strong relationship (highlighted), while absolute values between 0.3 and 0.7 indicates a moderate relationship. Negative values indicate a negative correlation.

	MDS			force-directed		
graph	$r(\mathrm{En}, \mathrm{Cr})$	$r(\mathrm{En}, \mathrm{Ang})$	$r(\mathrm{Cr}, \mathrm{Ang})$	$r(\mathrm{En}, \mathrm{Cr})$	$r(\mathrm{En}, \mathrm{Ang})$	$r(\mathrm{Cr}, \mathrm{Ang})$
GD	0.64	0.00	0.26	0.59	−0.02	−0.39
Recipes	**0.81**	−0.27	−0.15	0.61	−0.13	−0.13
Trade	**0.91**	**-0.82**	**-0.83**	0.62	0.02	−0.24
Universities	0.68	−0.53	−0.56	0.66	−0.09	−0.16
SODA	0.67	−0.69	−0.07	0.54	−0.16	0.10
IPL	**0.82**	−0.37	−0.12	**0.72**	−0.11	−0.04
TARJAN	0.62	−0.02	−0.08	0.54	−0.10	−0.04
SOCG	0.22	−0.64	−0.04	**0.72**	−0.61	−0.11
ALGO	0.41	−0.47	0.15	**0.78**	−0.64	−0.28

$i, j \in V$ by $\|p_i - p_j\|$. The energy of the graph layout is measured by

$$\sum_{i,j \in V} w_{ij}(\|p_i - p_j\| - d_{ij})^2, \tag{1}$$

where d_{ij} is the ideal distance between vertices i and j, and w_{ij} is a weight factor. Typically an ideal distance d_{ij} is defined as the length of the shortest path in G between i and j. Lower stress values correspond to a better layout. We use the conventional weighting factor of $w_{ij} = \frac{1}{d_{ij}^2}$.

We run the two algorithms fdp and neato on 9 graphs for 1,000 times on each graph. As in Section 3.2, we vary the initial layout to produce different drawings of the same graph. For each run, we measure stress, the number of edge crossings, and the average of all crossing angles of the layout. Note that Huang et al. [14] use the minimum crossing angle; in our dataset the minimum values range from 0.1 to 0.9 degrees and so the average angle provides a wider range. Then we consider the computed values for each graph as three random variables and compute the pairwise Pearson correlation coefficients; see Table 3.

The results indicate that there is a moderate positive correlation between the number of crossings and the energy of the layout for all 9 graphs processed with the force-directed algorithm and for 7 graphs processed with MDS. This means that there is a tendency for low-energy drawings to have fewer number of crossings (and vice versa). The effect is illustrated in Fig. 3, where crossings and energy are calculated for the Recipes dataset. We note here that the force-directed algorithm fdp (unlike neato) is not designed to reduce the energy function as defined by Equation (1). Yet the number of crossings is steadily correlated with the energy. This experimental evidence partially supports the observation of Dwyer et al. [5], who show that users prefer graph layouts with lower stress.

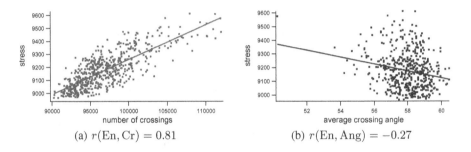

(a) $r(\text{En}, \text{Cr}) = 0.81$ (b) $r(\text{En}, \text{Ang}) = -0.27$

Fig. 3. Relationship between the energy of the drawing (stress) and (a) the number of crossings, (b) the average crossing angle. Dots represent values of the aesthetics computed for different layouts created by the multidimensional scaling algorithm for the Recipes graph.

On the other hand, there are no strong correlations between the other aesthetics. Our results indicate that the number of crossings and the crossing angles are independent in the layouts created by the two evaluated algorithms. We also note a negative correlation between the average crossing angle and the energy on 4 graphs processed with the MDS-based layout algorithm.

5   Conclusion and Future Work

All relevant materials for this study, including more detailed data analysis, are available at http://sites.google.com/site/gdpaper2014.

Our experimental results hopefully serve to inform designers of graph drawing algorithms that minimizing the number of edge crossings in large graphs is not as important as in small graphs. The correlation between low energy layouts and layouts with few crossings indicates that traditional energy-based methods might already result in some reduction in crossings. Although we attempted to be as diverse as possible, our results should be interpreted in the context of the specified graphs, sizes, densities, and tasks.

Due to natural limitations (e.g., length and complexity of experiments), we could not include graphs with more than 120 vertices and density greater than 2.5. Obtaining more results for larger range of the parameters would hopefully help provide a more complete picture. In our experiment we only considered relational reading of static graph drawings; results may be different in experiments that require an interpretive reading of graph drawings in the context of application domains. It would be also worthwhile to consider tasks beyond the network-topology category.

Another interesting direction would be to study in depth the effect of layout energy on understandability of graphs. Different energy function formulations (e.g., stress, distortion) likely have different impact. Evaluating such impact on a greater number of quantitatively measurable aesthetic criteria, as well as on actual tasks performance, is also a promising direction for future work.

Acknowledgements. The work supported in part by NSF grants CCF-1115971 and DEB-1053573.

References

1. Archambault, D., Purchase, C.H., Pinadu, B.: The readability of path-preserving clustering of graphs. EuroVis 29(3), 1173–1182 (2010)
2. Bateman, S., Mandryk, R.L., Gutwin, C., Genest, A., McDine, D., Brooks, C.: Useful junk? The effects of visual embellishment on comprehension and memorability of charts. In: CHI, pp. 2573–2582 (2010)
3. Buchheim, C., Chimani, M., Gutwenger, C., Jünger, M., Mutzel, P.: Crossings and planarization. In: Handbook of Graph Drawing and Visualization. CRC Press (2013)
4. Cabello, S., Mohar, B.: Adding one edge to planar graphs makes crossing number and 1-planarity hard. SIAM Journal on Computing 42(5), 1803–1829 (2013)
5. Dwyer, T., Lee, B., Fisher, D., Quinn, K.I., Isenberg, P., Robertson, G., North, C.: A comparison of user-generated and automatic graph layouts. IEEE Trans. Vis. Comput. Graphics 15(6), 961–968 (2009)
6. Ellson, J., Gansner, E.R., Koutsofios, L., North, S.C., Woodhull, G.: Graphviz - open source graph drawing tools. In: Mutzel, P., Jünger, M., Leipert, S. (eds.) GD 2001. LNCS, vol. 2265, pp. 483–484. Springer, Heidelberg (2002)
7. Fruchterman, T.M., Reingold, E.M.: Graph drawing by force-directed placement. Software: Practice and Experience 21(11), 1129–1164 (1991)
8. Gansner, E.R., Koren, Y., North, S.C.: Graph drawing by stress majorization. In: Pach, J. (ed.) GD 2004. LNCS, vol. 3383, pp. 239–250. Springer, Heidelberg (2005)
9. Garey, M.R., Johnson, D.S.: Crossing number is NP-complete. SIAM Journal on Algebraic Discrete Methods 4(3), 312–316 (1983)
10. Harel, D., Koren, Y.: A fast multi-scale method for drawing large graphs. J. Graph Algorithms Appl. 6(3), 179–202 (2002)
11. Hliněný, P.: Crossing number is hard for cubic graphs. J. Comb. Theory B 96(4), 455–471 (2006)
12. Hu, Y.: Efficient, high-quality force-directed graph drawing. Mathematica Journal 10(1), 37–71 (2005)
13. Huang, W., Huang, M.: Exploring the relative importance of number of edge crossings and size of crossing angles: A quantitative perspective. Advanced Intelligence 3(1), 25–42 (2014)
14. Huang, W., Eades, P., Hong, S.H.: Larger crossing angles make graphs easier to read. Visual Languages & Computing 1 (2014)
15. Huang, W., Hong, S.H., Eades, P.: Layout effects on sociogram perception. In: Healy, P., Nikolov, N.S. (eds.) GD 2005. LNCS, vol. 3843, pp. 262–273. Springer, Heidelberg (2006)
16. Jianu, R., Rusu, A., Hu, Y., Taggart, D.: How to display group information on node–link diagrams: an evaluation. IEEE Trans. Vis. Comput. Graphics (to appear, 2014)
17. Kamada, T., Kawai, S.: An algorithm for drawing general undirected graphs. Inf. Proc. Let. 31(1), 7–15 (1989)
18. Koren, Y., Çivril, A.: The binary stress model for graph drawing. In: Tollis, I.G., Patrignani, M. (eds.) GD 2008. LNCS, vol. 5417, pp. 193–205. Springer, Heidelberg (2009)
19. Lee, B., Plaisant, C., Parr, C., Fekete, J.D., Henry, N.: Task taxonomy for graph visualization. In: BELIV, pp. 81–85. ACM Press (2006)
20. Purchase, H.C.: Which aesthetic has the greatest effect on human understanding? In: DiBattista, G. (ed.) GD 1997. LNCS, vol. 1353, pp. 248–261. Springer, Heidelberg (1997)
21. Purchase, H., Cohen, R., James, M.: Validating graph drawing aesthetics. In: Brandenburg, F.J. (ed.) GD 1995. LNCS, vol. 1027, pp. 435–446. Springer, Heidelberg (1996)
22. Saket, B., Simonetto, P., Kobourov, S., Börner, K.: Node, node-link, and node-link-group diagrams: An evaluation. In: IEEE InfoVis (to appear, 2014)
23. Ware, C., Purchase, H.C., Colpoys, L., McGill, M.: Cognitive measurements of graph aesthetics. Information Visualization 1(2), 103–110 (2002)

The Importance of Being Proper
(In Clustered-Level Planarity and T-Level Planarity)

Patrizio Angelini[1], Giordano Da Lozzo[1], Giuseppe Di Battista[1],
Fabrizio Frati[2], and Vincenzo Roselli[1]

[1] Department of Engineering, Roma Tre University, Italy
{angelini,dalozzo,gdb,roselli}@dia.uniroma3.it
[2] School of Information Technologies, The University of Sydney, Australia
fabrizio.frati@sydney.edu.au

Abstract. In this paper we study two problems related to the drawing of level graphs, that is, T-LEVEL PLANARITY and CLUSTERED-LEVEL PLANARITY. We show that both problems are $\mathcal{NP}$-complete in the general case and that they become polynomial-time solvable when restricted to proper instances.

1 Introduction and Overview

A level graph is *proper* if every of its edges spans just two consecutive levels. Several papers dealing with the construction of level drawings of level graphs assume that the input graph is proper. Otherwise, they suggest to make it proper by "simply adding dummy vertices" along the edges spanning more than two levels. In this paper we show that this apparently innocent augmentation has dramatic consequences if, instead of constructing just a level drawing, we are also interested in representing additional constraints, like a clustering of the vertices or consecutivity constraints on the orderings of the vertices along the levels.

A *level graph* $G = (V, E, \gamma)$ is a graph with a function $\gamma : V \to \{1, 2, ..., k\}$, with $1 \leq k \leq |V|$ such that $\gamma(u) \neq \gamma(v)$ for each edge $(u, v) \in E$. The set $V_i = \{v | \gamma(v) = i\}$ is the i-th *level* of G. A level graph $G = (V, E, \gamma)$ is *proper* if for every edge $(u, v) \in E$, it holds $\gamma(u) = \gamma(v) \pm 1$. A *level planar drawing* of (V, E, γ) maps each vertex v of each level V_i to a point on the line $y = i$, denoted by L_i, and each edge to a y-monotone curve between its endpoints so that no two edges intersect. A level graph is *level planar* if it admits a level planar drawing. A linear-time algorithm for testing level planarity was presented by Jünger *et al.* in [10].

A *clustered-level graph* (*cl-graph*) (V, E, γ, T) is a level graph (V, E, γ) equipped with a *cluster hierarchy* T, that is, a rooted tree where each leaf is an element of V and each internal node μ, called *cluster*, represents the subset V_μ of V composed of the leaves of the subtree of T rooted at μ. A *clustered-level planar drawing* (*cl-planar drawing*) of (V, E, γ, T) is a level planar drawing of level graph (V, E, γ) together with a representation of each cluster μ as a simple closed region enclosing all and only the vertices in V_μ such that: (1) no edge intersects the boundary of a cluster more than once; (2) no two cluster boundaries intersect; and (3) the intersection of L_i with any cluster μ is a straight-line segment, that is, the vertices of V_i that belong to μ are consecutive along L_i. A cl-graph is *clustered-level planar* (*cl-planar*) if it admits a cl-planar drawing. CLUSTERED-LEVEL PLANARITY (CL-PLANARITY) is the problem

C. Duncan and A. Symvonis (Eds.): GD 2014, LNCS 8871, pp. 246–258, 2014.

of testing whether a given cl-graph is cl-planar. This problem was introduced by Forster and Bachmaier [9], who showed a polynomial-time testing algorithm for the case in which the level graph is a proper hierarchy and the clusters are level-connected.

A $\mathcal{T}$-*level graph* (also known as *generalized k-ary tanglegram*) $(V, E, \gamma, \mathcal{T})$ is a level graph (V, E, γ) equipped with a set $\mathcal{T} = T_1, \ldots, T_k$ of trees such that the leaves of T_i are the vertices of level V_i of (V, E, γ), for $1 \leq i \leq k$. A $\mathcal{T}$-*level planar drawing* of $(V, E, \gamma, \mathcal{T})$ is a level planar drawing of (V, E, γ) such that, for $i = 1, \ldots, k$, the order in which the vertices of V_i appear along L_i is *compatible* with T_i, that is, for each node w of T_i, the leaves of the subtree of T_i rooted at w appear consecutively along L_i. A $\mathcal{T}$-level graph is $\mathcal{T}$-*level planar* if it admits a $\mathcal{T}$-level planar drawing. T-LEVEL PLANARITY is the problem of testing whether a given $\mathcal{T}$-level graph is $\mathcal{T}$-level planar. This problem was introduced by Wotzlaw *et al.* [13], who showed a quadratic-time algorithm for the case in which the level graph is proper and the number of vertices of each level is bounded by a constant.

The definition of *proper* naturally extends to cl-graphs and $\mathcal{T}$-level graphs. Note that, given any non-proper level graph G it is easy to construct a proper level graph G' that is level planar if and only if G is level planar. However, as mentioned above, there exists no trivial transformation from a non-proper cl-graph (a non-proper $\mathcal{T}$-level graph) to an equivalent proper cl-graph (resp., an equivalent proper $\mathcal{T}$-level graph).

In this paper we show that CLUSTERED-LEVEL PLANARITY and T-LEVEL PLANARITY are $\mathcal{NP}$-complete for non-proper instances. Conversely, we show that both problems are polynomial-time solvable for proper instances. Our results have several consequences: (1) They narrow the gap between polynomiality and $\mathcal{NP}$-completeness in the classification of Schaefer [12] (see Fig. 1). The reduction of Schaefer between T-LEVEL PLANARITY and SEFE-2 holds for proper instances [12]. (2) They allow to partially answer a question from [12] asking whether a reduction exists from CL-PLANARITY to SEFE-2. We show that such a reduction exists for proper instances and that a reduction from general instances would imply the $\mathcal{NP}$-hardness of SEFE-2. (3) They improve on [9] and [13] by extending the classes of instances which are decidable in polynomial-time for CL-PLANARITY and T-LEVEL PLANARITY, respectively. (4) They provide the first, as far as we know, $\mathcal{NP}$-completeness for a problem that has all the constraints of the clustered planarity problem (and some more).

The paper is organized as follows. The $\mathcal{NP}$-completeness proofs are in Section 2, while the algorithms are in Section 3. We conclude with open problems in Section 4.

2 NP-Hardness

In this section we prove that the T-LEVEL PLANARITY and the CL-PLANARITY problems are $\mathcal{NP}$-complete. In both cases, the $\mathcal{NP}$-hardness is proved by means of a polynomial-time reduction from the $\mathcal{NP}$-complete problem BETWEENNESS [11], that takes as input a finite set A of n objects and a set C of m ordered triples of distinct elements of A, and asks whether a linear ordering $\mathcal{O}$ of the elements of A exists such that for each triple $\langle \alpha, \beta, \delta \rangle$ of C, we have either $\mathcal{O} = <\ldots, \alpha, \ldots, \beta, \ldots, \delta, \ldots>$ or $\mathcal{O} = <\ldots, \delta, \ldots, \beta, \ldots, \alpha, \ldots>$.

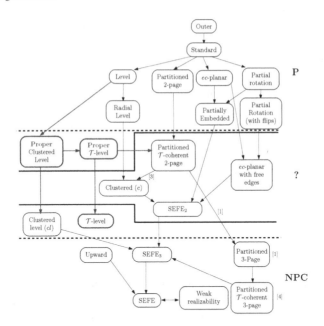

Fig. 1. Updates on the classification proposed by Schaefer in [12]. Dashed lines represent the boundaries between problems that were known to be polynomial-time solvable, problems that were known to be $\mathcal{NP}$-complete, and problems whose complexity was unknown before this paper. Solid lines represent the new boundaries according to the results of this paper. Reductions that can be transitively inferred are omitted. Results proved after [12] are equipped with references. The prefix "proper" has been added to two classes in [12] to better clarify their nature.

Theorem 1. T-LEVEL PLANARITY *is* $\mathcal{NP}$-complete.

Proof: The problem clearly belongs to $\mathcal{NP}$. We prove the $\mathcal{NP}$-hardness. Given an instance $\langle A, C \rangle$ of BETWEENNESS, we construct an equivalent instance $(V, E, \gamma, \mathcal{T})$ of T-LEVEL PLANARITY as follows. Let $A = \{1, \ldots, n\}$ and $m = |C|$. Graph (V, E) is a tree composed of n paths all incident to a common vertex v. Refer to Fig. 2(a). Initialize $V = \{v\}$, $E = \emptyset$, and $\gamma(v) = 0$. Let $T_0 \in \mathcal{T}$ be a tree with a single node v.

For each $j = 1, \ldots, n$, add a vertex v_j to V and an edge (v, v_j) to E, with $\gamma(v_j) = 1$. Also, let $T_1 \in \mathcal{T}$ be a star whose leaves are all the vertices of level V_1. Further, for each $j = 1, \ldots, n$, we initialize variable $last(j) = v_j$.

Then, for each $i = 1, \ldots, m$, consider the triple $t_i = \langle \alpha, \beta, \delta \rangle$. Add six vertices $u_\alpha(i), u'_\alpha(i), u_\beta(i), u'_\beta(i), u_\delta(i)$, and $u'_\delta(i)$ to V with $\gamma(u_\alpha(i)) = \gamma(u_\beta(i)) = \gamma(u_\delta(i)) = 2i$ and $\gamma(u'_\alpha(i)) = \gamma(u'_\beta(i)) = \gamma(u'_\delta(i)) = 2i+1$. Also, add edges $(last(\alpha), u_\alpha(i))$, $(last(\beta), u_\beta(i))$, $(last(\delta), u_\delta(i))$, $(u_\alpha(i), u'_\alpha(i))$, $(u_\beta(i), u'_\beta(i))$, and $(u_\delta(i), u'_\delta(i))$ to E. Further, set $last(\alpha) = u'_\alpha(i)$, $last(\beta) = u'_\beta(i)$, and $last(\delta) = u'_\delta(i)$. Let $T_{2i} \in \mathcal{T}$ be a binary tree with a root r_{2i}, an internal node x_{2i} and a leaf $u_\alpha(i)$ both adjacent to r_{2i}, and with leaves $u_\beta(i)$ and $u_\delta(i)$ both adjacent to x_{2i}. Moreover, let $T_{2i+1} \in \mathcal{T}$ be a binary tree with a root r_{2i+1}, an internal node x_{2i+1} and a leaf $u'_\delta(i)$ both adjacent to r_{2i+1}, and with leaves $u'_\alpha(i)$ and $u'_\beta(i)$ both adjacent to x_{2i+1}.

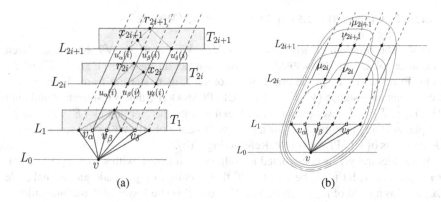

Fig. 2. Illustrations for the proof of (a) Theorem 1 and (b) Theorem 2

The reduction is easily performed in $O(n + m)$ time. We prove that $(V, E, \gamma, \mathcal{T})$ is $\mathcal{T}$-level planar if and only if $\langle A, C \rangle$ is a positive instance of BETWEENNESS.

Suppose that $(V, E, \gamma, \mathcal{T})$ admits a $\mathcal{T}$-level planar drawing Γ. Consider the left-to-right order $\mathcal{O}_1$ in which the vertices of level V_1 appear along L_1. Construct an order $\mathcal{O}$ of the elements of A such that $\alpha \in A$ appears before $\beta \in A$ if and only if $v_\alpha \in V_1$ appears before $v_\beta \in V_1$ in $\mathcal{O}_1$. In order to prove that $\mathcal{O}$ is a positive solution for $\langle A, C \rangle$, it suffices to prove that, for each triple $t_i = \langle \alpha, \beta, \delta \rangle \in C$, vertices v_α, v_β, and v_δ appear either in this order or in the reverse order in $\mathcal{O}_1$. Note that tree T_{2i} enforces $u_\alpha(i)$ not to lie between $u_\beta(i)$ and $u_\delta(i)$ along L_{2i}; also, tree T_{2i+1} enforces $u'_\alpha(i)$ not to lie between $u'_\alpha(i)$ and $u'_\beta(i)$ along L_{2i+1}. Since the three paths connecting $u'_\alpha(i)$, $u'_\beta(i)$, and $u'_\delta(i)$ with v are y-monotone, do not cross each other, and contain $u_\alpha(i)$ and v_α, $u_\beta(i)$ and v_β, and $u_\delta(i)$ and v_δ, respectively, we have that v_α, v_β, and v_δ appear either in this order or in the reverse order in $\mathcal{O}_1$.

Suppose that an ordering $\mathcal{O}$ of the elements of A exists that is a positive solution of BETWEENNESS for instance $\langle A, C \rangle$. In order to construct Γ, place the vertices of V_1 along L_1 in such a way that vertex $v_j \in V_1$ is assigned x-coordinate equal to s if j is the s-th element of $\mathcal{O}$, for $j = 1, \ldots, n$. Also, for $i = 1, \ldots, m$, let $t_i = \langle \alpha, \beta, \delta \rangle \in C$. Place vertices $u_\lambda(i)$ and $u'_\lambda(i)$, with $\lambda \in \{\alpha, \beta, \delta\}$, on L_{2i} and L_{2i+1}, respectively, in such a way that $u_\lambda(i)$ and $u'_\lambda(i)$ are assigned x-coordinate equal to s if λ is the s-th element of $\mathcal{O}$. Finally, place v at any point on L_0 and draw the edges of E as straight-line segments. We prove that Γ is a $\mathcal{T}$-level planar drawing of $(V, E, \gamma, \mathcal{T})$. First, Γ is a level planar drawing of (V, E, γ), by construction. Further, for each $i = 1, \ldots, m$, vertices $u_\alpha(i)$, $u_\beta(i)$, and $u_\delta(i)$ appear along L_{2i} either in this order or in the reverse order; in both cases, the order is compatible with tree T_{2i}. Analogously, vertices $u'_\alpha(i)$, $u'_\beta(i)$, and $u'_\delta(i)$ appear along L_{2i+1} either in this order or in the reverse order; in both cases, the order is compatible with tree T_{2i+1}. Finally, the order in which vertices of V_0 and V_1 appear along L_0 and L_1 are trivially compatible with T_0 and T_1, respectively. $\square$

Theorem 2. CLUSTERED-LEVEL PLANARITY *is $\mathcal{NP}$-complete.*

Proof: The problem clearly belongs to $\mathcal{NP}$. We prove the $\mathcal{NP}$-hardness. Given an instance $\langle A, C \rangle$ of BETWEENNESS, we construct an instance $(V, E, \gamma, \mathcal{T})$ of T-LEVEL PLANARITY as in the proof of Theorem 1; then, starting from $(V, E, \gamma, \mathcal{T})$, we construct an instance (V, E, γ, T) of CL-PLANARITY that is cl-planar if and only if $(V, E, \gamma, \mathcal{T})$ is $\mathcal{T}$-level planar. This, together with the fact that $(V, E, \gamma, \mathcal{T})$ is $\mathcal{T}$-level planar if and only if $\langle A, C \rangle$ is a positive instance of BETWEENNESS, implies the $\mathcal{NP}$-hardness of CL-PLANARITY. Refer to Fig. 2(b).

Cluster hierarchy T is constructed as follows. Initialize T with a root μ_{2m+1}. Next, for $i = m, \ldots, 1$, let $u'_\delta(i)$ be a leaf of T that is child of μ_{2i+1}; add an internal node ν_{2i+1} to T as a child of μ_{2i+1}; then, let $u'_\alpha(i)$ and $u'_\beta(i)$ be leaves of T that are children of ν_{2i+1}; add an internal node μ_{2i} to T as a child of ν_{2i+1}. Further, let $u_\alpha(i)$ be a leaf of T that is a child of μ_{2i}; add an internal node ν_{2i} to T as a child of μ_{2i}; then, let $u_\beta(i)$ and $u_\delta(i)$ be leaves of T that are children of ν_{2i}; add an internal node μ_{2i-1} to T as a child of ν_{2i}. Finally, let vertices $v \in V_0$ and $v_j \in V_1$, for $j = 1, \ldots, n$, be leaves of T that are children of μ_1.

We prove that (V, E, γ, T) is cl-planar if and only if $(V, E, \gamma, \mathcal{T})$ is $\mathcal{T}$-level planar.

Suppose that (V, E, γ, T) admits a cl-planar drawing Γ. Construct a $\mathcal{T}$-level planar drawing Γ^* of $(V, E, \gamma, \mathcal{T})$ by removing from Γ the clusters of T. The drawing of (V, E, γ) in Γ^* is level-planar, since it is level-planar in Γ. Further, for each $i = 1, \ldots, m$, vertex $u_\alpha(i)$ does not appear between $u_\beta(i)$ and $u_\delta(i)$ along L_{2i}, since $u_\beta(i), u_\delta(i) \in \nu_{2i}$ and $u_\alpha(i) \notin \nu_{2i}$; analogously, vertex $u'_\alpha(i)$ does not appear between $u'_\alpha(i)$ and $u'_\beta(i)$ along L_{2i+1}, since $u'_\alpha(i), u'_\beta(i) \in \nu_{2i+1}$ and $u'_\delta(i) \notin \nu_{2i+1}$. Hence, the order of the vertices of V_{2i} and V_{2i+1} along L_{2i} and L_{2i+1}, respectively, are compatible with trees T_{2i} and T_{2i+1}. Finally, the order in which the vertices of V_0 and V_1 appear along lines L_0 and L_1 are trivially compatible with T_0 and T_1, respectively.

Suppose that $(V, E, \gamma, \mathcal{T})$ admits a $\mathcal{T}$-level planar drawing Γ^*; we describe how to construct a cl-planar drawing Γ of (V, E, γ, T). Assume that Γ^* is a straight-line drawing, which is not a loss of generality [8]. Initialize $\Gamma = \Gamma^*$. Draw each cluster α in T as a convex region $R(\alpha)$ in Γ slightly surrounding the border of the convex hull of its vertices and slightly surrounding the border of the regions representing the clusters that are its descendants in T. Let j be the largest index such that V_j contains a vertex of α. Then, $R(\alpha)$ contains all and only the vertices that are descendants of α in T; moreover, any two clusters α and β in T are one contained into the other, hence $R(\alpha)$ and $R(\beta)$ do not cross; finally, we prove that no edge e in E crosses more than once the boundary of $R(\alpha)$ in Γ. First, if at least one end-vertex of e belongs to α, then e and the boundary of $R(\alpha)$ cross at most once, given that e is a straight-line segment and that $R(\alpha)$ is convex. All the vertices in $V_0 \cup \ldots \cup V_{j-1}$ and at least two vertices of V_j belong to α, hence their incident edges do not cross the boundary of $R(\alpha)$ more than once. Further, all the vertices in $V_{j+1} \cup \ldots \cup V_{2m+3}$ have y-coordinates larger than every point of $R(\alpha)$, hence edges between them do not cross $R(\alpha)$. It remains to consider the case in which e connects a vertex x_1 in V_j not in α (there is at most one such vertex) with a vertex x_2 in $V_{j+1} \cup \ldots \cup V_{2m+2}$; in this case e and $R(\alpha)$ do not cross given that x_1 is outside $R(\alpha)$, that x_2 has y-coordinate larger than every point of $R(\alpha)$, and that $R(\alpha)$ is arbitrarily close to the convex hull of its vertices. $\qquad\square$

The reductions described in Theorems 1 and 2 can be modified so that (V, E) consists of a set of paths (by removing levels V_0 and V_1), or that (V, E) is a 2-connected series-parallel graph (by introducing levels V_{2m+2} and V_{2m+3} "symmetric" to levels V_1 and V_0, respectively).

3   Polynomial-Time Algorithms

In this section we prove that both T-LEVEL PLANARITY and CL-PLANARITY are polyomial-time solvable problems if restricted to proper instances.

3.1 T-LEVEL PLANARITY

We start by describing a polynomial-time algorithm for T-LEVEL PLANARITY. The algorithm is based on a reduction to the *Simultanoues Embedding with Fixed Edges* problem for two graphs (SEFE-2), that is defined as follows.

A *simultanoues embedding with fixed edges* (SEFE) of two graphs $G_1 = (V, E_1)$ and $G_2 = (V, E_2)$ on the same set of vertices V consists of two planar drawings Γ_1 and Γ_2 of G_1 and G_2, respectively, such that each vertex $v \in V$ is mapped to the same point in both drawings and each edge of the *common graph* $G_\cap = (V, E_1 \cap E_2)$ is represented by the same simple curve in the two drawings. The SEFE-2 problem asks whether a given pair of graphs $\langle G_1, G_2 \rangle$ admits a SEFE [5]. The computational complexity of the SEFE-2 problem is unknown, but there exist polynomial-time algorithms for instances that respect some conditions [2,5,6,7,12]. We are going to use a result by Bläsius and Rütter [7], who proposed a quadratic-time algorithm for instances $\langle G_1, G_2 \rangle$ of SEFE-2 in which G_1 and G_2 are 2-connected and the common graph $G_\cap$ is connected.

In the analysis of the complexity of the following algorithms we assume that the internal nodes of the trees in $\mathcal{T}$ in any instance $(V, E, \gamma, \mathcal{T})$ of T-LEVEL PLANARITY and of tree T in any instance (V, E, γ, T) of CL-PLANARITY have at least two children. It is easily proved that this is not a loss of generality; also, this allows us to describe the size of the instances in terms of the size of their sets of vertices.

Lemma 1. *Let* $(V, E, \gamma, \mathcal{T})$ *be a proper instance of* T-LEVEL PLANARITY. *There exists an equivalent instance* $\langle G_1^*, G_2^* \rangle$ *of SEFE-2 such that* $G_1^* = (V^*, E_1^*)$ *and* $G_2^* = (V^*, E_2^*)$ *are 2-connected and the common graph* $G_\cap = (V^*, E_1^* \cap E_2^*)$ *is connected. Further, instance* $\langle G_1^*, G_2^* \rangle$ *can be constructed in linear time.*

Proof: We describe how to construct instance $\langle G_1^*, G_2^* \rangle$. Refer to Fig. 3.

Graph $G_\cap$ contains a cycle $\mathcal{C} = t_1, t_2, \ldots, t_k, q_k, p_k, q_{k-1}, p_{k-1}, \ldots, q_1, p_1$, where k is the number of levels of $(V, E, \gamma, \mathcal{T})$. For each $i = 1, \ldots, k$, graph $G_\cap$ contains a copy $\overline{T_i}$ of tree $T_i \in \mathcal{T}$, whose root is vertex t_i, and contains two stars P_i and Q_i centered at vertices p_i and q_i, respectively, whose number of leaves is determined as follows. For each vertex $u \in V_i$ such that an edge $(u, v) \in E$ exists connecting u to a vertex $v \in V_{i-1}$, star P_i contains a leaf $u(P_i)$; also, for each vertex $u \in V_i$ such that an edge $(u, v) \in E$ exists connecting u to a vertex $v \in V_{i+1}$, star Q_i contains a leaf $u(Q_i)$. We also denote by $u(\overline{T_i})$ a leaf of $\overline{T_i}$ corresponding to vertex $u \in V_i$.

Graph G_1^* contains $G_\cap$ plus the following edges. For $i = 1, \ldots, k$, consider each vertex $u \in V_i$. Suppose that i is even. Then, G_1^* has an edge connecting the leaf $u(\overline{T_i})$ of $\overline{T_i}$ corresponding to u with either the leaf $u(Q_i)$ of Q_i corresponding to u, if it exists,

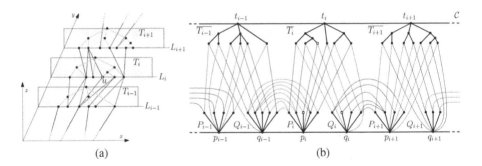

(a)   (b)

Fig. 3. Illustration for the proof of Lemma 1. Index i is assumed to be even. (a) A T-level planar drawing Γ of instance $(V, E, \gamma, \mathcal{T})$. (b) The SEFE $\langle \Gamma_1, \Gamma_2 \rangle$ of instance $\langle G_1^*, G_2^* \rangle$ of SEFE-2 corresponding to Γ. Correspondence between a vertex $u \in V_i$ and leaves $u(\overline{T_i}) \in \overline{T_i}$, $u(P_i) \in P_i$, and $u(Q_i) \in Q_i$ is highlighted by representing all such vertices as white boxes.

or with q_i, otherwise; also, for each edge in E connecting a vertex $u \in V_i$ with a vertex $v \in V_{i-1}$, graph G_1^* has an edge connecting the leaf $u(P_i)$ of P_i corresponding to u with the leaf $v(Q_{i-1})$ of Q_{i-1} corresponding to v (such leaves exist by construction). Suppose that i is odd. Then, graph G_1^* has an edge between $u(\overline{T_i})$ and either $u(P_i)$, if it exists, or p_i, otherwise. Graph G_2^* contains $G_\cap$ plus the following edges. For $i = 1, \ldots, k$, consider each vertex $u \in V_i$. Suppose that i is odd. Then, G_2^* has an edge connecting $u(\overline{T_i})$ with either the leaf $u(Q_i)$ of Q_i corresponding to u, if it exists, or with q_i, otherwise; also, for each edge in E connecting a vertex $u \in V_i$ with a vertex $v \in V_{i-1}$, graph G_2^* has an edge $(u(P_i), v(Q_{i-1}))$. Suppose that i is even. Then, graph G_2^* has an edge between $u(\overline{T_i})$ and either $u(P_i)$, if it exists, or p_i, otherwise.

Graph $G_\cap$ is clearly connected. We prove that G_1^* and G_2^* are 2-connected, that is, removing any vertex v disconnects neither G_1^* nor G_2^*. If v is a leaf of $\overline{T_i}$, P_i, or Q_i, with $1 \leq i \leq k$, then removing v disconnects neither G_1^* nor G_2^*, since $G_\cap$ remains connected. If v is an internal node (the root) of $\overline{T_i}$, P_i, or Q_i, say of $\overline{T_i}$, with $1 \leq i \leq k$, then removing v disconnects $G_\cap$ into one component $\overline{T_i}(v)$ containing all the vertices of $\mathcal{C}$ (resp. all the vertices of $\mathcal{C}$, except for v) and into some subtrees $\overline{T_{i,j}}$ of $\overline{T_i}$ rooted the children of v; however, by construction, each leaf $u(\overline{T_i})$ of $\overline{T_{i,j}}$ is connected to $\overline{T_i}(v)$ via an edge of G_1^*, namely either $(u(\overline{T_i}), u(P_i))$, $(u(\overline{T_i}), p_i)$, $(u(\overline{T_i}), u(Q_i))$, or $(u(\overline{T_i}), q_i)$ (and similar for G_2^*), hence G_1^* (and G_2^*) is connected after the removal of v.

Observe that, if $(V, E, \gamma, \mathcal{T})$ has $n_{\mathcal{T}}$ nodes in the trees of $\mathcal{T}$ (where $|V| < n_{\mathcal{T}}$), then $\langle G_1^*, G_2^* \rangle$ contains at most $3n_{\mathcal{T}}$ vertices. Also, the number of edges of $\langle G_1^*, G_2^* \rangle$ is at most $|E| + 2n_{\mathcal{T}}$. Hence, the size of $\langle G_1^*, G_2^* \rangle$ is linear in the size of $(V, E, \gamma, \mathcal{T})$; also, it is easy to see that $\langle G_1^*, G_2^* \rangle$ can be constructed in linear time.

We prove that $\langle G_1^*, G_2^* \rangle$ admits a SEFE if and only if $(V, E, \gamma, \mathcal{T})$ is $\mathcal{T}$-level planar. Suppose that $\langle G_1^*, G_2^* \rangle$ admits a SEFE $\langle \Gamma_1^*, \Gamma_2^* \rangle$. We show how to construct a drawing Γ of $(V, E, \gamma, \mathcal{T})$. For $1 \leq i \leq k$, let $\Theta(\overline{T_i})$ be the order in which the leaves of $\overline{T_i}$ appear in a pre-order traversal of $\overline{T_i}$ in $\langle \Gamma_1^*, \Gamma_2^* \rangle$; then, let the ordering $\mathcal{O}_i$ of the vertices of V_i along L_i be either $\Theta(\overline{T_i})$, if i is odd, or the reverse of $\Theta(\overline{T_i})$, if i is even.

We prove that Γ is $\mathcal{T}$-level planar. For each $i = 1, \ldots, k$, $\mathcal{O}_i$ is compatible with $T_i \in \mathcal{T}$, since the drawing of $\overline{T_i}$, that belongs to $G_\cap$, is planar in $\langle \Gamma_1^*, \Gamma_2^* \rangle$. Suppose, for a contradiction, that two edges $(u, v), (w, z) \in E$ exist, with $u, w \in V_i$ and $v, z \in V_{i+1}$, that intersect in Γ. Hence, either u appears before w in $\mathcal{O}_i$ and v appears after z in $\mathcal{O}_{i+1}$, or vice versa. Since i and $i + 1$ have different parity, either u appears before w in $\Theta(\overline{T_i})$ and v appears before z in $\Theta(\overline{T_{i+1}})$, or vice versa. We claim that, in both cases, this implies a crossing in $\langle \Gamma_1^*, \Gamma_2^* \rangle$ between paths $(q_i, u(Q_i), v(P_{i+1}), p_{i+1})$ and $(q_i, w(Q_i), z(P_{i+1}), p_{i+1})$ in $\langle G_1^*, G_2^* \rangle$. Since the edges of these two paths belong all to G_1^* or all to G_2^*, depending on whether i is even or odd, this yields a contradiction. We now prove the claim. The pre-order traversal $\Theta(Q_i)$ of Q_i (the pre-order traversal $\Theta(P_{i+1})$ of P_{i+1}) in $\langle \Gamma_1^*, \Gamma_2^* \rangle$ restricted to the leaves of Q_i (of P_{i+1}) is the reverse of $\Theta(\overline{T_i})$ (of $\Theta(\overline{T_{i+1}})$) restricted to the vertices of V_i (of V_{i+1}) corresponding to leaves of Q_i (of P_{i+1}). Namely, each leaf $x(Q_i)$ of Q_i ($y(P_{i+1})$ of P_{i+1}) is connected to leaf $x(\overline{T_i})$ of $\overline{T_i}$ ($y(\overline{T_{i+1}})$ of $\overline{T_{i+1}}$) in the same graph, either G_1^* or G_2^*, by construction. Hence, the fact that u appears before (after) w in $\Theta(\overline{T_i})$ and v appears before (after) z in $\Theta(\overline{T_{i+1}})$ implies that u appears after (before) w in $\Theta(Q_i)$ and v appears after (before) z in $\Theta(P_{i+1})$. In both cases, this implies a crossing in $\langle \Gamma_1^*, \Gamma_2^* \rangle$ between the two paths.

Suppose that $(V, E, \gamma, \mathcal{T})$ admits a $\mathcal{T}$-level planar drawing Γ. We show how to construct a SEFE $\langle \Gamma_1^*, \Gamma_2^* \rangle$ of $\langle G_1^*, G_2^* \rangle$. For $1 \leq i \leq k$, let $\mathcal{O}_i$ be the order of the vertices of level V_i along L_i in Γ. Since Γ is $\mathcal{T}$-level planar, there exists an embedding Γ_i of tree $T_i \in \mathcal{T}$ that is compatible with $\mathcal{O}_i$. If i is odd (even), then assign to each internal vertex of $\overline{T_i}$ the same (resp. the opposite) rotation scheme as its corresponding vertex in Γ_i. Also, if i is odd, then assign to p_i (to q_i) the rotation scheme in G_1^* (resp. in G_2^*) such that the paths that connect p_i (resp. q_i) to the leaves of $\overline{T_i}$, either with an edge or passing through a leaf of P_i (resp. of Q_i), appear in the same clockwise order as the vertices of V_i appear in $\mathcal{O}_i$; if i is even, then assign to p_i (to q_i) the rotation scheme in G_2^* (resp. in G_1^*) such that the paths that connect p_i (resp. q_i) to the leaves of $\overline{T_i}$ appear in the same counterclockwise order as the vertices of V_i appear in $\mathcal{O}_i$. Finally, consider the embedding $\Gamma_{i,i+1}$ obtained by restricting Γ to the vertices and edges of the subgraph induced by the vertices of V_i and V_{i+1}. If i is odd (even), then assign to the leaves of Q_i and P_{i+1} in G_1^* (in G_2^*) the same rotation scheme as their corresponding vertices have in $\Gamma_{i,i+1}$. This completes the construction of $\langle \Gamma_1^*, \Gamma_2^* \rangle$.

We prove that $\langle \Gamma_1^*, \Gamma_2^* \rangle$ is a SEFE of $\langle G_1^*, G_2^* \rangle$. Since the rotation scheme of the internal vertices of each $\overline{T_i}$ are constructed starting from an embedding of Γ_i of tree $T_i \in \mathcal{T}$ that is compatible with $\mathcal{O}_i$, the drawing of $\overline{T_i}$ is planar. Further, since the rotation schemes of p_i (of q_i) are also constructed starting from $\mathcal{O}_i$, there exists no crossing between two paths connecting t_i and p_i (t_i and q_i), one passing through a leaf $u(\overline{T_i})$ of $\overline{T_i}$ and, possibly, through a leaf $u(P_i)$ of P_i (through a leaf $u(Q_i)$ of Q_i), and the other passing through a leaf $v(\overline{T_i})$ of $\overline{T_i}$ and, possibly, through a leaf $v(P_i)$ of P_i (through a leaf $v(Q_i)$ of Q_i). Finally, since the rotation schemes of the leaves of Q_i and P_{i+1} are constructed from the embedding $\Gamma_{i,i+1}$ obtained by restricting Γ to the vertices and edges of the subgraph induced by the vertices of V_i and V_{i+1}, there exist no two crossing edges between leaves of Q_i and P_{i+1}.   □

We remark that a reduction from $\mathcal{T}$-LEVEL PLANARITY to SEFE-2 was described by Schaefer in [12]; however, the instances of SEFE-2 obtained from that reduction do not satisfy any conditions that make SEFE-2 known to be solvable in polynomial-time.

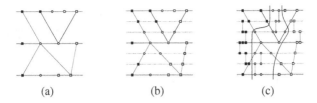

(a)	(b)	(c)

Fig. 4. Illustration for the proof of Lemma 2. (a) Instance (V, E, γ, T) with flat hierarchy containing clusters $\mu_\blacksquare$, $\mu_\square$, and $\mu_\circ$. (b) Insertion of dummy vertices in (V, E, γ, T) to obtain (V', E', γ', T'). (c) Level-connected instance $(V^*, E^*, \gamma^*, T^*)$ obtained from (V', E', γ', T').

Theorem 3. *There exists an $O(|V|^2)$-time algorithm that decides whether a proper instance $(V, E, \gamma, \mathcal{T})$ of $\mathcal{T}$-LEVEL PLANARITY is $\mathcal{T}$-level planar.*

Proof: The statement follows from Lemma 1 and from the existence of a quadratic-time algorithm [7] that decides whether an instance $\langle G_1, G_2 \rangle$ of SEFE-2 such that G_1 and G_2 are 2-connected and the common graph $G_\cap$ is connected admits a SEFE. □

3.2   CLUSTERED-LEVEL PLANARITY

In the following we show how to test in polynomial time the existence of a cl-planar drawing for a proper instance (V, E, γ, T) of CL-PLANARITY.

A proper cl-graph (V, E, γ, T) is *μ-connected between two levels* V_i and V_{i+1} if there exist two vertices $u \in V_\mu \cap V_i$ and $v \in V_\mu \cap V_{i+1}$ such that edge $(u, v) \in E$. For a cluster $\mu \in T$, let $\gamma_{\min}(\mu) = \min\{i | V_i \cap V_\mu \neq \emptyset\}$ and let $\gamma_{\max}(\mu) = \max\{i | V_i \cap V_\mu \neq \emptyset\}$. A proper cl-graph (V, E, γ, T) is *level-μ-connected* if it is μ-connected between levels V_i and V_{i+1} for each $i = \gamma_{\min}(\mu), \ldots, \gamma_{\max}(\mu) - 1$. A proper cl-graph (V, E, γ, T) is *level-connected* if it is μ-level-connected for each cluster $\mu \in T$.

Our strategy consists of first transforming a proper instance of CL-PLANARITY into an equivalent level-connected instance, and then transforming such a level-connected instance into an equivalent proper instance of T-LEVEL PLANARITY.

Lemma 2. *Let (V, E, γ, T) be a proper instance of CL-PLANARITY. An equivalent level-connected instance $(V^*, E^*, \gamma^*, T^*)$ of CL-PLANARITY whose size is quadratic in the size of (V, E, γ, T) can be constructed in quadratic time.*

Proof: The construction of $(V^*, E^*, \gamma^*, T^*)$ consists of two steps. See Fig. 4.

In the first step we turn (V, E, γ, T) into an equivalent instance (V', E', γ', T'). Initialize $V' = V$, $E' = E$, and $T' = T$. For each $i = 1, \ldots, k$ and for each vertex $u \in V_i$, set $\gamma'(u) = 3(i-1) + 1$. Then, for each $i = 1, \ldots, k - 1$, consider each edge $(u, v) \in E$ such that $\gamma(u) = i$ and $\gamma(v) = i + 1$; add two vertices d_u and d_v to V', and replace (u, v) in E' with edges (u, d_u), (d_u, d_v), and (d_v, v). Set $\gamma'(d_u) = 3(i-1) + 2$ and $\gamma'(d_v) = 3i$. Finally, add d_u (d_v) to T' as a child of the parent of u (of v) in T'.

We prove that (V', E', γ', T') is equivalent to (V, E, γ, T).

Suppose that (V, E, γ, T) admits a cl-planar drawing Γ; a cl-planar drawing Γ' of (V', E', γ', T') is constructed as follows. Initialize $\Gamma' = \Gamma$. Scale Γ' up by a factor of 3 and vertically translate it so that the vertices in V'_1 lie on line $y = 1$. After the two affine transformations have been applied (i) Γ' has no crossing, (ii) every edge is a y-monotone curve, (iii) for $i = 1, \ldots, k$, the vertices in $V_i = V'_{3(i-1)+1}$ are placed on line $y = 3(i-1) + 1$, that we denote by $L'_{3(i-1)+1}$, and (iv) the order in which the vertices in $V_i = V'_{3(i-1)+1}$ appear along $L'_{3(i-1)+1}$ is the same as the order in which they appear along L_i. For each $i = 1, \ldots, k-1$, consider each edge $(u, v) \in E$ such that $\gamma(u) = i$ and $\gamma(v) = i+1$. Place vertices d_u and d_v in Γ' on the two points of the curve representing (u, v) having y-coordinate equal to $3(i-1)+2$ and $3i$, respectively. Then, the curves representing in Γ' any two edges in E' are part of the curves representing in Γ' any two edges in E. Hence Γ' is a cl-planar drawing of (V', E', γ', T').

Suppose that (V', E', γ', T') admits a cl-planar drawing Γ'; a cl-planar drawing Γ of (V, E, γ, T) is constructed as follows. Initialize $\Gamma = \Gamma'$. For $i = 1, \ldots, k-1$, consider each path (u, d_u, d_v, v) such that $\gamma'(u) = 3(i - 1) + 1$ and $\gamma'(v) = 3i + 1$; remove vertices d_u and d_v, and their incident edges in E' from Γ; draw edge $(u, v) \in E$ in Γ as the composition of the curves representing edges (u, d_u), (d_u, d_v), and (d_v, v) in Γ'. Scale Γ down by a factor of 3 and vertically translate it so that the vertices of V_1 lie on line $y = 1$. After the two affine transformations have been applied (i) Γ has no crossing, (ii) every edge is a y-monotone curve, (iii) for $i = 1, \ldots, k$, the vertices of level V_i are placed on line $y = i$, and (iv) the order in which the vertices in $V_i = V'_{3(i-1)+1}$ appear along L_i is the same as the order in which they appear along $L'_{3(i-1)+1}$. Since Γ' is cl-planar, this implies that Γ is cl-planar, as well.

The goal of this transformation was to obtain an instance (V', E', γ', T') such that, if there exists a vertex $u \in V'_j$, with $1 \leq j \leq 3(k - 1) + 1$, that is adjacent to two vertices $v, w \in V'_h$, with $h = j \pm 1$, then u, v, and w have the same parent node $\mu \in T'$; hence, (V', E', γ', T') is μ-connected between levels V'_j and V'_h.

In the second step we transform (V', E', γ', T') into an equivalent level-connected instance $(V^*, E^*, \gamma^*, T^*)$. Initialize $(V^*, E^*, \gamma^*, T^*) = (V', E', \gamma', T')$. Consider each cluster $\mu \in T'$ according to a bottom-up visit of T'. If there exists a level V'_i, with $\gamma'_{\min}(\mu) \leq i < \gamma'_{\max}(\mu)$, such that no edge in E' connects a vertex $u \in V'_i \cap V'_\mu$ with a vertex $v \in V'_{i+1} \cap V'_\mu$, then add two vertices u^* and v^* to V^*, add an edge (u^*, v^*) to E^*, set $\gamma^*(u^*) = i$ and $\gamma^*(v^*) = i + 1$, and add u^* and v^* to T^* as children of μ.

Observe that, for each cluster $\mu \in T'$ and for each level $1 \leq i \leq 3k - 2$, at most two dummy vertices are added to $(V^*, E^*, \gamma^*, T^*)$. This implies that $|V^*| \in O(|V'|^2) \in O(|V|^2)$. Also, the whole construction can be performed in $O(|V|^2)$ time.

It remains to prove that $(V^*, E^*, \gamma^*, T^*)$ is equivalent to (V', E', γ', T').

Suppose that $(V^*, E^*, \gamma^*, T^*)$ admits a cl-planar drawing Γ^*; a cl-planar drawing Γ' of (V', E', γ', T') can be constructed as follows. Initialize $\Gamma' = \Gamma^*$ and remove from V', E', and Γ' all the vertices and edges added when constructing Γ^*. Since all the other vertices of V' and edges of E' have the same representation in Γ' and in Γ^*, and since Γ^* is cl-planar, it follows that Γ' is cl-planar, as well.

Suppose that (V', E', γ', T') admits a cl-planar drawing Γ'; a cl-planar drawing Γ^* of $(V^*, E^*, \gamma^*, T^*)$ can be constructed as follows. Initialize $\Gamma^* = \Gamma'$. Consider a level V'_i, with $1 \leq i \leq 3(k-1)$, such that vertices $u^*, v^* \in \mu$ with $\gamma'(u^*) = i$ and $\gamma'(v^*) = i + 1$, for some cluster $\mu \in T$, have been added to $(V^*, E^*, \gamma^*, T^*)$. By construction, (V', E', γ', T') is not μ-connected between levels V'_i and V'_{i+1}. As observed before, this implies that no vertex $u \in V'_i \cap V'_\mu$ exists that is connected to two vertices $v, w \in V'_{i+1}$, and no vertex $u \in V'_{i+1} \cap V'_\mu$ exists that is connected to two vertices $v, w \in V'_i$. Hence, vertices u^* and v^*, and edge (u^*, v^*), can be drawn in Γ^* entirely inside the region representing μ in such a way that u^* and v^* lie along lines L'_i and L'_{i+1} and there exists no crossing between edge (u^*, v^*) and any other edge.

This concludes the proof of the lemma. □

Lemma 3. *Let (V, E, γ, T) be a level-connected instance of* CL-PLANARITY. *An equivalent proper instance $(V, E, \gamma, \mathcal{T})$ of T-LEVEL PLANARITY whose size is linear in the size of (V, E, γ, T) can be constructed in linear time.*

Proof: We construct $(V, E, \gamma, \mathcal{T})$ from (V, E, γ, T) as follows. Initialize $\mathcal{T} = \emptyset$. For $i = 1, \ldots, k$, add to $\mathcal{T}$ a tree T_i that is the subtree of the cluster hierarchy T whose leaves are all and only the vertices of level V_i. Note that the set of leaves of the trees in $\mathcal{T}$ corresponds to the vertex set V. Since each internal node of the trees in $\mathcal{T}$ has at least two children, we have that the size of $(V, E, \gamma, \mathcal{T})$ is linear in the size of (V, E, γ, T). Also, the construction of $(V, E, \gamma, \mathcal{T})$ can be easily performed in linear time.

We prove that $(V, E, \gamma, \mathcal{T})$ is $\mathcal{T}$-level planar if and only if (V, E, γ, T) is cl-planar.

Suppose that $(V, E, \gamma, \mathcal{T})$ admits a $\mathcal{T}$-level planar drawing Γ^*; we show how to construct a cl-planar drawing Γ of (V, E, γ, T). Initialize $\Gamma = \Gamma^*$. Consider each level V_i, with $i = 1, \ldots, k$. By construction, for each cluster $\mu \in T$ such that there exists a vertex $v \in V_i \cap V_\mu$, there exists an internal node of tree $T_i \in \mathcal{T}$ whose leaves are all and only the vertices of $V_i \cap V_\mu$. Since Γ^* is $\mathcal{T}$-level planar, such vertices appear consecutively along L_i. Hence, in order to prove that Γ is a cl-planar drawing, it suffices to prove that there exist no four vertices u, v, w, z such that (i) $u, v \in V_i$ and $w, z \in V_j$, with $1 \leq i < j \leq k$; (ii) $u, w \in V_\mu$ and $v, z \in V_\nu$, with $\mu \neq \nu$; and (iii) u appears before v on L_i and w appears after z on L_j, or vice versa. Suppose, for a contradiction, that such four vertices exist. Note that, we can assume $j = i \pm 1$ without loss of generality, as (V, E, γ, T) is level-connected. Assume that u appears before v along L_i and w appears after z along L_j, the other case being symmetric. Since Γ^* is $\mathcal{T}$-level planar, all the vertices of V_μ appear before all the vertices of V_ν along L_i and all the vertices of V_μ appear after all the vertices of V_ν along L_j. Also, since (V, E, γ, T) is level-connected, there exists at least an edge (a, b) such that $a \in V_i \cap V_\mu$ and $b \in V_j \cap V_\mu$, and an edge (c, d) such that $c \in V_i \cap V_\nu$ and $d \in V_j \cap V_\nu$. However, under the above conditions, these two edges intersect in Γ and in Γ^*, hence contradicting the hypothesis that Γ^* is $\mathcal{T}$-level planar.

Suppose that (V, E, γ, T) admits a cl-planar drawing Γ; we show how to construct a $\mathcal{T}$-level planar drawing Γ^* of $(V, E, \gamma, \mathcal{T})$. Initialize $\Gamma^* = \Gamma$. Consider each level V_i, with $i = 1, \ldots, k$. By construction, for each internal node w of tree $T_i \in \mathcal{T}$, there exists a cluster $\mu \in T$ such that the vertices of $V_i \cap V_\mu$ are all and only the leaves of the subtree of T_i rooted at w. Since Γ is cl-planar, such vertices appear consecutively along L_i. Hence, Γ^* is $\mathcal{T}$-level planar. □

We get the following.

Theorem 4. *There exists an $O(|V|^4)$-time algorithm that decides whether a proper instance (V, E, γ, T) of* CLUSTERED-LEVEL PLANARITY *is cl-planar.*

Proof: By Lemma 2, it is possible to construct in $O(|V|^2)$ time a level-connected instance (V', E', γ', T') of CL-PLANARITY that is cl-planar if and only if (V, E, γ, T) is cl-planar, with $|V'| = O(|V|^2)$. By Lemma 3, it is possible to construct in $O(|V'|)$ time a proper instance $(V', E', \gamma', \mathcal{T}')$ of T-LEVEL PLANARITY that is $\mathcal{T}$-level planar if and only if (V', E', γ', T') is cl-planar. Finally, by Theorem 3, it is possible to test in $O(|V'|^2)$ time whether $(V', E', \gamma', \mathcal{T}')$ is $\mathcal{T}$-level planar. □

4   Open Problems

Several problems are opened by this research:

1. The algorithms for testing level planarity [10] and for testing cl-planarity for level-connected proper hierarchies [9] both have linear-time complexity. Although our algorithms solve more general problems than the ones above, they are less efficient. This leaves room for future research aiming at improving our complexity bounds.

2. Our $\mathcal{NP}$-hardness result on the complexity of CL-PLANARITY exploits a cluster hierarchy whose depth is linear in the number of vertices of the underlying graph. Does the $\mathcal{NP}$-hardness hold if the cluster hierarchy is flat?

3. The $\mathcal{NP}$-hardness of CL-PLANARITY is, to the best of our knowledge, the first hardness result for a variation of the clustered planarity problem in which none of the c-planarity constraints is dropped. Is it possible to use similar techniques to tackle the problem of determining the complexity of CLUSTERED PLANARITY?

Acknowledgments. Work partially supported by ESF EuroGIGA GraDR, by the Australian Research Council (grant DE140100708), by the MIUR project AMANDA "Algorithmics for MAssive and Networked DAta", prot. 2012C4E3KT_001, and by EU FP7 STREP "Leone: From Global Measurements to Local Management", no. 317647.

References

1. Angelini, P., Da Lozzo, G., Neuwirth, D.: On the complexity of some problems related to SEFE. CoRR abs/1207.3934 (2013)
2. Angelini, P., Di Battista, G., Frati, F., Patrignani, M., Rutter, I.: Testing the simultaneous embeddability of two graphs whose intersection is a biconnected or a connected graph. J. of Discrete Algorithms 14, 150–172 (2012)
3. Angelini, P., Da Lozzo, G.: Deepening the relationship between SEFE and C-planarity. CoRR abs/1404.6175 (2014)
4. Angelini, P., Da Lozzo, G., Neuwirth, D.: On some $\mathcal{NP}$-complete SEFE problems. In: Pal, S.P., Sadakane, K. (eds.) WALCOM 2014. LNCS, vol. 8344, pp. 200–212. Springer, Heidelberg (2014)
5. Blasiüs, T., Kobourov, S.G., Rutter, I.: Simultaneous embedding of planar graphs. In: Tamassia, R. (ed.) Handbook of Graph Drawing and Visualization. CRC Press (2013)
6. Bläsius, T., Rutter, I.: Disconnectivity and relative positions in simultaneous embeddings. In: Didimo, W., Patrignani, M. (eds.) GD 2012. LNCS, vol. 7704, pp. 31–42. Springer, Heidelberg (2013)
7. Bläsius, T., Rutter, I.: Simultaneous PQ-ordering with applications to constrained embedding problems. In: Khanna, S. (ed.) SODA, pp. 1030–1043. SIAM (2013)
8. Eades, P., Feng, Q.W., Lin, X., Nagamochi, H.: Straight-line drawing algorithms for hierarchical graphs and clustered graphs. Algorithmica 44(1), 1–32 (2006)

258 P. Angelini et al.

9. Forster, M., Bachmaier, C.: Clustered level planarity. In: Van Emde Boas, P., Pokorný, J., Bieliková, M., Štuller, J. (eds.) SOFSEM 2004. LNCS, vol. 2932, pp. 218–228. Springer, Heidelberg (2004)
10. Jünger, M., Leipert, S., Mutzel, P.: Level planarity testing in linear time. In: Whitesides, S.H. (ed.) GD 1998. LNCS, vol. 1547, pp. 224–237. Springer, Heidelberg (1999)
11. Opatrny, J.: Total ordering problem. SIAM J. Comput. 8(1), 111–114 (1979)
12. Schaefer, M.: Toward a theory of planarity: Hanani-Tutte and planarity variants. J. of Graph Alg. and Appl. 17(4), 367–440 (2013)
13. Wotzlaw, A., Speckenmeyer, E., Porschen, S.: Generalized k-ary tanglegrams on level graphs: A satisfiability-based approach and its evaluation. Discrete Applied Mathematics 160(16-17), 2349–2363 (2012)

Column Planarity and Partial Simultaneous Geometric Embedding

William Evans[1], Vincent Kusters[2], Maria Saumell[3], and Bettina Speckmann[4]

[1] University of British Columbia, Canada
will@cs.ubc.ca
[2] Department of Computer Science, ETH Zürich, Switzerland
vincent.kusters@inf.ethz.ch
[3] Department of Mathematics and European Centre of Excellence NTIS,
University of West Bohemia, Czech Republic
saumell@kma.zcu.cz
[4] Technical University Eindhoven, The Netherlands
b.speckmann@tue.nl

Abstract. We introduce the notion of *column planarity* of a subset R of the vertices of a graph G. Informally, we say that R is column planar in G if we can assign x-coordinates to the vertices in R such that any assignment of y-coordinates to them produces a partial embedding that can be completed to a plane straight-line drawing of G. Column planarity is both a relaxation and a strengthening of unlabeled level planarity. We prove near tight bounds for column planar subsets of trees: any tree on n vertices contains a column planar set of size at least $14n/17$ and for any $\epsilon > 0$ and any sufficiently large n, there exists an n-vertex tree in which every column planar subset has size at most $(5/6 + \epsilon)n$.

We also consider a relaxation of simultaneous geometric embedding (SGE), which we call partial SGE (PSGE). A PSGE of two graphs G_1 and G_2 allows some of their vertices to map to two different points in the plane. We show how to use column planar subsets to construct k-PSGEs in which k vertices are still mapped to the same point. In particular, we show that any two trees on n vertices admit an $11n/17$-PSGE, two outerpaths admit an $n/4$-PSGE, and an outerpath and a tree admit a $11n/34$-PSGE.

1 Introduction

A graph $G = (V, E)$ on n vertices is *unlabeled level planar (ULP)* if for all injections $\gamma : V \to \mathbb{R}$, there exists an injection $\rho : V \to \mathbb{R}$, so that embedding each $v \in V$ at $(\rho(v), \gamma(v))$ results in a plane straight-line embedding of

W. Evans is supported by an NSERC Discovery Grant. V. Kusters is partially supported by the ESF EUROCORES programme EuroGIGA, CRP GraDR and the Swiss National Science Foundation, SNF Project 20GG21-134306. M. Saumell is supported by the project NEXLIZ CZ.1.07/2.3.00/30.0038, which is co-financed by the European Social Fund and the state budget of the Czech Republic. B. Speckmann is supported by the Netherlands Organisation for Scientific Research (NWO) under project no. 639.023.208.

C. Duncan and A. Symvonis (Eds.): GD 2014, LNCS 8871, pp. 259–271, 2014.
© Springer-Verlag Berlin Heidelberg 2014

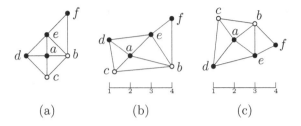

(a)   (b)   (c)

Fig. 1. (a) A graph $G = (V, E)$ with $R = \{a, d, e, f\}$ which is ρ-column planar for $\rho = \{d \mapsto 1, a \mapsto 2, e \mapsto 3, f \mapsto 4\}$. (b-c) Two assignments of y-coordinates to the vertices R and corresponding plane straight-line completions of G.

G. Estrella-Balderrama, Fowler and Kobourov [10] originally introduced ULP graphs and characterized ULP trees in terms of forbidden subgraphs. Fowler and Kobourov [12] extended this characterization to general ULP graphs. ULP graphs are exactly the graphs that admit a simultaneous geometric embedding with a monotone path: this was the original motivation for studying them.

In this paper we introduce the notion of *column planarity* of a subset R of the vertices V of a graph $G = (V, E)$. Informally, we say that R is column planar in G if we can assign x-coordinates to the vertices in R such that any assignment of y-coordinates to them produces a partial embedding that can be completed to a plane straight-line drawing of G. Column planarity is both a relaxation and a strengthening of unlabeled level planarity. It is a relaxation since it applies only to a subset R of the vertices and a strengthening since the requirements on R are more strict than in the case of unlabeled level planarity.

More formally, for $R \subseteq V$, we say that R is *column planar in* $G = (V, E)$ if there exists an injection $\rho : R \to \mathbb{R}$ such that for all ρ-*compatible* injections $\gamma : R \to \mathbb{R}$, there exists a plane straight-line embedding of G where each $v \in R$ is embedded at $(\rho(v), \gamma(v))$. Injection γ is ρ-compatible if the combination of ρ and γ does not embed three vertices on a line. Clearly, if R is column planar in G then any subset of R is also column planar in G. We say that R is ρ-*column planar* when we need to emphasize the injection ρ (see Fig. 1 for an example). If $R = V$ is column planar in G then G is ULP since column planarity implies the existence of one assignment of x-coordinates to vertices that will produce a planar embedding for all assignments of y-coordinates, while to be a ULP graph the x-coordinate assignment may depend on the y-coordinate assignment. In this sense, column planarity of V is strictly more restrictive than unlabeled level planarity of G. Di Giacomo et al. [8] study column planarity under a different name. Specifically, they define EAP graphs as the graphs $G = (V, E)$ where V is column planar in G. They consider a family of graphs called *fat caterpillars* and prove that these are exactly the EAP graphs.

As mentioned above, the study of ULP was originally motivated by simultaneous geometric embedding, a concept introduced by Brass et al. [4]. Formally, given two graphs $G_1 = (V, E_1)$ and $G_2 = (V, E_2)$ on the same set of n vertices, they defined a *simultaneous geometric embedding (SGE)* of G_1 and G_2 as an

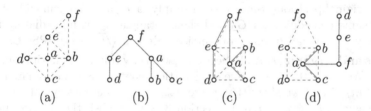

Fig. 2. (a-b) Two graphs on the same vertex set. (c) An SGE of these graphs. (d) A 3-PSGE of these graphs.

injection $\varphi : V \to \mathbb{R}^2$ such that the straight-line drawings of G_1 and G_2 induced by φ are both plane. With slight abuse of notation, we refer to these drawings as $\varphi(G_1)$ and $\varphi(G_2)$. Fig. 2c depicts an SGE of the graphs in Fig. 2a and Fig. 2b. Bläsius et al. [2] give an excellent survey of the subsequent papers on SGE with a comprehensive list of results. On the positive side, Brass et al. [4] prove that two paths, cycles or caterpillars always admit an SGE. Cabello et al. [5] prove that a matching and a tree or outerpath (a type of outerplanar graph) always admit an SGE. On the negative side, Brass et al. [4] prove that three paths sometimes do not admit an SGE. Erten and Kobourov [9] prove that a planar graph and a path may not admit an SGE. Frati, Kaufmann and Kobourov [13] strengthen this result to the case where the planar graph and the path do not share any edges. Geyer, Kaufmann and Kobourov [14] describe two trees that do not admit an SGE. Angelini et al. [1] close a long-standing open question by describing a tree and a path that admit no SGE. Finally, Estrella-Balderrama et al. [11] show that the decision problem for SGE is NP-hard.

In light of the restrictiveness of simultaneous geometric embedding, several other variations on the abstract problem have been studied. Cappos et al. [6] consider a version of SGE where edges are embedded as circular arcs or with bends. Di Giacomo et al. [7] consider *matched drawings*: a version of SGE where the location of a vertex in the drawing of G_1 need only have the same y-coordinate as its location in the drawing of G_2.

In this paper we consider a variant on SGE which we call *partial simultaneous geometric embedding* (PSGE). We do not require *every* vertex to map to a single point in the plane. Instead, some vertices can have a "split personality" and map to two different locations, one associated with G_1 and one associated with G_2. Specifically, given two graphs $G_1 = (V, E_1)$ and $G_2 = (V, E_2)$ on the same set of n vertices, a *k-partial simultaneous geometric embedding (k-PSGE)* of G_1 and G_2 is a pair of injections $\varphi_1 : V \to \mathbb{R}^2$ and $\varphi_2 : V \to \mathbb{R}^2$ such that (i) the straight-line drawings $\varphi_1(G_1)$ and $\varphi_2(G_2)$ are both plane; (ii) if $\varphi_1(v_1) = \varphi_2(v_2)$ then $v_1 = v_2$ and; (iii) $\varphi_1(v) = \varphi_2(v)$ for at least k vertices $v \in V$. An n-PSGE is simply an SGE. Fig. 2d depicts a 3-PSGE of the graphs in Fig. 2a and Fig. 2b.

PSGE is related to the notion of *planar untangling*: Given a straight-line drawing of a planar graph, change the embedding of as few vertices as possible in order to obtain a plane drawing. Goaoc et al. [15] describe an improvement of a result by Bose et al. [3] to show that $\sqrt[4]{(n + 1)/2}$ vertices can always be kept in their original positions. Since we can simply take any plane embedding of G_1, use the same embedding for G_2 and then untangle G_2, it immediately follows that every two planar graphs on n vertices admit a $\sqrt[4]{(n+1)/2}$-PSGE.

Results and Organization. In Section 2, we study column planarity for subsets of trees. We prove that every tree on n vertices contains a column planar subset of size $14n/17$ and we show that there exist trees where every column planar subset has size at most $5n/6$. In Section 3, we establish the relation between column planarity and PSGE. We show that every two trees admit an $11n/17$-PSGE, that every tree and ULP graph admit a $14n/17$-PSGE, that every two outerpaths admit an $n/4$-PSGE, and that every outerpath and a tree admit an $11n/34$-PSGE.

2   Column Planar Sets in Trees

In this section, we show how to find large column planar sets in trees. Let $p(v)$ be the parent of vertex v in a rooted tree T, and let $r(T)$ be the root of T. Given a subset R of the vertices of T, let $C_R(v)$ be the non-leaf children of v in R and let $C_R^+(v)$ be those vertices in $C_R(v)$ with at least one child in R. We first prove that subsets of T satisfying certain conditions are always column planar and next that every tree contains a large such subset.

Lemma 1. *For a rooted tree T, R is column planar in T if for all $v \in R$, either (1) $p(v) \in R$, the number of non-leaf children of v in R is at most two, and at most one of these children has a child in R (i.e. $C_R(v) \leq 2$ and $C_R^+(v) \leq 1$); or (2) $p(v) \notin R$, the number of non-leaf children of v in R is at most four, and at most two of these children have a child in R (i.e. $C_R(v) \leq 4$ and $C_R^+(v) \leq 2$).*

Proof. We will embed T recursively. The x-coordinates of V will be fixed in such a way that any assignment $\gamma : R \to \mathbb{R}$ of y-coordinates to R can be accommodated by embedding the vertices of $V \setminus R$ with y-coordinates much larger than $\max \gamma$ or much smaller than $\min \gamma$. Thus, the edges between $V \setminus R$ and R are embedded as near-vertical line segments. In the figures that accompany this proof, such edges will be drawn as curves.

For a subtree T' of T, let $p(T')$ be the parent of $r(T')$. If $r(T')$ is the root of T then $p(T')$, though it does not exist, is viewed as not in R. Our embedding will have the following properties for each subtree T': (i) if $r(T') \notin R$ or $\{r(T'), p(T')\} \subseteq R$, then $r(T')$ has either the smallest or largest x-coordinate among all vertices in T'; (ii) if $r(T') \notin R$, then $r(T')$ has either the smallest or largest y-coordinate among all vertices in T'; and (iii) no almost-vertical ray from $r(T')$ intersects any edge from T'.

Let T be the rooted tree we want to embed. Let $r = r(T)$. If $r \in R$, then recursively generate embeddings of all non-leaf children of r. Scale each such embedding horizontally to width 1. Suppose first that $p(T) \in R$. See Fig. 3a.

Embed r at $x = 1$ and its ℓ leaf children at $x = 2, \ldots, \ell + 1$. (Their y-coordinates are determined by γ.) Suppose $C_R(v) \subseteq \{r_1, s_1\}$ and $C_R^+(v) \subseteq \{r_1\}$. Embed r_1 and its subtree recursively and scale its x-coordinates to lie in $[\ell+3, \ell+$

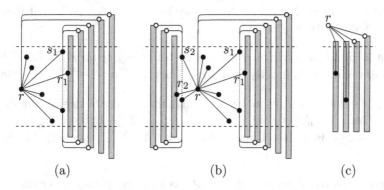

Fig. 3. Embedding a tree with a column planar set. The column planar vertices are black.

4]. By (i), and possibly after mirroring the embedding of the subtree rooted at r_1 horizontally, the edge $\{r, r_1\}$ does not cross edges in the subtree rooted at r_1.

Embed s_1 at $x = \ell + 2$. Let $T_1, \ldots, T_k$ be the child subtrees of s_1. Embed T_i recursively and scale its x-coordinates to lie in $[\ell + 3 + 2i, \ell + 4 + 2i]$ for all $1 \leq i \leq k$. Vertex s_1 will be above $\{r, r_1\}$ for some γ and below $\{r, r_1\}$ for other γ. If it is above, let $r(T_1), \ldots, r(T_k)$ have progressively larger y-coordinates (by scaling up and mirroring vertically if necessary). If it is below, let them have progressively smaller y-coordinates. Then none of the edges $\{s_1, r(T_i)\}$ cross $\{r, r_1\}$ and the edge $\{s_1, r(T_i)\}$ does not cross any edges in T_i by (i) and (ii).

Recursively, embed the remaining child subtrees $T'_1, \ldots, T'_t$ (none of whose roots are in R) with x-coordinates in $[\ell + 3 + 2k + 2i, \ell + 4 + 2k + 2i]$ for all $1 \leq i \leq t$ such that $r(T'_1), \ldots, r(T'_t)$ have progressively larger y-coordinates. The edge $\{r, r(T'_i)\}$ does not cross any edges in T'_i by (ii). In the completed drawing, note that r has the lowest x-coordinate, and thus (i) is satisfied. Properties (ii) and (iii) are trivially satisfied.

Suppose that $p(T) \notin R$. Proceed first as in the previous case. Suppose $C_R(v) \subseteq \{r_1, r_2, s_1, s_2\}$ and $C_R^+(v) \subseteq \{r_1, r_2\}$. Mirror the recursive embedding of the subtree rooted at r_2 horizontally and scale it to have x-coordinates in $[-3, -2]$. Embed the subtree rooted at s_1 as in the previous case. For s_2, proceed similarly but embed s_2 and its subtree to the left of r. See Fig. 3b. Properties (i)-(iii) are trivially satisfied.

Finally, suppose that $r = r(T) \notin R$. Embed its child subtrees $T_1, \ldots, T_t$ to have x-coordinates in $[2i, 2i + 1]$ for all $1 \leq i \leq t$, starting with the ones rooted at a vertex in R. Embed r sufficiently high on the line $x = 1$. For subtrees T_i with $r(T_i) \in R$, note that the edge $\{r, r(T_i)\}$ does not cross any edges of T_i due to (iii). For the other ones, $\{r, r(T_i)\}$ does not cross edges of T_i due to (i) and (ii). See Fig. 3c. Properties (i-iii) are satisfied. □

It remains to show that every tree contains a subset that satisfies the conditions imposed by Lemma 1. We show that every tree on n vertices contains such a subset of size at least $14n/17$ and that there are trees with no column planar subset of size larger than $5n/6$. Note that $14/17 \approx 5/6 - 0.01$, and thus our results are almost tight.

Lemma 2. *Let T be a tree on n vertices rooted at any vertex $r(T)$. Let c_i be the number of vertices with exactly i children. Then $c_0 = (n + 1 + \sum_{i=1}^{n-1}(i-2)c_i)/2$.*

Proof. The number of edges in T is $n - 1$ and also equals the degree sum divided by two. Thus, $\sum_{i=0}^{n-1} c_i(i + 1) = 2(n - 1) + 1 = 2n - 1$. Since $\sum_{i=0}^{n-1} c_i = n$, $\sum_{i=0}^{n-1} c_i(i - 2) + 3n = 2n - 1$, and $-2c_0 = -n - 1 - \sum_{i=1}^{n-1} c_i(i - 2)$. The lemma follows. $\square$

Theorem 1. *A tree T on n vertices contains a column planar set of size at least $14n/17$.*

Proof. Root T at an arbitrary non-leaf vertex $r(T)$. Orient every edge towards the root and topologically sort T to obtain an order $v_1, \ldots, v_n$. We will greedily add vertices to R in this order. More precisely, let $R_0 = \emptyset$ and let $R_i := R_{i-1} \cup \{v_i\}$ if $R_{i-1} \cup \{v_i\}$ satisfies Lemma 1 and let $R_i := R_{i-1}$ otherwise. Let $R = R_n$ be our final subset of T.

We say that a vertex is *marked* if it is in R. Consider a vertex $v = v_i \notin R$. The reason that v is not in R is that $R_{i-1} \cup \{v\}$ does not satisfy the condition in Lemma 1 for v or a child u of v (or both). More precisely, v is contained in exactly one of the following sets:

$$X_a = \{v \in T \setminus R : \qquad |C_R^+(v)| > 2\}$$
$$X_b = \{v \in T \setminus R \setminus X_a : \qquad |C_R(v)| > 4\}$$
$$X_c = \{v \in T \setminus R \setminus X_a \setminus X_b : \qquad |C_R^+(u)| > 1\}$$
$$X_d = \{v \in T \setminus R \setminus X_a \setminus X_b \setminus X_c : \qquad |C_R(u)| > 2\}.$$

We associate with each such v a witness tree $W(v)$ as follows (see Fig. 4). If $v \in X_a$, then let $W(v)$ be v, three vertices of $C_R^+(v)$ and a marked child of each of them (which must exist by definition of $C_R^+(v)$). If $v \in X_b$, then let $W(v)$ be v and five marked children of v. If $v \in X_c$, then let $W(v)$ be v, u, two vertices of $C_R^+(u)$ and a marked child of each of them. If $v \in X_d$, let $W(v)$ be v, u and three

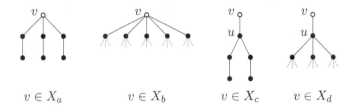

$$v \in X_a \qquad v \in X_b \qquad v \in X_c \qquad v \in X_d$$

Fig. 4. The witness tree $W(v)$ when v is in X_a, X_b, X_c or X_d. The marked vertices are black. Dotted line segments indicate that a vertex has at least one child.

marked children of u. Note that $W(v)$ and $W(v')$ are disjoint for $v, v' \in T \setminus R$ with $v \neq v'$. We have

$$|X_a| + |X_b| + |X_c| + |X_d| + |R| = n. \tag{1}$$

Let L_t and I_t be the set of marked vertices of $\bigcup_{v \in X_t} W(v)$ that are leaves and internal vertices in T, respectively, for $t = a, b, c, d$. We have

$$|I_a| + |L_a| = 6|X_a| \qquad\qquad |L_a| \leq 3|X_a| \tag{2}$$
$$|I_b| + |L_b| = 5|X_b| \qquad\qquad |L_b| = 0 \tag{3}$$
$$|I_c| + |L_c| = 5|X_c| \qquad\qquad |L_c| \leq 2|X_c| \tag{4}$$
$$|I_d| + |L_d| = 4|X_d| \qquad\qquad |L_d| = 0 \tag{5}$$

Since R always contains all leaves of T, we have

$$|R| \geq c_0 + |I_a| + |I_b| + |I_c| + |I_d|, \tag{6}$$

where c_i is the number of vertices with exactly i children in T. Note that $W(v)$ contains a vertex with at least three children if $v \in X_a \cup X_b \cup X_d$. Hence, by Lemma 2,

$$c_0 > \frac{n - c_1 + \sum_{i=3}^{n-1} c_i}{2} \geq \frac{n - c_1 + |X_a| + |X_b| + |X_d|}{2}. \tag{7}$$

In addition, we have

$$c_0 \geq |L_a| + |L_b| + |L_c| + |L_d|. \tag{8}$$

Before we bound $|R|$, consider the set S formed by all leaves and all vertices with one child. Then S is column planar by Lemma 1 and $|S| = c_0 + c_1$. Whenever the greedily chosen R has size less than $c_0 + c_1$, we choose $R = S$ instead. Thus, we may assume

$$|R| \geq c_0 + c_1. \tag{9}$$

Equations (7) and (9) yield

$$|R| > n - c_0 + |X_a| + |X_b| + |X_d|; \tag{10}$$

equations (2) and (8) yield

$$c_0 \geq 6|X_a| - |I_a| + |L_c|; \tag{11}$$

and equations (3), (4), (5), and (6) yield

$$|R| \geq c_0 + 5|X_b| + 5|X_c| + 4|X_d| - |L_c| + |I_a|. \tag{12}$$

To eliminate c_0, we combine equation (10) with two times (11) and three times (12) to obtain $4|R| > n + 13|X_a| + 16|X_b| + 15|X_c| + 13|X_d| - |L_c| + |I_a|$. With equation (4), this gives $4|R| > n + 13|X_a| + 16|X_b| + 13|X_c| + 13|X_d| + |I_a| \geq n + 13(|X_a| + |X_b| + |X_c| + |X_d|)$. Together with equation (1), this yields the desired bound of $|R| > 14n/17$. □

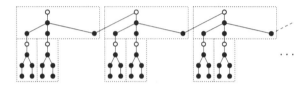

Fig. 5. A tree for which $|R| = |S| = 14n/17$. The set R is colored black.

The greedy algorithm achieves exactly this amount on the tree depicted in Fig. 5. Note that also $|S| = c_0 + c_1 = 14n/17$ in this tree. In general, Theorem 1 is close to best possible:

Theorem 2. *For any $\epsilon > 0$ and any $n > 2/\epsilon + 5$, there exists a tree T with n vertices in which every column planar subset in T has at most $(5/6 + \epsilon)n$ vertices.*

Proof. Let $p = \lfloor n/6 \rfloor$. Let T be p copies, $T_1, T_2, \ldots, T_p$, of the tree shown in Fig. 6a in which the root of T_{i+1} is made a child of the rightmost leaf of T_i, for $i = 1, \ldots, p - 1$. Suppose there is a column planar set R of marked vertices in T with $|R|/n > 5/6 + \epsilon$. Then in some sequence of at most $k = \lceil 1/(3\epsilon) \rceil$ subtrees $T_i, T_{i+1}, \ldots, T_j$ there must be at least two trees with 6 marked vertices and the other trees with 5 marked vertices. If not, since each subtree has 6 vertices, the average fraction of marked vertices per tree is less than $\frac{5k+2}{6k} < 5/6 + \epsilon$.

Let $T_i, T_{i+1}, \ldots, T_j$ be such a sequence. By possibly deleting a prefix of the sequence, we can assume that T_i has 6 marked vertices. Let $\ell > i$ be the smallest index such that the root of T_ℓ is marked. Since $T_i, T_{i+1}, \ldots, T_j$ contains at least two trees with 6 marked vertices, T_ℓ exists. Let H be the subtree induced by the root of T_ℓ and the vertices in $T_i \cup T_{i+1} \cup \cdots \cup T_{\ell-1}$. By definition, the unmarked vertices in H are exactly the roots of the subtrees $T_{i+1}, T_{i+2}, \ldots, T_{\ell-1}$. We claim that the marked vertices are not column planar in H.

To simplify notation, let $H_1, H_2, \ldots, H_{q-1}$ be the sequence of subtrees in H and let r_q be the (marked) root of T_ℓ. Label the vertices of H_i as in Fig. 6a subscripted by i. See Fig. 6b. Let R' be the marked vertices in H and suppose R' is ρ-column planar in H. For an edge $\{a, b\}$ in H with $a, b \in R'$, let $\rho(a, b) =$

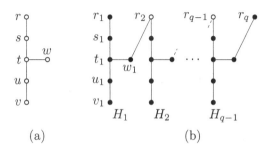

(a)   (b)

Fig. 6. (a) The tree T_i and (b) H used in the proof of Theorem 2

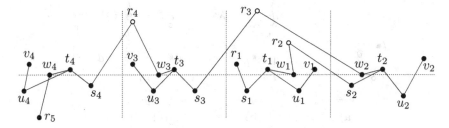

Fig. 7. An example of how γ is chosen in the proof of Theorem 2 where $q = 5$. Note that forcing r_5 (bottom left) below the x-axis causes the edge $\{w_4, r_5\}$ to intersect another edge.

$[\rho(a), \rho(b)]$ be the x-interval of edge $\{a, b\}$. For two edges $\{a, b\}$ and $\{c, d\}$ in H where a, b, c, and d are distinct vertices in R', $\rho(a, b) \cap \rho(c, d) = \emptyset$: otherwise, by choosing γ appropriately we can cause the edges to intersect within their shared x-interval. This implies, for example, that the x-interval spanned by marked vertices in one subtree does not intersect that of a different subtree.

For H_1, since $\rho(s_1, t_1) \cap \rho(u_1, v_1) = \emptyset$ and $\rho(t_1, u_1) \cap \rho(r_1, s_1) = \emptyset$, $\rho(t_1)$ is *between* $\rho(r_1, s_1)$ and $\rho(u_1, v_1)$ (meaning either $\rho(r_1, s_1) < \rho(t_1) < \rho(u_1, v_1)$ or $\rho(u_1, v_1) < \rho(t_1) < \rho(r_1, s_1)$, where $A < B$ if for all $a \in A$ and $b \in B$, $a < b$). By similar reasoning, $\rho(w_1)$ is between $\rho(t_1)$ and $\rho(u_1, v_1)$ or between $\rho(t_1)$ and $\rho(r_1, s_1)$. Let us assume, by renaming vertices if necessary, that $\rho(w_1)$ is between $\rho(t_1)$ and $\rho(u_1, v_1)$. See Fig. 7.

The basic idea is to choose γ so that vertices in R are close to the x-axis (with $\gamma(u_i) < \gamma(s_i) < 0 = \gamma(w_i) < \gamma(t_i) < \gamma(v_i)$ for all i except when mentioned otherwise) and so that unmarked vertices are forced to be above the x-axis. We set $\gamma(u_1)$ to be negative and $\gamma(v_1)$ to be positive (so w_1 lies in the triangle $t_1 u_1 v_1$). This, together with the fact that r_2 is connected to s_2, forces the edge from w_1 to r_2 to be upward and thus r_2 to be above the x-axis.

Consider the order of $\rho(s_2)$, $\rho(t_2)$ and $\rho(u_2, v_2)$. If $\rho(s_2)$ is between $\rho(t_2)$ and $\rho(u_2, v_2)$, then setting γ so that the path t_2, u_2, v_2 is above s_2 ($\gamma(t_2) < \gamma(v_2) < 0 < \gamma(s_2) < \gamma(u_2)$) causes the path to intersect $\{r_2, s_2\}$. Note that $\rho(u_2, v_2)$ cannot be between $\rho(t_2)$ and $\rho(s_2)$ since $\rho(u_2, v_2) \cap \rho(s_2, t_2) = \emptyset$. Hence, $\rho(t_2)$ is between $\rho(s_2)$ and $\rho(u_2, v_2)$. Now let us consider the possible positions of $\rho(w_2)$. If $\rho(s_2)$ is between $\rho(w_2)$ and $\rho(t_2)$, then setting γ so that the path u_2, t_2, w_2 is above s_2 ($\gamma(w_2) < \gamma(u_2) < 0 < \gamma(s_2) < \gamma(t_2)$) causes the path to intersect $\{r_2, s_2\}$. Note that $\rho(u_2, v_2)$ cannot be between $\rho(w_2)$ and $\rho(t_2)$ since $\rho(u_2, v_2) \cap \rho(t_2, w_2) = \emptyset$. Hence, $\rho(w_2)$ is between $\rho(s_2)$ and $\rho(t_2)$ or between $\rho(t_2)$ and $\rho(u_2, v_2)$. In the first case, we set $\gamma(s_2) < 0 = \gamma(w_2) < \gamma(t_2)$ so the edge from w_2 to r_3 is forced upward to avoid intersecting path r_2, s_2, t_2. In the second case, we set γ so that the path t_2, u_2, v_2 is below w_2 ($\gamma(u_2) < 0 = \gamma(w_2) < \gamma(t_2) < \gamma(v_2)$) and the edge from w_2 to r_3 is forced upward. By repeating this argument, we force all the unmarked vertices as well as r_q to be above the x-axis. Since r_q is marked, we derive a contradiction by setting $\gamma(r_q) < 0$. $\qquad\square$

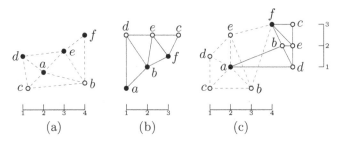

Fig. 8. (a) Graph G_1 with $R_1 = \{a, d, e, f\}$ and $\rho_1 = \{d \mapsto 1, a \mapsto 2, e \mapsto 3, f \mapsto 4\}$. (b) Graph G_2 with $R_2 = \{a, b, f\}$ and $\rho_2 = \{a \mapsto 1, b \mapsto 2, f \mapsto 3\}$. (c) A 2-PSGE of G_1 and G_2 where vertex set $R = R_1 \cap R_2 = \{a, f\}$ is shared.

3   Partial Simultaneous Geometric Embedding

The relation between column planarity and PSGE is expressed by the following theorem, which relates the size of column planar sets to PSGE.

Theorem 3. *Consider planar graphs $G_1 = (V, E_1)$ and $G_2 = (V, E_2)$ on n vertices. If R_1 is column planar in G_1, R_2 is column planar in G_2 and $|R_1| + |R_2| > n$, then G_1 and G_2 admit a $(|R_1| + |R_2| - n)$-PSGE.*

Proof. Fig. 8 illustrates the construction. The set $R = R_1 \cap R_2$ has size at least $|R_1| + |R_2| - n > 0$ and is column planar in both G_1 and G_2. More specifically, there exist injections $\rho_1 : R \to \mathbb{R}$ and $\rho_2 : R \to \mathbb{R}$ such that R is ρ_1-column planar in G_1 and ρ_2-column planar in G_2. By exchanging the roles of the x- and y-coordinates in the definition of column planar in G_2, we see that for all injections $\gamma : R \to \mathbb{R}$, there exists a plane straight-line embedding of G_2 that embeds each $v \in R$ at $(\gamma(v), \rho_2(v))$. In particular, we may choose $\gamma = \rho_1$.   □

Two trees. Combining Theorem 3 and Theorem 1 immediately yields the following lower bound on the size of a PSGE of two trees.

Corollary 1. *Every two trees on a set of n vertices admit an $11n/17$-PSGE.*

There are two trees T_1 and T_2 on 226 vertices that do not admit an SGE [14]. Thus, an upper bound on the size of the common set in a PSGE of T_1 and T_2 is 225. Root T_1 arbitrarily and let T_1^k be the result of taking k copies of T_1 and connecting their roots with a path. Define T_2^k similarly. Then an upper bound on the size of the common set in a PSGE of T_1^k and T_2^k is $225k$. It follows that there exist two trees on a set of n vertices that admit no k-PSGE for $k > 225n/226$.

Tree and ULP graph. If one of the two graphs in our PSGE is ULP, then the size of the common set depends only on how large a column planar set we can find in the other graph:

Lemma 3. *Consider a planar graph $G_1 = (V, E_1)$ and a ULP graph $G_2 = (V, E_2)$ on n vertices. If R is column planar in G_1, then G_1 and G_2 admit a $|R|$-PSGE.*

Proof. By exchanging the roles of x- and y-coordinates in the definition of column planar, we see that for all injections $\gamma : R \to \mathbb{R}$, there exists a plane straight-line embedding of G_1 with $v \in R$ at $(\gamma(v), \rho(v))$. Since G_2 is a ULP graph, for all injections $y : V \to \mathbb{R}$, there exists an injection $x : V \to \mathbb{R}$ such that placing $v \in V$ at $(x(v), y(v))$ results in a straight-line embedding of G_2. Thus, placing the vertices $v \in R$ at $(x(v), \rho(v))$ permits both a straight-line embedding of G_1 and G_2. □

Combining this with Theorem 1 yields

Corollary 2. *A tree and a ULP graph admit a $14n/17$-PSGE.*

Two outerpaths & outerpath and tree. An *outerplanar* graph is a planar graph that admits an embedding (called the *outerplane* embedding) that places all its vertices on the unbounded face. An *outerpath* is an outerplanar graph whose *weak dual* (the graph obtained from the dual graph by deleting the vertex corresponding to the unbounded face) is a path. A maximal outerpath has exactly two vertices of degree two: these vertices are on the faces that correspond to the terminal vertices of the dual path. Consider a maximal outerpath $G = (V, E)$. The *outer cycle* of G is the Hamiltonian cycle of G that bounds the unbounded face in the outerplane embedding of G. Denote by $C(G)$ the vertices of degree two in G. Deleting $C(G)$ from G partitions the outer cycle of G into two connected components whose vertices we refer to as $A(G)$ and $B(G)$. Note that $A(G) \cup B(G) \cup C(G) = V$. It is easy to see that:

Lemma 4. *Given a maximal outerpath $G = (V, E)$, the subsets $A(G) \cup C(G)$ and $B(G) \cup C(G)$ are column planar.*

Unlike in the tree setting, Theorem 3 does not immediately give a lower bound on the size of a PSGE of two outerpaths, since we might have $|A(G)| = |B(G)| = n/2 - 1$. Fortunately, this is easily resolved:

Theorem 4. *Every two outerpaths on a set of n vertices admit an $n/4$-PSGE.*

Proof. Consider outerpaths $G_1 = (V, E_1)$ and $G_2 = (V, E_2)$. Without loss of generality, G_1 and G_2 are maximal. Let $X_i^+ := X(G_i) \cup C(G_i)$ for $X = A, B$ and $i = 1, 2$. Then by Theorem 3 and Lemma 4, G_1 and G_2 admit a $\max\{|A_1^+ \cap A_2^+|, |A_1^+ \cap B_2^+|, |B_1^+ \cap A_2^+|, |B_1^+ \cap B_2^+|\}$-PSGE. Since the union of these four sets is again V, the maximum of their cardinalities must be at least $n/4$. □

Since $|C(G)| + \max\{|A(G)|, |B(G)|\} \geq n/2 + 1$, Theorem 1 and 3 yield:

Corollary 3. *An outerpath and a tree on n vertices admit a $11n/34$-PSGE.*

4 Discussion and Open Problems

Our results leave several directions for future research. The tree drawings produced by Theorem 1 may have exponential area. It would be interesting to see whether polynomial area is sufficient. Further research could be directed towards closing the gap between the lower and upper bound on the size of column planar sets for trees and on developing bounds for such sets in general planar graphs.

Acknowledgments. Research on the topic of this paper was initiated at the 1st International Workshop on Drawing Algorithms for Networks of Changing Entities (DANCE'2014) in Langbroek, The Netherlands, supported by the Netherlands Organisation for Scientific Research (NWO) under project no. 639.023.208. We wish to thank all participants, and in particular Csaba Tóth and Michael Hoffmann, for useful discussions on the topic of this paper.

References

1. Angelini, P., Geyer, M., Kaufmann, M., Neuwirth, D.: On a tree and a path with no geometric simultaneous embedding. In: Brandes, U., Cornelsen, S. (eds.) GD 2010. LNCS, vol. 6502, pp. 38–49. Springer, Heidelberg (2011)
2. Bläsius, T., Kobourov, S.G., Rutter, I.: Simultaneous embedding of planar graphs. In: Tamassia, R. (ed.) Handbook of Graph Drawing and Visualization (2013)
3. Bose, P., Dujmović, V., Hurtado, F., Langerman, S., Morin, P., Wood, D.R.: A polynomial bound for untangling geometric planar graphs. Disc. & Comp. Geom. 42(4), 570–585 (2009)
4. Brass, P., Cenek, E., Duncan, C.A., Efrat, A., Erten, C., Ismailescu, D.P., Kobourov, S.G., Lubiw, A., Mitchell, J.S.: On simultaneous planar graph embeddings. Comp. Geom. 36(2), 117–130 (2007)
5. Cabello, S., van Kreveld, M., Liotta, G., Meijer, H., Speckmann, B., Verbeek, K.: Geometric simultaneous embeddings of a graph and a matching. J. Graph Alg. Appl. 15(1), 79–96 (2011)
6. Cappos, J., Estrella-Balderrama, A., Fowler, J.J., Kobourov, S.G.: Simultaneous graph embedding with bends and circular arcs. Comp. Geom. 42(2), 173–182 (2009)
7. Di Giacomo, E., Didimo, W., van Kreveld, M., Liotta, G., Speckmann, B.: Matched drawings of planar graphs. In: Hong, S.-H., Nishizeki, T., Quan, W. (eds.) GD 2007. LNCS, vol. 4875, pp. 183–194. Springer, Heidelberg (2008)
8. Di Giacomo, E., Didimo, W., Liotta, G., Meijer, H., Wismath, S.: Planar and quasi planar simultaneous geometric embedding. In: Duncan, C., Symvonis, A. (eds.) GD 2014. LNCS, vol. 8871, pp. 52–63. Springer, Heidelberg (2014)
9. Erten, C., Kobourov, S.G.: Simultaneous embedding of planar graphs with few bends. In: Pach, J. (ed.) GD 2004. LNCS, vol. 3383, pp. 195–205. Springer, Heidelberg (2005)
10. Estrella-Balderrama, A., Fowler, J.J., Kobourov, S.G.: Characterization of unlabeled level planar trees. Comp. Geom. 42(6), 704–721 (2009)
11. Estrella-Balderrama, A., Gassner, E., Jünger, M., Percan, M., Schaefer, M., Schulz, M.: Simultaneous geometric graph embeddings. In: Hong, S.-H., Nishizeki, T., Quan, W. (eds.) GD 2007. LNCS, vol. 4875, pp. 280–290. Springer, Heidelberg (2008)

12. Fowler, J.J., Kobourov, S.G.: Characterization of unlabeled level planar graphs. In: Hong, S.-H., Nishizeki, T., Quan, W. (eds.) GD 2007. LNCS, vol. 4875, pp. 37–49. Springer, Heidelberg (2008)
13. Frati, F., Kaufmann, M., Kobourov, S.G.: Constrained simultaneous and near-simultaneous embeddings. In: Hong, S.-H., Nishizeki, T., Quan, W. (eds.) GD 2007. LNCS, vol. 4875, pp. 268–279. Springer, Heidelberg (2008)
14. Geyer, M., Kaufmann, M., Vrt'o, I.: Two trees which are self-intersecting when drawn simultaneously. Disc. Math. 309(7), 1909–1916 (2009)
15. Goaoc, X., Kratochvíl, J., Okamoto, Y., Shin, C.S., Spillner, A., Wolff, A.: Untangling a planar graph. Disc. & Comp. Geom. 42(4), 542–569 (2009)

Flat Foldings of Plane Graphs
with Prescribed Angles and Edge Lengths

Zachary Abel[1], Erik D. Demaine[2], Martin L. Demaine[2], David Eppstein[3],
Anna Lubiw[4], and Ryuhei Uehara[5]

[1] Department of Mathematics, MIT, Cambridge, USA
[2] MIT Computer Science and Artificial Intelligence Lab., Cambridge, USA
[3] Department of Computer Science, University of California, Irvine, USA
[4] David R. Cheriton School of Computer Science, University of Waterloo, Canada
[5] School of Information Science, Japan Advanced Institute of Science and
Technology, Ishikawa, Japan

Abstract. When can a plane graph with prescribed edge lengths and
prescribed angles (from among $\{0, 180°, 360°\}$) be folded flat to lie in
an infinitesimally thick line, without crossings? This problem generalizes
the classic theory of single-vertex flat origami with prescribed mountain-
valley assignment, which corresponds to the case of a cycle graph. We
characterize such flat-foldable plane graphs by two obviously necessary
but also sufficient conditions, proving a conjecture made in 2001: the
angles at each vertex should sum to 360°, and every face of the graph
must itself be flat foldable. .This characterization leads to a linear-time
algorithm for testing flat foldability of plane graphs with prescribed edge
lengths and angles, and a polynomial-time algorithm for counting the
number of distinct folded states.

1 Introduction

The modern field of origami mathematics began in the late 1980s with the goal
of characterizing flat-foldable crease patterns, i.e., which plane graphs form the
crease lines in a flat folding of a piece of paper [12]. This problem turns out to be
NP-complete in the general case, with or without an assignment of which folds
are mountains and which are valleys [6].

On the other hand, flat foldability can be solved in polynomial time for crease
patterns with just a single vertex (thus characterizing the local behavior of a
vertex in a larger graph). By slicing the paper with a small sphere centered at
the single vertex (the geometric *link* of the vertex), single-vertex flat foldability reduces to the 1D problem of folding a polygon (closed polygonal chain)
onto a line; see Figure 1. This problem can be solved by a greedy algorithm
that repeatedly folds both ends of a shortest edge with opposite fold directions
(mountain and valley)—either because such directions have already been pre-
assigned or, if the mountain-valley assignment is not given, by making such
an assignment [4,6,12,20]. The spherical, self-touching Carpenter's Rule Theorem [1,11,26] implies that any flat-folded single-vertex origami can be reached

C. Duncan and A. Symvonis (Eds.): GD 2014, LNCS 8871, pp. 272–283, 2014.
© Springer-Verlag Berlin Heidelberg 2014

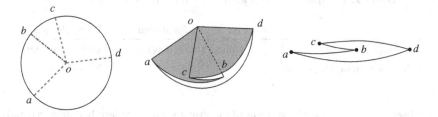

Fig. 1. Flat folding at a single vertex on a disc reduces to the problem of folding a polygon onto a line

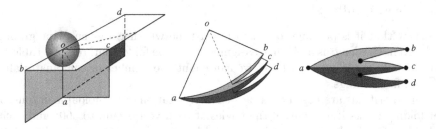

Fig. 2. Flat folding a two-dimensional cell complex with a single vertex reduces to the problem of folding a plane graph onto a line

from the unfolded piece of paper by a continuous motion that avoids bending or folding the uncreased parts of the paper.

In practical applications of folding beyond origami, the object being folded may not be a single flat sheet, but rather some 2D polyhedral cell complex with nonmanifold topology (more than two facets joined at an edge). Flat foldability of such complexes is no easier than the origami case, but again we can hope for reduced complexity when a complex has only a single vertex. As with one-vertex origami, we can reduce the problem to 1D by slicing with a small sphere centered at the vertex—now resulting in a general plane graph rather than a simple cycle—and asking whether this graph can be flattened onto a line [2]; see Figure 2. In this way, the problem of flat-folding single-vertex complexes can be reduced to finding embeddings of a given plane graph onto a line.

It is this problem that we study here: given a plane graph with specified edge lengths, does it have a straight-line plane embedding with all vertices arbitrarily close to a given line and with all edges arbitrarily close to their specified lengths? In the version of the problem we study, we are additionally given a specification of whether the angle between every two consecutive edges at each vertex is a *mountain fold* (the angle is arbitrarily close to 360°), a *valley fold* (the angle is arbitrarily close to 0), or *flat* (the angle is arbitrarily close to 180°). Without this information, the problem of testing whether a given plane graph can be folded flat with specified edge lengths (allowing angles of 180°) is weakly NP-complete, even for graphs that are just simple cycles, by a straightforward reduction from

Table 1. Complexity of flat folding a plane graph, by input model

	Flat angles forbidden	Flat angles allowed
Angle assignment given	Linear time (new)	Linear time (new)
Angle assignment unspecified	Open	NP-complete [2]

the subset sum problem. For general plane graphs, the problem becomes strongly NP-complete [2]. Therefore, we concentrate in this paper on the version of the problem with given angle assignments, posed as an open problem in [2].

1.1   New Results

We show that it is possible to test in linear time whether a given plane graph, with given edge lengths and angle assignment, can be folded flat; refer to Table 1. Additionally, in polynomial time, we can count the number of combinatorially distinct flat foldings.

Our algorithms are based on a new characterization of flat-foldable graphs: a flat folding exists if and only if the angles at each vertex sum to 360° and each individual face in the given graph can be folded flat. Even stronger, we show that independent flat foldings of the interior of each face can always be combined into a flat folding of the whole graph. Figure 3 shows an example of this combination of face foldings. A form of the theorem was conjectured in 2001 by Ilya Baran, Erik Demaine, and Martin Demaine, but not proved until now; it contradicts the intuitive (but false) idea that, for faces with ambiguous spiraling shapes, each face must be folded consistently with its neighboring faces. With this theorem in hand, our algorithms for constructing and counting folded states follow by using a greedy "crimping" strategy for flat-folding simple cycles [4,6,12] and by using dynamic programming to count cycles within each face.

Our characterization necessarily concerns flat folded states, not continuous folding motions from a given (nonflat) configuration. As shown by past work, even for trees, there exist locked states that cannot be continuously moved to a flat folded state [5,8], and testing the existence of a continuous motion between two states is PSPACE-complete [3].

We leave open the problem of finding a flat folded state for a graph in which the planar embedding and edge lengths are preassigned, and angles of 180° are forbidden, but the choice of which angles at each vertex are 0 and which are 360° is left free (bottom-left cell of Table 1). Even for trees, this open problem appears to be nontrivial; see Figure 4 and [15].

1.2   Related Work

There has been intensive study of straight-line drawings of graphs with specified edge lengths and/or specified angles between consecutive edges in a cyclic ordering of edges around each vertex. If only edge lengths are specified then—whether the drawing must be planar or not—the problem is NP-hard [9,24], or worse [25].

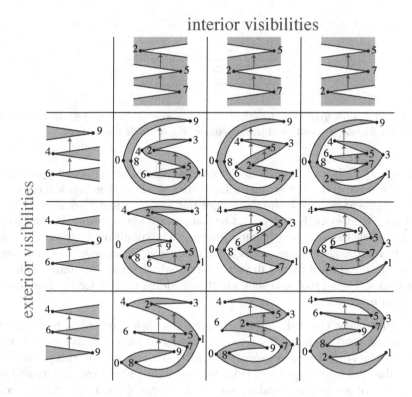

Fig. 3. A planar graph with two faces, each of which can be flat-folded to give three different patterns of vertical visibility (or "touching") within it. These patterns can be combined independently, giving nine flat-foldings of the whole graph.

It is also NP-hard to draw a plane graph with specified angles [16]. If both edge lengths and angles are specified then the drawing is uniquely determined and easy to construct, except in situations like ours where coincident edges give rise to ambiguities.

There are a number of results for special cases that have a similar flavor to ours, in that the whole plane graph can be realized if and only if each face can. We now describe some of these special cases, most of which arise as the prelude to finding an appropriate angle assignment.

Upward Planarity. A directed acyclic graph (DAG) is *upward planar* [17] if it has a planar drawing in which each edge is drawn as an increasing y-monotone curve. Recognizing upward planar graphs is NP-hard [18] but Bertolazzi et al. [7] gave a polynomial time algorithm for the special case of a plane graph whose cyclic order of edges around each vertex is prescribed. The main issue in their solution is to distinguish "small" versus "large" angles; if an upward planar drawing is flattened onto a vertical line, then its small and large angles correspond to our valley and mountain folds. The angle assignment is forced except at

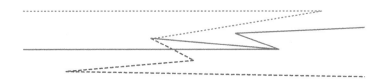

Fig. 4. A tree with fixed edge lengths that (when angles equal to 180° are forbidden) has no flat folding, regardless of planar embedding or angle assignment

vertices with only incoming or only outgoing edges, where exactly one angle should be large and the rest small. Bertolazzi et al. used network flows to determine these angles. To prove their algorithm's correctness, they showed that a graph with a given angle assignment has an upward planar drawing if and only if each face cycle has an upward planar drawing. The condition for drawing a single face, given an angle assignment, is particularly simple: an acyclic orientation of a cycle has an upward planar drawing if and only if it has two more small than large angles. Their proof also shows that embedding choices for the faces can be combined arbitrarily.

Level Planarity. Our flat folding problem differs from upward planarity in that we have assigned edge lengths as well as assigned angles. This makes it more similar to the problem of *level planarity* [13,19,21]. The input to this problem is a *leveled* directed acyclic graph: a DAG whose vertices have been partitioned into a sequence of levels (independent sets of its vertices), with all edges directed from earlier to later levels. The goal is to find an upward planar embedding that places the vertices of each level on a horizontal line [22]. This problem has a linear time solution [21] based on PQ-trees. When the cyclic order of edges around vertices is specified (and in fact for more general constraints) there is a quadratic time solution based on solving systems of binary equations [19].

The input to our folding problem may be interpreted as a leveled plane DAG. (Since our convention is to flatten to a horizontal line, we will map to a level planarity problem with levels progressing rightward rather than upward—this is a superficial difference.) Arbitrarily choose an x-coordinate for one vertex and a direction (left-to-right) for one edge incident to that vertex. These choices can be propagated to all the vertices and edges using the specified edge lengths and angles. The set of vertices at a given x-coordinate constitute a level, giving

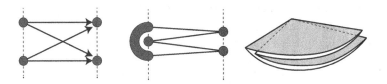

Fig. 5. A four-vertex cycle with vertices alternating between two levels is not level planar, but can be folded flat, representing a sheet of paper folded into quarters.

us an input to the level planarity problem with a prescribed plane embedding. However, the embeddings we seek in the folding problem are not the same as level planar embeddings. In a level planar embedding, vertices within a single level must be linearly ordered by the second coordinate value. In contrast, in the folding problem a vertex that has only incoming or only outgoing edges may be nested between two adjacent edges of another vertex at the same level. A four-vertex cycle, oriented with alternating edge directions, illustrates the difference between these two types of embedding: it is not level planar, but it still has a flat folding with three mountain folds and one valley fold, corresponding to the usual way of folding a square sheet of paper into quarters (Figure 5).

We have therefore been unable to apply level planarity algorithms to solve our flattening problem. On the other hand, our algorithm can be used to test level planarity of a plane graph (with a linear order of incoming and outgoing edges at each vertex) in linear time. Given an input to the level planarity problem, we assign increasing coordinates to the levels, and assign the length of an edge to be the difference in coordinates between the levels of its endpoints. Mountain/valley/flat angles are determined from the level assignment. Finally, to preclude the nesting of vertices that is allowed in flattening but not in level planarity, we add an extra edge incident to each vertex that has only incoming or only outgoing edges: if vertex v on level i has only outgoing edges, we add a new incoming edge from a new level just before i. The resulting plane graph has a flat-folding respecting the angles and edge-lengths if and only if the original has a level planar drawing. From this we also obtain the result that a leveled plane graph has a level planar drawing if and only if each cycle does.

Rectilinear Planarity. In our flat folding problem, the angles are multiples of 180°. When angles are multiples of 90°, we arrive at the important problems of orthogonal and rectilinear graph drawing [14]. A graph is *rectilinear planar* if it can be drawn in the plane so that every edge is a horizontal or vertical line segment. Coincident edges are forbidden, so the graph must have maximum degree 4. This problem is NP-complete in general [18] but—as with upward planarity—there is a polynomial time algorithm, due to Tamassia [27], if the cyclic order of edges around vertices is prescribed. Again, the main issue is to find an assignment of angles or, equivalently, a labeling of the edges incident to a vertex with distinct labels from the set U, D, L, R, where U stands for "Up", etc. Tamassia finds the angles using network flows (in fact, he solves a more general problem of minimizing the number of bends in the drawing). At the heart of this method is the result that, given an angle assignment that is locally consistent (i.e., the angles at every vertex sum to 360°), the graph has a rectilinear planar drawing if and only if each face cycle does [27]. As in the other cases we have discussed, the proof shows the stronger result that embedding choices for the faces can be combined arbitrarily. A cycle with an angle assignment has a rectilinear planar drawing if and only if the number of right turns minus the number of left turns is 4 in a clockwise traversal inside the cycle [27, 28].

Our result on flat folding can be used to prove an extension of the above result to rectilinear graph drawings with angles specified and with lengths assigned to the "horizontal" edges. (Note that the angle information allows us to distinguish the two classes of edges, although it is arbitrary which class is horizontal and which is vertical.) Given a rectilinear plane graph, contract all vertical edges, assigning angles of $0, 180°, 360°$ in the obvious way. Finally, in order to avoid "nested" vertices at the same coordinate (as in Figure 5), we use the same trick of adding an extra edge incident to each vertex that has only incoming or only outgoing edges. We claim that the resulting multi-graph has a flat-folding if and only if the original has a rectilinear planar drawing with horizontal edges of the specified lengths. For the non-trivial direction, suppose we have a flat folding of the constructed graph. We must expand each vertex to a vertical line segment with the horizontal edges touching the line segment in a way that is consistent with the original graph. This can easily be done in a left to right sweep.

From this reduction we obtain the following result.

Corollary 1. *If G is a plane graph of maximum degree 4 with specified angles that sum to $360°$ at each vertex and with lengths assigned to the horizontal edges, then G has a rectilinear planar drawing realizing these angles and edge lengths if and only if every cycle has a rectilinear planar drawing realizing the angles and lengths. Furthermore, we can combine any embedding choices for the faces that involve different vertical visibilities, so long as every vertical edge has the same length in its two cycles.*

2   Definitions

Following previous works in this area [10, 23] we formalize the notion of a flat folding using *self-touching configurations*. Intuitively, these are planar embeddings in which edges and vertices are allowed to be infinitesimally close to each other. A one-dimensional self-touching configuration of a graph G consists of a mapping from G to a path graph H that maps vertices of G to vertices of H and edges of G to paths in H, together with a *magnified view* of each vertex and edge of H that describes the local connectivity of the image of G at that point. In a self-touching configuration, the *multiplicity* of an edge in H is a positive integer, the number of different edges of G that map to it. The magnified view of an edge gives a linear ordering of the edges in its preimage. The magnified view of a vertex v of H is a planar embedding of a neighborhood of the preimage of v into a disk, consistent with the magnified views of the edges incident to v.

By replacing each vertex of H by its magnified view, and each edge of H by a corridor of finite width through which each edge passes, it is possible to transform a self-touching configuration into a conventional planar drawing (with edges that may curve or bend) of the given graph G. We call this the *expanded drawing* of a self-touching configuration (Figure 6).

We may now define a *flat folding* to be a self-touching configuration in which all edges of H lie on a single line. We consider two flat foldings to be equivalent if they have the same magnified views in the same order or in the reversed order as

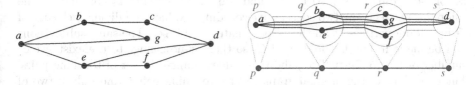

Fig. 6. A flat-folding of a seven-vertex graph G (left), described as a self-touching configuration in which G is mapped onto a four-vertex path H (right), shown with magnified views of its edges and vertices of H.

each other. A *face* of a flat folding or self-touching configuration is a cycle formed by a face of the expanded drawing. The *angle* formed by a pair of incident edges in a flat folding is one of three values, 0, 180°, or 360°, accordingly as the face lies between the two edges, the two edges extend in opposite directions from their common endpoint, or the two edges extend in the same direction with the face on both sides of them. An *angle assignment* to a plane graph is an assignment of the values 0, 180°, or 360° to each of its angles, regardless of whether this assignment is compatible with a flat-folding of the graph. An angle assignment is *consistent* if the angles sum to 360° at every vertex.

We define a *touching pair* of edges in a self-touching configuration of a graph G to be two edges e and f of G such that these two edges are consecutive in the magnified view of at least one edge in H. Each touching pair can be assigned to a single face of G, the face that lies between the two edges.

3   Local Characterization

In this section we show that for a plane graph with assigned lengths and consistent angles, being able to fold the whole graph flat is equivalent to being able to fold each of its faces flat.

Theorem 1. *Let G be a plane graph with given edge lengths and a consistent angle assignment. Then G has a flat folding if and only if every face cycle of G (with the induced assignment of lengths and angles) has a flat folding. More strongly, for every combination of flat foldings of the faces of G, there exists a flat folding of G itself whose touching pairs for each face are exactly the ones given in the folding of that face.*

Proof. One direction is straightforward: if G has a flat folding, then restricting to the faces of G gives flat foldings of the faces with the same touching pairs.

For the other direction, assume we have flat foldings of the faces of G. We will show that G has a flat folding with the same touching pairs. As described in Section 1.2, the assignment of lengths and angles given with G (together with an arbitrary choice of an x coordinate for one vertex and an orientation for one edge) gives us a unique assignment of x coordinates for the vertices of G in any possible flat folding. We will start by subdividing all the edges of G. Take the set of x-coordinates of vertices of G and add an extra "half" x-coordinate at the midpoint between any two consecutive coordinate values. Subdivide each edge of G by adding vertices at all the x-coordinates in this set. The same subdivisions can be made in any flat folding of G, so there is no change to the existence or number of flat foldings. The subdivision does change the set of touching pairs, but two edges of the original graph form a touching pair if and only if two of the edges in the paths they are subdivided into form a touching pair, so the correctness of the part of the theorem about touching pairs carries over.

With G subdivided in this way, we carry out the proof by induction on the number of face angles that are assigned to be 360° (*mountain folds*). The base case of the induction is the case in which G has only two such angles, on the outer face. In this case every cycle consists of two paths of increasing x-coordinates and has a unique flat folding, and it is easy to see that G has a flat folding with the same touching pairs. (Equivalently, the graph in this case is a directed st-plane graph so it is upward planar with each face drawn as two upward paths.)

If G contains a vertex v, and an interior face f in which v is a mountain fold, then let e be one of the two edges of f incident to v, the one that is uppermost in the magnified view of the flat folding edge corresponding to these two edges, and let e' be the edge immediately above that one. Edge e' must exist, because if e were the topmost edge in this magnified view, then f would necessarily be the exterior face. (For example, in Figure 6, vertex g is a mountain fold in a cycle; edge bg is the uppermost edge incident to g; and bc is the edge immediately above it.) Let v' be the endpoint of e' whose x-coordinate is the same as v. We form a graph G' by identifying v with v', ordering the edges of the combined super-vertex so that e' and e remain consecutive. This produces a graph, not a multigraph, because the other endpoints of e and e' are subdivision points at a "half" x-coordinate, and so cannot coincide with each other. (In the example, we would identify vertices g and c; the figure does not show the extra subdivision points.) This vertex identification reduces the number of mountain folds by one compared with G, and splits f into two simpler faces f_1 and f_2. The same split operation can be done to the flat folding of f, giving flat foldings of f_1 and f_2. Thus, G' meets the conditions of the theorem and has fewer mountain folds; by induction it has a flat folding realizing all the touching pairs of its face foldings, which are the same as the touching pairs of the face foldings of G. In this flat folding, the supervertex of G' formed from v and v' can be split back into the two separate vertices v and v', giving the desired flat folding of G.

The case when there exist three or more mountain folds on the exterior face is similar, but we must be more careful in our choice of v. Each mountain folded vertex on the exterior face is either a local minimum or local maximum of x-coordinates; because there are three or more of them, we may choose v to be a vertex that is not a unique global extremum. Then, as above, we find a vertex v' with the same coordinate, above or below v, and merge v and v' into a single vertex, giving a graph G' with fewer mountain folds in which the outer face has been split into two faces, one outer and one inner. As before, these two faces inherit a flat folded state from the given flat folding of the outer face of G, so by induction G' has a flat folding. And as before, v and v' may be split back into separate vertices in this flat folding, giving the desired flat folding of G.   □

4   Algorithm to Find a Folding

For completeness, we briefly describe a greedy "crimping" strategy for finding flat-folded states of simple cycles with pre-assigned fold angles. Bern and Hayes [6] used a similar strategy to flat-fold cycles without pre-assigned angles. Arkin et al. [4] applied this method to open polygonal chains with assigned angles. The version here for cycles with assigned angles is described by Demaine and O'Rourke [12]. We do not describe its (non-trivial) correctness proof.

First, remove any flat folds from the input by merging the edges on either side of the fold. Then, repeatedly find an edge e such that the two edges on either side of e are at least as long as e, with folds of opposite type at its ends. If no such edge e exists, the cycle has no folding. If an edge e that meets these conditions can be found, it is safe to perform both folds, merging e with its two neighboring edges into a single edge of a simpler polygon.

Maintaining a set of edges that are ready to be folded, and performing each fold, takes constant time per fold, so folding a cycle in this way, and recovering the covering relation of its ordered line embedding, may be done in linear time. Putting the characterization from Section 3 together with the algorithm for folding a single cycle described above gives us an algorithm for testing whether a given plane graph G with edge length and angle assignment is flat foldable:

Theorem 2. *We can test flat foldability of a plane graph with given edge lengths and given angle assignment in linear time.*

Proof. We partition the graph into its component faces, and apply the crimping algorithm to an Euler tour of each face. Each face takes time proportional to its size, so the total time is linear. For the correctness of forming simple cycles from each face by taking Euler tours, see the full version of this paper.   □

5   Counting Flat Foldings

We cannot use crimping to count the flat foldings of a cycle, because some flat foldings cannot be formed by a sequence of crimping steps (Figure 7). Instead, to count flat foldings in a single cycle, we use dynamic programming.

Lemma 1 (proof in the full version of this paper). *Given a single n-vertex cycle, with an assignment of edge lengths and angles, it is possible to count the flat foldings of the cycle in time $\tilde{O}(n^5)$.*

Theorem 3. *We can count the flat foldings of an n-vertex planar graph G with an assignment of edge lengths and angles in time $\tilde{O}(n^5)$.*

Proof. We apply Lemma 1 to the Euler tour of each face of G and return the product of the resulting numbers.   □

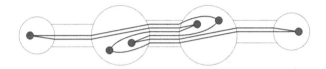

Fig. 7. Magnified view of a flat folding that cannot be obtained by crimping

Acknowledgements. This research was performed in part at the 29th Bellairs Winter Workshop on Computational Geometry. Erik Demaine thanks Ilya Baran and Muriel Dulieu, and the authors of [2], for many discussions attempting to solve this problem. We also thank Jason Ku for helpful comments on a draft of this paper. Erik Demaine was supported in part by NSF ODISSEI grant EFRI-1240383 and NSF Expedition grant CCF-1138967. David Eppstein was supported in part by NSF grant 1228639 and ONR grant N00014-08-1-1015.

References

1. Abbott, T.G., Demaine, E.D., Gassend, B.: arXiv:0901.1322 (January 2009), http://arxiv.org/abs/0901.1322

2. Abel, Z., Demaine, E.D., Demaine, M.L., Eisenstat, S., Lynch, J., Schardl, T.B., Shapiro-Ellowitz, I.: Folding equilateral plane graphs. Internat. J. Comput. Geom. Appl. 23(2), 75–92 (2013)

3. Alt, H., Knauer, C., Rote, G., Whitesides, S.: On the complexity of the linkage reconfiguration problem. In: Pach, J. (ed.) Towards a Theory of Geometric Graphs, Contemp. Math., vol. 342, pp. 1–13. Amer. Math. Soc., Providence (2004)

4. Arkin, E.M., Bender, M.A., Demaine, E.D., Demaine, M.L., Mitchell, J.S.B., Sethia, S., Skiena, S.S.: When can you fold a map? Comput. Geom. Th. Appl. 29(1), 23–46 (2004)

5. Ballinger, B., Charlton, D., Demaine, E.D., Demaine, M.L., Iacono, J., Liu, C.-H., Poon, S.-H.: Minimal locked trees. In: Dehne, F., Gavrilova, M., Sack, J.-R., Tóth, C.D. (eds.) WADS 2009. LNCS, vol. 5664, pp. 61–73. Springer, Heidelberg (2009)

6. Bern, M., Hayes, B.: The complexity of flat origami. In: Proc. 7th ACM-SIAM Symposium on Discrete Algorithms (SODA 1996), pp. 175–183 (1996)

7. Bertolazzi, P., Di Battista, G., Liotta, G., Mannino, C.: Upward drawings of tri-connected digraphs. Algorithmica 12(6), 476–497 (1994)

8. Biedl, T., Demaine, E.D., Demaine, M.L., Lazard, S., Lubiw, A., O'Rourke, J., Robbins, S., Streinu, I., Toussaint, G., Whitesides, S.: A note on reconfiguring tree linkages: trees can lock. Discrete Appl. Math. 117(1-3), 293–297 (2002)

9. Cabello, S., Demaine, E.D., Rote, G.: Planar embeddings of graphs with specified edge lengths. J. Graph Algorithms & Appl. 11(1), 259–276 (2007)

10. Connelly, R., Demaine, E.D., Rote, G.: Infinitesimally locked self-touching linkages with applications to locked trees. In: Physical Knots: Knotting, Linking, and Folding Geometric Objects in $\mathbb{R}^3$ (Las Vegas, NV, 2001). Contemp. Math., vol. 304, pp. 287–311. Amer. Math. Soc., Providence (2002)

11. Connelly, R., Demaine, E.D., Rote, G.: Straightening polygonal arcs and convexi-fying polygonal cycles. Discrete & Computational Geometry 30(2), 205–239 (2003)

12. Demaine, E.D., O'Rourke, J.: Geometric Folding Algorithms: Linkages, Origami, Polyhedra. Cambridge University Press (2007)
13. Di Battista, G., Nardelli, E.: Hierarchies and planarity theory. IEEE Trans. Systems Man Cybernet. 18(6), 1035–1046 (1988)
14. Duncan, C.A., Goodrich, M.T.: Planar orthogonal and polyline drawing algorithms. In: Tamassia, R. (ed.) Handbook of Graph Drawing and Visualization, ch. 7, pp. 223–246. Chapman and Hall/CRC (2013)
15. Estrella-Balderrama, A., Fowler, J.J., Kobourov, S.G.: Characterization of unlabeled level planar trees. Comput. Geom. Th. Appl. 42(6-7), 704–721 (2009)
16. Garg, A.: New results on drawing angle graphs. Comput. Geom. Th. Appl. 9(1), 43–82 (1998)
17. Garg, A., Tamassia, R.: Upward planarity testing. Order 12(2), 109–133 (1995)
18. Garg, A., Tamassia, R.: On the computational complexity of upward and rectilinear planarity testing. SIAM J. Comput. 31(2), 601–625 (2001)
19. Harrigan, M., Healy, P.: Practical level planarity testing and layout with embedding constraints. In: Hong, S.-H., Nishizeki, T., Quan, W. (eds.) GD 2007. LNCS, vol. 4875, pp. 62–68. Springer, Heidelberg (2008)
20. Hull, T.C.: The combinatorics of flat folds: a survey. In: Hull, T.C. (ed.) Origami[3] (Asilomar, CA, 2001), pp. 29–38. A K Peters, Natick (2002)
21. Jünger, M., Leipert, S., Mutzel, P.: Level planarity testing in linear time. In: Whitesides, S.H. (ed.) GD 1998. LNCS, vol. 1547, p. 224. Springer, Heidelberg (1999)
22. Pach, J., Tóth, G.: Monotone drawings of planar graphs. J. Graph Theory 46(1), 39–47 (2004)
23. Ribó Mor, A.: Realization and counting problems for planar structures. Ph.D. thesis, Free Univ. Berlin (2006)
24. Saxe, J.: Embeddability of weighted graphs in k-space is strongly NP-hard. In: Proc. 17th Allerton Conf. Commun. Control Comput., pp. 480–489 (1979)
25. Schaefer, M.: Realizability of graphs and linkages. In: Pach, J. (ed.) Thirty Essays on Geometric Graph Theory, pp. 461–482. Springer, New York (2013)
26. Streinu, I., Whiteley, W.: Single-vertex origami and spherical expansive motions. In: Akiyama, J., Kano, M., Tan, X. (eds.) JCDCG 2004. LNCS, vol. 3742, pp. 161–173. Springer, Heidelberg (2005), http://dx.doi.org/10.1007/11589440_17
27. Tamassia, R.: On embedding a graph in the grid with the minimum number of bends. SIAM J. Comput. 16(3), 421–444 (1987)
28. Vijayan, G., Wigderson, A.: Rectilinear graphs and their embeddings. SIAM J. Comput. 14(2), 355–372 (1985)

Disjoint Edges in Topological Graphs
and the Tangled-Thrackle Conjecture[*]

Andres J. Ruiz-Vargas[1], Andrew Suk[2], and Csaba D. Tóth[3]

[1] École polytechnique fédérale de Lausanne, Lausanne, Switzerland
andres.ruizvargas@epfl.ch
[2] University of Illinois at Chicago, Chicago, IL, USA
suk@math.uic.edu
[3] California State University Northridge, Los Angeles, CA, USA
cdtoth@acm.org

Abstract. It is shown that for a constant $t \in \mathbb{N}$, every simple topological graph on n vertices has $O(n)$ edges if the graph has no two sets of t edges such that every edge in one set is disjoint from all edges of the other set (i.e., the complement of the intersection graph of the edges is $K_{t,t}$-free). As an application, we settle the *tangled-thrackle* conjecture formulated by Pach, Radoičić, and Tóth: Every n-vertex graph drawn in the plane such that every pair of edges have precisely one point in common, where this point is either a common endpoint, a crossing, or a point of tangency, has at most $O(n)$ edges.

1 Introduction

A *topological graph* is a graph drawn in the plane such that its vertices are represented by distinct points and its edges are represented by Jordan arcs between the corresponding points satisfying the following (nondegeneracy) conditions: (a) no edge intersects any vertex other than its endpoints, (b) any two edges have only a finite number of interior points in common, (c) no three edges have a common interior point, and (d) if two edges share an interior point, then they properly cross at that point [7]. A topological graph is *simple* if every pair of edges intersect in at most one point. Two edges of a topological graph *cross* if their interiors share a point, and are *disjoint* if they neither share a common vertex nor cross.

In 2005, Pach and Tóth [10] conjectured that for every constant $t \geq 3$, an n-vertex simple topological graph has $O(n)$ edges if no t edges are pairwise disjoint. They gave an upper bound of $|E(G)| \leq O(n \log^{4t-8} n)$ for all such graphs. Despite much attention over the last 10 years (see related results in [3,11,15,16]), the conjecture is still open.

[*] Work on this paper began at the AIM workshop *Exact Crossing Numbers (Palo Alto, CA, 2014)*. Research by Ruiz-Vargas was supported by the Swiss National Science Foundation grants 200021-125287/1 and 200021-137574. Research by Suk was supported by an NSF Postdoctoral Fellowship and by the Swiss National Science Foundation grant 200021-125287/1. Research by Tóth was supported in part by the NSF award CCF 1423615.

C. Duncan and A. Symvonis (Eds.): GD 2014, LNCS 8871, pp. 284–293, 2014.
© Springer-Verlag Berlin Heidelberg 2014

The condition that *no t edges are pairwise disjoint* means that the intersection graph of the edges (Jordan arcs) contains no anti-clique of size t, or equivalently the complement of the intersection graph of the edges is K_t-free. In this paper, we consider a weaker condition that the complement of the intersection graph of the edges is $K_{t,t}$-free, where $t \in \mathbb{N}$ is a constant. This means that graph G has no set of t edges that are all disjoint from another set of t edges. Since no such graph G contains $2t$ pairwise disjoint edges, [10] implies $|E(G)| \leq O(n \log^{8t-8} n)$. Our main result improves this upper bound to $O(n)$.

Theorem 1. *Let $t \in \mathbb{N}$ be a constant. The maximum number of edges in a simple topological graph with n vertices that does not contain t edges all disjoint from another set of t edges is $O(n)$.*

Application to thrackles. More than 50 years ago, Conway asked what is the maximum number of edges in an n-vertex *thrackle*, that is, a simple topological graph G in which every two edges intersect, either at a common endpoint or at a proper crossing [1]. He conjectured that every n-vertex thrackle has at most n edges. The first linear upper bound was obtained by Lovász, Pach, and Szegedy [4], who showed that all such graphs have at most $2n$ edges. This upper bound was successively improved, and the current record is $|E(G)| \leq \frac{167}{117}n < 1.43n$ due to Fulek and Pach [2].

As an application of Theorem 1, we prove the tangled-thrackle conjecture recently raised by Pach, Radoičić, and Tóth [9]. A drawing of a graph G is a *tangled-thrackle* if it satisfies conditions (a)-(c) of topological graphs and every pair of edges have precisely one point in common: either a common endpoint, or a proper crossing, or a point of tangency. Note that such a drawing is not a topological graph due to tangencies. Pach, Radoičić, and Tóth [9] showed that every n-vertex tangled-thrackle has at most $O(n \log^{12} n)$ edges, and described a construction with at least $\lfloor 7n/6 \rfloor$ edges. They conjectured that the upper bound can be improved to $O(n)$. Here, we settle this conjecture in the affirmative.

Theorem 2. *Every tangled-thrackle on n vertices has $O(n)$ edges.*

2   Disjoint Edges in Topological Graphs

In this section, we prove Theorem 1. We start with reviewing a few graph theoretic results used in our argument. The following is a classic result in extremal graph theory due to Kővári, Sós, and Turán.

Theorem 3 (see [8]). *Let $G = (V, E)$ be a graph that does not contain $K_{t,t}$ as a subgraph. Then $|E(G)| \leq c_1 |V(G)|^{2-1/t}$, where c_1 is an absolute constant.*

Two edges in a graph are called *independent* if they do not share an endpoint. We define the *odd-crossing number* odd-cr(G) of a graph G to be the minimum number of unordered pairs of edges that are independent and cross an odd number of times over all topological drawings of G. The *bisection width* of a graph G, denoted by $b(G)$, is the smallest nonnegative integer such that there is a partition of the vertex set $V = V_1 \cup V_2$ with $\frac{1}{3}|V| \leq V_i \leq \frac{2}{3}|V|$ for $i = 1, 2$, and $|E(V_1, V_2)| = b(G)$. The following result, due to Pach and Tóth, relates the odd-crossing number of a graph to its bisection width.[1]

Theorem 4 ([10]). *There is an absolute constant c_2 such that if G is a graph with n vertices of vertex degrees $d_1, \ldots, d_n$, then*

$$b(G) \leq c_2 \log n \sqrt{\mathrm{odd\text{-}cr}(G) + \sum_{i=1}^{n} d_i^2}.$$

We also rely on the result due to Pach and Tóth [10] stated in the introduction.

Theorem 5 ([10]). *Let $G = (V, E)$ be an n-vertex simple topological graph, such that G does not contain t pairwise disjoint edges. Then $|E(G)| \leq c_3 n \log^{4t-8} n$, where c_3 is an absolute constant.*

From disjoint edges to odd crossings. Using a combination of Theorems 3–5, we establish the following lemma.

Lemma 1. *Let $G = (V, E)$ be simple topological bipartite graph on n vertices with vertex degrees $d_1, \ldots, d_n$, such that G does not contain a set of t edges all disjoint from another set of t edges. Then*

$$b(G) \leq c_4 n^{1-\frac{1}{2t}} \log^{8t-3} n + c_4 \log n \sqrt{\sum_{i=1}^{n} d_i^2}, \tag{1}$$

where c_4 is an absolute constant.

Proof: Since G does not contain $2t$ pairwise disjoint edges, Theorem 5 yields

$$|E(G)| \leq c_3 n \log^{8t-8} n. \tag{2}$$

Let V_a and V_b be the vertex classes of the bipartite graph G. Consider a simple curve γ that decomposes the plane into two parts, containing all points in V_a and V_b, respectively. By applying a suitable homeomorphism to the plane that maps γ to a horizontal line, G is deformed into a topological graph G' such that (refer to Fig. 1)

1. The vertices in V_a are above the line $y = 1$, the vertices in V_b are below the line $y = 0$,

---

[1] Pach and Tóth [10] defined the odd-crossing number of a graph G to be the minimum number of pairs of edges that cross an odd number of times (over all drawings of G), including pairs of edges with a common endpoint. However, since the number of pairs of edges with a common endpoint is at most $\sum_{i=1}^{n} d_i^2$, this effects Theorem 4 only by a constant factor.

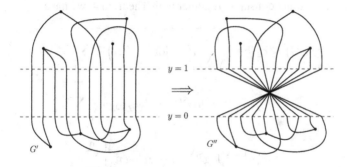

Fig. 1. Redrawing procedure

2. The part of any edge lying in the horizontal strip $0 \le y \le 1$ consists of vertical segments.

Since a homeomorphism neither creates nor removes intersections between edges, G and G' are isomorphic and their edges have the same intersection pattern.

Next we transform G' into a topological graph G'' by the following operations. Reflect the part of G' that lies above the $y = 1$ line about the y-axis. Replace the vertical segments in the horizontal strip $0 \le y \le 1$ by straight line segments that reconnect the corresponding pairs on the line $y = 0$ and $y = 1$, and perturb the segments if necessary to avoid triple intersections. Notice that if any two edges cross in G (and G'), then they must cross an even number of times in G''. Indeed, suppose the edges e_1 and e_2 cross in G. Since G is simple, they share exactly one point in common. Let k_i denote the number of times edge e_i crosses the horizontal strip for $i \in 1, 2$, and note that k_i must be odd since the graph is bipartite. These $k_1 + k_2$ segments within the strip pairwise cross in G'', creating $\binom{k_1+k_2}{2}$ crossings. Since edge e_i now crosses itself $\binom{k_i}{2}$ times in G'', there are

$$\binom{k_1 + k_2}{2} - \binom{k_1}{2} - \binom{k_2}{2} = k_1 k_2 \tag{3}$$

crossings between edges e_1 and e_2 within the strip, which is odd when k_1 and k_2 are odd. Since e_1 and e_2 had one point in common outside the strip in both G and G'', then e_1 and e_2 cross each other an even number of times in G''. (Note that one can easily eliminate self-intersections by local modifications around these crossings.)

Hence, the number of pairs of edges that are independent and cross an odd number of times in G'' is at most the number of disjoint pairs of edges in G, which is in turn at most $c_1 |E(G)|^{2-1/t}$ by Theorem 3. Combined with (2), we have

$$\text{odd-cr}(G) \le c_1 (c_3 n \log^{8t-8} n)^{2-\frac{1}{t}}$$
$$\le cn^{2-\frac{1}{t}} \log^{16t-8} n,$$

where c is an absolute constant. Together with Theorem 4, we have

$$b(G) \leq c_2 \log n \sqrt{\left(cn^{2-\frac{1}{t}} \log^{16t-8} n\right) + \sum_{i=1}^{n} d_i^2}$$

$$\leq c_2 \sqrt{cn}^{1-\frac{1}{2t}} \log^{8t-3} n + c_2 \log n \sqrt{\sum_{i=1}^{n} d_i^2}$$

$$\leq c_4 n^{1-\frac{1}{2t}} \log^{8t-3} n + c_4 \log n \sqrt{\sum_{i=1}^{n} d_i^2},$$

where c_4 is an absolute constant, as required.   □

If the maximum degree of G is relatively small, we obtain a sublinear bound on the bisection width.

Corollary 1. *Let $G = (V, E)$ be simple topological bipartite graph on n vertices with vertex degrees $d_1, \ldots, d_n \leq n^{1/5}$ such that G does not contain a set of t edges all disjoint from another set of t edges. Then*

$$b(G) \leq c_5 n^{1-\frac{1}{4t}}, \tag{4}$$

where c_5 is an absolute constant.

Proof: Substituting $d_i \leq n^{1/5}$ into (1), we have

$$b(G) \leq c_4 n^{1-\frac{1}{2t}} \log^{8t-3} n + c_4 \log n \sqrt{n \cdot n^{2/5}}$$
$$\leq c_4 n^{1-\frac{1}{2t}} \log^{8t-3} n + c_4 n^{7/10} \log n$$
$$\leq c_5 n^{1-\frac{1}{4t}}, \tag{5}$$

for a sufficiently large constant c_5.   □

Vertex splitting for topological graphs. Given a simple topological graph with n vertices, we reduce the maximum degree below $n^{1/5}$ by a standard vertex splitting operation. Importantly, this operation can be performed such that it preserves the intersection pattern of the edges.

Lemma 2. *Let G be a simple topological graph with n vertices and m edges; and let $\Delta \geq 2m/n$. Then there is a simple topological graph G' with maximum degree at most Δ, at most $n + 2m/\Delta$ vertices, and precisely m edges such that the intersection graph of its edges is isomorphic to that of G.*

Proof: We successively split every vertex in G whose degree exceeds Δ as follows. Refer to Fig. 2. Let v be a vertex of degree $d(v) = d > \Delta$, and let $vu_1, vu_2, \ldots, vu_d$ be the edges incident to v in counterclockwise order. In a small neighborhood around v, replace v by $\lceil d/\Delta \rceil$ new vertices, $v_1, \ldots, v_{\lceil d/\Delta \rceil}$ placed in counterclockwise order on a circle of small radius centered at v. Without introducing any new crossings, connect u_j to v_i if and only if $\Delta(i-1) < j \le \Delta i$ for $j \in \{1, \dots, d\}$ and $i \in \{1, \dots, \Delta\}$. Finally, we do a local change within the small circle by moving each vertex v_i across the circle, so that every edge incident to v_i crosses all edges incident to $v_{i'}$, for all $i \neq i'$. As a result, any two edges incident to some vertex $\{v_1, \dots, v_{\lceil d/\Delta \rceil}\}$ intersect precisely once: either at a common endpoint or at a crossing within the small circle centered at v.

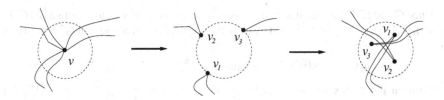

Fig. 2. Splitting a vertex v into new vertices v_1, v_2, v_3, such that each v_i has degree at most Δ. Moreover, we do not introduce any disjoint pairs of edges and our new graph remains simple.

After applying this procedure to all vertices G, we obtain a simple topological graph G' of maximum degree at most Δ. By construction, G' has m edges, and the intersection pattern of the edges is the same as in G. The number of vertices in G' is

$$|V(G')| \le \sum_{v \in V} \left\lceil \frac{d(v)}{\Delta} \right\rceil \le n + \sum_{v \in V} \frac{d(v)}{\Delta} \le n + \frac{2m}{\Delta},$$

as claimed. $\qquad\square$

Putting things together. Since all graphs have a bipartite subgraph with at least half of its edges, Theorem 1 immediately follows from the following.

Theorem 6. *Let G be an n-vertex simple topological bipartite graph such that G does not contain t edges all disjoint from another set of t edges. Then*

$$|E(G)| \le c_6(n - n^{1-\frac{1}{7t}}), \tag{6}$$

where $c_6 = c_6(t)$ depends only on t.

Proof: Let $t \in \mathbb{N}$ be a constant. We proceed by induction on n. Let $n_0 = n_0(t)$ be a sufficiently large constant (specified in (8) and (9) below) that depends only on t, and on the constants c_3 and c_5 defined in Theorem 5 and Corollary 1, respectively. Let c_6 be a sufficiently large constant such that $c_6 \ge 2c_5$ and for every positive integer $n \le n_0$, we have

$$c_3 n \log^{8t-8} n \le c_6(n - n^{1-\frac{1}{7t}}), \tag{7}$$

The choice of n_0 and c_6 ensures that (6) holds for all graphs with at most n_0 vertices. Now consider an integer $n > n_0$, and assume that (6) holds for all graphs with fewer than n vertices. Let G be a simple topological bipartite graph with n vertices such that G does not contain t edges all disjoint from another set of t edges.

By Theorem 5, G has $m \leq c_3 n \log^{8t-8} n$ edges. By Lemma 2, there is a simple topological graph G' of maximum degree at most $\Delta = n^{1/5}$, $n' \leq n + 2m/n^{1/5}$ vertices, and $m' = m$ edges, such that the intersection graph of its edges is isomorphic to that of G. Theorem 5 implies that $n' \leq n + 2m/n^{1/5} \leq n + 2c_3 n^{4/5} \log^{8t-8} n$. If $n \geq n_0$ for a sufficiently large constant n_0, then

$$n' \leq n + 2c_3 n^{4/5} \log^{8t-8} n \leq n + n^{5/6}. \tag{8}$$

Since G and G' have the same number of edges, it is now enough to estimate $|E(G')|$. Note that $\Delta = n^{1/5} \leq (n')^{1/5}$, and by Corollary 1, the bisection width of G' is bounded by

$$b(G') \leq c_5(n')^{1-\frac{1}{4t}} \leq c_5(n + n^{5/6})^{1-\frac{1}{4t}} \leq 2c_5 n^{1-\frac{1}{4t}}.$$

Partition the vertex set of G' as $V' = V_1 \cup V_2$ with $\frac{1}{3}|V'| \leq V_i \leq \frac{2}{3}|V'|$ for $i = 1, 2$, such that G' has $b(G')$ edges between V_1 and V_2. Denote by G_1 and G_2 the subgraphs induced by V_1 and V_2, respectively. Put $n_1 = |V_1|$ and $n_2 = |V_2|$, where $n_1 + n_2 = n' \leq n + n^{5/6}$.

Note that both G_1 and G_2 are simple topological graphs that do not contain t edges all disjoint from another set of t edges. By the induction hypothesis, $|E(G_i)| \leq c_6(n_i - n_i^{1-1/7t})$ for $i = 1, 2$. The total number of edges in G_1 and G_2 is

$$
\begin{aligned}
|E(G_1)| + |E(G_2)| &\leq c_6(n_1 - n_1^{1-\frac{1}{7t}}) + c_6(n_2 - n_2^{1-\frac{1}{7t}}) \\
&\leq c_6(n_1 + n_2) - c_6(n_1^{1-\frac{1}{7t}} + n_2^{1-\frac{1}{7t}}) \\
&\leq c_6(n') - c_6\left(\left(\frac{n_1}{n'}\right)^{1-\frac{1}{7t}} + \left(\frac{n_2}{n'}\right)^{1-\frac{1}{7t}}\right)(n')^{1-\frac{1}{7t}} \\
&\leq c_6(n + n^{5/6}) - c_6\left(\left(\frac{1}{3}\right)^{1-\frac{1}{7t}} + \left(\frac{2}{3}\right)^{1-\frac{1}{7t}}\right)n^{1-\frac{1}{7t}} \\
&= c_6(n + n^{5/6}) - c_6\alpha n^{1-\frac{1}{7t}} \\
&\leq c_6(n - n^{1-\frac{1}{7t}}) + c_6\left(n^{5/6} - (\alpha - 1)n^{1-\frac{1}{7t}}\right),
\end{aligned}
$$

where we write $\alpha = (1/3)^{1-1/7t} + (2/3)^{1-1/7t}$ for short. Note that for every $t \in \mathbb{N}$, we have $\alpha > 1$. Taking into account the edges between V_1 and V_2, the total number of edges in G' (and hence G) is

$$
\begin{aligned}
|E(G')| &= |E(G_1)| + |E(G_2)| + b(G') \\
&\leq c_6(n - n^{1-\frac{1}{7t}}) + c_6\left(n^{5/6} - (\alpha - 1)n^{1-\frac{1}{7t}}\right) + 2c_5 n^{1-\frac{1}{4t}} \\
&\leq c_6(n - n^{1-\frac{1}{7t}}) + c_6\left(n^{5/6} + n^{1-\frac{1}{4t}} - (\alpha - 1)n^{1-\frac{1}{7t}}\right) \\
&\leq c_6(n - n^{1-\frac{1}{7t}}), \tag{9}
\end{aligned}
$$

where the last inequality holds for $n \geq n_0$ if n_0 is sufficiently large (independent of c_6). This completes the induction step, hence the proof of Theorem 6.   □

3   Application: The Tangled-Thrackle Conjecture

Let G be tangled-thrackle with n vertices. By slightly modifying the edges (i.e., Jordan arcs) near the points of tangencies, we obtain a simple topological graph $\tilde{G}$ with the same number of vertices and edges such that every pair of tangent edges in G become disjoint in $\tilde{G}$ and all other intersection points between edges remain the same. In order to show that $|E(G)| \leq O(n)$, invoking Theorem 1, it suffices to prove the following.

Lemma 3. *For every tangled-thrackle G, the simple topological graph $\tilde{G}$ does not contain a set of 200 edges all disjoint from another set of 200 edges.*

Before proving the lemma, we briefly review the concept of Devenport-Schinzel sequences and arrangements of pseudo-segments.

A finite sequence $U = (u_1, \ldots, u_t)$ of symbols over a finite alphabet is a *Davenport-Schinzel sequence of order s* if it satisfies the following two properties:

- no two consecutive symbols in the sequence are equal to each other;
- for any two distinct letters of the alphabet, a and b, the sequence does not contain a (not necessarily consecutive) subsequence $(a, b, a, \ldots, b, a)$ consisting of $s + 2$ symbols alternating between a and b.

The maximum length of a Davenport-Schinzel sequence of order s over an alphabet of size n is denoted $\lambda_s(n)$. Sharp asymptotic bounds for $\lambda_s(n)$ were obtained by Nivasch [6] and Pettie [13]. However, to avoid the constants hidden in the big-Oh notation, we use simpler explicit bounds. Specifically, we use the following upper bound for $\lambda_3(n)$.

Theorem 7 (see Proposition 7.1.1 in [5]). $\lambda_3(n) < 2n \ln n + 3n$.

A set $\mathcal{L}$ of m Jordan arcs in the plane is called an *arrangement of pseudo-segment* if each pair of arcs intersects in at most one point (at an endpoint, a crossing, or a point of tangency), and no three arcs have a common interior point. An arrangement of pseudo-segments naturally defines a plane graph: The *vertices* of the arrangement are the endpoints and the intersection points of the Jordan arcs, and the *edges* are the portions of the Jordan arcs between consecutive vertices. The *faces* of the arrangement are the connected components of the complement of the union of the Jordan arcs. The vertices and edges are said to be *incident to* a face if they are contained in the (topological) closure of that face. The following theorem is a particular case of Theorem 5.3 of [14].

Theorem 8 (see [14]). *Let $\mathcal{L}$ be an arrangement of m pseudo-segments and F be a face of $\mathcal{L}$. Then number of edges incident to F is at most $\lambda_3(2m)$.*

Lemma 4. *Let $\mathcal{L}_1 \cup \mathcal{L}_2$ be an arrangement of pseudo-segments such that every arc in $\mathcal{L}_1$ is tangent to all arcs in $\mathcal{L}_2$; and $\mathcal{L}_1$ and $\mathcal{L}_2$ each form a connected arrangement. Then $\mathcal{L}_1$ or $\mathcal{L}_2$ contains at most 200 arcs.*

Proof: Suppose to the contrary, that both $\mathcal{L}_1$ and $\mathcal{L}_2$ contain at least 200 arcs. Without loss of generality, we may assume $|\mathcal{L}_1| = |\mathcal{L}_2| = 200$. Since no arc in $\mathcal{L}_1$ crosses any arc in $\mathcal{L}_2$, the arrangement $\mathcal{L}_1$ lies in the closure of a single face F_2 of the arrangement $\mathcal{L}_2$, and vice versa $\mathcal{L}_2$ lies in the closure of a single face F_1 of the arrangement $\mathcal{L}_1$. We construct a plane graph H representing the tangencies between the edges of the two arrangements: place a vertex on the relative interior of each edge of $\mathcal{L}_1$ incident to F_1 and each edge of $\mathcal{L}_2$ incident to F_2. Join two vertices, v and u, by an edge iff their corresponding edges of the arrangements, e_u and e_v, are tangent to each other. To see that H is indeed planar, note that each edge uv can be drawn closely following the arcs e_u and e_v to their intersection point in such a way that H has no crossings. As every arc in $\mathcal{L}_1$ is tangent to all arcs in $\mathcal{L}_2$, the graph H has exactly 200^2 edges. By Theorem 8, H has at most $2\lambda_3(400)$ vertices. However,

$$\frac{|E(H)|}{|V(H)|} \geq \frac{200^2}{2\lambda_3(400)} > \frac{200^2}{4 \cdot 400 \ln 400 + 6 \cdot 400} > 3.3,$$

which contradicts Euler's formula.   □

Note that Lemma 4 easily generalizes to the case when the arrangements $\mathcal{L}_1$ and $\mathcal{L}_2$ are not necessarily connected but they have the property that every pseudo-segment in $\mathcal{L}_1$ lies in the same face of $\mathcal{L}_2$, and vice versa.

It is now easy to see that Lemma 3 follows directly from Lemma 4: if 200 edges of the simple topological graph $\tilde{G}$ are disjoint from another set of 200 edges of $\tilde{G}$, then a set of the corresponding 200 edges of the tangled-thrackle G are tangent to corresponding other set of 200 edges of G.

Proof of Theorem 2: The statement follows by combining Theorem 1 and Lemma 3.□

We now show an analogue of Lemma 4 where we drop the condition that the arrangements $\mathcal{L}_1$ and $\mathcal{L}_2$ are connected. We find this interesting for its own sake.

Proposition 1. *Let $\mathcal{L}_1 \cup \mathcal{L}_2$ be an arrangements of pseudo-segments such that every arc in $\mathcal{L}_1$ is tangent to all arcs in $\mathcal{L}_2$. Then $\mathcal{L}_1$ or $\mathcal{L}_2$ contains at most 400 arcs.*

Proof: Suppose to the contrary, that both $\mathcal{L}_1$ and $\mathcal{L}_2$ contain at least 400 arcs. Without loss of generality, we may assume $|\mathcal{L}_1| = |\mathcal{L}_2| = 400$. The difference from Lemma 3 is that the arrangement $\mathcal{L}_1$ or $\mathcal{L}_2$ may not be connected, and so the arcs in $\mathcal{L}_1$ could be distributed in several faces of the arrangement $\mathcal{L}_2$.

Consider an arc $\ell \in \mathcal{L}_1$, and denote by s the number of other arcs in $\mathcal{L}_1$ that intersect ℓ, where $0 \leq s \leq 399$. Then ℓ contains precisely $s + 1$ edges of the arrangement $\mathcal{L}_1$. Partition ℓ into two Jordan arcs: $\ell_a \subset \ell$ consists of the first $\lfloor (s + 1)/2 \rfloor$ edges along ℓ, and $\ell_b = \ell \backslash \ell_a$. Recall that ℓ is tangent to all 400 arcs in $\mathcal{L}_2$. By the pigeonhole principle, we may assume w.l.o.g. that ℓ_a is tangent to at least 200 arcs in $\mathcal{L}_2$. Let $\mathcal{L}'_2 \subset \mathcal{L}_2$ be a set of 200 arcs in $\mathcal{L}_2$ that are tangent to ℓ_1. By construction, ℓ_a intersects at most $\lceil s/2 \rceil \leq 200$ arcs in $\mathcal{L}_1$. Consequently, there is a set $\mathcal{L}'_1 \subset \mathcal{L}_1$ of 200 arcs in $\mathcal{L}_1$ that do *not* intersect ℓ_a. Observe that ℓ_a lies in a single face of the arrangement $\mathcal{L}'_1$. Since every arc in $\mathcal{L}'_2$ intersects ℓ_a, all arcs in $\mathcal{L}'_2$ lie in the same face of the arrangement $\mathcal{L}'_1$.

We have found subsets $\mathcal{L}'_1 \subset \mathcal{L}_1$ and $\mathcal{L}'_2 \subset \mathcal{L}_2$ of size $|\mathcal{L}'_1| = |\mathcal{L}'_2| = 200$ such that every arc in $\mathcal{L}'_1$ is tangent to all arcs in $\mathcal{L}'_2$; and all arcs of $\mathcal{L}'_1$ lie in the same face of $\mathcal{L}'_2$ and vice versa. This contradicts Lemma 4 and the remark following its proof.   □

4   Concluding Remarks

1. We showed that for every integer t, the maximum number of edges in a simple topological graph with n vertices that does not contain t edges all disjoint from another set of t edges is cn, where $c = c(t)$. A careful analysis of the proof shows that $c = 2^{O(t \log t)}$. It would be interesting to see if one could improve the upper bound on c to $O(t)$.

2. We suspect that the bounds of 200 and 400 in Lemma 4 and Proposition 1 are not optimal. Since any constant bound yields a linear upper bound for the number of edges in tangled-thrackles, we have not optimized these values. However, finding the best possible constants, or shorter proofs for some arbitrary constant bounds, would be of interest.

Acknowledgments. We would like to thank Gábor Tardos for some suggestions on how to simplify the main proof.

References

1. Brass, P., Moser, W., Pach, J.: Research Problems in Discrete Geometry. Springer, New York (2005)
2. Fulek, R., Pach, J.: A computational approach to Conway's thrackle conjecture. Computational Geometry: Theory and Applications 44, 345–355 (2011)
3. Fulek, R., Ruiz-Vargas, A.J.: Topological graphs: empty triangles and disjoint matchings. In: Proc. Symposium on Computational Geometry, pp. 259–266. ACM Press (2013)
4. Lovász, L., Pach, J., Szegedy, M.: On Convway's trackle conjecture. Discrete & Compuational Geometry 18, 369–376 (1998)
5. Matousek, J.: Lectures on Discrete Geometry. Graduate Texts in Mathematics, vol. 212. Springer (2002)
6. Nivasch, G.: Improved bounds and new techniques for Davenport–Schinzel sequences and their generalizations. JACM 57(3), article 17 (2010)
7. Pach, J.: Geometric graph theory. In: Goodman, J., O'Rourke, J. (eds.) Handbook of Discrete and Computational Geometry, 2nd edn., ch. 10. CRC Press, Boca Rotan (2007)
8. Pach, J., Agarwal, P.K.: Combinatorial Geometry. Wiley, New York (1995)
9. Pach, J., Radoičić, R., Tóth, G.: Tangled thrackles. In: Márquez, A., Ramos, P., Urrutia, J. (eds.) EGC 2011. LNCS, vol. 7579, pp. 45–53. Springer, Heidelberg (2012)
10. Pach, J., Tóth, G.: Disjoint edges in topological graphs. In: Akiyama, J., Baskoro, E.T., Kano, M. (eds.) IJCCGGT 2003. LNCS, vol. 3330, pp. 133–140. Springer, Heidelberg (2005)
11. Pach, J., Sterling, E.: Conway's conjecture for monotone thrackles. American Mathematical Monthly 118(6), 544–548 (2011)
12. Pach, J., Tóth, G.: Which crossing number is it anyway? J. Comb. Theory Ser. B 80, 225–246 (2000)
13. Pettie, S.: Sharp bounds on Davenport-Schinzel sequences of every order. In: Proc. Symposium on Computational Geometry, pp. 319–328. ACM Press (2013)
14. Sharir, M., Agarwal, P.K.: Devenport-Schinzel Sequences and their Geometric Applications. Cambridge University Press (1995)
15. Suk, A.: Density theorems for intersection graphs of t-monotone curves. SIAM Journal on Discrete Mathematics 27, 1323–1334 (2013)
16. Suk, A.: Disjonit edges in complete toplogical graphs. Discrete & Computational Geometry 49, 280–286 (2013)

Morphing Schnyder Drawings
of Planar Triangulations

Fidel Barrera-Cruz[1], Penny Haxell[1], and Anna Lubiw[1]

University of Waterloo, Waterloo, Canada
{fbarrera,pehaxell,alubiw}@uwaterloo.ca

Abstract. We consider the problem of morphing between two planar drawings of the same triangulated graph, maintaining straight-line planarity. A paper in SODA 2013 gave a morph that consists of $O(n^2)$ steps where each step is a linear morph that moves each of the n vertices in a straight line at uniform speed [1]. However, their method imitates edge contractions so the grid size of the intermediate drawings is not bounded and the morphs are not good for visualization purposes. Using Schnyder embeddings, we are able to morph in $O(n^2)$ linear morphing steps and improve the grid size to $O(n) \times O(n)$ for a significant class of drawings of triangulations, namely the class of weighted Schnyder drawings. The morphs are visually attractive. Our method involves implementing the basic "flip" operations of Schnyder woods as linear morphs.

Keywords: algorithms, computational geometry, graph theory.

1 Introduction

Given a triangulation on n vertices and two straight-line planar drawings of it, Γ and Γ', that have the same unbounded face, it is possible to morph from Γ to Γ' while preserving straight-line planarity. This was proved by Cairns in 1944 [6]. Cairns's proof is algorithmic but requires exponentially many steps, where each step is a *linear morph* that moves every vertex in a straight line at uniform speed. Floater and Gotsman [13] gave a polynomial time algorithm using Tutte's graph drawing algorithm [19], but their morph is not composed of linear morphs so the trajectories of the vertices are more complicated, and there are no guarantees on how close vertices and edges may become. Recently, Alamdari et al. [1] gave a polynomial time algorithm based on Cairns's approach that uses $O(n^2)$ linear morphs, and this has now been improved to $O(n)$ by Angelini et al. [2]. The main idea is to contract (or almost contract) edges. With this approach, perturbing vertices to prevent coincidence is already challenging, and perturbing to keep them on a nice grid seems impossible.

In this paper we propose a new approach to morphing based on Schnyder drawings. We give a planarity-preserving morph that is composed of $O(n^2)$ linear morphs and for which the vertices of each of the $O(n^2)$ intermediate drawings are on a $6n \times 6n$ grid. Our algorithm works for *weighted Schnyder drawings* which are obtained from a Schnyder wood together with an assignment of positive weights

C. Duncan and A. Symvonis (Eds.): GD 2014, LNCS 8871, pp. 294–305, 2014.
© Springer-Verlag Berlin Heidelberg 2014

to the interior faces. A Schnyder wood is a special type of partition (colouring) and orientation of the edges of a planar triangulation into three rooted directed trees. Schnyder [15,16] used them to obtain straight-line planar drawings of triangulations in an $O(n) \times O(n)$ grid. To do this he defined barycentric coordinates for each vertex in terms of the number of faces in certain regions of the Schnyder wood. Dhandapani [7] noted that assigning any positive weights to the faces still gives straight-line planar drawings. We call these *weighted Schnyder drawings*— they are the drawings on which our morphing algorithm works.

Two weighted Schnyder drawings may differ in weights and in the Schnyder wood. We address these separately: we show that changing weights corresponds to a single planar linear morph; altering the Schnyder wood is more significant.

The set of Schnyder woods of a given planar triangulation forms a distributive lattice [5], [11], [14] possibly of exponential size [12]. The basic operation for traversing this lattice is a "flip" that reverses a cyclically oriented triangle and changes colours appropriately. It is known that the flip distance between two Schnyder woods in the lattice is $O(n^2)$ (see Section 2). Therefore, to morph between two Schnyder drawings in $O(n^2)$ steps, it suffices to show how a flip can be realized via a constant number of planar linear morphs. We show that flipping a facial triangle corresponds to a single planar linear morph, and that a flip of a separating triangle can be realized by three planar linear morphs.

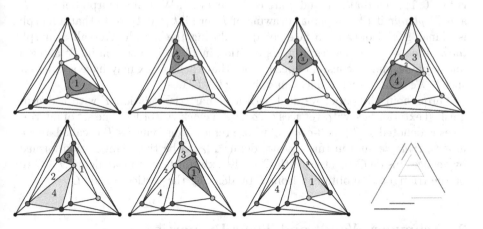

Fig. 1. A sequence of triangle flips, counterclockwise along the top row and clockwise along the bottom row. In each drawing the triangle to be flipped is darkly shaded, and the one most recently flipped is lightly shaded. The linear morph from each drawing to the next one is planar. Vertex trajectories are shown bottom right.

There is hope that our method will give good visualizations for morphing. See Figure 1. The edge-contraction method of Alamdari et al. [1] is not good for visualization purposes—at the end of the recursion, the whole graph has contracted to a triangle. The method of Floater and Gotsman [13] gives good visualizations, based on experiments and heuristic improvements developed by

Shurazhsky and Gotsman [17]. However, their method suffers the same draw-backs as Tutte's graph drawing method, namely that vertices and edges may come very close together. Our intermediate drawings lie on a $6n \times 6n$ grid where vertices are at least distance 1 apart and face areas are at least $\frac{1}{2}$.

Not all straight-line planar triangulations are weighted Schnyder drawings, but we can recognize those that are in polynomial time. The problem of extending our result to all straight-line planar triangulations remains open. There is partial progress in the first author's thesis [3].

This paper is structured as follows. Section 2 contains the relevant background on Schnyder woods. Section 3 contains the precise statement of our main result, and the general outline of the proof. In Section 4 we show that changing face weights corresponds to a linear morph. Flips of facial triangles are handled in Section 5 and flips of separating triangles are handled in Section 6. In Section 7 we explore which drawings are weighted Schnyder drawings.

1.1 Definitions and Notation

Consider two drawings Γ and Γ' of a planar triangulation T. A *morph* between Γ and Γ' is a continuous family of drawings of T, $\{\Gamma^t\}_{t \in [0,1]}$, such that $\Gamma^0 = \Gamma$ and $\Gamma^1 = \Gamma'$. We say a face xyz *collapses* during the morph $\{\Gamma^t\}_{t \in [0,1]}$ if there is $t \in (0,1)$ such that x, y and z are collinear in Γ^t. We call a morph between Γ and Γ' *planar* if Γ^t is a planar drawing of T for all $t \in [0,1]$. Note that a morph is planar if and only if no face collapses during the morph. We call a morph *linear* if each vertex moves from its position in Γ^0 to its position in Γ^1 along a line segment and at constant speed. Note that each vertex may have a different speed. We denote such a linear morph by $\langle \Gamma^0, \Gamma^1 \rangle$.

Throughout the paper we deal with a planar triangulation T with a distinguished exterior face with vertices a_1, a_2, a_3 in clockwise order. The set of interior faces is denoted $\mathcal{F}(T)$. A 3-cycle C whose removal disconnects T is called a *separating triangle*, and in this case we define $T|_C$ to be the triangulation formed by vertices inside C together with C as the exterior face, and we define $T \setminus C$ to be the triangulation obtained from T by deleting the vertices inside C.

2 Schnyder Woods and Their Properties

A *Schnyder wood* of a planar triangulation T with exterior vertices a_1, a_2, a_3 is an assignment of directions and colours 1, 2, and 3 to the interior edges of T such that the following two conditions hold.

(D1) Each interior vertex has three outgoing edges and they have colours 1, 2, 3 in clockwise order. All incoming edges in colour i appear between the two outgoing edges of colours $i - 1$ and $i + 1$ (index arithmetic modulo 3).

(D2) At the exterior vertex a_i, all the interior edges are incoming and of colour i.

The following basic concepts and properties are due to Schnyder [16]. For any Schnyder wood the edges of colour i form a tree T_i rooted at a_i. The path from internal vertex v to a_i in T_i is denoted $P_i(v)$.

(P1) If T_i^- denotes the tree in colour i with all arcs reversed, then $T_{i-1}^- \cup T_i \cup T_{i+1}^-$ contains no directed cycle. In particular, any two outgoing paths from a vertex v have no vertex in common, except for v, i.e., $P_i(v) \cap P_j(v) = \{v\}$ for $i \neq j$.

The *descendants* of vertex v in T_i, denoted $D_i(v)$, are the vertices that have paths to v in T_i. For any interior vertex v the three paths $P_i(v), i = 1, 2, 3$ partition the triangulation into three regions $R_i(v), i = 1, 2, 3$, where $R_i(v)$ is bounded by $P_{i+1}(v), P_{i-1}(v)$ and $a_{i+1}a_{i-1}$. Schnyder proved that every triangulation T has a Schnyder wood and that a planar drawing of T can be obtained from coordinates that count faces inside regions:

Theorem 1 (Schnyder [15,16]). *Let T be a planar triangulation on n vertices equipped with a Schnyder wood S. Consider the map $f : V(T) \rightarrow \mathbb{R}^3$, where $f(a_i) = (2n-5)e_i$, where e_i denotes the i-th standard basis vector in $\mathbb{R}^3$, and for each interior vertex v, $f(v) = (v_1, v_2, v_3)$, where v_i denotes the number of faces contained inside region $R_i(v)$. Then f defines a straight-line planar drawing.*

Dhandapani [7] noted that the above result generalizes to weighted faces. A *weight distribution* $\mathbf{w}$ is a function that assigns a positive weight to each internal face such that the weights sum to $2n - 5$. For any weight distribution, the above result still holds if v_i is defined as:

$$v_i = \sum \{\mathbf{w}(f) : f \in R_i(v)\}. \tag{1}$$

We call the resulting straight-line planar drawing the *weighted Schnyder drawing* obtained from $\mathbf{w}$ and S.

We now describe results of Brehm [5], Ossona De Mendez [14], and Felsner [11] on the flip operation that can be used to convert any Schnyder wood to any other. Let S be a Schnyder wood of planar triangulation T. A flip operates on a cyclically oriented triangle C of T. We use the following properties of such a triangle (proofs in the long version).

(S1) The triangle C has an edge of each colour in S. Furthermore, if C is oriented counterclockwise then the edges along C have colours $i, i-1, i+1$.

(S2) If C is a separating triangle, then the restriction of S to the interior edges of $T|_C$ is a Schnyder wood of $T|_C$.

Let $C = xyz$ be oriented counterclockwise with edges xy, yz, zx of colour $1, 3, 2$ respectively. A *clockwise flip* of C alters the colours and orientations of S as follows:

1. Edges on the cycle are reversed and colours change from i to $i - 1$. See triangle xyz in Figure 3.

2. Any interior edge of $T|_C$ remains with the same orientation and changes colour from i to $i + 1$. See edges incident to b in Figure 4.

Other edges are unchanged. The reverse operation is a *counterclockwise flip*, which Brehm calls a *flop*. Brehm [5, p. 44] proves that a flip yields another Schnyder wood. Consider the graph with a vertex for each Schnyder wood of T and a directed edge (S, S') when S' can be obtained from S by a clockwise flip. This graph forms a distributive lattice [5], [11], [14]. Ignoring edge directions, the distance between two Schnyder woods in this graph is called their *flip distance*.

Lemma 2 (Brehm (see the long version)). *In a planar triangulation on n vertices the flip distance between any two Schnyder woods is $O(n^2)$, and a flip sequence of that length can be found in linear time per flip.*

3   Main Result

Theorem 3. *Let T be a planar triangulation and let S and S' be two Schnyder woods of T . Let Γ and Γ' be weighted Schnyder drawings of T obtained from S and S' together with some weight distributions. There exists a sequence of straight-line planar drawings of T $\Gamma = \Gamma_0, \ldots, \Gamma_{k+1} = \Gamma'$ such that k is $O(n^2)$, the linear morph $\langle \Gamma_i, \Gamma_{i+1} \rangle$ is planar, $0 \leq i \leq k$, and the vertices of each Γ_i, $1 \leq i \leq k$, lie in a $(6n - 15) \times (6n - 15)$ grid. Furthermore, these drawings can be obtained in polynomial time.*

We now describe how the results in the upcoming sections prove the theorem. Lemma 4 (Section 4) proves that if we perform a linear morph between two weighted Schnyder drawings that differ only in their weight distribution then planarity is preserved. Thus, we may take Γ_1 and Γ_k to be the drawings obtained from the uniform weight distribution on S and S' respectively. By Schnyder's Theorem 1 these drawings lie on a $(2n - 5) \times (2n - 5)$ grid and we can scale them up to our larger grid. By Lemma 2 (Section 2) there is a sequence of k flips, $k \in O(n^2)$, that converts S to S'. Therefore it suffices to show that each flip in the sequence can be realized via a planar morph composed of a constant number of linear morphs. In Theorem 7 (Section 5) we prove that if we perform a linear morph between two weighted Schnyder drawings that differ only by a flip of a face then planarity is preserved. In Theorem 11 (Section 6) we prove that if two Schnyder drawings with the same uniform weight distribution differ by a flip of a separating triangle then there is a planar morph between them composed of three linear morphs. The intermediate drawings involve altered weight distributions (here Lemma 4 is used again), and lie on a grid of the required size. Putting these results together gives the final sequence $\Gamma_0, \ldots, \Gamma_{k+1}$. All the intermediate drawings lie in a $(6n - 15) \times (6n - 15)$ grid and each of them can be obtained in $O(n)$ time from the previous one. This completes the proof of Theorem 3 modulo the proofs in the following sections.

4   Morphing to Change Weight Distributions

Lemma 4. *(proof in the long version) Let T be a planar triangulation and let S be a Schnyder wood of T. Consider two weight distributions $\mathbf{w}$ and $\mathbf{w}'$ on the faces of T, and denote by Γ and Γ' the weighted Schnyder drawings of T obtained from $\mathbf{w}$ and $\mathbf{w}'$ respectively. Then the linear morph $\langle \Gamma, \Gamma' \rangle$ is planar.*

5   Morphing to Flip a Facial Triangle

In this section we prove that the linear morph from one Schnyder drawing to another one, obtained by flipping a cyclically oriented face and keeping the same weight distribution, preserves planarity. See Figure 2. We begin by showing how the regions for each vertex change during such a flip and then we use this to show how the coordinates change.

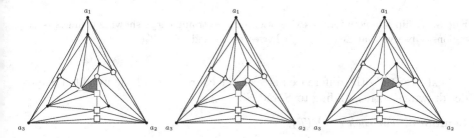

Fig. 2. Snapshots from a linear morph defined by a flip of the shaded face at times $t = 0$, $t = 0.5$ and $t = 1$. The trajectory of rectangular shaped vertices is parallel to $a_2 a_3$. Similar properties hold for triangular and pentagonal shaped vertices.

Let S and S' be Schnyder woods of triangulation T that differ by a flip on face xyz oriented counterclockwise in S with (x, y) of colour 1. Let (v_1, v_2, v_3) and (v_1', v_2', v_3') be the coordinates of vertex v in the weighted Schnyder drawings from S and S' respectively with respect to weight distribution $\mathbf{w}$. For an interior edge pq of T, let $\Delta_i(pq)$ be the set of faces in the region bounded by pq and the paths $P_i(p)$ and $P_i(q)$ in S, and we define $\delta_i(pq)$ to be the weight of that region, i.e., $\delta_i(pq) = \sum_{f \in \Delta_i(pq)} \mathbf{w}(f)$. We use notation $P_i(v), R_i(v)$, and $D_i(v)$ as defined in Section 2 and $\Delta_i(pq)$ as above and add primes to denote the corresponding structures in S'. Let us begin by identifying properties of S and S'. The following two lemmas are proved formally in the long version.

Lemma 5. *The following conditions hold (see Figure 3):*

1. $R_1(x) = R_1'(x)$, $R_3(y) = R_3'(y)$ and $R_2(z) = R_2'(z)$.
2. $R_2'(x) = R_2(x) \backslash (\Delta_1(yz) \cup \{f\})$, $R_3'(x) = R_3(x) \cup (\Delta_1(yz) \cup \{f\})$ *and similarly for y and z.*

3. $D_1(x) = D_1'(x)$, $D_2(z) = D_2'(z)$ and $D_3(y) = D_3'(y)$.
4. The interiors of $R_1(x)$, $R_2(z)$ and $R_3(y)$ are pairwise disjoint.
5. $D_1(x) \setminus \{x\}$ is contained in the interior of $R_1(x)$ and similarly for y and z. Consequently $D_1(x)$, $D_2(z)$ and $D_3(y)$ are pairwise disjoint.

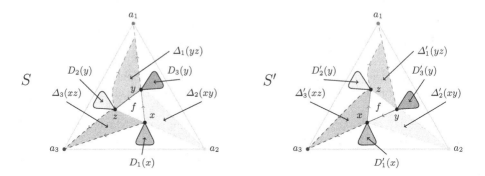

Fig. 3. A flip of a counterclockwise oriented face triangle xyz showing changes to the regions. Observe that $\Delta_1(yz) \cup \{f\}$ leaves $R_2(x)$ and joins $R_3'(x)$.

Next we study the difference between the coordinates of the weighted Schnyder drawings corresponding to S and S'.

Lemma 6. For each $v \in V(T)$,

$$(v_1', v_2', v_3') = \begin{cases} (v_1, v_2, v_3) & \text{if } v \notin D_1(x) \cup D_2(z) \cup D_3(y) \\ (v_1, v_2 - (\delta_1(yz) + \mathbf{w}(f)), v_3 + \delta_1(yz) + \mathbf{w}(f)) & \text{if } v \in D_1(x) \\ (v_1 + \delta_2(xy) + \mathbf{w}(f), v_2, v_3 - (\delta_2(xy) + \mathbf{w}(f))) & \text{if } v \in D_2(z) \\ (v_1 - (\delta_3(xz) + \mathbf{w}(f)), v_2 + \delta_3(xz) + \mathbf{w}(f), v_3) & \text{if } v \in D_3(y). \end{cases}$$

We are ready to prove the main result of this section. We express it in terms of a general weight distribution since we will need that in the next section.

Theorem 7. Let S be a Schnyder wood of a planar triangulation T that contains a face f bounded by a counterclockwise directed triangle xyz, and let S' be the Schnyder wood obtained from S by flipping f. Denote by Γ and Γ' the weighted Schnyder drawings obtained from S and S' respectively with weight distribution $\mathbf{w}$. Then $\langle \Gamma, \Gamma' \rangle$ is a planar morph.

Proof. If a triangle collapses during the morph, then it must be incident to at least one vertex that moves, i.e., one of $D_1(x)$, $D_2(y)$ or $D_3(z)$. By Lemma 5, apart from x, y, z these vertex sets lie in the interiors of regions $R_1(x)$, $R_2(y)$, $R_3(z)$ respectively. Thus it suffices to show that no triangle in one of these regions collapses, and that no triangle incident to x, y or z collapses.

Let t be a triangle such that $t \in R_1(x)$. (The argument for triangles in other regions is similar.) Any vertex of $R_1(x)$ that moves is in $D_1(x)$ and by Lemma 6 these vertices are all translated by the same amount. We argue that if triangle b, c, e in clockwise order collapses as we translate a subset of its vertices then the end result is triangle b, c, e in counterclockwise order. This contradicts the fact that Γ and Γ' have the same faces. A rigorous proof is in the long version. The same argument applies to a triangle in $\Delta_3(xz) \cup \Delta_2(xy)$ that is incident to x but not incident to either y or z.

It remains to prove that no triangle t incident to at least two vertices of x, y and z collapses. Here we only consider the case where $t = xyz$, the other case can be handled similarly. We will show that x never lies on the line segment yz during the morph. (The other two cases are similar.) Since (x, y) has colour 1 in S, it follows that $x \in R_1(y)$. Similarly, since (z, x) has colour 2 in S, we have that $x \in R_1(z)$. Therefore $x_1 < y_1, z_1$. Using a similar argument on S' we obtain that $x'_1 < y'_1, z'_1$. Finally, note that $x_1 = x'_1$. This implies that x never lies on the line segment yz during the morph. $\qquad\square$

6 Morphing to Flip a Separating Triangle

In this section we prove that there is a planar morph between any two weighted Schnyder drawings that differ by a separating triangle flip. Our morph will be composed of three linear morphs. Throughout this section we let S and S' be Schnyder woods of a planar triangulation T such that S' is obtained from S after flipping a counterclockwise oriented separating triangle $C = xyz$, with (x, y) coloured 1 in S. Let Γ and Γ' be two weighted Schnyder drawings obtained from S and S' respectively with weight distribution $\mathbf{w}$. For the main result of the section, it suffices to consider a uniform weight distribution because we can get to it via a single planar linear morph, as shown in Section 4. However, for the intermediate results of the section we need more general weight distributions.

We now give an outline of the strategy we follow. Morphing linearly from Γ to Γ' may cause faces inside C to collapse. An example is provided in the long version. However, we can show that there is a "nice" weight distribution that prevents this from happening. Our plan, therefore, is to morph linearly from Γ to a drawing $\overline{\Gamma}$ with a nice weight distribution, then morph linearly to drawing $\overline{\Gamma}'$ to effect the separating triangle flip. A final change of weights back to the uniform distribution gives a linear morph from $\overline{\Gamma}'$ to Γ'.

This section is structured as follows. First we study how the coordinates change between Γ and Γ'. Next we show that faces strictly interior to $T|_C$ do not collapse during a linear morph between $\overline{\Gamma}$ and $\overline{\Gamma}'$. We then give a similar result for faces of $T|_C$ that share a vertex or edge with C provided that the weight distribution satisfies certain properties. Finally we prove the main result by showing that there is a weight distribution with the required properties.

Let us begin by examining the coordinates of vertices. For vertex $b \in V(T)$ let (b_1, b_2, b_3) and (b'_1, b'_2, b'_3) denote its coordinates in Γ and Γ' respectively. For b an interior vertex of $T|_C$ let β_i be the i-th coordinate of b in $T|_C$ when considering the restriction of S to $T|_C$ with weight distribution $\mathbf{w}$. By analyzing

Figure 4, we can see that the coordinates for b in Γ are

$$
\begin{aligned}
(b_1, b_2, b_3) &= (x_1 + \delta_3(xz) + \beta_1, z_2 + \delta_1(yz) + \beta_2, y_3 + \delta_2(xy) + \beta_3) \\
&= (x_1, z_2, y_3) + (\delta_3(xz), \delta_1(yz), \delta_2(xy)) + (\beta_1, \beta_2, \beta_3).
\end{aligned}
\tag{2}
$$

We now analyze how the coordinates of vertices change from Γ to Γ'. We use $\mathbf{w}_C$ to denote the weight of faces inside C, i.e., $\mathbf{w}_C = \sum_{f \in \mathcal{F}(T|_C)} \mathbf{w}(f)$.

Lemma 8. *(proof in the long version) For each* $b \in V(T)$,

$$
(b_1', b_2', b_3') = \begin{cases}
(b_1, b_2 - (\delta_1(yz) + \mathbf{w}_C), b_3 + \delta_1(yz) + \mathbf{w}_C) & \text{if } b \in D_1(x) \\
(b_1 + \delta_2(xy) + \mathbf{w}_C, b_2, b_3 - (\delta_2(xy) + \mathbf{w}_C)) & \text{if } b \in D_2(z) \\
(b_1 - (\delta_3(xz) + \mathbf{w}_C), b_2 + \delta_3(xz) + \mathbf{w}_C, b_3) & \text{if } b \in D_3(y) \\
(x_1, z_2, y_3) + (\delta_2(xy), \delta_3(xz), \delta_1(yz)) + (\beta_3, \beta_1, \beta_2) & \text{if } b \in \mathcal{I} \\
(b_1, b_2, b_3) & \text{otherwise}
\end{cases}
$$

where $\mathcal{I}$ *is the set of interior vertices of* $T|_C$.

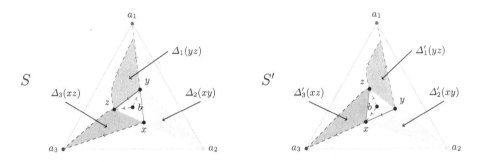

Fig. 4. A flip of a counter-clockwise oriented separating triangle xyz

We now examine what happens during a linear morph from Γ to Γ'. We first deal with faces strictly interior to C. The following two lemmas are proved formally in the long version.

Lemma 9. *For an arbitrary weight distribution no face formed by interior vertices of* $T|_C$ *collapses in the morph* $\langle \Gamma, \Gamma' \rangle$.

Proof sketch. Consider a face inside C formed by internal vertices b, c, e whose coordinates with respect to $T|_C$ are $\beta, \gamma, \varepsilon$, respectively. Examining (2) and Lemma 8 we see that the coordinates of b, c, e in Γ and Γ' depend in exactly the same way on the parameters from $T \setminus C$ and differ only in the parameters $\beta, \gamma, \varepsilon$. Therefore triangle bce collapses during the morph if and only if it collapses during the linear transformation on $\beta, \gamma, \varepsilon$ where we perform a cyclic shift of coordinates, viz., $(\beta_1, \beta_2, \beta_3)$ becomes $(\beta_3, \beta_1, \beta_2)$, etc. No triangle collapses during this transformation because it corresponds to moving each of the three outer vertices x, y, z in a straight line to its clockwise neighbour. $\square$

Next we consider the faces interior to C that share an edge or vertex with C. We show that no such face collapses, provided that the weight distribution $\mathbf{w}$ satisfies $\delta_1 = \delta_2 = \delta_3$ where we use δ_1, δ_2 and δ_3 to denote $\delta_1(yz)$, $\delta_2(xy)$ and $\delta_3(xz)$ respectively.

Lemma 10. *Let $\mathbf{w}$ be a weight distribution for the interior faces of T such that $\delta_1 = \delta_2 = \delta_3$. No interior face of $T|_C$ incident to an exterior vertex of $T|_C$ collapses during $\langle \Gamma, \Gamma' \rangle$.*

Proof sketch. We examine separately the case where the interior face is incident to the edge xy of C and the case where the interior face is only incident to the vertex x of C. The other cases follow by analogous arguments.

Consider the case of an interior face bxy incident to edge xy. Suppose by contradiction that at time $r \in [0,1]$ during the morph the face collapses with b^r lying on segment $x^r y^r$, say $b^r = (1-s)x^r + sy^r$ for some $s \in [0,1]$. We use formula (2) and Lemma 8 to re-write this equation. Some further algebraic manipulations (details in the long version) show that there is no solution for r. The case of a face involving vertex x and two interior vertices is similar. $\square$

We are now ready to prove the main result of this section.

Theorem 11. *Let T be a planar triangulation and let S and S' be two Schnyder woods of T such that S' is obtained from S by flipping a counterclockwise cyclically oriented separating triangle $C = xyz$ in S. Let Γ and Γ' be weighted Schnyder drawings obtained from S and S', respectively, with uniform weight distribution. Then there exist weighted Schnyder drawings $\overline{\Gamma}$ and $\overline{\Gamma}'$ on a $(6n-15) \times (6n-15)$ integer grid such that each of the following linear morphs is planar: $\langle \Gamma, \overline{\Gamma} \rangle$, $\langle \overline{\Gamma}, \overline{\Gamma}' \rangle$, and $\langle \overline{\Gamma}', \Gamma' \rangle$.*

Proof. Our aim is to define the planar drawings $\overline{\Gamma}$ and $\overline{\Gamma}'$. Each one will be realized in a grid that is three times finer than the $(2n-5) \times (2n-5)$ grid, i.e., in a $(6n-15) \times (6n-15)$ grid with weight distributions that sum to $6n-15$. Under this setup, the initial uniform weight distribution $\mathbf{u}$ takes a value of 3 in each interior face.

Drawings $\overline{\Gamma}$ and $\overline{\Gamma}'$ will be the weighted Schnyder drawings obtained from S and S' respectively with a new weight distribution $\overline{\mathbf{w}}$. We use Δ_1, Δ_2 and Δ_3 to denote the regions $\Delta_1(yz)$, $\Delta_2(xy)$ and $\Delta_3(xz)$ respectively, in S. We use δ_i and $\overline{\delta}_i$ to denote the weight of Δ_i, $i = 1, 2, 3$ with respect to the uniform weight distribution and the new weight distribution $\overline{\mathbf{w}}$, respectively.

We will define $\overline{\mathbf{w}}$ so that $\overline{\delta}_1, \overline{\delta}_2$, and $\overline{\delta}_3$ all take on the average value $\delta := (\delta_1 + \delta_2 + \delta_3)/3$. The idea is to remove weight from faces in a region of above-average weight, and add weight to faces in a region of below-average weight. The new face weights must be positive integers. Note first that δ is an integer. Note secondly that $\delta > \delta_i/3$ for any i since the other δ_j's are positive. Thus $\delta_i - \delta < \frac{2}{3}\delta_i$. This means that we can reduce δ_i to the average δ without removing more than 2 weight units from any face (of initial weight 3) in any region. There is more than one solution for $\overline{\mathbf{w}}$, but the morph might look best if $\overline{\mathbf{w}}$ is as uniform as possible. To be more specific, we can define new face weights $\overline{\mathbf{w}}$ via the following algorithm: Initialize $\overline{\mathbf{w}} = \mathbf{w}$. While some $\overline{\delta}_i$ is greater than the average δ, remove 1 from a maximum weight face of Δ_i and add 1 to a minimum weight face in a region Δ_j whose weight is less than the average.

This completes the description of $\overline{\Gamma}$ and $\overline{\Gamma}'$. It remains to show that the three linear morphs are planar. The morphs $\langle \Gamma, \overline{\Gamma} \rangle$ and $\langle \overline{\Gamma}', \Gamma' \rangle$ only involve changes to the weight distribution so they are planar by Lemma 4. Consider the linear morph $\langle \overline{\Gamma}, \overline{\Gamma}' \rangle$. The two drawings differ by a flip of a separating triangle. They have the same weight distribution $\overline{\mathbf{w}}$ which satisfies $\overline{\delta}_1 = \overline{\delta}_2 = \overline{\delta}_3$. By Lemmas 9, and 10 no interior face of $T|_C$ collapses during the morph. By Theorem 7 no face of $T \setminus C$ collapses during the morph. Thus $\langle \Gamma, \Gamma' \rangle$ defines a planar morph. $\square$

7 Identifying Weighted Schnyder Drawings

In this section we give a polynomial time algorithm to test if a given straight-line planar drawing Γ of triangulation T is a weighted Schnyder drawing. The first step is to identify the Schnyder wood. A recent result of Bonichon et al. [4] shows that, given a point set P with triangular convex hull, a Schnyder drawing on P is exactly the "half-Θ_6-graph" of P, which can be computed efficiently. Thus, given drawing Γ, we first ignore the edges and compute the half-Θ_6 graph of the points. If this differs from Γ, we do not have a weighted Schnyder drawing. Otherwise, the half-Θ_6 graph determines the Schnyder wood S. We next find the face weights. We claim that there exists a unique assignment of (not necessarily positive) weights $\mathbf{w}$ on the faces of T such that Γ is precisely the drawing obtained from S and $\mathbf{w}$ as described in (1). Furthermore, $\mathbf{w}$ can be found in polynomial time by solving a system of linear equations in the $2n - 5$ variables $\mathbf{w}(f), f \in \mathcal{F}(T)$. The equations are those from (1). The rows of the coefficient matrix are the characteristic vectors of $R_i(v), i \in \{1, 2, 3\}, v$ an interior vertex of T, and the system of equations has a solution because the matrix has rank $2n-5$. This was proved by Felsner and Zickfeld [10, Theorem 9]. (Note that their theorem is about coplanar orthogonal surfaces; however, their proof considers the exact same set of equations and Claims 1 and 2 give the needed result.)

8 Conclusions and Open Problems

We have made a first step towards morphing straight-line planar graph drawings with a polynomial number of linear morphs and on a well-behaved grid. Our method applies to weighted Schnyder drawings. There is hope of extending to all straight-line planar triangulations. The first author's thesis [3] gives partial progress: an algorithm to morph from any straight-line planar triangulation to a weighted Schnyder drawing in $O(n)$ steps—but not on a nice grid.

It might be possible to extend our results to general (non-triangulated) planar graphs using Felsner's extension [8,9] of Schnyder's results. The problem of efficiently morphing planar graph drawings to preserve convexity of faces is wide open—nothing is known besides Thomassen's existence result [18].

Acknowledgements. We thank Stefan Felsner for discussions, David Eppstein for suggestions, and an anonymous referee for pointing us to the work of Bonichon et al. [4]. F. Barrera-Cruz partially supported by Conacyt. P. Haxell and A. Lubiw partially supported by NSERC.

References

1. Alamdari, S., Angelini, P., Chan, T.M., Di Battista, G., Frati, F., Lubiw, A., Patrignani, M., Roselli, V., Singla, S., Wilkinson, B.T.: Morphing planar graph drawings with a polynomial number of steps. In: Proc. of the Twenty-Fourth Annual ACM-SIAM Symposium on Discrete Algorithms (SODA 2013), pp. 1656–1667. SIAM (2013)
2. Angelini, P., Da Lozzo, G., Di Battista, G., Frati, F., Patrignani, M., Roselli, V.: Morphing planar graph drawings optimally. In: Esparza, J., Fraigniaud, P., Husfeldt, T., Koutsoupias, E. (eds.) ICALP 2014. LNCS, vol. 8572, pp. 126–137. Springer, Heidelberg (2014)
3. Barrera-Cruz, F.: Morphing planar triangulations. Ph.D. thesis, University of Waterloo (2014)
4. Bonichon, N., Gavoille, C., Hanusse, N., Ilcinkas, D.: Connections between theta-graphs, Delaunay triangulations, and orthogonal surfaces. In: Thilikos, D.M. (ed.) WG 2010. LNCS, vol. 6410, pp. 266–278. Springer, Heidelberg (2010)
5. Brehm, E.: 3-orientations and Schnyder 3-tree-decompositions. Master's thesis, FB Mathematik und Informatik, Freie Universität Berlin (2000)
6. Cairns, S.S.: Deformations of plane rectilinear complexes. The American Mathematical Monthly 51(5), 247–252 (1944)
7. Dhandapani, R.: Greedy drawings of triangulations. Discrete & Computational Geometry 43(2), 375–392 (2010)
8. Felsner, S.: Convex drawings of planar graphs and the order dimension of 3-polytopes. Order 18(1), 19–37 (2001)
9. Felsner, S.: Geodesic embeddings and planar graphs. Order 20(2), 135–150 (2003)
10. Felsner, S., Zickfeld, F.: Schnyder woods and orthogonal surfaces. Discrete & Computational Geometry 40(1), 103–126 (2008)
11. Felsner, S.: Lattice structures from planar graphs. The Electronic Journal of Combinatorics 11(1), 15 (2004)
12. Felsner, S., Zickfeld, F.: On the number of α-orientations. In: Brandstädt, A., Kratsch, D., Müller, H. (eds.) WG 2007. LNCS, vol. 4769, pp. 190–201. Springer, Heidelberg (2007)
13. Floater, M.S., Gotsman, C.: How to morph tilings injectively. Journal of Computational and Applied Mathematics 101(1), 117–129 (1999)
14. Ossona de Mendez, P.: Orientations bipolaires. Ph.D. thesis, Ecole des Hautes Etudes en Sciences Sociales, Paris (1994)
15. Schnyder, W.: Planar graphs and poset dimension. Order 5, 323–343 (1989)
16. Schnyder, W.: Embedding planar graphs on the grid. In: Proc. of the First Annual ACM-SIAM Symposium on Discrete Algorithms, SODA 1990, pp. 138–148. SIAM, Philadelphia (1990)
17. Surazhsky, V., Gotsman, C.: Controllable morphing of compatible planar triangulations. ACM Trans. Graph. 20(4), 203–231 (2001)
18. Thomassen, C.: Deformations of plane graphs. Journal of Combinatorial Theory, Series B 34(3), 244–257 (1983)
19. Tutte, W.T.: How to draw a graph. Proc. London Math. Soc. 13(3), 743–768 (1963)

Trade-Offs in Planar Polyline Drawings

Stephane Durocher* and Debajyoti Mondal**

Department of Computer Science, University of Manitoba, Canada
{durocher,jyoti}@cs.umanitoba.ca

Abstract. Angular resolution, area and the number of bends are some important aesthetic criteria of a polyline drawing. Although trade-offs among these criteria have been examined over the past decades, many of these trade-offs are still not known to be optimal. In this paper we give a new technique to compute polyline drawings for planar triangulations. Our algorithm is simple and intuitive, yet implies significant improvement over the known results. We present the first smooth trade-off between the area and angular resolution for 2-bend polyline drawings of any given planar graph. Specifically, for any given n-vertex triangulation, our algorithm computes a drawing with angular resolution $r/d(v)$ at each vertex v, and area $f(n, r)$, for any $r \in (0, 1]$, where $d(v)$ denotes the degree at v. For $r < 0.389$ or $r > 0.5$, $f(n, r)$ is less than the drawing area required by previous algorithms; $f(n, r)$ ranges from $7.12n^2$ when $r \leq 0.3$ to $32.12n^2$ when $r = 1$.

1 Introduction

Polyline drawing is a classic style of drawing planar graphs, which has a wide range of applications in the area of software visualization [8,18] and layout of circuit diagrams [7]. Given an n-vertex planar graph G, a *polyline drawing* Γ of G maps each vertex to a distinct point in $\mathbb{R}^2$, and each edge to a simple polygonal chain between its endpoints such that no two edges intersect except possibly at their common end point. Γ is a *k-bend polyline drawing* if the number of line segments per edge is bounded by at most $k + 1$, i.e., each edge contains at most k bend points. Consequently, a k-bend polyline drawing can be considered as a $(k + \lambda)$-bend drawing for any $\lambda > 0$. Figures 1(a) and (b) illustrate a plane graph G and a 2-bend polyline drawing of G, respectively.

Researchers have examined the theoretical aspects of planar polyline drawings over a long time [2,4,9,10,13,17,20]. *Area* (i.e., the size of the smallest integer grid containing the drawing), *angular resolution* (i.e., the smallest angle formed at any vertex), number of bends per edge, edge separation and bend resolution are some examples of such aesthetic criteria. Even after decades of research effort, finding the optimal trade-off between the number of total bends and area still

---

* Work of the author is supported in part by the Natural Sciences and Engineering Research Council of Canada (NSERC).

** Work of the author is supported in part by a University of Manitoba Graduate Fellowship.

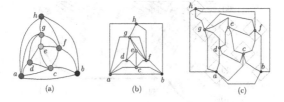

Fig. 1. (a) A planar graph G. (b)–(c) Two polyline drawings of G

seems to be an elusive goal. For example, every planar triangulation with n vertices admits a straight-line drawing (i.e., a 0-bend polyline drawing) in $O(n^2)$ area [9]. Several improvements on the constant hidden in $O(.)$ notation have been achieved [2,4,9,17], and the best known bound is $8n^2/9 = 0.89n^2$ [4]. Better upper bounds, i.e., $4n^2/9 < 0.45n^2$, can be attained if we use 1-bend polyline drawings, which takes at most $2n/3$ bends in total [20]. Although these drawings require small area, the compactness comes at the expense of bad angular resolution, i.e., $\Omega(1/n)$. Garg and Tamassia [19] showed that there exists planar graphs such that any of its straight-line drawing with angular resolution $\Omega(1/\rho)$ requires at least $\Omega(c^{\rho n})$ area, where $c > 1$, which suggests that drawings with angular resolution $\Omega(1/\Delta)$ and polynomial area may exist only if we allow the edges to have bends.

Allowing bends helps both to reduce area and to improve angular resolution, e.g., given an n-vertex planar graph with maximum degree Δ, one can construct a 3-bend polyline drawing with $2/\Delta$ radians of resolution and $3n^2$ area [13]. The angular resolution can be improved to $\Omega(1/d(v))$ radians (for each vertex v) with an expense of higher area [10,12], which also helps to reduce the number of bends per edge. Table 1 presents a brief summary of the related results.

Table 1. Angular resolution, area and total bends in k-bend polyline drawings, where $\alpha \in [1/4, 1/2]$, and $\beta \in [1/3, 1]$.

Graph Class	Area	Resolution	k-Bends	T. Bends	Reference
Maximal Planar	$7n^2/8$	$\Omega(1/n^2)$	0	0	[4]
Maximal Planar	$9n^2/2$	$\Omega(1/n)$	0	0	[16]
Maximal Planar	$12.5n^2$	$0.5/d(v)$	1	$3n$	[10]
Maximal Planar	$450n^2$	$1/d(v)$	1	$3n$	[5]
Maximal Planar	$4n^2/9$	$\Omega(1/n^2)$	1	$2n/3$	[20]
Maximal Planar	$200n^2$	$1/d(v)$	2	$6n$	[12]
Maximal Planar	$(6\alpha + 8/3)^2 n^2$	$\frac{\alpha}{(d(v)(\alpha^2 + 1/4)}$	2	**5.5n**	Theorem 2
Maximal Planar	$(6\beta + 2/3)^2 n^2$	$\frac{\beta}{(d(v)(\beta^2 + 1)}$	2	**5.5n**	Theorem 3
3-connected Planar	$6n^2$	$2/\Delta$	3	$5n - 15$	[15]
General Planar	$3n^2$	$2/\Delta$	3	$5n - 15$	[13]

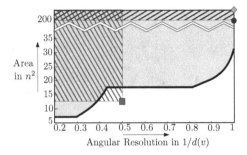

Fig. 2. Trade-off between angular resolution and area for 2-bend polyline drawings, where the bold line denotes the trade-off established in this paper. The square, circle and diamond denote the reference [10], [12] and [5], respectively.

Early polyline drawing algorithms were developed as a generalization of orthogonal drawings [1]. Before Duncan and Kobourov's algorithm [10], all the polyline drawing techniques with good angular resolution and $O(n^2)$ area were based on the idea of assigning an empty square surrounding each vertex (e.g., Figure 1(c)), which forced the constant factor in the $O(.)$ notation to be very large. The algorithm of Duncan and Kobourov [10] finds a drawing with smaller area, but loses the square-emptiness property around the vertices, as well as decreasing the angular resolution by a factor of 2. Observe that two solutions in a multi-objective optimization are comparable if and only if one of them dominates the other with respect to every optimization criteria. Hence although the drawing of [10] has smaller area than that of [5] (see Table 1), it is not an improvement over [5] because of its lower angular resolution.

Contributions. In this paper we examine the trade-offs between the angular resolution and area for 2-bend polyline drawings of planar triangulations. Figure 2 illustrates the solution space dominated by our algorithm in gray, which dominates all the previous 2-bend polyline drawing algorithms except Duncan and Kobourov's algorithm [10], which dominates our algorithm along a small interval of X-axis. Even under the model where each vertex v is surrounded by an empty square of size $d(v) \times d(v)$, we can construct a 2-bend polyline drawing with angular resolution $1/\Delta$ and area $32.12n^2$, where the best known bounds can achieve an $\Omega(1/d(v))$ angular resolution with an area at least $200n^2$ [5,12,14], or an $1/\Delta$ angular resolution with 3 bends per edge [13].

2   Technical Background

Let G be a *plane graph*, i.e., a planar graph with a fixed combinatorial embedding and a specified outerface. If every face of G including (respectively, excluding) the outer face is a cycle of length three, then G is called *triangulated* (respectively, *internally triangulated*). Let G be an n-vertex triangulated plane graph, where v_1, v_2 and v_n are the outer vertices of G in clockwise order, and let $\sigma = (v_1, v_2, ..., v_n)$ be an ordering of all the vertices of G. Then G_k, where $2 \leq k \leq n$, is the subgraph of G induced by $v_1 \cup v_2 \cup ... \cup v_k$, and P_k is the path (while walking clockwise) on the outer face of G_k that starts at v_1 and ends at v_2. The vertex-ordering σ is called a *canonical ordering* [9] with respect to the outer edge (v_1, v_2) if for each k, $3 \leq k \leq n$, the following conditions are satisfied: (a) G_k is 2-connected and internally triangulated. (b) If $k \leq n$, then v_k is an outer vertex of G_k and the neighbors of v_k in G_{k-1} appears consecutively on P_{k-1}. Figures 3(a)–(b) illustrate an example.

For some j, where $3 \leq j \leq n$, let P_j be the path $w_1(= v_1), \ldots, w_l, v_k(= w_{l+1}), w_r, \ldots, w_t(= v_2)$. We call the edges (w_l, v_j) and (v_j, w_r) the *l-edge* and the *r-edge* of v_j, respectively. The other edges incident to v_j in G_j are called the *m-edges* of v_j. For example, in Figure 3(c), the edges (v_6, v_4), (v_6, v_5), and (v_3, v_6) are the l-, r- and m-edges of v_6, respectively. By $d_l(v)$, $d_r(v)$ and $d_m(v)$ we denote the number of l, r and m-edges that are incoming to v, e.g., $d_l(v_6) = 0$, $d_r(v_6) = 1$ and $d_m(v_6) = 1$.

Let E_m be the set of all m-edges in G. Then the graph T_m induced by the edges in E_m is a tree with root v_n. Similarly, the graph T_l induced by all l-edges except (v_1, v_n) is a tree rooted at v_1 (Figure 3(d)), and the graph T_r induced by all r-edges except (v_2, v_n) is a tree rooted at v_2. These three trees form the *Schnyder realizer* [17] of G. A Schnyder realizer is called a *minimum realizer* if all the cyclic inner faces are oriented clockwise. By Δ_0 we denote the number of cyclic inner faces in the minimum realizer [21]. If $\{T_l, T_r, T_m\}$ is a minimum Schnyder realizer of G, then we have $\texttt{leaf}(T_l) + \texttt{leaf}(T_r) + \texttt{leaf}(T_m) = 2n - 5 - \Delta_0$ [3]. Hence we can observe the following property.

Remark 1. *Let $\{T_l, T_r, T_m\}$ be a minimum Schnyder realizer of an n-vertex triangulation. Then $\min\{\texttt{leaf}(T_l) + \texttt{leaf}(T_r), \texttt{leaf}(T_l) + \texttt{leaf}(T_m), \texttt{leaf}(T_r) + \texttt{leaf}(T_m)\} \leq (4n - 2\Delta_0 - 10)/3$.*

A non-root vertex in T_l is called a *primary vertex of T_l* if it is the first child of its parent in the clockwise order. Similarly, a non-root vertex in T_r is a *primary vertex of T_r* if it is the first child of its parent in the anticlockwise order. We now have the following lemma, whose proof is omitted due to space constraints.

Lemma 1. *Let n_l and n_r be the nonprimary vertices in T_l and T_r, respectively. Then $n_l + n_r \leq \texttt{leaf}(T_l) + \texttt{leaf}(T_r)$.*

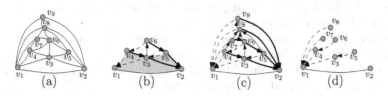

Fig. 3. (a) A canonical ordering of a plane triangulation G. (b) G_6. (c) The l-, r- and m- edges are shown in dashed, bold-solid, and thin-solid edges respectively. (d) T_l.

In a *plus-contact representation* of G, each vertex of G is represented as an axis-aligned plus shape (i.e., a shape consisting of two intersecting line segments) such that two plus shapes touch if and only if their corresponding vertices are adjacent in G [11]. Let Γ be a plus contact representation, and let v be any vertex in Γ. Then by $P(v)$ we denote the plus-shape that corresponds to v in Γ. By the *center* $C(v)$ of $P(v)$, we denote the intersection point of the vertical and horizontal straight line segments of $P(v)$. The four straight line segments that start at $C(v)$ and extend to the left, right, above and below $C(v)$ are the *left, right, up and down hands* of v, which we denote by $L(v), R(v), U(v)$ and $D(v)$, respectively. A *j-shift operation* on Γ with respect to an infinite horizontal line (respectively, vertical line) ℓ is performed as follows: Remove all the edges that are lying completely above (respectively, to the right of) ℓ. Increase the y-coordinate (respectively, x-coordinate) of every vertex lying above (respectively, to the right of) ℓ by j units. Draw the edges that were removed using the new vertex positions. Extend the edges intersected by ℓ upwards (respectively, to the right) until they reach to their other endpoint.

3   Polyline Drawing

Let G be an n-vertex maximal planar graph. We construct the drawing of G in three phases. In the first phase we construct a plus-contact representation of $G \setminus T_m$ on a rectangular grid. In the next phase we expand the drawing by inserting dummy grid lines, and in the third phase we use these grid lines to draw the edges of T_m, and route the l- and r-edges avoiding degeneracy.

Phase 1 (Plus-Contact): Let $\sigma = (v_1, v_2, \ldots, v_n)$ be a canonical ordering of G and let $\{T_l, T_r, T_m\}$ be the corresponding Schnyder realizer. Let Γ_k, where $2 \le k \le n$, be the drawing of all the edges of G_k except the m-edges. We first construct the drawing Γ_2 for G_2, as follows. Place $C(v_1)$ and $C(v_2)$ at coordinates $(1, 2)$ and $(2, 1)$, respectively. Then the horizontal and vertical unit-segments to the left and below $(1, 2)$ correspond to $L(v_1)$ and $D(v_1)$, respectively. Similarly, the horizontal and vertical unit-segments to the left and below $(2, 1)$ correspond to $L(v_2)$ and $D(v_2)$, respectively, as illustrated in Figure 4(b). We now insert the vertices in the canonical ordering maintaining the following invariants. While inserting a new vertex, we only draw the l and r-edges.

$\mathcal{I}_1$. The upper envelope of Γ_i is x-monotone, where the upper envelope is determined by the left and down hands of the vertices in P_i.

$\mathcal{I}_2$. The ray with slope $+1$ starting at any outer vertex of Γ_i can be extended towards infinity avoiding any edge crossing.

$\mathcal{I}_3$. Every l-edge starts as a left hand of some plus shape and ends either at a center or at a down hand of some other plus shape.

$\mathcal{I}_4$. Every r-edge starts as a down hand of some plus shape and ends either at a center or at a left hand of some other plus shape.

Since the upper envelope of G_2 forms a staircase, and does not contain any l- or r-edge, it is straightforward to verify the invariants for Γ_2. We now assume that invariants $\mathcal{I}_1$–$\mathcal{I}_4$ hold for $G_2, G_3, \ldots, G_{k-1}$, where $k-1 < n$, and consider the insertion of vertex v_k.

Insertion of v_k: Let $w_l, w_{l+1}, \ldots, w_{r-1}, w_r$ be the neighbors of v_k on P_{k-1}. Consider an infinite horizontal line ℓ_h that lies in between the horizontal grid line determined by $\mathrm{L}(w_l)$ and the horizontal grid line immediately below $\mathrm{L}(w_l)$. Similarly, let ℓ_v be an infinite vertical line that lies in between the vertical grid line determined by $\mathrm{D}(w_r)$ and the vertical grid line immediately to the left of $\mathrm{D}(w_r)$. We now add v_k considering the following cases. The case when $k = n$ is special, which is handled by Case 4.

Case 1 (v_k is a nonprimary vertex in both T_l and T_r): We first perform a 1-shift with respect to ℓ_h. This increases the number of horizontal lines by 1 and ensures that $\mathrm{D}(w_l)$ contains at least 1 grid point p that does not contain any vertex or contact point. Similarly, we perform a 1-shift with respect to ℓ_v, which increases the number of vertical lines by 1 and ensures that $\mathrm{L}(w_r)$ contains at least 1 grid point q that does not contain any vertex or contact point. We now consider the horizontal ray r_p that starts at p. Since the upper envelope of Γ_{k-1} is x monotone and p does not contain any vertex or contact point, r_p does not intersect Γ_{k-1} except at p. Similarly, we define a vertical ray r_q that starts at q, which does not intersect Γ_{k-1} except at q. We now place v_k at the intersection point of r_p and r_q, and draw the edges (v_k, w_l) and (v_k, w_r). Since r_p and r_q do not intersect Γ_{k-1} except at p and q, respectively, drawing of these edges does not introduce any crossing. Figure 4(c) illustrates such a scenario. We omit the proof that Γ_k respects the invariants $\mathcal{I}_1$–$\mathcal{I}_4$ due to space constraints.

Case 2 (v_k is a primary vertex in T_l but a nonprimary vertex in T_r): In this case we perform a 1-shift with respect to ℓ_v, which increases the number of vertical lines by 1 and ensures that $\mathrm{L}(w_r)$ contains at least 1 grid point q that does not contain any vertex or contact point. Assume that $p = \mathrm{C}(w_l)$. We now consider the horizontal ray r_p that starts at p. Since the upper envelope of Γ_{k-1} is x monotone and p does not contain any vertex or contact point, r_p does not intersect Γ_{k-1} except at p. Similarly, we define a vertical ray r_q starting at q, which does not intersect Γ_{k-1} except at q. We now place v_k at the intersection point of r_p and r_q, and draw the edges (v_k, w_l) and (v_k, w_r). Figure 4(e) illustrates such a scenario.

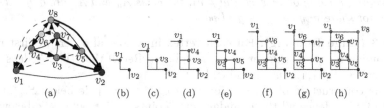

Fig. 4. (a) A plane graph G and a minimum Schnyder realizer of G. (b)–(h) Illustration for the drawing of $G \setminus T_m$.

Case 3 (v_k is a nonprimary vertex in T_l but a primary vertex in T_r):
This case is symmetric to Case 2, i.e., we perform a 1-shift with respect to ℓ_h
to obtain a new grid point p on $D(w_l)$ and assume that $q = C(w_r)$.
Case 4 (v_k is a primary vertex in both T_l and T_r): In this case we do
not perform any shift, and assume that $p = C(w_l)$ and $q = C(w_r)$.

We now have the following lemma whose proof is omitted due to space constraints.

Lemma 2. Γ_n *is a drawing on a* $(W + 2) \times (H + 2)$ *grid, where* $W + H \leq$
$\texttt{leaf}(T_l) + \texttt{leaf}(T_r)$.

Phase 2 (Expansion): For any plus-contact representation on an integer grid,
we define a *free grid line* as a grid line that does not contain any vertex-center
or contact points. We refer the reader to Figure 5.

Consider the horizontal grid lines from top to bottom. For every horizontal
grid line ℓ containing at least one vertex of Γ, we now perform two $\lfloor d(v)/2 \rfloor$-
shifts, where v is the vertex with the largest degree over all the vertices on ℓ. Let
ℓ_h (respectively, ℓ'_h) be an infinite horizontal line that lies in between the horizontal grid line ℓ and the horizontal grid line immediately below (respectively,
above) ℓ. Perform a $\lfloor d(v)/2 \rfloor$-shift with respect to ℓ_h, and then a $\lfloor d(v)/2 \rfloor$-shift
with respect to ℓ'_h. Observe that for each vertex w on ℓ, we now have a set of
$\lfloor d(v)/2 \rfloor$ free grid lines above w and a set of $\lfloor d(v)/2 \rfloor$ free grid lines below w.
We consider a corresponding set S_w that consists of these $2\lfloor d(v)/2 \rfloor$ free grid
lines along with the line ℓ. Furthermore, we assume that the grid lines of S_w are
ordered in the increasing order of y-coordinates. Figure 5(b) illustrates S_{v_4}.

Similarly, we consider the vertical grid lines from right to left, and for every
vertical grid line ℓ' containing at least one vertex of Γ, we perform two $\lfloor d(v)/2 \rfloor$-
shifts to the left and right side of ℓ', where v is the vertex with the largest degree
over all the vertices on ℓ'. We consider a corresponding set S'_w that contains
these $2\lfloor d(v)/2 \rfloor$ free vertical grid lines along with the line ℓ', where the lines are
ordered in the decreasing order of x-coordinates. Let the resulting drawing be Γ'_n,
as shown in Figure 5(c). The following property is a straightforward consequence
of the Expansion phase.

Remark 2. *For every vertex* v *in* Γ'_n, *the point* $C(v)$ *lies at the center of an
integer grid* A_v *of size* $(2\lfloor d(v)/2 \rfloor + 1) \times (2\lfloor d(v)/2 \rfloor + 1)$. *The grid* A_v *does
not contain any vertex, contact point, or edge of* Γ' *except the four hands of* v.
Furthermore, for any other vertex $u(\neq v)$, *the grids* A_u *and* A_v *are disjoint, i.e.,
they do not share any common grid point.*

Phase 3 (Edge Routing): For each vertex in canonical order, we first route
the incoming m-edges incident to v_k, as follows. Recall that the m-edges start
at the vertices $w_{l+1}, \ldots, w_{r-1}$ and ends at v_k.

By the construction of Γ'_n, the vertices $w_{l+1}, \ldots, w_{r-1}$ lie below S_{v_k} and to
the left of S'_{v_k}. Hence all the boundary grid points of A_{v_k}, which lie below S_{v_k}
and to the left of S'_{v_k}, are visible from the top-right corner c_{w_j} of A_{w_j}, for all
$l + 1 \leq j \leq r - 1$. Assume that $z = \lceil d_m(v)/2 \rceil$. Let M be the monotone chain

determined by the last line of S_w and first line of S'_w, where $w \in \{w_{l+1}, \ldots w_{r-1}\}$. Figure 5(d) illustrates M with a dotted line. For each $w \in \{w_{l+1}, \ldots w_z\}$, we now route the m-edge incident to w through the top-right corner c_w upto M, and then to a distinct grid point on the leftmost boundary of A_{v_k} below $L(v_k)$. Observe that $\lceil d_m(v_k)/2 \rceil \leq d_m(v_k)/2 + 1 \leq (d(v_k) - 3)/2 + 1 \leq (d(v_k) - 1)/2$. Since $(d(v_k) - 1)/2$ is at most $\lfloor d(v_k)/2 \rfloor$ (irrespective of the parity of $d(v_k)$), the grid points on the leftmost boundary of A_{v_k} below $L(v_k)$ are sufficient to route all the m-edges incident to $\{w_{l+1}, \ldots w_z\}$. Similarly, for each $w \in \{w_{z+1}, \ldots w_{r-1}\}$, we now route the m-edge incident to w through the top-right corner c_w upto M, and then to a distinct grid point to the left of $D(v_k)$ on the bottommost boundary of A_{v_k}. Since $\lfloor d_m(v_k)/2 \rfloor \leq \lfloor d(v_k)/2 \rfloor - 1$ (irrespective of the parity of $d(v_k)$), we have sufficient number of boundary points to route all the m-edges incident to $\{w_{z+1}, \ldots w_{r-1}\}$.

The l- and r-edges of Γ'_n contain edge overlapping on the left and down hands. From the Expansion phase it is straightforward to observe that the l-edges that are incoming to some vertex v in Γ'_n, are incident to $D(v)$, and properly intersects the first half of the S'_v. Let ℓ be the nearest vertical grid line to the right of S'_v, and remove the parts of these l-edges that lie in between $D(v)$ and ℓ (except for the l-edge incident to $C(v)$). Since all these l-edges lie below S_v, the points where these l-edges are incident to ℓ can see all the grid points on the rightmost boundary of A_v and on the right-half of the bottommost boundary of A_v. Consequently, we can route the l-edges to $C(v)$ through these boundary grid points, which removes the edge overlaps on $D(v)$. Figure 5(e) illustrates such a scenario. Symmetrically, we can remove the degeneracy of r-edges on $L(v)$. Remark 2 and the property that the lines in S_v and S'_v do not contain any vertex except v ensure that the above modifications do not introduce any edge crossing. Let the resulting drawing be Γ'', which is a planar polyline drawing of G, e.g., see Figure 5(e).

Area: By Lemma 2, the area before the Expansion phase was $(W+2) \times (H+2)$. For each i, where $1 \leq i \leq W + 2$, the Expansion phase increases the width of the drawing by $2\lfloor d(u_i)/2 \rfloor$, where u_i is the vertex with the largest degree on the ith column. Hence the total increase is at most $(\sum_{i=1}^{W+2} d(u_i)) - 3(n - W - 2) \leq (6n - 12) - 3(n - W - 2) = 3n + 3W - 6$. Similarly, the increase in

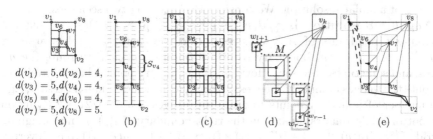

$$d(v_1) = 5, d(v_2) = 4,$$
$$d(v_3) = 5, d(v_4) = 4,$$
$$d(v_5) = 4, d(v_6) = 4,$$
$$d(v_7) = 5, d(v_8) = 5.$$

(a)   (b)   (c)   (d)   (e)

Fig. 5. Illustration for (a) Γ_n, (b) S_{v_k}, and (c) Γ'_n, where the grid A_v, for each vertex v, is shown in black squares. (d) Illustration for M. Note that A_ws are bounded by gray rectangles determined by S_w and S'_w. (e) Γ''.

height is at most $3n + 3H - 6$. Hence Γ'' is a drawing on an integer grid of size $(3n+4W-4)\times(3n+4H-4)$. Since $W+H \leq (4n-2\Delta_0-10)/3$ (see Remark 1), the area can be at most $(3n+4(2n-\Delta_0-5)/3)^2 = ((17n-4\Delta_0-20)/3)^2 \leq 32.12n^2$.

Bends per Edge: If (v, v') is an l-edge or r-edge in Γ_G, which starts at v and ends at v', then the edge has at most 2 bends: one before entering $A_{v'}$, and another at the boundary of $A_{v'}$. If (v, v') is an m-edge, then it contains one bend on M, and another bend on the boundary of $A_{v'}$. The l-and r-edges that connect a primary vertex to its parent, do not contain any bend. Since $\Delta_0 < n/2$ and $\text{leaf}(T_m) < n$, the drawing has at most $6n-\text{leaf}(T_l)-\text{leaf}(T_r) \leq 11n/2$ bends.

Angular Resolution: To compute the angular resolution, observe that the smallest possible angle θ at v is realized by a pair of consecutive integer grid points on the boundary of A_v where one of them is the corner of A_v, e.g., see Figure 6(a). Since A_v is a grid of size $(2\lfloor d(v)/2 \rfloor + 1) \times (2\lfloor d(v)/2 \rfloor + 1)$, the length of the line segment l connecting the center to any corner is $\sqrt{2}\lfloor d(v)/2 \rfloor$. Hence we have $\theta = \arctan\left(\frac{1/\sqrt{2}}{(\sqrt{2}\lfloor d(v)/2 \rfloor - 1/\sqrt{2}}\right) > 1/d(v)$, by the MacLaurin series expansion of $\arctan$ [12]. Observe that any edge e that intersects some grid A_v, where v does not correspond to any end vertex of e, must be an m-edge. We can avoid any such intersection by choosing for each vertex u, a rectangular grid of size $(2\lfloor d(u')/2 \rfloor + 1) \times (2\lfloor d(u'')/2 \rfloor + 1)$ (instead of A_u), where u' (respectively, u'') is the vertex with the largest degree over all the vertices on the horizontal (respectively, vertical) line through u. For example, see the gray rectangles in Figure 5(d). However, the angular resolution increases to $1/\Delta$.

Theorem 1. *Every n-vertex maximal planar graph admits a 2-bend polyline drawing Γ with angular resolution at least $1/d(v)$ for each vertex v, and area at most $(3n + 4W - 4) \times (3n + 4H - 4)$, where $W + H \leq (4n - 2\Delta_0 - 10)/3$. Within the same area, we can assign each vertex v in Γ a bounding box of size $(2\lfloor d(v)/2 \rfloor + 1) \times (2\lfloor d(v)/2 \rfloor + 1)$ that only intersect with the edges incident to v, but the angular resolution increases to $1/\Delta$.*

4   Trade-Offs between Angular Resolution and Area

In this section we show that one can significantly improve the area with an small expense of angular resolution. We consider the following two scenarios.

Angular Resolution is $\gamma/d(v)$, where $\gamma \in [0.8, 1]$: Observe that the bottom-left quadrants of A_v (with respect to the center $C(v)$) has at most $2\lfloor d(v)/2 \rfloor - 1 \geq d_m(v)$ boundary points, which are sufficient to route the m-edges, and sometimes necessary. However, the boundary points that are available to route the l-edges (similarly, r-edges) are significantly more than necessary, e.g., the number of boundary points to route the l-edges is $3\lfloor d(v)/2 \rfloor - 2$ (lying on the bottom-right quadrants and on the right-boundary of A_v). Hence assigning a grid of size $(\lfloor d(v)/2 \rfloor + 1 + \lceil d(v)/4 \rceil) \times (\lfloor d(v)/2 \rfloor + 1 + \lceil d(v)/4 \rceil)$ to each vertex v would be sufficient for routing the edges.

Observe that for each vertex v, the increase in width is at most $(\lfloor d(v)/2 \rfloor + \lceil d(v)/4 \rceil) \leq (3d(v)/4 + 1)$. Since one column may contain multiple vertices, and

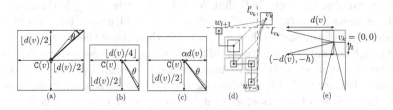

Fig. 6. Illustration for angular resolution

the degree of each vertex is at least three, we are overcounting the increase for $(n - W - 2)$ vertices. The amount of over computation for each such vertex v' is at least $\lfloor 3d(v')/4 \rfloor + 1 \geq 3$. Consequently, the total increase in the width in the Expansion phase is now bounded by $(\sum_{i=1}^{W+2}(3d(v_i)/4 + 1)) - 3(n - W - 2) \leq 3n/2 + 4W - 1$. Similarly, the increase in height is at most $3n/2 + 4H + 1$. Since $W + H \leq (4n - 2\Delta_0 - 10)/3$, the area can be at most $(3n/2 + 5(2n - \Delta_0 - 5)/3 + 5)^2 \leq 23.37n^2$. The number of bends remains the same, but the minimum angle θ is now at least $0.8/d(v)$, which is now determined by two consecutive points along the bottom-right corner, as shown in Figure 6(b).

We can parametrize the grid size with a parameter α, i.e., consider the grid assigned to v as $(\lfloor d(v)/2 \rfloor + 1 + \alpha d(v)) \times (\lfloor d(v)/2 \rfloor + 1 + \alpha d(v))$, where $\alpha \geq 1/4$. Then the increase in width is at most $(\sum_{i=1}^{W+2}((\alpha + 1/2)d(v_i) + 1)) - 3(n - W - 2) \leq (6(\alpha + 1/2)n - 3n + 4W + 8) \leq (6\alpha n + 4W + 8)$. Similarly, the increase in height is at most $(6\alpha n + 4H + 8)$, respectively. Hence the area is at most $(6\alpha n + 4(W + H)/2 + 10)^2 \leq (6\alpha n + 8n/3 + 10)^2 \approx (6\alpha + 8/3)^2 n^2$. The angular resolution is at least $\frac{\alpha/\sqrt{\alpha^2+1/4}}{d(v)\sqrt{\alpha^2+1/4}} > \frac{\alpha}{d(v)(\alpha^2+1/4)}$, as illustrated in Figure 6(c).

Theorem 2. *Every n-vertex maximal planar graph admits a 2-bend polyline drawing with angular resolution $\frac{\alpha}{d(v)(\alpha^2+1/4)}$ for each vertex v, and area $(6\alpha n + 4W + 10) \times (6\alpha n + 4H + 10)$. Here $\alpha \in [1/4, 1/2]$, and $W + H \leq (4n - 2\Delta_0 - 10)/3$.*

Angular Resolution is $\gamma/d(v)$, where $\gamma \in [0.3, 0.5]$: Recall that the new grid lines in the Expansion phase are inserted such that each vertex v has $h = \beta_v d(v)$ free grid lines, where $\beta_v \geq 1/d(v)$, in each of the four sides (above, below, left, right) around v, i.e., $C(v)$ is at the center of a free integer grid A_v of size $h \times h$. As in the Expansion phase, let S_v be the ordered set of horizontal free grid lines along with the horizontal line through v, and let S_v' be the ordered set of vertical free grid lines along with the vertical line through v. We now show that these free grids are sufficient for routing the l-, r- and m-edges.

Routing m-edges: Let l_{v_k} and l_{v_k}' be the grid lines that are immediately below and to the left of S_{v_k} and S_{v_k}', respectively. For each $w \in \{w_{l+1}, \ldots, w_{r-1}\}$, we now extend a line segment with slope $+1$ from $C(w)$ until we hit either l_{v_k} or l_{v_k}'. Let $B = \{b(w_{l+1}), \ldots, b(w_{r-1})\}$ be the set of points on l_{v_k} and l_{v_k}' reached by these extensions. We now extend these extensions further to reach $C(v_k)$, as follows:

- If the number of points of B that lie on l_{v_k} is z, where $z \leq h$, then we route the extensions of l_{v_k} through z consecutive grid points lying on the left side of A_{v_k} immediately below $L(v_k)$. We then route the extensions on l'_{v_k} through the next consecutive grid points along the same vertical line. Since there are at most $d_m(v_k)$ m-edges, we need at most $d(v)$ consecutive grid points below $L(v_k)$. Figure 6(d) illustrates such a scenario, where $h = 2$.
- If the number of points of B that lie on l'_{v_k} is at most z', where $z' \leq h$, then the drawing is symmetric to the case when $z < h$.
- Otherwise, both l_{v_k} and l'_{v_k} contains more than h extensions. In this case $\min\{z, z'\} > h$, and hence $\max\{z, z'\} \leq d_m(v) - h$. We first extend the extensions on l_{v_k} to the grid points that lie consecutively to the left of A_v (on the first line of S_{v_k}). We then extend the extensions on l'_{v_k} to the grid points that lie consecutively below of A_v (on the last line of S'_{v_k}). Finally, we connect all these new extensions directly to $C(v_k)$. Note that the maximum horizontal (respectively, vertical) distance between $C(v)$ and a bend point on l_{v_k} (respectively, l'_{v_k}) is at most $(d_m(v) - h) + h \leq d(v)$.

Routing l-edges: Let $u_1, u_2, \ldots, u_q$ be the vertices in top-to-bottom order that are incident to $D(v_k)$ by incoming l-edges. Let ℓ be the nearest vertical grid line to the right of S'_v, and remove the parts of these l-edges that lie in between $D(v_k)$ and ℓ (except for the l-edge incident to $C(v_k)$). We then connect these extensions to the q consecutive grid points on the first line of S'_{v_k} that lie immediately below the top-right corner of A_v. Finally, we connect all these new extensions directly to $C(v_k)$.

Routing r-edges: This scenario is symmetric for routing l-edges.

Angular Resolution and Area: In all the cases, the smallest angle θ at any vertex v is equal to the angle determined by the points $(-d(v), -h)$ and $(-d(v) + 1, -h)$ at $C(v) = (0,0)$, as illustrated in Figure 6(e). Here the angular resolution is at least $\frac{\beta_v}{d(v)(1+\beta_v^2)}$, where $1/d(v) \leq \beta_v \leq 1$, and the area is $(6\beta + 2/3)^2 n^2$. We omit the details due to space constraints.

Theorem 3. *Every n-vertex maximal planar graph admits a 2-bend polyline drawing with angular resolution $\frac{\beta}{d(v)(1+\beta^2)}$ for each vertex v, and area $(6n\beta + W + 2) \times (6n\beta + H + 2)$. Here $\beta \in [1/3, 1]$, and $W + H \leq (4n - 2\Delta_0 - 10)/3$.*

5   Conclusion

In this paper we have given the first smooth trade-off between the area and angular resolution for 2-bend polyline drawings of any given planar graph. Our algorithm dominates all the previous 2-bend polyline drawing algorithms except Duncan and Kobourov's algorithm [10], which uses 1-bend per edge and dominates our algorithm when the angular resolution is in the interval $[0.38/d(v), 0.5/d(v)]$. Similar to the previously known polyline drawing algorithms, one can implement our algorithm using standard techniques [6] such that the drawings are computed in linear time.

A natural open question is whether Duncan and Kobourov's algorithm could be modified (allowing 2-bends per edge) to achieve a better trade-off. Finding tight lower bounds would also be very interesting. Finally, we hope that the results in this paper will encourage the study of smooth trade-offs among different aesthetic criteria for other styles of drawing graphs.

References

1. Biedl, T.C., Kaufmann, M.: Area-efficient static and incremental graph drawings. In: Burkard, R.E., Woeginger, G.J. (eds.) ESA 1997. LNCS, vol. 1284, pp. 37–52. Springer, Heidelberg (1997)
2. Bonichon, N., Felsner, S., Mosbah, M.: Convex drawings of 3-connected plane graphs. Algorithmica 47(4), 399–420 (2007)
3. Bonichon, N., Le Saëc, B., Mosbah, M.: Wagner's theorem on realizers. In: Widmayer, P., Triguero, F., Morales, R., Hennessy, M., Eidenbenz, S., Conejo, R. (eds.) ICALP 2002. LNCS, vol. 2380, pp. 1043–1053. Springer, Heidelberg (2002)
4. Brandenburg, F.J.: Drawing planar graphs on $\frac{8}{9}n^2$ area. Electronic Notes in Discrete Mathematics 31, 37–40 (2008)
5. Cheng, C.C., Duncan, C.A., Goodrich, M.T., Kobourov, S.G.: Drawing planar graphs with circular arcs. Discrete & Computational Geometry 25(3), 405–418 (2001)
6. Chrobak, M., Payne, T.: A linear-time algorithm for drawing planar graphs. Information Processing Letters 54, 241–246 (1995)
7. CircuitLogix, https://www.circuitlogix.com/ (accessed June 03, 2014)
8. ConceptDraw: http://www.conceptdraw.com/ (accessed June 03, 2014)
9. De Fraysseix, H., Pach, J., Pollack, R.: How to draw a planar graph on a grid. Combinatorica 10(1), 41–51 (1990)
10. Duncan, C.A., Kobourov, S.G.: Polar coordinate drawing of planar graphs with good angular resolution. Journal of Graph Algorithms and Applications 7(4), 311–333 (2003)
11. Durocher, S., Mondal, D.: On balanced +-contact representations. In: Proceedings of GD. In: Wismath, S., Wolff, A. (eds.) GD 2013. LNCS, vol. 8242, pp. 143–154. Springer, Heidelberg (2013)
12. Goodrich, M.T., Wagner, C.G.: A framework for drawing planar graphs with curves and polylines. Journal of Algorithms 37(2), 399–421 (2000)
13. Gutwenger, C., Mutzel, P.: Planar polyline drawings with good angular resolution. In: Whitesides, S.H. (ed.) GD 1998. LNCS, vol. 1547, pp. 167–182. Springer, Heidelberg (1999)
14. Hong, S.H., Mader, M.: Generalizing the shift method for rectangular shaped vertices with visibility constraints. In: Tollis, I.G., Patrignani, M. (eds.) GD 2008. LNCS, vol. 5417, pp. 278–283. Springer, Heidelberg (2009)
15. Kant, G.: Drawing planar graphs using the canonical ordering. Algorithmica 16(1), 4–32 (1996)
16. Kurowski, M.: Planar straight-line drawing in an $o(n) \times o(n)$ grid with angular resolution $\omega(1/n)$. In: Vojtáš, P., Bieliková, M., Charron-Bost, B., Sýkora, O. (eds.) SOFSEM 2005. LNCS, vol. 3381, pp. 250–258. Springer, Heidelberg (2005)
17. Schnyder, W.: Embedding planar graphs on the grid. In: Proceedings of ACM-SIAM SODA, January 22-24, pp. 138–148. ACM (1990)

18. SmartDraw Software, LLC, http://www.smartdraw.com/ (accessed June 03, 2014)
19. Tamassia, R., Di Battista, G., Batini, C.: Automatic graph drawing and readability of diagrams. IEEE Transactions on Systems, Man and Cybernetics 18(1), 61–79 (1988)
20. Zhang, H.: Planar polyline drawings via graph transformations. Algorithmica 57(2), 381–397 (2010)
21. Zhang, H., He, X.: Canonical ordering trees and their applications in graph drawing. Discrete & Computational Geometry 33(2), 321–344 (2005)

Stress-Minimizing Orthogonal Layout of Data Flow Diagrams with Ports

Ulf Rüegg[1], Steve Kieffer[2],
Tim Dwyer[2], Kim Marriott[2], and Michael Wybrow[2]

[1] Department of Computer Science, Kiel University, Kiel, Germany
uru@informatik.uni-kiel.de
[2] Faculty of Information Technology, Monash University, NICTA Victoria, Australia
{Steve.Kieffer,Tim.Dwyer,Kim.Marriott,Michael.Wybrow}@monash.edu

Abstract. We present a fundamentally different approach to orthogonal layout of data flow diagrams with ports. This is based on extending constrained stress majorization to cater for ports and flow layout. Because we are minimizing stress we are able to better display global structure, as measured by several criteria such as stress, edge-length variance, and aspect ratio. Compared to the layered approach, our layouts tend to exhibit symmetries, and eliminate inter-layer whitespace, making the diagrams more compact.

Keywords: actor models, data flow diagrams, orthogonal routing, layered layout, stress majorization, force-directed layout.

1 Introduction

Actor-oriented data flow diagrams are commonly used to model movement of data between components in complex hardware and software systems [13]. They are provided in many widely used modeling tools including LabVIEW (National Instruments Corporation), Simulink (The MathWorks, Inc.), EHANDBOOK (ETAS), SCADE (Esterel Technologies), and Ptolemy (UC Berkeley). Complex systems are modeled graphically by composing *actors*, i. e., reusable block diagrams representing well-defined pieces of functionality. Actors can be *nested*— i. e., composed of other actors—or *atomic*. Fig. 1a shows an example of a data flow diagram with four nested actors. Data flow is shown by directed edges from the source port where the data is constructed to the target port where the data is consumed. By convention the edges are drawn orthogonally and the ports are fixed in position on the actors' boundaries. Automatic layout of data flow diagrams is important: Klauske and Dziobek [12] found that without automatic layout about 30 % of a modeler's time is spent manually arranging elements.

Current approaches to automatic layout of data flow diagram are modifications of the well-known Sugiyama layer-based layout algorithm [18] extended to handle ports and orthogonal edges. In particular Schulze et al. [16] have spent many years developing specialised layout algorithms that are used, for instance, in the EHANDBOOK and Ptolemy tools. However, their approach has a number

C. Duncan and A. Symvonis (Eds.): GD 2014, LNCS 8871, pp. 319–330, 2014.

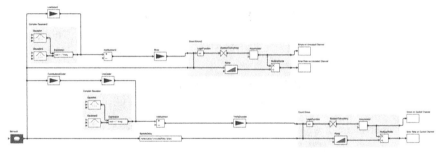

(a) Layout with layer-based algorithm KLay Layered by Schulze et al.

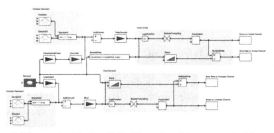

(b) Layout with the CoDaFlow algorithm presented here

Fig. 1. Two layouts of the same diagram. The result of our method, shown in (b), has less stress, lower edge length variance, less area, and better aspect ratio.

of drawbacks. First, it employs a strict layering which may result in layouts with poor aspect ratio and poor compactness, especially when large nodes are present. Furthermore, the diagrams often have long edges and the underlying structure and symmetries may not be revealed. A second problem with the approach of Schulze et al. is that it uses a recursive bottom-up strategy to compute a layout for nested actors independent of the context in which they appear. This can lead to bad arrangements with poor aspect ratio and a lack of compactness.

This paper presents a fundamentally different approach to the layout of actor-oriented data flow diagrams designed to overcome these problems. A comparison of our new approach with standard layer-based algorithm KLay Layered is shown in Fig. 1. Our starting point is *constrained stress majorization* [3]. Minimizing stress has been shown to improve readability by giving a better understanding of important graph structure such as cliques, chains and cut nodes [4]. However, stress-minimization typically results in a quite "organic" look with nodes placed freely in the plane that is quite different to the very "schematic" arrangement involving orthogonal edges, a left-to-right "flow" of directed edges, and precise alignment of node ports that practitioners prefer.

The main technical contribution of this paper is to extend constrained stress majorization to handle the layout conventions of data flow diagrams. In particular we: (1) augment the *P-stress* [7] model to handle ports that are constrained to node boundaries but are either allowed to float subject to ordering constraints or else are fixed to a given node boundary side, and (2) extend Adaptive Constrained Alignment (ACA) [10] for achieving grid-like layout to handle directed edges, orthogonal routing, ports, and widely varying node dimensions.

An empirical evaluation of the new approach (Sect. 4) shows it produces layouts of comparable quality to the method of Schulze et al. but with a different trade-off between aesthetic criteria. The layouts have more uniform edge length, better aspect ratio, and are more compact but have slightly more edge crossings and bends. Furthermore, our method is more flexible and requires far less implementation effort. The Schulze et al. approach took a team of developers and researchers several years to implement by extensively augmenting the Sugiyama method. While their infrastructure allows a flexible configuration of the existing functionality [16], it is very restrictive and brittle when it comes to extensions that affect multiple phases of the algorithm. The method described in this paper took about two months to implement and is also more extendible since it is built on modular components with well-defined work flows and no dependencies on each other.

Related Work. The most closely related work is the series of papers by Schulze et al. that show how to extend the layer-based approach to handle the layout requirements of data flow diagrams [16,17]. Their work presents several improvements over previous methods to reduce edge bend points and crossings in the presence of ports. While the five main phases (classically three) of the layer-based approach are already complex, they introduce between 10 and 20 *intermediate processes* in order to address additional requirements. The authors admit that their approach faces problems with unnecessary crossings of inter-hierarchy edges as they layout compound graphs bottom-up, i. e., processing the most nested actor diagrams first. Related work in the context of the layer-based approach has been studied thoroughly in [16,17]. Chimani et al. present methods to consider ports and their constraints during crossing minimization within the *upward planarization* approach [2]. While the number of crossings is significantly reduced, the approach eventually induces a layering, suffering from the same issues as above. There is no evaluation with real-world examples. Techniques from the area of VLSI design and other approaches that specifically target compound graphs have been discussed before and found to be insufficient to fulfil the layout requirements for data flow diagrams [17], especially due to lacking support for different port constraints.

2   CoDaFlow – The Algorithm

Data flow diagrams can be modeled as directed graphs $G = (V, E, P, \pi)$ where *nodes* or *vertices* $v \in V$ are connected by *edges* $e \in E \subseteq P \times P$ through *ports* $p \in P$—certain positions on a node's perimeter—and $\pi : P \to V$ maps each port p to the *parent* node $\pi(p)$ to which it belongs. An edge $e = (p_1, p_2)$ is directed, *outgoing* from port p_1 and *incoming* to p_2. A *hyperedge* is a set of edges where every pair of edges shares a common port.

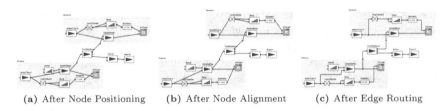

(a) After Node Positioning   (b) After Node Alignment   (c) After Edge Routing

Fig. 2. The results of pipeline stages (1), (2), (3) are shown in (a), (b), (c), respectively.

To better show flow it is preferable for sources of edges to be to the left of their targets and by convention edges are routed in an *orthogonal* fashion. Ports can—depending on the application—be restricted by certain constraints, e. g., all ports with incoming edges should be placed on the left border of the node. Spönemann et al. define five types of *port constraints* [17], ranging from ports being free to float arbitrarily on a node's perimeter, to ports having well-defined positions relative to nodes. Nodes that contain nested diagrams, i. e., child nodes, are referred to as *compound* or *nested nodes* (as opposed to *atomic nodes*); a graph that contains compound nodes is a *compound graph*. We refer to the ports of a compound node as *hierarchical ports*. These can be used to connect atomic nodes inside a compound node to atomic nodes on the outside.

The main additional requirements for layout of data flow diagrams on top of standard graph drawing conventions are therefore [17]: (R1) clearly visible flow, (R2) ports and port constraints, (R3) compound nodes, (R4) hierarchical ports, (R5) orthogonal edge routing, and (R6) orthogonalized node positions to emphasize R1 using horizontal edges.

The starting point for our approach is constrained stress majorization [3]. This extends the original stress majorization model [9] to support separation constraints that can be used to declaratively enforce node alignment, non-overlap of nodes, flow in directed graphs, and to cluster nodes inside non-overlapping regions. Brandes et al. [1] provide one method to orthogonalise an existing layout based on the topology-shape-metrics approach, but in order to handle requirements R1–6 we instead use the heuristic approach of Kieffer et al. [10] to apply alignment constraints within the stress-based model.

Our Constrained Data Flow (CoDaFlow) layout algorithm is a pipeline with three stages:

1. Constrained Stress-Minimizing Node Positioning
2. Grid-Like Node Alignment
3. Orthogonal Edge Routing

The intermediate results of this pipeline are depicted in Fig. 2. Single stages can be omitted, e. g., when no edge routing is required or initial node positions are given. In this section we restrict our attention to flat graphs, i. e., those without compound nodes, while Sect. 3 extends the ideas to compound graphs.

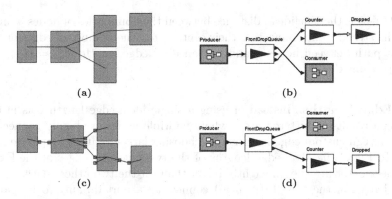

Fig. 3. Awareness of ports is important to achieve good node positioning. (a) and (c) show internal representations of what is passed to the layout algorithm, (b) and (c) show the resulting drawings. (a) is unaware of ports and yields node positions that introduce an edge crossing in (b). In (c) ports are considered and the unnecessary crossing is avoided in (d). Note, however, while the chance is higher that (c) is cross free, it is not guaranteed.

2.1 Constrained Stress-Minimizing Node Positioning

Traditional stress models for graph layout expect a simple graph without ports, so a key idea in order to handle data flow diagrams is to create a small node to represent each port, called a *port node* or *port dummy*, as in Fig. 3c. If D is the set of all these, and $\delta : P \to D$ maps each port to the dummy node that represents it, we construct a new graph $G' = (V', E')$ where $V' = V \cup D$, and

$$E' = \{(\delta(p_1), \delta(p_2)) : (p_1, p_2) \in E\} \cup \{(\pi(p), \delta(p)) : p \in P\}$$

includes one edge representing each edge of the original graph, and an edge connecting each port dummy to its parent node. We refer to the $v \in V$ as *proper nodes*.

Depending on the specified port constraints (R2) we restrict the position of each port dummy $\delta(p)$ relative to its parent node $\pi(p)$ using separation constraints. For instance, for a rigid relative position we use one separation constraint in each dimension, whereas we retain only the x-constraint if $\delta(p)$ need only appear on the left or right side of $\pi(p)$. The use of port nodes allows the constrained stress-minimizing layout algorithm to untangle the graph while being aware of relative port positions, resulting in fewer crossings, as illustrated in Fig. 3, and a better overall placement of nodes.

Our constrained stress-based layout uses the methods of Dwyer et al. [3] to minimize the *P-stress* function [7], a variant of *stress* [9] that does not penalise unconnected nodes being more than their desired distance apart:

$$\sum_{u < v \in V'} w_{uv} \left((\ell p_{uv} - b(u,v))^+ \right)^2 + \sum_{(u,v) \in E'} \ell^{-2} \left((b(u,v) - \ell)^+ \right)^2 \tag{1}$$

where $b(u,v)$ is the Euclidean distance between the boundaries of nodes u and v along the straight line connecting their centres, p_{uv} the number of edges on the shortest path between nodes u and v, ℓ an ideal edge length, $w_{uv} = (\ell p_{uv})^{-2}$, and $(z)^+ = \max(z, 0)$.

Ideal Edge Lengths. Instead of using a single ideal edge length ℓ as in (1), which can result in cluttered areas where multiple nodes are highly connected, we may assign custom edge lengths ℓ_{uv}, choosing larger values to separate such nodes. In Fig. 3 the ideal edge lengths of the two outgoing edges of the Front-DropQueue actor are chosen slightly larger than for the two other edges.

The length of the edge $(\pi(p), \delta(p))$ connecting a port dummy to its parent node is set to the exact distance from the node's center to the port's center.

Emphasizing Flow. A common requirement for data flow diagrams is that the majority of edges point in the same direction (here left-to-right). For this we introduce separation (*flow*) constraints for edges (u, v) of the form $x_u + g \leq x_v$, where $g > 0$ is a pre-defined spacing value, ensuring that u is placed left of v.

Special care has to be taken for cycles, as they would introduce contradicting constraints. We experimented with different strategies to handle this. 1) We introduced the constraints even though they were contradicting (and let the solver choose which one(s) to reject); 2) We did not generate *any* flow constraints for edges that are part of a strongly connected component; 3) We employed a greedy heuristic by Eades et al. [8] (known from the layer-based approach) to find the minimal feedback arc set, and withheld flow constraints for the edges in this set. Our experiments showed that the third strategy yields the best results.

Execution. We perform three consecutive layout runs, iteratively adding constraints: 1) Only port constraints are applied, allowing the graph to untangle and expose symmetry; 2) Flow constraints are added, but overlaps are still allowed so that nodes can float past each other, swapping positions where necessary; 3) Non-overlap constraints are applied to separate all nodes as desired.

2.2 Grid-Like Node Alignment

While yielding a good distribution of nodes overall, stress-minimization tends to produce an organic layout with paths splayed at all angles, which is inappropriate for data flow diagrams. The layout needs to be *orthogonalized*, i.e., connected nodes brought into alignment with one another so that where possible edges form straight horizontal lines, visually emphasizing horizontal flow.

For this purpose we apply the Adaptive Constrained Alignment (ACA) algorithm [10]. Since it respects existing flow constraints, it only attempts to align edges horizontally. However, our replacement of the given graph G by the auxiliary graph G' with port nodes tends to subvert the original intentions of ACA, so it requires some adaptation. Whereas the original ACA algorithm expected at

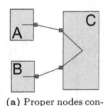

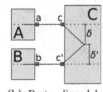

Fig. 4. In the new port model, two proper nodes may be connected to the same side of another via ports, as in (a). The systematic use of *offset alignments* between port nodes and their parents, i.e., constraints of the form $y_{\delta(p)} + \delta = y_{\pi(p)}$, $\delta \neq 0$ as shown in (b), creates a risk of node-edge and node-node overlaps far exceeding what was anticipated with the original ACA algorithm, as could have occurred here had node B been as tall as node C, for example. We have extended ACA to properly handle such cases.

(a) Proper nodes connected via port nodes

(b) Ports aligned by ACA

most one proper node to be aligned with another in a given compass direction, in our case (with ports) it will often be desirable to have more. See Fig. 4.

In order to adapt ACA to the new port model we made it possible to ignore certain edges—namely those connecting port nodes to their parents—and also generalised its overlap prevention methods significantly. Instead of the simple procedure for preventing multiple alignments in a single compass direction [10], we use the VPSC solver [5] for trial satisfaction of existing constraints, the new potential alignment, as well as non-overlap constraints between all nodes and a dummy node representing the potentially aligned edge.

Thus, while the ACA process continues to merely centre-align nodes—in this case port nodes $d \in D$—we have allowed it to *in effect* align several proper nodes $v_1, \ldots, v_k \in V$ with a single one $u \in V$ at port positions as in Fig. 4, meeting the requirement R6 of data flow diagrams.

2.3 Edge Routing

We now consider node positions to be fixed, and use the methods of Wybrow et al. [19] to route the edges orthogonally. We return from G' to G, using the final positions of the port nodes $d \in D$ to set *routing pins*, fixed port positions on the nodes $v \in V$ where the edges should connect.

3 Handling Compound Graphs

Graphs containing compound nodes can be handled using several different strategies. Schulze et al. employ a *bottom-up* strategy, treating every compound node as a separate graph, starting with the inner-most nodes. This allows application of different layout algorithms to each subgraph which reduces the size of the layout problem, and possibly the overall execution time. They remark, however, that the procedure can yield unsatisfying layouts since the surroundings of a compound node are not known; see Fig. 5a for an example where two unnecessary crossings are created inside the TM controllers actor and two separate

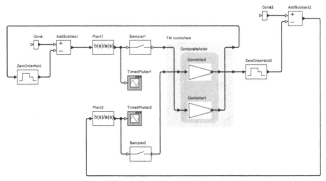

(a) Layout with layer-based methods by Schulze et al.

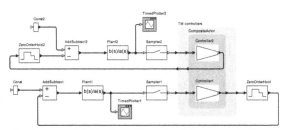

(b) Layout with the CoDaFlow algorithm presented here.

Fig. 5. Two layouts of the same Ptolemy diagram. While two distinct networks are interleaved in (a), they are clearly separated and the two crossings are avoided in (b).

networks are interleaved. A *global* approach would solve this issue, positioning all compound nodes along with their children at the same time.

Even though we focus our attention on a global approach in what follows, our methods are flexible in that we may choose between a bottom-up and a global strategy in each stage of our pipeline.

A compound graph G is transformed into G' as above, which is used to construct a *flat* graph $G'' = (A, E'')$ where $A \subseteq V'$ is the set of atomic nodes and their port nodes, and $E'' = U \cup H$ with

$$U = \{(\delta(p_1), \delta(p_n)) : \pi(p_1), \pi(p_n) \in A\}$$
$$H = \{(\delta(p_1), \delta(p_n)) : \exists (p_1, p_2), (p_2, p_3), \dots, (p_{n-2}, p_{n-1}), (p_{n-1}, p_n) \in E :$$
$$\pi(p_1), \pi(p_n) \in A \land \pi(p_2), \dots, \pi(p_{n-1}) \in V \backslash A\}$$

Intuitively, compound nodes are neglected along with their ports and only atomic nodes are retained. Sequences of edges that span hierarchy boundaries, e.g., the three edges between Sampler2 and Controller2 in Fig. 5b, are replaced by a single edge that directly connects the two atomic nodes. Note that for hyperedges multiple edges have to be created. *Cluster constraints* [3] guarantee that children of compound nodes are kept close together and are not interleaved with any other nodes. For instance, the CompositeActor in Fig. 5b yields a cluster containing Controller1 and Controller2.

To return to G, the clusters' dimensions, i.e., their rectangular bounding boxes, are applied to the compound nodes in $V \setminus A$. The edges in H are split into segments $s_1, \ldots, s_n$ based on the crossing points c_i with clusters. The route of s_i is applied to the corresponding edge $e \in E$ and the c_i determine the positions of the hierarchical ports.

4   Evaluation and Discussion

We evaluate our approach on a set of data flow diagrams that ship with the Ptolemy project[1], comparing with the KLay Layered algorithm of Schulze et al. Diagrams were chosen to be roughly the size Klauske found to be typical for real-world Simulink models from the automotive industry [11] (about 20 nodes and 30 edges per hierarchy level).

Metrics. Well established metrics to assess the quality of a drawing are *edge crossings* and *edge bends* [14], two metrics directly optimized by the layer-based approach. More recently, *stress* and *edge length variance* were found to have a significant impact on the readability of a drawing [4]. Additionally, we regard compactness in terms of *aspect ratio* and *area*.

So that comparisons of edge length and of layout area can be meaningful, we set the same value for KLay Layered's inter-layer distance and CoDaFlow's ideal separation between nodes.

The P-stress of a given (already layouted) diagram depends on the choice of the ideal edge length ℓ in (1), and the canonical choice $\bar{\ell}$ is that where the function takes its global minimum. If L is a list of all the individual ideal lengths $\ell_{uv} = b(u,v)/p_{uv}$, then $\bar{\ell}$ is equal to the *contraharmonic mean* $C(L_j)$ (i.e., the weighted arithmetic mean in which the weights equal the values) over a certain sublist $L_j \subseteq L$. Namely, if $L_E = \langle \ell_{uv} : (u,v) \in E \rangle$ and $L \setminus L_E = \langle \ell_1 \le \ell_2 \le \cdots \le \ell_\nu \rangle$, then $L_j = L_E \cup \langle \ell_1, \ell_2, \ldots, \ell_j \rangle$ for some $0 \le j \le \nu$. Since ν is finite, we can compute each $C(L_j)$ and take $\bar{\ell}$ to be that at which the P-stress is minimized, cf. [15].

Results. Table 1 and 2 show detailed results for layouts created by CoDaFlow and KLay Layered. We used two variations of the Ptolemy diagrams: small flat diagrams and compound diagrams (cf. [15] for further examples).

For flat diagrams CoDaFlow shows a better performance on stress, average edge length, and variance in edge length. CoDaFlow produced slightly more crossings, bends per edge, and slightly increased area.

More interesting are the results for the compound diagrams, which show more significant improvements. On average, CoDaFlow's diagram area was 88% that of KLay Layered, and edge length variance was only 29%. Also, the average aspect ratio shifts closer to that of monitors and sheets of paper. However, there is an increase in crossings. Currently our approach does not consider crossings

---

[1] http://ptolemy.eecs.berkeley.edu/

Table 1. Evaluations of 110 flat diagrams with 10–23 nodes (9–30 edges) and 10 compound diagrams with 12–38 nodes (12–52 edges). Figures for stress, average edge length, variance in edge length, and area are given as the ratio of CoDaFlow divided by KLay Layered. Values below 1 indicate a better performance of CoDaFlow. An average value shows the general tendency while minimal and maximal values show the best and worst performance.

	Stress			EL Variance			EL Average			Area		
	Min	Avg	Max	Min	Avg	Max	Min	Avg	Max	Min	Avg	Max
Comp.	0.27	0.75	0.97	0.11	0.29	0.61	0.39	0.57	0.79	0.50	0.88	1.28
Flat	0.34	0.77	1.13	0.03	0.60	1.92	0.34	0.84	1.10	0.62	1.11	2.01

Table 2. Results for the metrics aspect ratio, crossings, and average bends per edge. As opposed to Table 1, figures are absolute values.

		Aspect Ratio			Crossings			Bends/Edge		
		Min	Avg	Max	Min	Avg	Max	Min	Avg	Max
Comp.	CoDaFlow	1.27	1.83	2.51	0.00	3.40	10.0	0.92	1.25	1.56
	KLay	1.51	2.76	4.94	0.00	1.20	6.00	0.68	0.97	1.22
Flat	CoDaFlow	0.32	2.47	5.96	0.00	1.25	11.0	0.42	1.16	2.31
	KLay	0.37	2.77	9.00	0.00	1.02	7.00	0.22	1.04	1.73

at all, thus the increased average. As can be seen in Fig. 1, the small number of additional crossings are not ruinous to diagram readability, and they could be easily avoided by introducing further constraints, as discussed in Sect. 5.

As seen in Fig. 6, the current CoDaFlow implementation performs significantly slower than KLay Layered, but it still finishes in about half a second even on a larger diagram of 60 nodes. There is room for speedups, for instance, by avoiding re-initialization of internal data structures between pipeline stages. In addition, we plan to improve the incrementality of constraint solving in the ACA stage, as well as performing faster satisfiability checks wherever full projections are not required.

Compared to KLay Layered our approach is both easier to understand and implement, and more flexible in its application.

In addition to the five main phases of KLay Layered, about 10 to 20 *intermediate processes* of low to medium complexity are used during each layout run. Dependencies between these units have to be carefully managed and the phases have to be executed in strict order, e. g., the edge routing phase requires all previous phases.

CoDaFlow optimizes only one goal function and addresses the requirements of data flow diagrams by successively adding constraints to the optimization process. While we divide the algorithm into multiple stages, each stage merely introduces the required constraints. CoDaFlow's stages can be used independently of each other, e. g., to improve existing layouts. Also, users can fine-tune generated drawings using interactive layout [6] methods.

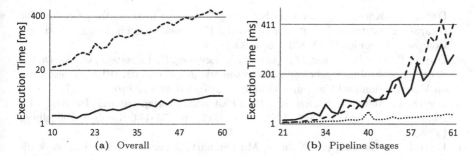

Fig. 6. Execution time plotted against the number of nodes n. For each n 10 graphs were generated randomly with an average of 1.5 outgoing edges per node. (a) Overall execution time of KLay Layered (solid line) and CoDaFlow (dashed line). (b) Execution time of the pipeline steps: Untangling (solid line), Alignment (dashed line), and Edge Routing (dotted line). Timings were conducted on an Intel i7 2.0 GHz with 8 GB RAM.

5   Conclusions

We present a novel approach to layout of data flow diagrams based on stress-minimization. We show that it is superior to previous approaches with respect to several diagram aesthetics. Also, it is more flexible and easier to implement.[2]

The approach can easily be extended to further diagram types with similar drawing requirements, such as the Systems Biology Graphical Notation (SBGN). To allow interactive browsing of larger diagram instances, however, execution time has to be reduced, e. g., by removing overhead from both the implementation and the pipeline steps. Avoiding the crossing in Fig. 3 is currently not guaranteed. We plan to detect such obvious cases via ordering constraints. In addition to ACA, the use of topological improvement strategies [7] could help to reduce the number of edge bends further where edges are almost straight.

Acknowledgements. Ulf Rüegg was funded by a doctoral scholarship (FIT-weltweit) of the German Academic Exchange Service. Michael Wybrow was supported by the Australian Research Council (ARC) Discovery Project grant DP110101390.

References

1. Brandes, U., Eiglsperger, M., Kaufmann, M., Wagner, D.: Sketch-driven orthogonal graph drawing. In: Goodrich, M.T., Kobourov, S.G. (eds.) GD 2002. LNCS, vol. 2528, pp. 1–11. Springer, Heidelberg (2002)
2. Chimani, M., Gutwenger, C., Mutzel, P., Spönemann, M., Wong, H.M.: Crossing minimization and layouts of directed hypergraphs with port constraints. In: Brandes, U., Cornelsen, S. (eds.) GD 2010. LNCS, vol. 6502, pp. 141–152. Springer, Heidelberg (2011)

---

[2] Author Ulf Rüegg has worked on both KLay Layered and CoDaFlow.

3. Dwyer, T., Koren, Y., Marriott, K.: IPSep-CoLa: An incremental procedure for separation constraint layout of graphs. IEEE Transactions on Visualization and Computer Graphics 12(5), 821–828 (2006)
4. Dwyer, T., Lee, B., Fisher, D., Quinn, K.I., Isenberg, P., Robertson, G., North, C.: A comparison of user-generated and automatic graph layouts. IEEE Transactions on Visualization and Computer Graphics 15(6), 961–968 (2009)
5. Dwyer, T., Marriott, K., Stuckey, P.J.: Fast node overlap removal. In: Healy, P., Nikolov, N.S. (eds.) GD 2005. LNCS, vol. 3843, pp. 153–164. Springer, Heidelberg (2006)
6. Dwyer, T., Marriott, K., Wybrow, M.: Dunnart: A constraint-based network diagram authoring tool. In: Tollis, I.G., Patrignani, M. (eds.) GD 2008. LNCS, vol. 5417, pp. 420–431. Springer, Heidelberg (2009)
7. Dwyer, T., Marriott, K., Wybrow, M.: Topology preserving constrained graph layout. In: Tollis, I.G., Patrignani, M. (eds.) GD 2008. LNCS, vol. 5417, pp. 230–241. Springer, Heidelberg (2009)
8. Eades, P., Lin, X., Smyth, W.F.: A fast and effective heuristic for the feedback arc set problem. Information Processing Letters 47(6), 319–323 (1993)
9. Gansner, E.R., Koren, Y., North, S.C.: Graph drawing by stress majorization. In: Pach, J. (ed.) GD 2004. LNCS, vol. 3383, pp. 239–250. Springer, Heidelberg (2005)
10. Kieffer, S., Dwyer, T., Marriott, K., Wybrow, M.: Incremental grid-like layout using soft and hard constraints. In: Wismath, S., Wolff, A. (eds.) GD 2013. LNCS, vol. 8242, pp. 448–459. Springer, Heidelberg (2013)
11. Klauske, L.K.: Effizientes Bearbeiten von Simulink Modellen mit Hilfe eines spezifisch angepassten Layoutalgorithmus. Ph.D. thesis, Technische Universität Berlin (2012)
12. Klauske, L.K., Dziobek, C.: Improving modeling usability: Automated layout generation for Simulink. In: Proceedings of the MathWorks Automotive Conference, MAC 2010 (2010)
13. Lee, E.A., Neuendorffer, S., Wirthlin, M.J.: Actor-oriented design of embedded hardware and software systems. Journal of Circuits, Systems, and Computers (JCSC) 12(3), 231–260 (2003)
14. Purchase, H.C.: Which aesthetic has the greatest effect on human understanding? In: DiBattista, G. (ed.) GD 1997. LNCS, vol. 1353, pp. 248–261. Springer, Heidelberg (1997)
15. Rüegg, U., Kieffer, S., Dwyer, T., Marriott, K., Wybrow, M.: Stress-minimizing orthogonal layout of data flow diagrams with ports. Technical report (August 2014), http://arxiv.org/abs/1408.4626
16. Schulze, C.D., Spönemann, M., von Hanxleden, R.: Drawing layered graphs with port constraints. Journal of Visual Languages and Computing, Special Issue on Diagram Aesthetics and Layout 25(2), 89–106 (2014)
17. Spönemann, M., Fuhrmann, H., von Hanxleden, R., Mutzel, P.: Port constraints in hierarchical layout of data flow diagrams. In: Eppstein, D., Gansner, E.R. (eds.) GD 2009. LNCS, vol. 5849, pp. 135–146. Springer, Heidelberg (2010)
18. Sugiyama, K., Tagawa, S., Toda, M.: Methods for visual understanding of hierarchical system structures. IEEE Transactions on Systems, Man and Cybernetics 11(2), 109–125 (1981)
19. Wybrow, M., Marriott, K., Stuckey, P.J.: Orthogonal connector routing. In: Eppstein, D., Gansner, E.R. (eds.) GD 2009. LNCS, vol. 5849, pp. 219–231. Springer, Heidelberg (2010)

Planar Octilinear Drawings
with One Bend Per Edge[*]

Michael A. Bekos[1], Martin Gronemann[2], Michael Kaufmann[1], and Robert Krug[1]

[1] Wilhelm-Schickard-Institut für Informatik, Universität Tübingen, Germany
{bekos,mk,krug}@informatik.uni-tuebingen.de
[2] Institut für Informatik, Universität zu Köln, Germany
gronemann@informatik.uni-koeln.de

Abstract. In *octilinear drawings* of planar graphs, every edge is drawn as an alternating sequence of horizontal, vertical and diagonal line-segments. In this paper, we study octilinear drawings of low edge complexity, i.e., with few bends per edge. A k-planar graph is a planar graph in which each vertex has degree less or equal to k. In particular, we prove that every 4-planar graph admits a planar octilinear drawing with at most one bend per edge on an integer grid of size $O(n^2) \times O(n)$. For 5-planar graphs, we prove that one bend per edge still suffices in order to construct planar octilinear drawings, but in super-polynomial area. However, for 6-planar graphs we give a class of graphs whose planar octilinear drawings require at least two bends per edge.

1 Motivation

Drawing edges as octilinear paths plays a central role in the design of metro-maps (see e.g., [9,18,19]), which dates back to the 1930's when Henry Beck, an engineering draftsman, designed the first schematic map of London Underground using mostly horizontal, vertical and diagonal segments. Laying out networks in such a way is called *octilinear graph drawing*, i.e., an *octilinear drawing* of a (planar) graph $G = (V, E)$ of maximum degree eight is a (planar) drawing $\Gamma(G)$ of G in which each vertex occupies a point on the integer grid and each edge is drawn as a sequence of horizontal, vertical and diagonal line-segments.

For drawings of (planar) graphs to be readable, special care is needed to keep the number of bends small. However, the problem of determining whether a given embedded 8-planar graph (that is, a planar graph of maximum degree eight with given combinatorial embedding) admits a bendless octilinear drawing is NP-hard [17]. This negative result motivated us to study octilinear drawings of low *edge complexity*, that is, with few bends per edge. Surprisingly enough, very few results relevant to this problem were known, even if the octilinear model has been well-studied in the context of metro-map visualization and map schematization (see e.g. [21]). As an immediate byproduct

---

[*] The work of M.A. Bekos is implemented within the framework of the Action "Supporting Postdoctoral Researchers" of the Operational Program "Education and Lifelong Learning" (Action's Beneficiary: General Secretariat for Research and Technology), and is co-financed by the European Social Fund (ESF) and the Greek State.

C. Duncan and A. Symvonis (Eds.): GD 2014, LNCS 8871, pp. 331–342, 2014.
© Springer-Verlag Berlin Heidelberg 2014

of a result of Keszegh et al. [13], it turns out that every d-planar graph, with $3 \leq d \leq 8$, admits a planar octilinear drawing with at most two bends per edge; see Section 1. On the other hand, every 3-planar graph on five or more vertices admits a planar octilinear drawing in which all edges are bendless [6,12].

In this paper, we bridge the gap between the two aforementioned results. We prove that every 4-planar graph admits a planar octilinear drawing with at most one bend per edge in cubic area. We also show that every 5-planar graph also admits a planar octilinear drawing with at most one bend per edge, but our construction may require super-polynomial area. Finally, we demonstrate an infinite class of 6-planar graphs whose planar octilinear drawings require at least two bends per edge.

Related Work. The research on the (planar) *slope number of graphs* focuses on minimizing the number of used slopes (see e.g., [10,13,14,15,16]). Octilinear drawings can be seen as a special case thereof, since only four slopes are used. Keszegh et al. [13] showed that any d-planar graph admits a planar drawing with one bend per edge, in which all edge-segments have at most $2d$ different slopes. So, for $d = 4$ and $d = 5$, we reduce the number of different slopes from 8 and 10 to 4. They also proved that d-planar graphs, $d \geq 3$, admit planar drawings with two bends per edge that require at most $\lceil \frac{d}{2} \rceil$ different slopes. One can transfer this technique to the octilinear model and show that any d-planar graph, with $3 \leq d \leq 8$, admits a planar octilinear drawing with two bends per edge. For $d = 3$, Di Giacomo et al. [6] recently proved that any 3-planar graph with $n \geq 5$ vertices has a bendless planar drawing with at most 4 different slopes and angular resolution $\pi/4$ (see also [12]); their approach also yields octilinear drawings.

Tamassia [20] showed that one can minimize the total number of bends in orthogonal drawings of embedded 4-planar graphs. However, minimizing the number of bends over all embeddings of a 4-planar graph is NP-hard [7]. The core of Tamassia's approach is a min-cost flow algorithm that specifies the angles and the bends of the drawing, producing an *orthogonal representation*, and then computes the actual drawing by specifying the drawing's exact coordinates. Tamassia's algorithm can be employed to produce a bend-minimum octilinear representation for any given embedded 8-planar graph. However, a bend-minimum octilinear representation may not be realizable by a corresponding planar octilinear drawing [4].

Biedl and Kant [2] showed that any 4-planar graph except the octahedron admits a planar orthogonal drawing with at most two bends per edge on an $O(n^2)$ integer grid. Hence, the octilinear drawing model allows us to reduce the number of bends per edge at the cost of an increased area. On the other hand, not all 4-planar graphs admit orthogonal drawings with one bend per edge; however, testing whether a 4-planar graph admits such a drawing can be done in polynomial time [3]. In the context of metro-map visualization, several approaches have been proposed to produce metro-maps using octilinear or nearly-octilinear polylines; see e.g., [9,18,19].

Preliminaries. In our algorithms, we incrementally construct the drawings similar to the method of Kant [11]. We first employ the canonical order to cope with triconnected graphs. Then, we extend them to biconnected graphs using the SPQR-tree and to simply connected graphs using the BC-tree. In this section we briefly recall them.

Definition 1 (Canonical order [11]). *For a given triconnected plane graph $G = (V, E)$ let $\Pi = (P_0, \ldots, P_m)$ be a partition of V into paths such that $P_0 = \{v_1, v_2\}$, $P_m = \{v_n\}$ and $v_2 \to v_1 \to v_n$ is a path on the outer face of G. For $k = 0, \ldots, m$ let G_k be the subgraph induced by $\cup_{i=0}^{k} P_i$ and assume it inherits its embedding from G. Partition Π is a canonical order of G if for each $k = 1, \ldots, m - 1$ the following hold: (i) G_k is biconnected, (ii) all neighbors of P_k in G_{k-1} are on the outer face, of G_{k-1} (iii) all vertices of P_k have at least one neighbor in P_j for some $j > k$. P_k is called a singleton if $|P_k| = 1$ and a chain otherwise.*

Definition 2 (BC-tree). *The BC-tree $\mathcal{B}$ of a connected graph G has a B-node for each biconnected component of G and a C-node for each cutvertex of G. Each B-node is connected with the C-nodes that are part of its biconnected component.*

An SPQR-tree [8,5] provides information about the decomposition of a biconnected graph into its triconnected components. Every triconnected component is associated with a node μ in the SPQR-tree $\mathcal{T}$. The triconnected component itself is referred to as the *skeleton* of μ, denoted by $G_\mu^{skel} = (V_\mu^{skel}, E_\mu^{skel})$. We refer to the degree of a vertex $v \in V_\mu^{skel}$ in G_μ^{skel} as $deg_\mu^{skel}(v)$. We say that μ is an *R-node*, if G_μ^{skel} is a simple triconnected graph. A bundle of at least three parallel edges classifies μ as a *P-node*, while a simple cycle of length at least three classifies μ as an *S-node*. By construction R-nodes are the only nodes of the same type that are allowed to be adjacent in $\mathcal{T}$. The leaves of $\mathcal{T}$ are formed by the *Q-nodes*. Their skeleton consists of two parallel edges; one of them corresponds to an edge of G and is referred to as *real edge*. The skeleton edges that are not real are referred to as *virtual edges*. A virtual edge e in G_μ^{skel} corresponds to a tree node μ' that is adjacent to μ in $\mathcal{T}$, more exactly, to another virtual edge e' in $G_{\mu'}^{skel}$. We assume that $\mathcal{T}$ is rooted at a Q-node. Hence, every skeleton (except the one of the root) contains exactly one virtual edge $e = (s, t)$ that has a counterpart in the skeleton of the parent node. We call this edge the *reference edge* of μ denoted by $ref(\mu)$. Its endpoints, s and t, are named the *poles* of μ denoted by $\mathcal{P}_\mu = \{s, t\}$. Every subtree rooted at a node μ of $\mathcal{T}$ induces a subgraph of G called the *pertinent graph* of μ that we denote by $G_\mu^{pert} = (V_\mu^{pert}, E_\mu^{pert})$. We abbreviate the degree of a node v in G_μ^{pert} with $deg_\mu^{pert}(v)$. The pertinent graph is the subgraph of G for which the subtree describes the decomposition. The following lemmata provide useful properties of SPQR-trees. Due to lack of space, their proofs are given in [1].

Lemma 1. *Let μ be a tree node that is not the root in the SPQR-tree $\mathcal{T}$ of a simple, biconnected, k-planar graph G and μ' its parent in $\mathcal{T}$. For $v \in \mathcal{P}_\mu$, it holds that $deg_\mu^{pert}(v) \leq k - 2$, if μ' is a P- or an R-node and $deg_\mu^{pert}(v) \leq k - 1$ otherwise, i.e. μ' is an S- or a Q-node.*

Lemma 2. *In the SPQR-tree $\mathcal{T}$ of a planar biconnected graph $G = (V, E)$ with $\deg(v) \geq 3$ for every $v \in V$, there exists at least one Q-node that is adjacent to a P- or an R-node.*

2   Octilinear Drawings of 4-Planar Graphs

In this section, we focus on octilinear drawings of 4-planar graphs. We first consider the triconnected case and then we extend it to biconnected and simply connected graphs.

The Triconnected Case. Let $G = (V, E)$ be a triconnected 4-planar graph and $\Pi = \{P_0, \ldots, P_m\}$ be a canonical order of G. We momentarily neglect the edge (v_1, v_2) of the first partition P_0 of Π and we start by placing the second partition, say a chain $P_1 = \{v_3, \ldots, v_{|P_1|+2}\}$, on a horizontal line from left to right. Since v_3 and $v_{|P_1|+2}$ are adjacent to v_1 and v_2, we place v_1 to the left of v_3 and v_2 to the right of $v_{|P_1|+2}$. So, they form a single chain where all edges are drawn using horizontal line-segments that are attached to the east and west port at their endpoints. The case where P_1 is a singleton is analogous. Having laid out the base of our drawing, we now place in an incremental manner the remaining partitions. Assume that we have already constructed a drawing for G_{k-1} and we now have to place P_k, for some $k = 2, \ldots, m - 1$.

In case where $P_k = \{v_i, \ldots, v_j\}$ is a chain of $j - i + 1$ vertices, we draw them from left to right along a horizontal line one unit above G_{k-1}. Since v_i and v_j are the only vertices that are adjacent to vertices in G_{k-1}, both only to one, we place the chain between those two as in Fig.1a. The port used at the endpoints of P_k in G_{k-1} depends on the following rule: Let v_i' (v_j', resp.) be the neighbor of v_i (v_j, resp.) in G_{k-1}. If the edge (v_i, v_i') ((v_j, v_j'), resp.) is the last to be attached to vertex v_i' (v_j', resp.), i.e., there is no vertex v in $P_l \in \Pi$, $l > k$ such that $(v_i', v) \in E$ ($(v_j', v) \in E$, resp.), then we use the northern port of v_i' (v_j', resp.). Otherwise, we choose the north-east port for (v_i, v_i') or the north-west port for (v_j, v_j').

In case of a singleton $P_k = \{v_i\}$, we can apply the previous rule if the singleton is of degree three. However, if v_i is of degree four, then we may have to deal with an additional third edge (v_i, v) that connects v_i with G_{k-1}. However, we may assume that v lies between the other two endpoints, thus, we place v_i such that $x(v_i) = x(v)$. This enables us to draw (v_i, v) as a vertical line-segment; see Fig.1b.

The above procedure is able to handle all chains and singletons except the last partition P_m, as v_n may have 4 edges pointing downwards. We exclude (v_n, v_1) and draw v_n as an ordinary singleton. Then, we shift v_1 to the left and up as in Fig.1c and draw (v_1, v_n) as a horizontal-vertical segment combination. Vertex v_2 is analogously moved. The drawings of the remaining edges incident to v_n are depicted in Fig.1c.

A *cut* is a y-monotone continuous curve that crosses only horizontal segments and divides the current drawing into a left and a right part. Since every edge, except the ones drawn as vertical line-segments, contains exactly one horizontal segment, we can shift the right part of the drawing that is defined by the cut further to the right while keeping the left part of the drawing on place and the result remains a valid octilinear drawing.

To compute the x-coordinates, we first assign consecutive x-coordinates to the first two partitions. We may have to stretch the drawing in two cases: (i) when we introduce a chain, say P_k, as it may not fit into the gap defined by its two adjacent vertices in G_{k-1}, and, (ii) when an edge that contains a diagonal segment is to be drawn, to prevent it from intersecting any horizontal-vertical combinations in the face below it. We can cope with both cases by horizontally stretching the drawing by a factor that is bounded by the current height of the drawing. Since the height of the resulting drawing is bounded by $|\Pi| = O(n)$, it follows that in the worst case its width is $O(n^2)$. Note that our algorithm produces drawings that have a linear number of bends in total (in particular, exactly $2|\Pi| = O(n)$ bends). One can prove that this bound is asymptotically tight (see [1]). We are now ready to state the main theorem of this subsection.

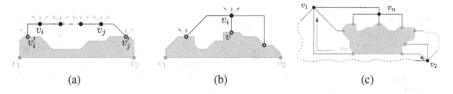

(a)   (b)   (c)

Fig. 1. (a) Horizontal placement of a chain $P_k = \{v_i, \ldots, v_j\}$. (b) Placement of a singleton $P_k = \{v_i\}$ with degree four. (c) Final layout after repositioning v_1 and v_2 (the shape of the dotted edges can be obtained by extending the stubs until they intersect).

Theorem 1. *Given a triconnected 4-planar graph G, we can compute in $O(n)$ time an octilinear drawing of G with at most 1 bend per edge on an $O(n^2) \times O(n)$ integer grid.*

Proof. In order to keep the time complexity of our algorithm linear, we employ a simple trick. We assume that any two adjacent points of the underlying integer grid are by n units apart in the horizontal direction and by one unit in the vertical direction. This a priori ensures that all edges that contain a diagonal segment will not be involved in crossings and simultaneously does not affect the total area of the drawing, which asymptotically remains cubic. On the other hand, the advantage of this approach is that we can use the shifting method of Kant [11] to cope with the introduction of chains in the drawing, that needs $O(n)$ time in total by keeping relative coordinates that can be efficiently updated and computing the absolute values only at the last step.   □

The Biconnected Case. Consider a node μ in the rooted SPQR-tree $\mathcal{T}$ of G with poles $\mathcal{P}_\mu = \{s, t\}$. In the drawing of G_μ^{pert}, s should be located at the upper-left and t at the lower-right corner of the drawing's bounding box with a port assignment as in Fig.2a. We also assume that the edges incident to s (t, resp.) use the western (eastern, resp.) port at their other endpoint, except of the northern (southern, resp.) most edge which may use the north (south, resp.) port instead. In that case we refer to s and t as *fixed*; see $\overline{e}_s, \overline{e}_t$ in Fig.2a. We maintain the following invariants:

IP-1: The width (height) of the drawing of μ is quadratic (linear) in the size of G_μ^{pert}. s is located at the upper-left; t at the lower-right corner of the drawing's bounding box.

IP-2: If $deg_\mu^{pert}(s) \geq 2$, s is fixed; t is fixed if $deg_\mu^{pert}(t) = 3$ and μ's parent is not the root.

IP-3: The edges that are incident at s and t in G_μ^{pert} use the south, south-east and east ports at s and the north, north-west and west port at t, resp. If s or t is not fixed, incident edges are attached at their other endpoints via the west and east port, respectively. If s or t is fixed, the northern-most edge at s and the southern-most edge at t may use the north (south, resp.) port at its other endpoint.

The port assignment, i.e. IP-3, guarantees the ability to stretch the drawing horizontally even in the case where both poles are fixed. Furthermore, IP-2 is *interchangeable* in the following sense: If $deg_\mu^{pert}(s) = 2$ and $deg_\mu^{pert}(t) = 1$, then s is fixed but t is

Fig. 2. (a) Schematic view of the layout requirements. (b) Creating a nose at t. (c) First P-node subcase without an (s, t)-edge but s might be fixed in a child μ_1. (d) Second P-node subcase with an (s, t)-edge where t might get fixed in a child μ_2.

not. But, if we relabel s and t such that $t' = s$ and $s' = t$, then $deg_\mu^{pert}(s') = 1$ and $deg_\mu^{pert}(t') = 2$. By IP-2, we can create a drawing where both s' and t' are not fixed and located in the upper-left and lower-right corner of the drawing's bounding box. Afterwards, we mirror the resulting layout vertically and horizontally to obtain one where s and t are in their respective corners and not fixed. For a non-fixed vertex, we introduce an operation referred to as forming or creating a *nose*; see Fig.2b, where t has been moved downwards at the cost of a bend. As a result, its west port is no longer occupied.

First consider the case where μ is a P-node. If there is no (s, t)-edge, then we draw the children of μ from top to bottom such that a possible child in which s is fixed, is drawn topmost (see μ_1 in Fig.2c). If there is an (s, t)-edge, then we draw it at the top and afterwards the remaining children of μ (see Fig.2d). This is possible only if s is not fixed in any of the other children. Let μ' be such a potential child where s is fixed, i.e., $deg_{\mu'}^{pert}(s) = 2$, and thus, the only child that remains to be drawn. Here, we use the property of interchangeability to "unfix" s in μ'. As a result s can form a nose, whereas t may now be fixed in μ' when $deg_{\mu'}^{pert}(t) = 2$ holds, as in Fig.2d. However, then $deg_\mu^{pert}(t) = 3$ follows. Notice that the presence of an (s, t)-edge implies that the parent of μ is not the root of $\mathcal{T}$, since this would induce a pair of parallel edges. Hence, by IP-2 we are allowed to fix t in μ. Port assignment and area requirements comply in both cases with our invariant properties.

In the case where μ is an S-node, we place the drawings of its children, say $\mu_1, \ldots, \mu_\ell$ in a "diagonal manner" such that their corners touch as in Fig.3a. In case of Q-nodes being involved, we draw their edges as horizontal segments (see, e.g., (v_3, v_4) in Fig.3a). s and t inherit their port assignment and pertinent degree from μ_1 and μ_ℓ, respectively. So, s (t, resp.) is fixed in μ, if it is fixed in μ_1 (μ_ℓ, resp.). By IP-2, t is not allowed to be fixed in the case where the parent of μ is the root of $\mathcal{T}$. However, from Lemma 2 we can choose the root such that t is not fixed in that case, and thus, complies with IP-2. Since we only concatenated the drawings of the children, IP-1 and IP-3 are satisfied.

For the case where μ is an R-node with poles $\mathcal{P}_\mu = \{s, t\}$, we follow the idea of the triconnected algorithm and describe only the required modifications. We assume the worst case where no child of μ is a Q-node. Let μ_{uv} be the child that is represented by the virtual edge $(u, v) \in E_\mu^{skel}$. Due to Lemma 1, $deg_{\mu_{uv}}^{pert}(u) \leq 2$ and $deg_{\mu_{uv}}^{pert}(v) \leq 2$ holds. By IP-2 we may assume that either u or v is fixed in μ_{uv} and choose the first partition in the canonical ordering to be $P_0 = \{s, t\}$.

In case of a chain, say $P_k = \{v_i, \ldots, v_j\}$ with two neighbors v_i' and v_j' in G_{k-1}, we have to replace two types of edges with the drawings of the corresponding children:

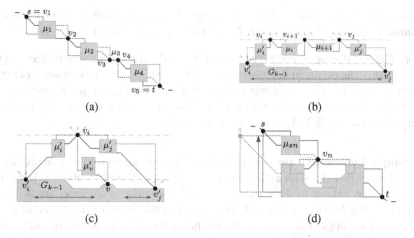

Fig. 3. (a) S-node with children $\mu_1, \ldots, \mu_4$; μ_3 is a Q-node representing the edge (v_3, v_4). Optional edges are drawn dotted. (b) Example for a chain $v_i, \ldots, v_j$ with virtual edges representing $\mu_i, \ldots, \mu_{j-1}$ in the R-node case. (c) Singleton v_i with possibly three incident virtual edges representing μ'_i, μ'_v, μ'_j. (d) Placing v_n and moving up s which might be fixed in μ_{sn}.

the edges $(v_i, v_{i+1}), \ldots, (v_{j-1}, v_j)$ representing the children $\mu_i, \ldots, \mu_{j-1}$ and (v'_i, v_i) $((v_j, v'_j)$ resp.) representing μ'_i (μ'_j resp.). We place the vertices of P_k on a horizontal line high enough above G_{k-1} so that every drawing fits in-between. Then, we insert the drawings aligned below the horizontal line and choose for $i \le l < j$, v_l to be the fixed node in μ_l, whereas in μ'_i (μ'_j resp.), we set v_i (v_j resp.) to be fixed. So, for $i \le l < j$, v_{l+1} may form a nose in μ_l pointing upwards while v'_i and v'_j form each one downwards as in Fig.3b. For the extra height and width, we stretch the drawing horizontally.

In case where $P_k = \{v_i\}$, $i \ne n$ is a singleton, we only outline the difference which is a possible third edge (v_i, v) to G_{k-1} representing say μ'_v. While the other two involved children, say μ'_i and μ'_j, are handled as in the chain-case, μ'_v requires extra height. We place v_i so that μ'_v fits below μ'_j as in Fig.3c. Notice that $deg^{pert}_{\mu'_v}(v_i) = 1$ and by IP-2 both v_i and v are not fixed in μ'_v. So, forming a nose at v_i and v is feasible.

For the last singleton $P_k = \{v_n\}$, observe that since $s, t \in P_0$, both have not been fixed. As in the triconnected algorithm we move $s = v_1$ above v_n as in Fig.3d to accommodate the drawing of the child μ_{sn} represented by the edge (s, v_n). Since we may require v_n to form a nose in μ_{sn} as in Fig.3d, we choose s to be fixed in μ_{sn}. By IP-2 we are allowed to fix s since t remains unfixed. Although some diagonal segments may force us to stretch the whole drawing by its height, the height of the drawing has been kept linear in the size of G^{pert}_μ. Since we increase the width by the height a constant number of times per step, the resulting width remains quadratic.

If there is a vertex $v \in V$ with $\deg(v) \le 3$, then we root $\mathcal{T}$ at a Q-node μ that represents one of its three incident edges and orient the poles $\{s, t\}$ such that $t = v$. So, for the child μ' of μ follows $deg^{pert}_{\mu'}(t) \le 2$. If $\deg(v) = 4$ for every $v \in V$, then we root $\mathcal{T}$ at a Q-node that is not adjacent to an S-node, which exists due to Lemma 2. In both cases, we may form a nose with t pointing downwards.

Theorem 2. *Given a biconnected 4-planar graph G, we can compute in $O(n)$ time an octilinear drawing of G with at most 1 bend per edge on an $O(n^2) \times O(n)$ integer grid.*

Proof. The SPQR-tree $\mathcal{T}$ can be computed in $O(n)$-time and its size is linear to the size of G [8]. The pertinent degrees of the poles at every node can be pre-computed by a bottom-up traversal of $\mathcal{T}$. Drawing a P-node requires constant time; S- and R-nodes require time linear to the size of the skeleton. However, the sum over all skeleton edges is linear, as every virtual edge corresponds to a tree node. □

The Simply Connected Case. The main idea of our algorithm is to root the BC-tree at some arbitrary B-node. With exception of the root, every B-node contains a designated cut vertex that links it to the parent. Similar to the biconnected case, we define an invariant for the drawing of a subtree. The cut vertex that links the subtree to the parent is located in the upper left corner of the bounding box. Due to lack of space, we only state the main result; its proof is given in [1].

Theorem 3. *Given a connected 4-planar graph G, we can compute in $O(n)$ time an octilinear drawing of G with at most 1 bend per edge on an $O(n^2) \times O(n)$ integer grid.*

3 Octilinear Drawings of 5-Planar Graphs

In this section, we focus on planar octilinear drawings of 5-planar graphs. As in Section 2, we first consider the case of triconnected 5-planar graphs and then we extend our approach first to biconnected and then to the simply connected graphs.

The Triconnected Case. Let $G = (V, E)$ be a triconnected 5-planar graph and $\Pi = \{P_0, \ldots, P_m\}$ be a canonical order of G. We place partitions P_0 and P_1 similar to the 4-planar case. Assume that we have already constructed a drawing for G_{k-1} which is stretchable in the following sense: If $e \in E(G_{k-1})$ is an edge incident to the outer face, then there is a cut which crosses e and can be utilized to horizontally stretch the drawing of G_{k-1}. In other words, one can define a cut through every edge incident to the outer face of G_{k-1} (*stretchability-invariant*).

If $P_k = \{v_i, \ldots, v_j\}$ is a chain, it is placed exactly as in the case of 4-planar graphs, but with different port assignment. Among the northern available ports of vertex v_i' (v_j', resp.), edge (v_i, v_i') ((v_j, v_j'), resp.) uses the eastern-most unoccupied port of v_i' (western-most unoccupied port of v_j', resp.); see Fig.4a. If P_k does not fit into the gap between v_i' and v_j' in G_{k-1}, then we horizontally stretch G_{k-1} between v_i' and v_j' to ensure that the horizontal distance between v_i' and v_j' is at least $|P_k| + 1$. This can be done due to the stretchability-invariant, as both v_i' and v_j' are on the outer face of G_{k-1}. Potential crossings introduced by edges of P_k containing diagonal segments can be eliminated by employing similar cuts to the ones presented in the 4-planar case. So, we may assume that G_k is plane. Also, G_k complies with the stretchability-invariant, as one can define a cut that crosses any of the newly inserted edges of P_k and then follows one of the cuts of G_{k-1} that crosses an edge between v_i' and v_j'.

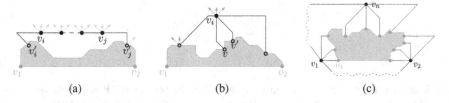

Fig. 4. (a) Horizontal placement of a chain $P_k = \{v_i, \ldots, v_j\}$. (b) Placement of a singleton $P_k = \{v_i\}$ of degree five. (c) Final layout (the shape of the dotted edges can be obtained by extending the stubs until they intersect).

In case of a singleton $P_k = \{v_i\}$ of degree 5, we have to deal with two additional edges (called *nested*) that connect v_i with G_{k-1}, say (v_i, v) and (v_i, v'); see Fig.4b. Such a pair of edges does not always allow vertex v_i to be placed along the next available horizontal grid line. A careful case analysis on the type of ports that are unoccupied at v and v' in conjunction with the fact that G_{k-1} is horizontally stretchable shows that we can find a feasible placement for v_i. Potential crossings due to the remaining edges incident to v_i are eliminated by employing similar cuts to the ones presented in the 4-planar case. So, G_k is planar. Also, G_k complies with the stretchability-invariant. The last partition $P_m = \{v_n\}$ is treated in the same way, even if v_n can be incident to three nested edges. Since v_1 and v_2 are along a common horizontal line, (v_1, v_2) can be drawn using two diagonal segments that form a bend pointing downwards; see Fig.4c. Note that our algorithm may result in drawings of super-polynomial area, as proven in [1].

Theorem 4. *Given a triconnected 5-planar graph G, we can compute in $O(n^2)$ time an octilinear drawing of G with at most one bend per edge.*

Proof. We can no longer use the shifting method of Kant [11], since the x- and y-coordinates are not independent. However, the computation of each cut can be done in linear time, which implies that our drawing algorithm needs $O(n^2)$ time in total.   □

The Biconnected Case. For the 4-planar case we defined several invariants in order to keep the area of the resulting drawings polynomial. Since we drop this requirement now we can define a simpler new invariant for the biconnected 5-planar case. When considering a node μ in $\mathcal{T}$ and its poles $\mathcal{P}_\mu = \{s, t\}$, then in the drawing of G_μ^{pert}, s and t are horizontally aligned at the bottom of the drawing's bounding box as in Fig.5a. If an (s, t)-edge is present, it can be drawn at the bottom. An (s, t)-edge only occurs in the pertinent graph of a P-node (and Q-node). We use the term *fixed* for a pole-node that is not allowed to form a nose. We maintain the following properties through the recursive construction process: In S- and R- nodes, s and t are not fixed. In P- and Q-nodes, only one of them is fixed, say s. But as in the 4-planar case, we may swap their roles.

If μ is a P-node, then it has at most 4 children; one of them might be a Q-node, i.e., an (s, t)-edge, which can be drawn at the bottom as a horizontal segment. Since P-nodes are not adjacent to each other in $\mathcal{T}$, the remaining children are S- or R-nodes. By our invariant we may form noses enabling us to stack them as in Fig.5b.

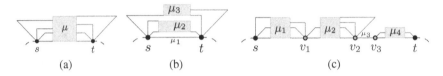

Fig. 5. (a) Layout specification; s and t are located at the bottom. (b) P-node with an (s,t)-edge from a Q-node μ_1. s and t form a nose in μ_2, μ_3. (c) S-node example with four children $\mu_1, \ldots, \mu_4$.

In case where μ is an S-node, we align its children $\mu_1, \ldots, \mu_l$ horizontally; see Fig.5c. The poles inherit their pertinent degree from the children. The same holds for the property of being fixed. However, by our new invariant this is forbidden, as it requires that s and t are not fixed. It is easy to see that when μ_1 is a P-node, s is fixed by the invariant in μ_1. In this case, we swap the roles of the poles in μ_1 such that s is not fixed. However, the other pole of μ_1, say v_1, is fixed now. Since the skeleton of an S-node is a cycle of length at least three, $v_1 \neq t$. So, s and t are not fixed in the resulting drawing.

To compute a layout of an R-node μ, we employ the triconnected algorithm (with $s = v_1$ and $t = v_2$). Let μ_e be a child of μ that corresponds to virtual edge $e = (u,v)$ in G_μ^{skel}. Then, $deg_{\mu_e}^{pert}(u), deg_{\mu_e}^{pert}(v) \leq 3$. When inserting the drawing of $G_{\mu_e}^{pert}$, we require at most three consecutive ports at u and v for the additional edges. As the triconnected algorithm assigns ports in a consecutive manner based on the relative position of the endpoints, we modify the port assignment so that an edge may have more than one port assigned. To do so, we assign each edge $e = (u,v)$ in G_μ^{skel} a pair $(deg_{\mu_e}^{pert}(u), deg_{\mu_e}^{pert}(v)) \in \{1,2,3\}^2$ that reflects the number of ports required by this edge at its endpoints. Then, we extend the triconnected algorithm such that when a port of u is assigned to an edge $e = (u,v)$, $deg_{\mu_e}^{pert}(u) - 1$ additional consecutive ports in clockwise or counterclockwise order are reserved. The direction depends on the different types of edges that we will discuss next.

The simplest type of edges are the ones among consecutive vertices v_i, v_{i+1} of a chain. For each such edge we reserve the additional ports at v_i in counterclockwise and at v_{i+1} in clockwise order; see Fig.6a. So, we can later plug the drawing of the children into the layout as in Fig.6b without forming noses. In the same manner, we reserve the ports for the edges that connect $P_k = \{v_i, \ldots, v_j\}$ to v_i' and v_j' in G_{k-1} (where P_k is singleton or chain), i.e., at v_i clockwise, (v_j counter-clockwise, resp.) and at v_i' counter-clockwise (v_j' clockwise); see Fig.6c. In case where (v_i, v_i') or (v_j, v_j') is a virtual edge, we choose the poles such that v_i (v_j resp.) is fixed in $\mu_{(v_i, v_i')}$ ($\mu_{(v_j, v_j')}$ resp.). Thus, we can create a nose with v_i' (v_j' resp.). Having exactly the ports required at both endpoints, we insert the drawing by replacing the bend with a nose as in Fig.6d. The remaining edges from P_k to G_{k-1} in case of a singleton $P_k = \{v_i\}$ are handled similarly; see Fig.6. During the replacement of the edges, the fixed vertex is always the upper one. The only exception are the horizontal drawn edges of a chain, for which it does not matter which one is fixed. Finally, we root $\mathcal{T}$ at an arbitrarily chosen Q-node representing a real edge (s,t). By our invariant we may construct a drawing with s and t at the bottom of the drawing's bounding box, so that (s,t) has a $90°$ bend downwards.

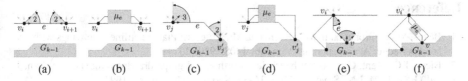

Fig. 6. (a) Virtual edge $e = (v_i, v_{i+1})$ connecting two consecutive vertices of a chain. At both endpoints the drawing of μ_e requires two ports. (b) Replacing e in (a) with the corresponding drawing of the child μ_e. (c) Example of an edge $e = (v_j, v'_j)$ that requires three ports at v_j and two at v'_j. (d) Inserting the drawing of μ_e into (c) with v_j being fixed and v'_j forming a nose. (e) Reserving ports for the nested edges. A single port for a real edge is reserved and then two ports for the virtual edge e $= (v_i, v)$. (f) Final layout after inserting the drawing of μ_e.

Theorem 5. *Given a biconnected 5-planar graph G, we can compute in $O(n^2)$ time an octilinear drawing of G with at most one bend per edge.*

Proof. The ability to rotate and scale suffices to extend the result from 4-planar to 5-planar at the expense of the area. Similar to the 4-planar case, computing $\mathcal{T}$ takes linear time. Hence, the overall runtime is governed by the triconnected algorithm.   □

The Simply Connected Case. Due to lack of space, we outline the differences in comparison to the 4-planar case in [1]. Here, we simply state the main theorem.

Theorem 6. *Given a connected 5-planar graph G, we can compute in $O(n^2)$ time an octilinear drawing of G with at most one bend per edge.*

4   A Note on Octilinear Drawings of 6-Planar Graphs

In this section, we show that it is not always possible to construct a planar octilinear drawing of a given 6-planar graph with at most one bend per edge.

Theorem 7. *There exists an infinite class of 6-planar graphs which do not admit planar octilinear drawings with at most one bend per edge.*

Sketch of Proof. Due to lack of space the detailed proof of this theorem is given in [1]. The main idea is to construct an infinite class of maximal 6-planar graphs, whose outer face is always delimited by exactly three vertices, say v, v' and v'', such that $deg(v) = deg(v') = 6$ and $5 \le deg(v'') \le 6$. Then, it is not difficult to prove that it is not feasible to draw all edges incident to the outer face with at most one bend per edge.   □

5   Conclusions

We presented algorithms for the construction of planar octilinear drawings with at most one bend per edge for 4- and 5-planar graphs. Our work raises several open problems: (i) Is it possible to construct planar octilinear drawings of 4-planar (5-planar) graphs with at most one bend per edge in $o(n^3)$ (polynomial, resp.) area? (ii) Does any triangle-free 6-planar graph admit a planar octilinear drawing with at most one bend per edge? (iii) What is the number of necessary slopes for bendless drawings of 4-planar graphs?

References

1. Bekos, M.A., Gronemann, M., Kaufmann, M., Krug, R.: Planar octilinear drawings with one bend per edge. Arxiv report arxiv.org/abs/1408.5920 (2014)
2. Biedl, T.C., Kant, G.: A better heuristic for orthogonal graph drawings. In: van Leeuwen, J. (ed.) ESA 1994. LNCS, vol. 855, pp. 24–35. Springer, Heidelberg (1994)
3. Bläsius, T., Krug, M., Rutter, I., Wagner, D.: Orthogonal graph drawing with flexibility constraints. Algorithmica 68(4), 859–885 (2014)
4. Bodlaender, H.L., Tel, G.: A note on rectilinearity and angular resolution. Journal of Graph Algorithms and Applications 8(1), 89–94 (2004)
5. Di Battista, G., Tamassia, R.: On-line graph algorithms with SPQR-trees. In: Paterson, M. (ed.) ICALP 1990. LNCS, vol. 443, pp. 598–611. Springer, Heidelberg (1990)
6. Di Giacomo, E., Liotta, G., Montecchiani, F.: The planar slope number of subcubic graphs. In: Pardo, A., Viola, A. (eds.) LATIN 2014. LNCS, vol. 8392, pp. 132–143. Springer, Heidelberg (2014)
7. Garg, A., Tamassia, R.: On the computational complexity of upward and rectilinear planarity testing. SIAM Journal on Computing 31(2), 601–625 (2001)
8. Gutwenger, C., Mutzel, P.: A linear time implementation of SPQR-trees. In: Marks, J. (ed.) GD 2000. LNCS, vol. 1984, pp. 77–90. Springer, Heidelberg (2001)
9. Hong, S.H., Merrick, D., do Nascimento, H.A.D.: Automatic visualisation of metro maps. Journal of Visual Languages and Computing 17(3), 203–224 (2006)
10. Jelínek, V., Jelínková, E., Kratochvíl, J., Lidický, B., Tesar, M., Vyskocil, T.: The planar slope number of planar partial 3-trees of bounded degree. Graphs and Combinatorics 29(4), 981–1005 (2013)
11. Kant, G.: Drawing planar graphs using the lmc-ordering. In: 33rd Annual Symposium on Foundations of Computer Science (FOCS 1992), pp. 101–110. IEEE (1992)
12. Kant, G.: Hexagonal grid drawings. In: Mayr, E.W. (ed.) WG 1992. LNCS, vol. 657, pp. 263–276. Springer, Heidelberg (1993)
13. Keszegh, B., Pach, J., Pálvölgyi, D.: Drawing planar graphs of bounded degree with few slopes. SIAM Journal of Discrete Mathematics 27(2), 1171–1183 (2013)
14. Keszegh, B., Pach, J., Pálvölgyi, D., Tóth, G.: Drawing cubic graphs with at most five slopes. Computational Geometry 40(2), 138–147 (2008)
15. Lenhart, W., Liotta, G., Mondal, D., Nishat, R.I.: Planar and plane slope number of partial 2-trees. In: Wismath, S., Wolff, A. (eds.) GD 2013. LNCS, vol. 8242, pp. 412–423. Springer, Heidelberg (2013)
16. Mukkamala, P., Pálvölgyi, D.: Drawing cubic graphs with the four basic slopes. In: van Kreveld, M., Speckmann, B. (eds.) GD 2011. LNCS, vol. 7034, pp. 254–265. Springer, Heidelberg (2011)
17. Nöllenburg, M.: Automated drawings of metro maps. Tech. Rep. 2005-25, Fakultät für Informatik, Universität Karlsruhe (2005)
18. Nöllenburg, M., Wolff, A.: Drawing and labeling high-quality metro maps by mixed-integer programming. IEEE Transactions on Visualization and Computer Graphics 17(5), 626–641 (2011)
19. Stott, J.M., Rodgers, P., Martinez-Ovando, J.C., Walker, S.G.: Automatic metro map layout using multicriteria optimization. IEEE Transactions on Visualization and Computer Graphics 17(1), 101–114 (2011)
20. Tamassia, R.: On embedding a graph in the grid with the minimum number of bends. SIAM Journal of Computing 16(3), 421–444 (1987)
21. Wolff, A.: Graph drawing and cartography. In: Tamassia, R. (ed.) Handbook of Graph Drawing and Visualization, ch. 23, pp. 697–736. CRC Press (2013)

On the Complexity of HV-rectilinear Planarity Testing*

Walter Didimo[1], Giuseppe Liotta[1], and Maurizio Patrignani[2]

[1] Dept. of Engineering, University of Perugia, Italy
[2] Dept. of Engineering, Roma Tre University, Italy

Abstract. An HV-restricted planar graph G is a planar graph with vertex-degree at most four and such that each edge is labeled either H (horizontal) or V (vertical). The *HV-rectilinear planarity testing* problem asks whether G admits a planar drawing where every edge labeled V is drawn as a vertical segment and every edge labeled H is drawn as a horizontal segment. We prove that HV-rectilinear planarity testing is NP-complete even for graphs having vertex degree at most three, which solves an open problem posed by both Manuch *et al.* (*GD 2010*) and Durucher *et al.* (*LATIN 2014*). We also show that HV-rectilinear planarity can be tested in polynomial time for partial 2-trees of maximum degree four, which extends a previous result by Durucher *et al.* (*LATIN 2014*) about HV-restricted planarity testing of biconnected outerplanar graphs of maximum degree three. When the test is positive, our algorithm returns an orthogonal representation of G that satisfies the given H- and V-labels on the edges.

1 Introduction

Let $G = (V, E)$ be a planar graph with vertex-degree at most four. A *rectilinear orthogonal drawing* Γ of G is a planar drawing of G where each vertex $v \in V$ corresponds to a distinct point p_v of the plane and each edge $(u, v) \in E$ corresponds to a horizontal or vertical segment between p_u and p_v. If G is a planar embedded graph, i.e., a graph with a given planar embedding (a planar embedding defines for each vertex $v \in V$ the circular order of the edges incident to v and it also specifies the external face), we assume that a rectilinear orthogonal drawing of G preserves its embedding. If v is a vertex of a planar embedded graph and if e_1 and e_2 are two (possibly coincident) edges of v that are consecutive in the clockwise order around v, we say that $a = \langle e_1, v, e_2 \rangle$ is an *angle at* v *of* G or simply an *angle* of G. Two rectilinear orthogonal drawings Γ and Γ' of the same planar embedded graph G are *shape equivalent* if for any angle a of G, the geometric angle corresponding to a is the same in Γ and Γ'. A *rectilinear orthogonal representation* $\mathcal{H}$ of a planar embedded graph G is a class of shape equivalent rectilinear orthogonal drawings of G; $\mathcal{H}$ can be described by the embedding of G equipped with a label for each angle of G; the labels for an angle can be R, F, L, or LL, corresponding to a geometric angle of 90, 180, 270, and 360 degrees, respectively.

The *rectilinear planarity testing* problem asks whether a planar graph G with vertex-degree at most four admits a rectilinear orthogonal drawing (or equivalently, a rectilinear orthogonal representation). The problem can be solved in polynomial time in the

---

* This research is supported in part by the Italian Ministry of Education, University, and Research (MIUR) under PRIN 2012C4E3KT national research project "AMANDA – Algorithmics for MAssive and Networked DAta"

C. Duncan and A. Symvonis (Eds.): GD 2014, LNCS 8871, pp. 343–354, 2014.

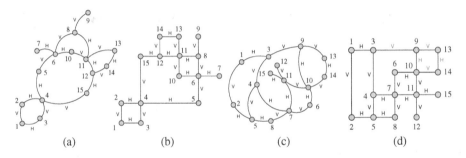

Fig. 1. (a) An HV-graph. (b) An HV-drawing of the HV-graph of Fig. 1(a). (c) An HV-graph that does not admit an HV-drawing. (d) A rectilinear orthogonal drawing of the graph of Fig. 1(c).

fixed embedding setting, i.e., when the rectilinear orthogonal representation must preserve a planar embedding of G given as part of the input [18]. The rectilinear planarity testing problem is however NP-complete in the *variable embedding setting*, i.e., over all planar embeddings of G [11]. Polynomial-time solutions exist if G is a biconnected series-parallel graph or if G has maximum vertex-degree three (see, e.g., [7,16,20]).

The rectilinear planarity testing problem is a classical subject of investigation in the graph drawing literature, and it can be regarded as a special case of the bend minimization problem for orthogonal drawings, probably one of the most explored topics in the field (see, e.g., [1,2,7,10,14,15,18]). Direction constrained versions of the rectilinear planarity testing problem also have a long tradition in the literature. An example is when every edge of G is labeled 'Up", or "Down", or "Left", or "Right" and one wants to test whether G has a rectlinear drawing where every edge has a direction consistent with its label (see, e.g. [12,19]); the 3D version of this problem has also been studied [5,6,8].

In this paper we study another direction constrained version of the rectilinear planarity testing problem that has been receiving increasing interest. An *HV-restricted planar graph* (*HV-graph* for short) $G = (V, E)$ is a planar graph with vertex-degree at most four such that each edge is labeled either H (horizontal) or V (vertical). Denote by $E_H \subset E$ and $E_V \subseteq E$ the subsets of H- and V-labeled edges, respectively. An *HV-drawing* of G is a rectilinear orthogonal drawing of G such that each edge $e \in E_H$ (resp. $e \in E_V$) corresponds to an orthogonal (resp. vertical) segment. An *HV-realization* of G is a rectilinear orthogonal representation $\mathcal{H}$ of G such that every drawing of $\mathcal{H}$ is an HV-drawing of G, up to a rotation of 90 degrees. The *HV-rectilinear planarity testing* problem asks whether an HV-graph admits an HV-realization. Figure 1 (b) shows an HV-drawing of the HV-graph in Fig. 1 (a). Figure 1 (c) shows an HV-graph that does not admit HV-realizations, although it admits a rectilinear orthogonal drawing (Fig. 1 (d)).

Manuch *et al.* [13] ask what is the time complexity of HV-rectilinear planarity testing both in the fixed and in the variable embedding setting. Durocher *et al.* [9] describe a polynomial-time testing algorithm in the fixed embedding setting; for the variable embedding setting they present a quadratic-time testing algorithm for biconnected outerplanar graphs of vertex-degree at most three. The authors leave as open problem both to characterize the outerplanar HV-graphs that admit an HV-realization and to establish the complexity of HV-rectilinear planarity testing in the variable embedding setting for general HV-graphs. We study these problems and establish the following results:

(i) HV-rectilinear planarity testing is NP-complete in the variable embedding setting even for HV-graphs with vertex-degree at most three. We recall, for a contrast, that Garg and Tamassia proved the NP-completeness of rectlinear planarity testing for planar graphs of vertex-degree at most four [11], but that rectlinear planarity testing can be solved in linear time for planar graphs of vertex-degree at most three [16].

(ii) There exists a polynomial-time algorithm to test HV-rectilinear planarity for HV-graphs that are partial 2-trees. In the affirmative case, the algorithm returns an HV-realization of the graph. Recall that biconnected outerplanar graphs are a sub-family of partial 2-trees, hence our result provides an algorithmic answer to the open problem of Durocher et al. [9], even for graphs with maximum vertex-degree four.

The remainder of the paper is organized as follows. Section 2 proves that HV- rectilinear planarity testing is NP-complete. Section 3 describes a polynomial-time algorithm for HV-rectilinear planarity testing of series-parallel graphs, which is used as a building-block for the design of an HV-rectilinear planarity testing algorithm for partial 2-trees presented in Sec. 4. Conclusions and open problems are in Sec. 5. For space reasons several proofs are sketched.

2   NP-completeness of HV-rectilinear Planarity Testing

It is easy to see that HV-rectilinear planarity testing is in NP, since the problem is polynomial when a planar embedding of the graph is given [9] and since all planar embeddings can be non-deterministically explored [1]. We show the hardness of this problem even on instances of maximum vertex-degree three by reducing SWITCH FLOW NETWORK to it. Hence, the following theorem holds.

Theorem 1. *HV-rectilinear planarity testing is NP-complete even for HV-graphs of maximum vertex-degree three.*

A *switch-flow network* $\mathcal{N}$ is an undirected graph where each edge e is labeled with a range $[c'...c'']$ of nonnegative integers, called the *capacity range* of e. For simplicity, the capacity range $[c...c]$ is denoted with $[c]$. A *flow* for a switch-flow network is an orientation of its edges and an assignment of integer values to them. A flow is *feasible* if it satisfies the following two properties: (i) the total flow entering a vertex from the incoming edges is equal to the total flow exiting the vertex from the outgoing edges, and (ii) the flow assigned to an edge is an integer within the capacity range of the edge. Given a switch-flow network $\mathcal{N}$, the SWITCH FLOW NETWORK problem is the problem of finding a feasible flow for $\mathcal{N}$.

The SWITCH FLOW NETWORK problem is trivially in NP, by assigning to the edges all possible flow values and orientations and computing the sum of the flows at each vertex. In [11] it is shown that SWITCH FLOW NETWORK is NP-hard even in a very restrictive setting, that is, if its instances are such that: (a) the lower bounds of the capacities ranges of the edges are either zero (as in $[0...c]$) or equal to the upper bounds (as in $[c]$), (b), edges with a proper capacity range (as in $[0...c]$) do not form a cut, and

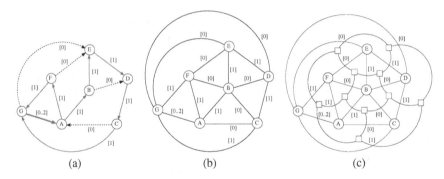

(a)   (b)   (c)

Fig. 2. (a) A feasible flow for an instance $\mathcal{N}$ of the SWITCH FLOW NETWORK problem; the instance is the underlying undirected graph. The thick arrow represents a flow of two units while the thinner arrows represent flows of one unit. (b) The maximal planar SWITCH FLOW NETWORK instance $\mathcal{N}^*$ obtained from $\mathcal{N}$. (b) Graph $\mathcal{N}^*$ and its dual $\mathcal{D}$.

(c) the network is planar. Starting from an instance $\mathcal{N}$ of the SWITCH FLOW NETWORK problem satisfying Properties (a), (b), and (c), we create an instance G of HV-planarity testing of maximum degree three as follows (refer to Figs 2 and 3):

Step 1. Construct a maximal planar instance $\mathcal{N}^*$ by inserting dummy edges with capacity range $[0]$ into $\mathcal{N}$. Observe that $\mathcal{N}^*$ admits a feasible flow if and only if $\mathcal{N}$ does.

Step 2. Compute the dual plane graph D of $\mathcal{N}^*$. Observe that, since $\mathcal{N}^*$ is a maximal triconnected graph, each vertex of D has degree three and D is also triconnected. We label the edges of D with the capacity range of the corresponding edge of $\mathcal{N}^*$.

Step 3. Compute an orthogonal drawing Γ_D of D with the linear-time algorithm in [17]. This algorithm takes as input a 4-plane biconnected graph and computes a drawing with at most $2n + 4$ bends and such that each edge has at least one vertical segment.

Step 4. Transform Γ_D into a positive instance F of HV-planarity testing by replacing orthogonal and vertical segments with rectangular boxes and by labeling each horizontal and vertical edge of F with labels H and V, respectively (see also Fig. 3(b)). Note that F has maximum vertex-degree three and, as D is triconnected, it has a unique HV-realization $\mathcal{H}_F$ up to horizontal and vertical flips.

Step 5. Build the instance G of HV-planarity testing. First, identify for each edge e of D with a label different from $[0]$ a rectangular box of F corresponding to a vertical segment of Γ_D. If the label of e is $[c]$, insert the HV-graph T_c, called *tendril*, in the rectangular box, attaching it with two edges e' and e'' called the *handles* of the tendril. See Fig 3(d) and 3(e) for T_1 and T_2. If e is labeled $[0...c]$ insert into the rectangular box the HV-graph W_c, called *wiggle*. Wiggles W_1 and W_2 are shown in Figs. 3(f) and 3(g).

Lemma 1. *An HV-realization of the HV-graph G corresponds to a feasible flow of the switch network $\mathcal{N}$ and vice versa.*

Proof sketch: First, we show that, starting from an HV-realization $\mathcal{H}_G$ of G, a feasible flow for $\mathcal{N}$ can be found. Observe that each tendril T_h necessarily has the HV-realization $\mathcal{H}_{T_h}$ or $\overline{\mathcal{H}}_{T_h}$, giving to its left and right faces f_l and f_r, $4h$ and $-4h$ (or

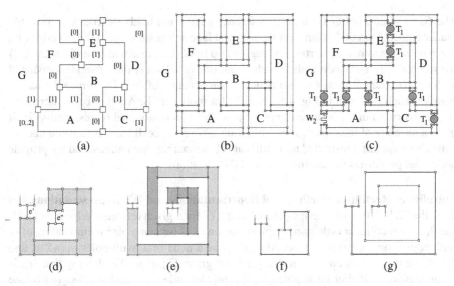

Fig. 3. (a) An orthogonal drawing Γ_D of the dual graph D. (b) The positive instance F of HV-planarity testing built by Step 4. (c) The instance G of HV-planarity testing corresponding to the SWITCH FLOW NETWORK instance $\mathcal{N}$ depicted in Fig. 2(a). (d) Tendril T_1. Edges e' and e'' are the handles of the tendril. (e) Tendril T_2. (f) Wiggle W_1. (g) Wiggle W_2.

$-4h$ and $4h$, respectively) right angles. Since in any HV-realization of G, the subgraph of G obtained by neglecting tendrils and wiggles is drawn as it is in F, and since in F each face is balanced in terms of left and right turns, it follows that the HV-realization chosen for tendrils and wiggles of G decides a flow traversing the faces of G. By construction, such a flow (divided by 4) gives a feasible flow for $\mathcal{N}$. Conversely, suppose a feasible flow exists for $\mathcal{N}$. By choosing the corresponding HV-realizations for tendrils and wiggles, each face is balanced in terms of right and left angles, yielding an HV-realization for G. □

3   Testing Algorithm for Series-Parallel Graphs

The decomposition of a biconnected graph G into its triconnected components is described by the *SPQR-tree* data structure, which implicitly represents all planar embeddings of G. We assume familiarity with $SPQR$-trees and related terminology [4]. If the $SPQR$-tree T of G has no R-nodes (corresponding to triconnected components that are triconnected graphs), G is a *series-parallel graph* and T is called an SPQ-tree: S-nodes represent series compositions, P-nodes parallel compositions, and Q-nodes single edges. Series-parallel graphs are a super-class of biconnected outerplanar graphs.

In [7] a polynomial-time algorithm is described that computes an orthogonal drawing of a series-parallel graph G, with the minimum number of bends over all planar embeddings of G. Our testing algorithm enhances the approach given in [7] to deal with H- and V-labels on the edges. As in [7] we use a variant of the SPQ-trees called SPQ^*-tree.

The SPQ^*-tree T of G for a given reference edge e implicitly describes all possible planar embeddings of G with e on the external face and is such that: (i) each edge or a chain of edges joined by vertices of degree two in G is represented by a Q^*-node; (ii) the root of T is a P-node with two children, one of which is the Q^*-node associated with e; (iii) each S-node has two children. An example of an SPQ^*-tree of a series-parallel graph is shown in Fig. 4(b). See [7] for more details. Another basic ingredient of the algorithm in [7] is the concept of *spirality* of an orthogonal representation $\mathcal{H}$; it gives a measure of how much $\mathcal{H}$ is "rolled-up". We now recall basic definitions and results about spirality, restricted to rectilinear orthogonal representations, and we provide additional properties for the spirality of HV-realizations.

Spirality of Rectilinear Orthogonal Representations and HV Representations. Let T be the SPQ^*-tree of a series-parallel graph G for a given reference edge $e = (s, t)$. Let $\mathcal{H}$ be a rectilinear orthogonal representation of G for some planar embedding of G with edge e on the external face. Also, let μ be a node of T with poles u and w, and let $\mathcal{H}_\mu$ be the restriction of $\mathcal{H}$ to the pertinent graph G_μ of μ. We also say that $\mathcal{H}_\mu$ is a *component* of $\mathcal{H}$. For each pole $v \in \{u, w\}$, let $\mathrm{indeg}_\mu(v)$ and $\mathrm{outdeg}_\mu(v)$ be the degree of v inside and outside $\mathcal{H}_\mu$, respectively. Define two (possibly coincident) *alias vertices* of v, denoted by v' and v'', as follows: (a) If $\mathrm{indeg}_\mu(v) = 1$, then $v' = v'' = v$; (b) If $\mathrm{indeg}_\mu(v) > 1$ and $\mathrm{outdeg}_\mu(v) = 1$, then $v' = v''$ is a dummy vertex that splits the edge of v that is outside $\mathcal{H}_\mu$; (c) If $\mathrm{indeg}_\mu(v) = \mathrm{outdeg}_\mu(v) = 2$, then v' and v'' are dummy vertices, each splitting a distinct edge of v that is outside $\mathcal{H}_\mu$.

Let A^v denote the set of distinct alias vertices of a pole v (if G has vertex-degree at most three, $|A^v| = 1$). Let P^{uw} be any simple path from u to w inside $\mathcal{H}_\mu$ and let u' and w' be alias vertices of u and of w, respectively. The path $S^{u'w'}$ obtained concatenating (u', u), P^{uw}, and (w, w') is called a *spine* of $\mathcal{H}_\mu$. Denote by $n(S^{u'w'})$ the number of right turns minus the number of left turns encountered moving along $S^{u'w'}$ from u' to w'. The *spirality* $\sigma(\mathcal{H}_\mu)$ of a component $\mathcal{H}_\mu$ with poles u and w is defined based on the following cases: (i) If $A^u = \{u'\}$ and $A^w = \{w'\}$, then $\sigma(\mathcal{H}_\mu) = n(S^{u'w'})$. (ii) If $A^u = \{u'\}$ and $A^w = \{w', w''\}$, then $\sigma(\mathcal{H}_\mu) = (n(S^{u'w'}) + n(S^{u'w''}))/2$. (iii) If $A^u = \{u', u''\}$ and $A^w = \{w'\}$, then $\sigma(\mathcal{H}_\mu) = (n(S^{u'w'}) + n(S^{u''w'}))/2$. (iv) If $A^u = \{u', u''\}$ and $A^w = \{w', w''\}$, assume w.l.o.g. that (u, u') immediately precedes (u, u'') counterclockwise around u and that (w, w') immediately precedes (w, w'') clockwise around w. Then $\sigma(\mathcal{H}_\mu) = (n(S^{u'w'}) + n(S^{u''w''}))/2$.

We now briefly recall the spirality properties used in this paper, restricted to our setting. We assume that G is a series-parallel graph equipped with an SPQ^*-tree T.

Substituting Components with the Same Spirality. Let $\mathcal{H}$ and $\mathcal{H}'$ be two rectilinear orthogonal representations of the same planar graph G and let μ be a node of T. We say that $\mathcal{H}$ and $\mathcal{H}'$ are μ-*different* if their planar embeddings may differ only for the clockwise orderings of the edges inside G_μ and if the labels of the external angles at the poles of μ stay the same in the two representations. See Figs. 4(c) and 4(d) for an example. Suppose that $\mathcal{H}$ and $\mathcal{H}'$ are μ-different and that $\sigma(\mathcal{H}_\mu) = \sigma(\mathcal{H}'_\mu)$. The next theorem claims that it is always possible to get a new rectilinear orthogonal representation $\mathcal{H}''$ by *substituting* $\mathcal{H}_\mu$ with $\mathcal{H}'_\mu$ in $\mathcal{H}$ (see, e.g., Fig. 4(e)).

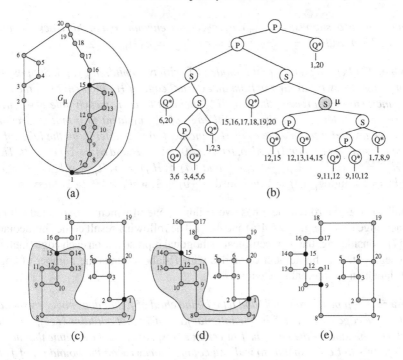

Fig. 4. (a) A planar graph G; (b) The SPQ^*-tree of G for the reference edge $(1, 20)$, and a high-lighted node μ (the pertinent graph G_μ of μ is highlighted in the graph); (c)-(d) Two rectilinear orthogonal representations $\mathcal{H}$ and $\mathcal{H}'$ that are μ-different; both $\mathcal{H}_\mu$ and $\mathcal{H}'_\mu$ have spirality 2. (e) The rectilinear orthogonal representation obtained by substituting $\mathcal{H}_\mu$ with $\mathcal{H}'_\mu$ in $\mathcal{H}$.

Theorem 2. [7] *Let $\mathcal{H}$ and $\mathcal{H}'$ be two rectilinear orthogonal representations of the same planar graph G such that $\mathcal{H}$ and $\mathcal{H}'$ are μ-different and $\sigma(\mathcal{H}_\mu) = \sigma(\mathcal{H}'_\mu)$. Then the embedded labeled graph $\mathcal{H}''$ obtained by substituting $\mathcal{H}_\mu$ with $\mathcal{H}'_\mu$ in $\mathcal{H}$ is a recti-linear orthogonal representation.*

Intuitively, Theorem 2 implies that in an orthogonal representation $\mathcal{H}$ we can always substitute a certain component $\mathcal{H}_\mu$ with a component $\mathcal{H}'_\mu$ having the same spirality, independently of the planar embedding of $\mathcal{H}_\mu$ and $\mathcal{H}'_\mu$.

Spirality Relationships for Series- and Parallel-Compositions. Let $\mathcal{H}$ be a rectilinear orthogonal representation of G. For an S- or a P-node μ of T, the spirality of $\mathcal{H}_\mu$ is related to the spiralities of the orthogonal representations of the pertinent graphs of the children of μ. For an S-node the relationship is given by Lemma 2. For a non-root P-node the relationship depends on the number of its children (see Lemmas 3 and 4).

Lemma 2. [7] *Let μ be an S-node of T with children μ_1 and μ_2. Then: $\sigma(\mathcal{H}_\mu) = \sigma(\mathcal{H}_{\mu_1}) + \sigma(\mathcal{H}_{\mu_2})$.*

Lemma 3. [7] *Let μ be a P-node of T with three children μ_1, μ_2, and μ_3. Let e_i be the edge of $\mathcal{H}_{\mu_i}$ incident to pole w ($i = 1, 2, 3$) and let e_{out} be the external edge of $\mathcal{H}_\mu$*

incident to w; also, suppose that e_{out}, e_1, e_2, e_3 are encountered in this clockwise order around w. Then: $\sigma(\mathcal{H}_\mu) = \sigma(\mathcal{H}_{\mu_1}) + 2 = \sigma(\mathcal{H}_{\mu_2}) = \sigma(\mathcal{H}_{\mu_3}) - 2$.

Lemma 4. [7] *Let μ be a non-root P-node of T with two children μ_1, μ_2. Suppose that while moving clockwise around w from an external edge of $\mathcal{H}_\mu$ incident to w, the edges of $\mathcal{H}_{\mu_1}$ incident to w precede those of $\mathcal{H}_{\mu_2}$ incident to w. For each pole $v \in \{u, w\}$ and for $i = 1, 2$, let e_{out} be the external edge of $\mathcal{H}_\mu$ incident to v that is circularly consecutive around v to an edge e_i of $\mathcal{H}_{\mu_i}$ incident to v, and let l_v^i be the label of the angle at v formed by e_{out} and e_i. Also, let $\alpha_v^i = 0$ if $l_v^i = F$ and $\alpha_v^i = 1$ if $l_v^i = R$. Then: (i) $\sigma(\mathcal{H}_\mu) = \sigma(\mathcal{H}_{\mu_1}) + k_u^i \alpha_u^1 + k_w^i \alpha_w^1$ and (ii) $\sigma(\mathcal{H}_\mu) = \sigma(\mathcal{H}_{\mu_2}) - k_u^2 \alpha_u^2 - k_w^i \alpha_w^2$, where $k_v^i = 1$ if $\text{indeg}_{\mu_i}(v) = 1$ and $\text{outdeg}_\mu(v) = 1$, while $k_v^i = 1/2$ otherwise.*

Finally, if μ is the root node it has two children, one of which is associated with the reference edge e (see, e.g., Figs. 4(a) and 4(b)). The following result comes immediately from [7], considering that in a rectilinear orthogonal representation each edge (included e) is a segment, and the number of angles with R-labels minus the number of angles with L-labels in an internal face equals 4.

Lemma 5. *Let μ be the root of T and let ν be the child of μ that does not correspond to the reference edge (s, t). Let f be the internal face of $\mathcal{H}$ that contains (s, t), and let l_s (resp. l_t) be the label of the angle in f at vertex s (resp. at vertex t). Assume that moving along e from s to t corresponds to walking counterclockwise on the boundary of f. For $v \in \{s, t\}$, let $\alpha_v = 0$ if $l_v = F$, $\alpha_v = 1$ if $l_v = R$, and $\alpha_v = -1$ if $l_v = L$. Then: $\sigma(\mathcal{H}_\nu) + k_s \alpha_s + k_t \alpha_t = 4$, where $k_v = 1$ if $\text{indeg}_\nu(v) = 1$ and $k_v = 0$ otherwise.*

Spirality and HV-realizations. Let G be an HV-graph and let T be an SPQ^*-tree of G for a reference edge $e = (s, t)$. W.l.o.g, from now on we assume that for any 2-degree vertex v the two edges incident to v have different labels (otherwise, they can be simply replaced by a single edge with the same label as the original edges). With this assumption, the chain associated with a Q^*-node always consists of an alternated sequence of H-labeled and V-labeled edges. The following lemma can be easily proved.

Lemma 6. *Let μ be a Q^*-node of T and let k be the number of edges of G_μ. If k is even (resp. odd), then there exists an HV-realization of G_μ with spirality j, for every odd (resp. even) $j \in [-k+1, k-1]$.*

Testing Algorithm. Let G be a series-parallel HV-graph and let T be the SPQ^*-tree for a given reference edge $e = (s, t)$. The testing is done by traversing T bottom-up; if the test fails, G does not admit an HV-realization with e on the external face, a new reference edge is chosen and the test is repeated. If the test succeed for some reference edge of G, the algorithm reconstructs an HV-realization of G with a top-down visit of T, exploiting suitable information stored in the nodes of T during the bottom-up traversal.

Theorem 3. *Let G be a series-parallel HV-graph with n vertices and maximum vertex-degree four. There exists an $O(n^4)$-time algorithm that tests whether G admits an HV-realization and, if so, it computes an HV-realization of G. Also, if G has vertex-degree at most 3, the time-complexity can be reduced to $O(n^3 \log n)$.*

Proof sketch: Let T be the SPQ^*-tree for a given reference edge e. We describe an algorithm that tests whether G admits an HV-realization with e on the external face. Let μ the current node of T when traversing T from bottom to top. Let u and w be the poles of μ. The algorithm associates with μ a set of *tuples*, each one corresponding to a distinct value of spirality that an HV-realization $\mathcal{H}_\mu$ of G_μ can have within an HV-realization $\mathcal{H}$ of G. The size of the set of tuples associated with μ is $O(n)$, since, by definition, the absolute value of spirality of $\mathcal{H}_\mu$ cannot exceed the length of the shortest path from u to w in G_μ plus one. A tuple contains a value of spirality σ_μ and an encoding of an embedding of G_μ for which an HV-realization of G_μ with spirality σ_μ exists; this encoding describes for a P-node the cyclic order of the edges of G_μ incident to u and w (i.e., the permutation of the edges of the skeleton of μ); if μ is either a Q^*-node or an S-node, the algorithm does not keep any embedding information for μ (because the embedding is uniquely defined). It is sufficient to keep only one representative tuple for each possible value of spirality, since by Theorem 2 rectilinear orthogonal representations with the same spirality are interchangeable.

If during the bottom-up visit we obtain an empty set of tuples for a node of T, the algorithm reports that G does not admit an HV-realization with e on the external face. Otherwise, let ν be the child of the root that does not correspond to edge e; if the set of tuples of ν has a tuple whose spirality verifies the condition of Lemma 5, then the algorithm reports that an HV-realization of G exists; otherwise, it reports again that G does not have an HV-realization with e on the external face. Note that, the values α_s and α_t in Lemma 5 are uniquely determined by the H- and V-labels once the circular ordering of the edges around s and t is decided. If the algorithm fails on T, a different edge e is chosen and it is executed again. The tuples of Q^*-nodes are computed using Lemma 6, and those of non-root P-nodes using Lemmas 3 and 4. For an S-node μ with children μ_1 and μ_2, based on Lemma 2, we consider all distinct values of spirality obtained by summing up the spiralities of a tuple of μ_1 and of a tuple of μ_2. However, if μ_1 and μ_2 share a pole v of degree 4, the H- and V-labels on the edges incident to v may not be compatible with some pairs of spirality values for μ_1 and μ_2, and these pairs must be discarded.

The tuple sets for all nodes of T are computed in $O(|S|n^2 + |P|n + |Q^*|n)$ time, where $|S|, |P|, |Q|$ denote the number of S-, P-, and Q^*-nodes in T, respectively. Hence, we have $O(n^3)$-time complexity for a specific reference edge and $O(n^4)$ over all possible reference edges for G. If the test is positive the algorithm reconstructs an HV-realization of G in $O(n^2)$ time, by visiting T top-down.

If G has maximum vertex-degree 3, there cannot be forbidden pairs of spirality values for the children of an S-node, and finding its possible spiralities corresponds to computing a Cartesian sum of two sets of integers, which takes $O(n \log n)$ time [3]. Hence, the overall time complexity for series-parallel HV-graphs of vertex-degree at most 3 is $O(n(|S|(n \log n) + |P|n + |Q^*|n)) = O(n^3 \log n)$.   □

4   Testing Algorithm for Partial 2-Trees

We extend the HV-realizability testing algorithm described in the proof of Theorem 3 to simply connected graphs that do not contain rigid components. Namely, a 2-*tree* is a graph obtained by starting from an edge and iteratively attaching a new vertex per time to two already adjacent vertices. A *partial 2-tree* is any subgraph of a 2-tree. We give a testing algorithm for the class of partial 2-trees, which includes series-parallel graphs.

Let G be an HV-graph that is a connected partial 2-tree (if G is not connected one can execute the test independently on each connected component). We assume that G does not contain three adjacent edges with the same label (H or V), because in this case it is trivial to conclude that G does not admit an HV-realization. The testing algorithm exploits a constrained version of the algorithm described in Sec. 3 for the biconnected components and the popular data structure known as the *block-cutvertex tree* $\mathcal{T}$ of G, which describes the decomposition of G into its biconnected components, also called *blocks*. A block consisting of a single edge is called a *trivial block*. $\mathcal{T}$ has a node v_B for each block B of G and a node v_c for each cutvertex c of G; there is an arc (v_B, v_c) in $\mathcal{T}$ if c belongs to B in G. A key-ingredient of our testing algorithm is the following; the proof is easy and is omitted for space reasons.

Lemma 7. *Let B_1 and B_2 be any two blocks of G and let Π be the path from v_{B_1} to v_{B_2} in $\mathcal{T}$. Let v_{c_1} and v_{c_2} be the cutvertex-nodes on Π adjacent to v_{B_1} and to v_{B_2}, respectively (c_1 and c_2 may coincide). In any planar embedding of G, either c_1 is on the external face of B_1 or c_2 is on the external face of B_2.*

Let B be a block of G and c a cutvertex of G that belongs to B. Suppose we want to construct an HV-realization of G for a planar embedding where B has c on the external face and such that some other block is attached to c in the external face of B. To do this, we need the angle at c in the external face of B to be greater than 90 degrees. We say that B is *HV-extrovert with respect to c* if B admits an HV-realization $\mathcal{H}_B$ such that: (i) c on the external face of $\mathcal{H}_B$; (ii) the external angle at c is greater than 90 degrees. We also say that $\mathcal{H}_B$ is *extrovert with respect to c*. The testing algorithm works as follows:

Step 1. Consider all degree-1 block-nodes of $\mathcal{T}$ and for each of these nodes v_B let v_c be its adjacent cutvertex-node; test if B is HV-extrovert with respect to c (we explain how to test it right after the description of Step 2); if so, store its extrovert HV-realization in a list L and remove v_B from $\mathcal{T}$, otherwise mark B as *not HV-extrovert*. At the end of this step remove from $\mathcal{T}$ all cutvertex-nodes of degree less than 2, previously attached to some degree-1 block-node.

Step 2. Check whether one of the following cases holds; if not repeat Step 1:

Case 1. Two blocks that are not HV-extrovert are found: in this case the test is negative, because the property of Lemma 7 cannot be satisfied in any HV-realization of G.

Case 2. $\mathcal{T}$ becomes empty. The test is positive. Indeed, in this case there exists a planar embedding of G and every block B has an extrovert HV-realization, stored in L, that is compatible with this embedding; an HV-realization of G can be easily obtained by suitably merging the extrovert HV-realizations stored in L.

Case 3. $\mathcal{T}$ consists of just one block-node v_B marked as not HV-extrovert. In this case the algorithm tests whether B admits any HV-realization $\mathcal{H}_B$, using the algorithm described in Sec. 3. In the affirmative case the test is positive and we can still construct an HV-realization of G by suitably merging the HV-realizations stored in L. Namely: (i) embed the HV-realization of each block that shares a cutvertex c with B inside a face of G with angle at c larger than 90 degrees (such a face always exists, because c has degree at most 3 in B); (ii) merge the extrovert HV-realizations of the other blocks as in Case 2. If B does not admit any HV-realization, then the test is clearly negative.

We now explain how to test whether a block B is HV-extrovert with respect to a desired cutvertex c. If B is trivial (i.e., a single edge) the test is clearly positive. If B is not trivial, then c has degree greater than 2 in G and we distinguish between two cases:

Vertex c has Degree 3 in G. In this case c has degree 2 in B. Let s and u be the two vertices adjacent to c in B, and let T be the SPQ^*-tree of B with reference edge (s, c). Execute an HV-realizability testing of B with respect to T, using the algorithm of Sec. 3. If the test is negative, B does not have an HV-realization with c on the external face, and we conclude that B is not HV-extrovert with respect to c. If the test is positive, there exists an HV-realization $\mathcal{H}_B$ with c on the external face, but we have to verify that we can get an HV-realization with the external angle at c greater than 90 degrees. If edges (s, c) and (u, c) have the same label (H or V), we do not need to do any additional check, as $\mathcal{H}_B$ surely has two angles of 180 degrees at c. If (s, c) and (u, c) have different labels, let ν be the child-node of the root of T that does not correspond to the reference edge; it suffices to check whether ν has a tuple whose spirality value satisfies the equation of Lemma 5, with the constraint that α_c (i.e., α_t of Lemma 5) is 1 (corresponding to an internal angle at c of 90 degrees, and hence to an external angle at c of 270 degrees).
Vertex c has degree 4 in G. If c has degree 2 in B, the algorithm applies the same check as in the previous case. If c has degree 3 in B, let $e = (s, c)$ and $e' = (u, c)$ be the edges of B incident to c with the same label (H or V), and let $e'' = (v, c)$ be the third edge of B incident to c. Let T be the SPQ^*-tree of B with reference edge $e = (s, c)$. Note that T has a P-node μ such that one of its poles is c and such that G_μ contains both e' and e''. Also, in any HV-realization that is extrovert with respect to c, e and e' must be both on the external face. Hence, to check whether B is HV-extrovert with respect to c, we can execute an HV-realizability testing of B on tree T, using the algorithm of Sec. 3, with the restriction that when we compute the tuple set of μ we only consider the arrangement of its children corresponding to having e' on the external face.

About the computational complexity of the test described above, let n be the number of vertices of G, let $B_1, B_2, \ldots, B_h$ be the biconnected components of G, and let n_i be the number of vertices of B_i $(i = 1, \ldots, h)$. Also, let $\mathcal{T}$ the block-cutvertex tree of G. For each block-node v_{B_i} of $\mathcal{T}$, the algorithm takes $O(n_i^3)$ to test whether B_i is HV-extrovert with respect to a desired cutvertex using the algorithm of Sec. 3, as it only needs to test the HV-realizability of B_i for a suitably chosen reference edge. Also, if one block B_i is not HV-extrovert, an additional HV-realizability testing on B_i is run over all possible reference edges, spending $O(n_i^4)$ time. Since $\sum_{i=1,\ldots,h} n_i = O(n)$, the overall time complexity of the algorithm is $O(n^4)$. The time complexity is reduced to $O(n^3 \log n)$ if G has vertex-degree at most 3, with the same argument as in Theorem 3.

Theorem 4. *Let G be an HV-graph that is a partial 2-tree with n vertices and maximum-vertex degree four. There exists an $O(n^4)$-time algorithm that tests whether G admits an HV-realization and, if so, it computes an HV-realization of G. Also, if G has vertex-degree at most 3, the time-complexity can be reduced to $O(n^3 \log n)$.*

5 Conclusions and Open Problems

We suggest the study of these problems: (i) Can the polynomial bound of Theorem 4 be improved? Recall that there is a linear-time algorithm for the rectilinear planarity testing of series-parallel graphs with vertex-degree at most three [20], but it relies on properties that do not apply for HV-graphs. (ii) Find a combinatorial characterization, e.g. in terms of forbidden substructures, for the HV-graphs that have an HV-realization.

References

1. Bertolazzi, P., Di Battista, G., Didimo, W.: Computing orthogonal drawings with the minimum number of bends. IEEE Transactions on Computers 49(8), 826–840 (2000)
2. Bläsius, T., Krug, M., Rutter, I., Wagner, D.: Orthogonal graph drawing with flexibility constraints. Algorithmica 68(4), 859–885 (2014)
3. Cormen, T.H., Leiserson, C.E., Rivest, R.L., Stein, C.: Introduction to Algorithms. MIT Press (2009)
4. Di Battista, G., Tamassia, R.: On-line planarity testing. SIAM J. on Comp. 25, 956–997 (1996)
5. Di Battista, G., Kim, E., Liotta, G., Lubiw, A., Whitesides, S.: The shape of orthogonal cycles in three dimensions. Discrete & Computational Geometry 47(3), 461–491 (2012)
6. Di Battista, G., Liotta, G., Lubiw, A., Whitesides, S.: Embedding problems for paths with direction constrained edges. Theoretical Computer Science 289(2), 897–917 (2002)
7. Di Battista, G., Liotta, G., Vargiu, F.: Spirality and optimal orthogonal drawings. SIAM J. on Comp. 27(6), 1764–1811 (1998)
8. Di Giacomo, E., Liotta, G., Patrignani, M.: A note on 3D orthogonal drawings with direction constrained edges. Inf. Proc. Lett. 90(2), 97–101 (2004)
9. Durocher, S., Felsner, S., Mehrabi, S., Mondal, D.: Drawing HV-restricted planar graphs. In: Pardo, A., Viola, A. (eds.) LATIN 2014. LNCS, vol. 8392, pp. 156–167. Springer, Heidelberg (2014)
10. Felsner, S., Kaufmann, M., Valtr, P.: Bend-optimal orthogonal graph drawing in the general position model. Computational Geometry 47(3), 460–468 (2014)
11. Garg, A., Tamassia, R.: On the computational complexity of upward and rectilinear planarity testing. SIAM J. on Comp. 31(2), 601–625 (2001)
12. Hoffman, F.: Embedding rectilinear graphs in linear time. Inf. Proc. Lett. 29(2), 75–79 (1988)
13. Maňuch, J., Patterson, M., Poon, S.-H., Thachuk, C.: Complexity of finding non-planar rectilinear drawings of graphs. In: Brandes, U., Cornelsen, S. (eds.) GD 2010. LNCS, vol. 6502, pp. 305–316. Springer, Heidelberg (2011)
14. Mutzel, P., Weiskircher, R.: Bend minimization in planar orthogonal drawings using integer programming. SIAM J. on Opt. 17(3), 665–687 (2006)
15. Rahman, M.S., Nakano, S.I., Nishizeki, T.: A linear algorithm for bend-optimal orthogonal drawings of triconnected cubic plane graphs. J. of Graph Alg. and Appl. 3(4), 31–62 (1999)
16. Rahman, M.S., Nishizeki, T., Naznin, M.: Orthogonal drawings of plane graphs without bends. J. of Graph Alg. and Appl. 7(4), 335–362 (2003)
17. Tamassia, R., Tollis, I.G.: Planar grid embedding in linear time. IEEE Transactions on Circuits Systems CAS-36(9), 1230–1234 (1989)
18. Tamassia, R.: On embedding a graph in the grid with the minimum number of bends. SIAM J. on Comp. 16(3), 421–444 (1987)
19. Vijayan, G., Wigderson, A.: Rectilinear graphs and their embeddings. SIAM J. on Comp. 14(2), 355–372 (1985)
20. Zhou, X., Nishizeki, T.: Orthogonal drawings of series-parallel graphs with minimum bends. SIAM J. on Discr. Math. 22(4), 1570–1604 (2008)

Embedding Four-Directional Paths
on Convex Point Sets*

Oswin Aichholzer[1], Thomas Hackl[1], Sarah Lutteropp[2],
Tamara Mchedlidze[2], and Birgit Vogtenhuber[1]

[1] Institute for Software Technology, Graz University of Technology, Austria
{oaich,thackl,bvogt}@ist.tugraz.at
[2] Institute of Theoretical Informatics, Karlsruhe Institute of Technology, Germany
sarah.lutteropp@student.kit.edu, mched@iti.uka.de

Abstract. A directed path whose edges are assigned labels "up", "down", "right", or "left" is called *four-directional*, and *three-directional* if at most three out of the four labels are used. A *direction-consistent embedding* of an n-vertex four-directional path P on a set S of n points in the plane is a straight-line drawing of P where each vertex of P is mapped to a distinct point of S and every edge points to the direction specified by its label. We study planar direction-consistent embeddings of three- and four-directional paths and provide a complete picture of the problem for convex point sets.

1 Introduction

In 1974, Rosenfeld proved that every tournament has a spanning *antidirected* path [17] and conjectured that there exists an integer n_0 such that every tournament with more than n_0 vertices contains every *oriented path* as a spanning subgraph. A tournament is a digraph whose underlying undirected structure is a complete graph and an oriented path is a digraph whose underlying undirected structure is a simple path. An oriented path is antidirected if the directions of its edges alternate. During the following decade several simplifications of Rosenfeld's conjecture had been shown to be true. Alspach and Rosenfeld [4] and Straight [18] settled the conjecture for oriented paths with either a single source or a single sink. Forcade [12] proved the conjecture to be true for every tournament whose size is a power of two. Reid and Wormald[16] showed that any tournament of size n contains every oriented path of size $2n/3$ and Zhang [20] improved this result to $n - 1$. Finally, in 1986, the conjecture was established by Thomason [19].

More than two decades later, with the expansion of Geometric Graph Theory and Graph Drawing, a geometric counterpart of Rosenfeld's conjecture was considered. The subject of this study is an *upward geometric* tournament, that is, a tournament drawn on the plane with straight-line edges so that each edge points in the upward direction. It was asked whether an upward geometric tournament contains a planar copy of any

---

* O.A. supported by the ESF EUROCORES programme EuroGIGA - ComPoSe, Austrian Science Fund (FWF): I 648-N18. T.H. supported by the Austrian Science Fund (FWF): P23629-N18 'Combinatorial Problems on Geometric Graphs'.

C. Duncan and A. Symvonis (Eds.): GD 2014, LNCS 8871, pp. 355–366, 2014.

oriented path [9]. Despite several independent approaches to attack the problem by different research groups, this question is still unsolved. However, it was answered in the affirmative for several special cases of paths and tournaments. We use the following definitions to list these results. A vertex of a digraph which is either a source or a sink is called a *switch*. An oriented path whose edges are all oriented in the same direction is called *monotone*. For the following cases it was shown that every upward tournament contains a planar copy of each oriented path: the vertices of the tournament are in convex position [9], the oriented path has at most 3 switches [9], the oriented path has at most 5 switches and at least two of its monotone subpaths contain a single edge [5], the oriented path where every sink is directly followed by a source [9]. It was also shown that each oriented path of size n is contained in any upward geometric tournament of size $n2^{k-2}$, where k is the number of switches [5]. This result was later improved to $(n-1)^2+1$ in [15]. Recently, with the help of a computer, we could verify that every upward geometric tournament of size 10 contains a planar copy of any oriented path as a spanning subgraph. This was done by exhaustive testing of all distinct directed order types, that is, all order types [2] with an additional combinatorial upward direction.

The question whether any upward geometric tournament contains a planar copy of any oriented path was originally stated in terms of so-called *point set embeddings*. Here we are given a set S of n points in the plane and a planar n-vertex graph G, and we are asked to determine whether G has a planar straight-line drawing where each vertex of G is mapped to a distinct point of S. This problem has been extensively studied and many exciting facts were established, see for example [6,8,10,11,13]. In the upward counterpart of point set embeddings, G is an upward planar digraph and the obtained drawing is additionally required to be upwards oriented. Such a drawing, if it exists, is called an *upward point set embedding*. Upward point set embeddings have been studied for different classes of digraphs [5,7,9,14]. Observe that the question whether any upward geometric tournament contains a planar copy of any oriented path is equivalent to asking whether any oriented path has an upward planar embedding on any set of n points. We will refer to the latter as the *oriented path question*.

The number of distinct plane embeddings of an (undirected) spanning path on a point set could provide us some additional evidence for the oriented path question. It is not difficult to see that if S is a set of n points in convex position, then it admits $n2^{n-3}$ distinct plane spanning path embeddings. Further it is known that this is the minimum number of distinct plane spanning path embeddings that a point set can admit, i.e., convex point sets minimize this number [3]. Comparing this lower bound with the number of distinct oriented paths, which is 2^{n-1}, it sounds even surprising that every oriented path has an upward planar embedding on every convex point set [9]. In order to approach the oriented path question in its general form, we aim to understand better how the nature of the problem changes when in addition to planarity of a path one requires its upwardness. To this end, we generalize the oriented path problem with respect to the number of considered directions (see Section 2 for a rigorous definition). Observe that, instead of considering an oriented path, one can consider a monotone path with labels on edges that declare whether an edge is required to point up or down. In this work we study monotone paths with four possible labels on the edges: up, down, left, and right. We call such paths *four-directional*, and *three-directional* if at most three out of the four labels are used. An embedding of such a path on a point set where each edge points into the direction specified by its label is called *direction-consistent*. We study planar direction-consistent embeddings of three- and four-directional paths on convex point sets. Recall that convex point sets are extremal in the sense that they minimize the number of plane embeddings of (undirected) spanning paths. We provide a complete picture regarding four-directional paths and convex point sets. Our results are as follows:

- Every three-directional path admits a planar direction-consistent embedding on any convex point set.
- There exists a four-directional path P and a one-sided[1] convex point set S such that P does not admit a planar direction-consistent embedding on S. On the other hand, a four-directional path always admits a planar direction-consistent embedding for special cases of one-sided point sets, namely so-called quarter-convex point sets.
- Given a four-directional path P and a convex point set S, we can decide in $O(n^2)$ time whether P admits a planar direction-consistent embedding on S.

Our study is also motivated by applications similar to those of upward point set embeddings, i.e., any situation where a hierarchical structure must be represented and additional constraints on the positions of vertices are given. Our scenario, where instead of two directions the edges can point into four directions, allows for a more detailed control over a drawing.

The remainder of the paper is organized as follows. In Section 2, we give the necessary definitions. In Section 3, we prove several preliminary results which are utilized in our main Section 4, where the existence of a planar direction-consistent embedding of a three-directional path on a convex point set is shown. All results on four-directional paths are concentrated in Section 5. Due to space constraints omitted proofs can be found in the full version [1].

2   Definitions

Graphs. The graphs we study in this paper are directed and we denote by (u, v) an edge directed from u to v. A directed edge when drawn as a straight-line segment is said to *point up* or being *upward*, if its source is below its sink. Similarly we define the notions of pointing *down*, *left*, and *right*. Our study concentrates on directed paths each edge of which is assigned one of four labels U, D, L, R, which means that (when the path is embedded on a point set) this edge is required to point up, down, left, or right, respectively. For simplicity, we will denote such a path containing vertices $v_1, \ldots, v_n$ by $P = d_1, \ldots, d_{n-1}$, where $d_i \in \{U, D, L, R\}, 1 \leq i \leq n-1$. Let $T \subseteq \{U, D, L, R\}$. If $d_i \in T, 1 \leq i \leq n-1$, then P is called T-*path* and $|T|$-*directional path* in order to emphasize the number of directions it contains. We denote by $P_{i,j} = d_i, \ldots, d_j$, $1 \leq i \leq j \leq n-1$, a subpath of P. In addition, we define $P_{i,i-1} = v_i$.

---

[1] A convex point set is called *one-sided* if all of its points lie on the same side of the line through its bottommost and topmost points.

Point Sets. We say that a set S of points in the plane is in *general position* if no three points are collinear and no two points have the same x- or y-coordinate. All point sets mentioned in this paper are in general position. Let S be a convex point set. We denote by $\ell(S), r(S), t(S), b(S)$ the leftmost, the rightmost, the topmost, and the bottommost point of S, respectively. A subset of points of S is called (*clockwise*) *consecutive* if its points appear consecutively as we (clockwise) traverse the convex hull of S.

A convex point set S is called *left-sided (resp. right-sided)* if $t(S)$ and $b(S)$ (resp. $b(S), t(S)$) are clockwise consecutive on S, and S is called *one-sided* if S is left-sided or right-sided. S is called *strip-convex* if (i) the points $b(S)$ and $\ell(S)$ are either consecutive or coincide, and (ii) the points $t(S)$ and $r(S)$ are either consecutive or coincide. For $p, q \in S$, the points of S which lie between the vertical lines through p and q (including them) are said to be *vertically between* p and q.

Embeddings. Let P be an n-vertex path (labeled) with vertex set $V(P)$ and S be a set of n points in general position. An *embedding* of P on S is an injective function $\mathcal{E}: V(P) \to S$. If the edges of P are drawn as straight-line segments connecting corresponding end-vertices, the embedding $\mathcal{E}$ yields a drawing of P. We say that the embedding $\mathcal{E}$ is *planar* if this drawing is planar. We say that $\mathcal{E}$ is *direction-consistent* if each edge points to the direction corresponding to its label. Planar direction-consistent embeddings are abbreviated by PDCE. During the construction of an embedding, a point p is called *used* if a vertex has already been mapped to it. Otherwise, p is called *free*. Throughout the paper we consider embeddings of n-vertex paths on sets of n points, unless explicitly stated differently.

Operations with Paths, Point Sets, and Embeddings. Let $T \subseteq \{U, D, R, L\}$ and consider a T-path $P = d_1 d_2 \ldots d_{n-1}$. Let S be a set of n points and let $\mathcal{E}$ be a direction-consistent embedding of P on S. Observe that $\mathcal{E}$ describes a direction-consistent embedding of another path $P^{\mathcal{I}}$ on the same point set S. Path $P^{\mathcal{I}}$ is called the *reverse* path of P, and is constructed by reversing the directions of the edges of P and changing the labels to their opposite. Thus, formally $P^{\mathcal{I}} = \mathcal{I}(d_{n-1}) \ldots \mathcal{I}(d_2)\mathcal{I}(d_1)$, where $\mathcal{I}(U) = D, \mathcal{I}(D) = U, \mathcal{I}(R) = L$, and $\mathcal{I}(L) = R$. This embedding of $P^{\mathcal{I}}$ on S is denoted by $\mathcal{E}^{\mathcal{I}}$. For example, if $P = UUDRL$, then $P^{\mathcal{I}} = RLUDD$. Observe also that $(P^{\mathcal{I}})^{\mathcal{I}} = P$.

Observation 1. *Let $\mathcal{E}$ be a PDCE of a path P on a point set S. Then $\mathcal{E}^{\mathcal{I}}$ is a PDCE of $P^{\mathcal{I}}$ on the same point set S.*

Let P, S, and $\mathcal{E}$ be as above. The embedding $\mathcal{E}$ yields a straight-line drawing Γ of P. Consider the rotation of Γ counterclockwise by $\pi/2$. This rotated drawing represents a direction-consistent embedding, denoted by $\mathcal{R}(\mathcal{E})$, of a new path, denoted by $\mathcal{R}(P)$, on the rotated point set, denoted by $\mathcal{R}(S)$. This new path $\mathcal{R}(P)$ is formally defined as follows: $\mathcal{R}(P) = \mathcal{R}(d_1)\mathcal{R}(d_2) \ldots \mathcal{R}(d_{n-1})$, where $\mathcal{R}(U) = L, \mathcal{R}(D) = R, \mathcal{R}(R) = U$, and $\mathcal{R}(L) = D$. We use the notation $\mathcal{R}^k$ for k applications of $\mathcal{R}$. Thus, $\mathcal{R}^4(P) = P$ and $\mathcal{R}^4(S) = S$. Also, if P is an $\{U, D, L\}$-path and S is a right-sided point set then $\mathcal{R}^2(P)$ is an $\{U, D, R\}$-path and $\mathcal{R}^2(S)$ is a left-sided point set. Note that $P^{\mathcal{I}} \neq \mathcal{R}^2(P)$.

Observation 2. *Let $\mathcal{E}$ be a PDCE of a path P on a point set S. Then $\mathcal{R}(\mathcal{E})$ is a PDCE of $\mathcal{R}(P)$ on the point set $\mathcal{R}(S)$.*

Finally, we define the operation of mirroring. Let P, S, $\mathcal{E}$, and Γ be as before. Consider a vertical mirroring of Γ through a vertical line not separating the points of S. This mirrored drawing represents a direction-consistent embedding, denoted by $\mathcal{M}(\mathcal{E})$, of a new path, denoted by $\mathcal{M}(P)$, on the mirrored point set, denoted by $\mathcal{M}(S)$. This new path $\mathcal{M}(P)$ is formally defined as follows: $\mathcal{M}(P) = \mathcal{M}(d_1)\mathcal{M}(d_2)\ldots\mathcal{M}(d_{n-1})$, where $\mathcal{M}(U) = U$, $\mathcal{M}(D) = D$, $\mathcal{M}(R) = L$, and $\mathcal{M}(L) = R$.

Observation 3. *Let $\mathcal{E}$ be a PDCE of a path P on a point set S. Then $\mathcal{M}(\mathcal{E})$ is a PDCE of $\mathcal{M}(P)$ on the point set $\mathcal{M}(S)$.*

3   Preliminaries

In this work we prove that every n-vertex three-directional path P admits a PDCE on any set of n points in convex position. As an overview, we sketch the basic idea of the proof. First, we show that it is possible to construct a PDCE of an $\{U, D, R\}$-path on a one-sided point set, while controlling the position of one of its end-points (Lemma 2 and Lemma 3). Then we show that we can embed a two-directional $\{U, R\}$-path on a strip-convex point set S while controlling the positions of both end vertices of the path (Lemma 4). We use these results to show that an $\{U, D, R\}$-path admits an embedding on any convex point set (Lemma 5). For this, we separate a given convex point set into one-sided point sets and a strip-convex point set and go through a case distinction on the labels of the edges which correspond to the separation of the point set. Finally, we show that an embedding of any three-directional path can be reduced to the embedding of an $\{U, D, R\}$-path (Theorem 1). We discuss the direction-consistency of constructed embeddings in detail in the flow of the proofs. However, the planarity of the embedding always follows from a single simple principle that is described by the following lemma and which is based on Lemma 3 of Binucci et al. [9].

Lemma 1. *An embedding of an n-vertex path on a convex point set is planar if and only if for each i, $1 < i < n$, path $P_{1,i}$ is mapped to a consecutive subset of S.*

We next show that Algorithm BACKWARD EMBEDDING is able to accomplish two tasks: to construct a PDCE of an $\{U, D, R\}$-path on a left-sided point set, and to construct a PDCE of an $\{U, R\}$-path on a strip-convex point set. The algorithm traverses the path backwards and places the vertex v_i, $1 < i \leq n$, so that, wherever vertex v_{i-1} is placed, edge (v_{i-1}, v_i) is guaranteed to be direction-consistent. The algorithm is a generalization of the algorithm constructing a PDCE of an $\{U, D\}$-path [9].

Lemma 2. *Let S be a left-sided point set and let $P = d_1, \ldots, d_{n-1}$ be an $\{U, D, R\}$-path. Algorithm BACKWARD EMBEDDING computes a PDCE $\mathcal{E}$ of P on S such that $\mathcal{E}(v_n)$ is $t(S)$, $b(S)$, or $r(S) \in \{t(S), b(S)\}$, dependent on whether d_{n-1} is U, D, or R, respectively.*

Algorithm 1. BACKWARD EMBEDDING

Input: $\{U, D, L, R\}$-path $P = d_1, \ldots, d_{n-1}$, convex point set S of size n
Output: Function $\mathcal{E} : V(P) \to S$
1 **for** $i \leftarrow n - 1$ *downto* 1 **do**
2   **switch** d_i **do**
3   **case** U: $\mathcal{E}(v_{i+1}) \leftarrow t(S)$ **case** D: $\mathcal{E}(v_{i+1}) \leftarrow b(S)$ **case** L: $\mathcal{E}(v_{i+1}) \leftarrow \ell(S)$
      case R: $\mathcal{E}(v_{i+1}) \leftarrow r(S)$
4    $S \leftarrow S \backslash \{\mathcal{E}(v_{i+1})\}$
5 $\mathcal{E}(v_1) \leftarrow v \in S$ // S contains only one element
6 **return** $\mathcal{E}$

Proof. Observe that the algorithm traverses the path backwards and decides the placement of vertex v_{i+1} based on the label of the edge (v_i, v_{i+1}), i.e., d_i. If $d_i = U$ (resp. D, L, R), vertex v_{i+1} is placed on the topmost (resp. bottommost, leftmost, rightmost) of the currently free points. Hence, when vertex v_i is placed at the next step on any other free point, edge (v_i, v_{i+1}) is guaranteed to be direction-consistent.

For the planarity, observe that the procedure picking the rightmost, the topmost, and the bottommost points of a left-sided point set, creates a consecutive subset of S. Thus, for any i, $1 \leq i \leq n - 1$, path $P_{i,n-1}$ (and therefore also $P_{1,i-1}$) is mapped to a consecutive subset of S. Hence, by Lemma 1, the created embedding is also planar. □

The following lemmas can be proven based on Lemma 2 and the operations of rotation of a point set and reverse of a path. See [1] for the missing proofs.

Lemma 3. *An $\{U, D, R\}$-path $P = d_1, \ldots, d_{n-1}$ admits a PDCE on any right-sided point set S such that $\mathcal{E}(v_1)$ is $b(S)$, $t(S)$, or $\ell(S) \in \{t(S), b(S)\}$, dependent on whether d_1 is U, D, or R, respectively.*

Lemma 4. *Let S be a strip-convex point set and let $P = d_1, \ldots, d_{n-1}$ be an $\{U, R\}$-path. Algorithm BACKWARD EMBEDDING computes a PDCE $\mathcal{E}$ of P on S such that (i) $\mathcal{E}(v_1)$ is $b(S)$ or $l(S)$, and (ii) $\mathcal{E}(v_n)$ is $t(S)$ or $r(S)$, dependent on whether d_{n-1} is U or R, respectively.*

4 Three-Directional Paths

The following lemma is the key ingredient for the proof of a main result of this paper. We postpone its proof until we have seen how the lemma is used.

Lemma 5. *Let S be a convex point set with the property that $t(S)$ is to the right of $b(S)$. Any $\{U, D, R\}$-path admits a PDCE on S.*

Theorem 1. *Any three-directional path admits a PDCE on a convex point set.*

Proof. **Case 1:** P is an $\{U, D, R\}$-path. Since S is in general position, $t(S)$ is either to the right or to the left of $b(S)$. In the former case a PDCE of P on S exists by

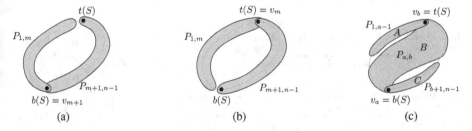

Fig. 1. Illustration of the construction in Cases 1-3

Lemma 5. For the latter case, observe that in $\mathcal{M}(S)$, point $t(\mathcal{M}(S))$ is to the right of $b(\mathcal{M}(S))$. Moreover, $P^{\mathcal{I}}$ is an $\{U, D, L\}$-path, and $\mathcal{M}(P^{\mathcal{I}})$ is again an $\{U, D, R\}$-path. By Lemma 5, there exists a PDCE $\mathcal{E}$ of $\mathcal{M}(P^{\mathcal{I}})$ on $\mathcal{M}(S)$. By Observation 3, $\mathcal{M}(\mathcal{E})$ is a PDCE of $P^{\mathcal{I}}$ on S. Due to Observation 1, $\mathcal{M}(\mathcal{E})^{\mathcal{I}}$ is a PDCE of P on S.

Case 2: P is an $\{U, D, L\}$-path. Observe that $P^{\mathcal{I}}$ is an $\{U, D, R\}$-path. Let $\mathcal{E}$ be a PDCE of $P^{\mathcal{I}}$ on S, which exists by Case 1. Then $\mathcal{E}^{\mathcal{I}}$ is a PDCE of P on S.

Case 3: P is an $\{U, L, R\}$-path. Thus, $\mathcal{R}(P)$ is an $\{U, D, L\}$-path. Due to Case 2, there exists a PDCE $\mathcal{E}$ of $\mathcal{R}(P)$ on $\mathcal{R}(S)$. By Observation 2, $\mathcal{R}(\mathcal{E})$ is a PDCE of P on S.

Case 4: P is a $\{D, L, R\}$-path. Notice that $\mathcal{R}(P)$ is an $\{U, D, R\}$-path. Thus, for a PDCE $\mathcal{E}$ of $\mathcal{R}(P)$ on $\mathcal{R}(S)$, which exists due to Case 1, $\mathcal{R}(\mathcal{E})$ is a PDCE of P on S. This concludes the proof of the theorem. □

Proof of Lemma 5. Let S_ℓ denote the subset of S containing all points on the left of the line through $b(S)$ and $t(S)$, and let $m = |S_\ell|$. We distinguish several cases based on the labels d_m and d_{m+1}.

Case 1: $d_m = D$, $d_{m+1} \in \{U, R\}$ (see Fig. 1(a) for an illustration). We embed $P_{1,m}$ on $S_l \cup \{b(S)\}$ using Algorithm BACKWARD EMBEDDING. By Lemma 2, vertex v_{m+1} is mapped to $b(S)$. Then, we embed $P_{m+1,n-1}$ on $S_r \cup \{t(S), b(S)\}$ in the way given by Lemma 3. Since $\ell(S_r \cup \{t(S), b(S)\}) = b(S_r \cup \{t(S), b(S)\}) = b(S)$ and $d_{m+1} \in \{U, R\}$, vertex v_{m+1} is mapped to $b(S)$. Thus, the union of these embeddings is a PDCE of P on S.

Case 2: $d_m \in \{U, R\}$, $d_{m+1} = D$ (see Fig. 1(b)). We embed $P_{1,m}$ on $S_l \cup \{t(S)\}$ using Algorithm BACKWARD EMBEDDING. By Lemma 2, vertex v_{m+1} is mapped to $t(S)$ since $r(S_l \cup \{t(S)\}) = t(S_l \cup \{t(S)\}) = t(S)$ and $d_m \in \{U, R\}$. Due to Lemma 3, we can embed $P_{m+1,n-1}$ on $S_r \cup \{t(S), b(S)\}$ such that vertex v_{m+1} is mapped to $t(S)$, since $t(S_r \cup \{t(S), b(S)\}) = t(S)$ and $d_{m+1} = D$. Thus, the union of these embeddings is a PDCE of P on S.

Case 3: $d_m = D$, $d_{m+1} = D$ (see Fig. 1(c)). Let $P_{a,b}$, $1 \le a \le m < m + 1 \le b \le n - 1$, be the maximal subpath of P containing d_m, d_{m+1} and only D labels. Let A be the a highest points of $S_l \cup \{t(S)\}$. Observe that A exists since $a \le m$. We embed $P_{1,a-1}$ on A using Algorithm BACKWARD EMBEDDING. By Lemma 2, vertex v_a is mapped to $t(S)$, since $d_{a-1} \in \{U, R\}$ and $r(A) = t(A) = t(S)$. Let C be the $n - b$ lowest points of $S_r \cup \{b(S)\}$. Since $|S_r \cup \{b(S)\}| = n - m - 1$, and $b \ge m + 1$, thus $n - b \le n - m - 1$, and therefore C exists. By Lemma 3, we can embed $P_{b+1,n-1}$ on

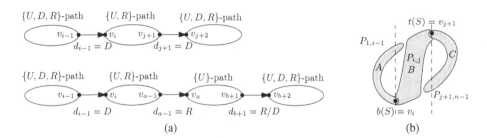

Fig. 2. (a) Structure of the path in Cases 4A (above) and 4B (below) (b) Construction in Case 4A

C such that v_{b+1} is mapped to $b(S)$ since $\ell(C) = b(C) = b(S)$ and $d_{b+1} \in \{U, R\}$. Let B be $(S \backslash (A \cup C)) \cup \{t(S), b(S)\}$. We embed the D-path $P_{a,b}$ on B, starting with v_a at $t(S)$ and ending with v_{b+1} at $b(S)$, by sorting the points of B by decreasing y-coordinate. Merging the PDCEs for $P_{1,a-1}, P_{a,b}$, and $P_{b+1,n-1}$, we obtain a PDCE of P on S.

Case 4: $d_m, d_{m+1} \in \{U, R\}$. Let $P_{i,j}$ where $1 \le i \le m < m+1 \le j \le n-1$ be the maximal subpath of P containing d_m, d_{m+1} and only U/R-labels. Thus $d_{i-1} = d_{j+1} = D$, if they exist. Let α (resp. β) denote the number of points of S lying to the left of $b(S)$ (resp. $t(S)$, including $t(S)$). We consider several cases based on how the indices i, j are related to the indices α, β. The intuition behind this is to distinguish whether or not the points that are vertically between $b(S)$ and $t(S)$ are enough to embed $P_{i,j}$.

Case 4A: $i > \alpha$ and $j < \beta$, i.e., the points vertically between $b(S)$ and $t(S)$ are enough to embed $P_{i,j}$ (see Fig. 2).
Let A be the i lowest points of $S_l \cup \{b(S)\}$; A exists since $i \le m$. By Lemma 2, we can embed $P_{1,i-1}$ on A such that v_i is mapped to $b(S)$. Let C be the $n-j$ highest points of $S_r \cup \{t(S)\}$; C exists since $n-j < n-m$. By Lemma 3, we can embed $P_{j+1,n-1}$ on C such that v_{j+1} is mapped to $t(S)$ since $d_{j+1} = D$. Let B be $(S \backslash (A \cup C)) \cup \{b(S), t(S)\}$. Since $i > \alpha$, $\ell(B) = b(B) = b(S)$, and since $j < \beta$, $r(B) = t(B) = t(S)$. Thus, B is a strip-convex point set. By Lemma 4, we can embed the $\{U, R\}$-path $P_{i,j}$ on B such that v_i lies on $b(S)$ and v_{j+1} lies on $t(S)$. By merging the constructed embeddings of $P_{1,i-1}, P_{i,j}$, and $P_{j+1,n-1}$, we obtain a PDCE of P on S.
Observe that if either $i-1 = \alpha$ and $d_\alpha = R$ or $j+1 = \beta$ and $d_\beta = R$ or both, then the embedding can be constructed identically. In case $d_\alpha = R$, vertex v_i is mapped to $r(A) = b(S)$. In case $d_\beta = R$, vertex v_{j+1} is mapped to $\ell(C) = t(S)$. Thus, these embeddings can be merged with the above embedding of $P_{i,j}$ on B.

Case 4B: $i > \alpha$ and $j \ge \beta$. In this case $d_\beta \in \{U, R\}$. If $d_\beta = R$ then the embedding is constructed as explained at the end of Case 4A. In the following we assume $d_\beta = U$. Let $P_{a,b}, i \le a \le \beta \le b \le j$ be the maximal subpath of P containing d_β and only U-edges; see Fig. 2(a)(below) for the structure of the constructed path. If $a > i, d_{a-1} = R$. Otherwise, if $a = i$ then $d_{a-1} = D$, i.e., the $\{U, R\}$-path $P_{i,a-2}$ is empty. Let A be the i lowest points of $S_l \cup \{b(S)\}$ (see Fig. 3(a)). Notice that A is a left-sided point set and $b(A) = b(S)$. We can embed $P_{1,i-1}$ on A by Lemma 2 such that vertex v_i is

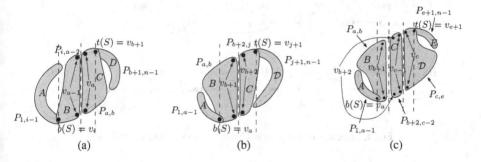

(a)   (b)   (c)

Fig. 3. Constructions for (a) Case 4B, (b) Case 4C, and (c) Case 4D when $P_{b+2,c-2}$ is non-empty (if $c = b + 2$ the set C is empty; if $a = c$ and $b = e$ the sets B and $\mathcal{D}$ are not distinguished).

mapped to $b(S)$. Let $\mathcal{D}$ be the $n - b$ highest points of $S_r \cup \{t(S)\}$. By Lemma 3, we can embed $P_{b+1,n-1}$ on $\mathcal{D}$ such that vertex v_{b+1} is mapped to $t(S)$. Let B be the $a - i$ leftmost points of $(S \backslash A) \cup \{b(S)\}$. If $a = i$ then B is empty. Otherwise, since $i > \alpha$, $\ell(B) = b(B) = b(S)$ and since $a \leq \beta$, the points $t(B)$ and $r(B)$ are consecutive in B. Thus, B is a strip-convex point set and by Lemma 4 we can embed the $\{U, R\}$-path $P_{i,a-2}$ on B such that vertex v_i is mapped to $b(S)$ and vertex v_{a-1} is mapped to either $t(B)$ or $r(B)$. Let $C = S \backslash (A \cup B \cup \mathcal{D}) \cup \{t(S)\}$. We embed $P_{a,b}$ on C by sorting the points by increasing y-coordinate. Thus, vertex v_a is mapped to $b(C)$ and vertex v_{b+1} is mapped to $t(S)$. If $a = i$, vertex $v_i = v_a$ is already mapped to $b(S)$, thus at this step we only embed the vertices of the $\{U\}$-path $P_{a+1,b}$.

Next we merge the constructed PDCEs of $P_{1,i-1}$, $P_{i,a-2}$, $P_{a,b}$, and $P_{b+1,n-1}$. If $a = i$, the edge d_i points upward since v_i is mapped to $b(S)$. Otherwise, since v_{a-1} is mapped to $t(B)$ or $r(B)$, v_a is mapped to $b(C)$, B and C are separable by a vertical line, and edge (v_{a-1}, v_a) points to the right and does not cross the remaining drawing.

Recall that this case considers the situation where $i > \alpha$. In case $i \leq \alpha$, we know that $d_\alpha \in \{U, R\}$. If it happens that $d_\alpha = R$, the construction can be accomplished identically by considering index $\alpha + 1$ everywhere in place of i. Here, Lemma 2 guarantees a mapping of $P_{1,\alpha}$ with $v_{\alpha+1}$ on $b(S)$ since it is the rightmost point of A and $d_\alpha = R$.

Case 4C: $i \leq \alpha$ and $j < \beta$. This case is symmetric to Case 4B. If $d_\alpha = R$ the embedding is constructed as explained at the end of Case 4A. Otherwise $d_\alpha = U$ and we again identify the maximal $\{U\}$-subpath $P_{a,b}$ of P containing d_α. The structure of the path in this case is shown in Fig. 4 and the embedding in Fig. 3(b).

Also, similar to Case 4B, we can use this construction to embed a path where $j \geq \beta$ and $d_\beta = R$. For that, consider the illustration of Fig. 3(b). We set $\mathcal{D}$ to contain only points to the right of $t(S)$ and $t(S)$, i.e., $|\mathcal{D}| = n - \beta + 1$. We embed $P_{\beta,n-1}$ on $\mathcal{D}$. By Lemma 3, we can map v_β to $t(S)$, since $d_\beta = R$ and $t(S)$ is the leftmost point of $\mathcal{D}$. The remaining construction is identical.

Case 4D: $i \leq \alpha$ and $j \geq \beta$, $d_\alpha = d_\beta = U$. Let $P_{a,b}$, $a \leq \alpha \leq b$, be the maximal $\{U\}$-subpath of P containing d_α. Similarly, let $P_{c,e}$, $c \leq \beta \leq e$, be the maximal $\{U\}$-subpath of P containing d_β. If there is no R-edge between d_α and d_β then $a = c$

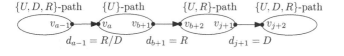

Fig. 4. Structure of the path in Cases 4C (above) and 4D (below)

and $b = e$. If there is a single R-edge between them then $c = b+2$. Otherwise, $P_{b+2,c-2}$ is a $\{U, R\}$-path containing at least one vertex; see Fig. 4 for this case.

We embed the $\{U, D, R\}$-path $P_{1,a-1}$ on the a lowest points, denoted by A, of $S_\ell \cup \{b(S)\}$. By Lemma 2, we can map v_a to $b(S)$, since the rightmost point of A is $b(S)$ and $d_{a-1} \in \{D, R\}$. By Lemma 3, we can embed $P_{e+1,n-1}$ on the $n - e - 1$ highest points, denoted by E, of $S_r \cup \{t(S)\}$, such that v_{e+1} is mapped to $t(S)$, since it is the leftmost point of E and $d_{e+1} \in \{D, R\}$. Fig. 3(c) shows the case where $P_{b+2,c-2}$ is non-empty. However, it presents the idea of the embedding in the remaining cases as well.

If $a = c$ and $b = e$ then $P_{a,e}$ is a $\{U\}$-path. We embed it on $S \setminus (A \cup E) \cup \{b(S), t(S)\}$, by sorting the points by increasing y-coordinate. This completes the construction of a PDCE of P on S. Otherwise, we let B (resp. $\mathcal{D}$) be the $b - a + 2$ leftmost (resp. $e - c + 2$ rightmost) points of $S \setminus (A \cup E) \cup \{b(S), t(S)\}$. We embed $P_{a,b}$ (resp. $P_{c,e}$) on B (resp. $\mathcal{D}$) by sorting its points by y-coordinates.

If $c = b + 2$, the $\{U\}$-paths $P_{a,b}$ and $P_{c,e}$ are joined by a single R-edge. Since v_{b+1} is to the left of $v_{b+2} = v_c$, the constructed embedding yields a direction-consistent embedding of the edge (v_{b+1}, v_{b+2}) and this completes the construction of a PDCE of P on S. Otherwise, $P_{b+2,c-2}$ is an $\{U, R\}$-path that contains at least one vertex and $d_{b-1} = d_{c-1} = R$. We embed $P_{b+2,c-2}$ on the remaining free points, i.e., on the point set $C = S \setminus (A \cup B \cup \mathcal{D} \cap E)$. By construction of B and $\mathcal{D}$, the set C is separated from the remaining points by vertical lines. Thus, $\ell(C)$ and $b(C)$ are either consecutive or coincide. Similarly, points $t(C)$ and $r(C)$ are either consecutive or coincide. Thus, C is a strip-convex point set. By Lemma 4, we can embed $P_{b+2,c-2}$ on C such that v_{b+2} is mapped to one of $\ell(C)$ or $b(C)$, and v_{c-1} to one of $t(C)$ or $r(C)$. As v_{b+2} is mapped to the highest point of B and v_c is mapped to the lowest point of $\mathcal{D}$, we infer that the obtained embedding of P on S is planar. Since $d_{b+1} = d_{c-1} = R$ and by the fact that C is separated from B and $\mathcal{D}$ by vertical lines, it is also direction-consistent. This concludes the proof of the lemma.   □

5   Four-Directional Paths

The proof of the following theorem, which can be found in the full version [1], is based on a counterexample showing that the path $P = LULRDR$ does not admit a PDCE on a left-sided point set.

Theorem 2. *There exists a one-sided point set S and an $\{U, D, L, R\}$-path P such that there is no PDCE of P on S.*

A one-sided point set S is a special case of a convex point set, such that $b(S)$ and $t(S)$ are consecutive. However, as Theorem 2 states, such a point set does not always admit a PDCE of every four-directional path. On the other hand, consider a one-sided convex point set S where one of the following pairs represents a clockwise consecutive subset of S: (i) $t(S)$ and $\ell(S)$, (ii) $r(S)$ and $t(S)$, (iii) $b(S)$ and $r(S)$, (iv) $\ell(S)$ and $b(S)$. Such a point set is called *quarter-convex*. It can be easily seen that every quarter-convex point set admits a PDCE of any four-directional path. Actually, in case (i) an edge pointing right always points up and an edge pointing left always points down. Thus, the problem of embedding a $\{U, D, R, L\}$-path is reduced to embedding a $\{U, D\}$-path, which always admits a PDCE on any convex point set [9]. Similar reductions can be made for any other type of a quarter-convex point set. Therefore, we state the following:

Observation 4. *Any $\{U, D, L, R\}$-path has a PDCE on any quarter-convex point set.*

Based on Lemma 1, it is easy to derive a dynamic programming algorithm to decide whether a four-directional path admits a PDCE on a convex point set. This is formalized in the following theorem. A similar algorithm, described in [14], tests whether an upward planar digraph admits an upward planar embedding on a convex point set.

Theorem 3. *Let P be an n-vertex four-directional path and S be a convex point set. It can be decided in $O(n^2)$ time whether P admits a PDCE on S.*

6   Conclusion

We investigated the question of finding a planar direction-consistent embedding on a convex point set for any given four-directional path. We have shown that this is always possible for paths that are restricted to at most three out of the four directions. To the contrary, we have provided an example showing that for paths using all four directions, this is not always possible. We also presented an $O(n^2)$ time algorithm to decide embeddability for a given four-directional path and convex point set.

The most challenging open problem is to determine whether any two- or three-directional path always admits a planar direction-consistent embedding on any point set in general position.

References

1. Aichholzer, O., Hackl, T., Lutteropp, S., Mchedlidze, T., Vogtenhuber, B.: Embedding four-directional paths on convex point sets. arXiv e-prints arXiv:1408.4933 [cs.CG] (2014)
2. Aichholzer, O., Krasser, H.: The point set order type data base: A collection of applications and results. In: 13th Annual Canadian Conference on Computational Geometry (CCCG 2001), pp. 17–20 (2001)
3. Aichholzer, O., Hackl, T., Huemer, C., Hurtado, F., Krasser, H., Vogtenhuber, B.: On the number of plane geometric graphs. Graphs and Comb. 23(1), 67–84 (2007)

4. Alspach, B., Rosenfeld, M.: Realization of certain generalized paths in tournaments. Discrete Math 34, 199–202 (1981)
5. Angelini, P., Frati, F., Geyer, M., Kaufmann, M., Mchedlidze, T., Symvonis, A.: Upward geometric graph embeddings into point sets. In: Brandes, U., Cornelsen, S. (eds.) GD 2010. LNCS, vol. 6502, pp. 25–37. Springer, Heidelberg (2011)
6. Bannister, M.J., Cheng, Z., Devanny, W.E., Eppstein, D.: Superpatterns and universal point sets. J. Graph Alg. Appl. 18(2), 177–209 (2014)
7. Bannister, M.J., Devanny, W.E., Eppstein, D.: Small superpatterns for dominance drawing. CoRR abs/1310.3770 (2013)
8. Biedl, T., Vatshelle, M.: The point-set embeddability problem for plane graphs. In: 28th Annual Symposium on Computational Geometry (SoCG 2012), pp. 41–50. ACM (2012)
9. Binucci, C., Di Giacomo, E., Didimo, W., Estrella-Balderrama, A., Frati, F., Kobourov, S., Liotta, G.: Upward straight-line embeddings of directed graphs into point sets. Computat. Geom. Th. Appl. 43, 219–232 (2010)
10. Cabello, S.: Planar embeddability of the vertices of a graph using a fixed point set is NP-hard. J. Graph Alg. Appl. 10(2), 353–366 (2006)
11. Durocher, S., Mondal, D.: On the hardness of point-set embeddability. In: Rahman, M.S., Nakano, S.-I. (eds.) WALCOM 2012. LNCS, vol. 7157, pp. 148–159. Springer, Heidelberg (2012)
12. Forcade, R.: Parity of paths and circuits in tournaments. Discrete Math. 6(2), 115 (1973)
13. Gritzmann, P., Mohar, B., Pach, J., Pollack, R.: Embedding a planar triangulation with vertices at specified points. The American Math. Monthly 98(2), 165–166 (1991)
14. Kaufmann, M., Mchedlidze, T., Symvonis, A.: On upward point set embeddability. Comput. Geom. 46(6), 774–804 (2013)
15. Mchedlidze, T.: Upward planar embedding of an n-vertex oriented path on $O(n^2)$ points. Comp. Geom.: Theory and Appl. 47(3), 493–498 (2014)
16. Reid, K., Wormald, N.: Embedding oriented n-trees in tournaments. Studia Sci. Math. Hungarica 18, 377–387 (1983)
17. Rosenfeld, M.: Antidirected hamiltonian circuits in tournaments. Journal of Comb. Theory, Ser. B 16(3), 234–242 (1974)
18. Straight, J.: The existence of certain type of semi-walks in tournaments. Congr. Numer. 29, 901–908 (1980)
19. Thomason, A.: Paths and cycles in tournaments. Trans. of the American Math. Society 296(1), 167–180 (1986)
20. Zhang, C.Q.: Some results on tournaments. J. Qufu Teachers College (1), 51–53 (1985)

Drawing Graphs within Restricted Area*

Maximilian Aulbach[1], Martin Fink[2], Julian Schuhmann[1], and Alexander Wolff[1]

[1] Lehrstuhl für Informatik I, Universität Würzburg, Germany
[2] Department of Computer Science, University of California, Santa Barbara, USA

Abstract. We study the problem of selecting a maximum-weight subgraph of a given graph such that the subgraph can be drawn within a prescribed drawing area subject to given non-uniform vertex sizes. We develop and analyze heuristics both for the general (undirected) case and for the use case of (directed) *calculation graphs* which are used to analyze the typical mistakes that high school students make when transforming mathematical expressions in the process of calculating, for example, sums of fractions.

1 Introduction

Our motivation for the problem that we study in this paper stems from so-called *calculation graphs*. Calculation graphs represent calculations starting from some initial task. They are used in studies [13] involving large numbers of high school students in order to analyze the students' typical mistakes in elementary mathematics. Even for relatively simple tasks such as evaluating the term "$3 \cdot (2 + 1/5)$", the different transformations performed by a large number of subjects can result in calculation graphs with hundreds of vertices. With the help of drawings of calculation graphs, human experts can analyze how students calculate and, in particular, which mistakes they frequently make. As Hennecke [13] suggests, such drawings are only useful if they are not too large, that is, if they fit into a relatively small drawing area. Hence, well-readable drawings of important parts of the graphs must be generated in an automated fashion. Certainly, the drawn subgraph should represent as much information of the original calculation graph as possible. Therefore, we consider the graphs to be edge- and vertex-weighted; see Fig. 1 for an example. The weight of an edge is the number of students who applied the respective calculation step; the weight of a vertex is the number of students who had the given term as an intermediate result in their calculation. Since we want the labels of the vertices to be readable, we assume that their sizes are fixed. Hence, often only a small fraction of the graph will fit into the prescribed drawing area. Note that a user of such a drawing must be made aware that only a subgraph is shown.

Certainly, vertices and edges that occur only in a single student's calculation can easily be dropped. For higher weights, however, this is not as easy. For instance, it is conceivable that dropping a single vertex of weight W makes it possible to include several other vertices into the drawing, each of weight slightly less than W; then the resulting drawing potentially allows for a better analysis. Hence, we need to select the subgraph to be drawn based on the graph structure and not only on the weights.

* M. Fink was partially supported by a fellowship within the Postdoc-Program of the German Academic Exchange Service (DAAD). A. Wolff acknowledges support by the ESF EuroGIGA project GraDR (DFG grant Wo 758/5-1).

C. Duncan and A. Symvonis (Eds.): GD 2014, LNCS 8871, pp. 367–379, 2014.

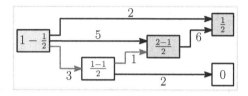

Fig. 1. Example of a calculation graph. Edge weights are indicated on the edges

Further Examples. There are many other scenarios in which most of the information of a large graph should be presented in a limited area. A first example are social networks, which are usually quite large. In this setting, vertices usually have labels and weights; the weight may describe the person's activity in the network. By drawing a heavy subgraph in a restricted area, a user can quickly get an overview over important actors in the network and connections between them. A second example, which we will use in the main part of the paper, are coauthor networks, where the weights represent numbers of publications. Here, we want to find a drawing of a subgraph that represents as many of the publications as possible, i.e., we prefer to keep authors with many publications.

Related Work. Surprisingly, the very natural problem of selecting a subgraph that can be drawn within a prescribed area seems to be new. Some work on related problems, however, is worth being mentioned. Rather than fixing the drawing area and maximizing the size of the subgraph that fits into this area, graph drawing research has focussed on drawing the whole input graph and minimizing the area needed for the drawing, which has also been the task in some graph drawing contests [6]. This problem is called compation and known to be NP-hard for orthogonal drawings with given orthogonal representation (that is, fixed bends) [14]. In constrained graph layout [12,9], the user can constrain the region into which a vertex must be placed. Dwyer et al. [8] presented the algorithm IPSep-CoLa where constraints demanding a vertical or horizontal separation of vertex labels can be specified. Such constraints make it possible to enforce that vertices are placed within a prescribed rectangular area. However, in contrast to our work, they do not consider dropping vertices. If the vertices do not fit into the area, their algorithm will not find a feasible drawing. The well-known algorithm of Fruchterman and Reingold [10] for force-directed graph drawing allows to specify a rectangular area within which the whole graph has to be drawn. However, in contrast to our work, the algorithm can easily achieve this since vertices are drawn as points without labels and edges can be made arbitrarily short, which we do not allow.

Dwyer et al. [7] developed a method for interactive exploration of graphs based on constrained graph layout. They use a fast heuristic for the overview drawing; for the detailed view of a smaller part of the graph, however, they can afford to use a slower constrained graph layout algorithm that yields better results. Constraints ensure consistency between the views and preserve the users's mental map. Da Lozzo et al. [5] considered the problem of drawing graphs on the very small display of a smartphone. They did not try to represent the whole graph on the display, but rather decided to provide only a local view around a focus vertex; their approach then offers interactive ways of navigating through the graph and exploring the graph based on the local view.

Our Contribution. While calculation graphs have quite specific characteristics and drawing requirements (see Section 3), the problem of graph drawing under area restrictions is also of interest for general, undirected graphs. For both cases, we present heuristics and evaluate them experimentally.

For the general case, we use the force-directed approach, reusing forces defined by Fruchterman and Reingold [10] and by Bertault [3]. We add extra phases in which the graph is pressed together; from time to time we remove vertices or edges from the graph if this is needed for further compaction. We use the desired edge length as parameter both for the usual iterations and our extra phases. In our tests, this proved to be effective for parametrizing the density of the output drawing. Furthermore, we experimentally improved the way of avoiding vertex-edge intersections and we developed a postprocessing that routes some of the edges as curves.

For calculation graphs, we chose the well-known Sugiyama framework [17] as basis for our algorithm since we want calculation paths to be readable from left to right. We add a method that successively removes the least important vertices and edges until the drawing fits into the given area. We also consider the weight of edges for crossing minimization so that important edges have few crossings. In our tests, it turned out that removing the lightest vertices as a preprocessing often improved the weight of the final subgraph. Furthermore, routing the edges as curves gave very nice results, also compared to the original orthogonal drawing style for calculation graphs.

2 General Graphs

We first present an algorithm for drawing arbitrary graphs. The input of our problem consists of an unweighted graph $G = (V, E)$ with a weight function $w \colon V \cup E \to \mathbb{R}^+$. For each vertex $v \in V$, we are given a geometric object $\ell(v)$ that will represent the vertex in the drawing. Vertices can be represented by different shapes, e.g., rectangles, disks, or ellipses. We will focus on rectangular vertices which are well-suited for text labels of vertices. We will denote the height and the width of $\ell(v)$ by h_v and w_v, respectively.

In addition to the graph input, we are given an axis-parallel rectangle of height H and width W, the *drawing area*. The task is to find a subgraph $G' = (V', E')$ of G with a *nice* drawing of G' within the given drawing area.

The hard constraints for the drawing are clear: Each vertex $v \in V'$ must be represented by $\ell(v)$, the vertices must not overlap, and each edge must connect its incident vertices. However, it needs to be clarified, what a nice drawing is. In our setting with a restricted drawing area, putting vertices close together can allow us to have more vertices—and thus more weight—in the final drawing. Certainly, a drawing with vertices that are very close is not nice; the same holds for very short edges. We will discuss these criteria and more later in detail.

By a straightforward reduction from SUBSET SUM, where we use height 1 for all vertices and the drawing area, we can easily observe that maximizing the weight of the subgraph that can be drawn in a prescribed area is NP-hard.

2.1   Our Algorithm

Our approach uses the force-directed framework; in this class of algorithms, the drawing is incrementally improved, starting with an arbitrary layout. Each improvement step is done by letting *forces*, defined using physical analogies, move the vertices.

In contrast to usual force-directed algorithms, we have to take both the dimensions of vertices and of the prescribed drawing area into account. Therefore, we add two important ingredients: We try to fit the drawing into a frame of decreasing size, and we remove vertices or edges from the graph in order to make the graph smaller so that the current drawing can fit into the current frame. While, as a first idea, fitting the drawing into the given area could be steered by a force pulling all vertices towards the center of the drawing region, this idea has some drawbacks. Therefore, we will introduce a more advanced approach. Similarly, removing vertices could simply be done by removing the lightest vertex in each step. However, this would take neither the structure of the graph nor the current drawing into account. Hence, we introduce a measure for the *stress* of vertices in the current situation; we will always remove the vertex with the highest stress value. In the following paragraphs, we will detail out our algorithm's individual steps.

Forces. We first define the forces used in our algorithm.

- We reuse existing forces from the algorithm of Fruchterman and Reingold [10]; that is, for any pair $u, v \in V$ of vertices, there is a force $F_r(u,v) = l^2_{unit}/(d(u,v)) \cdot \overrightarrow{uv}$ on v that repels v from u, where $\overrightarrow{uv}$ is the unit vector pointing from u towards v. If the vertices are adjacent, that is, if $uv \in E$, there is an additional force $F_a(u,v) = (d(u,v))^2/l_{unit} \cdot \overrightarrow{vu}$ that attracts v towards u. Both forces use a factor l_{unit} which describes the desired unit edge length. Since the desired edge length heavily influences the density of the drawing, and, hence, the number and weight of vertices and edges that can be placed within the given drawing area, the choice of l_{unit} is crucial for the results. While we allow the parameter to be set freely, we stress that the value must be chosen carefully, taking the sizes of vertices into account, so that one gets nice output drawings.

- Due to the high density of the input graphs and the given sizes of vertices, it may easily occur that vertices are intersected by nonincident edges, which reduces the readability significantly. As a first step to overcome this problem, we use a force that has been introduced by Bertault in his PrEd algorithm [3]: If an edge $\{u,w\}$ intersects a vertex v in its inner region, that is, close to the center of v, then we let a force $F_e(v, \{u,w\}) = (l_{unit} - d(v,i_v))^2 \cdot \overrightarrow{i_v v}$ repel v from $\{u,w\}$, where the point i_v is the orthogonal projection of v onto the straight-line segment $\overline{uw}$. Note that this does not guarantee that intersections between vertices and edges are avoided. However, such intersections become less likely; in Sec. 2.2, we will see how their number can be further reduced by routing edges as curves.

- For making the drawing more compact, we introduce a force $F_g(v) = d(v, p_c) \cdot \overrightarrow{vp_c}$ that attracts each vertex v to the center p_c of the drawing area. In our experiments it turned out that activating this force in later steps of the algorithm reduces the time for finding a final drawing, but also the quality of the output. Therefore, as a default, the force is only active when computing the very first equilibrium layout.

Handling the Frame. The forces described above yield a functional force-directed algorithm which can be applied for getting an initial layout. Once we have an initial drawing, we initialize the frame F as the bounding box of the drawing. Our algorithm iteratively reduces the size of the frame until it matches the prescribed drawing area.

In each step, we first uniformly reduce the height and the width of F by a small amount. It may happen that some vertices (partially) lie outside of the resulting new frame F'. If this is the case for a vertex v, we just place it at the closest position that lies completely within F'. This operation can result in intersections of vertices. Therefore, we compute a new equilibrium state which hopefully solves the intersections.

In all force-directed iterations in which there is a frame, we will always ensure that no vertex leaves the frame. This is done by cutting off the resulting movement vectors; Fruchterman and Reingold [10] did the same for ensuring that no vertex leaves the drawing area in their algorithm—with the difference that they did not shrink the frame but rather started with a very compressed drawing since in their setting vertices are points that can be arbitrarily close.

If there are intersections after computing a new equilibrium layout, we have a clear indication that the graph is still too large for the current frame and, hence, for the desired drawing area. In this case, and also in some more cases, we will remove vertices or edges as described in the next section.

Removing Vertices and Edges. Our indicator for the necessity of the removal of vertices or edges is, roughly speaking, the density of the drawing. If there are too many vertices in the graph for the current frame, then vertices will come very close. Therefore, we decide to remove a vertex or an edge if the minimum edge length is less than a value l_{adj} or if the minimum distance between two nonadjacent vertices is less than a value l_{nadj}. Note that this includes the case of two intersecting vertices.

Deciding what should be removed is more difficult. Since we want to keep as much weight as possible, the natural idea is to remove the lightest vertex. However, this is often not the best choice—even if it is unambiguous. A more advanced criterion should take also the degree of a vertex into account. The higher the number of neighbors of a vertex is, the more information on the graph is lost by removing the vertex.

Stress Calculation. However, we can still do better: So far, the current drawing is not taken into account, although it yields valuable information about the density of vertices in the vicinity of the vertex that should be removed. Thus, we suggest considering the forces in the last equilibrium layout. Even if the total movement vector of a vertex has length zero, this may actually result from strong forces that try to move the vertex to different directions, e.g., if the vertex is "trapped" between many other vertices.

These considerations lead us to a measure that we call the *pressure* of a vertex. Intuitively, the pressure is the maximum strength of forces in roughly opposite directions that act on a vertex. For formalizing this, we subdivide all force vectors applied on the vertex v into eight octants. For each octant, we sum up the force vector. For $i = 0, \ldots, 7$ let l_i be the length of the resulting force vector for octant i. Now, we first build the pressure p_i for octant i by comparing l_i with the force vectors in the three opposite directions, i.e., with l_{i+3}, l_{i+4}, and l_{i+5} (mod 8); see Fig. 2. The pressure then is the maximum over the pairwise minima, i.e., $p_i = \max\{\min\{p_i, p_{i+3}\}, \min\{p_i, p_{i+4}\}, \min\{p_i, p_{i+5}\}\}$. The total pressure on v is $p(v) = \max\{p_0, \ldots, p_7\}$.

Now, we must integrate the weight $w(v)$ and the degree $\deg(v)$—indicators of the vertex's resistance against pressure—with the pressure in order to get the *stress* $s(v)$. We do so by setting $s(v) = p(v)/(w(v) \cdot (\deg(v) + c_{\deg}))$. Here, $c_{\deg}$ is a small positive constant that ensures that we do not get problems for isolated vertices and that steers our preference for keeping isolated vertices in the drawing. With this definition of the stress, we can always choose the vertex with the highest stress for removal.

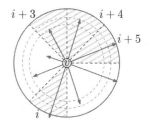

Fig. 2. Pressure computation for one of the octants

Boundary Vertices. There is a special case for vertices close to the boundary of the frame. We never move a vertex over the boundary although, in many cases, vertices are repelled in this direction by the inner vertices. This phenomenon is a cause of pressure on a vertex that is, so far, not covered by our definition. We therefore introduce a new "virtual" force that mimics the resistance against movements that push the vertex out of the frame. This force repels the vertex perpendicularly to the inside of the frame, away from the closest point p_f of the frame's boundary. More precisely, the virtual force is $F_f(v) = \overrightarrow{p_f v} \cdot l_{\text{unit}}^2/d(v, p_f)$. We stress that F_f is only taken into account for the stress computation and not actually applied.

Edges. In some cases, one may try to remove only an edge instead of a whole vertex; the hope is that after removing the edge, the graph becomes more flexible so that the available space can be used better. As an indicator for such a situation, we use the average edge length in the current drawing. If this length is larger than $l_{\text{unit}} \cdot c_{\text{len}}$, with a factor $c_{\text{len}} > 0$, we decide for removing an edge instead of a vertex. The intuition is that, on average, the edges are not very short, which means that by removing one of these longer edges we could allow more flexibility to the placement of vertices.

In order to determine which edge will be removed, we, again, use a definition of stress. To this end, we take both the weight $w(e)$ and the weights of the edges crossing e into account. Let E' be the set of these edges. Then, we set $s(e) = \sum_{e' \in E'} w(e') \cdot |E'|/w(e)$ to be the stress of e, and we remove the edge with the highest stress value.

2.2 Extensions

We developed and implemented two extensions that can help improve the runtime of the algorithm and the quality of the resulting drawings, respectively.

Preprocessing. In many input instances, there is a large number of vertices with very small weight, for which it is very unlikely to occur in the final drawing. To speed up the algorithm, we can remove all vertices that are lighter than a threshold value $w^\star$. Our choice of $w^\star$ is based on guessing a bound on the maximum number of vertices in the final drawing and depends on the height H and the weight W of the drawing area, the minimum height $h_{\min}$ and the minimum width $w_{\min}$ of a vertex as well as on the desired edge length l_{unit} and a factor $c_{\text{pre}} > 0$. We will make sure that we keep at least $(H \cdot W)/\big((l_{\text{unit}} c_{\text{pre}} + h_{\min}) \cdot (l_{\text{unit}} c_{\text{pre}} + w_{\min})\big)$ vertices in the graph.

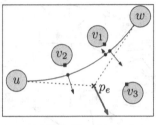
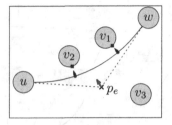

(a) Repelling forces   (b) Attracting forces; v_3 is too far away for an attraction.

Fig. 3. Repelling and attracting forces for Bézier curves

Postprocessing: Bézier Curves. In Sec. 2.1, we explained the force that aims at avoiding intersections between vertices and nonincident edges. However, we cannot guarantee that we do not have such intersections. We can do two things about this: (i) Intersections of an edge with the outer region of a vertex are relatively easy to distinguish from incidences and can, therefore, be tolerated. (ii) We can remove more edges if necessary.

Here, we present a third possibility: If we allow edges to be curves rather than only straight-line segments, we can avoid more intersections. In their improvement to Bertault's PrEd algorithm, Simonetto et al. [16] allowed polyline edges, where bends are introduced and removed based on the current drawing. However, this approach made the algorithm much slower; also, use a postprocessing in which edges are routed as Bézier curves around intersected nonadjacent vertices. We do this by representing edge $e = \{u,w\}$ by a quadratic Bézier curve, i.e., a parametric curve with a control point p_e in addition to the endpoints.

The computation of the curve is done as a postprocessing in the very last step of the algorithm. It is realized as an additional force-directed algorithm in which only the control point is moved, starting at the position in the middle between u and w. Each vertex v that is not far away from e causes a repelling force on p_e; see Fig. 3a. This force is parametrized by the width w_v and the height h_v of v as well as by the point p'_v of v that is closest to e and the point p'_e of e that is closes to v. The repelling force is defined as $F_{\text{rb}}(e,v) = \overrightarrow{p'_v p'_e} \cdot (w_v^2 + h_v^2)/d(p'_v, p'_e)$.

In order to avoid that the edge is curved too much, we also have an attracting force for vertices that have been (almost) intersected by e. If v is such a vertex, then the attracting force $F_{\text{ab}}(e,v) = \overrightarrow{p'_e p'_v} \cdot d(p'_v, p'_e)^2 / \sqrt{w_v^2 + h_v^2}$ is applied to p_e; see Fig. 3b.

2.3 Experiments and Evaluation

We implemented our heuristic in Java, using the graph library JUNG[1]. Our experiments were performed on a Core i5-2500K CPU with 8 GB of RAM. For the force-directed part of the algorithm, we also used a cooling factor that slows down the movement of vertices over the iterations in order to accelerate the computation of an equilibrium. We configured our algorithm such that there are always 25 steps of shrinking the frame around the current drawing. As input data, we primarily used (subgraphs of) the graph drawing collaboration graph from 1994 till 2012; in total, the graph has 950 vertices and 2559 edges. The weight of each vertex is the number of publications of

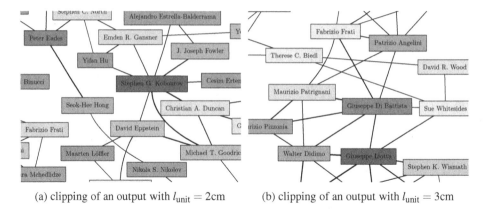

(a) clipping of an output with $l_{unit} = 2$cm   (b) clipping of an output with $l_{unit} = 3$cm

Fig. 4. Output examples with different choices for l_{unit}

the respective author and the weight of an edge connecting two authors is the number of their joint publications. We focused on a drawing area of $29.7cm \times 21cm$ (DIN A4) where the size of vertices was determined by the author's name (in 10pt font). For the complete publication graph, the runtimes were about 2 minutes, depending on the parameters.

The main results of our tests are the following.

- The preprocessing step described in Sec. 2.2 pays off; depending on the parameter c_{pre}, we may save runtime and get better results. In our tests, $c_{pre} = 0.7$ seemed a good choice; see Table 1 in the extended version of the paper [2].
- For testing whether something has to be removed, we set $l_{adj} = 0.1l_{unit}$ and $l_{nadj} = 0.15l_{unit}$. For the parameter c_{len} that determines whether an edge or a vertex will be removed, $c_{len} = 0.9$ gave a good compromise between the vertex and the edge weight in the final drawing; see also Table 2 in the extended version [2].
- We tested the effect of not always activating the force that repels vertices from edges; instead, we first computed an equilibrium without the force and then another one with it, so that we have more flexibility for vertices to cross edges. In our tests, this proved to have a significant impact on the total weight of edges in the final drawing, yielding an increase of almost 80%; see Table 3 in the extended version [2].
- The following way of computing the total force vector F for a vertex based on putting weights to the single forces showed the best results: $F = 0.01F_r + 0.01F_a + 0.005F_g + 0.0075F_e + 0.01F_f$. Note that not all forces are active at the same time.

A central parameter used in our algorithm is l_{unit}, describing the desired edge length. Figure 4 shows output examples that demonstrate that higher values lead to drawings that are less dense. Full outputs are available in the extended version of the paper [2]

3   Calculation Graphs

We now consider the initial problem of drawing calculation graphs in prescribed area. In our terminology, most of the input stays the same compared to general graphs, except that now our edges are directed since they represent calculation steps. Additionally, we are given a *start vertex* $s \in V$ that represents the task given to the students. Hence, s must be present in the output drawing and, furthermore, we insist that in the subgraph that is finally drawn, each vertex can be reached from s. To further improve the readability of the steps, we want as many edges as possible pointing from left to right. As a drawing convention for the edges, we will use the orthogonal drawing style with edges leaving vertices horizontally; we try to minimize the numbers of crossings and of bends and to optimize the gaps between edges and vertices. The hardness proof for general graphs can easily be adjusted. Hence, the problem remains NP-hard.

3.1 Our Algorithm

Due to the required readability from left to right, our approach is based on the Sugiyama framework [17] for hierarchical graph drawing. This framework consists of several steps that we will adjust to our problem. Our adjustments are optimized for our drawing style and especially for the task of removing vertices or edges from the graph, if necessary. We first briefly review the steps of the Sugiyama framework before going into detail and describing our adjustments.

In the first step, the graph is made acyclic by reverting some of the edges. Next, the vertices are assigned to layers from left to right. Then, based on the layer assignment, the number of edge crossings is minimized resulting in relative orders of the vertices of each layer. Eventually, final vertex coordinates are computed and the edges are routed.

Breaking the Cycles. In later steps of the Sugiyama framework, it is assumed that all edges are directed from left to right, i.e., the graph must be acyclic. Hence, we must revert some edges. For reverting the smallest number of edges, the NP-hard Feedback Set problem must be solved. We can either use existing heuristics, or even afford solving the problem optimally with the help of an MIP solver; in our tests, this worked quite fast and allows us also to minimize the weight of the reverted edges rather than their number.

Layer Assignment. Several approaches for the layer assignment in the Sugiyama framework exist, depending on the objective, e.g., of minimizing the number of layers or the number of vertices in a layer. Often, the number of layers is minimized subject to a prescribed maximum height of each layer. However, we will not use the height of our drawing area as the maximum height of a layer, although, at first, this may seem a good idea: If we do so, we would most probably have to remove many layers of vertices completely from the graph in later steps, which, subsequently, can also cause the removal of vertices of layers that are not removed, making these layers automatically smaller. Instead, we will set $n_{max} = \lceil |V|/k \rceil$, where k is the length of the longest path in the graph, so that we can hope for roughly equal numbers of vertices per layer. We mainly used the heuristic of Coffman and Graham [4] with the minor adjustment that preferably the leftmost layers have more vertices; we also tested Graham's list scheduling algorithm [11] and an assignment with the minimum number of layers.

Vertex Removal. After the layer assignment, the configuration usually does not fit into the drawing area. We now remove vertices until all vertices can be placed in the drawing area. We first remove vertices from each layer, so that the height of the layer is small enough. Afterwards, we remove whole layers until the width requirement is fulfilled. We first remove single vertices because this step can significantly influence the total weight of layers and, therefore, the choice of layers that will be deleted. The removal from the layers is done from left to right since the removal of a vertex from a layer can cause other vertices right of it to also be removed, if they become unreachable from s.

When removing from a layer, we should prefer light vertices. However, we must also take the heights of vertices into account: Removing a high vertex may save as much space as removing several lighter vertices whose weight sums up to a larger value. Hence, we measure the importance of a vertex v as $i(v) = w(v)/h_v$ and remove as many vertices of lowest importance as necessary so that the layer fits into the drawing area. We also tried other importance measures by taking the possible decrease of the width of the layer or the decrease of its area into account when removing a vertex. However, these measures did not perform better than the simpler height-based measure.

Once all layers have a feasible height, we will remove complete layers so that we are within the allowed total width. Note that we must always keep gaps between adjacent layers so that the edges can be drawn. The removal of layers is also done from left to right. We do this based on the importance of a layer L, which we define as $i(L) = \sum_{v \in L} w(v)/\text{width}(L)$, where the width of L is determined by its widest vertex.

Crossing Minimization. For the crossing minimization, we use the methods commonly used in the Sugiyama framework, which are based on considering only (parts of) edges between adjacent layers, but do so multiple times. For the adjacent exchange heuristic, we also considered the version where the weight of crossing edges is minimized. This heuristic just performs swaps of adjacent vertices in a layer—if this reduces the number (or weight) of crossings. Hence, weights can easily be integrated.

Edge Removal. Even after crossing minimization, there could still be too many crossings for the drawing to be well readable, if the graph is dense. Hence, we add a step in which edges are removed, if necessary. To this end, we introduce a measure for the importance of an edge e: If $E'(e)$ is the set of edges that cross e, then the importance of e is $i(e) = w(e)/(\sum_{e' \in E'(e)} w(e'))$. The result is that edges without a crossing are considered most important and will never be removed. Furthermore, an edge that crosses heavy edges—which are more valuable to us—will more likely be removed.

Gaps in Layers. Now, we know the orders of vertices in layers, where a layer also contains edges that are routed through several layers. We improve readability by using different gaps between the objects in a layer: Two edges are drawn closer together than two vertices, and edges that stay parallel until the next layer can be drawn even closer. While the different gaps make the drawings nicer, the consequence is also that only now we know the precise height each a layer. In some cases, this can make it necessary to remove another vertex of a layer, which, again, is done based on the importance measure.

Coordinate Assignment. For the final adjustment of the vertex positions, we still have some flexibility if the vertices (and edges) in the layer do not consume the total available height. We can use this flexibility and try to minimize the number of edge bends.

Therefore, we integrated a part of the heuristic of Sander [15]. However, due to the height constraint, we usually cannot save too many bends.

Edge Routing. Finally, only the edges need to be drawn, with several subproblems:
- We have to distribute the ports of the edges at the incident vertices or use a single port shared by all edges.
- We indicate the weight of edges by drawing them with different width. Since there are only few edges that are very heavy, it makes sense to not use a linear dependency between width and height but, e.g., a logarithmic dependency, or a dependency to the cube root (which gave the nicest results for our drawings).
- We have to distribute the vertical segments of the edges between consecutive layers such that both overlaps between segments and unnecessary (double) crossings are avoided. We first find a relative order of the segments from left to right for each pair of adjacent layers as follows: For any pair of edges between the layers that do not have to cross, there is at most one order with an unnecessary crossing. Using these orders, we build a directed graph of vertical segments. A topological sorting of this graph yields an order of the segments that avoids all unnecessary crossings.
  Once a relative order of the vertical segments is found, we assign the final coordinates. Small improvements are possible that locally optimize the spacing between the edge segments and we have put a lot of effort into implementing some of them. The most valuable optimization was using a force-directed algorithm with repelling forces between adjacent segments that optimizes the distances between the segments.

3.2   Extensions

Preprocessing. Similar to our algorithm for general graphs, we use a preprocessing step in which very light vertices and edges are removed in order to speed up the later steps.

Reinsertion of Removed Vertices. After crossing minimization, when the order of vertices is fixed, it is possible that we could safely reinsert some of the removed vertices so that the available area is used in a better way. We prefer the vertices with the highest importance as defined before and insert them in the leftmost available layer. Note that after reinserting vertices, we may have to reorder some of layers for crossing minimization.

Bézier Curves. While we try to avoid bends of the orthogonal edges, there still can be longer edges that have several bends, making them hard to follow. We suggest drawing the edges as smooth curves instead. To this end, we represent each edge segment between two adjacent layers as a cubic Bézier curve. As a simple version, this can be done by making the two bends of the orthogonal edge the two middle control points of the curve; this yields already quite nice results. We can further improve the drawings by adjusting the force-directed algorithm for the vertical segments (i.e., the control points): First, we can allow horizontal segments to overlap since they are not actual segments any more. Second, we can add a tendency to put the segments close to the middle between the layers in order to avoid sharp bends. We can, however, not place all vertical segments in the middle; doing so could result in unnecessary crossings of the respective curves.

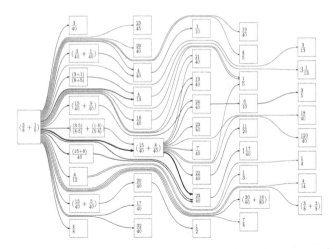

Fig. 5. An output example with edges drawn as Bézier curves (scaled down)

Weight Transfer. Suppose we delete a vertex v such that the edges (u,v) and (v,w) for vertices u and w exist. If both edges are heavy, it is possible that many students reached w from u with v as an intermediate step. Hence, after the removal, an edge (u,w) becomes more valuable to us because this edge can also partially represent the steps described above. We can model this by creating edge (u,w)—if it did not exist— and increasing its weight by $\min\{w(u,v),w(v,w)\}$ for the remainder of the algorithm. Similar weight transfers make sense also in more complicated situations.

3.3   Experiments and Evaluation

Also the algorithm for calculation graphs was implemented in Java. We used real-world data generated in user studies, with graphs of 107 and more vertices. The largest graph had 1031 vertices and 1549 edges. As for general graphs, we mainly used the A4 paper size as the prescribed drawing area. The tests were performed on a 3 GHz CPU with 4 GB RAM. Figure 5 shows an output example using Bézier curves, which, in our opinion, is the nicer and more interesting style compared to the version with orthogonal edges. Full examples can be found in the extended version [2]. Our main results are as follows.

- A preprocessing that removes the lightest vertices often improves the output, i.e., the drawn subgraph is heavier; see Table 4 in the extended version [2].
- There was no significant influence of the chosen layering algorithm on the weight of the final subgraph, especially when using the postprocessing for vertex reinsertion.
- Taking the weight of crossing edges into account in the adjacent exchange heuristic for crossing minimization reduces the weight of crossing edges significantly (factor > 2) and causes only few additional crossings; see Table 5 (extended version [2]).
- The computation for the largest graph with 1031 vertices took 3 to 4 seconds, de- pending on the parameters; most of the time was spent on the adjustment of seg- ments in the edge routing step and on writing the output file (about 2 seconds).

4 Conclusion

We have introduced the problem of drawing a heavy subgraph in a prescribed area. Both for general graphs without further constraints and for calculation graphs, we have developed and tested heuristics which yield quite nice results.

Acknowledgement. We thank Martin Hennecke for introducing the problem of drawing calculation graphs to us and for providing us with input data.

References

1. Java Universal Network/Graph Framework (JUNG), http://www.jung.sourceforge.net
2. Aulbach, M., Fink, M., Schuhmann, J., Wolff, A.: Drawing graphs within restricted area. CoRR (2014), ArXiv e-print http://arxiv.org/abs/1409.0499
3. Bertault, F.: A force-directed algorithm that preserves edge crossing properties. Inf. Proc. Letters 74(1-2), 7–13 (2000)
4. Coffman, E.G., Graham, R.L.: Optimal scheduling for two-processor systems. Acta Inform. 1(3), 200–213 (1972)
5. Da Lozzo, G., Di Battista, G., Ingrassia, F.: Drawing graphs on a smartphone. J. Graph Algorithms Appl. 16(1), 109–126 (2012)
6. Duncan, C.A., Gutwenger, C., Nachmanson, L., Sander, G.: Graph drawing contest report. In: Didimo, W., Patrignani, M. (eds.) GD 2012. LNCS, vol. 7704, pp. 575–579. Springer, Heidelberg (2013)
7. Dwyer, T., Marriott, K., Schreiber, F., Stuckey, P., Woodward, M., Wybrow, M.: Exploration of networks using overview+detail with constraint-based cooperative layout. IEEE Trans. Vis. Comput. Graph. 14(6), 1293–1300 (2008)
8. Dwyer, T., Koren, Y., Marriott, K.: IPSep-CoLa: An incremental procedure for separation constraint layout of graphs. IEEE Trans. Vis. Comput. Graph. 12(5), 821–828 (2006)
9. Dwyer, T., Marriott, K., Wybrow, M.: Topology preserving constrained graph layout. In: Tollis, I.G., Patrignani, M. (eds.) GD 2008. LNCS, vol. 5417, pp. 230–241. Springer, Heidelberg (2009)
10. Fruchterman, T.M.J., Reingold, E.M.: Graph drawing by force-directed placement. Softw. Pract. Exper. 21(11), 1129–1164 (1991)
11. Graham, R.L.: Bounds for certain multiprocessing anomalies. Bell Syst. Tech. J. 45(9), 1563–1581 (1966)
12. He, W., Marriott, K.: Constrained graph layout. Constraints 3(4), 289–314 (1998)
13. Hennecke, M.: Rechengraphen. Math. Didact. 30(1), 68–96 (2007)
14. Patrignani, M.: On the complexity of orthogonal compaction. Comput. Geom. Theory Appl. 19(1), 47–67 (2001)
15. Sander, G.: A fast heuristic for hierarchical Manhattan layout. In: Brandenburg, F.J. (ed.) GD 1995. LNCS, vol. 1027, pp. 447–458. Springer, Heidelberg (1996)
16. Simonetto, P., Archambault, D., Auber, D., Bourqui, R.: ImPrEd: An improved force-directed algorithm that prevents nodes from crossing edges. Comput. Graphics Forum 30(3), 1071–1080 (2011)
17. Sugiyama, K., Tagawa, S., Toda, M.: Methods for visual understanding of hierarchical system structures. IEEE Trans. Syst. Man Cyber. 11(2), 109–125 (1981)

Height-Preserving Transformations of Planar Graph Drawings

Therese Biedl[*]

David R. Cheriton School of Computer Science, University of Waterloo,
Waterloo, ON N2L 3G1, Canada
biedl@uwaterloo.ca

Abstract. There are numerous styles of planar graph drawings, such as straight-line drawings, poly-line drawings, orthogonal graph drawings and visibility representations. Given a planar drawing in one of these styles, can it be converted it to another style while keeping the height unchanged? This paper answers this question for (nearly) all pairs of these styles, as well as for related styles that additionally restrict edges to be y-monotone and/or vertices to be horizontal line segments. These transformations can be used to develop new graph drawing results, especially for height-optimal drawings.

Keywords: Planar graph drawing, poly-line drawing, straight-line drawing, orthogonal drawing, visibility representation, minimizing height.

1 Introduction

Let $G = (V, E)$ be a simple graph with $n = |V|$ vertices and $m = |E|$ edges. All graphs in this paper are *planar*, i.e., can be drawn without crossings. Some of the most commonly used drawings styles are the following: (1) A *straight-line drawing* is a drawing where vertices are points and edges are straight-line segments between their endpoints. Any planar graph has a planar straight-line drawing in an $O(n) \times O(n)$-grid [8][17]. (2) A *poly-line drawing* (called *mixed model* in [12]) is a drawing where vertices are points and edges are polygonal curves. Some results exist for poly-line drawings for which no equivalent straight-line drawing result is known; for example Kant gave drawings in an $O(n) \times O(n)$-grid with large minimum angle [12]. (3) A (2-directional) *visibility representation* is a drawing where vertices are axis-aligned boxes and edges are horizontal or vertical line segments. Every planar graph has a visibility representation in an $O(n) \times O(n)$-grid [16][20][21]. (4) An *orthogonal (box-)drawing* is a drawing where all vertices are axis-aligned boxes and edges are polygonals curves for which all line segments are horizontal or vertical. Every planar graph has an orthogonal drawing in an $O(n) \times O(n)$-grid [18].

---

[*] Supported by NSERC and the Ross and Muriel Cheriton Fellowship. The author would like to thank the referees of a preliminary version for helpful comments, and Fabrizio Frati and Géza Tóth for pointing out reference [15].

C. Duncan and A. Symvonis (Eds.): GD 2014, LNCS 8871, pp. 380–391, 2014.
© Springer-Verlag Berlin Heidelberg 2014

The research of this paper was driven by the following question: Does one really need to develop algorithms and lower bounds for these four drawing styles separately? Or is it possible to take a drawing in one style, and convert it directly to another style, preserving some of the features along the way? This would significantly simplify the development of algorithms and lower bounds.

This paper studies the existence of such transformations under the objective of maintaining the height of the drawing. It also includes some discussion on the width, and shows that transformations that maintain the height sometimes require an exponential blow-up in the width. Due to some prior work and applications, two variations of the drawing styles are included. First, a drawing is called *y-monotone* if each edge is drawn as a *y*-monotone curve. Every straight-line drawing and every visibility representation is *y*-monotone, but orthogonal drawings and poly-line drawings need not be. Second, a drawing is called *flat* if every vertex is represented by a horizontal line segment. Every straight-line drawing and every poly-line drawing is automatically flat, but orthogonal drawings and visibility representations need not be. Fig. 1 lists the results, which can be summarized as follows:

- Straight-line drawings, *y*-monotone poly-line drawings, flat visibility representations and flat *y*-monotone orthogonal drawings are equivalent, where "equivalent" means "can be transformed into each other such that the height of the drawings remains the same".
- Poly-line drawings and orthogonal drawings are also equivalent.
- *y*-monotone orthogonal drawings are strictly between orthogonal drawings and visibility representations.
- Visibility representations are between orthogonal *y*-monotone drawings and straight-line drawings, where the latter relationship may be an equality.

The transformations keep the width linearly bounded or better, with the notable exception of creating straight-line drawings: here we show that an exponential increase in width is required for some graphs if the height must stay the same.

These results have some applications given in Section 6. Most importantly, they allow to derive some height-bounds for which no direct proof appears known, and they can be used to formulate some NP-hard graph drawing problems as integer programs.

2   Preliminaries

Throughout this paper G denotes a planar graph, and Γ denotes a planar drawing of G that represents vertices of G as axis-aligned boxes (possibly degenerated into horizontal segments or points) and edges of G as polygonal curves (possibly straight line segments). The common end of two line segments in a polygonal curve is called a *bend*. Γ is called *y-monotone* if for all edges the *y*-coordinates monotonically increase while going from one end to the other; horizontal segments are allowed. Γ is called *flat* if all vertices are horizontal segments.

Call a drawing a *grid-drawing* if all corners of vertex-boxes and all endpoints of segments of polygonal curves have integer coordinates. A grid drawing is said

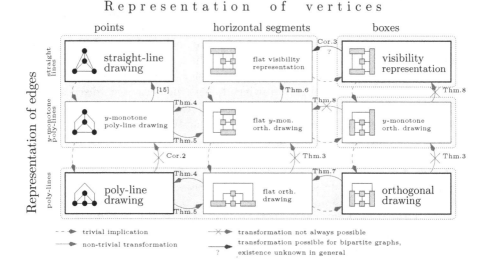

Fig. 1. Summary of height-preserving transformations

to have *width w* and *height h* if (possibly after translation) vertices and bends are placed on the $[1, w] \times [1, h]$-grid. The height is thus measured by the number of *rows*, i.e., horizontal lines with integer y-coordinates that intersect the drawing.[1] Drawings in this paper are required to be grid-drawings, with some exceptions for ease of description that will be pointed out as they occur.

The goal is to transform a planar drawing Γ into a different planar drawing Γ' which has the same height but uses a different drawing style. If Γ and Γ' are both flat, then these transformations will have two useful properties as follows. Γ' *preserves y-coordinates* if any vertex has the same y-coordinate in Γ and Γ'. This is well-defined since both drawings are flat. Γ' *preserves left-to-right-orders* if, scanning each row from left to right, one encounters the same edges and vertices, in the same order, in both Γ and Γ'. One can easily show that if Γ is planar and Γ' preserves y-coordinates and left-to-right-orders, then Γ' is also planar and has the same height. Furthermore, Γ' is y-monotone if Γ was.

3   Point-Drawings

This section considers transformations among drawing styles that represent vertices by points. One of the few existing results about height-preserving transformations is by Pach and Tóth:

Theorem 1. *[15] Any planar y-monotone poly-line drawing Γ can be transformed into a planar straight-line drawing Γ' with the same height.*

---

[1] In the pictures vertex representations are thickened for ease of readability, but this is not counted towards the height; e.g. the drawings in Fig. 1 have height 3.

3.1   Exponential Width

Pach and Tóth did not analyze what happens to the width during the transformation. In fact, they did not even worry about achieving integer coordinates, but it is clear that this can be done with minor modifications. One can show that the width must increase exponentially for some graphs:

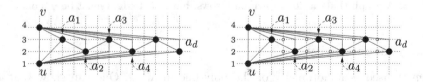

Fig. 2. (Left) A planar graph. (Right) Inserting vertices into inner faces.

Lemma 1. *Let Γ be the drawing in Fig. 2(left). Any planar straight-line drawing Γ' that preserves the y-coordinates and left-to-right-orders of Γ has width at least $\frac{1}{3}2^{d+1}$.*

Proof. Denote by $x(w)$ the x-coordinate of vertex w in Γ'. Assume that $x(v) \leq x(u)$; the other case is proved similar and in fact gives an even larger width bound. For ease of arithmetic operations, assume the drawing has been translated so that $x(v) = 0$. One can now show by induction on i (details are omitted) that

$$x(a_{2i-1}) \geq \frac{1}{3}(x(u) + 2^{2i}) - 1 \quad \text{and} \quad x(a_{2i}) \geq \frac{1}{3}(2x(u) + 2^{2i+1}) - 1$$

for $i \geq 1$. This implies the result.   □

Theorem 2. *There exists a graph H that has a planar straight-line drawing of height 4, but any such drawing has width at least $\frac{1}{3}2^{n/3}$.*

Proof. The graph H is obtained by taking the graph G from Fig. 2(left) with $d \geq 11$ and inserting into each inner face except $\{u, v, a_1\}$ a new vertex adjacent to the three vertices of the face. Note that H is triangulated and has $3d$ vertices. It has a y-monotone poly-line drawing on four rows (see Fig. 2(right)), and hence by Thm. 1 also a straight-line drawing on 4 rows.

Let Γ_H be an arbitrary planar straight-line drawing of H that uses four rows. Let Γ_G be the induced planar straight-line drawing of G. One can argue (details are omitted) that $\Gamma_G - a_d$ preserves, after possible horizontal and vertical flips, the y-coordinates and left-to-right-orders of the drawing of Fig. 2(left). Hence it requires width at least $\frac{1}{3}2^d$ by the previous lemma and Γ_H cannot be smaller.   □

One can easily show that H requires at least 4 rows in *any* straight-line drawing. Thm. 2 hence has the following consequence:

Corollary 1. *There exists an infinite class of planar graphs such that any planar straight-line drawing that has optimal height has exponential area.*

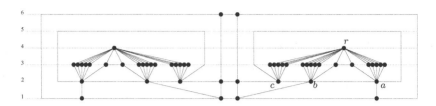

Fig. 3. A graph that can be drawn on 6 rows, but not if edges must be y-monotone

3.2 y-monotonicity

Thm. 1 used y-monotonicity of the poly-line drawing. One can show that y-monotonicity is required.

Theorem 3. *The graph in Fig. 3 has a planar poly-line drawing on 6 rows, but no planar y-monotone orthogonal drawing on 6 rows.*

Proof. (Sketch) Due to the $K_{2,5}$'s between r and each of $\{a, b, c\}$, vertices a, b, c must all be in the same row. This makes it impossible to draw cycle $a - b - c - a$ with y-monotone curves.   □

Corollary 2. *There exists a planar graph with a planar poly-line drawing on 6 rows that has no planar straight-line drawing on 6 rows.*

Proof. If the graph in Fig. 3 had a straight-line drawing on 6 rows, then by Thm. 5 (below) it would have a y-monotone orthogonal drawing on 6 rows, contradicting Thm. 3.   □

4   Flat Drawings

This section is devoted to drawings where vertices are horizontal segments or points, and in particular aims to show that poly-line drawings are equivalent to flat orthogonal drawings as far as height is concerned. Note that neither of these implications is trivial: A poly-line drawing allows slanted line segments while a flat orthogonal drawing does not, and a flat orthogonal drawing allows horizontal segments for vertices while a poly-line drawing does not.

Theorem 4. *Any planar flat orthogonal drawing Γ can be transformed into a planar poly-line drawing Γ' of the same height that preserves y-coordinates and left-to-right orders. Γ' has no more width than Γ. Moreover, if Γ is y-monotone then so is Γ'.*

Proof. First insert *pseudo-vertices* obtained by subdividing edges at bends and whenever a vertical segment of an edge crosses a row. Any pseudo-vertex is located on a grid-point since the edge segment was vertical, so this does not increase the width. Now within each row r enumerate the vertices and pseudo-vertices as $w_1, \ldots, w_k$ from left to right, and assign w_i to the point (r, i); clearly this preserves y-coordinates and left-to-right-orders. Since $w_1, \ldots, w_k$ had integer x-coordinates before, the x-coordinate of each can only decrease. Due to the pseudo-vertices edge-segments connect vertices in the same or two adjacent rows, so they can be drawn without crossings. Replacing the pseudo-vertices by bends gives the desired poly-line drawing Γ'. The claim on height and y-monotonicity holds since y-coordinates and left-to-right-orders are preserved. □

Theorem 5. *Any planar poly-line drawing Γ can be transformed into a planar flat orthogonal drawing Γ' of the same height that preserves y-coordinates and left-to-right orders. If Γ is y-monotone, then so is Γ'.*

Proof. Insert pseudo-vertices at bends and whenever a segment of an edge crosses a row, temporarily allowing non-integer x-coordinates. Now within each row r enumerate the vertices and pseudo-vertices as $w_1, \ldots, w_k$ from left to right. In Γ', replace each w_i by a box of width $\max\{1, \deg^{up}(w_i), \deg_{down}(w_i)\}$, where $\deg^{up}(w_i)$ and $\deg_{down}(w_i)$ are the number of neighbours above and below w_i, respectively. Place these boxes in row r in the same left-to-right order.

If an edge was drawn horizontally in Γ, then draw it horizontally in Γ' as well. Each non-horizontal edge ends in two adjacent rows due to the pseudo-vertices. Connect the edges between two adjacent rows using VLSI channel routing (see e.g. [13]), using two bends per edge and lots of new rows with non-integer y-coordinates that contain horizontal edge segments and nothing else.

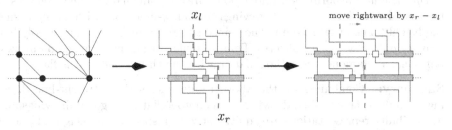

Fig. 4. Replace bends and row-crossings by pseudo-vertices (white). Replace points by boxes and route non-horizontal segments as zig-zags. Then remove zig-zags by shifting parts of the drawing rightward.

Each non-horizontal edge is now routed as a "zig-zag", ending vertically at its endpoints. It is well known (see e.g. [3]) that such a zig-zag can be removed as follows: Let $[x_l, x_r] \times y_s$ by the horizontal segment of a zig-zag. Let M be all bends and endpoints of vertices with x-coordinate exceeding x_r, or y-coordinate at least y_s and x-coordinate at least x_l. Then move all points in M rightwards, i.e., add $x_r - x_l$ to their x-coordinate. See also Fig. 4. Notice that this preserves y-coordinates and planarity, and eliminates the horizontal segment.

Repeating this for all zig-zags removes all horizontal segments with non-integer coordinates and hence gives the desired height and a flat visibility representation of the graph with pseudo-vertices. Any pseudo-vertex can now be removed and replaced by a bend or line-segment, if needed, to give an orthogonal drawing that satisfies all conditions.   □

A very similar construction converts y-monotone flat orthogonal drawings into flat visibility representations.

Theorem 6. *Any planar flat y-monotone orthogonal drawing Γ can be transformed into a planar flat visibility representation Γ' of the same height that preserves y-coordinates and left-to-right orders.*

Proof. Assume that there exists an edge e in Γ that attaches horizontally at one endpoint v and has bends. Since Γ is flat, v is a horizontal segment, which can expand until it covers the bend of e nearest to v. Afterwards e attaches vertically at v. Repeat this until all edges with bends attach vertically at both endpoints.

If there are now no bends then the claim holds. Otherwise, let e be an edge with bends; by the above it attaches vertically at both its endpoints. Since e is drawn y-monotone, it attaches at the top of one endpoint and the bottom of the other endpoint, and inbetween must have a zig-zag. As before (see Fig. 4) such a zig-zag can be removed by shifting parts of the drawing rightwards. This does not add height or bends. Applying this to all edges that have bends gives a planar visibility representation.   □

One height-transformation among point drawings and flat drawings remains to be studied: Can flat orthogonal drawings be made y-monotone without increasing the height? The answer here is "no" since the graph of Thm. 3 has a poly-line drawing on 6 rows, hence also (by Thm. 5) a flat orthogonal drawing on 6 rows, but it has no y-monotone orthogonal drawing, much less one that is flat.

Summary: Combining all the above results yields that straight-line drawings, y-monotone poly-line drawings, y-monotone flat orthogonal drawings and flat visibility representations are equivalent with respect to the height of planar drawings. [2] Poly-line drawings and flat orthogonal drawings are also equivalent, and the former group sometimes requires strictly more height than the latter.

Width Considerations: For Thm. 1 the width may have to increase exponentially (Thm. 3). For Thm. 4 the width did not increase. For Thm. 5 and 6 the width may increase, but after eliminating all zig-zags many columns are *redundant*: They contain no vertical edge, nor are they the only column of a vertex. In an orthogonal drawing redundant columns can simply be deleted. One can easily show the following:

---

[2] Historical note: At WAOA'12 [2], I claimed that any flat visibility representation can be transformed into a straight-line drawing of the same height directly, without using [15]. Unfortunately the transformation in [2] is incorrect since the resulting drawing may not be planar; details are in the full version.

Lemma 2. *Any visibility representation of a connected graph has width at most* max$\{m, n\}$ *after deleting redundant columns.*

Hence for Thm. 6 the width is max$\{m, n\}$. For Thm. 5 it is at most max$\{n, m\}$+ b, where b is the number of local minima and maxima of polygonal curves in the poly-line drawing, since all other pseudo-vertices are eliminated when removing zig-zags.

5 Box-Drawings Made Flat

Theorem 7. *Any planar orthogonal drawing Γ can be transformed into a planar flat orthogonal drawing Γ' of the same height.*

Proof. For any vertex v, let r be a row that intersects the box $B(v)$ representing v and set $B'(v) := B(v) \cap r$. Any edge that attached vertically at $B(v)$ can be extended to end at $B'(v)$ instead. Any edge that attached at $B(v)$ horizontally in row r also attaches at $B'(v)$. Any edge e that attached at $B(v)$ horizontally, but not in row r, is re-routed by inserting a bend where e attached at $B(v)$, and then going vertically towards $B'(v)$. Now vertical edge segments at v may overlap, but this can easily be remedied by replacing the leftmost and rightmost column of $B(v)$ by sufficiently many columns. See Fig. 5.   □

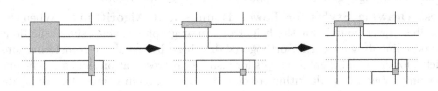

Fig. 5. Transforming orthogonal drawings into flat orthogonal drawings

Thm. 7 does not generally preserve y-monotonicity, and this is unavoidable.

Theorem 8. *The planar graph in Fig. 6 has a y-monotone orthogonal drawing on 6 rows, but it has no planar y-monotone flat orthogonal drawing on 6 rows, and also no planar visibility representation on 6 rows.*

Proof. (Sketch) The 4×2-grids force vertices v, w onto row 5 and v', w' onto row 2. The two cycles $\{v, w, c\}$ and $\{v', w', c'\}$ enclose x and x', respectively, which forces c to span both rows 3 and 4.   □

One transformation remains to be studied: Can every visibility representation be turned into a flat visibility representation? A variation of Thm. 7 shows that this is possible for bipartite graphs.

Corollary 3. *Any planar visibility representation of a bipartite graph G can be transformed into a planar flat visibility representation of the same height.*

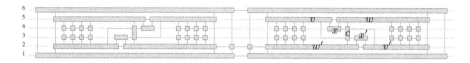

Fig. 6. A graph that has an orthogonal y-monotone on 6 rows, but not if vertices must be horizontal segments or edges have no bends

Proof. Since G is bipartite, it has a vertex-coloring with 2 colors. Proceed as in Thm. 7, but use the topmost row intersecting $B(v)$ for each vertex v in one color-class, and the bottommost row for each vertex in the other color-class. Each vertical edge remains vertical. Each horizontal edge becomes y-monotone since it connects two differently-colored vertices. So the result is a flat y-monotone orthogonal drawing, which can be converted into a flat visibility representation by Thm. 6. □

It remains open whether visibility representations of arbitrary graphs can also be made flat without increasing the height.

6   Applications

This section highlights some applications of the above results.

Best Drawing Styles for Lower Bounds and Algorithms: Whenever possible, lower bounds for the height of planar graph drawings should be done for poly-line drawings or for orthogonal drawings: by the above transformations such a lower bound then also holds for visibility representations and straight-line drawings. Vice versa, algorithms to create planar graph drawings of small height should ideally be for straight-line drawings; they then also hold for all other models. Alternative, algorithms could be given for flat visibility representations; the same height-bound then also holds for straight-line drawings and all other models, though the width does not transfer.

Drawings of Small Height: Two examples of using height-transformations to achieve new results shall suffice:

Theorem 9. *Every outer-planar graph G has a planar straight-line drawing of height $O(pw(G))$, which is in $O(\log n)$. Here $pw(G)$ is the so-called pathwidth of the graph.*[3]

Proof. Every outer-planar graph can be made 2-connected while increasing the pathwidth by at most a constant factor [1]. Every 2-connected outer-planar graph has a flat visibility representation of height $O(pw(G))$ [2]. This flat visibility representation is a flat y-monotone orthogonal drawing, which by Thms. 4 and 1 can be turned into a straight-line drawing of the same height. □

---

[3] A similar result was claimed in [2], but required the (incorrect) transformation in that paper.

Theorem 10. *Any 4-connected planar graph has a planar flat visibility representation of height at most $\lfloor n/2 \rfloor$.*

Proof. Any 4-connected planar graph has a straight-line drawing where the sum of the width and height is at most n [14]. After possible rotation, the height is at most $\lfloor n/2 \rfloor$, and with Thms. 5 and 6 one obtains a flat visibility representation of height at most $\lfloor n/2 \rfloor$.   □

The best previous bound on the height of visibility representations of 4-connected planar graphs was $\lceil \frac{3n}{4} \rceil$ [10].

Simplified Proofs: Some results are known about graphs that have planar straight-line drawings of height h. In particular, any such graph has pathwidth at most h [7], and there exists an algorithm that is fixed-parameter tractable in h to test whether a graph has a straight-line drawing of height at most h [6]. By the results in this paper a graph has a straight-line drawing of height h if and only if it has a flat visibility representation of height h. While "simplicity of proof" is a subjective matter, in my opinion the presentation of both of the above results can be simplified if one shows the properties for flat visibility representations of height h, rather than straight-line drawings.

HH-drawings: In a previous paper [19] we studied *HH-drawings*, where a planar graph G with a vertex partition $V = A \cup B$ should be drawn such that all vertices in A have positive y-coordinates and all vertices in B have negative y-coordinates. We gave a condition that is necessary for straight-line HH-drawings and sufficient for y-monotone poly-line HH-drawings. By the result by Pach and Tóth (Thm. 1) the condition is hence necessary and sufficient for straight-line HH-drawings. This proves:

Theorem 11. *Any planar bipartite graph has a planar straight-line HH-drawing. Testing whether a planar graph with a given partition has a planar straight-line HH-drawing can be done in linear time.*

Minimizing Heights Using Integer Programs: In a recent paper, we developed integer program (IP) formulations for many graph drawing problems where vertices and edges are represented by axis-aligned boxes [4]. In particular, we gave an integer program with $O(hn^2)$ variables and constraints to test whether a graph has a *bar-visibility representation* (i.e., a visibility representation where vertices are horizontal segments and all edges are vertical) that has height h. It is not hard to modify this IP so that horizontal edges are allowed as well; then it tests the existence of flat visibility representations of height h. Since straight-line drawings and flat visibility representations are equivalent with respect to height, therefore:

Theorem 12. *There exists an integer program with $O(hn^2)$ variables and constraints to test whether a graph has a planar straight-line drawing of height h.*

While an algorithm was already known to test whether G has a planar drawing of height at most h [6], its rather large run-time of $O(2^{32h^3} \text{poly}(n))$ means that solving the above integer program might well be faster in practice.

Upward Drawings: A directed acyclic graph has an *upward planar drawing* if it has a planar straight-line drawing such that for any directed edge $v \to w$ the y-coordinate of v is smaller than the y-coordinate of w. Testing whether a graph has an upward planar drawing is NP-hard [9]. There exists a way to formulate "G has an upward planar drawing" as either IP or as a Satisfiability-problem, using partial orders on the edges and vertices [5]. The transformations in this paper give a different way of testing this via IP:

Lemma 3. *A directed acyclic graph has an upward planar drawing if and only if it has a planar bar-visibility representation where for all edges the head is above the tail.*

Proof. Any straight-line upward planar drawing can be transformed into a flat visibility representation using Thm. 5 and 6. Since y-coordinates are unchanged, any edge is necessarily drawn vertical with the head above the tail, so this is a bar-visibility representation. Vice versa, given such a bar-visibility representation, it can be transformed into a y-monotone poly-line drawing (Thm. 4) and from there into a straight-line drawing (Thm. 1). Since y-coordinates and left-to-right-orders are preserved this gives an upward planar drawing. □

It is easy to express "the head of edge $v \to w$ must be above the tail" as constraints in the IP for bar-visibility representations defined in [4]. Therefore:

Corollary 4. *There exists an integer program with $O(n^3)$ variables and constraints to test whether a planar graph has an upward planar drawing. Moreover, the same integer program also finds the minimum-height upward planar drawing.*

7   Conclusion

This paper considered transformations of one type of planar drawings into another without increasing the height. In particular planar straight-line drawings are equivalent with respect to the height to flat visibility representations or y-monotone poly-line drawing, while they are more powerful than either orthogonal drawings or poly-line drawings. The latter two drawing styles are again equivalent with respect to drawing height.

The main gap left open concerns visibility representations. Is it possible to transform any visibility representation into a flat visibility representation of the same height?

As for future problems, it would be interesting to study other drawing objectives, and whether they can be preserved while changing the layout style. Hoffman et al. [11] recently gave some worst-case ratios, but for other graph drawing styles. Is it possible to transform any y-monotone poly-line drawing into a straight-line drawing of the same area? The example in Thm. 2 makes this unlikely, but can it be transformed into a straight-line drawing of asymptotically the same area?

References

1. Babu, J., Basavaraju, M., Chandran Leela, S., Rajendraprasad, D.: 2-connecting outerplanar graphs without blowing up the pathwidth. In: Du, D.-Z., Zhang, G. (eds.) COCOON 2013. LNCS, vol. 7936, pp. 626–637. Springer, Heidelberg (2013)
2. Biedl, T.: A 4-approximation for the height of drawing 2-connected outer-planar graphs. In: Erlebach, T., Persiano, G. (eds.) WAOA 2012. LNCS, vol. 7846, pp. 272–285. Springer, Heidelberg (2013)
3. Biedl, T., Lubiw, A., Petrick, M., Spriggs, M.J.: Morphing orthogonal planar graph drawings. ACM Transactions on Algorithms 9(4), 29 (2013)
4. Biedl, T., Bläsius, T., Niedermann, B., Nöllenburg, M., Prutkin, R., Rutter, I.: Using ILP/SAT to Determine Pathwidth, Visibility Representations, and other Grid-Based Graph Drawings. In: Wismath, S., Wolff, A. (eds.) GD 2013. LNCS, vol. 8242, pp. 460–471. Springer, Heidelberg (2013)
5. Chimani, M., Zeranski, R.: Upward planarity testing via SAT. In: Didimo, W., Patrignani, M. (eds.) GD 2012. LNCS, vol. 7704, pp. 248–259. Springer, Heidelberg (2013)
6. Dujmovic, V., Fellows, M., Kitching, M., Liotta, G., McCartin, C., Nishimura, N., Ragde, P., Rosamond, F., Whitesides, S., Wood, D.: On the parameterized complexity of layered graph drawing. Algorithmica 52, 267–292 (2008)
7. Felsner, S., Liotta, G., Wismath, S.: Straight-line drawings on restricted integer grids in two and three dimensions. Journal of Graph Algorithms and Applications 7(4), 335–362 (2003)
8. de Fraysseix, H., Pach, J., Pollack, R.: How to draw a planar graph on a grid. Combinatorica 10, 41–51 (1990)
9. Garg, A., Tamassia, R.: On the computational complexity of upward and rectilinear planarity testing. SIAM J. Comput. 31(2), 601–625 (2001)
10. He, X., Wang, J., Zhang, H.: Compact visibility representation of 4-connected plane graphs. Theor. Comput. Sci. 447, 62–73 (2012)
11. Hoffmann, M., van Kreveld, M., Kusters, V., Rote, G.: Quality ratios of measures for graph drawing styles. In: Canadian Conference on Computational Geometry, CCCG 2014 (to appear, 2014)
12. Kant, G.: Drawing planar graphs using the canonical ordering. Algorithmica 16, 4–32 (1996)
13. Lengauer, T.: Combinatorial Algorithms for Integrated Circuit Layout. Teubner/Wiley & Sons, Stuttgart/Chicester (1990)
14. Miura, K., Nakano, S., Nishizeki, T.: Convex grid drawings of four-connected plane graphs. Int. J. Found. Comput. Sci. 17(5), 1031–1060 (2006)
15. Pach, J., Tóth, G.: Monotone drawings of planar graphs. Journal of Graph Theory 46(1), 39–47 (2004)
16. Rosenstiehl, P., Tarjan, R.E.: Rectilinear planar layouts and bipolar orientation of planar graphs. Discrete Computational Geometry 1, 343–353 (1986)
17. Schnyder, W.: Embedding planar graphs on the grid. In: ACM-SIAM Symposium on Discrete Algorithms (SODA 1990), pp. 138–148 (1990)
18. Biedl, T., Kant, G.: A better heuristic for orthogonal graph drawings. Computational Geometry: Theory and Applications 9, 159–180 (1998)
19. Biedl, T., Kaufmann, M., Mutzel, P.: Drawing planar partitions II: HH-drawings. In: Hromkovič, J., Sýkora, O. (eds.) WG 1998. LNCS, vol. 1517, pp. 124–136. Springer, Heidelberg (1998)
20. Tamassia, R., Tollis, I.: A unified approach to visibility representations of planar graphs. Discrete Computational Geometry 1, 321–341 (1986)
21. Wismath, S.: Characterizing bar line-of-sight graphs. In: ACM Symposium on Computational Geometry (SoCG 1985), pp. 147–152 (1985)

Drawing Planar Graphs with Reduced Height

Stephane Durocher⋆ and Debajyoti Mondal⋆⋆

Department of Computer Science, University of Manitoba, Canada
{durocher,jyoti}@cs.umanitoba.ca

Abstract. A straight-line (respectively, polyline) drawing Γ of a planar graph G on a set L_k of k parallel lines is a planar drawing that maps each vertex of G to a distinct point on L_k and each edge of G to a straight line segment (respectively, a polygonal chain with the bends on L_k) between its endpoints. The height of Γ is k, i.e., the number of lines used in the drawing. In this paper we compute new upper bounds on the height of polyline drawings of planar graphs using planar separators. Specifically, we show that every n-vertex planar graph with maximum degree Δ, having a simple cycle separator of size λ, admits a polyline drawing with height $4n/9 + O(\lambda\Delta)$, where the previously best known bound was $2n/3$. Since $\lambda \in O(\sqrt{n})$, this implies the existence of a drawing of height at most $4n/9 + o(n)$ for any planar triangulation with $\Delta \in o(\sqrt{n})$. For n-vertex planar 3-trees, we compute straight-line drawings with height $4n/9 + O(1)$, which improves the previously best known upper bound of $n/2$. All these results can be viewed as an initial step towards compact drawings of planar triangulations via choosing a suitable embedding of the input graph.

1 Introduction

A *polyline drawing* of a planar graph G is a planar drawing of G such that each vertex of G is mapped to a distinct point in the Euclidean plane, and each edge is mapped to a polygonal chain between its endpoints. Let $L_k = \{l_1, l_2, \ldots, l_k\}$ be a set of k horizontal lines such that for each $i \leq k$, line l_i passes through the point $(0, i)$. A polyline drawing of G is called a *polyline drawing on L_k* if the vertices and bends of the drawing lie on the lines of L_k. The *height* of such a drawing is k, i.e., the number of parallel horizontal lines used by the drawing. Such a drawing is also referred to as a *k-layer drawing* in the literature [13,18]. Let Γ be a polyline drawing of G. We call Γ a *t-bend polyline drawing* if each of its edges has at most t bends. Thus a 0-bend polyline drawing is also known as a *straight-line drawing*. Drawing planar graphs on a small integer grid is an active research area in graph drawing [7,16], which is motivated by the need of compact layout of VLSI circuits and visualization of software architecture. Since simultaneously optimizing the

---

⋆ Work of the author is supported in part by the Natural Sciences and Engineering Research Council of Canada (NSERC).

⋆⋆ Work of the author is supported in part by a University of Manitoba Graduate Fellowship.

width and height of the drawing is very challenging, researchers have also focused their attention on optimizing one dimension of the drawing [6,11,13,17], while the other dimension is unbounded. In this paper we develop new techniques that can produce drawings with small height. We distinguish between the terms 'plane' and 'planar'. A *plane graph* is a planar graph with a fixed combinatorial embedding and a specified outer face. While drawing a planar graph, we allow the output to represent any planar embedding of the graph. On the other hand, while drawing a plane graph, the output is further constrained to respect the input embedding.

State-of-the-art algorithms that compute straight-line drawings of n-vertex plane graphs on an $(O(n) \times 2n/3)$-size grid imply an upper bound of $2n/3$ on the height of straight-line drawings [5,6]. This bound is tight for plane graphs, i.e., there exist n-vertex plane graphs such as plane nested triangles graphs and some plane 3-trees that require a height of $2n/3$ in any of their straight-line drawings. Recall that an n-vertex *nested triangles graph* is a plane graph formed by a sequence of $n/3$ vertex disjoint cycles, $C_1, C_2, \ldots, C_{n/3}$, where for each $i \in \{2, \ldots, n/3\}$, cycle C_i contains the cycles $C_1, \ldots, C_{i-1}$ in its interior, and a set of edges that connect each vertex of C_i to a distinct vertex in C_{i-1}. Besides, a *plane 3-tree* is a triangulated plane graph that can be constructed by starting with a triangle, and then repeatedly adding a vertex to some inner face of the current graph and triangulating that face.

The $2n/3$ upper bound on the height is also the currently best known bound for polyline drawings, even for planar graphs, i.e., when we are allowed to choose a suitable embedding for the output drawing. Frati and Patrignani [10] showed that in the variable embedding setting, an n-vertex nested triangles graph can be drawn with height at most $n/3 + O(1)$, which is significantly smaller than the lower bound of $2n/3$ in the fixed embedding setting. Similarly, Hossain et al. [11] showed that an *universal set* of $n/2$ horizontal lines can support all n-vertex planar 3-trees, i.e., every planar 3-tree admits a drawing with height at most $n/2$. They also showed that $4n/9$ lines suffice for some subclasses of planar 3-trees, and asked whether $4n/9$ is indeed an upper bound for planar 3-trees.

In the context of optimization, Dujmović et al.[9] gave fixed-parameter-tractable (FPT) algorithms, parameterized by pathwidth, to decide whether a planar graph admits a straight-line drawing on k horizontal lines. Drawings with minimum number of parallel lines have been achieved for trees [13]. Recently, Biedl [2] gave an algorithm to approximate the height of straight-line drawings of 2-connected outerplanar graphs within a factor of 4.

Contributions. In this paper we show that every n-vertex planar graph with maximum degree Δ, having a simple cycle separator of size λ, admits a drawing with height $4n/9 + O(\lambda\Delta)$, which is better than the previously best known bound of $2n/3$ for any $\lambda\Delta \in o(n)$. This result is an outcome of a new application of the planar separator theorem [8]. Although the technique is simple, it has the potential to be a powerful tool while computing compact drawings for planar triangulations in the variable embedding setting. If the input graphs are restricted to planar 3-trees, then we can improve the upper bound to $4n/9 + O(1)$, which settles the question of Hossain et al. [11]. Furthermore, the drawing we construct in this case is a straight-line drawing.

2 Preliminary Definitions and Results

Let G be an n-vertex triangulated plane graph. A simple cycle C in G is called a *cycle separator* if the interior and the exterior of C each contains at most $2n/3$ vertices. Let v_1, v_n and v_2 be the outer vertices of G in clockwise order on the outer face. Let $\sigma = (v_1, v_2, ..., v_n)$ be an ordering of all vertices of G. By G_k, $2 \leq k \leq n$, we denote the subgraph of G induced by $v_1 \cup v_2 \cup ... \cup v_k$. For each G_k, the notation P_k denotes the path (while walking clockwise) on the outer face of G_k that starts at v_1 and ends at v_2. We call σ a *canonical ordering* of G with respect to the outer edge (v_1, v_2) if for each k, $3 \leq k \leq n$, the following conditions are satisfied [7]:

(a) G_k is 2-connected and internally triangulated.
(b) If $k \leq n$, then v_k is an outer vertex of G_k and the neighbors of v_k in G_{k-1} are consecutive on P_{k-1}.

Let P_k, for some $k \in \{3, 4, ..., n\}$, be the path $w_1(= v_1), ..., w_l, v_k(= w_{l+1})$, $w_r, ..., w_t(= v_2)$. The edges (w_l, v_k) and (v_k, w_r) are the *l-edge* and *r-edge* of v_k, respectively. The other edges incident to v_k in G_k are called the *m-edges*. For example, in Figure 1(c), the edges (v_6, v_1), (v_6, v_4), and (v_5, v_6) are the *l*-, *r*- and *m*-edges of v_6, respectively. Let E_m be the set of all *m*-edges in G. Then the graph T_{v_n} induced by the edges in E_m is a tree with root v_n. Similarly, the graph T_{v_1} induced by all *l*-edges except (v_1, v_n) is a tree rooted at v_1 (Figure 1(b)), and the graph T_{v_2} induced by all *r*-edges except (v_2, v_n) is a tree rooted at v_2. These three trees form the *Schnyder realizer* [16] of G.

Lemma 1 (Bonichon et al. [4]). *The total number of leaves in all the trees in any Schnyder realizer of an n-vertex triangulation is at most $2n - 5$.*

Let G be a planar graph and let Γ be a straight-line drawing on k parallel lines. By $l(v)$, where v is a vertex of G, we denote the horizontal line in Γ that passes through v. We now have the following lemma that bounds the height of a straight-line drawing in terms of the number of leaves in a Schnyder tree. The

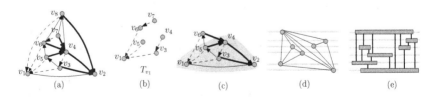

Fig. 1. (a) A plane triangulation G with a canonical ordering. The associated realizer, where the *l*-, *r*- and *m*- edges are shown in dashed, bold-solid, and thin-solid lines, respectively. (b) T_{v_1}. (c) Neighbors of v_6 in G_6. (d)–(e) Illustrating Lemma 3.

lemma can be derived from the known straight-line [5] and polyline drawing algorithms [4]. We omit the proof due to space constraints.

Lemma 2. *Let G be an n-vertex plane triangulation and let v_1, v_n, v_2 be the outer vertices of G in clockwise order on the outer face. Assume that T_{v_n} has at most p leaves. Then for any placement of v_n on line l_1 or l_{p+2}, there exists a straight-line drawing Γ of G on L_{p+2} such that v_2 and v_1 lie on lines l_{p+2} and l_1, respectively.*

Chrobak and Nakano [6] showed that every planar graph admits a straight-line drawing with height $2n/3$. We now observe some properties of Chrobak and Nakano's algorithm [6]. Let G be a plane triangulation with n vertices and let x, y be two user prescribed outer vertices of G in clockwise order. Let Γ be the drawing of G produced by the Algorithm of Chrobak and Nakano [6]. Then Γ has the following properties:

(CN$_1$) Γ is a drawing on L_q, where $q \leq 2n/3$.
(CN$_2$) For the vertices x and y, we have $l(x) = l_1$ and $l(y) = l_q$ in Γ. The remaining outer vertex z lies on either l_1 or l_q.

Note that the user cannot choose the placement of z, i.e., the algorithm may produce a drawing where $l(x) = l_1, l(y) = l_q$ and $l(z) = l_1$, however, this does not imply that there exists another drawing where $l(x) = l_1, l(y) = l_q$ and $l(z) = l_q$. We end this section with the following lemma.

Lemma 3. *Let G be a plane graph and let Γ be a straight-line drawing of G on k horizontal lines, but the lines are not necessarily equally spaced. Then there exists a drawing Γ' of G on a set of k horizontal lines that are equally spaced. Furthermore, for every $i \in \{1, 2, \ldots, k\}$, the left to right order of the vertices on the ith line in Γ coincides with that of Γ'.*

Proof (Outline). One can construct Γ' by first transforming Γ into a 'flat-visibility representation' on equally spaced horizontal lines, as shown in Figures 1(d)–(e), and then transforming this representation again into a straight-line drawing [1,3]. □

In the following sections we describe our drawing algorithms. Note that for simplicity we often omit the floor and ceiling functions while defining different parameters of the algorithms. One can describe a more careful computation using proper floor and ceiling functions, but that does not affect the asymptotic results discussed in this paper.

3   Drawing Triangulations with Small Height

Let $G = (V, E)$ be an n-vertex planar triangulation and let Γ be a planar drawing of G on the Euclidean plane. Let $C = (V_c, E_c)$ be a simple cycle separator of G of size λ. Let $G_i = (V_i, E_i)$ be the graph induced by the vertices that lie inside C and on the boundary of C. Similarly, let $G_o = (V_o, E_o)$ be the graph induced by the vertices that lie outside C and on the boundary of C. Specifically, $V = V_i \cup V_o$, $E = E_i \cup E_o$, $V_i \cap V_o = V_c$, and $E_i \cap E_o = E_c$. We now compute a polyline drawing of G.

3.1 Drawing Technique

If any edge $(a, b) \in E_c$ lies on the outer face of Γ, then we will draw G respecting the combinatorial embedding determined by Γ. Otherwise, there exists an edge $(a, b) \in E_c$ such that the face a, b, c with $c \in V_o$ does not lie interior to C. We redefine Γ as the embedding of G obtained by choosing a, b, c as the outer face, as illustrated in Figures 2(a)–(b).

Drawing G_i. Assume that $x = 4n/9 + 2\lambda/3 + 3$. Construct a plane graph G_i' by taking a copy of G_i from Γ, and then adding a vertex z to the outer face of G_i along with the edges (z, w), for all $w \in V_c$. Figure 2(c) illustrates G_i'. Since G_i has at most $(2n + 3\lambda)/3$ vertices, we now use the algorithm of Chrobak and Nakano [6] to compute a drawing Γ_i of G_i' on L_x, where a, b lie on l_1, l_x and z lies on either l_1 or l_x. Assume without loss of generality that z is in the right half-plane of the line through a, b.

Drawing G_o. Take a copy of G_o from Γ. Let u be any vertex in G_o. Then by $d_o(u)$ we denote the degree of vertex u in G_o. Let the cycle C be $a(= w_1), w_2, \ldots,$ $b(= w_\lambda)$. For each vertex $w_i \in V_c$, where $1 \leq i \leq \lambda$ and $w_{\lambda+1} = w_1$, if $d_o(w_i) > 3$, then replace (w_i, w_{i+1}) with a path $w_i, w_i^1, w_i^2, \ldots, w_i^{d_o(w_i)-3}, w_{i+1}$ of $d_o(w_i) - 3$ division vertices. Let $u_1, u_2, \ldots, u_{d_o(w_i)-2}$ be the neighbors of w_i in clockwise order outside of C. Then delete the edges from w_i to these neighbors, and add the edges $(w_i, u_1), (w_i^1, u_2), \ldots, (w_i^{d_o(w_i)-3}, u_{d_o(w_i)-2})$. Replace the edge $(w_1, w_\lambda^{d_o(w_\lambda)-3})$ by a path $w_1, w', w'', w_\lambda^{d_o(w_\lambda)-3}$, and redefine a and b such that $a = w''$ and $b = w'$. Let the resulting graph be H and let the newly constructed cycle be C'. Figure 2(d) illustrates H.

If z lies on l_1 in Γ_i, then we add the edges (a, w) to H, for each vertex w on C'. Otherwise, we add the edges (b, w) to H. Finally, we add a vertex z' on the outer face and triangulate H such that (a, b) remains an outer edge. Let the resulting graph be G_o'. Figure 2(e) illustrates G_o'. Observe that the number of vertices in G_o' is at most $2n/3 + \lambda\Delta + 3$. Hence we can use the algorithm of Chrobak and Nakano [6] to compute a drawing Γ_o of G_o' on L_y, where $y = (4n + 6\lambda\Delta + 18)/9$, such that a, b lie on l_1, l_y, respectively, and the segment ab is vertical. Assume without loss of generality that all the vertices of G_o' are in the right half-plane of the line through a, b.

Merging G_i and G_o. Without loss of generality assume that $l(z) = l_x$ in Γ_i, and recall that in this case b is adjacent to all the vertices on C' in Γ_o. Let ℓ_o be a vertical line to the right of segment ab in Γ_o such that all the other vertices of Γ_o are in the right half-plane of ℓ_o. Furthermore, ℓ_o must be close enough such that all the intersection points with the edges incident to b lie in between the horizontal line through b and the next horizontal line. For each intersection point, we insert a division vertex at that point and create a horizontal line through that vertex. We then delete vertex b from Γ_o, but not the division vertices. Figures 2(i)–(j) illustrate this scenario. By Lemma 3, we can modify Γ_o such that the horizontal lines are equally spaced. Since C' contains at most $\lambda\Delta$ vertices, Γ_o is a drawing on at most $y + \lambda\Delta$ horizontal lines. Similarly, we modify Γ_i, as follows.

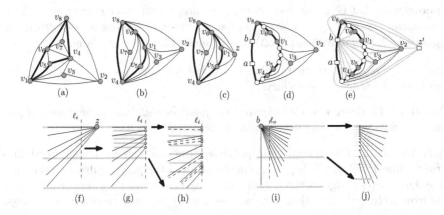

Fig. 2. (a) A plane triangulation G, where C is shown in bold. (b) Γ, where $v_4 = a$, $v_8 = b$ and $v_2 = c$. (c) G_i'. (d) H, where the division vertices are shown in white. (e) G_o', where the edges added to H are shown in gray. (f)–(j) Drawing G.

Let ℓ_i be a vertical line to the left of z in Γ_i such that all the other vertices of Γ_i are in the left half-plane of ℓ_i. Furthermore, ℓ_i must be close enough such that all the intersection points with the edges incident to z lie in between two consecutive parallel lines. For each intersection point, we insert a division vertex at that point and create a horizontal line through that vertex. Let $v_1, v_2, \ldots, v_\lambda$ be the division vertices on ℓ_i in the order of decreasing y-coordinate, where for each $i \in \{1, 2, \ldots, \lambda\}$, v_i is incident to the vertex w_i on C. Delete vertex z, but not the division vertices. For each vertex w_i, if $d_o(w_i) > 3$, then we place a set of division vertices $v_i^1, v_i^2, \ldots, v_i^{d_o(w_i)-3}$ below v_i and above the horizontal line closest to v_i. Besides, these new division vertices must be sufficiently close to v_i such that drawing of the edges (w_i, v_i^j), where $1 \le j \le d_o(w_i) - 3$, do not create any edge crossing. Figures 2(f)–(h) illustrate this scenario. Finally, by Lemma 3, we can modify Γ_i such that the horizontal lines are equally spaced. Note that Γ_i is a drawing on at most $x + \lambda\Delta$ horizontal lines.

Since the division vertices in Γ_i and Γ_o take a set of consecutive horizontal lines from their respective topmost lines, it is straightforward to merge these two drawings on a set of $\lambda\Delta + \max\{x, y\} = 4n/9 + O(\lambda\Delta)$ horizontal lines. We delete the edges on C', and consider all vertices of C' as division vertices. Since the division vertices correspond to the bends, each edge may contain at most six bends (two bends to enter Γ_o from Γ_i, two bends on C', and two bends to return to Γ_i from Γ_o). Since there are at most $\lambda\Delta$ edges that may have bends, the number of bends is at most $6\lambda\Delta$ in total. However, via 'flat-visibility representation' (similar to the proof of Lemma 3) one can reduce the number of bends to enter and exit Γ_o by one, we omit the details due to space constraints. Hence the number of bends reduces to $4\lambda\Delta$. The following theorem summarizes the result of this section.

Theorem 1. *Let G be an n-vertex planar graph with maximum degree Δ. If G contains a simple cycle separator of size λ, then G admits a 4-bend polyline drawing with height $4n/9 + O(\lambda\Delta)$ and at most $4\lambda\Delta$ bends in total.*

Since every planar triangulation has a simple cycle separator of size $O(\sqrt{n})$ [8], we obtain the following corollary.

Corollary 1. *Every n-vertex planar triangulation with maximum degree $o(\sqrt{n})$ admits a polyline drawing with height at most $4n/9 + o(n)$.*

Pach and Tóth [15] showed that polyline drawings can be transformed into straight-line drawings while preserving the height if the polyline drawing is monotone, i.e., if every edge in the polyline drawing is drawn as a y-monotone curve. Unfortunately, our algorithm does not necessarily produce monotone drawings.

4   Drawing Planar 3-Trees with Small Height

In this section we examine straight-line drawings of planar 3-trees.

4.1   Technical Background

Let G be an n-vertex planar 3-tree and let Γ be a straight-line drawing of G. Then Γ can be constructed by starting with a triangle, which corresponds to the outer face of Γ, and then iteratively inserting the other vertices into the inner faces and triangulating the resulting graph. Let a, b, c be the outer vertices of Γ in clockwise order. If $n > 3$, then Γ has a unique vertex p that is incident to all the outer vertices. This vertex p is called the representative vertex of G.

For any cycle i, j, k in G, let G_{ijk} be the subgraph induced by the vertices i, j, k and the vertices lying inside the cycle. Let G_{ijk}^* be the number of vertices in G_{ijk}. The following two lemmas describe some known results.

Lemma 4 (Mondal et al. [14]). *Let G be a plane 3-tree and let i, j, k be a cycle of three vertices in G. Then G_{ijk} is a plane 3-tree.*

Lemma 5 (Hossain et al. [11]). *Let G be an n-vertex plane 3-tree with the outer vertices a, b, c in clockwise order. Let D be a drawing of the outer cycle a, b, c on L_n, where the vertices lie on l_1, l_k and l_i with $k \le n$ and $i \in \{l_1, l_2, l_n, l_{n-1}\}$. Then G admits a straight-line drawing Γ on L_k, where the outer cycle of Γ coincides with D.*

Let G be a plane 3-tree and let a, b, c be the outer vertices of G. Assume that G has a drawing Γ on L_k, where a, b lie on lines l_1, l_k, respectively, and c lies on line l_i, where $1 \le i \le k$. Then the following properties hold for Γ [11].

Reshape. Let p, q and r be three different points on lines l_1, l_k and l_i, respectively. Then G has a drawing Γ' on L_k such that the outer face of Γ' coincides with triangle pqr (e.g., Figures 3(a)–(b)).

Stretch. For any integer $t \geq k$, G admits a drawing Γ' on L_t such that a, b, c lie on l_1, l_t, l_i, respectively (e.g., Figure 3(c)).

For any triangulation H with the outer vertices a, b, c, let $T_{a,H}, T_{b,H}, T_{c,H}$ be the Schnyder trees rooted at a, b, c, respectively. By $\mathtt{leaf}(T)$ we denote the number of leaves in T. The following lemma establishes a sufficient condition for a plane 3-tree G to have a straight-line drawing with height at most $4(n+3)/9+4$.

Lemma 6. *Let G be an n-vertex plane 3-tree with outer vertices a, b, c in clockwise order. Let $w_1, \ldots, w_k(= p), w_{k+1}(= q), \ldots, w_t(= c)$ be the maximal path P such that each vertex on P is adjacent to both a and b. Assume that $n' = n + 3$, and $x = 4n'/9$. If $G^*_{apq} \leq (n' + 2)/3$, $G^*_{bpq} \leq G^*_{abp} \leq n'/2$ and $\max_{\forall i > k+1}\{G^*_{aw_iw_{i-1}}, G^*_{bw_iw_{i-1}}\} \leq 4n'/9$, then G admits a drawing with height at most $4n'/9 + 4$.*

Proof. Let H be the subgraph of G induced by the vertices $\{a, b\} \cup \{w_k, \ldots, w_t\}$. The idea of the proof is first to construct a drawing of H on L_{x+4}, and then to extend it to the required drawing using Lemmas 2–5. We distinguish two cases depending on whether $\mathtt{leaf}(T_{p,G_{abp}}) \leq x$ or not.

Case 1 ($\mathtt{leaf}(T_{p,G_{abp}}) \leq x$). Since $G^*_{abp} \leq n'/2$, by Lemma 1, one of the trees in the Schnyder realizer of G_{bqp} has at most $n'/3 \leq x$ leaves. We now draw G_{abq} considering the following scenarios.

Case 1A ($\mathtt{leaf}(T_{p,G_{bqp}}) \leq x$). By Lemma 2 and the Stretch condition, G_{abp} admits a drawing Γ_{abp} on L_{x+2} such that the vertices a, b, p lie on l_1, l_{x+2}, l_{x+2}, respectively. Similarly, G_{bqp} admits a drawing Γ_{bpq} on L_{x+2} such that the vertices q, b, p lie on l_1, l_{x+2}, l_{x+2}, respectively, as shown in Figure 3(d). By the Stretch property, Γ_{abp} can be extended to a drawing Γ'_{abp} on L_{x+3}, where a, b, p lie on l_1, l_{x+3}, l_{x+2}, respectively. Similarly, Γ_{bpq} can be extended to a drawing Γ'_{bqp} on L_{x+3}, where q, b, p lie on l_1, l_{x+3}, l_{x+2}, respectively. Since $G^*_{apq} \leq (n'+2)/3$, by Lemma 5 and the Stretch condition, G_{apq} admits a drawing Γ_{apq} on $L_{(n'+2)/3}$. Finally, by the Stretch property Γ_{apq} can be extended to a drawing Γ'_{apq} on L_{x+2} such that a, p, q lie on l_1, l_{x+2}, l_1, respectively, and by the Reshape property we can merge these drawings to obtain a drawing of G_{abq} on L_{x+3}. Figure 3(e) depicts an illustration.

Case 1B ($\mathtt{leaf}(T_{b,G_{bqp}}) \leq x$). By Lemma 2 and the Stretch condition, G_{abp} admits a drawing Γ_{abp} on L_{x+2} such that the vertices a, b, p lie on l_1, l_{x+2}, l_1,

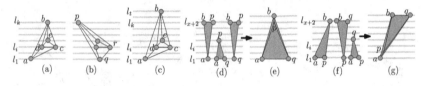

Fig. 3. (a)–(b) Illustrating Reshape. (c) Illustrating Stretch. (d)–(e) Illustration for Case 1A. (d)–(e) Illustration for Case 1B.

respectively. Similarly, G_{bqp} admits a drawing Γ_{bpq} on L_{x+2} such that the vertices p, b, q lie on l_1, l_{x+2}, l_{x+2}, respectively. By Lemma 5, G_{apq} admits a drawing Γ_{apq} on $L_{(n'+2)/3}$ such that a, p, q lie on l_1, l_1, l_{x+2}, respectively. Finally, by Stretch and Reshape we can merge these drawings to obtain a drawing of G_{abq} on L_{x+3}. Figures 3(f)–(g) show an illustration.

Case 1C (leaf($T_{q,G_{bqp}}$) $\leq x$). The drawing of this case is similar to Case 1B. The only difference is that we use $T_{q,G_{bqp}}$ while drawing G_{bqp}.

Observe that each of the Cases 1A–1C produces a drawing of G_{abq} such that a, b lie on l_1, l_{x+3}, respectively, and q lies on either l_1 or l_{x+3}. We stretch it to a drawing on L_{x+4} such that a, b lies on l_1, l_{x+4}, respectively, and q lies on either l_2 or l_{x+3}. If q lies on l_2, then we draw the path $w_{k+1}, \ldots, w_t(= c)$ in a zigzag fashion, placing the vertices on l_2 and l_{x+3} alternatively such that each vertex is visible to both a and b. Similarly, if q lies on l_{x+3}, then we place the vertices $w_{k+1}, \ldots, w_t(= c)$ on l_{x+3} and l_2 alternatively, as shown in Figure 4(a). For each $i > k + 1$, Lemma 4 ensures that the graphs $G_{aw_iw_{i-1}}$ and $G_{bw_iw_{i-1}}$ are plane 3-trees. Since $\max_{\forall i > k+1}\{G^*_{aw_iw_{i-1}}, G^*_{bw_iw_{i-1}}\} \leq x$, we can draw $G_{aw_iw_{i-1}}$ and $G_{bw_iw_{i-1}}$ using Lemma 5 inside their corresponding triangles.

Case 2 (leaf($T_{p,G_{abp}}$) $> x$). Since $G^*_{abp} \leq n'/2$, by Lemma 1, leaf($T_{a,G_{abp}}$) $+$ leaf($T_{b,G_{abp}}$) $\leq n' -$ leaf($T_{p,G_{abp}}$) $\leq 5n'/9$. Hence we draw G_{abq} considering the following scenarios.

Case 2A (leaf($T_{a,G_{abp}}$) $\leq x$ and leaf($T_{b,G_{abp}}$) $\leq x$). Since $G^*_{bqp} \leq n'/2$, by Lemma 1, one of the trees in the Schnyder realizer of G_{bqp} has at most $n'/3 \leq x$ leaves. If leaf($T_{p,G_{bpq}}$) $\leq x$, then we draw G_{abq} on L_{x+3}, where a, b, p, q lie on $l_1, l_{x+3}, l_{x+2}, l_1$, respectively, as in Figure 4(b). Otherwise, either leaf($T_{b,G_{bpq}}$) $\leq x$ or leaf($T_{q,G_{bpq}}$) $\leq x$. In this case we draw G_{abq} on L_{x+3}, where a, b, p, q lie on $l_1, l_{x+3}, l_2, l_{x+3}$, respectively, as in Figure 4(c).

Case 2B (leaf($T_{a,G_{abp}}$) $> x$ and leaf($T_{b,G_{abp}}$) $\leq n'/9$). If leaf($T_{p,G_{bpq}}$) $\leq n'/3$, then we first draw G_{bpq} using Lemma 2 such that b, p, q lie on $l_{n'/3+2}, l_{n'/3+2}, l_1$, respectively, and then use the Stretch condition to shift b to l_{x+3}. By Lemma 2 and the Stretch condition, there exists a drawing of G_{abp} on L_{x+3} with a, b, p lying on $l_1, l_{x+3}, l_{n'/3+2}$, respectively. Since $G^*_{apq} \leq (n' + 2)/3$, we can draw G_{apq} using Lemma 5 inside triangle apq. Figure 4(d) illustrates the scenario after applying Stretch and Reshape.

If leaf($T_{p,G_{bpq}}$) $> n'/3$, then by Lemma 1 either leaf($T_{b,G_{bpq}}$) $\leq n'/3$ or leaf($T_{q,G_{bpq}}$) $\leq n'/3$. Hence we can use Lemma 2 and the Stretch condition

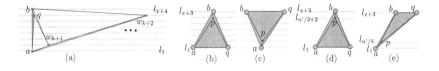

Fig. 4. (a) Illustrating Case 1. (b)–(c) Illustrating Case 2A. (d)–(e) Case 2B.

to draw G_{bpq} such that b, p, q lie on $l_{x+3}, l_{n'/9}, l_{x+3}$, respectively. On the other hand, we use Lemma 2 and the Stretch condition to draw G_{abp} such that a, b, p lie on $l_1, l_{x+3}, l_{n'/9}$, respectively. Since $G_{apq}^* \leq (n' + 2)/3$, we can draw G_{apq} using Lemma 5 inside triangle apq. Figure 4(e) illustrates the scenario.

Case 2C (leaf$(T_{a,G_{abp}}) \leq n'/9$ and leaf$(T_{b,G_{abp}}) > x$). Since each of the edges among (a, b) and (b, p) spans at least $n'/9 + 2$ parallel lines in Case 2B, the drawing in this case is analogous to Case 2B. The only difference is that we use $T_{a,G_{abp}}$ while drawing G_{abp}.

Each of the Cases 2A–2C produces a drawing of G_{abq} such that a, b lies on l_1, l_{x+3}, respectively, and q lies on either l_1 or l_{x+3}. Hence we can extend these drawings to draw G as in Case 1. □

4.2   Drawing Algorithm

Decomposition. Let G be an n-vertex plane 3-tree with the outer vertices a, b, c and the representative vertex p. A tree spanning the inner vertices of G is called the *representative tree* T if it satisfies the following conditions [14]:

(a) If $n = 3$, then T is empty.
(b) If $n = 4$, then T consists of a single vertex.
(c) If $n > 4$, then the root p of T is the representative vertex of G and the subtrees rooted at the three clockwise ordered children p_1, p_2 and p_3 of p in T are the representative trees of G_{abp}, G_{bcp} and G_{cap}, respectively.

Recall that every n-vertex tree T' has a vertex v' such that the connected components of $T' \setminus v'$ are all of size at most $n/2$ [12]. Such a vertex v in T corresponds to a decomposition of G into four smaller plane 3-trees G_1, G_2, G_3, and G_4, as follows.

- The plane 3-tree G_i, where $1 \leq i \leq 3$, is determined by the representative tree rooted at the ith child of v, and thus contains at most $(n + 3)/2$ vertices.
- The plane 3-tree G_4 is obtained by deleting v and the vertices from G that are descendent of v in T, and contains at most $(n + 3)/2$ vertices.

Drawing Technique. Without loss generality assume that $G_3^* \leq G_2^* \leq G_1^*$. If G_1 is incident to the outer face of G, then let (a, b) be the corresponding outer edge. Otherwise, G_1 does not have any edge incident to the outer face of G. In this case there exists an inner face f in G that is incident to G_1, but does not belong to G_1. We choose f as the outer face of G, and now we have an edge (a, b) of G_1 that is incident to the outer face. Let $P = (w_1, \ldots, w_k (= p), w_{k+1} (= q), \ldots, w_t)$ be the maximal path in G such that each vertex on P is adjacent to both a and b, where $\{a, b, p\}, \{a, p, q\}, \{b, q, p\}$ are the outer vertices of G_1, G_2, G_3, respectively. Assume that $n' = n + 3$ and $x = 4n'/9$. We draw G on L_{x+4} by distinguishing two cases depending on whether $G_4^* > x$ or not.

Case 1 ($G_4^* > x$). Observe that $G_2^* \leq G_1^* \leq n'/2$ and since $G_3^* + G_2^* + G_1^* \leq n' + 5 - G_4^*$, we have $G_3^* \leq 5n'/27 + 5/3 \leq n'/3$ for sufficiently large values of n.

If $\max_{\forall i > k+1}\{G^*_{aw_i w_{i-1}}, G^*_{bw_i w_{i-1}}\} \leq x$ holds, then G admits a drawing on L_{x+4} by Lemma 6. We may thus assume that there exists some $j > q$ such that either $G^*_{aw_j w_{j-1}} > x$ or $G^*_{bw_j w_{j-1}} > x$. Hence $\max_{\forall i > k+1, i \neq j}\{G^*_{aw_i w_{i-1}}, G^*_{bw_i w_{i-1}}\} \leq n'/9$.

We first show that G_{abq} can be drawn on L_{x+3} in two ways: One drawing Γ_1 contains the vertices a, b, q on l_1, l_{x+3}, l_2, respectively, and the other drawing Γ_2 contains a, b, q on l_1, l_{x+3}, l_{x+2}, respectively. We then extend these drawings to obtain the required drawing of G. Consider the following scenarios depending on whether $G^*_1 \leq x$ or not.

- If $G^*_1 \leq x$, then $G^*_3 \leq G^*_2 \leq G^*_1 \leq x$. Here we draw the subgraph G' induced by the vertices a, b, p, q such that they lie on $l_1, l_{x+3}, l_{x+2}, l_2$, respectively. Since $G^*_3 \leq G^*_2 \leq G^*_1 \leq x$, by Lemma 5, G_1, G_2 and G_3 can be drawn inside their corresponding triangles, which corresponds to Γ_1. Similarly, we can find another drawing Γ_2 of G_{abq}, where the vertices a, b, p, q lie on $l_1, l_{x+3}, l_2, l_{x+2}$, respectively.
- If $G^*_1 > x$, then $G^*_3 \leq G^*_2 \leq n'/9$. We use Chrobak and Nakano's algorithm [6] and the Stretch condition to draw G_1 on L_{x+3} layers such that a, b lie on l_1, l_{x+3}, respectively, and p lies either on l_2 or $l_{n'/3+2}$. If $l(p) = l_2$ (i.e., Γ_2), then we place q on l_{x+2}. Otherwise, $l(p) = l_{n'/3+2}$ (i.e., Γ_1), and we place q on l_2. Since $G^*_3 \leq G^*_2 \leq n'/9$, for each of these two placements we can draw G_2 and G_3 using Lemma 5 inside their corresponding triangles.

We now show how to extend the drawing of G_{abq} to compute the drawing of G. Consider two scenarios depending on whether $G^*_{aw_j w_{j-1}} > x$ or $G^*_{bw_j w_{j-1}} > x$.

- Assume that $G^*_{aw_j w_{j-1}} > x$. Shift b to l_{x+4}, and draw the path $w_{k+1}, \ldots, w_{j-1}$ in a zigzag fashion, placing the vertices on l_2 and l_{x+3} alternatively, such that $l(w_{k+1}) \neq l(w_{k+2})$, and each vertex is visible to both a and b. Choose Γ_1 or Γ_2 such that the edge (a, w_{j-1}) spans at least $x + 3$ lines. We now draw $G_{aw_j w_{j-1}}$ using Chrobak and Nakano's algorithm [6]. Since $x < G_{aw_j w_{j-1}} \leq n'/2$, we can draw $G_{aw_j w_{j-1}}$ on at most $n'/3$ parallel lines. By the Stretch and Reshape conditions, we merge this drawing with the current drawing such that w_j lies on either l_{x+3} or $l_{n'/9+2}$. Since $G^*_{bw_j w_{j-1}} \leq n'/9$, we can draw $G_{bw_j w_{j-1}}$ inside its corresponding triangle using Lemma 5. Since $\max_{\forall i > j}\{G^*_{aw_i w_{i-1}}, G^*_{bw_i w_{i-1}}\} \leq n'/9$, it is straightforward to extend the current drawing to a drawing of G on $x + 4$ parallel lines by continuing the path $w_j, \ldots, w_t$ in the zigzag fashion.
- Assume that $G^*_{bw_j w_{j-1}} > x$. The drawing in this case is similar to the case when $G^*_{aw_j w_{j-1}} > x$. The only difference is that while drawing the path $w_{k+1}, \ldots, w_{j-1}$, we choose Γ_1 or Γ_2 such that the edge (b, w_{j-1}) spans at least $x + 3$ lines.

Case 2 ($G^*_4 \leq x$). Observe that $G^*_2 \leq G^*_1 \leq n'/2$. Since $G^*_3 \leq G^*_2 \leq G^*_1$ and $G^*_3 + G^*_2 + G^*_1 = n + 5$, we have $G^*_3 \leq (n' + 2)/3$. Hence G admits a drawing on L_{x+4} by Lemma 6.

The following theorem summarizes the result of this section.

Theorem 2. *Every n-vertex planar 3-tree admits a straight-line drawing with height $4(n + 3)/9 + 4 = 4n/9 + O(1)$.*

Acknowledgement. We thank the anonymous reviewers for their detailed feedback to improve the presentation of the paper.

References

1. Biedl, T.: Height-preserving transformations of planar graph drawings. In: Duncan, C., Symvonis, A. (eds.) GD 2014. LNCS, vol. 8871, pp. 380–391. Springer, Heidelberg (2014)
2. Biedl, T.: A 4-approximation for the height of drawing 2-connected outer-planar graphs. In: Erlebach, T., Persiano, G. (eds.) WAOA 2012. LNCS, vol. 7846, pp. 272–285. Springer, Heidelberg (2013)
3. Biedl, T.C.: Transforming planar graph drawings while maintaining height. CoRR abs/1308.6693 (2013), http://arxiv.org/abs/1308.6693
4. Bonichon, N., Le Saëc, B., Mosbah, M.: Wagner's Theorem on Realizers. In: Widmayer, P., Triguero, F., Morales, R., Hennessy, M., Eidenbenz, S., Conejo, R. (eds.) ICALP 2002. LNCS, vol. 2380, pp. 1043–1053. Springer, Heidelberg (2002)
5. Brandenburg, F.J.: Drawing planar graphs on $\frac{8}{9}n^2$ area. Electronic Notes in Discrete Mathematics 31, 37–40 (2008)
6. Chrobak, M., Nakano, S.: Minimum width grid drawings of plane graphs. In: Tamassia, R., Tollis, I.G. (eds.) GD 1994. LNCS, vol. 894, pp. 104–110. Springer, Heidelberg (1995)
7. De Fraysseix, H., Pach, J., Pollack, R.: How to draw a planar graph on a grid. Combinatorica 10(1), 41–51 (1990)
8. Djidjev, H., Venkatesan, S.M.: Reduced constants for simple cycle graph separation. Acta Informatica 34(3), 231–243 (1997)
9. Dujmović, V., et al.: On the parameterized complexity of layered graph drawing. In: Meyer auf der Heide, F. (ed.) ESA 2001. LNCS, vol. 2161, pp. 488–499. Springer, Heidelberg (2001)
10. Frati, F., Patrignani, M.: A note on minimum-area straight-line drawings of planar graphs. In: Hong, S.-H., Nishizeki, T., Quan, W. (eds.) GD 2007. LNCS, vol. 4875, pp. 339–344. Springer, Heidelberg (2008)
11. Hossain, M.I., Mondal, D., Rahman, M.S., Salma, S.A.: Universal line-sets for drawing planar 3-trees. Journal of Graph Algorithms and Applications 17(2), 59–79 (2013)
12. Jordan, C.: Sur les assemblages de lignes. Journal für die reine und angewandte Mathematik 70(2), 185–190 (1869)
13. Mondal, D., Alam, M.J., Rahman, M.S.: Minimum-layer drawings of trees - (extended abstract). In: Katoh, N., Kumar, A. (eds.) WALCOM 2011. LNCS, vol. 6552, pp. 221–232. Springer, Heidelberg (2011)
14. Mondal, D., Nishat, R.I., Rahman, M.S., Alam, M.J.: Minimum-area drawings of plane 3-trees. Journal of Graph Algorithms and Applications 15(2), 177–204 (2011)
15. Pach, J., Tóth, G.: Monotone drawings of planar graphs. Journal of Graph Theory 46(1), 39–47 (2004)
16. Schnyder, W.: Embedding planar graphs on the grid. In: Proceedings of ACM-SIAM SODA, January 22-24, pp. 138–148. ACM (1990)
17. Suderman, M.: Pathwidth and layered drawing of trees. Journal of Computational Geometry & Applications 14(3), 203–225 (2004)
18. Suderman, M.: Pathwidth and layered drawings of trees. International Journal of Computational Geometry and Applications 14, 203–225 (2004)

Anchored Drawings of Planar Graphs*

Patrizio Angelini[1], Giordano Da Lozzo[1], Marco Di Bartolomeo[1],
Giuseppe Di Battista[1], Seok-Hee Hong[2], Maurizio Patrignani[1], and Vincenzo Roselli[1]

[1] Department of Engineering, Roma Tre University, Italy
{angelini,dalozzo,dibartolomeo,gdb,
patrigna,roselli}@dia.uniroma3.it
[2] School of Information Technologies, The University of Sydney, Australia
shhong@it.usyd.edu.au

Abstract. In this paper we study the ANCHORED GRAPH DRAWING (AGD) problem: Given a planar graph G, an initial placement for its vertices, and a distance d, produce a planar straight-line drawing of G such that each vertex is at distance at most d from its original position.

We show that the AGD problem is NP-hard in several settings and provide a polynomial-time algorithm when d is the uniform distance L_∞ and edges are required to be drawn as horizontal or vertical segments.

1 Introduction

Several applications require to draw graphs whose vertices are constrained to be not too much *distant* from specific points [1,9]. As an example, consider a graph whose vertices are cities and whose edges are relationships between cities. It is conceivable that the user wants to draw the graph on a geographic map where vertices have the coordinates of the corresponding cities. Unfortunately, depending on the local density of the cities, the drawing may be cluttered or may contain crossings between edges that might disappear if the vertices could move from their locations. Hence, the user may be interested to trade precision for quality of the drawing, accepting that the vertices move of a certain distance from the location of the cities, provided that the readability of the drawing increases. Problems in which the input consists of a set of imprecise points have also been studied in Computational Geometry [4,7].

In this paper we consider the following problem, that we call ANCHORED GRAPH DRAWING (AGD)[1]. Given a graph $G = (V, E)$, an initial placement for its vertices, and a distance δ, we ask whether there exists a planar drawing of G, according to a certain drawing convention, such that each vertex $v \in V$ can move at distance at most δ from its initial placement. Note that the problem can have different formulations depending on how the concepts of "readability" and "distance" are defined.

We consider both straight-line planar drawings and rectilinear planar drawings. Further, in addition to the traditional L_2 Euclidean distance, we consider the L_1 Manhattan

---

* Work partially supported by ESF EuroGIGA GraDR, by the MIUR project AMANDA "Algorithmics for MAssive and Networked DAta", prot. 2012C4E3KT_001, and by the EU FP7 STREP Project "Leone: From Global Measurements to Local Management", no. 317647.
[1] We remark that the term 'anchored graph' was used within a different setting in [3].

C. Duncan and A. Symvonis (Eds.): GD 2014, LNCS 8871, pp. 404–415, 2014.
© Springer-Verlag Berlin Heidelberg 2014

Table 1. The complexity of the ANCHORED GRAPH DRAWING problem depending on the metric and drawing style adopted when the areas of the vertices do not overlap

Metric	Distance	Region Shape	Straight-line	Rectilinear
L_1	Manhattan	◇	NP-hard	NP-hard
L_2	Euclidean	○	NP-hard	NP-hard
L_∞	Uniform	□	NP-hard	Polynomial

distance and the L_∞ 'uniform' distance. Note that, adopting L_2 distance is equivalent to allowing vertices to be placed into circular regions centered at their original positions, and adopting L_1 or L_∞ distances is equivalent to allowing vertices to be placed into diamond-shaped or square-shaped areas, respectively.

Observe that, if the regions of two vertices overlap, the positions of the two vertices can be swapped with respect to their initial placement, which may be confusing to a user of the drawing. Moreover, overlapping between vertex regions would make problem AGD as difficult as known Clustered Planarity variants, such as the Strip Planarity problem [2] in the straight-line setting, whose complexity is a non-trivial open problem. Hence, we restrict to instances such that the regions of the vertices do not overlap.

We remark that the version of the problem where each circle may have a different size was shown to be NP-hard in [6] by reducing Planar-$(3, 4)$-SAT with variable repetitions (where repeated occurrences of one variable in one clause are counted repeatedly). The proof in [6] uses disks with radius zero and disks with large radii. Also, the reduction relies on overlapping disks.

Furthermore, we observe that the NP-hardness of the problem with different distances and overlapping areas trivially follows from the NP-hardness of extending a planar straight-line drawing [10] by setting $\delta(v) = 0$ for each fixed vertex v and allowing suitably large distances for vertices that have to be planarly added to the drawing.

In this paper we show that the ANCHORED GRAPH DRAWING problem is NP-hard for any combination of metrics and drawing standards that we considered, with the exception of rectilinear drawings and uniform distance metric (square-shaped regions). These results, summarized in Table 1, were somehow unexpected, as computing a planar rectilinear drawing of a graph, without any further constraint, is NP-hard [5].

The paper is organized as follows. Section 2 contains basic definitions and terminology. Section 3 describes a polynomial-time algorithm when the considered distance is the uniform distance L_∞ and edges are required to be drawn as either horizontal or vertical segments. Section 4 is devoted to the NP-hardness proofs of all the other considered settings of the problems. Finally, Section 5 discusses some open problems.

2 Problem Definition and Instances Classification

A *straight-line planar drawing* of a graph G is a drawing of G where edges are straight-line segments that do not intersect except at common end-points. A *rectilinear planar drawing* is a straight-line planar drawing where edges are parallel to the axes.

Given two points p and q in the plane, denote by $dx(p, q)$ and $dy(p, q)$ the differences of their coordinates, i.e., $dx(p, q) = |x(p) - x(q)|$ and $dy(p, q) = |y(p) - y(q)|$, where

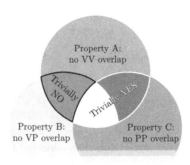

Fig. 1. Venn diagram describing the logical relationships among Properties A–C

$x(r)$ and $y(r)$ are the x- and y-coordinate of a point r, respectively. The *Euclidean distance* $d_2(p,q)$ of p and q is defined as $d_2(p,q) = (dx(p,q)^2 + dy(p,q)^2)^{\frac{1}{2}}$. The *Manhattan distance* is defined as $d_1(p,q) = dx(p,q) + dy(p,q)$. The *uniform distance* $d_\infty(p,q) = \lim_{i\to\infty}(dx(p,q)^i + dy(p,q)^i)^{\frac{1}{i}} = \max(dx(p,q), dy(p,q))$.

We define the ANCHORED GRAPH DRAWING problem parametrically in the metric L_k and the drawing style $\mathcal{X}$, which can be straight-line ($\mathcal{X} = \mathcal{S}$) or rectilinear ($\mathcal{X} = \mathcal{R}$). Hence, for any $L_k \in \{L_1, L_2, L_\infty\}$ and any $\mathcal{X} \in \{\mathcal{S}, \mathcal{R}\}$ we define: **Problem:** ANCHORED GRAPH DRAWING-L_k-$\mathcal{X}$ (AGD-L_k-$\mathcal{X}$). **Instance:** A graph $G = (V, E)$, an initial placement for its vertices $\alpha(v) : V \to \Re^2$, and a distance δ. **Question:** Does there exist a planar drawing of G according to the $\mathcal{X}$ drawing convention such that each vertex $v \in V$ is at distance L_k at most δ from $\alpha(v)$?

We define *anchored drawing* a planar drawing satisfying all the requirements of the particular version of problem ANCHORED GRAPH DRAWING.

Given an instance $\langle G, \alpha, \delta \rangle$ of the ANCHORED GRAPH DRAWING problem, each vertex v identifies a region $R(v)$ of the plane, called *vertex region*, that encloses the initial position of the vertex and whose shape depends on the metric adopted for computing the distance. In particular, for the Euclidean distance the vertex regions are circles, for the Manhattan distance they are diamonds, and for the uniform distance they are squares. Each edge (u, v) of the graph, instead, identifies a *pipe* $P(u, v)$, defined as follows. Consider the convex hull H of $R(u)$ and $R(v)$; pipe $P(u, v)$ is the closed region obtained by removing $R(u)$ and $R(v)$ from H.

Instances can be classified based on the intersections among vertex and pipe regions. Namely, we can have instances satisfying the following properties:

Property A. No overlap between two vertex regions (VV-overlaps);
Property B. No overlap between a vertex region and a pipe (VP-overlaps);
Property C. No overlap between pipes (PP-overlaps) not incident to the same vertex.

The Venn diagram in Fig. 1 shows the logical relationships between the three properties. The following observation is immediate.

Observation 1. *If Properties A, B, and C are all satisfied, then the instance is trivially positive, since choosing any point in the vertex region (including the initial placement of the vertex) yields an anchored drawing of the input graph.*

In this paper we always assume that Property A is satisfied. In fact, if vertex regions were allowed to overlap, then it would be possible to reduce to this problem a variant of the Clustered Planarity problem whose complexity is still unknown. In this variant, which includes Strip Planarity [2] as a special case, the cluster regions are already drawn and edges are straight-line.

Two further observations can be made which reduce the set of instances of interest.

Observation 2. *An instance satisfying Property B but not satisfying Property C (i.e., with PP-overlaps but without VP-overlaps) is trivially false, as in this case any PP-overlap would enforce a crossing between two edges for any placement of their end-vertices in the corresponding vertex regions.*

Observation 3. *An instance satisfying Property C but not satisfying Property B (i.e., with VP-overlaps but without PP-overlaps) is trivially true.*

Proof: Since Property C holds, no crossing can occur outside a vertex region. First, suppose that regions are diamonds or squares. If the center of region $R(v)$ of a vertex v lies inside a pipe $P(x,y)$, then at least two consecutive vertices, say a and b, delimiting $R(v)$ lie inside $P(x,y)$. This implies that v has degree at most 1, as otherwise there would be a PP-overlap between $P(x,y)$ and a pipe $P(v,w)$ delimited by either a or b.

As for the case in which regions are circles, if the center of $R(v)$ lies inside $P(x,y)$, then at least half of the circle delimiting $R(v)$ lies inside $P(x,y)$. Hence, a similar argument applies to prove that $\deg(v) \le 1$.

In all the three cases, since $\deg(v) \le 1$ and $R(v)$ is not completely contained into $P(x,y)$, v can be placed on any point of $R(v)$ outside $P(x,y)$. Hence, placing each other vertex at the center of its region yields an anchored drawing.   □

Due to the above properties and observations, the remaining part of this paper focuses on the instances for which Property A holds, while Properties B and C do not. These instances correspond to the blue region at the top of Fig. 1.

3   Polynomial-Time Algorithm

In this section we describe an algorithm, called **Algo-AGD-L_∞-$\mathcal{R}$**, that decides in poly-nomial time instances $\langle G, \alpha, \delta \rangle$ of problem AGD-L_∞-$\mathcal{R}$ such that G is connected.

For each vertex $v \in V$, denote by $x_l(v)$ and $x_r(v)$ the x-coordinate of the left and right side of $R(v)$, respectively. Similarly, denote by $y_t(v)$ and $y_b(v)$ the y-coordinate of the top and bottom side of $R(v)$, respectively. See region $R(u)$ in Fig. 2.

First note that, for each edge $(u, v) \in E$, the relative placement of $R(u)$ and $R(v)$ determines whether (u, v) has to be drawn as a vertical or a horizontal segment, or (u, v) cannot be drawn neither horizontal nor vertical with its endpoints lying inside their corresponding regions. In the latter case, instance I is negative. An edge that has to be drawn as a horizontal (vertical) segment is a *horizontal* (*vertical*) edge. In the fol-lowing we assume w.l.o.g. that any horizontal edge (u, v) is such that $x_r(u) < x_l(v)$, while any vertical edge (u, v) is such that $y_t(u) < y_b(v)$. A path composed only of horizontal (vertical) edges is a *horizontal* (*vertical*) path. Given that each edge (u, v)

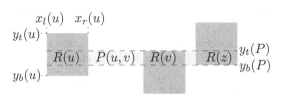

$x_l(u)$ $x_r(u)$

$y_t(u)$

$R(u)$ $P(u,v)$ $R(v)$ $R(z)$ $y_t(P)$
$y_b(P)$

$y_b(u)$

Fig. 2. Geometric description of a region $R(u)$ and of a pipe $P(u,v)$, after procedure PIPEE-QUALIZER has been applied

can be categorized as either horizontal or vertical, we can label its pipe $P(u,v)$ as either `horizontal` or `vertical` accordingly. Also, we can determine the minimum and maximum y-coordinate (x-coordinate) that a horizontal (vertical) edge (u,v) can assume while placing both its endvertices inside their regions. In the following we describe pipe $P(u,v)$ by means of these coordinates, which are denoted by $y_b(P)$ and $y_t(P)$ ($x_l(P)$ and $x_r(P)$), respectively. See `horizontal` pipe $P(u,v)$ in Fig. 2.

Also note that, if a vertex v of degree 2 is incident to two horizontal (vertical) edges (u,v) and (v,z), then replacing v and its incident edges with a horizontal (vertical) edge (u,z) yields an equivalent instance. Hence, we assume that, if there exists a vertex of degree 2, then it is incident to both a horizontal and a vertical edge.

As a preliminary step of the algorithm, we initialize the geometric description of each pipe $P(u,v)$ as follows. If P is `vertical`, then set $x_r(P) = min(x_r(u), x_r(v))$ and $x_l(P) = max(x_l(u), x_l(v))$. If P is `horizontal`, then set $y_t(P) = min(y_t(u), y_t(v))$ and $y_b(P) = max(y_b(u), y_b(v))$. Here and in the following, whenever a vertex region $R(w)$ (a pipe $P(u,v)$) is modified by the algorithm, we assume the pipes incident to w (the regions $R(u)$ and $R(v)$) to be modified accordingly.

In order to ensure that `horizontal` (`vertical`) pipes whose edges belong to the same horizontal (vertical) path have the same geometric description, we refine the pipes by applying the following procedure, that we call PIPEEQUALIZER. As long as there exist two `vertical` pipes $P'(u,v)$ and $P''(v,w)$ incident to the same vertex v such that $x_l(P') \neq x_l(P'')$ or $x_r(P') \neq x_r(P'')$, set $x_l(P') = x_l(P'') = max(x_l(P'), x_l(P''))$ and $x_r(P') = x_r(P'') = min(x_r(P'), x_r(P''))$. Analogously, as long as there exist two `horizontal` pipes $P'(u,v)$ and $P''(v,w)$ incident to the same vertex v such that $y_b(P') \neq y_b(P'')$ or $y_t(P') \neq y_t(P'')$, set $y_b(P') = y_b(P'') = max(y_b(P'), y_b(P''))$ and $y_t(P') = y_t(P'') = min(y_t(P'), y_t(P''))$. See pipe $P(u,v)$ in Fig. 2 after the application of PIPEEQUALIZER.

We then perform the following procedure, that we call PIPECHECKER. It first checks whether there exists a pipe P such that $x_r(P) < x_l(P)$ or $y_t(P) < y_b(P)$. Then, it checks whether there exists a PP-overlap between two pipes $P(u,v)$ and $P(w,z)$ such that: (i) neither of $R(u)$ or $R(v)$ has a VP-overlap with $P(w,z)$; and (ii) neither of $R(w)$ or $R(z)$ has a VP-overlap with $P(u,v)$. If one of the two checks succeeds, then we conclude that instance I is negative, otherwise we proceed with the algorithm.

In the following, every time a pipe is modified, we will apply procedure PIPEE-QUALIZER to extend this modification to other pipes, and procedure PIPECHECKER to test whether such modifications resulted in uncovering a negative instance.

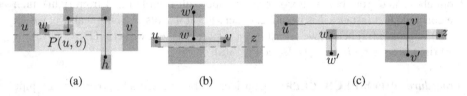

(a)   (b)   (c)

Fig. 3. Vertices exiting pipes. (a) Vertex w exits $P(u,v)$ from below. The cut of $P(u,v)$ applied by procedure PIPEBLOCKCHECKER is described by a dashed line. (b) Vertex v exits $P(w,z)$ through w. The cut of $R(w)$ and the consequent cut of $P(w,z)$ applied by procedure VERTEX-CHECKER is described by a dashed line. (c) A situation recognized by procedure PIPEINTER-LEAVECHECKER.

The general strategy of the main part of the algorithm is to progressively reduce the size of the pipes. In particular, at each step we consider the current instance I^i and modify it to obtain an instance I^{i+1} with smaller pipes than I^i that admits an anchored drawing if and only if I^i admits an anchored drawing. Eventually, such a process will lead either to an instance I^m for which it is easy to construct an anchored drawing or to conclude that instance $I = I^1$ is negative.

Let $P(u,v)$ be a `horizontal` pipe, and w be a vertex having a VP-overlap with $P(u,v)$. Refer to Fig. 3(a). We say that w *exits P from below* if there exists a vertex h such that: (i) $y_b(P) < y_b(w) < y_t(P)$ and $x_r(u) < x_l(w) < x_r(w) < x_l(v)$; (ii) $y_t(h) < y_b(P)$ and $x_r(u) < x_l(h) < x_r(h) < x_l(v)$; and (iii) there exists a path $\gamma = (w, \ldots, h)$ in G connecting w to h in which every internal vertex r is such that $R(r)$ intersects P. Symmetrically, we say that w *exits P from above* if there exists a vertex h with the same properties as before, except for the fact that $y_b(P) < y_t(w) < y_t(P)$, $y_b(h) > y_t(P)$. Otherwise, we say that w *exits P through a vertex*, either u or v. In Fig. 3(b), vertex v exits pipe $P(w,z)$ through w. Observe that, since G is connected and no VV-overlap occurs in I, there always exists a path $\gamma = (w, \ldots, h)$ in G connecting w to a vertex h such that h does not have any VP-overlap with P; hence, w always exits P, either from above or below, or through a vertex.

For the case of a `vertical` pipe $P(u,v)$, we assume analogous definitions of vertices exiting P *from left, right* or *through a vertex*, either u or v. As long as one of the following conditions is satisfied, we apply one of the procedures described hereunder.

***Procedure* VERTEXCHECKER:** Consider a vertex w having a VP-overlap with a `horizontal` (`vertical`) pipe $P(u,v)$ such that $y_b(w) \leq y_b(P) < y_t(P) \leq y_t(w)$ (resp., $x_l(w) \leq x_l(P) < x_r(P) \leq x_r(w)$). If w is incident to two `vertical` (`horizontal`) pipes, then we conclude that instance I is negative. Otherwise, if w is incident to a `vertical` (`horizontal`) pipe $P(w,w')$, then set $y_b(w) = max(y_b(w), y_b(P))$ (set $x_l(w) = max(x_l(w), x_l(P))$). See Fig. 3(b). Analogously, if w is incident to a `vertical` (`horizontal`) pipe $P(w',w)$, then set $y_t(w) = min(y_t(w), y_t(P))$ (set $x_r(w) = min(x_r(w), x_r(P))$).

***Procedure* PIPEBLOCKCHECKER:** Consider a pipe $P(u,v)$ having a VP-overlap with a vertex w such that w does not exit through a vertex. If w exits $P(u,v)$ both from above and from below (a vertical pipe both from left and from right), then we conclude that instance I is negative. Otherwise, if w exits P from (i) above, we set $y_t(P) = y_t(w)$; (ii) below, we set $y_b(P) = y_b(w)$; (iii) left, we set $x_l(P) = x_l(w)$; or (iv) right, we set $x_r(P) = x_r(w)$. See Fig. 3(a).

Procedure PIPESIDECHECKER: Consider a horizontal (vertical) pipe $P(u, v)$ and a vertex w exiting $P(u, v)$ both through vertex u and through vertex v. If u and v are incident to vertical (horizontal) pipes, either $P(u, u')$ and $P(v', v)$, or $P(u', u)$ and $P(v, v')$, respectively, then we conclude that instance I is negative.

Procedure PIPEINTERLEAVECHECKER: Suppose that there exist two horizontal (vertical) pipes $P(u, v)$ and $P(w, z)$ such that v and $P(w, z)$ have a VP-overlap, and w and $P(u, v)$ have a VP-overlap. If either v is incident to a vertical (horizontal) pipe $P(v, v')$ and w is incident to a vertical (horizontal) pipe $P(w, w')$, or v is incident to a vertical (horizontal) pipe $P(v', v)$ and w is incident to a vertical (horizontal) pipe $P(w', w)$, then we conclude that instance I is negative. See Fig. 3(c).

If none of the above procedures can be applied, then we conclude that I is a positive instance.

Theorem 1. *Let $I = \langle G, \alpha, \delta \rangle$ be an instance of* AGD-L_∞-$\mathcal{R}$ *such that G is connected. Algorithm* **Algo-AGD-L_∞-$\mathcal{R}$** *decides in polynomial time whether $\langle G, \alpha, \delta \rangle$ admits an anchored drawing.*

Proof: The initialization of the pipes and their refinement operated by procedure PIPEEQUALIZER, both after the initialization and after each further modification, is trivially necessary to meet the requirements that vertices are placed inside their regions and edges are drawn as either horizontal or vertical segments.

Suppose that procedure PIPECHECKER concluded that instance I is negative at some point of the algorithm. If $x_r(P) < x_l(P)$ (if $y_t(P) < y_b(P)$), then there exist two vertical (horizontal) pipes sharing a vertex that cannot be placed inside its region while drawing both its incident edges as rectilinear segments. Otherwise, there exists a PP-overlap between two pipes $P(u, v)$ and $P(w, z)$ not overlapping with regions $R(u)$, $R(v)$, $R(w)$, and $R(z)$. By Observation 2, the instance is negative.

We prove that the modifications operated by VERTEXCHECKER, when a vertex w has a VP-overlap with a horizontal (vertical) pipe $P(u, v)$ and w is incident to a vertical (horizontal) $P(w, w')$, do not restrict the possibility of constructing an anchored drawing of $\langle G, \alpha, \delta \rangle$. Refer to Fig. 3(b). In fact, in this case, in any anchored drawing of $\langle G, \alpha, \delta \rangle$, edge (w, w') cannot traverse $P(u, v)$ from top to bottom. As for the fact that an instance in which w is incident to two vertical (horizontal) pipes is correctly recognized as negative, observe that in this case one of the two vertical edges incident to w necessarily crosses edge (u, v).

We prove that the modifications operated by PIPEBLOCKCHECKER, when a vertex w overlaps a pipe $P(u, v)$ and does not exit through one of its vertices, do not restrict the possibility of constructing an anchored drawing of $\langle G, \alpha, \delta \rangle$. Suppose that w exits $P(u, v)$ from below, the other cases being analogous. Refer to Fig. 3(a). The statement

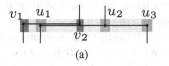

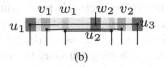

(a)   (b)

Fig. 4. Construction of the drawing when none of the procedures can be applied. (a) Two maximal horizontal paths (u_1, u_2, u_3) and (v_1, v_2) whose pipes have the same y-coordinates. Path (v_1, v_2) is assigned a y-coordinate slightly larger than (u_1, u_2, u_3). (b) Three maximal horizontal paths (u_1, u_2, u_3), (v_1, v_2), and (w_1, w_2) whose pipes have the same y-coordinates. Path (v_1, v_2) is assigned a y-coordinate slightly larger than (w_1, w_2).

follows from the fact that, in any anchored drawing of $\langle G, \alpha, \delta \rangle$, the drawing of path $\gamma = (w, \ldots, h)$ blocks visibility from $R(u)$ to $R(v)$ inside $P(u, v)$ at least for all the y-coordinates in the range between the point where w is placed and the point where h is placed. Since $y_t(h) < y_b(P)$, the point where w is placed determines a new lower bound for the value of $y_b(P)$. Since such a point cannot be below $y_b(w)$, the statement follows. As for the fact that an instance containing a vertex w that exits a horizontal pipe both from above and from below (a vertical pipe both from left and from right) is correctly recognized as negative, observe that in this case the two paths starting from w completely block visibility between from $R(u)$ to $R(v)$ inside $P(u, v)$.

We prove that an instance containing a vertex w that exits $P(u, v)$ through both of its vertices, and such that u and v are also incident to pipes $P(u, u')$ and $P(v', v)$, or vice versa, reaching them from different sides, is correctly recognized as negative by procedure PIPESIDECHECKER. Namely, observe that in this case path (u', u, v, v') necessarily crosses one of the two paths starting from w.

Finally, suppose that procedure PIPEINTERLEAVECHECKER concluded that instance I is negative. Refer to Fig. 3(c). It is easy to observe that the fact that v and w are reached from the same side is not compatible with an anchored drawing of $\langle G, \alpha, \delta \rangle$.

We conclude the proof of the theorem by showing that, when none of the described procedures can be applied, it is always possible to draw every edge (u, v) inside its pipe $P(u, v)$, as follows.

Consider every maximal horizontal path $(u_1, \ldots, u_r)$. Note that, each vertex u_i, with $1 \le i \le r$, is incident to at least a vertical pipe, either (u_i, u'_i) or (u'_i, u_i), as otherwise edges (u_{i-1}, u_i) and (u_i, u_{i+1}) would have been replaced with edge (u_{i-1}, u_{i+1}). If all the vertices u_i are incident to a vertical (u_i, u'_i), then assign y-coordinate equal to $y_t(u_i)$ to u_i, for $i = 1, \ldots, r$; if all the vertices u_i are incident to a vertical (u'_i, u_i), then assign y-coordinate equal to $y_b(u_i)$ to u_i, for $i = 1, \ldots, r$; finally, if there exists at least a vertex u_i incident to a vertical (u_i, u'_i) and at least a vertex u_j incident to a vertical (u'_j, u_j), then assign y-coordinate equal to $\frac{y_b(u_i) + y_t(u_i)}{2}$ to u_i, for $i = 1, \ldots, r$. Assign x-coordinates to vertices of every maximal vertical path equal to $x_l(u_i)$, to $x_r(u_i)$, or to $\frac{x_l(u_i) + x_r(u_i)}{2}$, in an analogous way.

With a straightforward case analysis, it is possible to observe that, since none of the conditions activating the described procedures is satisfied, there exists no crossing in the drawing, apart from possible overlaps between edges belonging to different maximal horizontal (vertical) paths whose pipes have the same bottom and top y-coordinates (the same left and right x-coordinates). However, these overlaps can be always eliminated by increasing (decreasing) of an arbitrarily small amount the coordinates of the overlapping paths (see two examples in Fig. 4(a) and 4(b)), again due to the fact that none of the conditions activating the described procedures is satisfied.   □

4   Hardness Results

In this section we prove the hardness of the ANCHORED GRAPH DRAWING problem in different settings. In particular, Theorem 2 is devoted to the hardness of the AGD-L_2-$\mathcal{S}$ problem, i.e., the problem of generating planar straight-line drawings of the input graph where the vertex regions are circles of radius δ. Theorem 3, instead, is devoted to the hardness of the AGD-L_2-$\mathcal{R}$ problem, where the regions are circles of radius δ and edges are required to be drawn as horizontal or vertical segments.

The proofs of hardness for the remaining variants of the problem listed in Table 1 can be derived from these two and, thus, will not be explained in detail. Namely, the reduction to AGD-L_1-$\mathcal{R}$ is very similar to that used for AGD-L_2-$\mathcal{R}$, and can be obtained by suitably replacing circles with diamond-shaped regions that ensure analogous geometric visibility and obstruction properties. The same holds for the hardness proof of AGD-L_∞-$\mathcal{S}$, that can be obtained from AGD-L_2-$\mathcal{S}$ with small adaptations of the gadgets. Finally, the reduction to AGD-L_1-$\mathcal{S}$ is the same as the one for AGD-L_∞-$\mathcal{S}$ where all the geometric constructions are rotated by 45°, transforming the square-shaped regions of AGD-L_∞-$\mathcal{S}$ into the diamond-shaped regions of AGD-L_1-$\mathcal{S}$.

All our proofs are based on a reduction from the NP-complete problem PLANAR 3-SATISFIABILITY [8], defined as follows. **Problem:** PLANAR 3-SATISFIABILITY (P3SAT). **Instance:** A planar bipartite graph $G = (V_v, V_c, E)$ where: (i) V_v is a set of variables; (ii) V_c is a set of clauses, each consisting of exactly three literals representing variables in V_v; and (iii) E is a set of edges connecting each variable $v \in V_v$ to all the clauses containing a literal representing v. **Question:** Does there exist a truth assignment to the variables so that each clause has at least one true literal?

For each of our problems, we describe gadgets that, given an instance ϕ of P3SAT, can be combined to construct an instance γ of the considered problem. Namely, we describe a gadget for each of the following: *variable*, *not*, *turn*, *split*, and *clause*.

The variable gadget has two families of planar drawings, corresponding to the two truth values. The not gadget admits planar drawings that invert its input truth value. The turn gadget admits planar drawings that propagate its input truth value in a direction that is orthogonal to the original one. The split gadget admits planar drawings that propagate its input truth value to two different directions. Finally, the clause gadget is planar if and only if at least one of its input literals is true. The gadgets are combined following the structure of a planar drawing of ϕ, so that any planar drawing of γ corresponds to a truth assignment for the variables satisfying ϕ. Similarly, given a truth assignment for the variables that satisfies ϕ, the gadgets for variables can be drawn accordingly to obtain a planar drawing of γ.

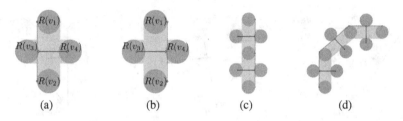

Fig. 5. *Variable* gadget for the reduction to the AGD-L_2-$\mathcal{S}$ problem in its false (a) and true (b) configurations. (c) Propagation of the true configuration of a variable gadget. (d) *Turn* gadget in its false configuration.

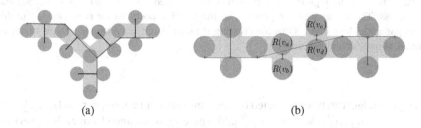

Fig. 6. (a) *Split* gadget in its true configuration. (b) *Not* gadget.

Theorem 2. AGD-L_2-$\mathcal{S}$ *is NP-hard.*

Proof: To prove hardness we reduce problem P3SAT to AGD-L_2-$\mathcal{S}$, under the hypothesis that Property A is satisfied (no overlap among vertex regions).

Let ϕ be an instance of P3SAT with n variables and m clauses. We describe how to construct an equivalent instance γ of AGD-L_2-$\mathcal{S}$. For each variable x_i, $i = 1, \ldots, n$, we create a *variable* gadget, whose two families of planar drawings are depicted in Figs. 5(a) and 5(b), consisting of four vertices v_1, v_2, v_3, and v_4 and edges (v_1, v_2) and (v_3, v_4). The regions assigned to the vertices are placed as follows: (i) the centers of $R(v_1)$ and of $R(v_2)$ lie on the same vertical line; (ii) the centers of $R(v_3)$ and of $R(v_4)$ lie on the same horizontal line; (iii) pipe $P(v_1, v_2)$ has an intersection of arbitrarily small area with both $R(v_3)$ and $R(v_4)$; and (iv) pipe $P(v_3, v_4)$ intersects neither $R(v_1)$ nor $R(v_2)$. Hence, in any anchored drawing of γ, edge (v_1, v_2) is drawn either to the left of v_3 (as in Fig. 5(a)) or to the right of v_4 (as in Fig. 5(b)). In both cases, edge (v_1, v_2) is drawn almost vertical. We call these two configurations *false* and *true* configurations for the variable gadget, respectively, and associate them with the `false` and `true` values for the corresponding variable x_i. The truth value of a variable can be propagated by concatenating a sequence of variable gadgets $\mu_1, \ldots, \mu_k$ in which $R(v_1)$ of μ_i is identified with $R(v_2)$ of μ_{i+1}, for each $i = 1, \ldots, k-1$. See Fig. 5(c).

The *turn* gadget can be constructed by concatenating three variable gadgets, μ_1, μ_2 and μ_3, as depicted in Fig. 5(d), in such a way that μ_2 has a clockwise rotation of $45°$ with respect to μ_1, and μ_3 has a clockwise rotation of $90°$ with respect to μ_1.

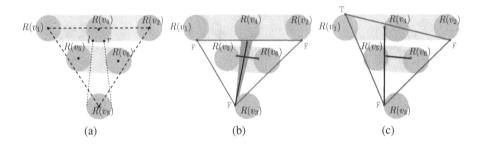

Fig. 7. Clause for the reduction to the AGD-L_2-S problem. Vertices v_1, v_2, and v_3 represent the three literals of the clause. For readability, we show only pipes $P(v_1, v_2)$ and $P(v_5, v_6)$. (a) Arrangement of the regions. (b) All literals are assigned false, and edges (v_5, v_6) and (v_3, v_4) cross. The darker wedge represents all the possible positions for edge (v_3, v_4) in this truth assignment, which implies that the crossing is unavoidable. (c) Assigning true to any of the literals allows for a planar drawing.

The *split* gadget can be constructed by combining two turn gadgets $\tau_L = \langle \mu_1^L, \mu_2^L, \mu_3^L \rangle$ and $\tau_R = \langle \mu_1^R, \mu_2^R, \mu_3^R \rangle$, with $\mu_1^L = \mu_1^R$ and where τ_L is obtained from τ_R by a vertical mirroring. See Fig. 6(a).

The *not* gadget is constructed as follows. Consider two horizontally (vertically) aligned variable gadgets μ_1 and μ_2. Add an edge connecting v_2 of μ_1 to v_1 of μ_2, as in Fig. 6(b). Place between μ_1 and μ_2 two pairs of adjacent vertices (v_a, v_b) and (v_c, v_d) whose regions are placed in such a way that: (i) any drawing of edge (v_a, v_b) blocks the visibility between the true configurations of μ_1 and μ_2; (ii) any drawing of edge (v_c, v_d) blocks the visibility between the false configurations of μ_1 and μ_2; and (iii) edges (v_a, v_b) and (v_c, v_d) can be drawn in such a way that there exists visibility between different truth configurations of μ_1 and μ_2. Hence, in any anchored drawing of γ, the configurations of μ_1 and μ_2 are different.

The *clause* gadget is constructed as follows. Refer to Fig. 7(a). Consider three vertices v_1, v_2, and v_3 whose regions are placed in such a way that their centers induce a non-degenerated triangle $\mathcal{T}$ and the centers of $R(v_1)$ and of $R(v_2)$ lie on the same horizontal line. These three vertices represent the three literals of the clause. While $R(v_2)$ and $R(v_3)$ maintain the usual convention to encode the truth value of the represented variable, for $R(v_1)$ it is inverted. It can be easily realized by negating the value of the variable. The gadget contains three more vertices: v_4, v_5, and v_6, and edges (v_1, v_2), (v_2, v_3), (v_1, v_3), (v_3, v_4), and (v_5, v_6). The center of $R(v_i)$, with $i = 4, 5, 6$, lies inside $\mathcal{T}$. Region $R(v_4)$ is completely contained in pipe $P(v_1, v_2)$, except for an arbitrarily small part Π, which lies inside $\mathcal{T}$. Consider the two points l and r in which the boundary of $R(v_4)$ intersects $P(v_1, v_2)$. The boundary of $R(v_5)$ is tangent to the leftmost segment of the convex hull H of $\{l, r\} \cup R(v_3)$. Region $R(v_6)$ completely lies to the right of the rightmost segment of H, except for an arbitrarily small part. Neither $R(v_5)$ nor $R(v_6)$ intersects $P(v_1, v_2)$.

If all the literals are set to false, then v_4 must lie below edge (v_1, v_2) (and hence in Π). However, visibility between Π and $R(v_3)$ is prevented by edge (v_5, v_6) (Fig. 7(b)).

Otherwise, if at least one of the literals is set to `true`, then an anchored drawing of γ can be realized (see Fig. 7(c) for an example). □

Theorem 3. AGD-L_2-$\mathcal{R}$ is NP-hard.

Proof sketch: To prove hardness we reduce problem P3SAT to AGD-L_2-$\mathcal{R}$, under the hypothesis that Property A is satisfied (no overlap among vertex regions). The adopted gadgets are similar to those used in the proof of Theorem 2 with the exception of the clause gadget. That gadget is based on creating three horizontal strips that are the only possible containers of a specific edge. If all the literals are `false`, then suitable edges obstruct such strips and make it not possible to construct an anchored drawing. □

5   Conclusions and Open Problems

We considered the ANCHORED GRAPH DRAWING problem in several settings, showing that, provided that the input instance do not have overlaps between vertex regions (Property A), the problem of producing planar drawings is NP-hard in most of the settings. The only exception is for the case with rectilinear drawings and uniform distances (square-shaped regions), for which a polynomial-time algorithm is provided in Section 3.

We leave open the following questions: (i) Does problem AGD belong to class NP? (ii) The instances in our NP-hardness proofs can be augmented to equivalent instances whose graphs are biconnected (we omit details for space reasons). In these instances, different truth values correspond to different embeddings. What is the complexity of AGD when the input graph is triconnected or has at least a fixed embedding? (iii) What if we allow the vertex regions to (at least partially) overlap?

References

1. Abellanas, M., Aiello, A., Hernández, G., Silveira, R.I.: Network drawing with geographical constraints on vertices. In: Actas XI Encuentros de Geom. Comput., pp. 111–118 (2005)
2. Angelini, P., Da Lozzo, G., Di Battista, G., Frati, F.: Strip planarity testing. In: Wismath, S., Wolff, A. (eds.) GD 2013. LNCS, vol. 8242, pp. 37–48. Springer, Heidelberg (2013)
3. Cabello, S., Mohar, B.: Adding one edge to planar graphs makes crossing number and 1-planarity hard. SIAM J. Comput. 42(5), 1803–1829 (2013)
4. Dumitrescu, A., Mitchell, J.S.B.: Approximation algorithms for TSP with neighborhoods in the plane. J. Algorithms 48(1), 135–159 (2003)
5. Garg, A., Tamassia, R.: On the computational complexity of upward and rectilinear planarity testing. SIAM J. Comput. 31(2), 601–625 (2001)
6. Godau, M.: On the difficulty of embedding planar graphs with inaccuracies. In: Tamassia, R., Tollis, I.G. (eds.) GD 1994. LNCS, vol. 894, pp. 254–261. Springer, Heidelberg (1995)
7. Löffler, M., van Kreveld, M.J.: Largest and smallest convex hulls for imprecise points. Algorithmica 56(2), 235–269 (2010)
8. Lichtenstein, D.: Planar formulae and their uses. SIAM J. Comput. 11, 185–225 (1982)
9. Lyons, K.A., Meijer, H., Rappaport, D.: Algorithms for cluster busting in anchored graph drawing. J. Graph Algorithms Appl. 2(1) (1998)
10. Patrignani, M.: On extending a partial straight-line drawing. International Journal of Foundations of Computer Science (IJFCS) 17(5), 1061–1069 (2006)

Advances on Testing C-Planarity
of Embedded Flat Clustered Graphs*

Markus Chimani[1], Giuseppe Di Battista[2], Fabrizio Frati[3], and Karsten Klein[3]

[1] Theoretical Computer Science, University Osnabrück, Germany
markus.chimani@uni-osnabrueck.de
[2] Dipartimento di Ingegneria, University Roma Tre, Italy
gdb@dia.uniroma3.it
[3] School of Information Technologies, The University of Sydney, Australia
{fabrizio.frati,karsten.klein}@sydney.edu.au

Abstract. We show a polynomial-time algorithm for testing c-planarity of embedded flat clustered graphs with at most two vertices per cluster on each face.

1 Introduction

A *clustered graph* $C(G, T)$ consists of a graph $G(V, E)$, called *underlying graph*, and of a rooted tree T, called *inclusion tree*, representing a cluster hierarchy on V. The vertices in V are the leaves of T, and the inner nodes of T, except for the root, are called *clusters*. The vertices that are descendants of a cluster α in T *belong to* α or *are in* α. A *c-planar drawing* of C is a planar drawing of G together with a representation of each cluster α as a simple connected region R_α enclosing all and only the vertices that are in α; further, the boundaries of no two such regions R_α and R_β intersect; finally, only the edges connecting vertices in α to vertices not in α cross the boundary of R_α, and each does so only once. A clustered graph is *c-planar* if it admits a c-planar drawing.

Clustered graphs find numerous applications in computer science [22], thus theoretical questions on clustered graphs have been deeply investigated. From the visualization perspective, the most intriguing question is to determine the complexity of testing c-planarity of clustered graphs. Unlike for other planarity variants [21], like *upward planarity* [14] and *partial embedding planarity* [1], the complexity of testing c-planarity remains unknown since the problem was posed nearly two decades ago [13].

Polynomial-time algorithms to test the c-planarity of a clustered graph C are known if C belongs to special classes of clustered graphs [7–11, 13, 15, 16, 18, 19], including *c-connected clustered graphs*, that are clustered graphs $C(G, T)$ in which, for each cluster α, the subgraph $G[\alpha]$ of G induced by the vertices in α is connected [8, 10, 13]. Effective ILP formulations and FPT algorithms for testing c-planarity have been presented [5, 6]. Generalizations of the c-planarity testing problem have also been considered [2, 12].

An important variant of the c-planarity testing problem is the one in which the clustered graph $C(G, T)$ is *flat* and *embedded*. That is, every cluster is a child of the root of T and a planar embedding for G (an order of the edges incident to each vertex) is fixed

---

* Research partially supported by the Australian Research Council (grant DE140100708).

in advance; then, the c-planarity testing problem asks whether a c-planar drawing exists in which G has the prescribed planar embedding. This setting can be highly regarded for several reasons. First, several NP-hard graph drawing problems are polynomial-time solvable in the fixed embedding scenario, e.g., *upward planarity testing* [3, 14] and *bend minimization in orthogonal drawings* [14, 23]. Second, testing c-planarity of embedded flat clustered graphs generalizes testing c-planarity of triconnected flat clustered graphs. Third, testing c-planarity of embedded flat clustered graphs is strongly related to a seemingly different problem, that we call *planar set of spanning trees in topological multigraphs* (PSSTTM): Given a non-planar topological multigraph A with k connected components $A_1, \ldots, A_k$, do spanning trees $S_1, \ldots, S_k$ of $A_1, \ldots, A_k$ exist such that no two edges in $\bigcup_i S_i$ cross? Starting from an embedded flat clustered graph $C(G, T)$, an instance A of the PSSTTM problem can be constructed that admits a solution if and only if $C(G, T)$ is c-planar: A is composed of the edges that can be inserted inside the faces of G between vertices of the same cluster, where each cluster defines a multigraph A_i. The PSSTTM problem is NP-hard, even if $k = 1$ [20].

Testing c-planarity of an embedded flat clustered graph $C(G, T)$ is a polynomial-time solvable problem if G has no face with more than five vertices and, more in general, if C is a *single-conflict* clustered graph [11], i.e., the instance A of the PSSTTM problem associated with C is such that each edge has at most one crossing. A polynomial-time algorithm is also known for testing c-planarity of embedded flat clustered graphs such that the graph induced by each cluster has at most two connected components [17]. Finally, the c-planarity of clustered cycles with at most three clusters [9] or with each cluster containing at most three vertices [19] can be tested in polynomial time.

Our Contribution. In this paper we show how to test c-planarity in polynomial time for embedded flat clustered graphs $C(G, T)$ such that at most two vertices of each cluster are incident to any face of G. While this setting might seem unnatural at a first glance, its study led to a deep (in our opinion) exploration of some combinatorial properties of highly non-planar topological graphs. Namely, every instance A of the PSSTTM problem arising from our setting is such that there exists no sequence $e_1, e_2, \ldots, e_h$ of edges in A with e_1 and e_h in the same connected component of A and with e_i crossing e_{i+1}, for every $1 \leq i \leq h - 1$; these instances might contain a quadratic number of crossings, which is not the case for single-conflict clustered graphs [11]. Within our setting, performing all the "trivial local" tests and simplifications results in the rise of nice global structures, called α-*donuts*, whose study was interesting to us.

Refer to the full version of the paper [4] for complete proofs.

2 Saturators, Con-Edges, and Spanning Trees

A natural approach to test c-planarity of a clustered graph $C(G(V, E), T)$ is to search for a *saturator* for C. A set $S \subseteq V \times V$ is a saturator for C if $C'(G'(V, E \cup S), T)$ is a c-connected c-planar clustered graph. Determining the existence of a saturator for C is equivalent to testing the c-planarity of C [13]. Thus, the core of the problem consists of determining S so that $G'[\alpha]$ is connected, for each $\alpha \in T$, and so that G' is planar. For embedded flat clustered graphs (see Fig. 1(a)), the problem of finding saturators

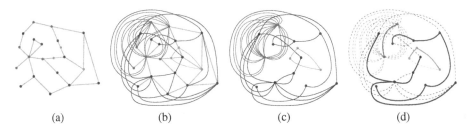

Fig. 1. (a) A clustered graph C. (b) Con-edges in C. (c) Multigraph A. (d) A planar set S of spanning trees for A. Edges in S are thick and solid, while edges in $A \setminus S$ are thin and dashed.

becomes seemingly simpler. Since the embedding of G is fixed and since G' has to be planar, edges in S can only be embedded inside faces of G. This implies that, for any two edges e_1 and e_2 that can be inserted inside a face f of G, it is known *a priori* whether e_1 and e_2 can be both in S, namely only if their end-vertices do not alternate along the boundary of f. Also, S can be assumed to contain only edges between vertices in distinct connected components of $G[\alpha]$, for each cluster α, as other types of edges do not help to connect any cluster.

Consider a face f of G and let $(o_1, \ldots, o_k)$ be the clockwise order of the occurrences of vertices along the boundary of f, where o_i and o_j might be occurrences of the same vertex u (this might happen if u is a cut-vertex of G). A *con-edge* (short for *connectivity-edge*) is a pair of occurrences (o_i, o_j) of distinct vertices both belonging to a cluster α, both incident to f, and belonging to different connected components of $G[\alpha]$ (see Fig. 1(b)). If there are ℓ distinct pairs of occurrences of vertices u and v along a single face f, then there are ℓ con-edges connecting u and v in f, one for each pair of occurrences. A *con-edge for* α is a con-edge connecting vertices in a cluster α. Two con-edges e and e' in f *have a conflict* or *cross* (we write $e \otimes e'$) if the occurrences in e alternate with the occurrences in e' along the boundary of f.

The *multigraph A of the con-edges* is an embedded multigraph that is defined as follows. Starting from G, insert all the con-edges inside the faces of G; then, for each cluster α and for each connected component $G_i[\alpha]$ of $G[\alpha]$, contract $G_i[\alpha]$ into a single vertex; finally, remove all the edges of G. See Fig. 1(c). With a slight abuse of notation, we denote by A both the multigraph of the con-edges and the set of its edges. For each cluster α, we denote by $A[\alpha]$ the subgraph of A induced by the con-edges for α. A *planar set of spanning trees for A* is a set $S \subseteq A$ such that: (i) for each cluster α, the subset $S[\alpha]$ of S induced by the con-edges for α is a tree that spans the vertices belonging to α; and (ii) there exist no two edges in S that have a conflict. See Fig. 1(d). The PSSTTM problem asks whether a planar set of spanning trees for A exists.

The following lemma relates the c-planarity problem for embedded flat clustered graphs to the PSSTTM problem.

Lemma 1 (**[11]**). *An embedded flat clustered graph $C(G, T)$ is c-planar if and only if: (1) G is planar; (2) there exists a face f in G such that when f is chosen as outer face for G no cycle composed of vertices of the same cluster encloses a vertex of a different cluster; and (3) a planar set of spanning trees for A exists.*

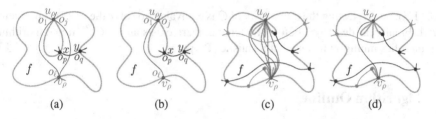

Fig. 2. Illustration for the reduction to a multigraph of the con-edges satisfying Property 1

We now introduce the concept of *conflict graph* K_A, which is defined as follows. Graph K_A has a vertex for each con-edge in A and has an edge (e, e') if $e \otimes e'$. In the remainder of the paper we will show how to decide whether a set of planar spanning trees for A exists by assuming that the following property holds for A.

Property 1. *No two con-edges for the same cluster belong to the same connected component of K_A.*

We now show that A can be assumed to satisfy Property 1, since $C(G, T)$ has at most two vertices per cluster on each face. Consider any face f of G and any cluster ϱ that has two vertices u_ϱ and v_ϱ incident to f. No con-edge for ϱ that connects a pair of vertices different from (u_ϱ, v_ϱ) is in the connected component of K_A containing (u_ϱ, v_ϱ), given that no vertex of ϱ different from u_ϱ and v_ϱ is incident to f. However, it might be the case that several con-edges (u_ϱ, v_ϱ) are in the same connected component of K_A, which happens if u_ϱ, or v_ϱ, or both have several occurrences on the boundary of f. We show a simple reduction that gets rid of these multiple con-edges.

Let $(o_1, \ldots, o_k)$ be the clockwise order of the occurrences of vertices along f. Let o_1, o_j, and o_ℓ be occurrences of u_ϱ, u_ϱ, and v_ϱ, respectively, with $1 < j < \ell \le k$. Suppose that o_p and o_q are occurrences of vertices x and y in a cluster $\tau \ne \varrho$, for some $1 < p < j < q < \ell$, as in Fig. 2(a). Then, all the con-edges (x, y) have a conflict with con-edge $e_\varrho = (o_j, o_\ell)$; moreover, the con-edges (x, y) form a separating set for $A[\tau]$, hence any planar set S of spanning trees for A contains one of them. Thus, $e_\varrho \notin S$ and e_ϱ can be removed from A, as in Fig. 2(b). Similar reductions can be performed if $\ell < q \le k$ and by exchanging the roles of u_ϱ and v_ϱ. If no two occurrences o_p and o_q as above exist, then all the con-edges (u_ϱ, v_ϱ) left cross the same set of con-edges for clusters different from ϱ (see Fig. 2(c)). Hence, a single edge (u_ϱ, v_ϱ) can be kept in A, and all the other con-edges (u_ϱ, v_ϱ) can be removed from A (see Fig. 2(d)). After repeating this reduction for all the con-edges in A, an equivalent instance A is eventually obtained in which Property 1 is satisfied. Observe that the described simplification can be easily performed in $O(|C|^2)$ time. Thus, we get the following:

Lemma 2. *Assume that the PSSTTM problem can be solved in $f(|A|)$ time for instances satisfying Property 1. Then the c-planarity of any embedded flat clustered graph C with at most two vertices per cluster on each face can be tested in $O(f(|A|) + |C|^2)$ time.*

Proof. Consider any embedded flat clustered graph C with at most two vertices per cluster on each face. Conditions (1) and (2) in Lemma 1 can be tested in $O(|C|)$ time (see [11]); hence, testing the c-planarity of C is equivalent to solve the PSSTTM problem for A. Finally, as described before the lemma, there exists an $O(|C|^2)$-time algorithm that modifies multigraph A so that it satisfies Property 1.   □

3   Algorithm Outline

In this section we give an outline of our algorithm for testing the existence of a planar set S of spanning trees for A. We assume that A satisfies Property 1.

Our algorithm repeatedly tries to detect certain substructures in A. When it does find one of such substructures, the algorithm either "simplifies" A or concludes that A does not admit any planar set of spanning trees. For example, if a cluster α exists such that $A[\alpha]$ is not connected, then the algorithm concludes that no planar set of spanning trees exists and terminates; as another example, if conflicting con-edges e_α and e_β for clusters α and β exist in A such that e_α is a bridge for $A[\alpha]$, then the algorithm determines that e_α has to be in S and that e_β can be assumed not to be in S.

If the algorithm determines that certain edges have to be in S or can be assumed not to be in S, these edges are contracted or removed, respectively. Given a set $A' \subseteq A$, the operation of *removing* A' from A consists of updating $A := A \setminus A'$. Given a set $A' \subseteq A$, the operation of *contracting* the edges in A' consists of identifying the end-vertices of each con-edge e in A' (all the con-edges different from e and incident to the end-vertices of e remain in A), and of updating $A := A \setminus A'$.

Edges are removed or contracted only when this does not alter the possibility of finding a planar set of spanning trees for A. Also, contractions are only applied to con-edges that cross no other con-edges; hence, after any contraction, graph K_A only changes for the removal of the isolated vertices corresponding to the contracted edges.

As a consequence of a removal or of a contraction operation, the number of edges in A decreases, that is, A is "simplified". After any simplification due to the detection of a certain substructure in A, the algorithm will run again all previous tests for the detection of the other substructures. In fact, it is possible that a certain substructure arises from performing a simplification on A (e.g., a bridge might be present in A after a set of edges has been removed from A). Since detecting each substructure that leads to a simplification in A can be performed in quadratic time, and since the initial size of A is linear in the size of C, the algorithm has a cubic running time.

If none of the four tests (called TEST 1–4) and none of the eight simplifications (called SIMPLIFICATION 1–8) described in Section 4 applies to A, then A is a *single-conflict* multigraph. That is, each con-edge in A crosses at most one con-edge in A. A linear-time algorithm for deciding the existence of a planar set of spanning trees in a single-conflict multigraph A is known [11]. Hence, our algorithm uses that algorithm [11] to conclude the test of the existence of a planar set of spanning trees in A.

4   Algorithm

To ease the reading and avoid text duplication, when introducing a new lemma we always assume, without making it explicit, that all the previously defined simplifications do not apply, and that all the previously defined tests fail. Also, we do not make explicit the removal and contraction operations that we perform, as they straight-forwardly follow from the statement of each lemma. We start with the following test.

Lemma 3 (TEST 1). *Let α be a cluster such that $A[\alpha]$ is disconnected. Then, there exists no planar set S of spanning trees for A.*

Proof. No set $S \subseteq A$ is such that $S[\alpha]$ induces a tree spanning the vertices in α.   □

We continue with the following simplification.

Lemma 4 (SIMPLIFICATION 1). *Let e be a bridge of $A[\alpha]$. Then, for every planar set S of spanning trees for A, we have $e \in S$.*

Proof. Graph $A[\alpha] \setminus \{e\}$ is disconnected; hence, by Lemma 3, no planar set of spanning trees for A exists with $e \notin S$.   □

The following lemma is used massively in the remainder of the paper.

Lemma 5. *Let $e_\alpha, e_\beta \in A$ be con-edges such that $e_\alpha \otimes e_\beta$. Let S be a planar set of spanning trees for A and suppose that $e_\alpha \notin S$. Then, $e_\beta \in S$.*

Proof sketch. If S contains neither e_α nor e_β, then the two paths in S connecting the end-vertices of e_α and connecting the end-vertices of e_β cross, a contradiction.   □

The algorithm continues with the following test.

Lemma 6 (TEST 2). *If the conflict graph K_A is not bipartite, then there exists no planar set S of spanning trees for A.*

Proof sketch. If an odd cycle $\mathcal{C}$ exists in K_A, then by repeated applications of Lemma 5 and of the fact that S does not contain two conflicting edges, we get that any edge of $\mathcal{C}$ simultaneously should be in S and should not be in S, a contradiction. □

The contraction of con-edges chosen to be in S might lead to self-loops in A, a situation that is handled in the following.

Lemma 7 (SIMPLIFICATION 2). *Let $e \in A$ be a self-loop. Then, for every planar set S of spanning trees for A, we have $e \notin S$.*

Proof. Since a tree does not contain any self-loop, the lemma follows.   □

Con-edges that do not cross any other con-edge can be safely chosen to be in S.

Lemma 8 (SIMPLIFICATION 3). *Let e be any con-edge in A that does not have a conflict with any other con-edge in A. Then, there exists a planar set S of spanning trees for A if and only if there exists a planar set S' of spanning trees for A such that $e \in S'$.*

Proof sketch. Let S be any planar set of spanning trees for A. If $e \notin S$, then $S \cup \{e\}$ contains a cycle $\mathcal{C}$ of con-edges for the same cluster. Let e' be any edge of $\mathcal{C}$ different from e. Then, $S' = S \cup \{e\} \setminus \{e'\}$ is a planar set of spanning trees for A.   □

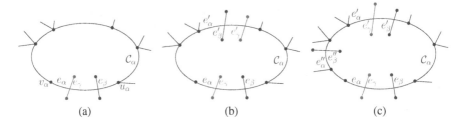

Fig. 3. The setting for (a) Lemma 9, (b) Lemma 10, and (c) Lemma 11

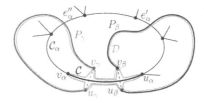

Fig. 4. Illustration for the proof of Lemma 9

In the next three lemmata we deal with the following setting. Assume that there exist con-edges $e_\alpha, e_\beta, e_\gamma \in A$ for distinct clusters α, β, and γ, respectively, such that $e_\alpha \otimes e_\beta$ and $e_\alpha \otimes e_\gamma$. Since TEST 2 fails on A, e_β does not cross e_γ. Let $\mathcal{C}_\alpha$ be any of the two facial cycles of $A[\alpha]$ incident to e_α, where a facial cycle of $A[\alpha]$ is a simple cycle all of whose edges appear on the boundary of a single face of $A[\alpha]$. Assume w.l.o.g. that e_α is crossed first by e_β and then by e_γ when $\mathcal{C}_\alpha$ is traversed clockwise. See Fig. 3(a). The next lemma presents a condition in which we can delete e_α from A.

Lemma 9 (SIMPLIFICATION 4). *Suppose that there exists no con-edge of $\mathcal{C}_\alpha$ different from e_α that has a conflict with both a con-edge for β and a con-edge for γ. Then, for every planar set S of spanning trees for A, we have $e_\alpha \notin S$.*

Proof sketch. See Fig. 4. Let u_α and v_α (u_β and v_β, u_γ and v_γ) be the end-vertices of e_α (resp. of e_β, of e_γ). By Property 1, a closed simple curve $\mathcal{C}$ can be drawn through u_α, u_β, u_γ, v_α, v_γ, and v_β, with e_α, e_β, and e_γ in its interior and every other con-edge for α, β, and γ in its exterior. If $e_\alpha \in S$, then $e_\beta, e_\gamma \notin S$. Then, the path P_β in S connecting u_β and v_β and the path P_γ in S connecting u_γ and v_γ cross $\mathcal{C}_\alpha$ at different edges e'_α and e''_α. Hence, the end-vertices of e'_α are on different sides of the cycle $\mathcal{D}$ composed of P_β, of P_γ, and of the paths in $\mathcal{C}$ between u_β and u_γ and between v_β and v_γ. However, no con-edge for α in S crosses $\mathcal{D}$, hence S does not connect α.   □

The next two lemmata state conditions in which no planar set of spanning trees for A exists. Their statements are illustrated in Figs. 3(b) and 3(c), respectively; further, they can be proved with arguments similar to the ones in the proof of Lemma 9.

Lemma 10 (TEST 3). *Suppose that there exist con-edges $e'_\alpha, e'_\beta, e'_\gamma \in A$ for clusters α, β, and γ, respectively, such that $e'_\alpha \neq e_\alpha$, e'_α belongs to $\mathcal{C}_\alpha$, and $e'_\alpha \otimes e'_\beta$ as well*

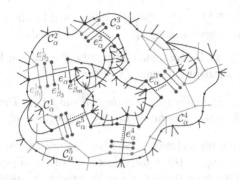

Fig. 5. A partial representation of the α-donut for e_α

as $e'_\alpha \otimes e'_\gamma$. Assume that e'_α is crossed first by e'_β and then by e'_γ when C_α is traversed clockwise. Then, no planar set of spanning trees for A exists.

Lemma 11 (TEST 4). *Suppose that con-edges $e'_\alpha, e''_\alpha \in A$ for α exist in C_α, and such that e_α, e''_α, and e'_α occur in this order along C_α, when clockwise traversing C_α. Suppose also that there exist con-edges $e'_\beta, e''_\beta \in A$ for β and $e'_\gamma \in A$ for γ such that $e'_\alpha \otimes e'_\beta$, $e'_\alpha \otimes e'_\gamma$, and $e''_\alpha \otimes e''_\beta$. Then, no planar set of spanning trees for A exists.*

If SIMPLIFICATIONS 1–4 do not apply to A and TESTS 1–4 fail on A, then the con-edges for a cluster α that are crossed by con-edges for (at least) two other clusters have a nice structure, that we call α-*donut* (see Fig. 5). Consider a con-edge $e_\alpha \in A$ for α crossing con-edges $e_{\beta_1}, \ldots, e_{\beta_m}$ for clusters $\beta_1, \ldots, \beta_m$, with $m \geq 2$. An α-donut for e_α consists of a sequence $e^1_\alpha, \ldots, e^k_\alpha, e^{k+1}_\alpha$ of con-edges for α with $k \geq 2$, called *spokes* of the α-donut, of a sequence $C^1_\alpha, \ldots, C^k_\alpha, C^{k+1}_\alpha = C^1_\alpha$ of facial cycles in $A[\alpha]$, and of sequences $e^1_{\beta_j}, \ldots, e^k_{\beta_j}$ of con-edges for β_j, for each $1 \leq j \leq m$, such that the following hold for every $1 \leq i \leq k$: (a) $e_\alpha = e^1_\alpha$; (b) $e^i_\alpha \otimes e^i_{\beta_j}$, for every $1 \leq j \leq m$; (c) C^i_α and C^{i+1}_α share edge e^i_α; (d) edge e^i_α is crossed by $e^i_{\beta_1}, \ldots, e^i_{\beta_m}$ in this order when C^i_α is traversed clockwise; (e) all the con-edges of C^{i+1}_α encountered when clockwise traversing C^{i+1}_α from e^i_α to e^{i+1}_α do not cross any con-edge for $\beta_2, \ldots, \beta_m$; and (f) all the con-edges of C^{i+1}_α encountered when clockwise traversing C^{i+1}_α from e^{i+1}_α to e^i_α do not cross any con-edge for $\beta_1, \ldots, \beta_{m-1}$. We have the following.

Lemma 12. *For every con-edge $e_\alpha \in A$ for α, there exists an α-donut for e_α.*

Proof sketch. Let $e^1_\alpha = e_\alpha$ and let C^1_α and C^2_α be the facial cycles incident to e^1_α. Since SIMPLIFICATION 1 does not apply to A, $C^1_\alpha \neq C^2_\alpha$. Let $e^1_{\beta_1}, \ldots, e^1_{\beta_m}$ be the con-edges for $\beta_1, \ldots, \beta_m$, respectively, ordered as they cross e^1_α when clockwise traversing C^1_α. Since SIMPLIFICATION 4 does not apply to A and TESTS 3-4 fail on A, a con-edge e^2_α exists in C^2_α that is crossed by con-edges $e^2_{\beta_1}, \ldots, e^2_{\beta_m}$ for $\beta_1, \ldots, \beta_m$, respectively, in this order when clockwise traversing C^2_α; further, all the con-edges of C^2_α encountered when clockwise traversing C^2_α from e^1_α to e^2_α (from e^2_α to e^1_α) do not cross any con-edges for $\beta_2, \ldots, \beta_m$ (resp. for $\beta_1, \ldots, \beta_{m-1}$). This argument is repeated for $i = 3, \ldots, k$ to determine a facial cycle C^i_α containing e^{i-1}_α and to determine edges $e^i_\alpha, e^i_{\beta_1}, \ldots, e^i_{\beta_m}$.

Eventually, facial cycle $\mathcal{C}_\alpha^{k+1} = \mathcal{C}_\alpha^1$ of $A[\alpha]$ is considered in which the two con-edges that are crossed by con-edges for all of $\beta_1, \ldots, \beta_m$ are e_α^k and e_α^1. □

The α-donut for e_α can be computed efficiently. Further, we have the following lemma, whose proof is similar to the one of Lemma 9.

Lemma 13. *Let $e_\alpha^1, \ldots, e_\alpha^k$ be the spokes of the α-donut for e_α. Then, if a planar set S of spanning trees for A exists, it contains exactly one of $e_\alpha^1, \ldots, e_\alpha^k$.*

Consider a con-edge e for a cluster α. The *conflicting structure* $M(e)$ of e is a sequence of sets $H_0(e), L_1(e), H_1(e), L_2(e), H_2(e), \ldots$ of con-edges which correspond to the layers of a BFS traversal starting at e of the connected component of K_A containing e. That is: $H_0(e) = \{e\}$; then, for $i \geq 1$, $L_i(e)$ is the set of con-edges that cross con-edges in $H_{i-1}(e)$ and that are not in $L_{i-1}(e)$, and $H_i(e)$ is the set of con-edges that cross con-edges in $L_i(e)$ and that are not in $H_{i-1}(e)$.

We now study the conflicting structures of the spokes $e_\alpha^1, \ldots, e_\alpha^k$ of the α-donut of a con-edge e_α for α. No two edges in a set $H_i(e_\alpha)$ or in a set $L_i(e_\alpha)$ have a conflict, as otherwise TEST 2 would succeed. Also, by Lemma 5, any planar set S of spanning trees for A contains either all the edges in $\bigcup_i H_i(e_\alpha)$ or all the edges in $\bigcup_i L_i(e_\alpha)$.

Assume that e_α has a conflict with at least two con-edges for other clusters. For any $1 \leq i \leq k$, we say that e_α^i and e_α^{i+1} have *isomorphic conflicting structures* if they belong to isomorphic connected components of K_A and if the vertices of these components that are in correspondence under the isomorphism represent con-edges for the same cluster. Formally, e_α^i and e_α^{i+1} have isomorphic conflicting structures if there exists a bijective mapping δ from the edges in $M(e_\alpha^i)$ to the edges in $M(e_\alpha^{i+1})$ such that: (1) e is a con-edge for a cluster ϱ if and only if $\delta(e)$ is a con-edge for ϱ, for every $e \in M(e_\alpha^i)$; (2) $e \in H_j(e_\alpha^i)$ if and only if $\delta(e) \in H_j(e_\alpha^{i+1})$, for every $e \in M(e_\alpha^i)$; (3) $e \in L_j(e_\alpha^i)$ if and only if $\delta(e) \in L_j(e_\alpha^{i+1})$, for every $e \in M(e_\alpha^i)$; and (4) $e \otimes f$ if and only if $\delta(e) \otimes \delta(f)$, for every $e, f \in M(e_\alpha^i)$. Observe that the isomorphism of two conflicting structures can be tested efficiently.

We will prove in the following four lemmata that by examining the conflicting structures for the spokes of the α-donut for e_α, a decision on whether some spoke is or is not in S can be taken without loss of generality. We start with the following:

Lemma 14 (SIMPLIFICATION 5). *Suppose that spokes e_α^i and e_α^{i+1} have isomorphic conflicting structures. Then, there exists a planar set S of spanning trees for A if and only if there exists a planar set S' of spanning trees for A such that $e_\alpha^i \notin S'$.*

Proof sketch. Suppose that a planar set S of spanning trees for A exists with $e_\alpha^i \in S$. By Lemma 5, $\bigcup_j H_j(e_\alpha^i) \subseteq S$ and $S \cap \bigcup_j L_j(e_\alpha^i) = \emptyset$. By Lemma 13, $e_\alpha^{i+1} \notin S$, hence $\bigcup_j L_j(e_\alpha^{i+1}) \subseteq S$ and $S \cap \bigcup_j H_j(e_\alpha^{i+1}) = \emptyset$. Let S' be the set of con-edges obtained from S by removing $\bigcup_j H_j(e_\alpha^i)$ and $\bigcup_j L_j(e_\alpha^{i+1})$ and by adding $\bigcup_j L_j(e_\alpha^i)$ and $\bigcup_j H_j(e_\alpha^{i+1})$. The lemma follows from the claim that S' is a planar set of spanning trees for A. The proof for this claim consists of two parts. In the first one, it is shown that no two con-edges in S' cross, by exploiting the absence of crossings in S and the properties of $M(e_\alpha^i)$ and $M(e_\alpha^{i+1})$. In the second one, it is shown that, for each cluster μ, the graph induced by the con-edges in $S'[\mu]$ is a tree that spans the vertices in μ; this proof uses topological arguments to establish that the only con-edge for μ in $S' \setminus S$ has its end-vertices in different connected components of the graph obtained from $S[\mu]$ by removing the only con-edge for μ in $S \setminus S'$. $\qquad\square$

Next, we study non-isomorphic spokes. Let e^i_α be a spoke of the α-donut for e_α. Assume that $L_1(e^i_\alpha)$ contains a con-edge e^i_β for a cluster β, and that $H_1(e^i_\alpha)$ contains a con-edge e^i_γ for a cluster γ, where $e^i_\alpha \otimes e^i_\beta$ and $e^i_\beta \otimes e^i_\gamma$. By Property 1, since e^i_γ and e^i_α belong to the same connected component of K_A and do not cross (as otherwise TEST 2 would succeed), it follows that e^i_γ does not cross any con-edge for α, hence it lies in one of the two faces f^i_α and f^{i+1}_α of $A[\alpha]$ that e^i_α shares with spokes e^{i-1}_α and e^{i+1}_α, respectively. Assume w.l.o.g. that e^i_γ lies in f^{i+1}_α. By Lemma 12, $L_1(e^{i+1}_\alpha)$ contains a con-edge e^{i+1}_β for β, where $e^{i+1}_\alpha \otimes e^{i+1}_\beta$.

The next two lemmata discuss the case in which $M(e^{i+1}_\alpha)$ contains a con-edge for γ that has a conflict with e^{i+1}_β and the case in which it does not. We start with the latter.

Lemma 15 (SIMPLIFICATION 6). *Suppose that no con-edge e^{i+1}_γ for γ exists such that $e^{i+1}_\gamma \otimes e^{i+1}_\beta$, and that a planar set S of spanning trees for A exists. Then, $e^i_\alpha \in S$.*

Proof sketch. If a planar set S of spanning trees for A exists with $e^i_\alpha \notin S$, then by Lemma 5 we have $e^i_\beta \in S$ and $e^i_\gamma \notin S$. Then, the paths P^i_α and P^i_γ connecting the end-vertices of e^i_α and e^i_γ, together with a closed simple curve $\mathcal{C}$ surrounding e^i_α, e^i_β, and e^i_γ form a closed curve $\mathcal{D}$ that contains vertices in β on both sides. However, $\mathcal{D}$ cannot be crossed by any con-edge for β in S, thus S does not connect β. $\qquad\square$

Lemma 16 (SIMPLIFICATION 7). *Suppose that a con-edge e^{i+1}_γ for γ exists with $e^{i+1}_\gamma \otimes e^{i+1}_\beta$. If a planar set S of spanning trees for A exists, then either $e^i_\alpha \in S$ or $e^{i+1}_\alpha \in S$.*

Proof sketch. By Lemma 13, at most one out of e^i_α and e^{i+1}_α belongs to S. To prove that at least one out of e^i_α and e^{i+1}_α belongs to S, by Lemma 5 it suffices to prove that at most one out of e^i_β and e^{i+1}_β belongs to S. This is accomplished again by Lemma 13 and by proving that e^{i+1}_β is a spoke of the β-donut for e^i_β. $\qquad\square$

Observe that Simplification 7 can be applied in the case in which the α-donut for e_α has at least three spokes. Namely, in that case, by Lemmata 13 and 16 all the spokes different from e^i_α and e^{i+1}_α can be removed from A.

Next, assume that there exists an α-donut with exactly two spokes e^1_α and e^2_α. Consider the smallest $j \geq 1$ such that one of the following holds:

(1) there exist con-edges $e_\mu \in L_j(e^a_\alpha)$ and $e_\nu \in H_{j-1}(e^a_\alpha)$ for clusters μ and ν, resp., such that $e_\mu \otimes e_\nu$, and there exists no con-edge $g_\mu \in L_j(e^b_\alpha)$ for μ such that $g_\mu \otimes g_\nu$ with g_ν con-edge for ν in $H_{j-1}(e^b_\alpha)$, for some $a, b \in \{1, 2\}$ with $a \neq b$; or

(2) there exist con-edges $e_\mu \in H_j(e^a_\alpha)$ and $e_\nu \in L_j(e^a_\alpha)$ for clusters μ and ν, resp., such that $e_\mu \otimes e_\nu$, and there exists no con-edge $g_\mu \in H_j(e^b_\alpha)$ for μ such that $g_\mu \otimes g_\nu$ with g_ν con-edge for ν in $L_j(e^b_\alpha)$, for some $a, b \in \{1, 2\}$ with $a \neq b$. We have the following.

Lemma 17 (SIMPLIFICATION 8). *Assume that a planar set S of spanning trees for A exists. Then, $e_\mu \in S$.*

Proof sketch. The proof uses topological arguments to establish the following claim: If $e_\mu \notin S$, then the end-vertices of e_μ are on diffent sides of a cycle composed of con-edges for ν that cannot be crossed by con-edges for μ in S, hence S does not connect μ, a contradiction.   □

We now prove that our simplifications form a "complete set".

Lemma 18. *Suppose that* SIMPLIFICATIONS *1–8 do not apply to* A *and that* TESTS *1–4 fail on* A. *Then, every con-edge in* A *crosses exactly one con-edge in* A.

Proof sketch. Since SIMPLIFICATIONS 2-3 do not apply to A, every con-edge crosses at least one con-edge. Suppose, for a contradiction, that there exists a con-edge for a cluster α that has a conflict with at least two con-edges. Since SIMPLIFICATIONS 1–4 do not apply to A and TESTS 1–4 fail on A, by Lemma 12 there exists an α-donut with spokes $e_\alpha^1, \ldots, e_\alpha^k$. If the conflicting structures of e_α^1 and e_α^2 are isomorphic, then SIMPLIFICATION 5 applies to A. Otherwise, if $k \geq 3$ and the conflicting structures of e_α^1 and e_α^2 are isomorphic (not isomorphic) when restricted to sets $H_0(e_\alpha^j)$, $L_1(e_\alpha^j)$, and $H_1(e_\alpha^j)$, then SIMPLIFICATION 7 (resp. SIMPLIFICATION 6) applies to A. If $k = 2$ and the conflicting structures of e_α^1 and e_α^2 are not isomorphic, then SIMPLIFICATION 8 applies to A. This provides a contradiction.   □

A linear-time algorithm to determine whether a planar set S of spanning trees exists for a single-conflict graph is known [11]. We thus finally get:

Theorem 1. *There exists an* $O(|C|^3)$*-time algorithm to test the c-planarity of an embedded flat clustered graph* C *with at most two vertices per cluster on each face.*

Proof. Multigraph A can be easily constructed in $O(|C|^2)$ time, so that A has $O(|C|)$ vertices and edges and satisfies Property 1. By Lemma 2, it suffices to show how to solve the PSSTTM problem for A in $O(|C|^3)$ time. Algorithm 1 correctly determines whether a planar set S of spanning trees for A exists, by Lemmata 3–18. It can be easily tested in $O(|A|^2)$ time whether the pre-conditions of each of SIMPLIFICATIONS 1–8 and TESTS 1–4 are satisfied; also, the actual simplifications, that is, removing and contracting edges in A, can be performed in $O(|A|)$ time. Furthermore, the algorithm in [11] runs in $O(|A|)$ time. Since the number of performed tests and simplifications is in $O(|A|)$, the total running time is in $O(|A|^3)$, and hence in $O(|C|^3)$.   □

5   Conclusions

We presented a polynomial-time algorithm for testing c-planarity of embedded flat clustered graphs with at most two vertices per cluster on each face. An interesting extension of our results would be to devise an FPT algorithm to test c-planarity of embedded flat clustered graphs, where the parameter is the maximum number k of vertices of the same cluster on any face. Several key lemmata (e.g. Lemmata 5 and 6) do not apply if $k > 2$, hence even an algorithm with running time $n^{O(f(k))}$ seems to be an elusive goal.

References

1. Angelini, P., Di Battista, G., Frati, F., Jelínek, V., Kratochvíl, J., Patrignani, M., Rutter, I.: Testing planarity of partially embedded graphs. In: SODA 2010, pp. 202–221. ACM (2010)

2. Angelini, P., Frati, F., Patrignani, M.: Splitting clusters to get C-planarity. In: Eppstein, D., Gansner, E.R. (eds.) GD 2009. LNCS, vol. 5849, pp. 57–68. Springer, Heidelberg (2010)
3. Bertolazzi, P., Di Battista, G., Liotta, G., Mannino, C.: Upward drawings of triconnected digraphs. Algorithmica 12(6), 476–497 (1994)
4. Chimani, M., Di Battista, G., Frati, F., Klein, K.: Advances on testing c-planarity of embedded flat clustered graphs. CoRR, abs/1408.2595 (2014)
5. Chimani, M., Gutwenger, C., Jansen, M., Klein, K., Mutzel, P.: Computing maximum c-planar subgraphs. In: Tollis, I.G., Patrignani, M. (eds.) GD 2008. LNCS, vol. 5417, pp. 114–120. Springer, Heidelberg (2009)
6. Chimani, M., Klein, K.: Shrinking the search space for clustered planarity. In: Didimo, W., Patrignani, M. (eds.) GD 2012. LNCS, vol. 7704, pp. 90–101. Springer, Heidelberg (2013)
7. Cornelsen, S., Wagner, D.: Completely connected clustered graphs. J. Discrete Algorithms 4(2), 313–323 (2006)
8. Cortese, P.F., Di Battista, G., Frati, F., Patrignani, M., Pizzonia, M.: C-planarity of c-connected clustered graphs. J. Graph Algorithms Appl. 12(2), 225–262 (2008)
9. Cortese, P.F., Di Battista, G., Patrignani, M., Pizzonia, M.: Clustering cycles into cycles of clusters. J. Graph Alg. Appl. 9(3), 391–413 (2005)
10. Dahlhaus, E.: A linear time algorithm to recognize clustered planar graphs and its parallelization. In: Lucchesi, C.L., Moura, A.V. (eds.) LATIN 1998. LNCS, vol. 1380, pp. 239–248. Springer, Heidelberg (1998)
11. Di Battista, G., Frati, F.: Efficient c-planarity testing for embedded flat clustered graphs with small faces. J. Graph Alg. Appl. 13(3), 349–378 (2009)
12. Didimo, W., Giordano, F., Liotta, G.: Overlapping cluster planarity. J. Graph Algorithms Appl. 12(3), 267–291 (2008)
13. Feng, Q.W., Cohen, R.F., Eades, P.: Planarity for clustered graphs. In: Moore, W., Luk, W. (eds.) FPL 1995. LNCS, vol. 975, pp. 213–226. Springer, Heidelberg (1995)
14. Garg, A., Tamassia, R.: On the computational complexity of upward and rectilinear planarity testing. SIAM Journal on Computing 31(2), 601–625 (2001)
15. Goodrich, M.T., Lueker, G.S., Sun, J.Z.: C-planarity of extrovert clustered graphs. In: Healy, P., Nikolov, N.S. (eds.) GD 2005. LNCS, vol. 3843, pp. 211–222. Springer, Heidelberg (2006)
16. Gutwenger, C., Jünger, M., Leipert, S., Mutzel, P., Percan, M., Weiskircher, R.: Advances in c-planarity testing of clustered graphs. In: Goodrich, M.T., Kobourov, S.G. (eds.) GD 2002. LNCS, vol. 2528, pp. 220–235. Springer, Heidelberg (2002)
17. Jelínek, V., Jelínková, E., Kratochvíl, J., Lidický, B.: Clustered planarity: Embedded clustered graphs with two-component clusters. In: Tollis, I.G., Patrignani, M. (eds.) GD 2008. LNCS, vol. 5417, pp. 121–132. Springer, Heidelberg (2009)
18. Jelínek, V., Suchý, O., Tesař, M., Vyskočil, T.: Clustered planarity: Clusters with few outgoing edges. In: Tollis, I.G., Patrignani, M. (eds.) GD 2008. LNCS, vol. 5417, pp. 102–113. Springer, Heidelberg (2009)
19. Jelínková, E., Kára, J., Kratochvíl, J., Pergel, M., Suchý, O., Vyskocil, T.: Clustered planarity: Small clusters in cycles and Eulerian graphs. J. Graph Alg. Appl. 13(3), 379–422 (2009)
20. Kratochvíl, J., Lubiw, A., Nesetril, J.: Noncrossing subgraphs in topological layouts. SIAM J. Discrete Math. 4(2), 223–244 (1991)
21. Schaefer, M.: Toward a theory of planarity: Hanani-Tutte and planarity variants. J. Graph Algorithms Appl. 17(4), 367–440 (2013)
22. Schaeffer, S.E.: Graph clustering. Computer Science Review 1(1), 27–64 (2007)
23. Tamassia, R.: On embedding a graph in the grid with the minimum number of bends. SIAM J. Comput. 16(3), 421–444 (1987)

Clustered Planarity Testing Revisited

Radoslav Fulek[1,4,*], Jan Kynčl[1,**], Igor Malinović[2], and Dömötör Pálvölgyi[3,* * *]

[1] Department of Applied Mathematics and Institute for Theoretical Computer Science,
Faculty of Mathematics and Physics, Charles University, Malostranské nám. 25,
118 00 Praha 1, Czech Republic
radoslav.fulek@gmail.com, kyncl@kam.mff.cuni.cz
[2] Faculté Informatique et Communications, École Polytechnique Fédérale de Lausanne,
1015 Lausanne, Switzerland
igor.malinovic@epfl.ch
[3] Institute of Mathematics, Eötvös University, Pázmány Péter sétány 1/C,
H-1117 Budapest, Hungary
domotorp@gmail.com
[4] Department of Industrial Engineering and Operations Research, Columbia University,
New York City, NY, USA

Abstract. The Hanani–Tutte theorem is a classical result proved for the first time in the 1930s that characterizes planar graphs as graphs that admit a drawing in the plane in which every pair of edges not sharing a vertex cross an even number of times. We generalize this classical result to clustered graphs with two disjoint clusters, and show that a straightforward extension of our result to flat clustered graphs with three or more disjoint clusters is not possible.

We also give a new and short proof for a related result by Di Battista and Frati based on the matroid intersection algorithm.

1 Introduction

Investigation of graph planarity can be traced back to the 1930s and developments accomplished at that time by Hanani [21], Kuratowski [26], Whitney [38] and others. Forty years later, with the advent of computing machinery, a linear time algorithm for graph planarity was discovered [23]. Nowadays, a polynomial time algorithm for testing whether a graph admits a crossing-free drawing in the plane could almost be considered a folklore result.

Nevertheless, many variants of planarity are still only poorly understood. As a consequence of this state of affairs, the corresponding decision problem for these variants has neither been shown to be polynomial nor NP-hard. *Clustered planarity* is one of

---

* The author gratefully acknowledges support from the Swiss National Science Foundation Grant No. 200021-125287/1 and ESF Eurogiga project GraDR as GAČR GIG/11/E023.
** Supported by the ESF Eurogiga project GraDR as GAČR GIG/11/E023 and by the grant SVV-2013-267313 (Discrete Models and Algorithms).
* * * Supported by Hungarian National Science Fund (OTKA), under grant PD 104386 and under grant NN 102029 (EUROGIGA project GraDR 10-EuroGIGA-OP-003) and the János Bolyai Research Scholarship of the Hungarian Academy of Sciences.

C. Duncan and A. Symvonis (Eds.): GD 2014, LNCS 8871, pp. 428–439, 2014.
© Springer-Verlag Berlin Heidelberg 2014

the most prominent [5] of such planarity notions. Roughly speaking, an instance of this problem is a graph whose vertices are partitioned into clusters. The question is, then, whether the graph can be drawn in the plane so that the vertices from the same cluster belong to the same region and no edge crosses the boundary of a particular region more than once. The aim of the present work is to offer novel perspectives on clustered planarity, which seem to be worth pursuing in order to better our understanding of the problem.

More precisely, a *clustered graph* is a pair (G, T) where $G = (V, E)$ is a graph and T is a rooted tree whose set of leaves is the set of vertices of G. The non-leaf vertices of T represent the clusters. For $\nu \in V(T)$, let T_ν denote the subtree of T rooted at ν. The *cluster* $V(\nu)$ is the set of leaves of T_ν. The subgraph of G induced by $V(\nu)$ is denoted by $G(\nu)$.

A *drawing* of G is a representation of G in the plane where every vertex is represented by a unique point and every edge $e = uv$ is represented by a simple arc joining the two points that represent u and v. If it leads to no confusion, we do not distinguish between a vertex or an edge and its representation in the drawing and we use the words "vertex" and "edge" in both contexts. We assume that in a drawing no edge passes through a vertex, no two edges touch and every pair of edges cross in finitely many points. A drawing of a graph is an *embedding* if no two edges cross.

A clustered graph (G, T) is *clustered planar* (or briefly *c-planar*) if G has an embedding in the plane such that

(i) for every $\nu \in V(T)$, there is a topological disc $d(\nu)$ containing all the leaves of T_ν and no other vertices of G such that if $\mu \in T_\nu$ then $d(\mu) \subseteq d(\nu)$;

(ii) if μ_1 and μ_2 are children of ν in T, then $d(\mu_1)$ and $d(\mu_2)$ are internally disjoint;

(iii) every edge of G intersects the boundary of $d(\nu)$ at most once for every $\nu \in V(T)$.

A *clustered drawing (or embedding)* of a clustered graph (G, T) is a drawing (or embedding, respectively) of G satisfying (i)–(iii). See Fig. 1 for an illustration. We will be using the word "cluster" for both the topological disc $d(\nu)$ and the subset of vertices $V(\nu)$.

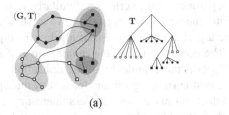

(a)   (b)

Fig. 1. (a) A clustered embedding of a clustered graph (G, T) and its tree T; (b) A clustered graph with two non-trivial clusters, which is not c-planar

The notion of clustered planarity was introduced in the work of Feng, Cohen and Eades [13,14] under the name of c-planarity and a similar problem was considered already in [28]. Since then an efficient algorithm for c-planarity testing or embedding has been discovered only in some special cases. The general problem whether the c-planarity of a clustered graph can be tested in polynomial time is wide open, already when we restrict ourselves to three pairwise disjoint clusters and the case when the embedding of G is a part of the input!

A clustered graph (G, T) is *c-connected* if every cluster of (G, T) induces a connected subgraph. In order to test a c-connected clustered graph (G, T) for c-planarity, it is enough to test whether there exists an embedding of G in which for every $\nu \in V(T)$ all vertices $u \in V(G)$ such that $u \notin V(\nu)$ are drawn in a single face of the subgraph $G(\nu)$ [14]. Cortese et al. [6] gave a structural characterization of c-planarity for c-connected clustered graphs and provided a linear-time algorithm. The extended abstract of Gutwenger et al. [19] contains an efficient algorithm in a more general case of *almost connected* clustered graphs, which can be also used in the case of two clusters. Biedl [2] is usually credited for giving the first polynomial time algorithm for c-planarity with two clusters, including the case of straight-line or y-monotone drawing. An alternative approach to the problem is given in [22]. On the other hand, only very little is known in the case of three clusters, where already clustered cycles are non-trivial to test for c-planarity [7] in a polynomial time.

The Hanani–Tutte theorem [21,37] is a classical result that provides an algebraic characterization of planarity with interesting (and not only algorithmic) consequences; see Section 2. The (strong) Hanani–Tutte theorem says that a graph is planar as soon as it can be drawn in the plane so that no pair of independent edges crosses an odd number of times. Moreover, its variant known as the weak Hanani–Tutte theorem [3,30,33] states that if we have a drawing $\mathcal{D}$ of a graph G where every pair of edges cross an even number of times then G has an embedding that preserves the cyclic order of edges at vertices from $\mathcal{D}$. Note that the weak variant does not directly follow from the strong Hanani–Tutte theorem. For sub-cubic graphs, the weak variant implies the strong variant. Other variants of the Hanani–Tutte theorem were proved for surfaces of higher genus [32,34], x-monotone drawings [17,31], partially embedded planar graphs, and several special cases of simultaneously embedded planar graphs [36]. See [35] for a (not too recent) survey on applications of the Hanani–Tutte theorem and related results.

We prove a variant of the Hanani–Tutte theorem for clustered graphs consisting only of two non-trivial clusters forming a partition of the vertex set. Similarly, as in the case of other variants of the Hanani–Tutte theorem, as a byproduct of our result, we immediately obtain a polynomial-time algorithm based on linear algebra for c-planarity testing in the corresponding case. The downside is that the running time of the algorithm is in $O(|V(G)|^{2\omega})$, where $O(n^\omega)$ is the complexity of multiplication of square $n \times n$ matrices; see Section 2. The best current algorithms for matrix multiplication give $\omega < 2.3729$ [18,39]. This fact does not make our work less interesting, since the purpose of our results lies more in theoretical foundations than in its immediate consequences. Also the worst case running time analysis often gives an unfair perspective on the performance of algebraic algorithms, e.g., the simplex method.

We remark that there exist more efficient algorithms for planarity testing using the Hanani–Tutte theorem such as the one in [9,10], which runs in linear time, see also [35, Section 1.4.1]. Moreover, in the case of x-monotone drawings a computation study [4] showed that the Hanani–Tutte approach [17] performs really well in practice. This should come as no surprise, since Hanani–Tutte theory seems to provide solid theoretical foundations for graph planarity that brings together its combinatorial, algebraic, and computational aspects [36].

Notation. In the present paper we assume that $G = (V, E)$ is a (multi)graph. We use a shorthand notation $G - v$ and $G \cup E'$ for $(V \setminus \{v\}, E \setminus \{vw \mid vw \in E\})$, and $(V, E \cup E')$, respectively. The *rotation* at a vertex v is the clockwise cyclic order of the end pieces of edges incident to v. The *rotation system* of a graph is the set of rotations at all its vertices. We say that two embeddings of a graph are the *same* if they have the same rotation system up to switching the orientations of all the rotations simultaneously. We say that a pair of edges in a graph are *independent* if they do not share a vertex. An edge in a drawing is *even* if it crosses every other edge an even number of times. A drawing of a graph is *even* if all edges are even.

Hanani–Tutte for Clustered Graphs. A clustered graph (G, T) is *two-clustered* if the root of T has exactly two children and only leaves as grandchildren. In other words, a two-clustered graph has exactly two non-trivial clusters, which form a partition of the vertex set.

We extend the strong version of the Hanani–Tutte theorem as follows. A drawing of a graph is *independently even* if every pair of independent edges in the drawing cross an even number of times.

Theorem 1. *If a two-clustered graph (G, T) admits an independently even clustered drawing then (G, T) is c-planar.*

The weak variant of Theorem 1 is a special case of the result obtained recently by the first author [15]. On the other hand, we exhibit examples of clustered graphs with more than two disjoint clusters that are not c-planar, but admit an even clustered drawing.

Theorem 2. *For every $r \geq 3$ there exists a flat clustered cycle with r clusters that is not c-planar and admits an even clustered drawing.*

Gutwenger et al. [20] recently showed that by using the reduction from [36] our counter-examples can be turned into counter-examples for [36, Conjecture 1.2] and for a variant of the Hanani–Tutte theorem for two simultaneously embedded planar graphs [36, Conjecture 6.20]. Our counter-examples show that a straightforward extension of Theorem 1 or its weak variant to flat clustered graphs with more than two clusters is not possible. Nevertheless, interesting extensions are still possible, e.g., in the c-connected case, as shown in the full version of this extended abstract [16], or for strip clustered graphs [15].

A clustered graph (G, T) is *flat* if no non-root cluster of (G, T) has a non-trivial subcluster; that is, if every root-leaf path in T has at most three vertices. A pair $(\mathcal{D}(G), T)$ is an *embedded clustered graph* if (G, T) is a clustered graph and $\mathcal{D}(G)$ is an embedding of G in the plane, not necessarily a clustered embedding. The embedded clustered graph $(\mathcal{D}(G), T)$ is *c-planar* if it can be extended to a clustered embedding of (G, T), by choosing a topological disc for each cluster.

We give an alternative polynomial time algorithm for deciding c-planarity of embedded flat clustered graphs with small faces, reproving a result of Di Battista and Frati [11]. Our algorithm is based on the matroid intersection theorem. Its running time is $O(|V(G)|^{3.5})$ by [8], so it does not outperform the linear algorithm from [11], and similarly as for our other results we see its purpose more in mathematical foundations than in giving an efficient algorithm. We find it quite surprising that by using completely different techniques we obtained an algorithm for exactly the same case. Our approach is very similar to a technique used [25] for deciding the global connectivity of switch graphs.

Theorem 3. [11] *Let G denote an embedded planar graph such that all its faces are incident to at most five vertices. Let (G, T) denote a flat clustered graph. We can decide in polynomial time whether (G, T) admits a c-planar embedding, in which G keeps its given embedding.*

The rest of the paper is organized as follows. In Section 2 we describe an algorithm for c-planarity testing of clustered graphs belonging to classes for which the corresponding variant of the strong Hanani-Tutte theorem holds. In Section 3, we prove Theorem 1. In Section 4 we prove Theorem 2. In Section 5 we prove Theorem 3. We conclude with some remarks in Section 6.

2 Algorithm

Let (G, T) belong to a class of clustered graph for which the corresponding variant of the strong Hanani-Tutte theorem holds, i.e., an independently even clustered drawing of (G, T) implies that (G, T) is c-planar.

Our algorithm for c-planarity testing is an adaption of the algorithm for planarity testing from [35, Section 1.4.2]. The algorithm tests whether we can continuously deform a given clustered drawing $\mathcal{D}$ of (G, T) into an independently even clustered drawing $\mathcal{D}'$ of (G, T). By the corresponding variant of the strong Hanani–Tutte theorem, the existence of such a drawing is equivalent to c-planarity of (G, T).

During the deformation the parity of crossings between a pair of edges is affected only when an edge e passes over a vertex v, in which case we change the parity of crossings of e with all the edges adjacent to v. We call such an event an *edge-vertex switch*. Note that every edge-vertex switch can be performed independently of others, for any initial drawing: we can always deform a given edge to pass close to a given vertex, while introducing new crossings only in pairs. Thus, for our purpose the continuous deformation of $\mathcal{D}$ can be represented by a set S of edge-vertex switches. In S, an edge-vertex switch of an edge e with a vertex v is represented as the ordered pair (e, v).

A drawing of (G, T) can then be represented as a vector $\mathbf{v} \in \mathbb{Z}_2^M$, where M denotes the number of unordered pairs of independent edges. The component of $\mathbf{v}$ corresponding to a pair $\{e, f\}$ is 1 if e and f cross an odd number of times and 0 otherwise. An edge-vertex switch (e, v) is represented as a vector $\mathbf{w}_{(e,v)} \in \mathbb{Z}_2^M$ such that its only components equal to 1 are those indexed by pairs $\{e, f\}$ where f is incident to v. The set of all drawings that can be obtained from (G, T) by the switches from S then corresponds to an affine subspace $\mathbf{v} + W$, where W is the subspace generated by the set

$\{\mathbf{w}_{(e,v)}; (e,v) \in S\}$. The algorithm tests whether $0 \in \mathbf{v} + W$, which is equivalent to the solvability of a system of linear equations over $\mathbb{Z}_2$.

The difference between the original algorithm for planarity testing and our version for c-planarity testing is the following. To keep the drawing of (G, T) clustered after every deformation, for every edge $e = v_1 v_2$, we allow only those edge-vertex switches (e, v) such that v is a child of some vertex of the shortest path between v_1 and v_2 in T. We also include *edge-cluster switches* (e, C), where C is a child of some vertex of the shortest path between v_1 and v_2 in T, that move e over all vertices of C simultaneously. The corresponding vector $\mathbf{w}_{(e,C)}$ is the sum of all $\mathbf{w}_{(e,v)}$ for $v \in C$. Therefore, the set of allowed switches generates a subspace W_c of W. Our algorithm then tests whether $0 \in \mathbf{v} + W_c$.

Before running the algorithm, we first remove any loops and parallel edges and check whether $|E(G')| < 3|V(G')|$ for the resulting graph G'. Then we run our algorithm on (G', T). This means solving a system of $O(|E(G')||V(G')|) = O(|V(G)|^2)$ linear equations in $O(|E(G')|^2) = O(|V(G)|^2)$ variables. This can be performed in $O(|V(G)|^{2\omega}) \leq O(|V(G)|^{4.752})$ time using the algorithm by Ibarra, Moran and Hui [24].

3   Two Clusters

Let (G, T) be a two-clustered graph. Let A and B denote the two clusters of (G, T) forming a partition of $V = V(G)$. For a subset $V' \subseteq V$, let $G[V']$ denote the subgraph of G induced by V'. By the assumption of Theorem 1 and the strong Hanani–Tutte theorem, G has an embedding. However, in this embedding, $G[B]$ does not have to be contained in a single face of $G[A]$ and vice-versa. Hence, we cannot guarantee that a clustered embedding of (G, T) exists so easily.

For an induced subgraph H of G, the *boundary* of H is the set of vertices in H that have a neighbor in $G - H$. We say that an embedding $\mathcal{D}(H)$ of H is *exposed* if all vertices from the boundary of H are incident to the outer face of $\mathcal{D}(H)$.

The following lemma is an easy consequence of the strong Hanani–Tutte theorem. It helps us to find an exposed embedding of each connected component X of $G[A]$ and $G[B]$. Later in the proof of Theorem 1 this allows us to remove non-essential parts of each such component X and concentrate only on a subgraph G' of G in which both $G[A]$ and $G[B]$ are outerplanar.

Lemma 1. *Suppose that (G, T) admits an independently even clustered drawing. Then every connected component of $G[A] \cup G[B]$ admits an exposed embedding.*

Our proof of Theorem 1 proceeds by reducing the problem to an application of the following lemma.

Lemma 2. *Let (G, T) denote a two-clustered bipartite graph in which the two nontrivial clusters induce independent sets. If G admits an even drawing then (G, T) is c-planar. Moreover, there exists a clustered embedding of (G, T) with the same rotation system as in the given even drawing of G.*

3.1 Proof of Theorem 1

The proof is inspired by the proof of the strong Hanani–Tutte theorem from [33] and its outline is as follows. First we obtain a subgraph G' of G containing the boundary of each component of $G[A]$ and $G[B]$ and such that each of $G'[A]$ and $G'[B]$ is a *cactus forest*, that is, a graph where every two cycles are edge disjoint. Equivalently, a cactus forest is a graph with no subdivision of $K_4 - e$. A connected component of a cactus forest is called a *cactus*.

Then we apply the strong Hanani–Tutte theorem on a graph which is constructed from G' by turning all cycles in $G'[A]$ and $G'[B]$ into wheels, and by splitting certain vertices of G' into edges. The wheels in G' guarantee that everything that has been removed from G in order to obtain G' can be inserted back.

Let $X_1, \ldots, X_k$ denote the connected components of $G[A]$ and $G[B]$. By Lemma 1 we find an exposed embedding $\mathcal{D}(X_i)$ of each X_i. Let X_i' denote the subgraph of X_i obtained by deleting from X_i all the vertices and edges not incident to the outer face of $\mathcal{D}(X_i)$. Observe that X_i' is a cactus.

Let $G' = (\bigcup_{i=1}^{k} X_i') + E(A, B)$. That is, G' is subgraph of G that consists of all X_i'-s and all edges between the two clusters. Let $\mathcal{D}'$ denote the drawing of G' obtained from the initial independently even drawing of G by deleting the edges and vertices of G not belonging to G'. Thus, $\mathcal{D}'$ is independently even.

In what follows we process the cycles of $G'[A]$ and $G'[B]$ one by one. We will be modifying G' and therefore also the drawing $\mathcal{D}'$. At each stage of this process some cycles in $G'[A]$ and $G'[B]$ will be labeled as processed and the rest will be labeled as unprocessed. We will maintain the property that all processed cycles are vertex disjoint and that all their edges are even. We start with all the cycles in $G'[A]$ and $G'[B]$ being labeled as unprocessed. Let C denote an unprocessed cycle in $G'[A]$. For cycles in $G'[B]$, the procedure is analogous. We consider two cases.

a) C Shares no Vertex with an Already Processed Cycle. We two-color the connected regions in the complement of C so that two regions sharing a non-trivial part of the boundary receive opposite colors. We say that a point not lying on C is "outside" of C if it is contained in the region with the same color as the unbounded region. Otherwise, such a point is "inside" of C.

We locally modify the drawing $\mathcal{D}'$ at the vertices of C so that all the edges of C cross every other edge an even number of times [33]. Since $\mathcal{D}'$ is a clustered drawing of G', all vertices of B are "outside" of C. Therefore, every path joining C with a vertex in B internally vertex disjoint from C is attached to its endpoint on C from the "outside" of C.

Now we fill the cycle C with a wheel. More precisely, we add a vertex v_C into A and place it very close to an arbitrary vertex of C "inside" of C. We connect v_C with all the vertices of C by edges that closely follow the closed curve representing C either from the left or from the right, and attach to their endpoints on C from "inside". Portions of these new edges may lie "outside" of C due to self-crossings of C, but not in the neighborhood of vertices of C. Therefore, the new edges can introduce an odd crossing pair only with an edge e attached to a vertex v of C from the "inside" of C.

Since $G'[A]$ is a cactus forest, it follows that such a vertex v is a cut vertex in $G'[A]$ and that the endpoint of e different from v belongs to a connected component K of

$G'[A] - v$, which is also a connected component of $G' - v$. Thus, we shrink the drawing of $G'[V(K) \cup v]$ so that $G'[V(K) \cup v]$ is drawn very close to v and none of its edges crosses an edge in the rest of the graph. In particular, by shrinking $G'[V(K) \cup v]$ we do not introduce a pair of edges crossing an odd number of times. We label all the cycles in $G'[V(K) \cup v]$ as processed. By repeating this for all the troublesome cut-vertices of C we modify $\mathcal{D}'$ so that none of the edges incident to v_C crosses another edge an odd number of times. Finally, we label C as processed.

b) C Shares a Vertex with an Already Processed Cycle. Let v be a vertex on C belonging to an already processed cycle C_p. Since processed cycles are vertex disjoint, the cycle C_p is unique. Since the edges of C_p are even, the edges $v_1 v$ and $v_2 v$ of C_p adjacent to v are attached to v both from the "inside" or both from the "outside" of C. Suppose the latter. (The other case is analogous.) We split the vertex v by replacing it with two new vertices v' and v'' connected by an edge. Every edge uv attaching to v from the "outside" of C is replaced by an edge uv' (including the edges $v_1 v$ and $v_2 v$). All other edges uv are replaced by an edge uv''. The cycle that is obtained from C_p by replacing v with v' is then labeled as processed. Note that we can do such vertex-splitting in $\mathcal{D}'$ without introducing any pair of edges crossing an odd number of times by drawing v' and v'' very close to v. After performing all necessary vertex splits for vertices of C, we may apply the procedure in case a) to the modified cycle C.

It is easy to see that the algorithm terminates after a finite number of steps a) or b) with all cycles processed. Let G'' denote the graph we obtain from G' after processing all the cycles of $G'[A]$ and $G'[B]$. By applying the strong Hanani–Tutte theorem on G'' we obtain an embedding which can be easily modified so that the only vertices of G'' not incident to the outer face of $G''[A]$ or $G''[B]$ are the vertices v_C that form the centers of the wheels. In particular, $G''[A]$ is drawn in the outer face of $G''[B]$ and vice-versa. In the resulting embedding we delete all the vertices v_C and contract the edges between the pairs of vertices v', v'' that were obtained by vertex-splits.

Thus, we obtain an embedding of G' in which for every component X of $G'[A] \cup G'[B]$, all vertices of $G' - X$ are drawn in the outer face of X. By inserting the removed parts of G back to G' we obtain an embedding of G in which for every component X of $G[A] \cup G[B]$, all vertices of $G - X$ are drawn in the outer face of X. The theorem follows by contracting each component of $G[A] \cup G[B]$ to a point and applying Lemma 2.

4 Proof of Theorem 2

In this section we construct a family of even clustered drawings of flat clustered cycles on more than two clusters that are not clustered planar. Thus, the straightforward generalization of the Hanani–Tutte theorem for graphs with three or more clusters is not possible. For any $r \geq 3$ and any odd $k \geq 3$, our counterexample is a clustered cycle with kr vertices and r clusters. In our clustered drawing the clusters are drawn as regions bounded by a pair of rays emanating from the same vertex p. We call p the *center* of our drawing.

Topologically our construction is equivalent to a cylindrical drawing, where c clusters are separated by vertical lines. For every odd integer $k > 0$, we can describe the curve representing the cycle analytically as a height function $f(\alpha) = \sin\left(\frac{rk+1}{k}\alpha\right)$ on a standing cylinder taking the angle as the parameter. The vertices of the cycle are at $\left(i\frac{2k}{rk+1}\pi, 0\right)$, where $i = 0, \ldots, rk - 1$, and the lines separating clusters at $\frac{2ki+1}{rk+1}\pi$, for $i = 0, \ldots r - 1$, see Fig. 2. By [7], for any clustered drawing of any of our examples, the curve representing the cycle has winding number k around p, and therefore, it is not c-planar when $k > 1$.

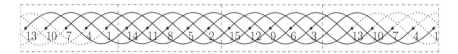

Fig. 2. A counter-example to the variant of the Hanani–Tutte theorem with parameters $r = 3$ and $k = 5$, and hence, the underlying graph is a cycle on 15 vertices. The vertices are labeled by positive integers in correspondence with their appearance on the cycle. The leftmost and the rightmost cluster need to be identified in the actual cylindrical drawing.

5   Small Faces

This section reproves a result of Di Battista and Frati [11] that c-planarity can be decided in polynomial time for embedded flat clustered graphs if all faces are incident to at most five vertices. Our approach seems quite different from theirs, as we use (a corollary of) the matroid intersection theorem [12,27], which says that the largest common independent set of two matroids can be found in polynomial time. See e.g. [29] for further references.

In this section, we will use a shorthand notation (G, T) instead of $(\mathcal{D}(G), T)$ for an embedded clustered graph. Let (G, T) be a embedded flat clustered graph. A *saturator* of (G, T) is a subset F of $\binom{V}{2}$ disjoint from $E(G)$ such that $(G \cup F, T)$ is planar, every cluster of $(G \cup F, T)$ is connected, and the edges in F can be embedded so that every cluster of $(G \cup F, T)$ is in the outer face of every other cluster. We have the following simple fact regarding saturators already observed by Feng, Cohen and Eades [14].

Observation 1. *An embedded flat clustered graph* (G, T) *is c-planar if and only if* (G, T) *has a saturator.*

In order to model our problem by matroids we need to avoid two saturating edges in one face coming from two different clusters (even if they do not cross). In general, this is not possible if the face is not a simple cycle. Thus, we first augment the graph by adding some edges inside the faces.

Lemma 3. *An embedded flat clustered graph* (G, T), *all of whose faces are incident to at most five vertices, can be augmented by adding edges into an embedded flat clustered graph* (G', T') *such that* (G, T) *is c-planar if and only if* (G', T') *is c-planar, and the following holds for* (G', T'). *If* (G', T') *is c-planar then* (G', T') *has a saturator* F *whose edges can be embedded so that each face of* G' *contains at most one edge of* F.

Proof of Theorem 3

We give an algorithm for deciding c-planarity for embedded flat clustered graphs satisfying the hypothesis of the claim. By an algorithmic version of Lemma 3, from the given embedded flat clustered graph (G, T) we obtain a new embedded graph (G', T') such that every minimal saturator of (G', T') has at most one edge inside each face and (G', T') is c-planar if and only if (G, T) is c-planar. This can be done easily in linear time in the number of vertices. Thus, it is enough to show that we can decide c-planarity of (G', T') in polynomial time.

By Observation 1, it is enough to decide whether we can saturate G' so that all the clusters are connected and every cluster is drawn in the outer face of every other cluster. The latter can be tested in quadratic time in the number of vertices. In order to test the existence of a saturator we define two matroids for which we will use the matroid intersection algorithm. The ground set of each matroid is the multiset $\overline{E'}$ of non-edges of G' defined as the union $\bigcup_f E_f$, over faces of G', where E_f is the set of diagonals of the face f.

The first matroid, M_1, is the direct sum of graphic matroids constructed for each cluster. More precisely, denote the clusters by C_i, $i = 1, \ldots, k$, and let $v \sim_i u$ if u and v are connected in $G'[C_i]$. Denote by G_i the multigraph obtained from $\overline{G'} = (V, \overline{E'})$ by deleting the vertices not in C_i, contracting the $\sim_i$-equivalent vertices into new vertices, and deleting all loops. Now, the ground set of the graphic matroid $M(G_i)$ can be identified with the set of edges from $\overline{E'}$ that go between two vertices from C_i belonging to distinct connected components of C_i. The rank of $M(G_i)$ is the number of vertices of G_i minus one. Since the matroids $M(G_i)$, $i = 1, \ldots, k$, are pairwise disjoint, their direct sum, M_1, is also a matroid and its rank is the sum of the ranks of $M(G_i)$-s. The second matroid, M_2, is a partition matroid. A subset of $\overline{E'}$ is independent in M_2 if it has at most one edge in every face of G'.

Let M be the intersection of M_1 and M_2. If M has the same rank as M_1 then there exists a saturator of (G', T') that has at most one edge inside each face. Thus, (G', T') is c-planar by Observation 1, and that in turn implies that (G, T) is c-planar as well. On the other hand, if (G, T), and hence (G', T'), is c-planar then there exists a minimal saturator F of G' that has at most one edge inside each face by the property of G' guaranteed by Lemma 3. Thus, F witnesses the fact that the rank of M_1 and the rank of M are the same. Hence, M has the same rank as M_1 if and only if (G', T') is c-planar and the theorem follows by the matroid intersection algorithm.

6   Concluding Remarks

By the construction in Section 4 we cannot hope for the fully general variant of the Hanani–Tutte theorem for clustered graphs. Nevertheless, it is still interesting to ask whether the weak or the strong Hanani–Tutte theorem for the case of flat clustered graphs holds if the graph obtained by contracting the clusters is acyclic (after deleting loops and multiple edges). More formally, given a flat clustered graph (G, T), let G_T denote the simple graph whose vertices correspond to clusters of (G, T) and two distinct vertices μ and ν are joined by an edge if and only if there exists an edge in G between the clusters $V(\mu)$ and $V(\nu)$.

Conjecture 1. *If G_T is acyclic and (G, T) admits an independently even clustered drawing then (G, T) is c-planar.*

A variant of the conjecture when G_T is a path would provide a polynomial time algorithm for c-planarity testing for strip clustered graph, which is an open problem stated in [1]. Note that our proof from Section 5 fails if the graph has hexagonal faces. We wonder if this difficulty can be overcome or rather could lead to NP-hardness.

Acknowledgements. We are grateful to the numerous anonymous reviewers for many valuable comments.

References

1. Angelini, P., Da Lozzo, G., Di Battista, G., Frati, F.: Strip planarity testing. In: Wismath, S., Wolff, A. (eds.) GD 2013. LNCS, vol. 8242, pp. 37–48. Springer, Heidelberg (2013)
2. Biedl, T.C.: Drawing planar partitions III: Two constrained embedding problems. Technical report, RUTCOR, Rutgers University (1998)
3. Cairns, G., Nikolayevsky, Y.: Bounds for generalized thrackles. Discrete Comput. Geom. 23(2), 191–206 (2000)
4. Chimani, M., Zeranski, R.: Upward planarity testing: A computational study. In: Wismath, S., Wolff, A. (eds.) GD 2013. LNCS, vol. 8242, pp. 13–24. Springer, Heidelberg (2013)
5. Cortese, P.F., Di Battista, G.: Clustered planarity (invited lecture). In: Twenty-First Annual Symposium on Computational Geometry (Proc. SoCG 2005), pp. 30–32. ACM (2005)
6. Cortese, P.F., Di Battista, G., Frati, F., Patrignani, M., Pizzonia, M.: C-planarity of c-connected clustered graphs. J. Graph Algorithms Appl. 12(2), 225–262 (2008)
7. Cortese, P.F., Di Battista, G., Patrignani, M., Pizzonia, M.: Clustering cycles into cycles of clusters. J. Graph Algorithms Appl. 9(3), 391–413 (2005)
8. Cunningham, W.H.: Improved bounds for matroid partition and intersection algorithms. SIAM Journal on Computing 15(4), 948–957 (1986)
9. de Fraysseix, H., de Mendez, P.O., Rosenstiehl, P.: Trémaux trees and planarity. International Journal of Foundations of Computer Science 17(05), 1017–1029 (2006)
10. de Fraysseix, H., Rosenstiehl, P.: A characterization of planar graphs by Trémaux orders. Combinatorica 5(2), 127–135 (1985)
11. Di Battista, G., Frati, F.: Efficient c-planarity testing for embedded flat clustered graphs with small faces. In: Hong, S.-H., Nishizeki, T., Quan, W. (eds.) GD 2007. LNCS, vol. 4875, pp. 291–302. Springer, Heidelberg (2008)
12. Edmonds, J.: Submodular functions, matroids, and certain polyhedra. In: Jünger, M., Reinelt, G., Rinaldi, G. (eds.) Combinatorial Optimization - Eureka, You Shrink! LNCS, vol. 2570, pp. 11–26. Springer, Heidelberg (2003)
13. Feng, Q.-W., Cohen, R.F., Eades, P.: How to draw a planar clustered graph. In: Li, M., Du, D.-Z. (eds.) COCOON 1995. LNCS, vol. 959, pp. 21–30. Springer, Heidelberg (1995)
14. Feng, Q.W., Cohen, R.F., Eades, P.: Planarity for clustered graphs. In: Spirakis, P.G. (ed.) ESA 1995. LNCS, vol. 979, pp. 213–226. Springer, Heidelberg (1995)
15. Fulek, R.: Towards Hanani–Tutte theorem for clustered graphs. In: 40th International Workshop on Graph-Theoretic Concepts in Computer Science (accepted)
16. Fulek, R., Kynčl, J., Malinović, I., Pálvölgyi, D.: Efficient c-planarity testing algebraically. arXiv:1305.4519

17. Fulek, R., Pelsmajer, M., Schaefer, M., Štefankovič, D.: Hanani-Tutte, monotone drawings and level-planarity. In: Pach, J. (ed.) Thirty Essays in Geometric Graph Theory, pp. 263–288 (2012)
18. Gall, F.L.: Powers of tensors and fast matrix multiplication. CoRR abs/1401.7714 (2014)
19. Gutwenger, C., Jünger, M., Leipert, S., Mutzel, P., Percan, M., Weiskircher, R.: Advances in c-planarity testing of clustered graphs. In: Goodrich, M.T., Kobourov, S.G. (eds.) GD 2002. LNCS, vol. 2528, pp. 220–236. Springer, Heidelberg (2002)
20. Gutwenger, C., Mutzel, P., Schaefer, M.: Practical experience with Hanani-Tutte for testing c-planarity. In: 2014 Proceedings of the Sixteenth Workshop on Algorithm Engineering and Experiments (ALENEX), pp. 86–97 (2014)
21. Hanani, H.: Über wesentlich unplättbare Kurven im drei-dimensionalen Raume. Fundamenta Mathematicae 23, 135–142 (1934)
22. Hong, S., Nagamochi, H.: Two-page book embedding and clustered graph planarity. Technical report, Dept. of Applied Mathematics and Physics, University of Kyoto (2009)
23. Hopcroft, J., Tarjan, R.: Efficient planarity testing. J. ACM 21(4), 549–568 (1974)
24. Ibarra, O.H., Moran, S., Hui, R.: A generalization of the fast LUP matrix decomposition algorithm and applications. J. Algorithms 3(1), 45–56 (1982)
25. Katz, B., Rutter, I., Woeginger, G.: An algorithmic study of switch graphs. In: Paul, C., Habib, M. (eds.) WG 2009. LNCS, vol. 5911, pp. 226–237. Springer, Heidelberg (2010)
26. Kuratowski, K.: Sur le problème des courbes gauches en topologie. Fund. Math. 15, 271–283 (1930)
27. Lawler, E.L.: Matroid intersection algorithms. Mathematical Programming 9, 31–56 (1975)
28. Lengauer, T.: Hierarchical planarity testing algorithms. J. ACM 36(3), 474–509 (1989)
29. Oxley, J.: Matroid Theory. Oxford University Press (2011)
30. Pach, J., Tóth, G.: Which crossing number is it anyway? J. Combin. Theory Ser. B 80(2), 225–246 (2000)
31. Pach, J., Tóth, G.: Monotone drawings of planar graphs. J. Graph Theory 46(1), 39–47 (2004), updated version: arXiv:1101.0967
32. Pelsmajer, M.J., Schaefer, M., Stasi, D.: Strong Hanani–Tutte on the projective plane. SIAM Journal on Discrete Mathematics 23(3), 1317–1323 (2009)
33. Pelsmajer, M.J., Schaefer, M., Štefankovič, D.: Removing even crossings. J. Combin. Theory Ser. B 97(4), 489–500 (2007)
34. Pelsmajer, M.J., Schaefer, M., Štefankovič, D.: Removing even crossings on surfaces. European Journal of Combinatorics 30(7), 1704–1717 (2009)
35. Schaefer, M.: Hanani-Tutte and related results. To appear in Bolyai Memorial Volume
36. Schaefer, M.: Toward a theory of planarity: Hanani-tutte and planarity variants. J. Graph Algorithms Appl. 17(4), 367–440 (2013)
37. Tutte, W.T.: Toward a theory of crossing numbers. J. Combin. Theory 8, 45–53 (1970)
38. Whitney, H.: Non-separable and planar graphs. Trans. Amer. Math. Soc. 34, 339–362 (1932)
39. Williams, V.V.: Multiplying matrices faster than Coppersmith-Winograd. In: Proceedings of the Forty-Fourth Annual ACM Symposium on Theory of Computing, STOC 2012, pp. 887–898 (2012)

A New Perspective on Clustered Planarity as a Combinatorial Embedding Problem*

Thomas Bläsius and Ignaz Rutter

Faculty of Informatics, Karlsruhe Institute of Technology (KIT), Germany
{blaesius,rutter}@kit.edu

Abstract. The clustered planarity problem (c-planarity) asks whether a hierarchically clustered graph admits a planar drawing such that the clusters can be nicely represented by regions. We introduce the cd-tree data structure and give a new characterization of c-planarity. It leads to efficient algorithms for c-planarity testing in the following cases. (i) Every cluster and every co-cluster has at most two connected components. (ii) Every cluster has at most five outgoing edges.

Moreover, the cd-tree reveals interesting connections between c-planarity and planarity with constraints on the order of edges around vertices. On one hand, this gives rise to a bunch of new open problems related to c-planarity, on the other hand it provides a new perspective on previous results.

1 Introduction

When visualizing graphs whose nodes are structured in a hierarchy, one usually has two objectives. First, the graph should be drawn nicely. Second, the hierarchical structure should be expressed by the drawing. Regarding the first objective, we require drawings without edge crossings, i.e., *planar drawings*. A natural way to represent a cluster is a simple region containing exactly the vertices in the cluster. To express the hierarchical structure, the boundaries of two regions must not cross and edges of the graph can cross region boundaries at most once (if only one of its endpoints lies inside the cluster). Such a drawing is called *c-planar*; see Sec. 2 for a formal definition. Testing a clustered graph for c-planarity is a fundamental open problem in the field of Graph Drawing.

C-planarity was first considered by Lengauer [21] (in a different context). He gave an efficient algorithm for the case that every cluster is connected. Feng et al. [13], who coined the name c-planarity, rediscovered the problem and gave a similar algorithm. Cornelsen and Wagner [7] showed that c-planarity is equivalent to planarity when additionally every co-cluster is connected.

Relaxing the condition that every cluster must be connected, makes testing c-planarity surprisingly difficult. Efficient algorithms are known only for very restricted cases and many of these algorithms are very involved. One example is the efficient algorithm by Jelínek et al. [17, 18] for the case that every cluster consists of at most two connected components while the planar embedding of the graph is fixed. Another algorithm by Jelínek et al. [19] solves the case that every cluster has at most four outgoing edges.

---

* Partly done within GRADR – EUROGIGA project no. 10-EuroGIGA-OP-003. Supported by a fellowship within the Postdoc-Program of the German Academic Exchange Service (DAAD).

C. Duncan and A. Symvonis (Eds.): GD 2014, LNCS 8871, pp. 440–451, 2014.
© Springer-Verlag Berlin Heidelberg 2014

A popular restriction is to require a *flat* hierarchy, i.e., every pair of clusters has empty intersection. For example, Di Battista and Frati [12] solve the case where the clustering is flat, the graph has a fixed embedding and the faces have size at most 5. Sec. 4.1 and Sec. 4.2 contain additional related work viewed from the new perspective.

Contribution and Outline. We first present the cd-tree data structure (Sec. 3) and use it to characterize c-planarity in terms of combinatorial embeddings of planar graphs. This provides a useful new perspective and significantly simplifies some previous results.

In Sec. 4 we define different constrained-planarity problems. We show in Sec. 4.1 that they are equivalent to different variants of the c-planarity problem of flat-clustered graphs. We also discuss which cases of the constrained embedding problems are solved by previous results on c-planarity. Based on these insights we derive a generic algorithm for testing c-planarity in Sec. 4.2 and discuss previous work in this context.

In Sec. 5, we show how the cd-tree characterization together with results on the problem SIMULTANEOUS PQ-ORDERING [4] lead to efficient algorithms for the cases that (i) every cluster and every co-cluster consists of at most two connected components; or (ii) every cluster has at most five outgoing edges. The latter extends the result by Jelínek et al. [19], where every cluster has at most four outgoing edges.

2 Preliminaries

We denote graphs by G with vertex set V and edge set E. We implicitly assume graphs to be *simple* (no multiple edges or loops). We use the prefix *multi-* to indicate that a graph may have multiple edges (but no loops), e.g., a multi-cycle is obtained from a cycle by multiplying edges. A (multi-)graph G is *planar* if it admits a planar drawing (no edge crossings). The *edge-ordering* of a vertex v is the clockwise cyclic order of its incident edges in a planar drawing of G. An *embedding* of G consists of an edge-ordering for every vertex such that G has a planar drawing with these edge-orderings.

A *PQ-tree* [5] is an unrooted tree T with leaves L such that every inner node is either a *P-node* or a *Q-node*. When embedding T, one can choose the edge-orderings of P-nodes arbitrarily, whereas the edge-orderings of Q-nodes are fixed up to reversal. Every such embedding of T defines a cyclic order on the leaves L. The PQ-tree T *represents* the orders one can obtain in this way. A set of orders is *PQ-representable* if it can be represented by a PQ-tree. The valid edge-orderings of non-cutvertices in planar graphs are PQ-representable (e.g., [4]). Conversely, replace each Q-node of a PQ-tree T by a wheel (to fix its edge-ordering) and connect all leaves to a new vertex v. Then T represents the edge-orderings of v in embeddings of the resulting graph (e.g., [21]).

C-Planarity. A *clustered graph* (G, T) is a graph G together with a rooted tree T whose leaves are the vertices of G. Let μ be a node of T. The tree T_μ is the subtree of T consisting of the root μ and all its successors. The graph induced by the leaves of T_μ is a *cluster* in G. We identify this cluster with the node μ. A cluster is *proper* if it is neither the whole graph (root cluster) nor a single vertex (leaf cluster).

A *c-planar drawing* of (G, T) is a planar drawing of G together with a *simple* (= simply-connected) region R_μ for every cluster μ satisfying the following properties.

(i) Every region R_μ contains exactly the vertices of the cluster μ. (ii) Two regions have non-empty intersection only if one contains the other. (iii) Edges cross the boundary of a region at most once. A clustered graph is *c-planar* if it admits a c-planar drawing. This definition relies on embeddings in the plane using terms like "outside" and "inside". Instead, one can consider drawings on the sphere by unrooting T, using cuts instead of clusters and simple closed curves instead of simple regions. Removing an edge ε of T splits T in two components. As the leaves of T are the vertices of G, this induces a *corresponding cut* $(V_\varepsilon, V_\varepsilon')$ with $V_\varepsilon' = V \setminus V_\varepsilon$ on G. For a c-planar drawing of G on the sphere, we require a planar drawing of G together with a simple closed curve C_ε for every cut $(V_\varepsilon, V_\varepsilon')$ with the following properties. (i) The curve C_ε separates V_ε from V_ε'. (ii) No two curves intersect. (iii) Edges of G cross C_ε at most once.

Using clusters instead of cuts corresponds to orienting the cuts, using one side as cluster and the other side as the cluster's complement (*co-cluster*). C-planarity on the sphere and in the plane are equivalent; one simply has to choose an appropriate point on the sphere to lie in the outer face. We use the rooted and unrooted view interchangeably.

3   The CD-Tree

The *cd-tree (cut- or cluster-decomposition-tree)* of a clustered graph (G, T) is the tree T together with a multi-graph associated with each node of T that represents the decomposition of G along its cuts corresponding to edges in T; see Fig. 1a and b for an example. Lengauer [21] uses a similar structure. Our notation is inspired by SPQR-trees.

Let μ be a node of T with neighbors $\mu_1, \ldots, \mu_k$ and incident edges $\varepsilon_i = \{\mu, \mu_i\}$. Removing μ separates the leaves of T into k subsets and thus partitions the vertices of G into $V_1, \ldots, V_k \subseteq V$. The *skeleton* skel($\mu$) of μ is the multi-graph obtained from G by contracting each subset V_i into a *virtual vertex* v_i (we keep multiple edges but remove loops). Note that skeletons of inner nodes of T contain only virtual vertices, while skeletons of leaves consist of one virtual and one non-virtual vertex. The node μ_i is the neighbor of μ *corresponding* to v_i and the virtual vertex in skel(μ_i) corresponding to μ is the *twin* of v_i, denoted by twin(v_i). Note that twin(twin(v_i)) = v_i.

The edges incident to v_i are exactly the edges of G crossing the cut corresponding to the tree edge ε_i. Thus, the same edges of G are incident to v_i and twin(v_i). This gives a bound on the total size c of the cd-tree's skeletons (which we shortly call the *size of the cd-tree*). The total number of edges in skeletons of T is twice the total size of all cuts represented by T. Since T represents $O(n)$ cuts, each of size $O(n)$, it is $c \in O(n^2)$.

Assume the cd-tree is rooted. Recall that in this case every node μ represents a cluster of G. The *pertinent graph* pert(μ) of the node μ is the cluster represented by μ. Note that one could also define the pertinent graph recursively, by removing the virtual vertex corresponding to the parent of μ (the *parent vertex*) from skel(μ) and replacing each remaining virtual vertex by the pertinent graph of the corresponding child of μ. Clearly, the pertinent graph of a leaf of T is a single vertex and the pertinent graph of the root is the whole graph G. A similar concept, also defined for unrooted cd-trees, is the *expansion graph*. The expansion graph exp(v_i) of a virtual vertex v_i in skel(μ) is the pertinent graph of its corresponding neighbor μ_i of μ, when rooting T at μ. One can think of the expansion graph exp(v_i) as the subgraph of G represented by v_i in skel(μ).

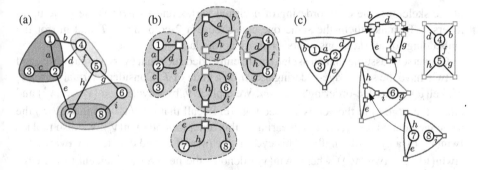

Fig. 1. (a) A c-planar drawing of a clustered graph. (b) The corresponding (rooted) cd-tree (without leaves). The skeletons are drawn inside their corresponding (gray) nodes. Every pair of twins has the same edge-ordering. (c) Construction of a c-planar drawing from the cd-tree.

The leaves of a cd-tree represent singleton clusters that exist only due to technical reasons. It is often more convenient to consider cd-trees with all leaves removed as follows. Let μ be a node with virtual vertex v in $\mathrm{skel}(\mu)$ that corresponds to a leaf. The leaf contains $\mathrm{twin}(v)$ and a non-virtual vertex $v \in V$ in its skeleton (with an edge between $\mathrm{twin}(v)$ and v for each edge incident to v in G). We replace v in $\mathrm{skel}(\mu)$ with the non-virtual vertex v and remove the leaf containing v. Clearly, this preserves all clusters except for the singleton cluster. Moreover, the graph G represented by the cd-tree remains unchanged as we replaced the virtual vertex v by its expansion graph $\exp(v) = v$. In the following we always assume the leaves of cd-trees to be removed.

The CD-Tree Characterization. We show that c-planarity testing can be expressed in terms of edge-orderings in embeddings of the skeletons of T.

Theorem 1. *A clustered graph is c-planar if and only if the skeletons of all nodes in its cd-tree can be embedded such that every virtual vertex and its twin have the same edge-ordering.*

Proof. Assume G admits a c-planar drawing Γ on the sphere. Let μ be a node of T with incident edges $\varepsilon_1, \ldots, \varepsilon_k$ connecting μ to its neighbors $\mu_1, \ldots, \mu_k$, respectively. Let further v_i be the virtual vertex in $\mathrm{skel}(\mu)$ corresponding to μ_i and let V_i be the nodes in the expansion graph $\exp(v_i)$. For every cut (V_i, V_i') (with $V_i' = V \setminus V_i$), Γ contains a simple closed curve C_i representing it. Since the V_i are disjoint, we can choose a point on the sphere to be the outside such that V_i lies inside C_i for $i = 1, \ldots, k$. Since Γ is a c-planar drawing, the C_i do not intersect and only the edges of G crossing the cut (V_i, V_i') cross C_i exactly once. Thus, one can contract the inside of C_i to a single point while preserving the embedding of G. Doing this for each of the curves C_i yields $\mathrm{skel}(\mu)$ together with a planar embedding. Moreover, the edge-ordering of v_i is the same as the order in which the edges cross the curve C_i. Applying the same construction for the neighbor μ_i corresponding to v_i yields a planar embedding of $\mathrm{skel}(\mu_i)$ in which the edge-ordering of $\mathrm{twin}(v_i)$ is the same as the order in which these edges cross the curve C_i, when traversing C_i in counter-clockwise direction. Thus, in the resulting embeddings of the skeletons, the edge-ordering of a virtual vertex and its twin is the same up to reversal. To make them the same one can choose a 2-coloring of T and mirror the embeddings of all skeletons of nodes in one color class.

Conversely, assume that all skeletons are embedded such that every virtual vertex and its twin have the same edge-ordering. Let μ be a node of T. Consider a virtual vertex v_i of $\mathrm{skel}(\mu)$ with edge-ordering $e_1, \ldots, e_\ell$. We replace v_i by a cycle $C_i = (v_i^1, \ldots, v_i^\ell)$ and attach the edge e_j to the vertex v_i^j; see Fig. 1c. Recall that $\mathrm{twin}(v_i)$ has in $\mathrm{skel}(\mu_i)$ the same incident edges $e_1, \ldots, e_\ell$ appearing in this order around $\mathrm{twin}(v_i)$. We also replace $\mathrm{twin}(v_i)$ by a cycle of length ℓ. This cycle is the *twin* of C_i and denote it by $\mathrm{twin}(C_i) = (\mathrm{twin}(v_i^1), \ldots, \mathrm{twin}(v_i^\ell))$ where $\mathrm{twin}(v_i^j)$ denotes the new vertex incident to the edge e_j. As the interiors of C_i and $\mathrm{twin}(C_i)$ are empty, we can glue the skeletons $\mathrm{skel}(\mu)$ and $\mathrm{skel}(\mathrm{twin}(\mu))$ together by identifying the vertices of C_i with the corresponding vertices in $\mathrm{twin}(C_i)$ (one of the embeddings has to be flipped). Applying this replacement for every virtual vertex and gluing it with its twin leads to an embedded planar graph G^+ with the following properties. First, G^+ contains a subdivision of G. Second, for every cut corresponding to an edge $\varepsilon = \{\mu, \mu_i\}$ in T, G^+ contains the cycle C_i with exactly one subdivision vertex of an edge e of G if the cut corresponding to ε separates the endpoints of e. Third, no two of these cycles share a vertex. The planar drawing of G^+ gives a planar drawing of G. Moreover, the drawings of the cycles can be used as curves representing the cuts, yielding a c-planar drawing of G.   □

Cutvertices in Skeletons. We show that cutvertices in skeletons correspond to different connected components in a cluster or in a co-cluster.

Lemma 1. *Let v be a virtual vertex that is a cutvertex in its skeleton. The expansion graphs of virtual vertices in different blocks incident to v belong to different connected components in $\exp(\mathrm{twin}(v))$.*

Proof. Let μ be the node whose skeleton contains v. Recall that one can obtain the graph $\exp(\mathrm{twin}(v))$ by removing v from $\mathrm{skel}(\mu)$ and replacing all other virtual vertices of $\mathrm{skel}(\mu)$ with their expansion graphs. Clearly, this yields (at least) one connected component for each of the blocks incident to v.   □

Lemma 2. *Every cluster in a clustered graph is connected if and only if in every node μ of the rooted cd-tree the parent vertex is not a cutvertex in $\mathrm{skel}(\mu)$.*

Proof. By Lemma 1, the existence of a cutvertex implies a disconnected cluster. Conversely, let $\mathrm{pert}(\mu)$ be disconnected and assume without loss of generality that $\mathrm{pert}(\mu_i)$ is connected for every child $\mu_1, \ldots, \mu_k$ of μ in the cd-tree. One obtains $\mathrm{skel}(\mu)$ without the parent vertex v by contracting in $\mathrm{pert}(\mu)$ the child clusters $\mathrm{pert}(\mu_i)$ to virtual vertices v_i. As the contracted graphs $\mathrm{pert}(\mu_i)$ are connected while the initial graph $\mathrm{pert}(\mu)$ is not, the resulting graph must be disconnected. Thus, v is a cutvertex in $\mathrm{skel}(\mu)$.   □

4   Clustered and Constrained Planarity

We first describe several constraints on planar embeddings, each restricting the edge-orderings of vertices. We then show the relation to c-planarity.

Consider a finite set S (e.g., edges incident to a vertex). Denote the set of all cyclic orders of S by O_S. An *order-constraint* on S is simply a subset of O_S (only the orders in the subset are *allowed*). A *family of order-constraints* for the set S is a set of different order constraints, i.e., a subset of the power set of O_S. We say that a family of order-constraints has a *compact representation*, if one can specify every order-constraint in this family with polynomial space (in $|S|$). In the following we describe families of order-constraints with compact representations.

A *partition-constraint* is given by partitioning S into subsets $S_1 \cup \ldots \cup S_k = S$. It requires that no two partitions *alternate*, i.e., elements $a_i, b_i \in S_i$ and $a_j, b_j \in S_j$ must not appear in the order a_i, a_j, b_i, b_j. A *PQ-constraint* requires that the order of elements in S is represented by a given PQ-tree with leaves S. A *full-constraint* contains only one order, i.e., the order of S is completely fixed.

A *partitioned full-constraint* restricts the orders of elements in S according to a partition constraint (partitions must not alternate) and additionally completely fixes the order within each partition. Similarly, *partitioned PQ-constraints* require the elements in each partition to be ordered according to a PQ-constraint. Clearly, this notion of partitioned order-constraints generalizes to arbitrary order-constraints.

Consider a planar graph G. By *constraining* a vertex v of G, we mean that there is an order-constraint on the edges incident to v. We then only allow planar embeddings of G where the edge-ordering of v is allowed by the order-constraint. By constraining G, we mean that several (or all) vertices of G are constrained.

4.1 Flat-Clustered Graph

Consider a flat-clustered graph, i.e., a clustered graph where the cd-tree is a star. We choose the center μ of the star to be the root. Let $v_1, \ldots, v_k$ be the virtual vertices in $\text{skel}(\mu)$ corresponding to the children $\mu_1, \ldots, \mu_k$ of μ. Note that $\text{skel}(\mu_i)$ contains exactly one virtual vertex, namely $\text{twin}(v_i)$. The possible ways to embed $\text{skel}(\mu_i)$ restrict the possible edge-orderings of $\text{twin}(v_i)$ and thus, by the characterization in Theorem 1, the edge-orderings of v_i in $\text{skel}(\mu)$. Hence, the graph $\text{skel}(\mu_i)$ essentially yields an order constraint for v_i in $\text{skel}(\mu)$. We consider c-planarity with differently restricted instances leading to different families of order-constraints. To show that testing c-planarity is equivalent to testing whether $\text{skel}(\mu)$ is planar with respect to order-constraints of a specific family, we have to show two directions. First, the embeddings of $\text{skel}(\mu_i)$ only yield order-constraints of the given family. Second, we can get every possible order-constraint of the given family by choosing an appropriate graph for $\text{skel}(\mu_i)$.

Theorem 2. *Testing c-planarity of flat-clustered graphs* (i) where each proper cluster consists of isolated vertices; (ii) where each cluster is connected; (iii) with fixed planar embedding; (iv) without restriction *is linear-time equivalent to testing planarity of a multi-graph with* (i) partition-constraints; (ii) PQ-constraints; (iii) partitioned full-constraints; (iv) partitioned PQ-constraints, *respectively.*

Proof. We start with case (i); see Fig. 2. Consider a flat-clustered graph G and let μ_i be one of the leaves of the cd-tree. As $\text{pert}(\mu_i)$ is a proper cluster, it consists of isolated vertices. Thus, $\text{skel}(\mu_i)$ is a set of vertices $v_1, \ldots, v_\ell$, each connected (with multiple

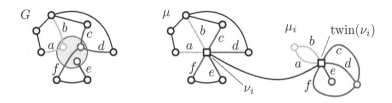

Fig. 2. A graph G with a single cluster consisting of isolated vertices together with an illustration of its cd-tree. An edge-ordering of $\text{twin}(v_i)$ corresponds to a planar embedding of $\text{skel}(\mu_i)$ if and only if no two partitions of $\{\{a,b\},\{c,d,f\},\{e\}\}$ alternate.

edges) to the virtual vertex $\text{twin}(v_i)$. The vertices $v_1,\dots,v_\ell$ partition the edges incident to $\text{twin}(v_i)$ into ℓ subsets. Clearly, in every planar embedding of $\text{skel}(\mu_i)$ no two partitions alternate. Moreover, every edge-ordering of $\text{twin}(v_i)$ in which no two partitions alternate gives a planar embedding of $\text{skel}(\mu_i)$. Thus, the edges incident to v_i in $\text{skel}(\mu)$ are constrained by a partition-constraint, where the partitions are determined by the incidence of the edges to the vertices $v_1,\dots,v_\ell$. One can easily construct the resulting instance of planarity with partition-constraints problem in linear time in the size of the cd-tree, which is linear in the size of G for flat-clustered graphs.

Conversely, given a planar graph H with partition-constraints, we set $\text{skel}(\mu) = H$. For every vertex of H we have a virtual vertex v_i in $\text{skel}(\mu)$ with corresponding child μ_i. We can simulate every partitioning of the edges incident to v_i by connecting edges incident to $\text{twin}(v_i)$ (in $\text{skel}(\mu_i)$) with vertices such that two edges are connected with the same vertex if and only if they belong to the same partition.

Consider case (ii). By Lemma 2 the condition of connected clusters is equivalent to requiring that the virtual vertex $\text{twin}(v_i)$ in the skeleton of any leaf μ_i of the cd-tree is not a cutvertex. The statement follows from the fact that the possible edge-orderings of non-cutvertices is PQ-representable and that any PQ-tree can be achieved by choosing an appropriate planar graph in which $\text{twin}(v_i)$ is not a cutvertex (see Sec. 2).

Consider case (iii). As in case (i), the blocks incident to $\text{twin}(v_i)$ in $\text{skel}(\mu_i)$ partition the edges incident to v_i in $\text{skel}(\mu)$ such that two partitions must not alternate. The fixed embedding of G fixes the edge-ordering of non-virtual vertices and thus fixes the embeddings of the blocks in $\text{skel}(\mu_i)$. Hence, we get partitioned full-constraints for v_i. Conversely, we can construct an arbitrary partitioned full-constraint as in case (i).

For case (iv) the arguments from case (iii) show that we again get partitioned order-constraints, while the arguments from case (ii) show that these order-constraints (for the blocks) are PQ-constraints.   □

Related Work. Biedl [1] proposes different drawing-models for graphs whose vertices are partitioned into two subsets. The model matching the requirements of c-planar drawings is called *HH-drawings*. Biedl et al. [2] show that one can test for the existence of HH-drawings in linear time. Hong and Nagamochi [16] rediscovered this result in the context of 2-page book embeddings. These results solve c-planarity for flat-clustered graphs if the skeleton of the root node contains two virtual vertices. This is equivalent to testing planarity with partitioned PQ-constraints for multi-graphs with only two vertices (Theorem 2). Thus, to solve c-planarity for flat-clustered graphs, one needs to solve an embedding problem on general planar multi-graphs that is so far only solved on a set of parallel edges (with absolutely non-trivial algorithms). This indicates that we are still far away from solving the c-planarity problem even for flat-clustered graphs.

Cortese et al. [9] give a linear-time algorithm for testing c-planarity of a flat-clustered cycle (i.e., G is a simple cycle) if the skeleton of the cd-tree's root is a multi-cycle. Requiring that G is a cycle implies that the skeleton of each non-root node in T has the property that the blocks incident to the parent vertex are simple cycles. Thus, in terms of constrained planarity, they show how to test planarity of multi-cycles with partition-constraints where each partition has size two. The result can be extended to a special case of c-planarity where the clustering is not flat. However, the cd-tree fails to have easy-to-state properties in this case, which shows that the cd-tree perspective of course has some limitations. Later, Cortese et al. [10] extended this result to the case where G is still a cycle, while the skeleton of the root can be an arbitrary planar multi-graph that has a fixed embedding up to the ordering of parallel edges. This is equivalent to testing planarity of such a graph with partition-constraints where each partition has size two.

Jelínková et al. [20] consider the case where each cluster contains at most three vertices (with additional restrictions). Consider a cluster containing only two vertices u and v. If u and v are connected, then the region representing the cluster can be always added and we can omit the cluster. Otherwise, the region representing the cluster in a c-planar drawing implies that one can add the edge uv to G, yielding an equivalent instance. Thus, one can assume that every cluster has size exactly 3, which yields flat-clustered graphs. In this setting they give efficient algorithms for the cases that G is a cycle and G is 3-connected. Moreover, they give an FPT-algorithm for the case that G is an *Eulerian graph* with k nodes, i.e., a graph obtained from a 3-connected graph of size k by multiplying and then subdividing edges.

In case G is 3-connected, its planar embedding is fixed and thus the edge-ordering of non-virtual vertices is fixed. Thus, one obtains partitioned full-constraints with the restriction that there are only three partitions. Clearly, the requirement that G is 3-connected also restricts the possible skeletons of the root of the cd-tree. It is an interesting open question whether planarity with partitioned full-constraints with at most three partitions can be tested efficiently for arbitrary planar graphs. In case G is a cycle, one obtains partition constraints with only three partitions and each partition has size two. Note that this in particular restricts the skeleton of the root to have maximum degree 6. Although these kind of constraints seem pretty simple to handle, the algorithm by Jelínková et al. is pretty involved. It seems like one barrier where constrained embedding becomes difficult is when there are partition constraints with three or more partitions (see also Theorem 4). The result about Eulerian graphs in a sense combines the cases where G is 3-connected and a cycle. A vertex has either degree two and thus yields a partition of size two or it is one of the constantly many vertices with higher degree for which the edge-ordering is partly fixed.

4.2 General Clustered Graphs

Expressing c-planarity for general clustered graphs (not necessarily flat) in terms of constrained planarity problems is harder for the following reason. Consider a leaf μ in the cd-tree. The skeleton of μ is a planar graph yielding (as in the flat-clustered case) partitioned PQ-constraints for its parent μ'. This restricts the possible embeddings of $\text{skel}(\mu')$ and thus the order-constraints one obtains for the parent of μ' are not necessarily again partitioned PQ-constraints.

One can express this issue in the following, more formal way. Let G be a planar multi-graph with vertices $v_1, \ldots, v_n$ and designated vertex $v = v_n$. The map φ_G^v maps a tuple $(C_1, \ldots, C_n)$ where C_i is an order-constraint on the edges incident to v_i to an order-constraint C on the edges incident to v. The order-constraint $C = \varphi_G^v(C_1, \ldots, C_n)$ contains exactly those edge-orderings of v that one can get in a planar embedding of G that respects $C_1, \ldots, C_n$. Note that C is empty if and only if there is no such embedding. Note further that testing planarity with order-constraints is equivalent to deciding whether φ_G^v evaluates to the empty set. We call such a map φ_G^v a *constrained-embedding operation*.

The issue mentioned above (constraints iteratively handed to parents) boils down to the fact that partitioned PQ-constraints are not closed under constrained-embedding operations. On the positive side, we obtain a general algorithm for solving c-planarity as follows. Assume we have a family of order-constraints C with compact representations that is closed under constrained-embedding operations. Assume further that we can evaluate the constrained embedding operations in polynomial time on order-constraints in C. Then one can simply solve c-planarity by traversing the cd-tree bottom-up, evaluating for a node μ with parent vertex v the constrained-embedding operation $\varphi_{\text{skel}(\mu)}^v$ on the constraints one computed in the same way for the children of μ.

Clearly, when restricting the skeletons of the cd-tree or requiring properties for the parent vertices in these skeletons, these restrictions carry over to the constrained-embedding operations one has to consider. More precisely, let $\mathcal{R}$ be a set of pairs (G, v), where v is a vertex in G. We say that a clustered graph is $\mathcal{R}$-restricted if $(\text{skel}(\mu), v) \in \mathcal{R}$ holds for every node μ in the cd-tree with parent vertex v. Moreover, the $\mathcal{R}$-restricted constrained-embedding operations are those operations φ_G^v with $(G, v) \in \mathcal{R}$. The following theorem directly follows.

Theorem 3. *One can solve c-planarity of $\mathcal{R}$-restricted clustered graphs in polynomial time if there is a family C of order-constraints such that*

- *C has a compact representation,*
- *C is closed under $\mathcal{R}$-restricted constrained-embedding operations,*
- *every $\mathcal{R}$-restricted constrained-embedding operation on order-constraints in C can be evaluated in polynomial time.*

When dropping the requirement that C has a compact representation the algorithm becomes super-polynomial only in the maximum degree d of the virtual vertices (the number of possible order-constraints for a set of size d depends only on d). Moreover, if φ_G^v has only k order constraints (whose sizes are bounded by a function of d) as input, then φ_G^v can be evaluated by iterating over all combinations of orders, applying a planarity test in every step. This gives an FPT-algorithm with parameter $d + k$ (running time $O(f(d+k)p(n))$, where f is a computable function depending only on $d + k$ and p is a polynomial). In other words, we obtain an FPT-algorithm where the parameter is the sum of the maximum degree of the tree T and the maximum number of edges leaving a cluster. Note that this generalizes the FPT-algorithm by Chimani and Klein [6] with respect to the total number of edges connecting different clusters.

Moreover, Theorem 3 has the following simple implication. Consider a clustered graph where each cluster is connected. This restricts the skeletons of the cd-tree such that none of the parent vertices is a cutvertex (Lemma 1). Thus, we have $\mathcal{R}$-restricted clustered graphs where $(G, v) \in \mathcal{R}$ implies that v is not a cutvertex in G. PQ-constraints are closed under $\mathcal{R}$-restricted constrained-embedding operations as the valid edge-ordering of non-cutvertices is PQ-representable and planarity with PQ-constraints is basically equivalent to planarity (one can model a PQ-tree with a simple gadget; see Sec. 2). Thus, Theorem 3 directly implies that c-planarity can be solved in polynomial time if each cluster is connected.

Related Work. The above algorithm resulting from Theorem 3 is more or less the one described by Lengauer [21]. The algorithm was later rediscovered by Feng et al. [13] who coined the term "c-planarity". The algorithm runs in $O(c) \subseteq O(n^2)$ time (recall that c is the size of the cd-tree). Dahlhaus [11] improves the running time to $O(n)$. Cortese et al. [8] give a characterization that also leads to a linear-time algorithm.

Goodrich et al. [14] consider the case where each cluster is either connected or *extrovert*. Let μ be a node in the cd-tree with parent μ'. The cluster pert(μ) is extrovert if the parent cluster pert(μ') is connected and every connected component in pert(μ) is connected to a vertex not in the parent pert(μ'). They show that one obtains an equivalent instance by replacing the extrovert cluster pert(μ) with one cluster for each of its connected components while requiring additional PQ-constraints for the parent vertex in the resulting skeleton. In this instance every cluster is connected and the additional PQ-constraints clearly do no harm.

Another extension to the case where every cluster must be connected is given by Gutwenger et al. [15]. They give an algorithm for the case where every cluster is connected with the following exception. Either, the disconnected clusters form a path in the tree or for every disconnected cluster the parent and all siblings are connected. This has basically the effect that at most one order-constraint in the input of a constrained-embedding operation is not a PQ-tree.

Jelínek et al. [17, 18] assume each cluster to have at most two connected components and the underlying (connected) graph to have a fixed planar embedding. Thus, they consider $\mathcal{R}$-restricted clustered graphs where $(G, v) \in \mathcal{R}$ implies that v is incident to at most two different blocks. The fixed embedding of the graph yields additional restrictions that are not so easy to state within this model.

5   Cutvertices with Two Non-trivial Blocks

The input of the SIMULTANEOUS PQ-ORDERING problem consists of several PQ-trees together with child-parent relations between them (the PQ-trees are the nodes of a directed acyclic graph) such that the leaves of every child form a subset of the leaves of its parents. SIMULTANEOUS PQ-ORDERING asks whether one can choose orders for all PQ-trees *simultaneously* in the sense that every child-parent relation implies that the order of the leaves of the parent are an extension of the order of the leaves of the child.

In this way one can represent orders that cannot be represented by a single PQ-tree. For example, adding one or more children to a PQ-tree T restricts the set of orders represented by T by requiring the orders of different subsets of leaves to be represented by some other PQ-tree. Moreover, one can synchronize the orders of different trees that share a subset of leaves by introducing a common child containing these leaves.

SIMULTANEOUS PQ-ORDERING is NP-hard but efficiently solvable for so-called 2-fixed instances [4]. For every biconnected planar graph G, there exists an instance of SIMULTANEOUS PQ-ORDERING, the *PQ-embedding representation*, that represents all planar embeddings of G [4]. It has the following properties.

- For every vertex v in G there is a PQ-tree $T(v)$, the *embedding tree*, that has the edges incident to v as leaves.
- For every solution of the PQ-embedding representation, setting the edge-ordering of every vertex v to the order given by $T(v)$ yields a planar embedding. Moreover, one can obtain every embedding of G in this way.
- The instance remains 2-fixed when adding up to one child to each embedding tree.

A PQ-embedding representation still exists if every cutvertex in G is incident to at most two *non-trivial blocks* (blocks that are not just bridges) [3].

Theorem 4. *C-planarity can be tested in $O(c^2) \subseteq O(n^4)$ time if every virtual vertex in the skeletons of the cd-tree is incident to at most two non-trivial blocks.*

Proof. Let G be a clustered graph with cd-tree T. For the skeleton of each node in T, we get a PQ-embedding representation with the above-mentioned properties. Let μ be a node of T and let v be a virtual vertex in $\text{skel}(\mu)$. Let μ' be the node whose skeleton contains $\text{twin}(v)$. The embedding representations of $\text{skel}(\mu)$ and $\text{skel}(\mu')$ contain the embedding trees $T(v)$ and $T(\text{twin}(v))$ representing the edge-orderings of v and $\text{twin}(v)$, respectively. To ensure that v and $\text{twin}(v)$ have the same edge-ordering, one can simply add a PQ-tree as common child of $T(v)$ and $T(\text{twin}(v))$. We do this for every virtual node in the skeletons of T. Due to the last property of the PQ-embedding representations, the resulting instance remains 2-fixed and can thus be solved efficiently.

Every solution of this SIMULTANEOUS PQ-ORDERING instance D yields planar embeddings of the skeletons such that every virtual vertex and its twin have the same edge-ordering and vice versa. By Theorem 1, testing c-planarity is equivalent to solving D. The size of D is linear in the size c of T. Moreover, solving SIMULTANEOUS PQ-ORDERING for 2-fixed instances can be done in quadratic time [4], yielding the running time $O(c^2)$. □

Theorem 4 includes the following interesting cases. The latter extends the result by Jelínek et al. [19] from four to five outgoing edges per cluster.

Corollary 1. *C-planarity can be tested in $O(c^2) \subseteq O(n^4)$ time if every cluster and every co-cluster has at most two connected components.*

Corollary 2. *C-planarity can be tested in $O(n^2)$ time if every cluster has at most five outgoing edges.*

References

1. Biedl, T.: Drawing planar partitions I: LL-drawings and LH-drawings. In: SoCG 1998, pp. 287–296. ACM (1998)
2. Biedl, T., Kaufmann, M., Mutzel, P.: Drawing planar partitions II: HH-drawings. In: Hromkovič, J., Sýkora, O. (eds.) WG 1998. LNCS, vol. 1517, pp. 124–136. Springer, Heidelberg (1998)
3. Bläsius, T., Rutter, I.: Simultaneous PQ-ordering with applications to constrained embedding problems. CoRR abs/1112.0245, 1–46 (2011)
4. Bläsius, T., Rutter, I.: Simultaneous PQ-ordering with applications to constrained embedding problems. In: SODA 2013. SIAM (2013)
5. Booth, K.S., Lueker, G.S.: Testing for the consecutive ones property, interval graphs, and graph planarity using PQ-tree algorithms. J. Comput. System Sci. 13(3), 335–379 (1976)
6. Chimani, M., Klein, K.: Shrinking the search space for clustered planarity. In: Didimo, W., Patrignani, M. (eds.) GD 2012. LNCS, vol. 7704, pp. 90–101. Springer, Heidelberg (2013)
7. Cornelsen, S., Wagner, D.: Completely connected clustered graphs. J. of Disc. Alg. 4(2), 313–323 (2006)
8. Cortese, P.F., Di Battista, G., Frati, F., Patrignani, M., Pizzonia, M.: C-planarity of c-connected clustered graphs. J. Graph Alg. Appl. 12(2), 225–262 (2008)
9. Cortese, P.F., Di Battista, G., Patrignani, M., Pizzonia, M.: Clustering cycles into cycles of clusters. J. Graph Alg. Appl. 9(3), 391–413 (2005)
10. Cortese, P.F., Di Battista, G., Patrignani, M., Pizzonia, M.: On embedding a cycle in a plane graph. Disc. Math. 309(7), 1856–1869 (2009)
11. Dahlhaus, E.: A linear time algorithm to recognize clustered planar graphs and its parallelization. In: Lucchesi, C.L., Moura, A.V. (eds.) LATIN 1998. LNCS, vol. 1380, pp. 239–248. Springer, Heidelberg (1998)
12. Di Battista, G., Frati, F.: Efficient C-planarity testing for embedded flat clustered graphs with small faces. In: Hong, S.-H., Nishizeki, T., Quan, W. (eds.) GD 2007. LNCS, vol. 4875, pp. 291–302. Springer, Heidelberg (2008)
13. Feng, Q.W., Cohen, R.F., Eades, P.: Planarity for clustered graphs. In: Spirakis, P.G. (ed.) ESA 1995. LNCS, vol. 979, pp. 213–226. Springer, Heidelberg (1995)
14. Goodrich, M.T., Lueker, G.S., Sun, J.Z.: C-planarity of extrovert clustered graphs. In: Healy, P., Nikolov, N.S. (eds.) GD 2005. LNCS, vol. 3843, pp. 211–222. Springer, Heidelberg (2006)
15. Gutwenger, C., Jünger, M., Leipert, S., Mutzel, P., Percan, M., Weiskircher, R.: Advances in c-planarity testing of clustered graphs. In: Goodrich, M.T., Kobourov, S.G. (eds.) GD 2002. LNCS, vol. 2528, pp. 220–235. Springer, Heidelberg (2002)
16. Hong, S.H., Nagamochi, H.: Two-page book embedding and clustered graph planarity. Tech. Rep. 2009-004, Kyoto University, Depart. Appl. Math. & Phys. (2009)
17. Jelínek, V., Jelínková, E., Kratochvíl, J., Lidický, B.: Clustered planarity: Embedded clustered graphs with two-component clusters (2009), http://kam.mff.cuni.cz/~bernard/pub/flat.pdf (manuscript)
18. Jelínek, V., Jelínková, E., Kratochvíl, J., Lidický, B.: Clustered planarity: Embedded clustered graphs with two-component clusters (extended abstract). In: Tollis, I.G., Patrignani, M. (eds.) GD 2008. LNCS, vol. 5417, pp. 121–132. Springer, Heidelberg (2009)
19. Jelínek, V., Suchý, O., Tesař, M., Vyskočil, T.: Clustered planarity: Clusters with few outgoing edges. In: Tollis, I.G., Patrignani, M. (eds.) GD 2008. LNCS, vol. 5417, pp. 102–113. Springer, Heidelberg (2009)
20. Jelínková, E., Kára, J., Kratochvíl, J., Pergel, M., Suchý, O., Vyskočil, T.: Clustered planarity: Small clusters in cycles and eulerian graphs. J. Graph Alg. Appl. 13(3), 379–422 (2009)
21. Lengauer, T.: Hierarchical planarity testing algorithms. J. ACM 36(3), 474–509 (1989)

Mapsets: Visualizing Embedded and Clustered Graphs

Alon Efrat[1], Yifan Hu[2], Stephen G. Kobourov[1], and Sergey Pupyrev[1,3]

[1] Department of Computer Science, University of Arizona, Tucson, Arizona, USA
[2] Yahoo Labs, New York, USA
[3] Institute of Mathematics and Computer Science, Ural Federal University, Russia

Abstract. We describe MapSets, a method for visualizing embedded and clustered graphs. The proposed method relies on a theoretically sound geometric algorithm, which guarantees the contiguity and disjointness of the regions representing the clusters, and also optimizes the convexity of the regions. A fully functional implementation is available online and is used in a comparison with related earlier methods.

1 Introduction

In many real-world examples of relational datasets, groups of objects (clusters) are an inherent part of the input. For example, scientists belong to specific research communities, politicians are affiliated with specific parties, and living organisms are divided into biological species in the tree of life. Such clusters are often visualized with regions in the plane that enclose related objects. By explicitly defining the boundary and coloring the regions, the cluster information becomes evident. In many instances the data objects are often associated with fixed or relative positions in the plane. In geo-referenced data, for example, the positions of the objects might be based on their geographic coordinates. Thus a natural problem arises: How to best visualize graphs in which vertices are divided into clusters and embedded with fixed positions in the plane?

Several existing visualization approaches seem suitable. For example, methods for visualizing set relations over existing embedded pointsets, such as BubbleSets [6] and LineSets [2] use colored shapes to connect objects that belong to the same set. Alternatively, a geographic map metaphor can be used to represent such data. With self-organizing maps [22] or geometry-based GMaps [9], objects become cities and cluster information is captured by uniquely colored countries. While both approaches can produce compelling visualizations, we argue that neither is perfectly suited to the problem of visualizing embedded and clustered graphs.

As the number of sets increases, set-based methods generate complex and sometimes ambiguous results. More recent methods, such as KelpDiagrams [7] and Kelp-Fusion [15], reduce visual clutter and guarantee unambiguous visualization. But more importantly, all of these methods result in overlapping regions for the sets, even when the input sets are disjoint. This unnecessarily increases visual complexity and might mislead the viewer about the disjointness of the sets. The geographic map approach suffers from a different problem. A country in the map, that represents a given cluster of vertices, might not be a contiguous region in the plane. Even though each cluster is colored with

C. Duncan and A. Symvonis (Eds.): GD 2014, LNCS 8871, pp. 452–463, 2014.
© Springer-Verlag Berlin Heidelberg 2014

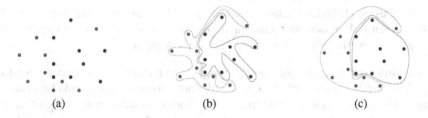

Fig. 1. (a) An embedded and clustered (red/blue) pointset. (b-c) Two different ways to construct contiguous shapes bounding points of the same color.

a unique color, such fragmented maps are difficult to read as human perception of color changes based on surrounding colors [19] and can be misinterpreted [11].

We want to combine the advantages of existing methods, while attempting to avoid their problems. That is, we are interested in visualizing embedded and clustered graphs with non-fragmented and non-overlapping regions. While constructing such representations is easy in theory, in practice the regions may still have high visual complexity; see Fig. 1. Ideally the regions should be as *convex* as possible, as the convex hull best captures cohesive grouping according to Gestalt theory [12].

With this in mind, we describe MapSets, a method for creating non-fragmented, non-overlapping regions that are as convex as possible, from a given embedded and clustered graph. We consider several criteria for measuring convexity of a shape, and propose a novel geometric problem aiming at optimizing convexity. We present a theoretical analysis of the problem in Section 3. Next, in Section 4, we describe a practical method for visualizing clustered graphs. A comparison of the method with existing techniques is provided in Section 5.

2   Related Work

Set Visualization. Graph clusters can be viewed as sets over graph vertices. In Venn diagrams and their generalization, Euler diagrams, closed curves correspond to (possibly overlapping) sets, and overlaps between the curves indicate intersections. Simonetto et al. [21] automatically generate Euler-like diagrams, by allowing disconnected regions, which can be complex and non-convex. Riche and Dwyer [20] propose a way to avoid the visual complexity problem by drawing simplified rectangular Euler-like diagrams, that do not depict the intersections between the sets explicitly, by duplicating objects that belong to multiple sets. In a user study, they found that it is beneficial to show intersections using simple set regions and strict containment, enabled by the duplication. For the setting where the positions of the objects are fixed, Collins et al. [6] present BubbleSets, a method based on isocontours to overlay such an arrangement with enclosing set regions. The readability of these visualizations suffer when there are many overlapping regions. LineSets [2] aim to improve the readability of complex set intersections and to minimize the overall visual clutter by reducing set regions to simple curved lines drawn through set elements. KelpDiagrams [7] incorporate classic graph-drawing "bubble and stick" style graph or tree spanners over the member points in a set. KelpFusion [15] adds filled-in regions to provide a stronger sense of grouping for close elements. A significant limitation of all these set visualization techniques is that they produce overlapping regions even when the sets are disjoint.

Visualizing Graphs as Maps. The geographic map metaphor is utilized as visual interface for relational data, where objects, relations between objects, and clustering are captured by cities, roads, and countries. Using maps to visualize non-cartographic data has been considered in the context of spatialization [22]. Maps of science showing groups of scientific disciplines are used by a wide range of professionals to grasp developments in science and technology [4].

The geographic map metaphor is used in the Graph-to-Map approach (GMap) [9]. GMap combines graph layout and graph clustering, together with appropriate coloring of the clusters and creating boundaries based on clusters and connectivity in the original graph. However, since layout and clustering are two separate steps, a region representing a cluster may often be fragmented; see Fig. 7(b). Such fragmentation makes it difficult to identify the correct regions and can result in misinterpretation of the map [11]. Note that in the setting when either an input embedding or clustering can be modified, the GMap approach can be improved to achieve contiguous regions [13].

Colored Spanning Trees. From an algorithmic perspective, our geometric approach of optimizing convexity of regions that cover points in the plane is related to several problems in which the input is a multicolored point set [1, 3]. The group Steiner tree problem deals with a graph with colored vertices, and the objective is to find a minimum weight subtree covering all colors [16]. Also related is the problem of computing spanning graphs for multicolored point set [10]. The problem is motivated by optimizing the amount of "ink" needed to connect monochromatic points that arise when visualizing sets using the KelpFusion technique. These trees cannot be directly used as "skeletons" of regions in the plane as they can result in overlapping regions.

3   Constructing Contiguous Non-overlapping Regions

We assume that the input instance consists of a set of objects P with fixed positions $p_i \in \mathbb{R}^2$ for all $i \in P$, for example, cities and their geographic locations. In practical applications labels are often associated with the objects. In this case, we assume that non-overlapping bounding boxes for the labels are given. The input also specifies a clustering $C = \{C_1, \ldots, C_k\}$ of the objects with $\cup_{i=1}^k C_i = P$ and $C_i \cap C_j = \emptyset$ for $i \neq j$. We wish to enclose all objects of the same cluster by a single contiguous region so that regions corresponding to different clusters do not overlap.

On one hand, simply overlaying each cluster with a convex region (e.g., bounding box or convex hull) is not always a valid solution, as it might cover elements in other clusters. On the other hand, representing clusters by some minimal regions (e.g., spanning or Steiner trees) is also not always valid, as it might result in intersecting regions.

We require regions that are contiguous and disjoint, and it is not difficult to see that such regions can be easily computed. We can begin by computing a crossing-free spanning tree of points belonging to some cluster. Once the tree is constructed, its vertices and edges become "obstacles" that should be avoided by subsequent trees. Note that

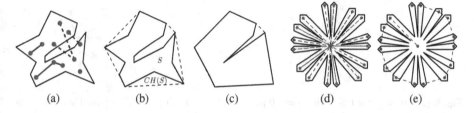

Fig. 2. Convexity measures for a shape S enclosing red points. (a) Solid segments are within S, while dashed ones are not. (b) A shape and its convex hull (dashed). (c) Area-based measure ignores boundary defects. (d-e) Ink needed to connect the points is much bigger than the length of the minimum spanning tree. The shape is enclosed in solid black, while the tree is dashed red.

all the clusters will be processed as the trees do not separate the plane into more than one region. Finally, contiguous non-overlapping regions can be grown, starting from these disjoint trees. However, this procedure often generates "octopus"-like shapes that are neither aesthetically pleasant nor practically useful for visualization; see Fig. 1. Hence, we require a method for creating regions that are as convex as possible. In order to design such a method, a quality criterion for measuring the convexity of regions is needed. Next we review and formalize several convexity measures.

3.1 Convexity Measures

A shape S is said to be convex if it has the following property: If points $p, q \in \mathbb{R}$ belong to S then all points from the line segment $[pq]$ belong to S as well. The definition allows for several different ways to measure the convexity of non-convex shapes.

Point/Vertex Visibility. For a given shape S, this convexity measure is defined as the probability that for points p and q, chosen uniformly at random from S, all points from the line segment $[pq]$ also belong to S [24]. The result is a real number from $[0, 1]$, with 1 corresponding to convex shapes. A problem with this definition is that it is difficult to compute, even if S is a polygon. Hence, we consider its discrete variant, taking into account that the input of our problem specifies points in the plane; see Fig. 2(a).

This vertex-based measure takes into account how many segments $[pq]$ are completely in S for pairs of input points $p, q \in P$ of the cluster corresponding to S. The measure is defined as $\frac{\sum_{p,q \in P} \delta(p,q)}{|P|^2}$, where the sum is over all pairs of input points P and $\delta(p, q) = 1$ if $[pq]$ lies inside S and $\delta(p, q) = 0$, otherwise.

Convex Hull Area/Perimeter. Recall that the smallest convex set which includes a shape S is called the *convex hull*, $CH(S)$, of S; see Fig. 2(b). The area-based convexity measure is defined as $\frac{Area(S)}{Area(CH(S))}$; it is frequently used and appears in textbooks [23]. The result is a real number from $[0, 1]$, with 1 corresponding to convex shapes. Unlike visibility-based measures, the convex hull-based one is very easy to calculate efficiently and is robust with respect to noise. However, the definition does not allow to detect defects on boundary that have a relatively small impact on the shape area; see Fig. 2(c). The perimeter-based definition attempts to remedy this: $\frac{Perimeter(S)}{Perimeter(CH(S))}$.

(a)   (b)

Fig. 3. (a) An input for CST with $n = 10$ points and $k = 3$ colors. (b) An optimal solution with minimum ink containing Steiner points.

If a shape S is convex, then there exists a minimum spanning tree on the given point set such that every edge of the tree lies completely in S; non-convex shapes do not necessarily admit such a spanning tree. Hence, the length of a shortest curve that belongs to S and connects all the input points is an indicator of convexity of S. In the following measure, we compare the length of such a curve (or equivalently, the amount of "ink" needed to connect all the points) with the length of a minimum spanning tree on the same point set; see Figs. 2(d)-2(e).

Minimum Ink. Let $|\,\mathrm{INK}(P)|$ be the length of the shortest curve connecting all vertices of V lying in S, and let $|\,\mathrm{MST}(P)|$ be the length of the minimum spanning tree of V. The measure is defined as $\frac{|\,\mathrm{MST}(P)|}{|\,\mathrm{INK}(P)|}$. Again, 1 indicates the best possible value (though, it does not always correspond to a convex shape); smaller values are worse.

There are advantages and disadvantages of all of the proposed convexity measures, and there are also many other ways to define convexity of shapes or polygons. In an attempt to balance theoretical and practical considerations, we focus on visibility-based and the ink-based measures. Similar ink-based criteria are used for constructing LineSets and KelpDiagrams. By minimizing the ink needed for drawing, all of these techniques aim to reduce visual clutter and increase the readability of the representation.

3.2 Algorithm for Ink Minimization

Here we study a problem motivated by computing contiguous regions with minimum ink. The input consists of n points in the plane, and each point is associated with one of k colors. The CST (COLORED SPANNING TREES) problem is to connect points of the same color by mutually non-intersecting curves of shortest total length. In an optimal solution each curve forms a tree spanning points of the corresponding color. The trees may use additional (Steiner) points that do not belong to the original pointset; see Fig. 3.

Computing an optimal solution for CST is NP-hard. This follows from the observation that the known NP-complete MINIMUM STEINER TREE problem is a special case of CST, in which the input consists of monochromatic points. Next we present a heuristic for CST and prove that it is an approximation algorithm in the theoretical sense.

We refer to the minimum spanning and Steiner trees of a set of points P as $\mathrm{MST}(P)$ and $\mathrm{SMT}(P)$, respectively; their lengths are denoted by $|\,\mathrm{MST}(P)|$ and $|\,\mathrm{SMT}(P)|$. We use the Steiner ratio, denoted by ρ, which is the supremum of the ratio of the length of a minimum spanning tree to the length of a minimum Steiner tree. It is conjectured that $\rho = \frac{2}{\sqrt{3}} \approx 1.15$, and ≈ 1.21 is the best-known upper bound on ρ [5].

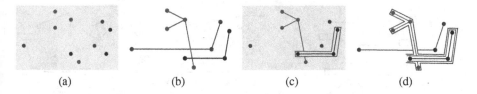

Fig. 4. Steps of the algorithm for the CST problem. (a) An input with $n = 10$ points and $k = 3$. (b) Computing minimum spanning trees. (c) Bounding the tree having the shortest length, and removing red-blue crossings. (d) Merging with the green tree.

We begin with the description of our algorithm in the setting when the input consists of blue and red points. First, we compute a minimum spanning tree of the blue points (ignoring the red ones), and a minimum spanning tree of the red points; see Fig.4(b). If the trees do not intersect, then they form a solution for CST. Otherwise, we create a red "shell" bounding the blue tree; see Fig.4(c). Now red-blue crossings appear inside the constructed shell, and they can be eliminated by removing all portions of the red tree inside the shell. Finally, the red curve, consisting of the original spanning tree and the constructed shell, can be transformed to a tree by disconnecting its cycles; see Fig.4(d).

The general algorithm works in the following steps. First, create a minimum tree $\mathrm{MST}(C_i)$ spanning the set of points C_i for $1 \le i \le k$, ignoring points of the other colors. Sort the colors with respect to the length of the corresponding spanning trees. Without loss of generality, we may assume that the resulting order is $C_1, \ldots, C_k$ and $|\mathrm{MST}(C_1)| \le \cdots \le |\mathrm{MST}(C_k)|$. Then the resulting curve for C_1 is the tree $\mathrm{MST}(C_1)$. A curve for each successive color C_i is constructed by adding a "shell" bounding the curve corresponding to C_{i-1}. The length of the shell is exactly $2 \sum_{j<i} |\mathrm{MST}(C_j)|$, since it bounds all the spanning trees corresponding to already processed colors; see Fig. 4. The length of a curve for C_i is then $|\mathrm{MST}(C_i)| + 2 \sum_{j<i} |\mathrm{MST}(C_j)|$.

In order to analyze the algorithm, we denote the amount of ink in the optimal solution by OPT, and the total length of the constructed solution by ALG. An optimal solution induces a curve connecting all points of the same cluster, that is, the solution is a Steiner tree for the set of points (but not necessarily the minimum one). Hence, $\mathrm{OPT} \ge \sum_i |\mathrm{SMT}(C_i)| \ge \sum_i |\mathrm{MST}(C_i)|/\rho$. On the other hand,

$$\mathrm{ALG} \le \sum_{i=1}^{k} \left(|\mathrm{MST}(C_i)| + 2 \sum_{j=1}^{i-1} |\mathrm{MST}(C_j)| \right) = \sum_{i=1}^{k} (2k - 2i + 1)|\mathrm{MST}(C_i)|, \text{ and}$$

$$\frac{\mathrm{ALG}}{\mathrm{OPT}} \le \frac{\sum_{i=1}^{k}(2k - 2i + 1)|\mathrm{MST}(C_i)|}{\sum_{i=1}^{k} |\mathrm{MST}(C_i)|/\rho} =$$

$$= \rho \frac{\sum_{i=1}^{\lfloor k/2 \rfloor} (2k - 2i + 1)|\mathrm{MST}(C_i)| + \sum_{i=1}^{\lceil k/2 \rceil} (2i - 1)|\mathrm{MST}(C_{k-i+1})|}{\sum_{i=1}^{k} |\mathrm{MST}(C_i)|} \le k\rho.$$

Hence, our algorithm is a $(k\rho)$-approximation for the CST problem for any $k \ge 1$.

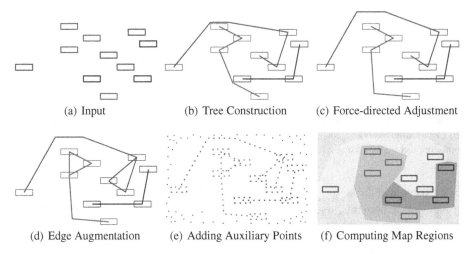

(a) Input   (b) Tree Construction   (c) Force-directed Adjustment

(d) Edge Augmentation   (e) Adding Auxiliary Points   (f) Computing Map Regions

Fig. 5. Algorithmic pipeline of MapSets

4   MapSets

Here we describe MapSets, starting with a high-level overview; see Fig. 5. We assume that the input is a set of rectangular shapes (bounding boxes of labels) embedded in the plane along with a clustering. In the first step, we compute spanning mutually non-crossing trees interconnecting centers of rectangles corresponding to the same cluster, while minimizing the total ink needed to draw the trees. In the second step, we modify the trees by adding buffers of free space around the segments of the trees, using a force-directed heuristic. In the third step, we try to optimize the convexity of the resulting regions based on the vertex visibility measure, by adding edges between vertices in the same cluster, while ensuring that edges of different clusters do not cross. In the fourth step, we use the modified trees and the added edges to build contiguous non-overlapping boundaries for all clusters.

Tree Construction. In order to construct the trees, we employ the approximation algorithm described in Section 3.2. For each cluster, we first compute a minimum tree spanning the set of rectangle centers, ignoring other clusters. The clusters are then sorted in non-decreasing order by the length of the computed trees and processed in this order. At each step we consider all the precomputed trees as obstacles that should be avoided when constructing the current tree. The rectangles are also treated as obstacles. We compute a sparse visibility graph on the set of obstacles, where the vertices are all the centers and corners of the rectangles, and there is an edge between two vertices if one can draw a straight-line segment without crossing the obstacles. The sparse visibility graph (unlike the full visibility graph) has a linear number of edges and can be constructed efficiently [8]. We then compute shortest paths (of the visibility graph) between every pair of rectangles of the current cluster. From these shortest paths, we compute a minimum spanning tree for the current cluster. We add the tree to the set of obstacles and proceed with the next cluster.

Force-directed Adjustment. This step improves the constructed trees. Our goal is to provide some free space around the edges of the trees so as to avoid (1) narrow channels between parts of the same region and (2) region borders lying too close to the input vertex labels. To accomplish this, we consider an adjustment graph G^{adj} in which vertices are the end points and bends of the constructed trees and edges are maximal straight-line segments of the trees. We then build a force system moving the vertices of G^{adj} that correspond to the bends of the tree. The system relies on the following forces.

- **Vertex-vertex Attraction.** We would like to keep the ink of the drawing low. Therefore, for every vertex of G^{adj}, there is a force pushing the vertex towards its neighbor vertices in G^{adj}.
- **Edge-edge Repulsion.** This repulsive force attempts to push the edges of G^{adj} apart to provide enough space to draw the regions. In order to compute the force, it is convenient to replace edges of G^{adj} with cylinders of a specified thickness. Then, if two cylinders corresponding to different trees intersect, the force repels them away from each other. This force also ensures that the trees do not overlap and do not intersect during the adjustment process.
- **Edge-label Repulsion.** This force prevents edges from being routed too close to the input text labels. Again, it is convenient to consider the edges of G^{adj} as cylinders. If a cylinder occludes a label, then we introduce a repulsive force moving the corresponding vertices of G^{adj} away from the label.

We use iterative refinement similar to that used in drawing graphs with edge bundles [18] to adjust the positions of the vertices of G^{adj} under these three forces: repulsive forces have equal priorities, and the attractive force is weaker. In our experiments, the force system provides the desired buffer of free space around the trees and converges quickly; see Fig. 8.

Edge Augmentation. In this step we try to optimize the convexity of the regions using the vertex visibility metric. Consider all possible straight-line segments connecting centers of rectangles corresponding to the same cluster. Our goal is to select and add as many of these segments as possible, subject to the condition that they do not cross each other. To this end, we construct a graph H in which vertices are the straight-line segments. A segment is added to H only if it does not intersect the trees found in the previous step. Two vertices of H are connected by an edge if the corresponding straight-line segments cross each other. Notice that now the problem reduces to the problem of finding a maximum non-crossing (independent) set of segments in the plane. The problem can be solved optimally in polynomial time for two clusters, that is, if $k = 2$. Indeed, in this setting the graph H is bipartite, and the size of a maximum independent set in a bipartite graph equals to the number of edges in a minimum edge covering by König's theorem. The latter can be found using a maximum matching algorithm. Unfortunately, the general variant is NP-hard even for $k = 3$ [14]. Therefore, unless $k = 2$, we use a greedy strategy to solve the problem. At every step, we choose the minimum degree vertex in H and remove its neighbors. It is well-known that this strategy guarantees an approximation ratio of $(\Delta + 2)/3$ on graphs with maximum degree Δ.

Adding Auxiliary Points and Computing Map Regions. Given the initial placement of the labels and curves connecting the labels from the previous steps, we need explicit

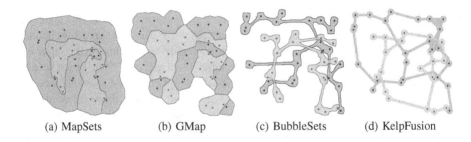

(a) MapSets   (b) GMap   (c) BubbleSets   (d) KelpFusion

Fig. 6. The senator voting graph (the part of the U.S. west of Mississippi). The vertices are senators (red republicans and blue democrats) positioned according to their home-cities.

regions grouping together labels and curves in the same cluster. As in GMap, we generate boundaries by adding dummy points to the current embedding. There are three types of the dummy points: (a) random points, sufficiently far away from the set of the input labels, lead to more rounded and thus more realistic region boundaries; (b) random points along bounding boxes of the labels help ensure that the labels are drawn inside the regions; (c) auxiliary points along all the edges constructed on the previous step, that keep the regions connected. The distance between consecutive points on an edge is chosen to be less than the distance to any other point of a different color. After adding the dummy points, we compute the Voronoi diagram of the set of all points and merge the Voronoi cells that belong to the points of the same color.

Time Complexity. Now we discuss the complexity of our algorithm on an input with n points and k clusters, assuming we can compute distances and intersections between geometric primitives (points, line-segments, rectangles) in constant time. The sparse visibility graph can be constructed in $O(n \log n)$ time and it contains $O(n)$ edges [8]. Therefore, computing all pairwise distances takes $O(n^2)$ time and finding a minimum spanning tree for one cluster takes $O(n^2 + n \log n)$ time. Summing over all clusters, we get $O(kn^2)$. In the iterative force-directed heuristic we compute forces between pairs of edges, which can take $O(n^2)$ in the worst case. Hence, the time complexity of the force-directed heuristic is $O(cn^2)$, where c is the maximum number of iterations in the adjustment ($c = 10$ in our implementation). The complexity of the edge augmentation step is $O(n^3)$, as we may add quadratic number of edges in the greedy process. Finally, computing the boundaries takes $O(n \log n)$ time. Therefore, the overall time complexity is $O(kn^2 + n^3)$. More details and actual running times are given in the next section.

5   Experiments

Here we compare our new algorithm, MapSets, with the existing approaches for map-like visualizations: GMap [9], BubbleSets [6], and KelpFusion [15]. A fully functional implementation of MapSets, GMap, and BubbleSets, together with a complete dataset, is available in an online system at http://gmap.cs.arizona.edu.

Our first example is the senator voting graph; see Fig. 6. The vertices in the graph are the U.S. senators in 2010 positioned according to their home-cities in the U.S. The

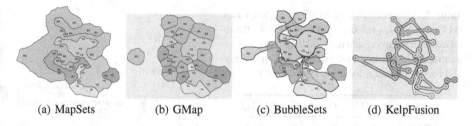

| (a) MapSets | (b) GMap | (c) BubbleSets | (d) KelpFusion |

Fig. 7. The graph of genetic similarities between 50 individuals in Europe. The layout is computed using the principal component analysis, while the clusters correspond to the countries of origin of the individuals.

clustering is based on the political party they represent, red for republicans and blue for democrats. Clearly, both clustering and geographic information of the vertices are fixed and cannot be changed. One can see that GMap produces fragmented clusters, while BubbleSets and KelpFusion compute overlapping regions. On the other hand, the result of MapSets is contiguous and non-overlapping, which makes it easier to analyze the distribution of senators over the map.

The second example shows the population structure within Europe [17]. The original points correspond to genetic data from $1,387$ Europeans (but we sampled only 50 vertices corresponding to Eastern Europe for illustration purposes). The positions of the vertices come from the original principal component analysis, based on the similarity matrix. As the authors point out, the PCA plot (appropriately rotated) closely matches the geographic outlines of Europe; hence, it is undesirable to change the node positions. The clusters are extracted independently and corresponds to the countries of origin of the individuals. Again, only MapSets constructs non-fragmented disjoint regions; see Fig. 7. Arguably, this is easier to analyze than the overlapping regions produced by BubbleSets and KelpFusion.

We next analyze the performance of our ink minimization heuristic. To this end, we utilize a collection of 9 real-world networks, that are embedded and clustered using the GMap tool with the default setting. Table 1 gives details about the graphs and measurements of our ink saving algorithm. Here, ALG shows the ratio of the total ink of the computed trees to the total length of the minimum spanning trees computed individually for every cluster. In other words, this is an approximation factor achieved by our algorithm on the test cases. Although we can only guarantee factor $k\rho$, in practice the algorithm performs very well, always producing a solution at most 1.6 times worse than the optimal. Our experiments indicate that ink minimization strategy often results in aesthetically more pleasant map visualizations.

Similarly, ALG_{fd} indicates the utilized ink after the force-directed adjustments. As expected, the ink increases after the step, but the increase is not significant. On the other hand, the adjustments improve the quality of the resulting regions.

Table 1. Measurements of MapSets on test cases: ALG and ALG$_{fd}$ stand for the ratio between the total ink of the drawing and the total length of the minimum spanning trees after the steps *Tree Construction* and *Force-directed Adjustment*, respectively.

graph	$|P|$	k	ALG	ALG$_{fd}$
Colors	50	6	1.002	1.012
GD	506	23	1.582	1.612
Recipes	381	15	1.356	1.502
Trade	211	8	1.101	1.259
Universities	161	8	1.366	1.443
SODA	316	11	1.204	1.296
IPL	336	11	1.337	1.414
SOCG	500	11	1.492	1.601
TARJAN	252	16	1.150	1.197
ALGO	500	5	1.547	1.650

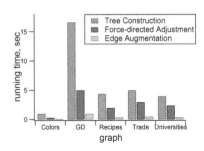

Fig. 8. Running times of the different steps of MapSets on some of the test cases.

The algorithm is implemented in C++. We use a machine with Intel i5 3.2GHz and 8GB RAM for measuring running time; see Fig. 8. The last two steps, *Adding Auxiliary Points* and *Computing Regions*, are very efficient taking few milliseconds for the largest graphs, and hence are not included in the chart. The first step, *Tree Construction*, is usually the most time consuming; it is more efficient for nearly contiguous clusters (e.g, Colors) and less efficient for graphs with many fragments (e.g., GD). Although *Edge Augmentation* theoretically has cubic time complexity, it is among the fastest steps in practice, because there are usually not many edges added. Overall, our algorithm processes all the graphs (most with hundreds of vertices) in less than a minute. This is slower than the GMap and LineSets but comparable to BubbleSets. Since our algorithm extensively utilizes many primitive geometric operations (e.g., testing for segment intersections), using a specialized geometric library will likely improve the performance.

6   Conclusion and Future Work

We designed and implemented a new approach for visualizing embedded and clustered graphs. Unlike existing techniques, our MapSets method always produces contiguous and non-overlapping regions. Results of the initial evaluations seem promising. We also presented a simple approximation algorithm for the geometric problem of ink minimization motivated by the method. A natural future direction is to improve the approximation factor. It would be also worthwhile to carefully evaluate different convexity measures and select one that offers the best balance between ease of computation and visual quality of the resulting regions. Similarly interesting would be in-depth user study comparing map-based visualizations constructed with different approaches considered in the paper.

Acknowledgements. The work supported in part by NSF grants CCF-1115971 and DEB 1053573. We thank the authors of [17] for the DNA dataset. The drawings of KelpFusion are courtesy of the authors of [15].

References

1. Agarwal, P.K., Edelsbrunner, H., Schwarzkopf, O., Welzl, E.: Euclidean minimum spanning trees and bichromatic closest pairs. Discrete & Comput. Geom. 6(1), 407–422 (1991)
2. Alper, B., Riche, N.H., Ramos, G., Czerwinski, M.: Design study of LineSets, a novel set visualization technique. IEEE Trans. Vis. Comput. Graphics 17(12), 2259–2267 (2011)
3. Arora, S., Chang, K.: Approximation schemes for degree-restricted MST and red–blue separation problems. Algorithmica 40(3), 189–210 (2004)
4. Boyack, K.W., Klavans, R., Börner, K.: Mapping the backbone of science. Scientometrics 64, 351–374 (2005)
5. Chung, F., Graham, R.: A new bound for Euclidean Steiner minimal trees. Annals of the New York Academy of Sciences 440(1), 328–346 (1985)
6. Collins, C., Penn, G., Carpendale, S.: Bubble sets: Revealing set relations with isocontours over existing visualizations. IEEE Trans. Vis. Comput. Graphics 15(6), 1009–1016 (2009)
7. Dinkla, K., van Kreveld, M.J., Speckmann, B., Westenberg, M.A.: Kelp diagrams: Point set membership visualization. Comput. Graph. Forum 31(3, pt1), 875–884 (2012)
8. Dwyer, T., Nachmanson, L.: Fast edge-routing for large graphs. In: Eppstein, D., Gansner, E.R. (eds.) GD 2009. LNCS, vol. 5849, pp. 147–158. Springer, Heidelberg (2010)
9. Hu, Y., Gansner, E.R., Kobourov, S.G.: Visualizing graphs and clusters as maps. IEEE Comput. Graphics and Appl. 30(6), 54–66 (2010)
10. Hurtado, F., Korman, M., van Kreveld, M., Löffler, M., Sacristán, V., Silveira, R.I., Speckmann, B.: Colored spanning graphs for set visualization. In: Wismath, S., Wolff, A. (eds.) GD 2013. LNCS, vol. 8242, pp. 280–291. Springer, Heidelberg (2013)
11. Jianu, R., Rusu, A., Hu, Y., Taggart, D.: How to display group information on node-link diagrams: An evaluation. IEEE Trans. Vis. Comput. Graphics 20(11), 1530–1541 (2014)
12. Kanizsa, G., Gerbino, W.: Convexity and symmetry in figure-ground organization. Vision and Artifact, 25–32 (1976)
13. Kobourov, S.G., Pupyrev, S., Simonetto, P.: Visualizing graphs as maps with contiguous regions. Comput. Graph. Forum (2014)
14. Kratochvíl, J., Nešetřil, J.: Independent set and clique problems in intersection-defined classes of graphs. Commentationes Math. Univ. Carolinae 31(1), 85–93 (1990)
15. Meulemans, W., Riche, N., Speckmann, B., Alper, B., Dwyer, T.: KelpFusion: A hybrid set visualization technique. IEEE Trans. Vis. Comput. Graphics 19(11), 1846–1858 (2013)
16. Mitchell, J.S.: Geometric shortest paths and network optimization. Handbook of Computational Geometry 334, 633–702 (2000)
17. Novembre, et al.: Genes mirror geography within Europe. Nature 456(7218), 98–101 (2008)
18. Pupyrev, S., Nachmanson, L., Bereg, S., Holroyd, A.E.: Edge routing with ordered bundles. In: van Kreveld, M., Speckmann, B. (eds.) GD 2011. LNCS, vol. 7034, pp. 136–147. Springer, Heidelberg (2011)
19. Purves, D., Lotto, R.B.: Why we see what we do: An empirical theory of vision. Sinauer Associates (2003)
20. Riche, N.H., Dwyer, T.: Untangling Euler diagrams. IEEE Trans. Vis. Comput. Graphics 16(6), 1090–1099 (2010)
21. Simonetto, P., Auber, D., Archambault, D.: Fully automatic visualisation of overlapping sets. Comput. Graph. Forum 28(3), 967–974 (2009)
22. Skupin, A., Fabrikant, S.I.: Spatialization methods: a cartographic research agenda for non-geographic information visualization. Cartogr. Geogr. Inform. 30, 95–119 (2003)
23. Sonka, M., Hlavac, V., Boyle, R.: Image Processing, Analysis, and Machine Vision. Thomson-Engineering (2007)
24. Zunic, J., Rosin, P.L.: A convexity measurement for polygons. IEEE Trans. Pattern Anal. Mach. Intell. 26, 173–182 (2002)

Increasing-Chord Graphs On Point Sets*

Hooman Reisi Dehkordi[1], Fabrizio Frati[2], and Joachim Gudmundsson[2]

[1] School of Information Technologies, Monash University, Australia
hooman.dehkordi@monash.edu
[2] School of Information Technologies, The University of Sydney, Australia
{fabrizio.frati,joachim.gudmundsson}@sydney.edu.au

Abstract. We tackle the problem of constructing increasing-chord graphs spanning point sets. We prove that, for every point set P with n points, there exists an increasing-chord planar graph with $O(n)$ Steiner points spanning P. Further, we prove that, for every convex point set P with n points, there exists an increasing-chord graph with $O(n \log n)$ edges (and with no Steiner points) spanning P.

1   Introduction

A *proximity graph* is a geometric graph that can be constructed from a point set by connecting points that are "close", for some local or global definition of proximity. Proximity graphs constitute a topic of research in which the areas of graph drawing and computational geometry nicely intersect. A typical graph drawing question in this topic asks to characterize the graphs that can be represented as a certain type of proximity graphs. A typical computational geometry question asks to design an algorithm to construct a proximity graph spanning a given point set.

Euclidean minimum spanning trees and Delaunay triangulations are famous examples of proximity graphs. Given a point set P, a *Euclidean minimum spanning tree* (MST) of P is a geometric tree with P as vertex set and with minimum total edge length; the *Delaunay triangulation* of P is a triangulation T such that no point in P lies inside the circumcircle of any triangle of T. From a computational geometry perspective, given a point set P with n points, an MST of P with maximum degree five exists [12] and can be constructed in $O(n \log n)$ time [4]; also, the Delaunay triangulation of P exists and can be constructed in $O(n \log n)$ time [4]. From a graph drawing perspective, every tree with maximum degree five admits a representation as an MST [12] and it is NP-hard to decide whether a tree with maximum degree six admits such a representation [7]; also, characterizing the class of graphs that can be represented as Delaunay triangulations is a deeply studied question, which still eludes a clear answer; see, e.g., [5,6]. Refer to the excellent survey by Liotta [10] for more on proximity graphs.

While proximity graphs have constituted a frequent topic of research in graph drawing and computational geometry, they gained a sudden peak in popularity even outside these communities in 2004, when Papadimitriou *et al.* [14] devised an elegant routing protocol that works effectively in all the networks that can be represented as a certain type of proximity graphs, called *greedy graphs*. For two points p and q in the plane,

* Work partially supported by the Australian Research Council (grant DE140100708).

C. Duncan and A. Symvonis (Eds.): GD 2014, LNCS 8871, pp. 464–475, 2014.

denote by $\overline{pq}$ the straight-line segment having p and q as end-points, and by $|\overline{pq}|$ the length of $\overline{pq}$. A geometric path $(v_1, \ldots, v_n)$ is *greedy* if $|\overline{v_{i+1}v_n}| < |\overline{v_iv_n}|$, for every $1 \le i \le n - 1$. A geometric graph G is *greedy* if, for every ordered pair of vertices u and v, there exists a greedy path from u to v in G. A result related to our paper is that, for every point set P, the Delaunay triangulation of P is a greedy graph [13].

In this paper we study *self-approaching* and *increasing-chord graphs*, that are types of proximity graphs defined by Alamdari *et al.* [2]. A geometric path $\mathcal{P} = (v_1, \ldots, v_n)$ is *self-approaching* if, for every three points a, b, and c in this order on $\mathcal{P}$ from v_1 to v_n (possibly a, b, and c are internal to segments of $\mathcal{P}$), it holds that $|\overline{bc}| < |\overline{ac}|$. A geometric graph G is *self-approaching* if, for every ordered pair of vertices u and v, G contains a self-approaching path from u to v; also, G is *increasing-chord* if, for every pair of vertices u and v, G contains a path between u and v that is self-approaching both from u to v and from v to u; thus, an increasing-chord graph is also self-approaching. The study of self-approaching and increasing-chord graphs is motivated by their relationship with greedy graphs (a self-approaching graph is also greedy), and by the fact that such graphs have a small geometric dilation, namely at most 5.3332 [9] (self-approaching graphs) and at most 2.094 [15] (increasing-chord graphs).

Alamdari *et al.* showed: (i) how to test in linear time whether a path in $\mathbb{R}^2$ is self-approaching; (ii) a characterization of the class of self-approaching trees; and (iii) how to construct, for every point set P with n points in $\mathbb{R}^2$, an increasing-chord graph that spans P and uses $O(n)$ Steiner points.

In this paper we focus our attention on the problem of constructing increasing-chord graphs spanning given point sets in $\mathbb{R}^2$. We prove two main results.

- We show that, for every point set P with n points, there exists an increasing-chord planar graph with $O(n)$ Steiner points spanning P. This answers a question of Alamdari *et al.* [2] and improves upon their result (iii) above, since our increasing-chord graphs are planar and contain increasing-chord paths between every pair of points, including the Steiner points (which is not the case for the graphs in [2]). It is interesting that our result is achieved by studying Gabriel triangulations, which are proximity graphs strongly related to Delaunay triangulations (a Gabriel triangulation of a point set P is a subgraph of the Delaunay triangulation of P). It has been proved in [2] that Delaunay triangulations are not, in general, self-approaching.
- We show that, for every convex point set P with n points, there exists an increasing-chord graph that spans P and that has $O(n \log n)$ edges (and no Steiner points).

2   Definitions and Preliminaries

A *geometric graph* (P, S) consists of a point set P in the plane and of a set S of straight-line segments (called *edges*) between points in P. A geometric graph is *planar* if no two of its edges cross. A planar geometric graph partitions the plane into connected regions called *faces*. The bounded faces are *internal* and the unbounded face is the *outer face*. A geometric planar graph is a *triangulation* if every internal face is delimited by a triangle and the outer face is delimited by a convex polygon.

Let p, q, and r be points in the plane. We denote by $\angle pqr$ the angle defined by a clockwise rotation around q bringing $\overline{pq}$ to coincide with $\overline{qr}$.

Fig. 1. A convex point set that is one-sided with respect to a directed straight line d

A *convex combination* of a set of points $P = \{p_1, \ldots, p_k\}$ is a point $\sum \alpha_i p_i$ where $\sum \alpha_i = 1$ and $\alpha_i \geq 0$ for each $1 \leq i \leq k$. The *convex hull* $\mathcal{H}_P$ of P is the set of points that can be expressed as a convex combination of the points in P. A *convex point set* P is such that no point is a convex combination of the others. Let P be a convex point set and d be a directed straight line not orthogonal to any line through two points of P. Order the points in P as their projections appear on d; then, the *minimum point* and the *maximum point* of P with respect to d are the first and the last point in such an ordering. We say that P is *one-sided with respect to d* if the minimum and the maximum point of P with respect to d are consecutive along the border of $\mathcal{H}_P$. See Fig. 1. A *one-sided convex point set* is a convex point set that is one-sided with respect to some directed straight line d. The proof of our first lemma shows an algorithm to construct an increasing-chord planar graph spanning a one-sided convex point set.

Lemma 1. *Let P be any one-sided convex point set with n points. There exists an increasing-chord planar graph spanning P with $2n - 3$ edges.*

Proof. Assume that P is one-sided with respect to the positive x-axis x. Such a condition can be met after a suitable rotation of the Cartesian axes. Let $\{p_1, p_2, \ldots, p_n\}$ be the points in P, ordered as their projections appear on x.

We show by induction on n that an increasing-chord planar graph G spanning P exists, in which all the edges on the border of $\mathcal{H}_P$ are in G. If $n = 2$ then the graph with a single edge $\overline{p_1 p_2}$ is an increasing-chord planar graph spanning P. Next, assume that $n > 2$ and let p_j be a point with largest y-coordinate in P (possibly $j = 1$ or $j = n$). Point set $Q = P \setminus \{p_j\}$ is convex, one-sided with respect to x, and has $n - 1$ points. By induction, there exists an increasing-chord planar graph G' spanning Q in which all the edges on the border of $\mathcal{H}_Q$ are in G'. Let G be the graph obtained by adding vertex p_j and edges $\overline{p_{j-1} p_j}$ and $\overline{p_j p_{j+1}}$ to G'. We have that G is planar, given that G' is planar and that edges $\overline{p_{j-1} p_j}$ and $\overline{p_j p_{j+1}}$ are on the border of $\mathcal{H}_P$. Further, all the edges on the border of $\mathcal{H}_P$ are in G. Moreover, G contains an increasing-chord path between every pair of points in Q, by induction; also, G contains an increasing-chord path between p_j and every point p_i in Q, as one of the two paths on the border of $\mathcal{H}_P$ connecting p_j and p_i is both x- and y-monotone, and hence increasing-chord by the results in [2]. Finally, G is a maximal outerplanar graph, hence it has $2n - 3$ edges. □

The *Gabriel graph* of a point set P is the geometric graph that has an edge $\overline{pq}$ between two points p and q if and only if the closed disk whose diameter is $\overline{pq}$ contains no point of $P \setminus \{p, q\}$ in its interior or on its boundary. A *Gabriel triangulation* is a triangulation that is the Gabriel graph of its point set P. We say that a point set P *admits* a Gabriel triangulation if the Gabriel graph of P is a triangulation. A triangulation is a Gabriel triangulation if and only if every angle of a triangle delimiting an internal face is acute [8]. See [8,10,11] for more properties about Gabriel graphs.

In Section 3 we will prove that every Gabriel triangulation is increasing-chord. A weaker version of the converse is also true, as proved in the following.

Lemma 2. *Let P be a set of points and let $G(P, S)$ be an increasing-chord graph spanning P. Then all the edges of the Gabriel graph of P are in S.*

Proof. Suppose, for a contradiction, that there exists an increasing-chord graph $G(P, S)$ and an edge $\overline{uv}$ of the Gabriel graph of P such that $\overline{uv} \notin S$. Then, consider any increasing-chord path $\mathcal{P} = (u = w_1, w_2, \ldots, w_k = v)$ in G. Since $\overline{uv} \notin S$, it follows that $k > 2$. Assume w.l.o.g. that w_1, w_2, and w_k appear in this clockwise order on the boundary of triangle (w_1, w_2, w_k). Since the closed disk with diameter $\overline{uv}$ does not contain any point in its interior or on its boundary, it follows that $\angle w_k w_2 w_1 < 90°$. If $\angle w_2 w_1 w_k \geq 90°$, then $|w_1 w_k| < |w_2 w_k|$, a contradiction to the assumption that $\mathcal{P}$ is increasing-chord. If $\angle w_2 w_1 w_k < 90°$, then the altitude of triangle (w_1, w_2, w_k) incident to w_k hits $\overline{w_1 w_2}$ in a point h. Hence, $|hw_k| < |w_2 w_k|$, a contradiction to the assumption that $\mathcal{P}$ is increasing-chord which proves the lemma.   □

3 Planar Increasing-Chord Graphs with Few Steiner Points

We show that, for any point set P, one can construct an increasing-chord planar graph $G(P', S)$ such that $P \subseteq P'$ and $|P'| \in O(|P|)$. Our result has two ingredients. The first one is that Gabriel triangulations are increasing-chord graphs. The second one is a result of Bern *et al.* [3] stating that, for any point set P, there exists a point set P' such that $P \subseteq P'$, $|P'| \in O(|P|)$, and P' admits a Gabriel triangulation. Combining these two facts proves our main result. The proof that Gabriel triangulations are increasing-chord graphs consists of two parts. In the first one, we prove that geometric graphs having a θ-*path* between every pair of points are increasing-chord. In the second one, we prove that in every Gabriel triangulation there exists a θ-path between every pair of points.

We introduce some definitions. The *slope* of a straight-line segment $\overline{uv}$ is the angle spanned by a clockwise rotation around u that brings $\overline{uv}$ to coincide with the positive x-axis. Thus, if θ is the slope of $\overline{uv}$, then $\theta + k \cdot 360°$ is also the slope of $\overline{uv}$, $\forall k \in \mathbb{Z}$. A straight-line segment $\overline{uv}$ is a θ-*edge* if its slope is in the interval $[\theta - 45°; \theta + 45°]$. Also, a geometric path $\mathcal{P} = (p_1, \ldots, p_k)$ is a θ-*path* from p_1 to p_k if $\overline{p_i p_{i+1}}$ is a θ-edge, for every $1 \leq i \leq k - 1$. Consider a point a on a θ-path $\mathcal{P}$ from p_1 to p_k. Then, the subpath $\mathcal{P}_a$ of $\mathcal{P}$ from a to p_k is also a θ-path. Moreover, denote by $W_\theta(a)$ the closed wedge with an angle of $90°$ incident to a and whose delimiting lines have slope $\theta - 45°$ and $\theta + 45°$; then $\mathcal{P}_a$ is contained in $W_\theta(a)$ (see Fig. 2). We have the following:

Lemma 3. *Let $\mathcal{P}$ be a θ-path from p_1 to p_k. Then, $\mathcal{P}$ is increasing-chord.*

Proof. Lemma 3 in [9] states the following (see also [1]): A curve $\mathcal{C}$ with end-points p and q is self-approaching from p to q if and only if, for every point a on $\mathcal{C}$, there exists a closed wedge with an angle of $90°$ incident to a and containing the part of $\mathcal{C}$

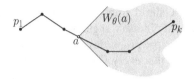

Fig. 2. Wedge $W_\theta(a)$ contains path $\mathcal{P}_a$

between a and q. By definition of θ-path, for every point a on $\mathcal{P}$, the closed wedge $W_\theta(a)$ with an angle of $90°$ incident to a and whose delimiting lines have slope $\theta - 45°$ and $\theta + 45°$ contains the subpath $\mathcal{P}_a$ of $\mathcal{P}$ from a to p_k. Hence, by Lemma 3 in [9], $\mathcal{P}$ is self-approaching from p_1 to p_k. An analogous proof shows that $\mathcal{P}$ is self-approaching from p_k to p_1, given that $\mathcal{P}$ is a $(\theta + 180°)$-path from p_k to p_1. ☐

We now prove that Gabriel triangulations contain θ-paths.

Lemma 4. *Let G be a Gabriel triangulation on a point set P. For every two points $s, t \in P$, there exists an angle θ such that G contains a θ-path from s to t.*

Proof. Consider any two points $s, t \in P$. Clockwise rotate G of an angle ϕ so that $y(s) = y(t)$ and $x(s) < x(t)$. Observe that, if there exists a θ-path from s to t after the rotation, then there exists a $(\theta + \phi)$-path from s to t before the rotation.

A θ-path $(p_1, \ldots, p_k)$ in G is *maximal* if there is no $z \in P$ such that $\overline{p_k z}$ is a θ-edge. For every maximal θ-path $\mathcal{P} = (p_1, \ldots, p_k)$ in G, p_k lies on the border of $\mathcal{H}_P$. Namely, assume the converse, for a contradiction. Since G is a Gabriel triangulation, the angle between any two consecutive edges incident to an internal vertex of G is smaller than $90°$, thus there is a θ-edge incident to p_k. This contradicts the maximality of $\mathcal{P}$. A maximal θ-path $(s = p_1, \ldots, p_k)$ is *high* if either (a) $y(p_k) > y(t)$ and $x(p_k) < x(t)$, or (b) $\overline{p_i p_{i+1}}$ intersects the vertical line through t at a point above t, for some $1 \leq i \leq k - 1$. Symmetrically, a maximal θ-path $(s = p_1, \ldots, p_k)$ is *low* if either (a) $y(p_k) < y(t)$ and $x(p_k) < x(t)$, or (b) $\overline{p_i p_{i+1}}$ intersects the vertical line through t at a point below t, for some $1 \leq i \leq k - 1$. High and low $(\theta + 180°)$-paths starting at t can be defined analogously. The proof of the lemma consists of two main claims.

Claim 1. If a maximal θ-path $\mathcal{P}_s$ starting at s and a maximal $(\theta + 180°)$-path $\mathcal{P}_t$ starting at t exist such that $\mathcal{P}_s$ and $\mathcal{P}_t$ are both high or both low, for some $-45° \leq \theta \leq 45°$, then there exists a θ-path in G from s to t.

Claim 2. For some $-45° \leq \theta \leq 45°$, there exist a maximal θ-path $\mathcal{P}_s$ starting at s and a maximal $(\theta + 180°)$-path $\mathcal{P}_t$ starting at t that are both high or both low.

Observe that Claims 1 and 2 imply the lemma.

We now prove Claim 1. Suppose that G contains a maximal high θ-path $\mathcal{P}_s$ starting at s and a maximal high $(\theta + 180°)$-path $\mathcal{P}_t$ starting at t, for some $-45° \leq \theta \leq 45°$. If $\mathcal{P}_s$ and $\mathcal{P}_t$ share a vertex $v \in P$, then the subpath of $\mathcal{P}_s$ from s to v and the subpath of $\mathcal{P}_t$ from v to t form a θ-path in G from s to t. Thus, it suffices to show that $\mathcal{P}_s$ and $\mathcal{P}_t$ share a vertex. For a contradiction assume the converse. Let p_s and p_t be the end-vertices of $\mathcal{P}_s$ and $\mathcal{P}_t$ different from s and t, respectively. Recall that p_s and p_t

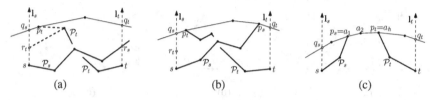

Fig. 3. Paths $\mathcal{P}_s$ and $\mathcal{P}_t$ intersect if: (a) $x(p_s) \geq x(t)$, (b) $x(s) < x(p_t) < x(p_s) < x(t)$, and (c) $x(s) < x(p_s) < x(p_t) < x(t)$

lie on the border of $\mathcal{H}_P$. Denote by l_s and l_t the vertical half-lines starting at s and t, respectively, and directed towards increasing y-coordinates; also, denote by q_s and q_t the intersection points of l_s and l_t with the border of $\mathcal{H}_P$, respectively. Finally, denote by Q the curve obtained by clockwise following the border of $\mathcal{H}_P$ from q_s to q_t.

Assume that $x(p_s) \geq x(t)$, as in Fig. 3(a). Path $\mathcal{P}_s$ starts at s and passes through a point r_s on l_t (possibly $r_s = q_t$), given that $x(p_s) \geq x(t)$. Path $\mathcal{P}_t$ starts at t and either passes through a point r_t on l_s, or ends at a point p_t on Q, depending on whether $x(p_t) \leq x(s)$ or $x(p_t) > x(s)$, respectively. Since $\mathcal{P}_s$ is x-monotone and lies in $\mathcal{H}_P$, it follows that r_t and p_t are above or on $\mathcal{P}_s$; also, t is below $\mathcal{P}_s$ given that $\mathcal{P}_s$ is a high path. It follows $\mathcal{P}_s$ and $\mathcal{P}_t$ intersect, hence they share a vertex given that G is planar.

Analogously, if $x(p_t) \leq x(s)$, then $\mathcal{P}_s$ and $\mathcal{P}_t$ share a vertex.

If $x(p_t) = x(p_s)$, then $\mathcal{P}_s \cup \mathcal{P}_t$ is a θ-path from s to t.

Next, if $x(s) < x(p_t) < x(p_s) < x(t)$, as in Fig. 3(b), then the end-points of $\mathcal{P}_s$ and $\mathcal{P}_t$ alternate along the boundary of the region R that is the intersection of $\mathcal{H}_P$, of the half-plane to the right of l_s, and of the half-plane to the left of l_t. Since $\mathcal{P}_s$ and $\mathcal{P}_t$ are x-monotone, they lie in R, thus they intersect, and hence they share a vertex.

Finally, assume that $x(s) < x(p_s) < x(p_t) < x(t)$, as in Fig. 3(c). Let $a_1, \dots, a_h$ be the clockwise order of the points along Q, starting at $p_s = a_1$ and ending at $a_h = p_t$. By the assumption $x(p_s) < x(p_t)$ we have $h \geq 2$. We prove that $\overline{a_1 a_2}$ is a θ-edge. Suppose, for a contradiction, that $\overline{a_1 a_2}$ is not a θ-edge. Since the slope of $\overline{a_1 a_2}$ is larger than $-90°$ and smaller than $90°$, it is either larger than $\theta + 45°$ and smaller than $90°$, or it is larger than $-90°$ and smaller than $\theta - 45°$. First, assume that the slope of $\overline{a_1 a_2}$ is larger than $\theta + 45°$ and smaller than $90°$, as in Fig. 4(a). Since the slope of $\overline{s a_1}$ is between $\theta - 45°$ and $\theta + 45°$, it follows that a_1 is below the line composed of $\overline{s a_2}$ and $\overline{a_2 t}$, which contradicts the assumption that a_1 is on Q. Second, if the slope of $\overline{a_1 a_2}$ is larger than $-90°$ and smaller than $\theta - 45°$, then we distinguish two further cases. In the first case, represented in Fig. 4(b), the slope of $\overline{a_1 t}$ is larger than $\theta - 45°$, hence a_2 is below the line composed of $\overline{s a_1}$ and $\overline{a_1 t}$, which contradicts the assumption that a_2 is on Q. In the second case, represented in Fig. 4(c), the slope of $\overline{a_1 t}$ is in the interval $[-90°; \theta - 45°]$. It follows that the slope of $\overline{t a_1}$ is in the interval $[90°; \theta + 135°]$; since the slope of $\overline{t a_h}$ is smaller than the one of $\overline{t a_1}$, we have that $\mathcal{P}_t$ is not a $(\theta + 180°)$-path. This contradiction proves that $\overline{a_1 a_2}$ is a θ-edge. However, this contradicts the assumption that $\mathcal{P}_s$ is a maximal θ-path, and hence concludes the proof of Claim 1.

We now prove Claim 2. First, we prove that, for *every* θ in the interval $[-45°; 45°]$, there exists a maximal θ-path starting at s that is low or high. Indeed, it suffices to prove

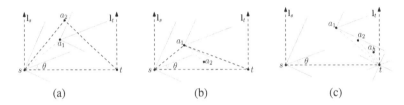

(a)   (b)   (c)

Fig. 4. Illustration for the proof that $\overline{a_1 a_2}$ is a θ-edge

that there exists a θ-edge incident to s, as such an edge is also a θ-path starting at s, and the existence of a θ-path starting at s implies the existence of a maximal θ-path starting at s. Consider a straight-line segment e_θ that is the intersection of a directed half-line incident to s with slope θ and of a disk of arbitrarily small radius centered at s. If e_θ is internal to $\mathcal{H}_P$, then consider the two edges e_1 and e_2 of G that are encountered when counter-clockwise and clockwise rotating e_θ around s, respectively. Then, e_1 or e_2 is a θ-edge, as the angle spanned by a clockwise rotation bringing e_1 to coincide with e_2 is smaller than $90°$, given that G is a Gabriel triangulation, and e_θ is encountered during such a rotation. If e_θ is outside $\mathcal{H}_P$, which might happen if s on the boundary of $\mathcal{H}_P$, then assume that the slope of e_θ is in the interval $[0°; 45°]$ (the case in which the slope of e_θ is in the interval $[-45°; 0°]$ is analogous). Then, the angle spanned by a clockwise rotation bringing e_θ to coincide with $\overline{st}$ is at most $45°$. Since $\overline{st}$ is in interior or on the boundary of $\mathcal{H}_P$, an edge e_1 of G is encountered during such a rotation, hence e_1 is a θ-edge. An analogous proof shows that, for *every* θ in the interval $[-45°; 45°]$, there exists a maximal $(\theta + 180°)$-path starting at t that is low or high.

Second, we prove that, for *some* $\theta \in [-45°; 45°]$, there exist a maximal low θ-path *and* a maximal high θ-path both starting at s. All the maximal $(-45°)$-paths (all the maximal $(45°)$-paths) starting at s are low (resp. high), given that every edge on these paths has slope in the interval $[-90°; 0°]$ (resp. $[0°; 90°]$). Thus, let θ be the smallest constant in the interval $[-45°; 45°]$ such that a maximal high θ-path exists. We prove that there also exists a maximal low θ-path starting at s. Consider an arbitrarily small $\epsilon > 0$. By assumption, there exists no high $(\theta - \epsilon)$-path. Hence, from the previous argument there exists a low $(\theta - \epsilon)$-path $\mathcal{P}$. If ϵ is sufficiently small, then no edge of $\mathcal{P}$ has slope in the interval $[\theta - 45° - \epsilon; \theta - 45°)$. Thus every edge of $\mathcal{P}$ has slope in the interval $[\theta - 45°; \theta + 45° - \epsilon)$, hence $\mathcal{P}$ is a maximal low θ-path starting at s.

Since there exist a maximal high θ-path starting at s, a maximal low θ-path starting at s, and a maximal $(\theta + 180°)$-path starting at t that is low or high, it follows that there exist a maximal θ-path $\mathcal{P}_s$ starting at s and a maximal $(\theta + 180°)$-path $\mathcal{P}_t$ starting at t that are both high or both low. This proves Claim 2 and hence the lemma.   □

Lemma 3 and Lemma 4 immediately imply the following.

Corollary 1. *Any Gabriel triangulation is increasing-chord.*

We are now ready to state the main result of this section.

Theorem 1. *Let P be a point set with n points. One can construct in $O(n \log n)$ time an increasing-chord planar graph $G(P', S)$ such that $P \subseteq P'$ and $|P'| \in O(n)$.*

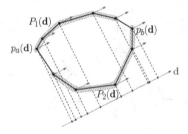

Fig. 5. Subsets $P_1(d)$ and $P_2(d)$ of a point set P determined by a directed straight line d

Proof. Bern, Eppstein, and Gilbert [3] proved that, for any point set P, there exists a point set P' with $P \subseteq P'$ and $|P'| \in O(n)$ such that P' admits a Gabriel triangulation G. Both P' and G can be computed in $O(n \log n)$ time [3]. By Corollary 1, G is increasing-chord, which concludes the proof.   □

We remark that $o(|P|)$ Steiner points are not always enough to augment a point set P to a point set that admits a Gabriel triangulation. Namely, consider any point set B with $O(1)$ points that admits no Gabriel triangulation. Construct a point set P out of $|P|/|B|$ copies of B placed "far apart" from each other, so that any triangle with two points in different copies of B is obtuse. Then, a Steiner point has to be added inside the convex hull of each copy of B to obtain a point set that admits a Gabriel triangulation.

4   Increasing-Chord Convex Graphs with Few Edges

In this section we prove the following theorem;

Theorem 2. *For every convex point set P with n points, there exists an increasing-chord geometric graph $G(P, S)$ such that $|S| \in O(n \log n)$.*

The main idea behind the proof of Theorem 2 is that any convex point set P can be decomposed into some one-sided convex point sets $P_1, \ldots, P_k$ (which by Lemma 1 admit increasing-chord spanning graphs with linearly many edges) in such a way that every two points of P are part of some P_i and that $\sum |P_i|$ is small. In order to perform such a decomposition, we introduce the concept of *balanced (d_1, d_2)-partition*.

Let P be a convex point set and let d be a directed straight line not orthogonal to any line through two points of P. See Fig. 5. Let $p_a(d)$ and $p_b(d)$ be the minimum and maximum point of P with respect to d, respectively. Let $P_1(d)$ be composed of those points in P that are encountered when clockwise walking along the boundary of $\mathcal{H}_P$ from $p_a(d)$ to $p_b(d)$, where $p_a(d) \in P_1(d)$ and $p_b(d) \notin P_1(d)$. Analogously, let $P_2(d)$ be composed of those points in P that are encountered when clockwise walking along the boundary of $\mathcal{H}_P$ from $p_b(d)$ to $p_a(d)$, where $p_b(d) \in P_2(d)$ and $p_a(d) \notin P_2(d)$.

Let d_1 and d_2 be two directed straight lines not orthogonal to any line through two points of P, where the clockwise rotation that brings d_1 to coincide with d_2 is at most $180°$. The (d_1, d_2)-*partition* of P partitions P into subsets $P_a = P_1(d_1) \cap P_1(d_2)$,

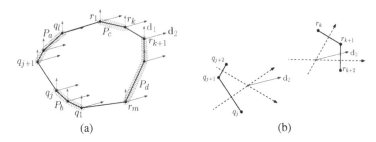

(a)   (b)

Fig. 6. (a) Sets P_a, P_b, P_c, and P_d at a certain time instant during the rotation of d_2. (b) The slope of d_2 with respect to the slopes of the lines orthogonal to $\overline{q_j q_{j+1}}$, to $\overline{q_{j+1} q_{j+2}}$, to $\overline{r_k r_{k+1}}$, and to $\overline{r_{k+1} r_{k+2}}$.

$P_b = P_1(d_1) \cap P_2(d_2)$, $P_c = P_2(d_1) \cap P_1(d_2)$, and $P_d = P_2(d_1) \cap P_2(d_2)$. Note that every point in P is contained in one of P_a, P_b, P_c, and P_d. A (d_1, d_2)-partition of P is *balanced* if $|P_a| + |P_d| \leq \frac{|P|}{2} + 1$ and $|P_b| + |P_c| \leq \frac{|P|}{2} + 1$. We now argue that, for every point set P, a balanced (d_1, d_2)-partition of P always exists, even if d_1 is arbitrarily prescribed.

Lemma 5. *Let P be a convex point set and let d_1 be a directed straight line not orthogonal to any line through two points of P. Then, there exists a directed straight line d_2 that is not orthogonal to any line through two points of P such that the (d_1, d_2)-partition of P is balanced.*

Proof. Denote by $q_1 = p_a(d_1), q_2, \ldots, q_l, q_{l+1} = p_b(d_1)$ the points of P encountered when clockwise walking on the boundary of $\mathcal{H}_P$ from $p_a(d_1)$ to $p_b(d_1)$. Also, denote by $r_1 = p_b(d_1), r_2, \ldots, r_m, r_{m+1} = p_a(d_1)$ the points of P encountered when clockwise walking on the boundary of $\mathcal{H}_P$ from $p_b(d_1)$ to $p_a(d_1)$.

Initialize d_2 to be a directed straight line coincident with d_1. When $d_2 = d_1$, we have $P_a = \{q_1, q_2, \ldots, q_l\}$, $P_d = \{r_1, r_2, \ldots, r_m\}$, $P_b = \emptyset$, and $P_c = \emptyset$. We now clockwise rotate d_2 until it is opposite to d_1 (that is, parallel and pointing in the opposite direction). As we rotate d_2, sets $P_1(d_2)$ and $P_2(d_2)$ change, hence sets P_a, P_b, P_c, and P_d change as well. When d_2 is opposite to d_1, we have $P_a = \emptyset$, $P_d = \emptyset$, $P_b = \{q_1, q_2, \ldots, q_l\}$, and $P_c = \{r_1, r_2, \ldots, r_m\}$. We will argue that there is a moment during such a rotation of d_2 in which the corresponding (d_1, d_2)-partition of P is balanced. Assume that at any time instant during the rotation of d_2 the following hold (see Figs. 6(a)–(b)):

– $P_b = \{q_1, q_2, \ldots, q_j\}$ (possibly P_b is empty);
– $P_a = \{q_{j+1}, q_{j+2}, \ldots, q_l\}$ (possibly P_a is empty);
– $P_c = \{r_1, r_2, \ldots, r_k\}$ (possibly P_c is empty);
– $P_d = \{r_{k+1}, r_{k+2}, \ldots, r_m\}$ (possibly P_d is empty); and
– q_{j+1} and r_{k+1} are the minimum and maximum point of P w.r.t. d_2, respectively.

The assumption is indeed true when d_2 starts moving, with $j = 0$ and $k = 0$.

As we keep on clockwise rotating d_2, at a certain moment d_2 becomes orthogonal to $\overline{q_{j+1}q_{j+2}}$ or to $\overline{r_{k+1}r_{k+2}}$ (or to both if $\overline{q_{j+1}q_{j+2}}$ and $\overline{r_{k+1}r_{k+2}}$ are parallel). Thus, as we keep on clockwise rotating d_2, sets P_a, P_b, P_c, and P_d change. Namely:

If d_2 becomes orthogonal first to $\overline{q_{j+1}q_{j+2}}$ and then to $\overline{r_{k+1}r_{k+2}}$, then as d_2 rotates clockwise after the position in which it is orthogonal to $\overline{q_{j+1}q_{j+2}}$, we have

- $P_b = \{q_1, q_2, \ldots, q_j, q_{j+1}\}$;
- $P_a = \{q_{j+2}, q_{j+3}, \ldots, q_l\}$ (possibly P_a is empty);
- $P_c = \{r_1, r_2, \ldots, r_k\}$ (possibly P_c is empty);
- $P_d = \{r_{k+1}, r_{k+2}, \ldots, r_m\}$ (possibly P_d is empty); and
- q_{j+2} and r_{k+1} are the minimum and maximum point of P w.r.t. d_2, respectively.

If d_2 becomes orthogonal first to $\overline{r_{k+1}r_{k+2}}$ and then to $\overline{q_{j+1}q_{j+2}}$, then as d_2 rotates clockwise after the position in which it is orthogonal to $\overline{r_{k+1}r_{k+2}}$, we have that P_a and P_b stay unchanged, that r_{k+1} passes from P_d to P_c, and that q_{j+1} and r_{k+2} are the minimum and maximum point of P w.r.t. d_2, respectively.

If d_2 becomes orthogonal to $\overline{q_{j+1}q_{j+2}}$ and $\overline{r_{k+1}r_{k+2}}$ simultaneously, then as d_2 rotates clockwise after the position in which it is orthogonal to $\overline{q_{j+1}q_{j+2}}$, we have that q_{j+1} passes from P_a to P_b, that r_{k+1} passes from P_d to P_c, and that q_{j+2} and r_{k+2} are the minimum and maximum point of P w.r.t. d_2, respectively.

Observe that:

1. whenever sets P_a, P_b, P_c, and P_d change, we have that $|P_a| + |P_d|$ and $|P_b| + |P_c|$ change at most by two;
2. when d_2 starts rotating we have that $|P_a| + |P_d| = |P|$, and when d_2 stops rotating we have that $|P_a| + |P_d| = 0$;
3. when d_2 starts rotating we have that $|P_b| + |P_c| = 0$, and when d_2 stops rotating we have that $|P_b| + |P_c| = |P|$; and
4. $|P_a| + |P_b| + |P_c| + |P_d| = |P|$ holds at any time instant.

By continuity, there is a time instant in which $|P_a|+|P_d| = \lfloor |P|/2 \rfloor$ and $|P_b|+|P_c| = \lceil |P|/2 \rceil$, or in which $|P_a| + |P_d| = \lfloor |P|/2 \rfloor + 1$ and $|P_b| + |P_c| = \lceil |P|/2 \rceil - 1$. This completes the proof of the lemma.   □

We now show how to use Lemma 5 in order to prove Theorem 2.

Let P be any point set. Assume that no two points of P have the same y-coordinate. Such a condition is easily met after rotating the Cartesian axes. Denote by l a vertical straight line directed towards increasing y-coordinates. Each of $P_1(l)$ and $P_2(l)$ is convex and one-sided with respect to l. By Lemma 1, there exist increasing-chord graphs $G_1 = (P_1(l), S_1)$ and $G_2 = (P_2(l), S_2)$ with $|S_1| < 2|P_1(l)|$ and $|S_2| < 2|P_2(l)|$. Then, graph $G(P, S_1 \cup S_2)$ has less than $2(|P_1(l)|+|P_2(l)|) = 2|P|$ edges and contains an increasing-chord path between every pair of vertices in $P_1(l)$ and between every pair of vertices in $P_2(l)$. However, G does not have increasing-chord paths between any pair (a, b) of vertices such that $a \in P_1(l)$ and $b \in P_2(l)$.

We now present and prove the following claim. Consider a convex point set Q and a directed straight line d_1 not orthogonal to any line through two points of Q. Then, there exists a geometric graph $H(Q, R)$ that contains an increasing-chord path between every point in $Q_1(d_1)$ and every point in $Q_2(d_1)$, such that $|R| \in O(|Q| \log |Q|)$.

The application of the claim with $Q = P$ and $d_1 = l$ provides a graph $H(P, R)$ that contains an increasing-chord path between every pair (a, b) of vertices such that $a \in P_1(l)$ and $b \in P_2(l)$. Thus, the union of G and H is an increasing-chord graph with $O(|P| \log |P|)$ edges spanning P. Therefore, the above claim implies Theorem 2.

We show an inductive algorithm to construct H. Let $f(Q, d_1)$ be the number of edges that H has as a result of the application of our algorithm on a point set Q and a directed straight-line d_1. Also, let $f(n) = \max\{f(Q, d_1)\}$, where the maximum is among all point sets Q with $n = |Q|$ points and among all the directed straight-lines d_1 that are not orthogonal to any line through two points of Q.

Let Q be any convex point set with n points and let d_1 be any directed straight line not orthogonal to any line through two points of Q. By Lemma 5, there exists a directed straight line not orthogonal to any line through two points of Q and such that the (d_1, d_2)-partition of Q is balanced.

Let $Q_a = Q_1(d_1) \cap Q_1(d_2)$, let $Q_b = Q_1(d_1) \cap Q_2(d_2)$, let $Q_c = Q_2(d_1) \cap Q_1(d_2)$, and let $Q_d = Q_2(d_1) \cap Q_2(d_2)$.

Point set $Q_a \cup Q_c$ is convex and one-sided with respect to d_2. By Lemma 1 there exists an increasing-chord graph $H_1(Q_a \cup Q_c, R_1)$ with $|R_1| < 2(|Q_a| + |Q_c|)$ edges. Analogously, by Lemma 1 there exists an increasing-chord graph $H_2(Q_b \cup Q_d, R_2)$ with $|R_2| < 2(|Q_b| + |Q_d|)$ edges.

Hence, there exists a graph $H_3(Q, R_1 \cup R_2)$ with $|R_1 \cup R_2| < 2(|Q_a| + |Q_c| + |Q_b| + |Q_d|) = 2|Q| = 2n$ edges containing an increasing-chord path between every point in Q_a and every point in Q_c, and between every point in Q_b and every point in Q_d. However, G does not have an increasing-chord path between any point in Q_a and any point in Q_d, and does not have an increasing-chord path between any point in Q_b and any point in Q_c.

By Lemma 5, it holds that $|Q_a| + |Q_d| \leq \frac{n}{2} + 1$ and $|Q_b| + |Q_d| \leq \frac{n}{2} + 1$. By definition, we have $f(Q_a \cup Q_d, d_1) \leq f(|Q_a| + |Q_d|) \leq f(\frac{n}{2} + 1)$. Analogously, it holds that $f(Q_b \cup Q_c, d_1) \leq f(|Q_b| + |Q_c|) \leq f(\frac{n}{2} + 1)$. Hence, $f(n) \leq 2n + 2f(\frac{n}{2} + 1) \in O(n \log n)$. This proves the claim and hence Theorem 2.

5   Conclusions

We considered the problem of constructing increasing-chord graphs spanning point sets. We proved that, for every point set P, there exists a planar increasing-chord graph $G(P', S)$ with $P \subseteq P'$ and $|P'| \in O(|P|)$. We also proved that, for every convex point set P, there exists an increasing-chord graph $G(P, S)$ with $|S| \in O(|P| \log |P|)$.

Despite our research efforts, the main question on this topic remains open:

Problem 1. Is it true that, for every (convex) point set P, there exists an increasing-chord planar graph $G(P, S)$?

One of the directions we took in order to tackle this problem is to assume that the points in P lie on a constant number of straight lines. While a simple modification of the proof of Lemma 1 allows us to prove that an increasing-chord planar graph always exists spanning a set of points lying on two straight lines, it is surprising and disheartening that we could not prove a similar result for sets of points lying on three straight lines.

The main difficulty seems to lie in the construction of planar increasing-chord graphs spanning sets of points lying on the boundary of an acute triangle. Gabriel graphs naturally generalize to higher dimensions, where empty balls replace empty disks. In Section 3 we showed that, for points in $\mathbb{R}^2$, every Gabriel triangulation is increasing-chord. Can this result be generalized to higher dimensions?

Problem 2. Is it true that, for every point set P in $\mathbb{R}^d$, any Gabriel triangulation of P is increasing-chord?

Finally, it would be interesting to understand if increasing-chord graphs with few edges can be constructed for any (possibly non-convex) point set:

Problem 3. Is it true that, for every point set P, there exists an increasing-chord graph $G(P, S)$ with $|S| \in o(|P|^2)$?

References

1. Aichholzer, O., Aurenhammer, F., Icking, C., Klein, R., Langetepe, E., Rote, G.: Generalized self-approaching curves. Discr. Appl. Math. 109(1-2), 3–24 (2001)
2. Alamdari, S., Chan, T.M., Grant, E., Lubiw, A., Pathak, V.: Self-approaching graphs. In: Didimo, W., Patrignani, M. (eds.) GD 2012. LNCS, vol. 7704, pp. 260–271. Springer, Heidelberg (2013)
3. Bern, M.W., Eppstein, D., Gilbert, J.R.: Provably good mesh generation. J. Comput. Syst. Sci. 48(3), 384–409 (1994)
4. de Berg, M., Cheong, O., van Kreveld, M., Overmars, M.: Computational Geometry: Algorithms and Applications, 3rd edn. Springer, Heidelberg (2008)
5. Di Battista, G., Vismara, L.: Angles of planar triangular graphs. SIAM J. Discrete Math. 9(3), 349–359 (1996)
6. Dillencourt, M.B., Smith, W.D.: Graph-theoretical conditions for inscribability and Delaunay realizability. Discrete Mathematics 161(1-3), 63–77 (1996)
7. Eades, P., Whitesides, S.: The realization problem for Euclidean minimum spanning trees is NP-hard. Algorithmica 16(1), 60–82 (1996)
8. Gabriel, K.R., Sokal, R.R.: A new statistical approach to geographic variation analysis. Systematic Biology 18, 259–278 (1969)
9. Icking, C., Klein, R., Langetepe, E.: Self-approaching curves. Math. Proc. Camb. Phil. Soc. 125(3), 441–453 (1999)
10. Liotta, G.: Chapter 4 of Handbook of Graph Drawing. CRC Press (2014); Tamassia, R. (ed.)
11. Matula, D.W., Sokal, R.R.: Properties of Gabriel graphs relevant to geographic variation research and clustering of points in the plane. Geographical Analysis 12(3), 205–222 (1980)
12. Monma, C.L., Suri, S.: Transitions in geometric minimum spanning trees. Discrete & Computational Geometry 8, 265–293 (1992)
13. Papadimitriou, C.H., Ratajczak, D.: On a conjecture related to geometric routing. Theoretical Computer Science 344(1), 3–14 (2005)
14. Rao, A., Papadimitriou, C.H., Shenker, S., Stoica, I.: Geographic routing without location information. In: Johnson, D., Joseph, A., Vaidya, N. (eds.) MOBICOM 2003, pp. 96–108 (2003)
15. Rote, G.: Curves with increasing chords. Math. Proc. Camb. Phil. Soc. 115(1), 1–12 (1994)

On Self-Approaching and Increasing-Chord Drawings of 3-Connected Planar Graphs

Martin Nöllenburg, Roman Prutkin, and Ignaz Rutter

Institute of Theoretical Informatics, Karlsruhe Institute of Technology, Germany

Abstract. An st-path in a drawing of a graph is self-approaching if during a traversal of the corresponding curve from s to any point t' on the curve the distance to t' is non-increasing. A path has increasing chords if it is self-approaching in both directions. A drawing is self-approaching (increasing-chord) if any pair of vertices is connected by a self-approaching (increasing-chord) path.

We study self-approaching and increasing-chord drawings of triangulations and 3-connected planar graphs. We show that in the Euclidean plane, triangulations admit increasing-chord drawings, and for planar 3-trees we can ensure planarity. Moreover, we give a binary cactus that does not admit a self-approaching drawing. Finally, we show that 3-connected planar graphs admit increasing-chord drawings in the hyperbolic plane and characterize the trees that admit such drawings.

1 Introduction

Finding a path between two vertices is one of the most fundamental tasks users want to solve when considering graph drawings. Empirical studies have shown that users perform better in path-finding tasks if the drawings exhibit a strong geodesic-path tendency [10, 17]. Not surprisingly, graph drawings in which a path with certain properties exists between every pair of vertices have become a popular research topic. Over the last years a number of different drawing conventions implementing the notion of strong geodesic-path tendency have been suggested, namely *greedy drawings* [18], *(strongly) monotone drawings* [2], and *self-approaching* and *increasing-chord drawings* [1]. Note that throughout this paper, all drawings are straight-line and vertices are mapped to distinct points.

The notion of greedy drawings came first and was introduced by Rao et al. [18]. Motivated by greedy routing schemes, e.g., for sensor networks, one seeks a drawing, where for every pair of vertices s and t there exists an st-path, along which the distances to t decrease in every vertex. This ensures that greedily sending a message to a vertex that is closer to the destination guarantees delivery. Papadimitriou and Ratajczak conjectured that every 3-connected planar graph admits a greedy embedding into the Euclidean plane [16]. This conjecture has been proved independently by Leighton and Moitra [13] and Angelini et al. [5]. Kleinberg [12] showed that every connected graph has a greedy drawing in the hyperbolic plane. Eppstein and Goodrich [7] showed how to construct such an embedding, in which the coordinates of each vertex are represented using only $O(\log n)$ bits, and Goodrich and Strash [9] provided a corresponding *succinct* representation for greedy embeddings of 3-connected planar graphs in $\mathbb{R}^2$. Angelini et al. [3] showed that some graphs require exponential area for a greedy drawing

C. Duncan and A. Symvonis (Eds.): GD 2014, LNCS 8871, pp. 476–487, 2014.

in $\mathbb{R}^2$. Wang and He [21] used a custom distance metric to construct planar, convex and succinct greedy embeddings of 3-connected planar graphs using Schnyder realizers [20]. Nöllenburg and Prutkin [14] characterized trees admitting a Euclidean greedy embedding. However, a number of interesting questions remain open, e.g., whether every 3-connected planar graph admits a planar and convex Euclidean greedy embedding (strong Papadimitriou-Ratajczak conjecture [16]). Regarding planar greedy drawings of triangulations, the only known result is an existential proof by Dhandapani [6].

While getting closer to the destination, a greedy path can make numerous turns and may even look like a spiral, which hardly matches the intuitive notion of geodesic-path tendency. To overcome this, Angelini et al. [2] introduced monotone drawings, where one requires that for every pair of vertices s and t there exists a *monotone path*, i.e., a path that is monotone with respect to some direction. Ideally, the monotonicity direction should be $\overrightarrow{st}$. This property is called *strong monotonicity*. Angelini et al. showed that biconnected planar graphs admit monotone drawings [2] and that plane graphs admit monotone drawings with few bends [4]. The existence of strongly monotone planar drawings remains open, even for triangulations.

Both greedy and monotone paths may have arbitrarily large *detour*, i.e., the ratio of the path length and the distance of the endpoints can, in general, not be bounded by a constant. Motivated by this fact, Alamdari et al. [1] recently initiated the study of *self-approaching* graph drawings. Self-approaching curves, introduced by Icking [11], are curves where for any point t' on the curve, the distance to t' decreases continuously while traversing the curve from the start to t'. Equivalently, a curve is self-approaching if, for any three points a, b, c in this order along the curve, it is $\text{dist}(a, c) \geq \text{dist}(b, c)$, where dist denotes the Euclidean distance. An even stricter requirement are so-called *increasing-chord* curves, which are curves that are self-approaching in both directions. The name is motivated by the characterization of such curves, which states that a curve has increasing chords if and only if for any four distinct points a, b, c, d in that order, it is $\text{dist}(b, c) \leq \text{dist}(a, d)$. Self-approaching curves have detour at most 5.333 [11] and increasing-chord curves have detour at most 2.094 [19]. Alamdari et al. [1] studied the problem of recognizing whether a given graph drawing is self-approaching as well as connecting given points to a self-approaching drawing. They also gave a complete characterization of trees admitting self-approaching drawings.

We note that every increasing chord drawing is self-approaching and strongly monotone [1]. The converse is not true. A self-approaching drawing is greedy, but not necesserily monotone, and a greedy drawing is generally neither self-approaching nor monotone. For trees, the notions of self-approaching and increasing-chord drawing coincide.

Contribution. We obtain the following results on constructing self-approaching or increasing-chord drawings.

1. We show that every triangulation has an increasing-chord drawing (answering an open question of Alamdari et al. [1]) and construct a *binary cactus* that does not admit a self-approaching drawing (Sect. 3). The latter is a notable difference to greedy drawings since both constructions of greedy drawings for 3-connected planar graphs [5, 13] essentially show that every binary cactus has a greedy drawing.

2. We show how to construct plane increasing-chord drawings for *planar 3-trees* (a special class of triangulations) using Schnyder realizers (Sect. 4). To the best of our knowledge, this is the first construction for this graph class, even for greedy and strongly monotone plane drawings, which addresses an open question of Angelini et al. [2].

3. We show that, similarly to the greedy case, the hyperbolic plane $\mathbb{H}^2$ allows representing a broader class of graphs than $\mathbb{R}^2$ (Sect. 5). We prove that a tree has a self-approaching or increasing-chord drawing in $\mathbb{H}^2$ if and only if it either has maximum degree 3 or is a subdivision of $K_{1,4}$ (this is not the case in $\mathbb{R}^2$; see the characterization by Alamdari et al. [1]), implying every 3-connected planar graph has an increasing-chord drawing. We also show how to construct planar increasing-chord drawings of binary cactuses in $\mathbb{H}^2$.

2 Preliminaries

For points $a, b, c, d \in \mathbb{R}^2$, let ray$(a, b)$ denote the ray with origin a and direction $\overrightarrow{ab}$ and ray$(a, \overrightarrow{bc})$ the ray with origin a and direction $\overrightarrow{bc}$. Let dir(ab) be the vector $\overrightarrow{ab}$ normalized to unit length. Let $\angle(\overrightarrow{ab}, \overrightarrow{cd})$ denote the smaller angle formed by the two vectors $\overrightarrow{ab}$ and $\overrightarrow{cd}$. For an angle $\alpha \in [0, 2\pi]$, let R_α denote the rotation matrix $\left(\begin{smallmatrix} \cos\alpha & -\sin\alpha \\ \sin\alpha & \cos\alpha \end{smallmatrix}\right)$.

For vectors $\overrightarrow{v_1}, \overrightarrow{v_2}$ with dir$(\overrightarrow{v_2}) = R_\alpha \cdot \text{dir}(\overrightarrow{v_1})$, $\alpha \in [0, 2\pi)$, we write $\angle_{\text{ccw}}(\overrightarrow{v_1}, \overrightarrow{v_2}) := \alpha$. Further, let $[\overrightarrow{v_1}, \overrightarrow{v_2}]$ denote the *cone of directions* $\{\overrightarrow{v} \mid \text{dir}(\overrightarrow{v}) = R_\beta \cdot \text{dir}(\overrightarrow{v_1}), \beta \in [0, \alpha]\}$. Let $|[\overrightarrow{v_1}, \overrightarrow{v_2}]| := \alpha$ be its *size*. For a set of directions D, let $\overline{D}$ denote a minimum cone of directions containing D, and let $|D| = |\overline{D}|$. Note that if $|D| < 180°$, $\overline{D}$ is unique.

We reuse some notation from the work of Alamdari et al. [1]. For points $p, q \in \mathbb{R}^2$, $p \neq q$, let l_{pq}^+ denote the halfplane not containing p bounded by the line through q orthogonal to the segment pq. A piecewise-smooth curve is self-approaching if and only if for each point a on the curve, the line perpendicular to the curve at a does not intersect the curve at a later point [11]. This leads to the following characterization of self-approaching paths.

Fact 1 (Corollary 2 in [1]). *Let $\rho = (v_1, v_2, \ldots, v_k)$ be a directed path embedded in $\mathbb{R}^2$ with straight-line segments. Then, ρ is self-approaching if and only if for all $1 \leq i < j \leq k$, the point v_j lies in $l_{v_i v_{i+1}}^+$.*

We shall denote the reverse of a path ρ by ρ^{-1}. Let $\rho = (v_1, v_2, \ldots, v_k)$ be a self-approaching path. Define front$(\rho) = \bigcap_{i=1}^{k-1} l_{v_i v_{i+1}}^+$, see also Fig. 1. Using Fact 1, we can decide whether a concatenation of two paths is self-approaching.

Fact 2. *Let $\rho_1 = (v_1, \ldots, v_k)$ and $\rho_2 = (v_k, v_{k+1}, \ldots, v_m)$ be self-approaching paths. The path $\rho_1.\rho_2 := (v_1, \ldots, v_k, v_{k+1}, \ldots, v_m)$ is self-approaching if and only if $\rho_2 \subseteq$ front(ρ_1).*

Fig. 1. self-approaching path ρ and front(ρ)

A path ρ has *increasing chords* if for any points a, b, c, d in this order along ρ, it is dist$(b, c) \leq$ dist(a, d). A path has increasing chords if and only if it is self-approaching in both directions. The following result is easy to see.

Lemma 1. *Let $\rho = (v_1, \ldots, v_k)$ be a path such that for any $i < j$, $i, j \in \{1, \ldots, k-1\}$, it is $\angle(\overrightarrow{v_i v_{i+1}}, \overrightarrow{v_j v_{j+1}}) \leq 90°$. Then, ρ has increasing chords.*

Let $G = (V, E)$ be a connected graph. A *separating k-set* is a set of k vertices whose removal disconnects the graph. A vertex forming a separating 1-set is called *cutvertex*. A graph is *c-connected* if it does not admit a separating k-set with $k \leq c - 1$; 2-connected graphs are also called *biconnected*. A connected graph is biconnected if and only if it does not contain a cutvertex. A *block* is a maximal biconnected subgraph. The *block-cutvertex tree* (or *BC-tree*) T_G of G has a *B-node* for each block of G, a *C-node* for each cutvertex of G and, for each block ν containing a cutvertex v, an edge between the corresponding B- and C-node. We associate B-nodes with their corresponding blocks and C-nodes with their corresponding cutvertices.

The following notation follows the work of Angelini et al. [5]. Let T_G be rooted at some block ν containing a non-cutvertex (such a block ν always exists). For each block $\mu \neq \nu$, let $\pi(\mu)$ denote the *parent block* of μ, i.e., the grandparent of μ in T_G. Let $\pi^2(\mu)$ denote the parent block of $\pi(\mu)$ and, generally, $\pi^{i+1}(\mu)$ the parent block of $\pi^i(\mu)$. Further, we define the *root* $r(\mu)$ of μ as the cutvertex contained in both μ and $\pi(\mu)$. Note that $r(\mu)$ is the parent of μ in T_G. In addition, for the root node ν of T_G, we define $r(\nu)$ to be some non-cutvertex of ν. Let $\text{depth}_B(\mu)$ denote the number of B-nodes on the $\nu\mu$-path in T_G minus 1, and let $\text{depth}_C(r(\mu)) = \text{depth}_B(\mu)$. If μ is a leaf of T_G, we call it a *leaf block*.

If every block of G is outerplanar, G is called a *cactus*. In a *binary* cactus every cutvertex is part of exactly two blocks. For a binary cactus G with a block μ containing a cutvertex v, let G_μ^v denote the maximal connected subgraph containing v but no other vertex of μ. We say that G_μ^v is a *subcactus* of G.

A cactus is *triangulated* if each of its blocks is internally triangulated. A *triangular fan with vertices* $V_t = \{v_0, v_1, \ldots, v_k\}$ and *root* v_0 is a graph on V_t with edges $v_i v_{i+1}$, $i = 1, \ldots, k - 1$, as well as $v_0 v_i$, $i = 1, \ldots, k$. Let us consider a special kind of triangulated cactuses, each of whose blocks μ is either an edge or a *triangular fan* with root $r(\mu)$. We call such a cactus *downward-triangulated* and every edge of a block μ incident to $r(\mu)$ a *downward* edge.

For a fixed straight-line drawing of a binary cactus G, we define sets $U(G) = \{\overrightarrow{r(\mu)v} \mid \mu \text{ is a block of } G \text{ containing } v, v \neq r(\mu)\}$ and $D(G) = \{\overrightarrow{uv} \mid \overrightarrow{vu} \in U(G)\}$, i.e. the sets of *upward* and *downward* directions of G.

3 Graphs with Self-Approaching Drawings

A natural approach to construct (not necessarily plane) self-approaching drawings is to construct a self-approaching drawing of a spanning subgraph. For instance, to draw a graph G containing a Hamiltonian path H with increasing chords, we simply draw H consecutively on a line. In this section, we consider 3-connected planar graphs and the special case of triangulations, which addresses an open question of Alamdari et al. [1]. These graphs are known to have a spanning binary cactus [5, 13]. Angelini et al. [5] showed that every triangulation has a spanning downward-triangulated binary cactus.

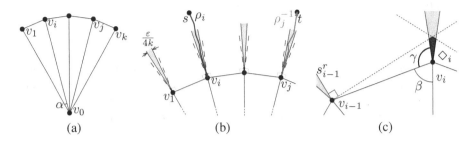

Fig. 2. Drawing a triangulated binary cactus with increasing chords inductively. The drawings $\Gamma_{i,\varepsilon'}$ of the subcactuses, $\varepsilon' = \frac{\varepsilon}{4k}$, are contained inside the gray cones. It is $\beta = 90° - \varepsilon'$, $\gamma = 90° + \varepsilon'/2$.

3.1   Increasing-Chord Drawings of Triangulations

We show that every downward-triangulated binary cactus has an increasing-chord drawing. The construction is similar to the one of the greedy drawings of binary cactuses in the two proofs of the Papadimitriou-Ratajczak conjecture [5, 13]. Our proof is by induction on the height of the BC-tree. We show that G can be drawn such that all downward edges are almost vertical and the remaining edges almost horizontal. For vertices s, t, an st-path with increasing chords goes downwards to some block μ, then sideways to another cutvertex of μ and, finally, upwards to t. Let $\vec{e_1}, \vec{e_2}$ be vectors $(1,0)^\top, (0,1)^\top$.

Theorem 1. *Let $G = (V, E)$ be a downward-triangulated binary cactus. For any $0° < \varepsilon < 90°$, there exists an increasing-chord drawing Γ_ε of G, such that for each vertex v contained in some block μ, $v \neq r(\mu)$, the angle formed by $\overrightarrow{r(\mu)v}$ and $\vec{e_2}$ is at most $\frac{\varepsilon}{2}$.*

Proof. Let G be rooted at block ν. As our base case, let $\nu = G$ be a triangular fan with vertices $v_0, v_1, \ldots, v_k$ and root $v_0 = r(\nu)$. We draw v_0 at the origin and distribute $v_1, \ldots, v_k$ on the unit circle, such that $\angle(\vec{e_2}, \overrightarrow{v_0v_1}) = k\alpha/2$ and $\angle(\overrightarrow{v_0v_i}, \overrightarrow{v_0v_{i+1}}) = \alpha$, $\alpha = \varepsilon/k$; see Fig. 2a. By Lemma 1, path $(v_1, \ldots, v_k)$ has increasing chords.

Now let G have multiple blocks. We draw the root block ν, $v_0 = r(\nu)$, as in the previous case, but with $\alpha = \frac{\varepsilon}{2k}$. Then, for each $i = 1, \ldots, k$, we choose $\varepsilon' = \frac{\varepsilon}{4k}$ and draw the subcactus G_i rooted at v_i inductively, such that the corresponding drawing $\Gamma_{i,\varepsilon'}$ is aligned at $\overrightarrow{v_0v_i}$ instead of $\vec{e_2}$; see Fig 2b. Note that ε' is the angle of the cones (gray) containing $\Gamma_{i,\varepsilon'}$. Obviously, all downward edges form angles at most $\frac{\varepsilon}{2}$ with $\vec{e_2}$.

We must be able to reach any t in any G_j from any s in any G_i via an increasing-chord path ρ. To achieve this, we make sure that no normal on a downward edge of G_i crosses the drawing of G_j, $j \neq i$. Let Λ_i be the cone with apex v_i and angle ε' aligned with $\overrightarrow{v_0v_i}$, $v_0 \notin \Lambda_i$ (gray regions in Fig. 2b). Let s_i^l and s_i^r be the left and right boundary rays of Λ_i with respect to $\overrightarrow{v_0v_i}$, and h_i^l, h_i^r the halfplanes with boundaries containing v_i and orthogonal to s_i^l and s_i^r respectively, such that $v_0 \in h_i^l \cap h_i^r$. Define $\Diamond_i = \Lambda_i \cap h_{i-1}^r \cap h_{i+1}^l$ (thin blue quadrangle in Fig. 2c), and analogously $\Diamond_j$ for $j \neq i$. It holds $\Diamond_j \subseteq h_i^r \cap h_i^l$ for each $i \neq j$. We now scale each drawing $\Gamma_{i,\varepsilon'}$ such that it is contained in $\Diamond_i$. In particular, for any downward edge uv in $\Gamma_{i,\varepsilon'}$, we have $\Gamma_{j,\varepsilon'} \subseteq \Diamond_j \subseteq l_{uv}^+$ for $j \neq i$. We claim that the resulting drawing of G is an increasing-chord drawing.

Consider vertices s,t of G. If s and t are contained in the same subgraph G_i, an increasing-chord st-path in G_i exists by induction. If s is in G_i and t is v_0, let ρ_i be the sv_i-path in G_i that uses only downward edges. By Lemma 1, path ρ_i is increasing-chord and remains so after adding edge $v_i v_0$.

Finally, assume t is in G_j with $j \neq i$. Let ρ_j be the tv_j-path in G_j that uses only downward edges. Due to the choice of ε', $h_i^r \cap h_i^l \subseteq \mathrm{front}(\rho_i)$ contains $v_1, \dots, v_k$ in its interior. Consider the path $\rho' = (v_i, v_{i+1}, \dots, v_j)$. It is self-approaching by Lemma 1; also, $\rho' \subseteq \mathrm{front}(\rho_i)$ and $\rho_j \subseteq \mathrm{front}(\rho')$. It also holds $\rho_j \subseteq \Diamond_j \subseteq \mathrm{front}(\rho_i)$. Fact 2 lets us concatenate ρ_i, ρ' and ρ_j^{-1} to a self-approaching path. By a symmetric argument, it is also self-approaching in the opposite direction and, thus, is increasing-chord.   □

Since every triangulation has a spanning downward-triangulated binary cactus [5], this implies that planar triangulations admit increasing-chord drawings.

Corollary 1. *Every planar triangulation admits an increasing-chord drawing.*

3.2 Non-triangulated Cactuses

The above construction fails if the blocks are not triangular fans since we now cannot just use downward edges to reach the common ancestor block. Consider the family of rooted binary cactuses $G_n = (V_n, E_n)$ defined as follows. Graph G_0 is a single 4-cycle, where an arbitrary vertex is designated as the root. For $n \geq 1$, consider two disjoint copies of G_{n-1} with roots a_0 and c_0. We create G_n by adding new vertices r_0 and b_0 both adjacent to a_0 and c_0; see Fig. 3a. For the new block ν containing r_0, a_0, b_0, c_0, we set $r(\nu) = r_0$. We select r_0 as the root of G_n and ν as its root block. For a block μ_i with root r_i, let a_i, b_i, c_i be its remaining vertices, such that $b_i r_i \notin E_n$. For a given drawing, due to the symmetry of G_n, we can rename the vertices a_i and c_i such that $\angle_{\mathrm{ccw}}(\overrightarrow{r_i c_i}, \overrightarrow{r_i a_i}) \leq 180°$. We now prove the following negative result.

Theorem 2. *For $n \geq 9$, G_n has no self-approaching drawing.*

The outline of the proof is as follows. We show that every self-approaching drawing Γ of G_9 contains a self-approaching drawing of G_3 with the following properties.

1. If μ_i is contained in the subcactus rooted at c_j, each self-approaching $b_i a_j$-path uses edge $b_i a_i$, and analogously for the symmetric case; see Lemma 5.
2. Each block is drawn significantly smaller than its parent block; see Lemma 6(i).
3. If the descendants of block μ form subcactuses G_k with $k \geq 2$ on both sides, the parent block of μ must be drawn smaller than μ; see Lemma 6(ii).

Obviously, the second and third conditions are contradictory. The following lemmas will be used to show that the drawings of certain blocks must be relatively thin, i.e., their downward edges have similar directions; see the full version for the omitted proofs [15].

Lemma 2. *For cactus $G = (V, E)$ and $s, t \in V$, consider cutvertices $v_1, \dots, v_k$ lying on any st-path in G in this order. Then, the path $(s, v_1, \dots, v_k, t)$ is drawn greedily, i.e., each of its subpaths is greedy. In particular, $\mathrm{ray}(v_1, s)$ and $\mathrm{ray}(v_k, t)$ diverge.*

Obviously, this divergence property also holds for a self-approaching drawing of any cactus. From now on, we consider a fixed self-approaching drawing Γ of G_9. For

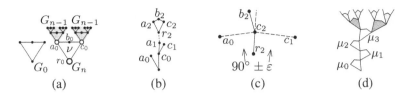

Fig. 3. (a) cactuses G_n; (b),(c) construction for Lemma 5; (d) subcactus G_5 providing the contradiction in the proof of Theorem 2

a block μ of G_9 with root $r = r(\mu)$, we write G^r for $(G_9)_\mu^r$, i.e., the binary cactus subgraph of G_9 rooted at r. We write U^r for the set of directions of the upward edges of G^r and define $I^r = \overline{U^r}$. Using Lemma 2, we can show that vectors in $U^{a_i} \cup U^{c_i}$ have the following circular order: first vectors in U^{a_i}, then vectors in U^{c_i}. It follows easily: $\min\{|I^{a_i}|, |I^{c_i}|\} < |I^{r_i}|/2$. Thus, we can provide a bound for the smallest of the cones of a subcactus depending on the depth of its root.

Lemma 3. *Every self-approaching drawing of G_9 contains a cutvertex $\bar{r}$ with* $\operatorname{depth}_C(\bar{r}) = 4$ *and* $|I^{\bar{r}}| < 22.5°$.

Let $\bar{r}$ be a cutvertex from Lemma 3 in the fixed drawing, and let $\varepsilon := |I^{\bar{r}}|$. Then, $G^{\bar{r}}$ is isomorphic to G_6. From now on, we only consider non-leaf blocks μ_i and vertices r_i, a_i, b_i, c_i in $G^{\bar{r}}$. We shall sometimes name the points a instead of a_i etc. for convenience. We assume $\angle(\overrightarrow{e_2}, \overrightarrow{ra}), \angle(\overrightarrow{e_2}, \overrightarrow{rc}) < \varepsilon/2$. The following lemma is proved using basic trigonometric arguments.

Lemma 4. *It holds:* (i) $\angle abc \geq 90°$; (ii) $G^a \subseteq l_{ba}^+$, $G^c \subseteq l_{bc}^+$; (iii) $\angle bar \leq 90° + \varepsilon$, $\angle bcr \leq 90° + \varepsilon$. (iv) *For vertices u in G^a, v in G^c of degree 4 it is* $\angle(\overrightarrow{uv}, \overrightarrow{e_1}) \leq \varepsilon/2$

We can now describe block angles at a_i, c_i more precisely and characterize certain self-approaching paths in $G^{\bar{r}}$. We show that a self-approaching path from b_i *downwards and to the left*, i.e., to an ancestor block μ_j of μ_i, such that μ_i is in G^{c_j}, must use a_i. Similarly, a self-approaching path *downwards and to the right* must use c_i. Since for several ancestor blocks of μ_i the roots lie on both of these two kinds of paths, we can bound the area containing them and show that it is relatively small. This implies that the ancestor blocks are small as well, providing a contradiction.

Lemma 5. *Consider non-leaf blocks μ_0, μ_1, μ_2, such that $r(\mu_1) = c_0$ and μ_2 in G^{a_1}; see Fig. 3b.* (i) *It is $\angle r_2 a_2 b_2, \angle r_2 c_2 b_2 \in [90°, 90° + \varepsilon]$, b_2 lies to the right of $\operatorname{ray}(r_2, a_2)$ and to the left of $\operatorname{ray}(r_2, c_2)$.* (ii) *Each self-approaching $b_2 a_0$-path uses a_2; each self-approaching $b_2 c_1$-path uses c_2.*

Proof. (i) Assume $\angle r_2 a_2 b_2 < 90°$. Then, all self-approaching $b_2 a_0$ and $b_2 c_1$-paths must use c_2. By Lemma 4(iv), the lines through $a_0 c_2$ and $c_2 c_1$ are "almost horizontal", i.e., $\angle(\overrightarrow{a_0 c_2}, \overrightarrow{e_1}), \angle(\overrightarrow{c_2 c_1}, \overrightarrow{e_1}) \leq \varepsilon/2$. Since $r_2 c_2$ is "almost vertical", r_2 must lie below these lines and it is $\angle a_0 c_2 r_2, \angle c_1 c_2 r_2 \in [90° - \varepsilon, 90° + \varepsilon]$; see Fig. 3c. First, let b_2 lie to the left of $\operatorname{ray}(r_2, c_2)$. Then, b_2 is above $a_0 c_2$, and it is $\angle r_2 c_2 b_2 = \angle a_0 c_2 r_2 + \angle a_0 c_2 b_2 \geq$

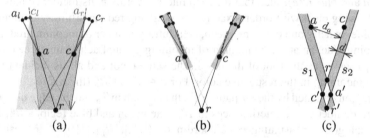

Fig. 4. Showing the contradiction in Theorem 2

$(90° − ε) + 90°$. Now let b_2 lie to the right of ray(r_2, c_2). Then, b_2 is above c_2c_1, and it is $\angle r_2c_2b_2 = \angle c_1c_2r_2 + \angle c_1c_2b_2 \geq (90° − ε) + 90°$. Since $ε < 22.5°$, this contradicts Lemma 4(iii). Similarly, $\angle r_2c_2b_2 \geq 90°$. Thus, by Lemma 4(iii), $\angle r_2c_2b_2, \angle r_2c_2b_2 \in [90°, 90° + ε]$. Since $\angle a_2b_2c_2 \geq 90°$, b_2 lies to the right of ray(r_2, a_2) and to the left of ray(r_2, c_2). (If b_2 lies to the left of both rays, it is $\angle a_2b_2c_2 = \angle(\overrightarrow{a_2b_2}, \overrightarrow{c_2b_2}) \leq 2ε < 90°$.) (ii) Similarly, if a self-approaching b_2a_0-path uses c_2 instead of a_2, it is $\angle r_2c_2b_2 \geq 180° − ε$. The last part follows analogously. □

From now on, let μ_0 be the root block of $G^{\bar{r}}$ and μ_1, μ_2, μ_3 its descendants such that $r(\mu_1) = c_0, r(\mu_2) = a_1, r(\mu_3) \in \{a_2, c_2\}$; see Fig. 3d. Light gray blocks are the subject of Lemma 6(i), which shows that several ancestor roots lie inside a cone with a small angle. Dark gray blocks are the subject of Lemma 6(ii), which considers the intersection of the cones corresponding to a pair of sibling blocks and shows that some of their ancestor roots lie inside a narrow strip; see Fig. 4a for a sketch.

Lemma 6. *Let μ be a block in G^{c_2} with vertices a, b, c, $r(\mu)$. (i) Let μ have depth 5 in $G^{\bar{r}}$. Then, the cone $l^+_{ba} \cap l^+_{bc}$ contains $r(\mu)$, $r(\pi(\mu))$, $r(\pi^2(\mu))$ and $r(\pi^3(\mu))$. (ii) Let μ have depth 4 in $G^{\bar{r}}$. There exist u in G^a and v in G^c of degree 4 and a strip S containing $r(\mu)$, $r(\pi(\mu))$, $r(\pi^2(\mu)) = r(\mu_2)$, such that u and v lie on the different boundaries of S.*

Again, we consider two siblings and the intersection of their corresponding strips, which forms a small diamond containing the root of the ancestor block; see Fig. 4b, 4c.

Lemma 7. *Consider block $\mu = \mu_3$ containing $r = r(\mu)$, a, b, c, and let $r_\pi := r(\pi(\mu_3))$. It holds: (i) $|r_\pi r| \leq \frac{(1 + 2\tan ε)(\tan ε)^2}{\cos ε}(|ra| + |rc|)$; (ii) $|ra|, |rc| \leq |rr_\pi|(\tan ε)^2$.*

For $ε \leq 22.5°$, the two claims of Lemma 7 contradict each other. This concludes the proof of Theorem 2.

4   Planar Increasing-Chord Drawings of 3-Trees

In this section, we show how to construct planar increasing-chord drawings of 3-trees. We make use of *Schnyder labelings* [20] and drawings of triangulations based on them. For a plane triangulation $G = (V, E)$ with external vertices r, g, b, its Schnyder labeling is an orientation and partition of the interior edges into three trees T_r, T_g, T_b (called *red, green* and *blue tree*), such that for each internal vertex v, its incident edges appear in the following clockwise order: exactly one outgoing red, an arbitrary number of incoming blue, exactly one outgoing green, an arbitrary number of incoming red, exactly one outgoing blue, an arbitrary number of incoming green. Each of the three outer vertices r, g, b serves as the root of the tree in the same color and all its incident interior edges are incoming in the respective color. For $v \in V$, let R_v^r (the *red region* of v) denote the region bounded by the vg-path in T_g, the vb-path in T_b and the edge gb. Let $|R_v^r|$ denote the number of interior faces in R_v^r. The green and blue regions R_v^g, R_v^b are defined analogously. Assigning v the coordinates $(|R_v^r|, |R_v^g|, |R_v^b|) \in \mathbb{R}^3$ results in a plane straight-line drawing of G in the plane $\{x = (x_1, x_2, x_3) \mid x_1 + x_2 + x_3 = f - 1\}$ called *Schnyder drawing*. Here, f denotes the number of faces of G. For a thorough introduction to this topic, see the book of Felsner [8].

For $\alpha, \beta \in [0°, 360°]$, let $[\alpha, \beta]$ denote the corresponding counterclockwise cone of directions. We consider drawings satisfying the following constraints.

Definition 1. *Let $G = (V, E)$ be a plane triangulation graph with a Schnyder labeling. For $0° \le \alpha \le 60°$, we call an arbitrary planar straight-line drawing of G α-Schnyder if for each internal vertex $v \in V$, its outgoing red edge has direction in $[90° - \frac{\alpha}{2}, 90° + \frac{\alpha}{2}]$, blue in $[210° - \frac{\alpha}{2}, 210° + \frac{\alpha}{2}]$ and green in $[330° - \frac{\alpha}{2}, 330° + \frac{\alpha}{2}]$ (see Fig. 5a).*

According to Def. 1, classical Schnyder drawings are $60°$-Schnyder. The next lemma shows an interesting connection between α-Schnyder and increasing-chord drawings.

Lemma 8. *$30°$-Schnyder drawings are increasing-chord drawings.*

Proof. Let $G = (V, E)$ be a plane triangulation with a given Schnyder labeling and Γ a corresponding $30°$-Schnyder drawing. Let r, g, b be the red, green and blue external vertex, respectively, and T_r, T_g, T_b the directed trees of the corresponding color.

Consider vertices $s, t \in V$. First, note that monochromatic directed paths in Γ have increasing chords by Lemma 1. Assume s and t are not connected by such a path. Then, they are both internal and s is contained in one of the regions R_t^r, R_t^g, R_t^b. Without loss of generality, we assume $s \in R_t^r$. The sr-path in T_r crosses the boundary of R_t^r, and we assume without loss of generality that it crosses the blue boundary of R_t^r in $u \neq t$; see Fig. 5b. The other cases are symmetric.

Let ρ_r be the su-path in T_r and ρ_b the tu-path in T_b; see Fig. 5c. On the one hand, the direction of a line orthogonal to a segment of ρ_r is in $[345°, 15°] \cup [165°, 195°]$. On the other hand, ρ_b is contained in a cone $[15°, 45°]$ with apex u. Thus, $\rho_b^{-1} \subseteq \text{front}(\rho_r)$, and $\rho_r . \rho_b^{-1}$ is self-approaching by Fact 2. By a symmetric argument it is also self-approaching in the other direction, and hence has increasing chords. $\square$

Planar 3-trees are the graphs obtained from a triangle by repeatedly choosing a (triangular) face f, inserting a new vertex v into f, and connecting v to each vertex of f.

Lemma 9. *Planar 3-trees have α-Schnyder drawings for any $0° < \alpha \le 60°$.*

Proof. We describe a recursive construction of an α-Schnyder drawing of a planar 3-tree. We start with an equilateral triangle and put a vertex v in its center. Then, we align

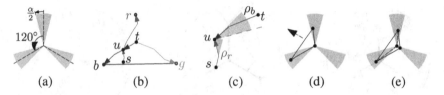

Fig. 5. (a)–(c) 30°-Schnyder drawings are increasing-chord; (d),(e) special case of planar 3-trees.

the pattern from Fig. 5a at v. For the induction step, consider a triangular face xyz and assume that the pattern is centered at one of its vertices, say x, such that the other two vertices are in the interiors of two distinct cones; see Fig. 5d. It is now possible to move the pattern inside the triangle slightly, such that the same holds for all three vertices x, y, z; see Fig. 5e. We insert the new vertex at the center of the pattern and again get the situation as in Fig. 5d.   □

Lemmas 8 and 9 provide a constructive proof for the following theorem.

Theorem 3. *Every planar 3-tree has a planar increasing-chord drawing.*

5   Self-Approaching Drawings in the Hyperbolic Plane

Kleinberg [12] showed that any tree can be drawn greedily in the hyperbolic plane $\mathbb{H}^2$. This is not the case in $\mathbb{R}^2$. Thus, $\mathbb{H}^2$ is more powerful than $\mathbb{R}^2$ in this regard. Since self-approaching drawings are closely related to greedy drawings, it is natural to investigate the existence of self-approaching drawings in $\mathbb{H}^2$.

We shall use the *Poincaré disk* model for $\mathbb{H}^2$, in which $\mathbb{H}^2$ is represented by the unit disk $D = \{x \in \mathbb{R}^2 : |x| < 1\}$ and lines are represented by circular arcs orthogonal to the boundary of D. For an introduction to the Poincaré disk model, see e.g. Kleinberg [12] and the references therein.

First, let us consider a tree $T = (V, E)$. A drawing of T in $\mathbb{R}^2$ is self-approaching if and only if no normal on an edge of T in any point crosses another edge [1]. The same condition holds in $\mathbb{H}^2$; see full version for the proof [15]. According to the characterization by Alamdari et al. [1], some binary trees have no self-approaching drawings in $\mathbb{R}^2$. We show that this is no longer the case in $\mathbb{H}^2$.

Theorem 4. *Let $T = (V, E)$ be a tree, such that each node of T has degree either 1 or 3. Then, T has a self-approaching drawing in $\mathbb{H}^2$, in which every arc has the same hyperbolic length and every pair of incident arcs forms an angle of $120°$.*

Proof. For convenience, we subdivide each edge of T once. We shall show that both pieces are collinear in the resulting drawing Γ and have the same hyperbolic length.

First, consider a regular hexagon $\bigcirc = p_0 p_1 p_2 p_3 p_4 p_5$ centered at the origin o of D; see Fig. 6a. In $\mathbb{H}^2$, it can have angles smaller than $120°$. We choose them to be $90°$ (any angle between $0°$ and $90°$ would work). Next, we draw a $K_{1,3}$ with center v_0 in o and the leaves v_1, v_2, v_3 in the middle of the arcs $p_0 p_1$, $p_2 p_3$, $p_4 p_5$ respectively.

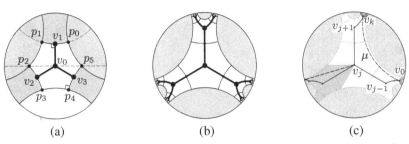

(a)   (b)   (c)

Fig. 6. Constructing increasing-chord drawings of binary trees and cactuses in $\mathbb{H}^2$

For each such building block of the drawing consisting of a $K_{1,3}$ inside a regular hexagon with $90°$ angles, we add its copy mirrored at an arc of the hexagon containing a leaf node of the tree constructed so far. For example, in the first iteration, we add three copies of $\circ$ mirrored at p_0p_1, p_2p_3 and p_4p_5, respectively, and the corresponding inscribed $K_{1,3}$ subtrees. The construction after two iterations is shown in Fig. 6b. This process can be continued infinitely to construct a drawing Γ_∞ of the infinite binary tree. However, we stop after we have completed Γ for the tree T.

We now show that Γ_∞ (and thus also Γ) has the desired properties. Due to isometries and the aforementioned sufficient condition, it suffices to consider edge $e = v_0v_1$ and show that a normal on e does not cross Γ_∞ in another point. To see this, consider Fig. 6a. Due to the choice of the angles of $\circ$, all the other hexagonal tiles of Γ_∞ are contained in one of the three blue quadrangular regions $\square_i := l_{v_0v_i}^+ \setminus (l_{v_ip_{2i-1}}^+ \cup l_{v_ip_{2i-2}}^+)$, $i = 1, 2, 3$. Thus, the regions $l_{v_1p_1}^+$ and $l_{v_1p_0}^+$ (gray) contain no point of Γ_∞. Therefore, since each normal on v_0v_1 is contained in the "slab" $D \setminus (l_{v_0v_1}^+ \cup l_{v_1v_0}^+)$ bounded by the diameter through p_2, p_5 and the line through p_0, p_1 (dashed) and is parallel to both of these lines, it contains no other point of Γ_∞. $\qquad\square$

We note that our proof is similar in spirit to the one by Kleinberg [12], who also used tilings of $\mathbb{H}^2$ to prove that any tree has a greedy drawing in $\mathbb{H}^2$.

As in the Euclidean case, it can be easily shown that if a tree T contains a node v of degree 4, it has a self-approaching drawing in $\mathbb{H}^2$ if and only if T is a subdivision of $K_{1,4}$ (apply an isometry, such that v is in the origin of D). This completely characterizes the trees admitting a self-approaching drawing in $\mathbb{H}^2$. Further, it is known that every binary cactus and, therefore, every 3-connected planar graph has a binary spanning tree [5, 13].

Corollary 2. (i) *A tree T has an increasing-chord drawing in $\mathbb{H}^2$ if and only if T either has maximum degree 3 or is a subdivision of $K_{1,4}$.* (ii) *Every binary cactus and, therefore, every 3-connected planar graph has an increasing-chord drawing in $\mathbb{H}^2$.*

Again, note that this is not the case for binary cactuses in $\mathbb{R}^2$; see the example in Theorem 2. We use the above construction to produce *planar* self-approaching drawings of binary cactuses in $\mathbb{H}^2$. We show how to choose a spanning tree and angles at vertices of degree 2, such that non-tree edges can be added without introducing crossings; see Fig. 6c for a sketch and the full version [15] for the proof.

Corollary 3. *Every binary cactus has a planar increasing-chord drawing in $\mathbb{H}^2$.*

References

1. Alamdari, S., Chan, T.M., Grant, E., Lubiw, A., Pathak, V.: Self-approaching Graphs. In: Didimo, W., Patrignani, M. (eds.) GD 2012. LNCS, vol. 7704, pp. 260–271. Springer, Heidelberg (2013)
2. Angelini, P., Colasante, E., Di Battista, G., Frati, F., Patrignani, M.: Monotone drawings of graphs. J. Graph Algorithms Appl. 16(1), 5–35 (2012)
3. Angelini, P., Di Battista, G., Frati, F.: Succinct greedy drawings do not always exist. Networks 59(3), 267–274 (2012)
4. Angelini, P., Didimo, W., Kobourov, S., Mchedlidze, T., Roselli, V., Symvonis, A., Wismath, S.: Monotone drawings of graphs with fixed embedding. In: Speckmann, B. (ed.) GD 2011. LNCS, vol. 7034, pp. 379–390. Springer, Heidelberg (2012)
5. Angelini, P., Frati, F., Grilli, L.: An algorithm to construct greedy drawings of triangulations. J. Graph Algorithms Appl. 14(1), 19–51 (2010)
6. Dhandapani, R.: Greedy drawings of triangulations. Discrete Comput. Geom. 43, 375–392 (2010)
7. Eppstein, D., Goodrich, M.T.: Succinct greedy geometric routing using hyperbolic geometry. IEEE Trans. Computers 60(11), 1571–1580 (2011)
8. Felsner, S.: Geometric Graphs and Arrangements. Vieweg+Teubner Verlag (2004)
9. Goodrich, M.T., Strash, D.: Succinct greedy geometric routing in the Euclidean plane. In: Dong, Y., Du, D.-Z., Ibarra, O. (eds.) ISAAC 2009. LNCS, vol. 5878, pp. 781–791. Springer, Heidelberg (2009)
10. Huang, W., Eades, P., Hong, S.-H.: A graph reading behavior: Geodesic-path tendency. In: IEEE Pacific Visualization Symposium (PacificVis 2009), pp. 137–144 (2009)
11. Icking, C., Klein, R., Langetepe, E.: Self-approaching curves. Math. Proc. Camb. Phil. Soc. 125, 441–453 (1999)
12. Kleinberg, R.: Geographic routing using hyperbolic space. In: Computer Communications (INFOCOM 2007), pp. 1902–1909. IEEE (2007)
13. Moitra, A., Leighton, T.: Some Results on Greedy Embeddings in Metric Spaces. Discrete Comput. Geom. 44, 686–705 (2010)
14. Nöllenburg, M., Prutkin, R.: Euclidean greedy drawings of trees. In: Bodlaender, H.L., Italiano, G.F. (eds.) ESA 2013. LNCS, vol. 8125, pp. 767–778. Springer, Heidelberg (2013)
15. Nöllenburg, M., Prutkin, R., Rutter, I.: On Self-Approaching and Increasing-Chord Drawings of 3-Connected Planar Graphs. CoRR arXiv:1409.0315 (2014)
16. Papadimitriou, C.H., Ratajczak, D.: On a conjecture related to geometric routing. Theor. Comput. Sci. 344(1), 3–14 (2005)
17. Purchase, H.C., Hamer, J., Nöllenburg, M., Kobourov, S.G.: On the usability of Lombardi graph drawings. In: Didimo, W., Patrignani, M. (eds.) GD 2012. LNCS, vol. 7704, pp. 451–462. Springer, Heidelberg (2013)
18. Rao, A., Ratnasamy, S., Papadimitriou, C., Shenker, S., Stoica, I.: Geographic routing without location information. In: Mobile Computing and Networking (MobiCom 2003), pp. 96–108. ACM (2003)
19. Rote, G.: Curves with increasing chords. Math. Proc. Camb. Phil. Soc. 115, 1–12 (1994)
20. Schnyder, W.: Embedding planar graphs on the grid. In: Discrete Algorithms (SODA 1990), pp. 138–148. SIAM (1990)
21. Wang, J.-J., He, X.: Succinct strictly convex greedy drawing of 3-connected plane graphs. Theor. Comput. Sci. 532, 80–90 (2014)

On Monotone Drawings of Trees[*]

Philipp Kindermann[1], André Schulz[2], Joachim Spoerhase[1], and Alexander Wolff[1]

[1] Lehrstuhl für Informatik I, Universität Würzburg, Germany
http://www1.informatik.uni-wuerzburg.de/en/staff
[2] Institut für Mathematische Logik und Grundlagenforschung, Universität Münster, Germany
andre.schulz@uni-muenster.de

Abstract. A crossing-free straight-line drawing of a graph is *monotone* if there is a monotone path between any pair of vertices with respect to *some* direction. We show how to construct a monotone drawing of a tree with n vertices on an $O(n^{1.5}) \times O(n^{1.5})$ grid whose angles are close to the best possible angular resolution. Our drawings are *convex*, that is, if every edge to a leaf is substituted by a ray, the (unbounded) faces form convex regions. It is known that convex drawings are monotone and, in the case of trees, also crossing-free.

A monotone drawing is *strongly monotone* if, for every pair of vertices, the direction that witnesses the monotonicity comes from the vector that connects the two vertices. We show that every tree admits a strongly monotone drawing. For biconnected outerplanar graphs, this is easy to see. On the other hand, we present a simply-connected graph that does not have a strongly monotone drawing in any embedding.

1 Introduction

A natural requirement for the layout of a connected graph is that between any source vertex and any target vertex, there should be a source–target path that approaches the target according to some distance measure. A large body of literature deals with problems of this type; various measures have been studied. For example, in a *greedy drawing* it is possible to decide locally where to go, by selecting in the current vertex any neighbor closer to the target. In a *monotone* drawing, the distance between vertices (on the desired source-target path) is measured with respect to their projections on *some* line, which may be different for any source–target pair. In *strongly monotone* drawings, that line is always the line from source to target, and in *upward* drawings, the line is always the vertical line, directed upwards.

In this paper, we focus on monotone and strongly monotone drawings of trees with additional aesthetic properties such as convexity or small area. Given a tree, we call the edges incident to the leaves *leaf edges* and all other edges *interior edges*. We direct all edges away from the root. Given a straight-line drawing of a tree, we substitute each leaf edge by a ray whose initial part coincides with the edge. The embedding of the tree defines a combinatorial embedding of the tree, that is the order of the edges around every vertex. The faces are then specified by this combinatorial embedding as leaf-to-leaf paths. If the faces of the augmented drawing are realized as convex nonoverlapping

---

[*] This research was supported by the ESF EuroGIGA project GraDR (DFG grant Wo 758/5-1).

(unbounded) polygonal regions, then we call the original drawing a *convex drawing*. If every region is *strictly convex* (that is, all interior angles are strictly less than π), we also call the drawing *strictly convex*. Note that a convex drawing is also monotone [4,2], but a monotone drawing is not necessarily convex. Strict convexity forbids vertices of degree 2. In this paper, when we talk about (strongly) monotone drawings, this always includes the planarity requirement. Otherwise, as Angelini et al. [2] observed, drawing any spanning tree of the given graph in a (strongly) monotone way and inserting the remaining edges would yield a (strongly) monotone drawing.

Previous Work. While any 3-connected plane graph has a greedy drawing in the Euclidean plane [10] (even without crossing [7]), this is, unfortunately, not true for trees. Nöllenburg and Prutkin [11] gave a complete characterization for the tree case, which shows that no tree with a vertex of degree 6 or more admits a greedy drawing. Alamdari et al. [1] have studied a subclass of greedy drawings, so-called *self-approaching drawings* which require that there always is a source–target path such that the distance decreases for any triplet of intermediate points on the *edges*, not only for the vertices.

Carlson and Eppstein [6] study convex drawings of trees. They give linear-time algorithms that optimize the angular resolution of the drawings, both for the fixed- and the variable-embedding case. They observe that convexity allows them to pick edge lengths arbitrarily, without introducing crossings.

For monotone drawings, Angelini et al. [2] studied the variable-embedding case. They showed that any n-vertex tree admits a straight-line monotone drawing on a grid of size $O(n^{1.6}) \times O(n^{1.6})$ (using a BFS-based algorithm) or $O(n) \times O(n^2)$ (using a DFS-based algorithm). They also showed that any biconnected planar graph has a monotone drawing (using exponential area). Further, they observed that not every planar graph admits a monotone drawing if its embedding is fixed. They introduced the concept of *strong monotonicity* and showed that there is a drawing of a planar triangulation that is not strongly monotone. Hossain and Rahman [9] improve some of the results of Angelini et al. by showing that every connected planar graph admits a monotone drawing of size $O(n) \times O(n^2)$ and that such a drawing can be computed in linear time.

Both the BFS- and the DFS-based algorithms of Angelini et al. precompute a set of $n - 1$ vectors in decreasing order of slope. For this, they use two different partial traversals of the so-called Stern–Brocot tree, an infinite tree whose vertices are in bijection with the irreducible positive rational numbers. Such numbers can be seen as *primitive vectors* in 2d, that is, as vectors with pairwise different slopes. Then both algorithms do a depth-first (pre-order) traversal of the input tree. Whenever they hit a new edge, they assign to it the steepest unused vector. They place the root of the input tree at the origin and draw each edge (u, v) by adding the vector assigned to (u, v) to the position of u. They call such tree drawings *slope-disjoint*. We won't formally define this notion here, but it is not hard to see that it implies monotonicity.

Angelini, with a different set of co-authors [3], investigated the fixed-embedding case. They showed that, on the $O(n) \times O(n^2)$ grid, every connected plane graph admits a monotone drawing with two bends per edge and any outerplane graph admits a straight-line monotone drawing.

Our contribution. We present two main results. First, we show that any n-vertex tree admits a strictly convex and, hence, monotone drawing on the $O(n^{1.5}) \times O(n^{1.5})$ grid (see Section 3). As the drawings of Angelini et al. [2], our drawings are slope-disjoint, but we use a different set of primitive vectors (based on Farey sequences), which slightly decreases the grid size. (This also works for the BFS-based algorithm of Angelini et al.) Instead of pre-oder, we use a kind of in-oder traversal (first child – root – other children) of the input tree, which helps us to achieve convexity. Our ideas can be applied to modify the optimal angular resolution algorithm of Carlson and Eppstein [6] such that a drawing on an $O(n^{1.5}) \times O(n^{1.5})$ grid is constructed at the expense of missing the optimal angular resolution by a constant factor. Second, we show that any tree admits a *strongly* monotone drawing (see Section 4). So far, no positive results have been known for strongly monotone drawings.

In the case of proper binary trees, our drawings are additionally strictly convex. For biconnected outerplanar graphs, it is easy to construct strongly monotone drawings. On the other hand, we present a simply-connected planar graph that does not have a strongly monotone drawing in any embedding.

We leave it as an open question whether trees admit strongly monotone drawings on a grid of polynomial size (see Section 5).

2   Building Blocks: Primitive Vectors

The following algorithms require a set of integral vectors with distinct directed slopes and bounded length. In particular, we ask for a set of *primitive vectors* $P_d = \{(x, y) \mid \gcd(x, y) \in \{1, d\}, 0 \le x \le y \le d\}$. Our goal is to find the right value of d such that P_d contains at least k primitive vectors, where k is a number that we determine later. We can then use the reflections on the lines $x = y$, $y = 0$ and $x = 0$ to get a sufficiently large set of integer vectors with distinct directed slopes. The edges of the monotone drawings in Section 3 are translates of these vectors; each edge uses a different vector.

Assume that we have fixed d and want to generate the set P_d. If we consider each entry (x, y) of P_d to be a rational number x/y and order these numbers by value, we get the *Farey sequence* $\mathcal{F}_d$ (see, for example, Hardy and Wright's book [8]). The Farey sequence is well understood. In particular, it is known that $|\mathcal{F}_d| = 3d^2/\pi^2 + O(d \log d)$ [8, Thm. 331]. Furthermore, the entries of $\mathcal{F}_d$ can be computed in time $O(|\mathcal{F}_d|)$. We remark that the set $\bigcup_d \mathcal{F}_d$ coincides with the entries of the Stern–Brocot tree. However, collecting the latter level by level is not the most effective method to build a set of primitive vectors for our purpose.

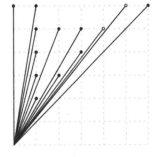

Fig. 1. The 13 primitive vectors obtained from $\mathcal{F}_6$. The smallest angle of $\approx 1.14°$ is realized between the vectors $(4, 5)$ and $(5, 6)$ marked with white dots; the best possible angular resolution in this case is $45°/12 = 3.75°$.

To get access to a set of k primitive vectors, we use the first k entries of the Farey sequence $\mathcal{F}_d$, for $d := 4\lceil\sqrt{k}\rceil$, replacing each rational by its corresponding 2d vector. By selecting k vectors form this set we get a set of exactly k primitive vectors, which we denote by V_k; see Fig. 1.

If we wish to have more control over the aspect ratio in our final drawing, we can pick a set of primitive vectors contained inside a triangle spanned by the grid points $(0, 0)$, $(m_x, 0), (m_x, m_y)$. By stretching the triangle and keeping its area fixed, we may end up with fewer primitive vectors. This will result in an (only slightly) smaller constant compared to the case $m_x = m_y$. As proven by Bárány and Rote [5, Thm. 2], any such triangular domain contains at least $m_x m_y/4$ primitive vectors. This implies that we can adapt the algorithm easily to control the aspect ratio by selecting the box for the primitive vectors accordingly. For the sake of simplicity, we detail our algorithms only for the most interesting case $(m_x = m_y)$.

Lemma 1. *Let $P \subseteq P_d$ be a set of $k = |P_d|/c$ primitive vectors with no coordinate greater than d for some constant $c \geq 1$. Then any two primitive vectors of P are separated by an angle of $\Omega(1/k)$.*

Proof. Since $|P_d| = 3d^2/\pi^2 + O(d \log d)$ we have that $2d^2 \approx 2\pi^2 ck/3$. Any line with slope m encloses an angle α with the x-axis, such that $\tan(\alpha) = m$. Let m_1 and m_2 be the slopes of two lines and let α_1 and α_2 be the corresponding angles with respect to the x-axis. By the trigonometric addition formulas we have that the separating angle of these two lines equals

$$\tan \phi := \tan(\alpha_1 - \alpha_2) = \frac{\tan \alpha_1 - \tan \alpha_2}{1 + \tan \alpha_1 \tan \alpha_2} = \frac{m_1 - m_2}{1 + m_1 m_2}.$$

For any two neighboring entries p/q and r/s in the Farey sequence, it holds that $qr - ps = 1$ [8, Thm. 3.1.2], and therefore p/q and r/s differ by exactly $(qr - ps)/(qs) = 1/(qs)$. As a consequence, $\tan \phi = 1/(pr + qs)$. The angle ϕ is minimized if $pr + qs$ is maximized. Clearly, we have that $pr + qs < 2d^2 \approx 2\pi^2 ck/3$. By the Taylor expansion, $\arctan(x) = x - x^2\xi/(1 + \xi^2)^2$ for some value $0 \leq \xi \leq x$. Substituting x with $3/(2\pi^2 ck)$ yields, for $k \geq 2$, that

$$\phi \geq \frac{3}{2\pi^2 ck} - \frac{9\xi}{4\pi^4 c^2 k^2 (1 + \xi^2)^2} > \frac{3}{2\pi^2 ck} - \frac{9}{4\pi^4 c^2 k^2} = \Omega(1/k). \qquad \square$$

Since the best possible resolution for a set of k primitive vectors is $2\pi/k$, Lemma 1 shows that the resolution of our set differs from the optimum by at most a constant.

3 Monotone Grid Drawings with Good Angles

We start by ensuring that convex tree drawings are crossing-free. This has already been stated by Carlson and Eppstein [6].

Lemma 2. *Any convex straight-line drawing of a tree is crossing-free.*

We now present a simple method for drawing a tree on a grid in a strictly convex, and therefore monotone and, by Lemma 2, crossing-free way. We name our strategy the *inorder-algorithm*. The algorithm first computes a reasonable large set of primitiv vectors, then selects a subset of these vectors and finally assigns the slopes to the edges.

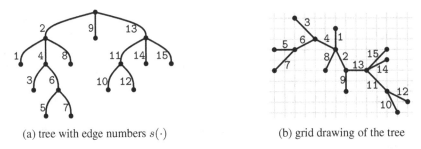

(a) tree with edge numbers $s(\cdot)$ (b) grid drawing of the tree

Fig. 2. A strictly convex drawing of a tree

The drawing is then generated by translating the selected primitive vectors. In the following, an *extended* subtree will refer to a subtree including the edge leading into the subtree (if the subtree is not the whole tree).

Every edge e will be assign with a number $s(e)$. This number will refer to the rank of the edge's slope (in circular order) in the final assignment. The rank assignment is done in a recursive fashion. At any time, let $\hat{s}$ be 1 plus the maximum rank $s(e)$ assigned so far. Initially, $\hat{s} = 1$. Let $e = uv$ be an edge (directed away from the root), and let $v_1, v_2, \ldots, v_\ell$ be the children of v ordered from left to right. We recursively set the ranks of all edges in the extended subtree rooted at v_1. Then we set $s(e) = \hat{s}$ (which increases $\hat{s}$ by one). Finally, we set, for $i = 2, \ldots, \ell$, the ranks of the edges in the extended subtree rooted at v_i. For an example of a tree with its edge ranks, see Fig. 2a.

Second, we assign actual slopes to the edges. Let e be an edge with $s(e) = j$. Then we assign to e some vector $s_j \in \mathbb{Z}^2$ and draw e as a translate of s_j. We pick the vectors $s_1, s_2, \ldots, s_{n-1}$ by selecting a sufficiently large set of primitive vectors and their reflections in counterclockwise order, see Section 2. Our drawing algorithm has the following requirements, which can be fulfilled as the following lemma shows:

(R1) Edges that are incident to the root and consecutive in circular order are assigned to vectors that together span an angle less than π.

(R2) In every extended subtree hanging off the root, the edges are assigned to a set of vectors that spans an angle less than π.

Lemma 3. *We can select $n - 1$ vectors with distinct directed slopes from a $[-d, d] \times [-d, d]$ grid with $d = 4\lceil\sqrt{n}\rceil$ such that the requirements (R1) and (R2) are fulfilled.*

Proof. We first preprocess our tree by adding temporary edges at some leaves. These edges will receive slopes, but are immediately discarded after the assignment.

First, our objective is to ensure that the tree can be split up into three parts that all have n edges. In particular, we adjust the sizes of the extended subtrees hanging off the root by adding temporary edges such that we can partition them into three sets of consecutive extended subtrees which all contain n edges. Note that we have to add $2n + 1$ edges to achieve this.

Second, we define three cones C_1, C_2, and C_3 (see Fig. 3). Each cone has its apex at the origin and spans an angle of $\pi/4$. The angular ranges are $C_1 = [0, \pi/4]$, $C_2 = [3\pi/4, \pi]$, and $C_3 = [3\pi/2, 7\pi/4]$; angles are measured from the x-axis pointing in positive direction. Note that C_2 is separated from the two other cones by an angle of $\pi/2$. As mentioned in Section 2 the set V_n contains n primitive vectors in the $[0, d] \times [0, d]$ grid. When reflected on the $x = y$ line these vectors lie in C_1. Reflecting the vectors in C_1 further we generate n vectors in C_2 and n vectors in C_3. In every cone we "need" at most $n - 3$ edges, hence we can remove the vectors on the boundary of each cone. After removing the temporary edges, the number of vectors will drop from $3n$ to $n - 1$.

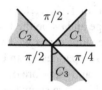

Fig. 3. The cones that contain the slopes used in the algorithm

Now, we observe the following. Every two consecutive edges incident to the root lie in the *interiors* of our cones. This yields requirement (R1) given the sizes and angular distances of the cones. Furthermore, any extended subtree is assigned slopes from a single cone. This yields (R2). □

For the example tree of Fig. 2a, it suffices to pick the 16 vectors that one gets from reflecting the primitive vectors from the $[0, 2] \times [0, 2]$ grid. These vectors already fulfill requirements (R1) and (R2). Hence, we did not have to apply the more involved slope selection as described in Lemma 3. The resulting drawing is shown in Fig. 2b.

Every face in the drawing contains two leaves. The leaves are ordered by their appearance in some DFS-sequence $\mathcal{D}$ respecting some rooted combinatorial embedding of T. For a face f, we call the leaf that comes first in $\mathcal{D}$ the *left leaf* and the other leaf of f the *right leaf* of f. The only exception is the face whose leaves are the first and last child of $\mathcal{D}$. Here we call the first vertex in $\mathcal{D}$ the right leaf and the last vertex in $\mathcal{D}$ the left leaf.

Lemma 4. *Let u be the left leaf, and let v be the right leaf of a face of T. Further, let w be the lowest common ancestor of u and v. The above assignment of slope ranks s to the tree edges implies the following.*
(a) If edge e_1 is on the w–u path and edge e_2 is on the w–v path, then $s(e_1) < s(e_2)$.
(b) The ordered sequence of edges on the path $w \rightarrow u$ is increasing in $s(\cdot)$.
(c) The ordered sequence of edges on the path $w \rightarrow v$ is decreasing in $s(\cdot)$.

The proof is omitted because of space constraints. We now prove the correctness of our algorithm.

Theorem 1. *Given an embedded tree with n vertices (none of degree 2), the inorder-algorithm produces a strictly convex and crossing-free drawing with angular resolution $\Omega(1/n)$ on a grid of size $O(n^{1.5}) \times O(n^{1.5})$. The algorithm runs in $O(n)$ time.*

Proof. We first show that in the drawing no face is incident to an angle larger than π. Let f be a face, let e and e' be two consecutive edges on the boundary of f, and let α be the angle formed by e and e' in the interior of f. If e and e' are incident to the root, requirement (R1) implies $\alpha < \pi$. If both edges contain the lowest common ancestor of the leafs belonging to f, then by requirement (R2) also $\alpha < \pi$. In the remaining case, e and e' both lie on a path to the left leaf of f, or both lie on a path to the right leaf of f. At vertex v we have at least two outgoing edges. Let e_1 be the first outgoing edge and e_2 be the last outgoing edge at v – one of the edges is e'. By the selection of the slope ranks we have $s(e_1) < s(e) < s(e_2)$. As a consequence, the supporting line of e separates e_1 and e_2, and hence both faces containing e have an angle less than π at v and therefore $\alpha < \pi$.

Next, we show that the edges and rays of a face do not intersect. Then, by Lemma 2, no edges will cross. Assume that there are two edges/rays ℓ and r in a common face that have an intersection in some point x. Let t be the lowest common ancestor of ℓ and r and assume that ℓ lies on the path to the left leaf and r on the path to the right leaf. We define a closed polygonal chain P of as follows. The chain starts with the path from t to ℓ, continues via x to r, and finally returns to t. We direct the edges according to this walk (for measuring the directed slopes) and call them $e_1, e_2, \ldots, e_k$. We may assume that P is simple, otherwise we find another intersection point. By Lemma 4, the slopes are monotone when we traverse P. For $i = 1, \ldots, k - 1$, let α_i be the difference between the directed slopes of the edges e_i and e_{i+1}. Then the sum $\sum_{i<k} \alpha_i$ equals the angle between the slopes of e_1 and e_k. Due to requirement (R2), this angle is less than π. Let $\beta_i = \pi - \alpha_i$ be the angle between e_i and e_{i+1} in P, and let $\beta_0 > 0$ be the "interior" angle at t. We have that

$$\sum_{0 \le i < k} \beta_i = \beta_0 + \sum_{1 \le i < k} (\pi - \alpha_i) > 0 + (k - 1)\pi - \pi = (k - 2)\pi.$$

This, however, contradicts the fact that the angle sum of the polygon with boundary P is $(k - 2)\pi$. Thus, our assumption that two edges/rays cross was wrong.

Since the drawing is assembled from $n - 1$ vectors whose absolute coordinates are at most $O(\sqrt{n})$, the complete drawing uses a grid of dimension $O(n^{1.5}) \times O(n^{1.5})$. Since all vectors are reflections of (a subset of) vectors defined by a Farey sequence with at most n entries, Lemma 1 yields that the angular resolution is bounded by $\Omega(1/n)$. □

We conclude this section with comparing our result with the drawing algorithm of Carlson and Eppstein [6]. Their algorithm produces a drawing with optimal angular resolution. It draws trees convex, but, in contrast to our algorithm, not necessarily strictly convex. Allowing parallel leaf edges can have a great impact on the angular resolution. However, our ideas can be applied to modify the algorithm of Carlson and Eppstein. For the leaf edges, their algorithm uses a set of k slopes and picks the slopes such that they are separated by an angle of $2\pi/k$. The slopes of interior edges have either one of the slopes of the leaf edges, or are chosen such that they bisect the wedge spanned by their outermost child edges. However, it suffices to assure that the slope of an interior edge differs from the extreme slopes in the following subtree by at least $2\pi/(2k)$.

We can now modify the algorithm as follows. We pick $2k/8$ primitive vectors and reflect them such that they fill the whole angular space with $2k$ distinct integral vectors. We use every other vector of this set for the leaf edges. For an interior edge we take any vector from our preselected set whose slope lies in between the extreme slopes of the edges in its subtree. We can always find such a vector, since we have sufficiently spaced out our set of primitive vectors. By this we obtain a drawing on the $O(n^{1.5}) \times O(n^{1.5})$ grid. Clearly, the drawing doesn't have optimal angular resolution. However, since we use $2k$ integral vectors having, by Lemma 1, an angular resolution of $\Omega(1/k)$, we differ from the best possible angular resolution $2\pi/k$ only by a constant factor.

4 Strongly Monotone Drawings

We first show how to draw any proper binary tree (that is, any internal vertex has exactly two children). We name our strategy the *disk-algorithm*. Then, we generalize our result to arbitrary trees. Further, we show that connected planar graphs do not necessarily have a strongly monotone drawing. Finally, we show how to draw biconnected outerplanar graphs in a strongly monotone fashion.

Let T be a proper binary tree, let D be any disk with center c, and let C be the boundary of D. Recall that a strictly convex drawing cannot have a vertex of degree 2. Thus, we consider the root of T a dummy vertex and ensure that the angle at the root is π. We draw T inside D. We start by mapping the root of T to c. Then, we draw a horizontal line h through c and place the children of the root on $h \cap \text{int}(D)$ such that they lie on opposite sites relative to c. We cut off two circular segments by dissecting D with two vertical lines running through points representing the children of the root. We inductively draw the right subtree of T into the right circular segment and the left subtree into the left circular segment.

In any step of the inductive process, we are given a vertex v of T, its position in D (which we also denote by v) and a circular segment D_v; see Fig. 4a. The preconditions for our construction are that

(i) v lies in the relative interior of the chord s_v that delimits D_v, and

(ii) D_v is empty, that is, the interiors of D_v and D_u are disjoint for any vertex u that does not lie on a root–leaf path through v.

In order to place the two children l and r of v (if any), we shoot a ray v from v perpendicular to s_v into D_v. Let v' be the point where v hits C. Consider the chords that connect the endpoints of s_v to v'. The chords and s_v form a triangle with height vv'. The height is contained in the interior of the triangle and splits it into two right subtriangles. The chords are the hypotenuses of the subtriangles. We contruct l and r by connecting v to these chords perpendicularly. Note that, since the subtriangles are right triangles, the heights lie inside the subtriangles. Hence, l and r lie in the relative interiors of the chords. Further, note that the circular segments D_l and D_r delimited by the two chords are disjoint and both are contained in D_v. Hence, D_l and D_r are empty, and the preconditions for applying the above inductive process to r and l with D_l and D_r are fulfilled. See Fig. 4b for the output of our algorithm for a tree of height 3.

Lemma 5. *For a proper binary tree rooted in a dummy vertex, the disk-algorithm yields a strictly convex drawing.*

Proof. Let T be a proper binary tree and let f be a face of the drawing generated by the algorithm described above. Clearly, f is unbounded. Let a and b be the leaves of T that are incident to the two unbounded edges of f, and let v be the lowest common ancestor of a and b; see Fig. 4b. Consider the two paths from v to a and b. We assume that the path from v through its left child ends in a and the path through its right child ends in b.

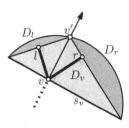
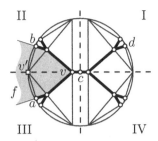

(a) sketch of the inductive step   (b) output of our algorithm for a tree of height 3

Fig. 4. Strongly monotone drawings of proper binary trees

Due to our inductive construction that uses disjoint disk sections for different sub-trees, it is clear that the two paths do not intersect. Moreover, each vertex on the two paths is convex, that is, the angle that such a vertex forms inside f is less than π. This is due to the fact that we always turn right when we go from v to a, and we always turn left when we go to b. Vertex v is also convex since the two edges from v to its children lie in the same half-plane (bounded by s_v).

It remains to show that the two rays a and b (defined analogously to v above) don't intersect. To this end, recall that $v' = v \cap C$. By our construction, a and b are orthogonal to two chords of C that are both incident to v'. Clearly, the two chords form an angle of less than π in v'. Hence, the two rays diverge, and the face f is strictly convex.   □

For the proof that the algorithm described above yields a strongly monotone drawing, we need the following tools. Let v_1 and v_2 be two vectors. We say that v_3 *lies between* v_1 and v_2 if v_3 is a positive linear combination of v_1 and v_2. For two vectors v and w, we define $\langle v, w \rangle = |v||w| \cos(v, w)$ as the scalar product of v and w.

Lemma 6. *If a path p is monotone with respect to two vectors v_1 and v_2, then it is monotone with respect to any vector v_3 between v_1 and v_2.*

Proof. Let $v_3 = \lambda_1 v_1 + \lambda_2 v_2$ with $\lambda_1, \lambda_2 > 0$. Assume that the path p is given by the sequence of vectors $w_1, w_2, \ldots, w_k$. Since p is monotone with respect to vectors v_1 and v_2, we have that $\langle v_1, w_i \rangle > 0$ and $\langle v_2, w_i \rangle > 0$ for all $i \leq k$. This yields, for all $i \leq k$,

$$\langle v_3, w_i \rangle = \langle \lambda_1 v_1 + \lambda_2 v_2, w_i \rangle = \lambda_1 \langle v_1, w_i \rangle + \lambda_2 \langle v_2, w_i \rangle > 0,$$

since $\lambda_1, \lambda_2 > 0$. It follows that p is monotone with respect to v_3.   □

Lemma 7. *For a proper binary tree rooted in a dummy vertex, the disk-algorithm yields a strongly monotone drawing.*

Proof. We split the drawing into four sectors: I, II, III and IV; see Fig. 4b. Let a and b be two vertices in the graph. We will show that the path that connects a and b in the drawing output by our algorithm is strongly monotone. Let c be the root of the tree. W.l.o.g., assume that a lies in sector III. Then, we distinguish three cases.

Case 1: a and b lie on a common root–leaf path; see a and v in Fig. 4b. Obviously, b lies in sector III. W.l.o.g., assume that b lies on a path between a and c. By construction, all edges in sector III, seen as vectors directed towards c, lie between $\boldsymbol{x} = (0, 1)$ and $\boldsymbol{y} = (1, 0)$. Thus, all edges on the path from a to b, and in particular ab, lie between $\boldsymbol{x}$ and $\boldsymbol{y}$. Since $\boldsymbol{x}$ is perpendicular to $\boldsymbol{y}$, the path from a to b is monotone with respect to $\boldsymbol{x}$ and $\boldsymbol{y}$. Following Lemma 6, the path between a and b is monotone with respect to $\overrightarrow{ab}$, and thus strongly monotone.

Case 2: b lies in sector I; see a and d in Fig. 4b. In Case 1, we have shown that the all edges on the path from a to c lie between $\boldsymbol{x} = (0, 1)$ and $\boldsymbol{y} = (1, 0)$. Analogously, the same holds for the path from c to b. Thus, the path between a and b is monotone with respect to $\boldsymbol{x}$ and $\boldsymbol{y}$ and, following Lemma 6, strongly monotone.

Case 3: a and b do not lie on a common root–leaf path, and b does not lie in sector I; see a and b in Fig. 4b. Let d be the lowest common ancestor of a and b. Let $a_0, a_1, \ldots, a_{k-1}, a_k$ be the path from d to a where $a_0 = d$ and $a_k = a$. Now, let $b_0, b_1, \ldots, b_{m-1}, b_m$ be the path from d to b where $d = b_0$ and $b = b_m$. Finally, let $p = a_k, a_{k-1}, \ldots, a_0, b_1, \ldots, b_{m-1}, b_m$ be the path from a to b.

Below, we describe how to rotate and mirror the drawing such that the any vector $\overrightarrow{a_i, a_{i-1}}$, $1 \le i \le k$ lies between $\boldsymbol{x} = (0, 1)$ and $\boldsymbol{y} = (1, 0)$, and any vector $\overrightarrow{b_{j-1}, b_j}$, $1 \le j \le m$ lies between $\boldsymbol{x}$ and $-\boldsymbol{y}$. This statement is equivalent to $x(a_i) < x(d) < x(b_j), 1 \le i \le k, 1 \le j \le m$ and $y(a_k) < \ldots < y(a_1) < y(d) > y(b_1) > \ldots > y(b_m)$;

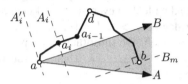

Fig. 5. Illustration of case 3 in the proof of Lemma 7

see Fig. 5. If b lies in sector IV, then $d = c$ and this statement is true by construction. If b lies in sector II, then d is a child of c. We rotate the drawing by $\pi/2$ in counterclockwise direction and then mirror it horizontally. If b lies in sector III, let $p(d)$ be the parent of d. We rotate the drawing such that the edge $(p(d), d)$ is drawn vertically. Recall that, by construction, the ray from d in direction $\overrightarrow{p(d)d} = -\boldsymbol{y}$ separates the subtrees of the two children of d; see Fig. 4a. Further, the angle between any edge (directed away from d) in the subtree of d and $\overrightarrow{p(d)d} = -\boldsymbol{y}$ is at most $\pi/2$, i.e., they are directed downwards.

Let $A_i, 1 \le i \le k$ be the straight line through a_i and perpendicular to $\overrightarrow{a_{i-1}a_i}$. Let A_i' be the parallel line to A_i that passes through a. Due to the x-monotonicity of p the point a lies below A_i. During the construction of the tree, the line A_i defined a circular sector in which the subtree rooted at a_i including a was exclusively drawn. It follows that a and b lie on opposite sites of A_i. Thus, b lies above A_i and also above A_i'. Let $B_j, 1 \le j \le m$ be the straight line through b_j and perpendicular to $\overrightarrow{b_{j-1}b_j}$. Let B_j' be the parallel line to B_j that passes through a. By construction, b lies below B_j and a lies above B_j. Thus, b lies below B_j'.

Let A be the line A_i' with maximum slope and let B be the line B_j' with minimum slope. First, we will show that the path is monotone with respect to the unit vector $\boldsymbol{A}$ on A directed to the right. By choice of A, the angle between any edge $(a_i, a_{i-1}), 1 \le i \le k$ and $\boldsymbol{A}$ is at most $\pi/2$. Recall that any vector $\overrightarrow{a_i, a_{i-1}}$, $1 \le i \le k$ lies between $\boldsymbol{x}$ and $\boldsymbol{y}$. Since $\boldsymbol{A}$ is perpendicular to one of these edges and directed to the right, it lies between x and $-y$. Since any vector $\overrightarrow{b_{j-1},b_j}$, $1 \leq j \leq m$ also lies between x and $-y$, the angle between A and these edges is at most $\pi/2$. Because the angle between A and any edge on the path from a to b is at most $\pi/2$, the path is monotone with respect to A.

Analogously, it can be shown that the path is monotone with respect to B. Recall that b lies above A and below B. So the vector $\overrightarrow{ab}$ lies between A and B. Following Lemma 6, the path is monotone with respect to $\overrightarrow{ab}$ and thus strongly monotone.   □

Lemmas 5 and 7 immediately imply the following.

Theorem 2. *Any proper binary tree rooted in a dummy vertex has a strongly monotone and strictly convex drawing.*

Next, we (partially) extend this result to arbitrary trees.

Theorem 3. *Any tree has a strongly monotone drawing.*

Proof. Let T be a tree. If T has a vertex of degree 2, we root T in this vertex. Otherwise, we subdivide any edge by creating a vertex of degree 2, which we pick as root. Then, we add a leaf to every vertex of degree 2, except the root. Now, let v be a vertex with out-degree $k > 2$. Let $(v, w_1), \ldots, (v, w_k)$ be the outgoing edges of v ordered from right to left. We substitute v by a path $\langle v = v_1, \ldots, v_{k+1} \rangle$, where v_{i+1} is the left child of v_i, for $i = 1, \ldots, k$. Then, we substitute the edges (v, w_i) by $(v_i, w_i), i = 2, \ldots, k$; see Fig. 6.

Let T' be the resulting proper binary tree. Clearly, all vertices of T', except its root, have degree 1 or 3, so T' is a proper binary tree. We use Theorem 2 to get a strongly monotone drawing $\Gamma_{T'}$ of T'. Then, we remove the dummy vertices inserted above and draw the edges that have been subdivided by a path as a straight-line. This yields a drawing Γ_T of T that is crossing-free since the only new edges form a set of stars that are drawn in disjoint areas of the drawing.

Now, we show that Γ_T is strongly monotone. Let (v, w) be an edge in T. Let $p = \langle v = v_1, \ldots, v_m = w \rangle$ be the path in T' between v and w. Suppose p is monotone with respect to some direction d. Thus, $\angle\{\overrightarrow{v_i v_{i+1}}, d\} < \pi/2$ for $1 \leq i \leq m - 1$. Clearly, $\overrightarrow{vw} = \sum_{i=1}^{m-1} \overrightarrow{v_i v_{i+1}}$ is a positive linear combination of $\overrightarrow{v_i v_{i+1}}, i = 1, \ldots, m$ and hence $\angle\{\overrightarrow{vw}, d\} < \pi/2$. It follows that the path between two vertices a and b is monotone to a direction d in Γ_T if the path between a and b is monotone to d in $\Gamma_{T'}$. With $d = \overrightarrow{ab}$, it follows that Γ_T is strongly monotone.   □

We add to this another positive result concerning biconnected outerplanar graphs.

Theorem 4. *Any biconnected outerplanar graph has a strongly monotone and strictly convex drawing.*

Proof. Let G be a biconnected outerplanar graph with outer cycle $\langle v_1, \ldots, v_n, v_1 \rangle$. We place the vertices $v_2, \ldots, v_{n-1}$ in counterclockwise order on a quarter circle C that has $v_1 = (0,0)$ and $v_n = (1,1)$ as its endpoints; see Fig. 7. Since the outer cycle is drawn strictly convex, the drawing is planar and strictly convex. Clearly, the path $\langle v_1, \ldots, v_n \rangle$ is x- and y-monotone. Also, every vector $\overrightarrow{v_i v_j}, j > i$ lies between $x = (0,1)$ and $y = (1,0)$. Thus, by Lemma 6, the drawing is strongly monotone.   □

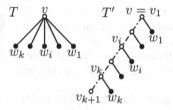

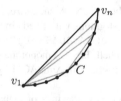

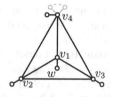

Fig. 6. Subdivision of a vertex v with k outgoing edges

Fig. 7. A strongly monotone drawing of a biconnected outerplanar graph

Fig. 8. A planar graph without any strongly monotone drawing

We close with a negative result. Note that the graphs in the family that we construct are neither outerplanar nor biconnected.

Theorem 5. *There is an infinite family of connected planar graphs that do not have a strongly monotone drawing in any combinatorial embedding.*

Proof. Let C be the graph that arises by attaching to each vertex of K_4 a "leaf"; see Fig. 8. Let $v_1, \ldots, v_4$ be the vertices of K_4. For K_4 to be crossing-free, one of its vertices, say v_1, lies in the interior. Let w be the leaf incident to v_1. Because of planarity, w has to be placed inside a triangular face incident to v_1. W.l.o.g., assume that w is placed in the face (v_1, v_2, v_3). If the drawing is strongly monotone, then $\angle(\overrightarrow{wv_2}, \overrightarrow{wv_1}) < \pi/2$ and $\angle(\overrightarrow{wv_1}, \overrightarrow{wv_3}) < \pi/2$ and thus $\angle(\overrightarrow{wv_3}, \overrightarrow{wv_2}) > \pi$. However, this means that w does not lie inside the triangle (v_1, v_2, v_3), which is a contradiction to the assumption. Thus, C does not have a strongly monotone drawing in any combinatorial embedding. We create an infinite family from C by adding more leaves to the vertices of K_4. □

5   Conclusion and Open Problems

We have shown that any tree has a monotone drawings on a grid with area $O(n^3)$ and a strongly monotone drawing, but can we combine the two features, that is, does any tree have a strongly monotone drawing on a grid of polynomial size?

Angelini et al. [2, Fig. 18(b)] have constructed a drawing of a triangulation that is not strongly monotone. But is there a triconnected (or biconnected) planar graph that does not have any strongly monotone drawing? If yes, can this be tested efficiently?

If we could show that our drawings for general trees are not just strongly monotone but also convex (as in the proper binary case), then all Halin graphs would automatically have convex and strictly monotone drawings, too.

References

1. Alamdari, S., Chan, T.M., Grant, E., Lubiw, A., Pathak, V.: Self-approaching graphs. In: Didimo, W., Patrignani, M. (eds.) GD 2012. LNCS, vol. 7704, pp. 260–271. Springer, Heidelberg (2013)
2. Angelini, P., Colasante, E., Battista, G.D., Frati, F., Patrignani, M.: Monotone drawings of graphs. J. Graph Algorithms Appl. 16(1), 5–35 (2012)

3. Angelini, P., Didimo, W., Kobourov, S., Mchedlidze, T., Roselli, V., Symvonis, A., Wismath, S.: Monotone drawings of graphs with fixed embedding. To appear in Algorithmica, http://dx.doi.org/10.1007/s00453-013-9790-3

4. Arkin, E.M., Connelly, R., Mitchell, J.S.: On monotone paths among obstacles with applications to planning assemblies. In: Proc. 5th Ann. ACM Symp. Comput. Geom. (SoCG 1989), pp. 334–343 (1989)

5. Bárány, I., Rote, G.: Strictly convex drawings of planar graphs. Doc. Math. 11, 369–391 (2006)

6. Carlson, J., Eppstein, D.: Trees with convex faces and optimal angles. In: Kaufmann, M., Wagner, D. (eds.) GD 2006. LNCS, vol. 4372, pp. 77–88. Springer, Heidelberg (2007)

7. Dhandapani, R.: Greedy drawings of triangulations. Discrete Comput. Geom. 43(2), 375–392 (2010)

8. Hardy, G., Wright, E.M.: An Introduction to the Theory of Numbers, 5th edn. Oxford University Press (1979)

9. Hossain, M. I., Rahman, M. S.: Monotone Grid Drawings of Planar Graphs. In: Chen, J., Hopcroft, J.E., Wang, J. (eds.) FAW 2014. LNCS, vol. 8497, pp. 105–116. Springer, Heidelberg (2014)

10. Leighton, T., Moitra, A.: Some results on greedy embeddings in metric spaces. Discrete Comput. Geom. 44(3), 686–705 (2010)

11. Nöllenburg, M., Prutkin, R.: Euclidean greedy drawings of trees. In: Bodlaender, H.L., Italiano, G.F. (eds.) ESA 2013. LNCS, vol. 8125, pp. 767–778. Springer, Heidelberg (2013)

Graph Drawing Contest Report

Carsten Gutwenger[1], Maarten Löffler[2], Lev Nachmanson[3], and Ignaz Rutter[4]

[1] Technische Universität Dortmund, Germany
carsten.gutwenger@tu-dortmund.de
[2] Utrecht University, The Netherlands
m.loffler@uu.nl
[3] Microsoft, USA
levnach@microsoft.com
[4] Karlsruhe Institute of Technology, Germany
rutter@kit.edu

Abstract. This report describes the 21st Annual Graph Drawing Contest, held in conjunction with the 2014 Graph Drawing Symposium in Würzburg, Germany. The purpose of the contest is to monitor and challenge the current state of graph-drawing technology.

1 Introduction

This year, the Graph Drawing Contest was divided into an *offline contest* and an *online challenge*. The offline contest had two categories: the first one dealt with creating a metro map layout from a given bus and tram network, and the second one was a composer's network. The data sets for the offline contest were published months in advance, and contestants could solve and submit their results before the conference started. The submitted drawings were evaluated according to aesthetic appearance, domain specific requirements, and how well the data was visually represented.

The online challenge took place during the conference in a format similar to a typical programming contest. Teams were presented with a collection of challenge graphs and had approximately one hour to submit their highest scoring drawings. This year's topic was the same as last year, namely to minimize the area for orthogonal grid layouts, where we allowed crossings (the number of crossings was not judged, only the area counted).

Overall, we received 24 submissions: 5 submissions for the offline contest and 19 submissions for the online challenge.

2 Metro Map Layout

In this category, the task was to visualize the bus and tram network of Würzburg in a metro map style layout. The data for the network included information about the stops like the name of the stops and their geographic locations, as well as the bus/tram lines with their stops and the distances between stops. We asked for a visualization of the whole network, presenting the connections in a clear way for a possible user of public transport in Würzburg. The data had been kindly provided by the WVV[1].

---

[1] http://www.wvv.de

C. Duncan and A. Symvonis (Eds.): GD 2014, LNCS 8871, pp. 501–506, 2014.
© Springer-Verlag Berlin Heidelberg 2014

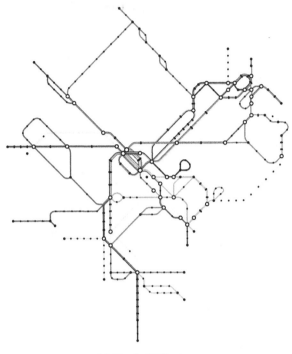

(a) Martin Nöllenburg

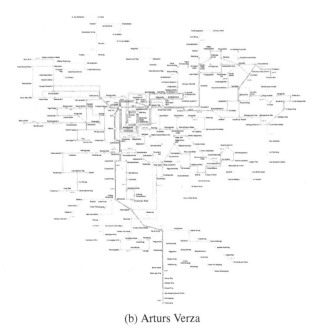

(b) Arturs Verza

Fig. 1. Metro map layout of Würzburg's public transport network

We received two submissions in this category, both presenting the network in a very nice way. Martin Nöllenburg's submission (see Fig. 1(a)) is a typical metro map drawing with a very nice routing of lines, created using their ILP-based metro map layout algorithm [1]. Fig. 1(b) shows Arturs Verza's submission, which gives a clear picture of the cluttered city center.

The winner in this category was Martin Nöllenburg from the Karlsruhe Institute of Technology, since we preferred the nicer global layout of his submission, which allows a user of the map to easily figure out possible connections.

3 Composers Graph

For this category, we used a data set that was already a contest graph in 2011. The *composers graph* is a large directed graph, where its 3,405 nodes represent Wikipedia articles about composers, and its 13,382 edges represent links between these articles. This graph has too many nodes and edges to be effectively presented in a straightforward way. Therefore, this time the task was to select the about 150 most important nodes and to create a drawing of a subgraph containing these nodes. Part of the task was to define *important* in a suitable way. The criterion should only depend on the given graph, not on any other sources or knowledge. It was also allowed to filter out some edges between important nodes using a reasonable criterion for filtering.

We received three submissions for this graph. Fig. 2(a) shows the submission from Remus Zelina et al.; they divided the composers into *influencers* and *influencees* (a composer could appear twice) and then used Girvan Newman modularization to obtain a set of modules. For selecting the most important composers, they used the corresponding factor in the modularity formula as well as the page rank algorithm. They also categorized the edges with respect to the module structure and selected only the most important ones. The final layout was then obtained by applying a layered approach that emphasized the module structure. The submission by Ulf Rüegg (see Fig. 2(b)) used the notion of betweenness to select the most important nodes in the graph; the edges were then selected as a maximum spanning tree, where the edges were weighed using edge betweenness. The resulting tree was laid out with a stress minimization approach. The third submission came again from Arturs Verza. He used centrality for selecting the top 150 composers, removed transitive edges in the subgraph, and finally applied a circular layout algorithm (due to lack of space we omit the drawing; it can be found on the contest web page).

The winner in this category was the team Remus Zelina, Sebastian Bota, Siebren Houtman, and Radu Balaban from Meurs, Romania, for their clear representation of global as well as local structure.

4 Online Challenge

The online challenge, which took place during the conference, dealt with minimizing the area in an orthogonal grid drawing. The challenge graphs were not necessarily planar and had at most four incident edges per node. Edge crossings were allowed and their number did not affect the score of a layout. Since typical drawing systems first try

(a) Remus Zelina et al.

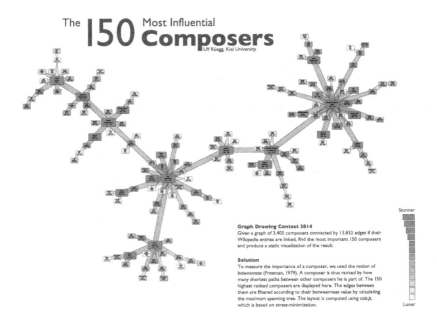

(b) Ulf Rüegg

Fig. 2. Composers graph

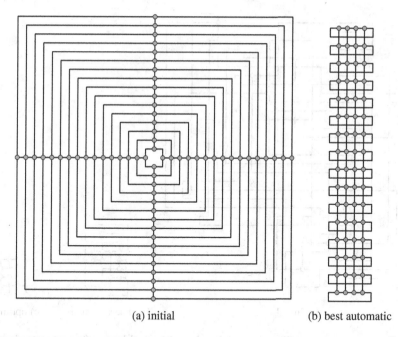

|(a) initial|(b) best automatic|

Fig. 3. Challenge graph with 64 nodes and 124 edges: (a) initial layout and (b) best automatic result by the team of Mchedlidze et al

to minimize the number of crossings, which might result in long edges increasing the required area, we were in particular interested in the effect of allowing crossings on the quality of layouts when trying to reduce the area.

The task was to place nodes, edge bends, and crossings on integer coordinates so that the edge routing is orthogonal and the layout contains no overlaps. At the start of the one-hour on-site competition, the contestants were given five graphs with an initial legal layout with a large area. The goal was to rearrange the layout to reduce the area, defined as the number of grid points in the smallest rectangle enclosing the layout. Only the area was judged; other aesthetic criteria, such as the number of crossings or edge bends, were ignored.

The contestants could choose to participate in one of two categories: *automatic* and *manual*. To determine the winner in each category, the scores of each graph, determined by dividing the area of the best submission in this category by the area of the current submission, were summed up. If no legal drawing of a graph was submitted (or a drawing worse than the initial solution), the score of the initial solution was used.

In the automatic category, contestants received six graphs ranging in size from 20 nodes / 29 edges to 100 nodes / 182 edges and were allowed to use their own sophisticated software tools with specialized algorithms. Manually fine-tuning the automatically obtained solutions was allowed. Fig. 3 shows a challenge graph from the automatic category with 64 nodes, 124 edges, and a very bad initial layout. The best obtained result improved the area from 1089 to 192. With a score of 4.964, the winner in the automatic category was the team of Tamara Mchedlidze, Martin Nöllenburg and their

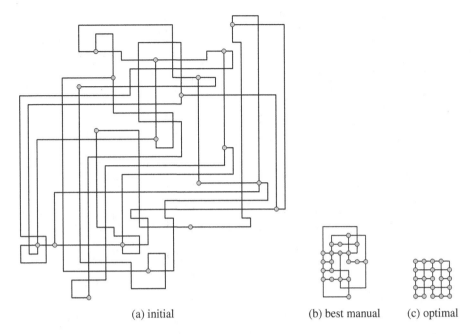

(a) initial   (b) best manual   (c) optimal

Fig. 4. Challenge graph with 20 nodes and 29 edges: (a) initial layout, (b) best manual result obtained by the team of Will and Jawaherul, and (c) optimal solution

Graph Drawing lecture students Igor, Alexander, and Denis from the Karlsruhe Institute of Technology, who found the best results for four of the five contest graphs.

The 19 manual teams solved the problems by hand using IBM's *Simple Graph Editing Tool* provided by the committee. They received five graphs ranging in size from 6 nodes / 8 edges to 20 nodes / 29 edges. The largest input graph was also in the automatic category. For this graph, both the best automatic and the best manual team could improve the area from initially 1056 to 54, whereas the optimal solution has an area of 25; see Fig. 4. With a score of 4.425, the winner in the manual category was the team of Philipp Kindermann, Fabian Lipp and Wadim Reimche from the University of Würzburg, who found the best results for three of the five contest graphs.

Acknowledgments. The contest committee would like to thank the generous sponsors of the symposium and all the contestants for their participation. Further details including winning drawings and challenge graphs can be found at the contest website:

http://www.graphdrawing.de/contest2014/results.html

References

1. Nöllenburg, M., Wolff, A.: Drawing and labeling high-quality metro maps by mixed-integer programming. IEEE Trans. Vis. Comput. Graph. 17(5), 626–641 (2011)

A User Study on the Visualization of Directed Graphs*

Walter Didimo[1], Fabrizio Montecchiani[1], Evangelos Pallas[2], and Ioannis G. Tollis[2]

[1] Dip. di Ingegneria, Università degli Studi di Perugia, Italy
[2] University of Crete and Institute of Computer Science-FORTH, Greece

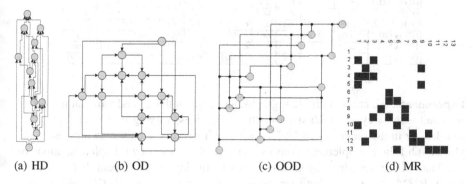

(a) HD   (b) OD   (c) OOD   (d) MR

Fig. 1. Different drawing styles for directed graphs

In a node-link visualization of a directed acyclic graph it is desirable that edges flow in a common direction (say upward) according to their orientations, as in *Hierarchical Drawings (HD)* (see, e.g., [3]). More general requirements are that the number of edge crossings, the number of edge bends, the drawing area, etc., are kept low. These are well addressed by *Orthogonal Drawings (OD)*, where edges are chains of horizontal and vertical segments (see, e.g., [4]). Unfortunately, OD algorithms do not control the flow of the edges in a desired direction. *Overloaded Orthogonal Drawing (OOD)* is a recent visualization paradigm conceived for drawing directed graphs, that merges and enforces the benefits of HD and of OD [2]. Indeed, edges are still represented using only horizontal and vertical segments, and if the digraph is acyclic, any directed edge (u, v) is drawn as an upward-rightward polyline consisting of one bend point (v is above and to the right of u). Also, adjacent edge segments can partially overlap to draw graphs with arbitrary vertex degree. We present the results of a user study aimed at measuring the usability of OOD on directed graphs against HD and OD. Moreover, we consider *Matrix-based Representations (MR)*. Indeed, a user study on undirected graphs [1] suggests that MR are more suitable than node-link diagrams computed with force-directed algorithms to perform some simple tasks when the size of the graph increases, but remains relatively small. Thus, we also aim to understand whether OOD maintains the same readability properties as MR with respect to user's tasks similar to those considered in [1], but tailored to directed graphs. See Figure 1 for an illustration.

---

* An extended abstract of this work has been presented at the 5*th* Conference on Information, Intelligence, Systems & Applications (IISA 2014). Research supported in part by the MIUR project AMANDA: Algorithmics for MAssive and Networked DAta, prot. 2012C4E3KT_001.

C. Duncan and A. Symvonis (Eds.): GD 2014, LNCS 8871, pp. 507–508, 2014.
© Springer-Verlag Berlin Heidelberg 2014

Table 1. The mean values and the pairwise significance between each pair of drawing paradigms are shown for error rate and for response time considering all the tasks (Overall) and single tasks

	Error rate					Response time (s)				
	Overall	PA	DE	CA	CY	Overall	PA	DE	CA	CY
mean OOD	0.119	0.190	0.047	0.154	0.083	58	78	36	69	48
mean HD	0.199	0.250	0.142	0.190	0.214	55	69	41	55	56
mean OD	0.369	0.547	0.095	0.369	0.452	67	114	17	71	66
mean MR	0.423	0.559	0.047	0.285	0.809	129	187	27	145	158
OOD vs HD	.003	n.s.	n.s.	n.s.	0.008	n.s.	n.s.	n.s.	.002	n.s.
OOD vs OD	< .001	.001	n.s.	.003	< .001	n.s.	n.s.	< .001	n.s.	n.s.
OOD vs MR	< .001	< .001	n.s	n.s.	< .001	< .001	.001	n.s.	< .001	< .001
HD vs OD	< .001	.002	n.s.	.003	.001	n.s.	.003	< .001	.006	n.s.
HD vs MR	< .001	.001	n.s.	n.s.	< .001	< .001	.001	.001	< .001	< .001
OD vs MR	n.s.	n.s.	n.s.	n.s.	< .001	< .001	.002	.001	< .001	< .001

Experiments. We chose 4 different graphs, modeling both real and artificial networks, with and without cycles, with size (number of vertices) varying in the range $[77, 122]$ and density in the range $[2.5, 3.5]$; for each graph we computed 4 drawings using the yEd Graph Editor[1] implementations of HD and OD and our own implementations of OOD and MR. After a training session, for each drawing the participants had to solve 4 tasks: *(PA)* "Is there a path between the two highlighted vertices?"; *(DE)* "What is the out-degree of the highlighted vertex?"; *(CA)* "Do the two highlighted vertices have any common adjacent vertex?"; *(CY)* "Is there a cycle including the highlighted vertex?". We compared the performance of all the drawing paradigms in terms of error rate and response time. 21 volunteering students participated in the experiments. We performed a non parametric analysis, the results are summarized in Table 1.

Conclusions. The results show a clear advantage in terms of accuracy in the reading of the displayed graphs when using the OOD paradigm, over all tasks and in particular for the tasks involving paths (PA) and cycles (CY). In terms of response time, the performance on the node-link representations (OOD, HD, OD) are comparable, although most tasks are executed slightly faster using HD. MR led to slower performance, except for task DE. In addition, node-link representations outperformed the matrix-based representation, both in terms of error rate and response time, especially for task CY.

References

1. Ghoniem, M., Fekete, J.-D., Castagliola, P.: A comparison of the readability of graphs using node-link and matrix-based representations. In: INFOVIS, pp. 17–24. IEEE (2004)
2. Kornaropoulos, E.M., Tollis, I.G.: Overloaded orthogonal drawings. In: Speckmann, B. (ed.) GD 2011. LNCS, vol. 7034, pp. 242–253. Springer, Heidelberg (2011)
3. Sugiyama, K., Tagawa, S., Toda, M.: Methods for visual understanding of hierarchical system structures. IEEE Tran. on Sys., Man, and Cyb. 11(2), 109–125 (1981)
4. Tamassia, R.: On embedding a graph in the grid with the minimum number of bends. SIAM J. on Comp. 16(3), 421–444 (1987)

---

[1] http://www.yworks.com/en/index.html

GraphBook: Making Graph Paging Real*

Alessio Arleo[1], Felice De Luca[1], Giuseppe Liotta[1],
Fabrizio Montecchiani[1], and Ioannis G. Tollis[2]

[1] Dip. di Ingegneria, Università degli Studi di Perugia, Italy
[2] University of Crete and Institute of Computer Science-FORTH, Greece

In the last years we observed an impressive growth of different mobile devices, like tablets and smartphones, that allow people to access and share a large variety of contents. These devices are often used to access social networks, route networks, or other kinds of networks that for example may deal with the job of the user. Although designing drawing algorithms and visualization systems tailored for mobile devices would be the best choice in this case (see, e.g., [1]), this is often not possible, as for example the drawing of the network might be computed by some diagram server in a collaborative environment, or attached to an e-mail, or returned by a web application that does not consider limited display capabilities. We present a system whose goal is to facilitate the reading of a given drawing on a mobile device by exploiting an analogy with electronic books[1]. It takes as input a drawing Γ of a graph $G=(V, E)$ and computes a *graph book* $\mathcal{B}(\Gamma)$. The idea is to adapt the drawing's aspect-ratio to the device display and distribute the visual complexity of the drawing in different pages that are browsed by using standard gestures. This is done in three steps. We describe two possible implementations for each step, depending on whether Γ is a straight-line drawing or an overloaded orthogonal drawing (OOD) [4]. In the first case we can assume that Γ has been computed by some force-directed algorithm (see, e.g., [3]). In the second case we recall that OOD is a graph visualization paradigm conceived for digraphs [4], which has similarities with hierarchical and orthogonal drawings. The presence of an edge (u, v) is conveyed by an e-point (a small circle) placed at the intersection point of the vertical segment passing through u and the horizontal segment passing through v. The information in an OOD is enriched by computing the transitive closure of G and by conveying paths as p-points.

Sizing. This step takes as input Γ and returns a resized drawing Γ_R that matches the device display resolution and that guarantees all the drawing conventions of the input drawing Γ. Also, Γ_R should preserve as much as possible some of the drawing original prominent features. If Γ is straight-line, a technique like the one described in [5] can be used. If Γ is OOD, it is enough to properly adjust the horizontal and vertical grid unit.

Paging. In this step the set of edges E is partitioned into subsets $E_1, E_2, \ldots, E_k$, such that the subdrawing $p_i = \Gamma_R[E_i]$ ($1 \leq i \leq k$), called *page*, guarantees some desired property (e.g., planarity). In each subdrawing, the vertices remain fixed in their original positions as in Γ_R. The goal is to distribute the drawing visual complexity along the pages, which can be explored separately. On the negative side, the user has to face

---

* Research supported in part by the MIUR project AMANDA: Algorithmics for MAssive and Networked DAta, prot. 2012C4E3KT_001.
[1] A demo video is available at http://youtu.be/Toi9lnkbzlo

C. Duncan and A. Symvonis (Eds.): GD 2014, LNCS 8871, pp. 509–510, 2014.

510   A. Arleo et al.

Fig. 1. Two screenshots of the GraphBook system. Left: The input is an overloaded orthogonal drawing [4] of a DBLP coauthorship network and the displayed page shows the paths of length one in the network. Right: The input is a straight-line drawing of a social network and the readability property used to compute the displayed page is planarity (see also [2]).

the difficulty of making sense of a distributed information, thus it is important to have as few pages as possible. If Γ_R is straight-line, then a user could prefer to see each page p_i without any edge crossing, i.e., as planar, or that any two crossing edges form a sufficiently large angle. Computing a minimum set of pages that cover all the edges of G and such that each page guarantees a prescribed readability property can be heuristically computed in polynomial time [2]. If Γ_R is an OOD, a natural choice to build the set of pages is to assign to page p_i all edges that represent shortest paths of length i.

Binding. Binding all the pages together is done by a set of interactive operations that can be used to browse the pages of the book and to explore a single page. To go from one page to the next one or to the previous one, it is enough to swipe the screen from left to right or from right to left, respectively: The edges in the current page will fade out, while the edges in the new page will fade in. Zooming in and out is done by the pinch and stretch gesture. The edges incident on a vertex, even those assigned to different pages, can be highlighted by tapping on the vertex. If tapping on two vertices the system shows the edge that connects the two vertices (if any). By selecting a region of the screen with a circle gesture the subdrawing induced by the vertices inside the region is highlighted.

References

1. Da Lozzo, G., Di Battista, G., Ingrassia, F.: Drawing graphs on a smartphone. JGAA 16(1), 109–126 (2012)
2. Di Giacomo, E., Didimo, W., Liotta, G., Montecchiani, F., Tollis, I.G.: Techniques for edge stratification of complex graph drawings. JVLC 25(4), 533–543 (2014)
3. Eades, P.A.: A heuristic for graph drawing. In: Congr. Num., vol. 42, pp. 149–160 (1984)
4. Kornaropoulos, E.M., Tollis, I.G.: Overloaded orthogonal drawings. In: van Kreveld, M.J., Speckmann, B. (eds.) GD 2011. LNCS, vol. 7034, pp. 242–253. Springer, Heidelberg (2011)
5. Wu, Y., Liu, X., Liu, S., Ma, K.-L.: Visizer: A visualization resizing framework. IEEE TVCG 19(2), 278–290 (2013)

Circular Tree Drawing by Simulating Network Synchronisation Dynamics and Scaling

Farshad Ghassemi Toosi* and Nikola S. Nikolov

CSIS Department, University of Limerick, Ireland
{farshad.toosi,nikola.nikolov}@ul.ie

Abstract. We present an algorithm which produces circular-shape layouts of trees by simulating synchronisation dynamics on the tree. Our approach consists of evolving scalar dynamical values assigned to the nodes. Then the dissimilarities between the values of each pair of nodes are utilised to calculate the coordinates of the nodes by using a lower bound on dissimilarities and scaling up the lower bound per iteration.

1 Introduction

We propose a new algorithm for tree drawing with a circular shape; nodes with higher betweenness centrality are placed near the center of the circle and leaves towards the periphery. A force-directed algorithm with a similar result has been proposed by Bannister et al. [2]; it utilises a third *gravity force* in addition to forces of attraction and repulsion between nodes. Our algorithm produces similar results without introducing gravity, and the layouts take an exact circular shape. We achieve this by simulating network synchronisation dynamics on the tree according to a modified version of the Kuramoto model [1]. That is, we evolve scalar (*dynamical*) variables assigned to the tree nodes according to a rule that brings variables of connected nodes close to each other. We then use the dissimilarities between pairs of dynamical values to compute the tree layout. Our experimental results demonstrate that we are able to find tree layouts with either no or a small number of edge crossings.

2 Algorithm

Consider tree T with a node set V rooted at node v_r with the highest betweenness centrality. Let w_i be a dynamic variable assigned to node $v_i \in V$ and let $w_i(t)$ be the value of w_i at time step $t \in \{0, 1, \ldots\}$. We choose to have dynamic variables with values in the interval $[0, 2\pi)$, so that they can be interpreted as angular coordinates of the nodes. The tree drawing algorithm we propose consists of the following four steps.

Step 0: $t \leftarrow 0$; let $w_r(t) = 0$ for any t; for each $v_i \neq v_r$ assign a random value in the interval $[0, 2\pi)$ to $w_i(0)$; assign random x- and y-coordinates to all nodes; set a lower bound L on the dissimilarity of a pair of dynamical variables.

---

* Supported by the Irish Research Council (IRC) under project no. GOIPG/2014/938.

C. Duncan and A. Symvonis (Eds.): GD 2014, LNCS 8871, pp. 511–512, 2014.

Step 1: $t \leftarrow t+1$; for each $v_i \neq v_r$ with parent v_k calculate $w_i(t)$ according to the equation $\frac{dw_i(t)}{dt} = \sin(w_k(t)) - w_i(t) + \phi_i(t)$ which models network synchronisation dynamics. The term $\phi_i(t)$ is an adjustment which allows larger drawing space for subtrees rooted at nodes with a relatively high degree.

Step 2: For each pair of nodes $v_i \neq v_j$, calculate the dissimilarity of their dynamic variables $dis_{ij}(t) = \max\{L, ((\cos(w_i(t)) - \cos(w_j(t)))^2 + (\sin(w_i(t)) - \sin(w_j(t)))^2)^{\frac{1}{2}}\}$.

Step 3: Update the coordinates of all nodes in force-directed manner. If v_i and v_j are adjacent then the force of attraction between them is $q \times \sqrt[p]{(dis_{ij}(t))}$; if they are not adjacent then $-1 \times \sqrt{(dis_{ij}(t))}$ is the force of repulsion between them. The values of $q \geq 1$ and $0 < p \leq 1$ grow with the size of the tree.

Step 4: Scaling up the value of L at each iteration gradually and continuing this process until layout gets stable (typically $|V|$ steps).

Figure 1 shows two trees drawn with our algorithm. The one in the left is one of the example trees which is also drawn by Bannister et al. [2]. We have observed that our algorithm achieves more even distribution of nodes within a circular drawing area compared to other tree drawing algorithms. Our algorithm has the same time complexity as the classical force-directed graph drawing algorithms, i.e. $O(m|V|^2)$, where m is the number of iterations.

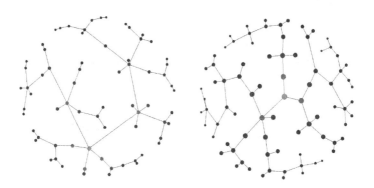

Fig. 1. A tree with 70 nodes (right) and a tree with 82 nodes (left). Nodes have different size and color depending on the distance to the root.

References

1. Arenas, A., Daz-Guilera, A., Prez-Vicente, C.J.: Synchronization reveals topological scales in complex networks. Phys. Rev. Lett. 96(11), 114102-1–114102-4 (2006)
2. Bannister, M.J., Eppstein, D., Goodrich, M.T., Trott, L.: Force-directed graph drawing using social gravity and scaling. In: Didimo, W., Patrignani, M. (eds.) GD 2012. LNCS, vol. 7704, pp. 414–425. Springer, Heidelberg (2013)

PiGra– A Tool for Pixelated Graph Representations

Thomas Bläsius, Fabian Klute, Benjamin Niedermann, and Martin Nöllenburg

Karlsruhe Institute of Technology (KIT), Germany

At GD 2013 Biedl et al. presented a simple and versatile formulation of grid-based graph representation problems as integer linear programs (ILPs) and corresponding SAT-instances [1]. In a grid-based representation each vertex and each edge of a graph is represented by a set of grid cells, which we also call *pixels*. Biedl et al. described a general ILP model, where each object (vertex or edge) corresponds to a set of variables that determine which pixels represent the object. They introduced constraints that restrict the shapes of objects (e.g., requiring the pixels of a vertex to form a 2D box) and how the representations of different objects can intersect. In this way, one can solve a variety of NP-hard graph problems, including pathwidth, bandwidth, optimum st-orientation, area-minimal (bar-k) visibility representation, boxicity-k graphs and others. For example, in a grid-based drawing of a visibility representation, each vertex is represented by a horizontal box of height 1 and each edge is represented by a vertical box of width 1. Moreover, two boxes overlap if and only if they represent a vertex and an incident edge.

Biedl et al. [1] implemented and evaluated the ILP-models for the above problems. The experiments showed that their models successfully solve NP-hard problems within few minutes on small to medium-size graphs. They further provided their C++ implementation as the framework GDSAT[1], which can be freely downloaded and adapted.

With GDSAT it requires little effort to solve problems that are already modeled within GDSAT for a given graph. However, adapting the models to solve different grid-based graph drawing problems requires deeper insights and adaptions. Moreover, there is no graphical output of the result. Hence, it is not an easy-to-use tool for tasks like running initial experiments to explore new grid-based graph drawing problems.

To make the framework of Biedl et al. [1] more widely accessible and useful to the community, we developed the GUI-based tool PiGra (**pi**xelated **gra**phs) that allows to easily combine pre-defined general constraints in order to model the above mentioned problems as well as other grid-based layout problems. Additionally, in case the pre-defined constraints are not sufficient, the user may adapt existing or define new constraints using the simple, mathematically-oriented language PGL (**p**ixelated **g**raphs **l**anguage), which we have introduced for this purpose.

In PiGra the typical workflow consists of the following three steps; see Fig. 1.

1. The user defines the problem as a generic ILP-formulation in PiGra, using combinations of pre-defined and custom constraints formulated in PGL.
2. PiGra instantiates the ILP-formulation for a user-specified graph and solves it with GUROBI[2] (an automatic conversion to an equivalent SAT-instance is planned).
3. The resulting grid-based representation is graphically displayed and can be interactively explored.

---

[1] i11www.iti.kit.edu/gdsat – The name GDSAT comes from the fact that solving an equivalent SAT-instance performs better than solving the ILP itself.

[2] www.gurobi.com

C. Duncan and A. Symvonis (Eds.): GD 2014, LNCS 8871, pp. 513–514, 2014.

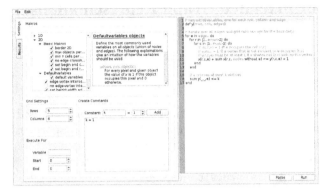 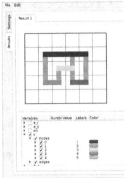

Fig. 1. Screenshots of a typical workflow in PIGRA using the example of bar-k visibility representations. Model formulation using a combination of pre-defined constraints and custom constraints expressed in PGL (left). Graphical output of the ILP solution for a small example graph (right).

PIGRA provides a variety of pre-defined ILP-constraints that are useful to formulate many custom grid-based graph drawing problems and can simply be selected by ticking the corresponding check-boxes. Among others, the following constraints are available: (a) vertices or edges are represented by boxes, (b) boxes have a certain height or width, (c) certain types of boxes may not overlap, (d) boxes of edges overlap exactly the boxes of their incident vertices. Additional constraints on shapes and intersections can be specified directly in PIGRA using PGL. For example, vertices could be modeled as the union of two boxes with non-empty intersection, including the special case of L-shapes. Or a constraint could be added so that each vertex box may only intersect a single incident edge, which models selecting a set of matching edges. In summary, the main features of PIGRA are:

- A macro system with pre-defined ILP-constraints for grid-based graph layouts.
- The simple, mathematically-oriented language PGL providing the capability to formulate ILP constraints with low overhead.
- A simple editor for writing constraints in PGL. Since all pre-defined constraints are also written in PGL, the user may adapt those constraints.
- A well-structured graphical user interface for presenting the result of the ILP as grid-based graph drawing.
- Support of GML-format for loading graphs (we use OGDF[3] to parse GML-files).
- Implemented in C++ and soon available for download[4] under the GPL.

References

1. Biedl, T., Bläsius, T., Niedermann, B., Nöllenburg, M., Prutkin, R., Rutter, I.: Using ILP/SAT to determine pathwidth, visibility representations, and other grid-based graph drawings. In: Wismath, S., Wolff, A. (eds.) GD 2013. LNCS, vol. 8242, pp. 460–471. Springer, Heidelberg (2013)

---

[3] www.ogdf.net

[4] i11www.iti.kit.edu/pigra

Simultaneous Drawing of Planar Graphs
with Right-Angle Crossings and Few Bends*

Michael A. Bekos[1], Thomas C. van Dijk[2], Philipp Kindermann[2],
and Alexander Wolff[2]

[1] Wilhelm-Schickard-Institut für Informatik, Universität Tübingen, Germany
bekos@informatik.uni-tuebingen.de
[2] Lehrstuhl für Informatik I, Universität Würzburg, Germany
http://www1.informatik.uni-wuerzburg.de/en/staff

1 Introduction

A simultaneous embedding of two planar graphs embeds each graph in a planar way—using the same vertex positions for both embeddings. Edges of one graph are allowed to intersect edges of the other graph. There are two versions of the problem: In the first version, called *Simultaneous Embedding with Fixed Edges* (SEFE), edges that occur in both graphs must be embedded in the same way in both graphs (and hence, cannot be crossed by any other edge). In the second version, these edges can be drawn differently for each of the graphs. Both versions of the problem have a geometric variant where edges must be drawn using straight-line segments.

When actually drawing these simultaneous embeddings, a natural choice is to use straight-line segments. Only very few graphs can be drawn in this way, however, and some existing results need exponential area. We suggest a new approach that overcomes these problems. We insist that crossings occur at right angles, thereby "taming" them. We do this while drawing on a grid of size $O(n) \times O(n)$ for n-vertex graphs, and we can still draw any pair of planar graphs simultaneously. We do not consider the problem of fixed edges. In a way, our results give a measure for the geometric complexity of simultaneous embeddability for various pairs of graph classes, some of which can be combined more easily (with fewer bends) and some not as easily (needing more bends).

More formally, in this paper we study the *RAC simultaneous drawing problem* (RACSIM *drawing problem*). Let $G_1 = (V, E_1)$ and $G_2 = (V, E_2)$ be two planar graphs on the same vertex set. We say that G_1 and G_2 admit a RACSIM drawing if we can place the vertices on the plane such that (i) each edge is drawn as a polyline, (ii) each graph is drawn planar, (iii) there are no edge overlaps and (iv) crossings between edges in E_1 and E_2 occur at right angles.

Argyriou et al. [1] introduced and studied the geometric version of RACSIM drawing. In particular, they proved that it is always possible to construct a geometric RACSIM drawing of a cycle and a matching in quadratic area, while there exists a pair of graphs (a wheel and a cycle) which which do not admit a geometric RACSIM drawing. The problem that we study was left as an open problem.

---

* This research was supported by the ESF EuroGIGA project GraDR (DFG grant Wo 758/5-1).

C. Duncan and A. Symvonis (Eds.): GD 2014, LNCS 8871, pp. 515–516, 2014.

Table 1. A short summary of our results

Graph classes			Number of bends	Grid size
Planar	+	Planar	$6+6$	$(14n - 26) \times (14n - 26)$
2-page book embed.	+	2-page book embed.	$4+4$	$(11n - 32) \times (11n - 32)$
Outerplanar	+	Outerplanar	$3+3$	$(7n - 10) \times (7n - 10)$
Cycle	+	Cycle	$1+1$	$2n \times 2n$
Caterpillar	+	Cycle	$1+1$	$(2n - 1) \times 2n$
Four Matchings			$1+1+1+1$	$2n \times 2n$
Tree	+	Matching	$1+0$	$n \times (n - 1)$
Wheel	+	Matching	$2+0$	$(1.5n - 1) \times (n + 2)$
Outerpath	+	Matching	$2+1$	$(3n - 2) \times (3n - 2)$

2   Our Contribution

First, we look at the most general version of the problem, where the input is a pair of planar graphs. (In a simultaneous drawing, certainly both graphs must—individually—be planar.) We give a linear-time algorithm for this case, which produces a drawing in quadratic area with at most six bends per edge. For 2-page book embeddable graphs and outerplanar graphs, we give algorithms that guarantee four and three bends respectively. Then we turn our attention to graph classes that are more restricted, but for which we can give algorithms that use very few bends. See Table 1 for a full list of results. The main approach in these algorithms is to find linear orders on the vertices of the two graphs and then to compute coordinates for the vertices based on these orders. All drawings can be computed in linear time on a grid whose size is quadratic in the number of vertices. The proofs of these claims can be found in the full version of the paper [2].

3   Open Problems

The results presented in this paper raise several questions that remain open, such as the following. (i) As a variant of the problem, it might be possible to reduce the required number of bends per edge by relaxing the strict constraint that intersections must be right-angle and instead ask for drawings that have close to optimal crossing resolution. (ii) The computational complexity of the general problem remains open: Given two or more planar graphs on the same set of vertices and an integer k, is there a RACSIM drawing in which each graph is drawn with at most k bends per edge and the crossing are at right angles?

References

1. Argyriou, E.N., Bekos, M.A., Kaufmann, M., Symvonis, A.: Geometric RAC simultaneous drawings of graphs. J. Graph Algorithms Appl. 17(1), 11–34 (2013)
2. Bekos, M.A., van Dijk, T.C., Kindermann, P., Wolff, A.: Simultaneous drawing of planar graphs with right-angle crossings and few bends. Arxiv report (August 2014), http://arxiv.org/abs/1408.3325

Touching Triangle Represenations in a k-gon of Biconnected Outerplanar Graphs

Nieke Aerts

Technische Universität Berlin, Institut für Mathematik, Berlin, Germany
aerts@math.tu-berlin.de

Introduction. A side-contact representation by triangles, without holes, is called a touching triangle representation. Gansner et al. have shown that every biconnected outerplanar graph has a touching triangle representation [2]. In their construction the boundary is a polygon with concave and convex angles. Fowler characterized the strongly-connected outerplanar graphs that have a proper touching triangle representation [1], i.e., the outer face is a triangle as well. An outerplanar graph is strongly-connected if it is biconnected and the graph induced by the interior edges is connected. Here we expand this characterization to biconnected outerplanar graphs. Moreover, the characterization allows for deciding precisely how many corners the boundary polygon needs to have. A touching triangle representation in a k-gon is called a kTTR.

Definitions. First we construct the graph H such that G is the weak dual of H. Start with the weak dual of G. Add an edge and a new endpoint for every boundary edge of G. The newly added points are cyclically connected. Every boundary edge, whose contraction does not induce a 2-face, is contracted.

Let $\text{VEIN}(G)$ be the graph consisting of all strictly interior edges of G and their endpoints. The venation graph of $\text{VEIN}(G)$, denoted by $\text{VENATION}(G)$, is the graph that has as vertices the components of $\text{VEIN}(G)$ and the faces between two components. There is an edge between a component and a face if and only if the face has a chord of this component on its boundary. There are no other edges.

The vertices of the venation graph can be divided into five classes, the components without interior face C_0, the components with precisely one interior face C_1, the components with precisely two interior faces C_2, the components with

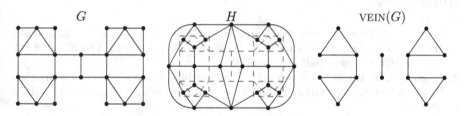

Fig. 1. A biconnected outerplanar graph G, its auxiliary graph H and $\text{VEIN}(G)$

C. Duncan and A. Symvonis (Eds.): GD 2014, LNCS 8871, pp. 517–518, 2014.

more than two interior faces C_3 and the connecting faces F. An orientation of the edges of VENATION(G) is called *valid* if all edges are oriented and:
- every vertex in C_2 has only incoming arcs,
- every vertex in C_1 has at most one outgoing arc,
- every vertex in C_0 has at most two outgoing arcs,
- every vertex in F has at precisely one outgoing arc.

We need one more definition, which will make it possible to decide how many suspensions are needed.

Dividing Path. Let c a component of VEIN(G) with two interior faces, f and f'. An edge on the boundary of G with one end on f, is called a *petiole* of f. A simple path P, between two boundary vertices of a biconnected outerplanar graph is said to be a *dividing path* for c if
[D1] there is at most one edge of c in P, and,
[D2] the interior faces of c are on opposite sides of P, and,
[D3] each face has at least one petiole that is not completely in P.

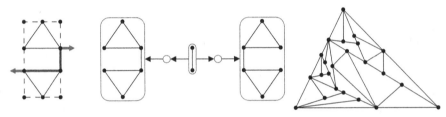

Fig. 2. A dividing path (red) and petioles (dashed), a valid orientation and a 3TTR

Theorem 1. *Let G be a biconnected outerplanar graph and v_2 the number of degree two vertices in of G. Let k be an integer such that $2 < k \leq v_2$ if $v_2 \geq 3$ and let $k = 3$ otherwise. A biconnected outerplanar graph has a kTTR iff*
[K1] *Each component of* VEIN(G) *has at most two interior faces.*
[K2] *The graph* VENATION(G) *admits a valid orientation.*
[K3] *There is a way to select k vertices of degree 2 in G, such that, for every component $c \in C_2$, there are two representatives in this set, v_i, v_j, and between the representatives there is a dividing path for c.*

References

1. Fowler, J.J.: Strongly-Connected Outerplanar Graphs with Proper Touching Triangle Representations. In: Wismath, S., Wolff, A. (eds.) GD 2013. LNCS, vol. 8242, pp. 155–160. Springer, Heidelberg (2013)
2. Gansner, E.R., Hu, Y., Kobourov, S.G.: On Touching Triangle Graphs. In: Brandes, U., Cornelsen, S. (eds.) GD 2010. LNCS, vol. 6502, pp. 250–261. Springer, Heidelberg (2011)

3D Graph Visualization with the Oculus Rift*

Farshad Barahimi and Stephen Wismath

University of Lethbridge, Canada

Abstract. Visualization of large graphs in 3D has been hampered by
the expense and inconvenience of virtual reality equipment. The Oculus
Rift is an affordable head-mounted VR system that is becoming popular
in the gaming and education markets. The GLuskap software package for
creating and editing graphs in 3D has been extended to include support
on the Oculus Rift for stereographic 3D viewing, and navigation.

Introduction. There are many theoretical results in the graph drawing literature regarding three-dimensional (3D) representations of graphs. However, the
actual usefulness of such 3D drawings has always been questioned; in particular, the effectiveness of visualization has been a main concern, as ultimately the
graph must be presented to the user, commonly:

- projected onto a (2D) monitor, possibly with a stereographic effect such as
  shutter glasses, or anaglyphically,
- displayed in a virtual reality environment such as a CAVE, or
- printed as a physical model with a 3D printer.

Early studies on the effectiveness of head-mounted VR systems vs. fish tank
VR noted that the hardware suffered from latency and resolution problems [1].
However, more recent HCI experiments indicate that users can improve performance on graph tasks (such as determining paths) for large graphs in a 3D VR
environment [3].

The GLuskap [4] software package allows for the creation and editing of graphs
in 3D and has been extended over the past decade to include each of the above
three output techniques; interactive visualization of 3D graphs has always been
a priority. The Oculus Rift [2] headset provides an affordable, individual, 3D
stereo experience with head tracking and display of appropriate left and right
eye information (Fig. 1). The low latency and accurate orientational tracking
allow an acceptable immersive experience.

Implementation. The Oculus Rift Software Development Kit provides pro-
grammable support in a C environment. Although GLuskap is written in python,
suitable wrappers and libraries permit its use as the main engine for creating the
virtual (graph) world. Head tracking on the first generation (DK1) rift is only
rotational and does not detect translational movements. For more effective and
convenient navigation, a simple game controller was added, allowing the user to

* Research supported by N.S.E.R.C.

C. Duncan and A. Symvonis (Eds.): GD 2014, LNCS 8871, pp. 519–520, 2014.

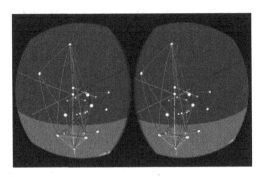

Fig. 1. Left and right eye information provided by the rift

efficiently and intuitively move through the displayed graph. For visualization purposes, when a user discovers an appropriate viewpoint and viewing angle, a *snapshot* or picture of the visible portion of the graph can be made. Note that the left/right eye information provided by the rift does not provide an acceptable 2D picture and thus the scene is totally recreated and supplied to separate rendering software (POVRay) for processing and generation.

The rift headset prevents the wearer from easily accessing the keyboard, mouse, and computer screen and thus the graph editing features of GLuskap (such as adding, deleting, and moving vertices and edges) are unavailable – only navigation, scrolling through stored graphs, and snapshots are provided in this mode, which required about 700 additional lines of python code.

Graph Drawing 2014. The poster presentation at GD14 will include a demo of navigating and viewing 3D graph drawings with an Oculus Rift.

References

1. Arthur, K.W., Booth, K.S., Ware, C.: Evaluating 3d task performance for fish tank virtual worlds. ACM Trans. Inf. Syst. 11(3), 239–265 (1993)
2. Oculus, http://www.oculusvr.com/rift/
3. Ware, C., Mitchell, P.: Visualizing graphs in three dimensions. TAP 5(1) (2008)
4. Wismath, S.: Website: **GLuskap** software, people.uleth.ca/~wismath/gluskap/

Force-Directed 3D Arc Diagrams

Michael J. Bannister, Michael T. Goodrich, and Peter Sampson

University of California, Irvine, USA

Abstract. We discuss a force-directed algorithm for constructing 3D arc diagrams. We introduce forces that allow curves in a 2D force directed graph to bow out and away from each other in the third dimension in order to achieve better angular resolution.

1 Introduction

In a 2D *arc diagram*[1], a graph is drawn by placing its vertices on a line and drawing its edges as circular arcs. Goodrich and Pszona [4] extend this definition to 3D arc diagrams, where vertices are placed in the xy-plane and edges are drawn as circular arcs that may bow out of that plane in the third dimension, and they show how to use graph coloring methods to determine tangent angles. Their approach works well for some types of graphs, but not all. In this poster, we present a general force-directed method for constructing a 3D arc diagram for any graph.

2 Our Algorithm

We start with a 2D force-directed layout produced with Fruchterman-Reingold [3] forces, that is, where all vertices repel each other based on electromagnetic forces and vertices connected by an edge are attracted to each other by a force that views the edge as a spring. Vertex placements are refined iteratively, where each iteration brings the 2D graph closer to a low energy state where nodes are experiencing the same forces in every direction. Our implementation starts with the standard 2D forces, without any kind of distortion, allowing vertices to move freely. Thus, in the xy-plane, all vertices exert a force inversely related to their distance, similar to an electromagnetic force, which tends to push away vertices that are not related to each other. Vertices that share an edge are pulled toward each other with a spring force, positively related to their distance, so that nodes that are related can be closer spatially. These two forces are tempered by a cooling function that gradually reduces the impact of the forces to avoid any case where nodes are pulled forward and backward indefinitely. This phase of the algorithm is similar to the first phase in the "dummy node" approach of Chemobelskiy *et al.* [2] for force-directed Lombardi drawings.

After a small number of iterations, our algorithm enters its second phase, where we allow intersecting edges to "pop" out of the xy-plane. We also allow edges that share common node to bow away from each other in the third dimension, based on their

---

[1] See http://en.wikipedia.org/wiki/Arc_diagram

C. Duncan and A. Symvonis (Eds.): GD 2014, LNCS 8871, pp. 521–522, 2014.

proximity and the angle between them. This is implemented by viewing edges as Bezier curves with 1 control point. The bowing force is exerted on the control points of two edges sharing a common node and is constructed by the current distance between the control points and the angle formed by the shared node and the location of each control point. Specifically, we compute a force that 2 control points will exert on each other as

$$F = c/(r^a \cdot d^b),$$

where r is the angle of the control points and their shared node and d is the distance between the two control points. In our case, we found that the values, $c = .5$ and $b = 2$, worked well, with $r = 1.5$ initially.

This force is applied in the x- and y-directions, leading the two control points away from each other and in the positive z-direction on both. It is then reduced so that the xy-force is perpendicular to the original straight line edge before any curve bowing force is applied. The control points are then pulled back to their original, unmoved location laying on the xy-plane by a spring force with power proportional to how far away the control point currently is from its origin. The spring force for a direction (x in this case) is given by

$$-Fx = kFx,$$

where $k = .5$.

When edges intersect, as detected by an orientation algorithm [1], we repel their control points in the z-direction only. This lifts up the higher edge and lowers the other. The magnitude of the force exerted on each control point is based on their distance in the xy-plane.

In the initial state, where all edges are lying on the xy-plane and have a z-value of 0, we exert a force to only one of the edges (at random), leaving the other edge still on the xy-plane.

3   Conclusion

We have observed that by adding another dimension to a graph drawing and allowing control points to enter this new dimension with forces that cause edges to bow out of the plane in a fashion dictated by intersections, angles, and proximity, we can improve the angular resolution of the graph as a whole. For almost all cases, using additional forces will provide better angular resolution than if we were in the xy-plane alone.

References

1. Battista, G.D., Eades, P., Tamassia, R., Tollis, I.G.: Graph Drawing: Algorithms for the Visualization of Graphs (1998)
2. Chernobelskiy, R., Cunningham, K.I., Goodrich, M.T., Kobourov, S.G., Trott, L.: Force-directed lombardi-style graph drawing. In: Speckmann, B. (ed.) GD 2011. LNCS, vol. 7034, pp. 320–331. Springer, Heidelberg (2011)
3. Fruchterman, T.M.J., Reingold, E.M.: Graph drawing by force-directed placement. Software: Practice and Experience 21(11), 1129–1164 (1991), doi:10.1002/spe.4380211102
4. Goodrich, M.T., Pszona, P.: Achieving good angular resolution in 3D arc diagrams. In: Wismath, S., Wolff, A. (eds.) GD 2013. LNCS, vol. 8242, pp. 161–172. Springer, Heidelberg (2013)

People Prefer Less Stress and Fewer Crossings[*]

Markus Chimani[1], Patrick Eades[2], Peter Eades[2], Seok-Hee Hong[2], Weidong Huang[3],
Karsten Klein[2], Michael Marner[4], Ross T. Smith[4], and Bruce H. Thomas[4]

[1] University of Osnabrück, Germany
Markus.Chimani@uni-osnabrueck.de
[2] University of Sydney, Australia
patrick.f.eades@gmail.com,
{peter.eades,seokhee.hong,karsten.klein}@sydney.edu.au
[3] University of Tasmania, Australia
tony.huang@utas.edu.au
[4] University of South Australia, Adelaide, Australia
{michael.marner,ross.smith,bruce.thomas}@unisa.edu.au

Human experiments in Graph Drawing, such as the seminal papers of Purchase et al. [1], have concentrated on *tasks*, such as path-tracing. Here we report the first results in an ongoing series of experiments to investigate the geometric properties of graph drawings that users *prefer*. The specific experiment described in this poster investigates the impact of edge crossings and stress in straight-line drawings on human preference. Our results suggest a positive preference for drawings with less stress and fewer crossings.

Experiment. Seventy nine subjects were recruited; subjects were about 50% Computer Science students and 50% Cognitive Science at the University of Osnabrück. After standard introductory material, subjects were shown a sequence of ten "instances". Each instance consisted of a screen containing a pair of graph drawings. The subject was asked to click on which one of the pair they prefer. A typical screenshot is in Fig. 1. The data set used in the experiment comes from publicly available graph sets, including the Hachul library, Walshaw's Graph Partitioning Archive, and sets of randomly generated biconnected and scale free graphs (using the OGDF generators). We used a total of 118 graphs, ranging in size from small (25 nodes and 29 edges) to moderately large (8000 nodes and 15580 edges). For each graph in the data set, four drawings were computed using a force-directed method; these four drawings varied in stress and crossings. For each instance of the experiment, a graph was randomly selected from the data set, and two drawings of the four drawings of that graph were randomly selected and displayed as in Fig. 1. The subject was then asked to click on the drawing that he/she prefers.

Results. The subject preferred the drawing with less (scaled[1]) stress in 57% of all (790) instances, and preferred the drawing with fewer crossings in 65% of all instances. We found that preference for lower stress drawings increases with *(scaled) stress ratio*, that is, the ratio of the stress between the two drawings presented to the user, to a maximum of 70% of instances when stress ratio is greater than 4. Curiously, the percentage

---

[*] Research supported by the Australian Research Council and Tom Sawyer Software.
[1] Scaled by average edge length

C. Duncan and A. Symvonis (Eds.): GD 2014, LNCS 8871, pp. 523–524, 2014.

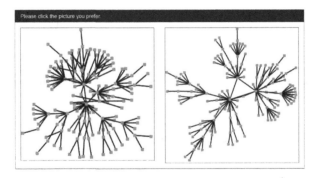

Fig. 1. *Example of a typical "instance" (a graph pair shown to participants)*

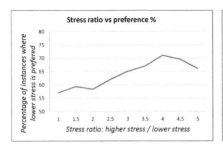
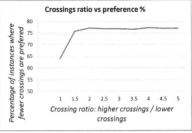

Fig. 2. *Stress ratios, crossing ratios, and preferences*

decreases when the stress ratio is between 4 and 5. Further, the user's preference for lower crossings increases from crossing ratio 1 to 1.5, where *crossing ratio* is the ratio of the crossings between the two drawings presented to the user. For 76% of instances where the crossing ratio was at least 2, the human preferred the drawing with fewer crossings. The relationships between stress ratios, crossing ratios, and preferences are illustrated in Fig. 2.

Remarks. Our results suggest that people prefer less stress; further, this effect increases as the stress decreases. A similar result can be inferred for crossings. However, this initial experiment raises a number of questions: In particular, crossings and stress are weakly correlated in our data set, in that graphs with higher stress tend to have more crossings. It is unclear whether the preference is for less stress or less crossings (or both).

Our experimental software is web-deployable. We welcome more researchers to collaborate in this project.

References

[1] Purchase, H.C.: Which aesthetic has the greatest effect on human understanding? In: DiBattista, G. (ed.) GD 1997. LNCS, vol. 1353, pp. 248–261. Springer, Heidelberg (1997)

A New Approach to Visualizing General Trees Using Thickness-Adjustable Quadratic Curves

M. Ali Rostami[1], Azin Azadi[2], and H. Martin Bücker[1]

[1] Friedrich Schiller University Jena, Germany
{a.rostami,martin.buecker}@uni-jena.de
[2] Jovoto Company, Berlin, Germany
aazadi@gmail.com

In this abstract, we present a new algorithm for visualization of trees. This algorithm illustrates the hierarchical data by using curve segments for each edge and how edges evolve on a path from the root to the leaves. More clearly, we visualize a tree based on the shape of a botanical tree. The software tool Graph-Tea [3,4] implements this algorithm where the adjustable thicknesses of the curve segments reflect the expansion [2] of botanical trees. Here, we consider a simple expansion definition for each vertex as the sum over the number of children and the children of its children up to the some specific level.

Given a tree and a layout of this tree, the algorithm computes a set of curve segments as follows. A quadratic parameter curve segment is drawn for each node starting from the tree root moving toward all leaves. Each curve segment is associated to three consecutive nodes on a path. Suppose that these nodes are located at the positions P_0, P_1 and P_2 in the layout. Then, the quadratic parameter curve segment

$$B(t) = (1-t)^2 P_0 + 2(1-t)t P_1 + t^2 P_2, \quad t \in [0,1],$$

is computed. After generating these curve segments for each node, a general shape is generated from them by interpolating the segments. This approach generates a tree like Fig. 1 (a) with a fixed thickness.

Although this visualization with a fixed thickness gives an overall image of the tree structure, we can visualize more information using thickness. To control the thickness, we visualize each curve segment by two so-called boundary curves. As a result, we can generate different thicknesses by changing the start and end positions of these boundary curves. More clearly, suppose we are given points P_0, P_1, P_2 of the curve segment and the angles $\theta_0, \theta_1, \theta_2$, between their position vectors. Also, let w_0, w_1, and w_2 denote the starting, the middle, and the end thicknesses of the curve segment, i.e., the distance between the boundary curves in their different parts. First, we compute three values: the start width S, the middle width M, and the end width E,

$$S = \frac{w_0 + w_1}{2}, \quad M = w_1, \quad \text{and} \quad E = \frac{w_1 + w_2}{2}.$$

These widths specify the distances between the two boundary curves at three positions.

C. Duncan and A. Symvonis (Eds.): GD 2014, LNCS 8871, pp. 525–526, 2014.
© Springer-Verlag Berlin Heidelberg 2014

Then, the two boundary curves are generated using the two sets of points

$$\{P_0 - SR(\theta_0), P_1 - MR(\theta_1), P_2 - ER(\theta_2)\},$$
$$\{P_0 + SR(\theta_0), P_1 + MR(\theta_1), P_2 + ER(\theta_2)\}.$$

where $R(\theta) = (\cos(\theta), \sin(\theta))^T$. This new definition controls the shape of each boundary curve segment. Fig. 1 (b) shows a tree generated by this new method. Here, all curve segments have different thicknesses. However, the tree is drawn smoothly and there is no inconsistency.

Fig. 1. (a) A general path is generated from quadratic curves with a fixed thickness. (b) A tree is drawn by different thicknesses for branches. (c,d) The thicknesses are changing by adding new nodes as a result of extending by the feature of adjustable thickness. (e) A tree is visualized using the adjustable thickness and adding circles for fruits.

Holton [1] was one of the first who proposed a strand model to investigate the tree drawing. Based on this model, we adjust the thicknesses automatically. Suppose α is a basic default thickness and l_0 is the number of levels we want to consider. Hence, the thickness of a node is computed by the number of its children and the children of its children up to the level l_0, multiplied by α. Figs. 1 (c,d) illustrate this model in the evolution of the tree. Computing this thickness for each node through the tree and interpolating the connection of two curve segments produce Fig. 1 (e). In this figure, circles are added as fruits only in favour of a better visualization.

References

1. Holton, M.: Strands, gravity and botanical tree imagery. Computer Graphics Forum 13(1), 57–67 (1994)
2. Hoory, S., Linial, N., Wigderson, A., Overview, A.: Expander graphs and their applications. Bull. Amer. Math. Soc. (N.S.) 43, 439–561 (2006)
3. Rostami, M.A., Azadi, A., Seydi, M.: GraphTea: Interactive Graph Self-Teaching Tool. In: Proc. 2014 Int. Conf. Edu. & Educat. Technol. II, pp. 48–52 (2014)
4. Rostami, M.A., Bücker, H.M., Azadi, A.: Illustrating a graph coloring algorithm based on the principle of inclusion and exclusion using GraphTea. In: Rensing, C., de Freitas, S., Ley, T., Muñoz-Merino, P.J. (eds.) EC-TEL 2014. LNCS, vol. 8719, pp. 514–517. Springer, Heidelberg (2014)

Minimum Representations
of Rectangle Visibility Graphs

John S. Caughman[1], Charles L. Dunn[2], Joshua D. Laison[3],
Nancy Ann Neudauer[4], and Colin L. Starr[3]

[1] Department of Mathematics and Statistics, Portland State University,
Box 751, Portland, OR 97202, USA
[2] Department of Mathematics, Linfield College, 900 SE Baker Street,
McMinnville, OR 97128, USA
[3] Department of Mathematics, Willamette University, Salem, OR 97301, USA
[4] Department of Mathematics & Computer Science, Pacific University,
Forest Grove, OR 97116, USA

Let S be a set of nonintersecting open rectangles in the plane with horizontal
and vertical sides. Two rectangles R_1 and R_2 are **visible** if there exists a **line
of sight** between them, a horizontal or vertical line segment that intersects
both R_1 and R_2 but no other object in S. We construct a graph G with a
vertex for each rectangle in S, and an edge between two vertices if and only
if their corresponding rectangles are visible. Note that we require rectangles to
have positive width and height, but two rectangles may share a part of their
boundary, so the distance between two rectangles may be 0. For a given graph
G, if such a representation of G with rectangles exists then G is a **rectangle
visibility graph** (**RVG**) and S is a **rectangle visibility representation** of
G [1].

Suppose the corners of the rectangles in S have integer coordinates. For a
given RVG G, we ask how small its rectangle visibility representation can be.
We think of the size of a rectangle visibility representation as the area of the
smallest axis-parallel rectangle that encloses it (the **area** of S), or as the length
of the shorter side of this rectangle (the **height** of S). Kant, Liotta, Tamassia,
and Tollis [2] found the area of a rectangle visibility representation of a tree
up to a multiplicative constant. In this work, we find the height of a rectangle
visibility representation of a tree within an additive constant of 1. We also ask
for the RVG with n vertices and largest area (i.e. the maximum over all graphs
G on n vertices of the minimum area of any rectangle visibility representation
of G). We begin answering this question for small values of n. Specifically, we
prove the following theorems.

Theorem 1. *Among graphs with n vertices, $1 \leq n \leq 6$, the empty graphs E_n
have largest area, n^2.*

Theorem 2. *Among graphs with n vertices, $n \geq 7$, the empty graphs E_n do not
have largest area.*

To prove Theorem 2, we show first that the graphs K_7 and K_8 must be
enclosed by axis-parallel rectangles of dimensions at least 7×8 and 10×10, and

C. Duncan and A. Symvonis (Eds.): GD 2014, LNCS 8871, pp. 527–528, 2014.

so have area 56 and 100, respectively. Then note that if $n \geq 9$ the disjoint union of K_8 and $n - 8$ isolated vertices has area at least $(n - 8 + 10)^2$, greater than the area of E_n.

Theorem 3. *A tree T with ℓ leaves has representation height $\lceil \ell/2 \rceil$ or $\lceil \ell/2 \rceil + 1$. Further, if T has no vertices of degree 2 then T has representation height $\lceil \ell/2 \rceil$ exactly.*

We prove Theorem 3 by converting a rectangle visibility representation of a tree T into an orienting path cover of T. An ***orienting path cover*** of T is a set of directed paths in T containing all of its vertices, such that whenever two paths share an edge, they point the same direction. We prove the following lemmas:

Lemma 1. *A tree has a representation with height k if and only if it has an orienting path cover with k paths.*

Lemma 2. *A tree with ℓ leaves has an orienting path cover with at most $\lceil \ell/2 \rceil + 1$ paths.*

Note that every orienting path cover must include a path ending at every leaf. This is at least $\lceil \ell/2 \rceil$ paths. So equivalently, Lemma 2 says that T has an orienting path cover using at most one extra path. We find a family of examples for which this extra path is necessary.

References

1. Dean, A.M., Hutchinson, J.P.: Rectangle-visibility representations of bipartite graphs. Discrete Appl. Math. 75(1), 9–25 (1997)
2. Kant, G., Liotta, G., Tamassia, R., Tollis, I.G.: Area requirement of visibility representations of trees. Inform. Process. Lett. 62(2), 81–88 (1997)

Author Index